Christmas 1999
To Ben
Love, mom

MODERN PRACTICAL JOINERY

MODERN PRACTICAL JOINERY

A TREATISE ON

THE PRACTICE OF JOINER'S WORK BY HAND AND MACHINE

FOR THE USE OF

WORKMEN, ARCHITECTS, BUILDERS, AND MACHINISTS

CONTAINING A FULL DESCRIPTION OF

TOOLS AND THEIR USES, WORKSHOP PRACTICE, ALL KINDS OF HOUSE JOINERY, BANK, OFFICE, CHURCH, MUSEUM, AND SHOP FITTINGS, AIR-TIGHT CASES, AND SHAPED WORK

WITH CHAPTERS ON

STAIRBUILDING AND HANDRAILING, FOREMEN'S WORK, FIXING, &c.

THIRD EDITION, REVISED AND ENLARGED

INCLUDING A CONCISE TREATISE ON JOINERY MACHINES, MACHINE SHOP PRACTICE, AND PREPARATION OF WORK FOR MACHINING

BY

GEORGE ELLIS

AUTHOR OF "MODERN PRACTICAL CARPENTRY"

Vice-President of the British Institute of Certified Carpenters,
Lecturer on Carpentry and Joinery and Hand-Railing
at the London County Council School of Building

LINDEN PUBLISHING COMPANY

First published in 1902
Reprinted 1903, 1904, 1907

This edition reprinted 1987 from
the third edition of London 1908

Linden Publishing Co.
336 W. Bedford, Suite 107
Fresno, CA 93711

Library of Congress Cataloguing-in-Publication Data
Ellis, George, fl. 1902—
Modern practical joinery.
Reprint. Originally pub. 3rd ed. London:
Batsford; New York: Scribner's, 1908.
Includes index.
1. Joinery. I Title.
TH5662.E49 1987 694'.6 87-22844

ISBN 0-941936-08-2

PUBLISHER'S INTRODUCTION

Over many years Ellis' books on carpentry, joinery, stairbuilding and handrailing have been keenly sought throughout the out-of-print and rare books market. The underlying reason for this continued demand is the depth and scope of coverage of the basic principles of the subject arising from the undisputed authority of the author.

Obviously, when a reprint is made from an early book such as this, many of the machines used then would now be updated with newer materials and with new and more stringent methods of guarding; there is also revised legislation covering the practical use of machining and a keener awareness of all safety matters in both hand and machine tool use. It is, therefore, prudent for the publishers to ask present readers to ensure that current legislation and safety aspects are observed whenever following suggested working methods in the book.

This reprint is a facsimile of the third edition revised and enlarged in 1908, with the exception of the following minor changes: many photographs were printed on one side only (the other side blank); this reprint has omitted all such blank pages. The page size has been reduced slightly to enable a more economical cover price to be set. Trade advertisements appearing in the original have been dispensed with; finally, fold-outs or double-page drawings have been re-arranged to suit modern book production methods. The original photographs, which now date back over 80 years to the first 1902 edition, have been retained to preserve the period of the book as a whole, although, to some extent, this has meant a sacrifice in quality in terms of present day production standards.

The publishers hope to reproduce Ellis' other books in due course and subscriptions to these new reprints are invited.

PREFACE TO THIRD EDITION.

THE first edition and subsequent reprints of this book having been exhausted, an opportunity has been afforded the author of revising and enlarging the book in accordance with suggestions made to him by reviewers and correspondents. He has especially introduced several chapters describing the chief types of machines, their use and working, and machine workshop practice. The subject cannot, of course, be dealt with exhaustively, or even fully, in the given space, but it is hoped that the treatment will not be found altogether inadequate, and that this new feature, which has entailed much research, will prove of value to craftsmen and builders.

Various additions have been made to the chapters upon "Doors," "Frames," "Windows," "Shop Fronts," and "Miscellaneous Fittings," mainly in the shape of further plates of modern examples. Some few drawings have been removed, and others substituted to render clearer the explanations. The Glossary and Index have been considerably enlarged; in the latter case the author believes that every important item in the book is now indexed at least twice, and this he trusts will render it much more serviceable as a work of reference than hitherto.

In conclusion, the author desires to express his gratitude for the friendly and appreciative reception accorded to his labours by his critics and the public, and trusts this new edition will prove as useful as its predecessors.

GEORGE ELLIS.

PREFACE TO FIRST EDITION.

THIS book has been written principally for the guidance and instruction of those learning the trade of joinery, by one who has practised the craft in all its branches for upwards of twenty-five years, and the result of whose personal experience is herein set down. Every endeavour has been made by the copious use of illustrations to render clear the explanations given in the text, which are themselves as full and comprehensive as the wide range of the subject will allow. The language used is that of the workshop, the author's desire being to make himself easily understood rather than to aim at any literary style. The methods of construction described are those generally adopted by skilful workmen, and are such as are in constant use in the best London workshops at the present time. Although handwork has throughout been given the primary place, the more important variations required for machine production have also been dealt with.

The author's experience as a teacher has led him to believe that there is a very general want for a work which does not omit or slur over the more elementary parts of the subject, but which, dealing comprehensively with this stage, shall proceed in a progressive manner. Commencing, then, with a description of the tools in common use, he has followed this by a chapter dealing fully with their application to the general purposes of the craft, and has then advanced step by step to the most difficult and elaborate examples, making each chapter, however, complete in itself.

A glossary of terms bearing on the subject of the book, the result of considerable research, and including many items not before given in print, is introduced, and will, it is hoped, prove of service to all interested in building. A short chapter on wood, giving in a condensed form the latest and most reliable information upon the subject, concludes the book; the necessity of keeping it within reasonable limits has confined it chiefly to the woods used by joiners.

Whilst no attempt has been made to furnish a "Text-book" in the narrower acceptance of the term, there will be found within these pages a far larger amount of practical information relative to the subjects included under the head of Joinery in the syllabuses of the several examining bodies in London than can be met with elsewhere. It was inevitable that some amongst the great number of illustrations given should resemble those that have already appeared in other works in which the subject is dealt with, but in no instance has either illustration or explanation been copied from other books; in every case the author has relied upon his own experience.*

He also desires to acknowledge the very valuable assistance he has received from Mr Batsford in the arrangement and preparation of the work for the press, and concludes with the earnest wish that his labours will prove of some little assistance to his brethren in the craft of joinery.

GEORGE ELLIS.

February 1902.

---

* Four of the plates illustrating mediæval church fittings are exceptions to the above general statement.

CONTENTS.

LIST OF PHOTOGRAPHS AND DOUBLE-PAGE PLATES.

[*Note.*—(P.) = Photograph.]

MODERN PRACTICAL JOINERY.

[*NOTE.—Throughout this work the letters "f." signify figure, and "p." page. When the figure referred to is on the same page, no page number is given. When the same or several figures on the same page are referred to close together, the page number is given only with the first reference.*]

CHAPTER I.

INTRODUCTORY.

Definition of Joinery—First Principles—Uses of Drawing—Hints on Drawing—Purchase and Use of Instruments—Drawing Boards, Paper, &c.—Method of Making a Drawing, Inking In—Sharpening the Ruling Pen.

JOINERY may be defined as the art of preparing, constructing, and fixing the internal and external wood fittings of a building, as distinguished from carpentry, which in like manner deals with the constructive woodwork.

Principles of Joinery.—The essentials of good joinery are:—(1) Thoroughly sound, seasoned, and dry material; (2) great accuracy in the preparation, setting out, fitting, and fixing of the various parts; (3) carefulness and neatness in the finishing or "cleaning off." Neglect of any one of these items will inevitably result in inferior or scamped work, which will be a continual reproach to its producer and a source of vexation to the users. Carefulness and accuracy are attributes which none who aspire to become skilful joiners can afford to dispense with, and the young tradesman should cultivate these virtues from the commencement of his career.

A knowledge of geometrical drawing, whilst not absolutely indispensable to the acquirement of skill as a craftsman, is such a great aid in developing the intelligence and resourcefulness of its possessor, rendering his daily tasks intelligible and interesting, that he is strongly urged to acquire the power of making conventional representations of his work himself, so that he may be the better able to read and understand these representations when made by others; for without this power he must ever be content to remain in inferior positions, the humble imitator of his more skilled brethren, a mere rule-of-thumb mechanic. It is not the primary purpose of this work to teach drawing, space alone forbids, and the art can readily be acquired at any of the numerous classes which now abound throughout the country; but a few hints are here given that will be of service to the beginner and the self-instructed student in making use of the examples furnished throughout this book.

Technical Drawings are distinguished from pictorial or perspective drawings by the fact that no allowance is made for distance in the former. Every part is drawn as though it were equally near to the observer, or as if every part in a particular direction could be seen at once. It is this difference from what we are accustomed to see in a photograph, for instance, that makes it somewhat difficult for the uneducated eye to understand a technical drawing. It is, however, necessary to make technical drawings in this manner; every line in the drawing in the same direction as the edge of the solid it represents, irrespective of the appearance it presents to the eye, or otherwise measurements could not be taken from them or the drawing "read." For definitions of the various parts of a drawing, refer to Chapter XXIII., under Reading of Plans.

Technical drawings are best made upon fine cartridge paper, that kind termed "engineers' cartridge"; the most useful size being imperial, measuring about 30 in. by 22 in. This can be used whole for large drawings, and cut through the middle for smaller. The sheets are secured to the drawing board with four drawing pins at the corners. To fix the sheet, first pin it near the middle of top and bottom edges, then spread it out equally from the centre and fix the corners.

The Drawing Board should be of American pine, with two battens fixed on the back by means of screws in slotted holes. In size, the board should show a margin of about $\frac{3}{8}$ in. all round the paper it is intended to use.

Pencils.—H. or F. pencils are used by draughtsmen for the general construction lines, to be afterwards "lined in" with H.B. when finishing in pencil (all drawings worth preserving should be finished in ink); but much depends upon the "hand" of the student, if he has, as most beginners have, a tendency to press heavily on the pencil, he will find it a corrective to use a soft pencil in preference to a hard one; the intense black lines he obtains will cause him to instinctively press lighter, with a corresponding improvement in his "touch." H.H. pencils may be used for fine or small scale work. For the drawing of straight lines the pencil is better sharpened to a flat or chisel edge, but a sharp round point is required for curves and setting off dimensions. A sharp chisel or piece of fine glass-paper glued to a strip of wood may be used for sharpening pencils. Use soft rubber for erasing pencil marks, and a sharp penknife for ink marks. After inking in, clean off the drawing with bread crumbs a day old, rubbed lightly over the surface with the palm of the hand.

Drawing Instruments.—A beginning can be made with a **T** square, which should have a blade within $\frac{1}{2}$ in. of the length of the board; two Set squares of vulcanite, 45 deg. and 60 deg. respectively; a combination compass with divider, pen and pencil legs; a protractor for measuring angles; and a set of cardboard scales. For advanced work an additional pair of dividers, and a set of spring bows similar to the above but smaller, a ruling pen, French curves, and a parallel rule. In buying instruments, purchase from a mathematical instrument dealer, and buy the best the pocket can afford. A cheap instrument is worse than useless, because its faulty action will often cause the beginner to throw up the subject in disgust at the result. Good English-made compasses, as mentioned above, can be obtained from mathematical instrument dealers at from 5s. to 7s. 6d., and will last a lifetime with care. In taking measurements with the dividers, apply them to the copy and the drawing sideways, not upon their points upright, but where possible use the Scale direct, rather than the dividers, to set off a dimension. In repeating small distances with the dividers, start and transfer all from a fixed point, not stepping one from the other, as any error is intensified thus. In using the compasses for describing curves, &c., avoid puncturing the paper. Endeavour to press lightly with both instruments and pencils. Draw horizontal lines with the T square and vertical lines with the set square, moving them both upwards or downwards as the length of line may require. Use the T square with the left hand. Parallel lines at any angle are drawn, by arranging the edge of a set square on the first or pattern line and then placing the T square or another set square at one of the free edges, and moving the first square as may be desired against the second.

When commencing a drawing, "balance" it upon the sheet, *i.e.*, roughly block out the space required, and place it so that an equal margin will be shown all round. Title, date, and scale all drawings. The scale should be at the bottom edge, and is required for future reference. If the drawing is at all symmetrical—that is, has two sides alike—lay down a centre line, and work equally on each side of this. Generally, plans are drawn first, then vertical sections, and the elevation projected from these two, by drawing straight lines (projectors) from all the points visible on the corresponding sides in the two drawings. Draw in the chief constructive lines first throughout, then fill in the details, and finally the sectioning. The latter should always be left until last, as the less lead there is on the paper the better. In the passage to and fro of the squares, much is taken up on them, which soils the paper elsewhere. Technical drawings are not now shaded. Graining or sectioning is used, to indicate that the parts so marked are cut by, or are in the line of section of the drawing upon which they occur, and all the other parts shown are farther away or out of the line of section. Parts "out of the picture," that is, above or behind the line of section, may be indicated by dotted lines. Drawings that are to be coloured should not be sectioned in ink, but distinguished by a deeper tint of colour. See remarks on colouring drawings, Chapter XXIII.

When **Inking In** a drawing, hold the pen upright with the screw away from you, use a straight-edge with the arris taken off the under edge, to prevent the ink clinging to it. When a line has been commenced, finish it without stopping or lifting the pen. Mark "stopping points" previously with a pencil, or carry all intersecting lines beyond the points of intersection. Ink in curved lines first, and join straight ones to them unbrokenly. Use "Indian ink," which can now be obtained in liquid form from artists' colourmen. That made with pure carbon is the best.

The Ruling Pen occasionally requires sharpening, as the nibs wear thick in places. The nibs should be exactly alike all round, with a slightly rounded end, the edges so fine as to be scarcely discernible. They are sharpened upon an oilstone. Draw the nibs close together with the screw, then holding the pen between finger and thumb, draw it slowly along the stone sideways, at the same time making the free end describe a semicircle. When the end is nicely rounded, rub each side to and fro on the stone with a slightly revolving motion until the edge is of equal thickness throughout, then remove the screw and give the faces a slight rub to remove the burr.

CHAPTER II.

HAND TOOLS.

Action of the Wedge—The Half-Wedge Type, difference of Action—Why the Chisel produces a Smooth Surface—Planes, the Use of the Stock, Influence of wide and narrow "Mouth" upon Character of Shavings, Influence of Angle, or Pitch of Iron, Object of High Pitch, Value of Double Irons, Reasons for reversing Single Irons—Action of Saws, Reason for gulleting Ripping Saws, Use of Set on a Saw—Description, Sizes, and Uses of various Tools, Advantages of certain Types, Points to be observed in Purchasing, &c.—Saws—Planes—Abrading Tools—Chisels and Gouges—Marking or Dimensioning Tools—Boring Tools—Testing Tools, the Steel Square—Miscellaneous Tools—Care of Tools—A New Saw Set.

Theory of Action of Wood-Cutting Tools.—As a preliminary to the description of the various hand tools used by the joiner, which will tend to make that description of more service by indicating how the various shapes arise out of the peculiarities of the materials dealt with, it is proposed to make a short investigation into the principles of action of the chief types of cutting tools. It will be found on analysis of the action of all the numerous forms of tools that their cutting actions may be resolved into three elementary kinds, viz., wedging, tearing, and abrading. The first of these actions separates the wood in layers, the cutting edge of the tool entering under and lifting up the fibres in comparatively thin slices; the second removes it in short strips or shreds by a combined crushing down and shearing off of the fibres; the third action reduces the surface of the material to powder, breaking up the fibre by numberless scores or fine grooves crossing each other at an angle.

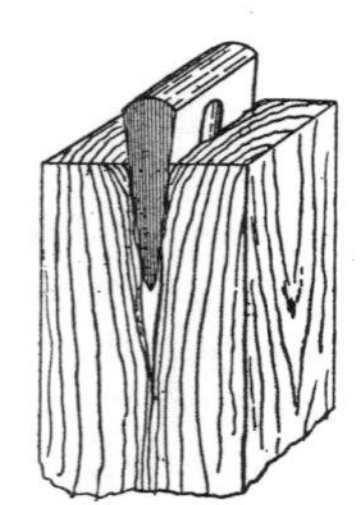

Cleaving Action of Wedge.

In the first group may be placed all axes, chisels, gouges, planes, draw-knives, spokeshaves, boring-bits, &c.

The second group comprises saws, scratch tools, and revolving machine cutters.

The third group, toothing planes, scrapers, files, glass-paper, and radial machine cutters.

We will first consider the action of the Wedge, which is the fundamental form of all the tools in the first group, and of which the chisel in its various shapes may be taken to typify the perfected form. When cleaving a block of wood into two parts, an iron wedge is driven in at one end, as shown in sketch, and the edge entering between the fibres of the wood, first crushes them together on each side, until the resistance offered by the outer layers to crushing is greater than the lateral cohesion of the parts in advance of the edge, when separation below the wedge takes place; the split continually extending as the wedge is driven in. An inspection of the surfaces thus split will show that the wood has been torn asunder irregularly, in consequence of the various components of the wood exhibiting varying cohesive strength, and also because the extreme edge of the wedge, after its first entry into the wood, has no further action upon it, the dividing force being exerted near the heel of the wedge. It is to avoid this loss of action by the front edge, that all cutting tools except the axe—and in this case the same effect is produced by the inclination given to the tool in use—are made wedge shape or bevelled on one side only, and the utility of this formation will be apparent on examination of f. 1, which represents a chisel in the act of paring longitudinally. Upon pressure being applied to the handle, the chisel edge insinuates itself under a layer of the wood, and drives it off on the bevelled side in the same manner that the wedge does in the previous example; but the face side of the chisel, lying precisely in the same plane as the cutting edge, keeps the plane of severance or "line of cut" in the same direction as the chisel face, and this will be so, at whatever angle the chisel may be applied to the surface of the wood. In addition to the advantage thus gained of being able to direct the line of split, independent of the amount of cohesion in the fibres, the face of the chisel bearing equally over the entire surface as its edge advances, prevents the weaker fibres yielding unequally with the stronger, and the result is that a comparatively smooth surface is left after the passage of the chisel, and the greater the pressure exerted through the chisel to keep the fibres down, the smoother will be the surface produced. This will be more clearly understood when we examine the action of the plane. This restraint upon the splitting action marks the difference between the crude form of wedge shown on p. 4, acting by expansion alone, and the improved chisel form having a keen cutting edge acting in any desired direction, and separating only those fibres that are contiguous to its edge. The same action takes place when the chisel pares across the grain, as shown in f. 2. To localise the cut, the fibres must be severed across with a knife, otherwise they will be torn out beyond the sides of the chisel, as their longitudinal adhesion is much greater than the lateral. This is exemplified by a piece of board splitting more readily in the direction of the grain than it will break across the grain. The reason that the surface produced in this direction of cut is not so smooth as in the longitudinal is, that the edge of the tool has a tendency to dip into the softer tissue, which in that direction alternates regularly with the harder, as will be found fully explained in the chapter on timber; but this tendency can to a great extent be overcome by increasing the bearing pressure on the chisel. In the plane it is met by setting the back iron "fine," which is a mechanical device to counteract this digging-in tendency.

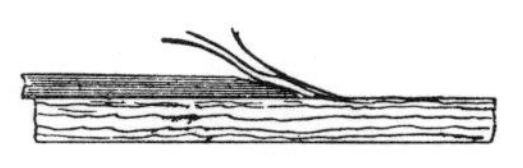
1. Paring Action of Chisel.

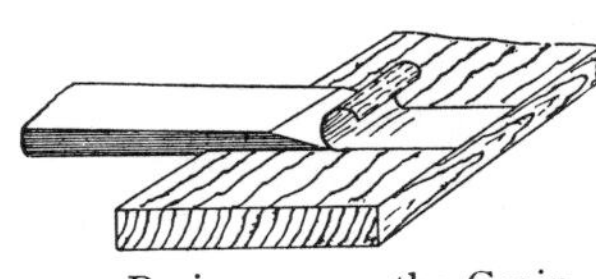
2. Paring across the Grain.

A Plane Iron or cutter is merely a chisel fixed securely in a block of wood or metal for the purpose of ensuring that the depth of cut shall be equal throughout the stroke—in other words, that the shaving removed shall be of uniform thickness. Incidentally the weight of the stock adds to the momentum of the stroke, but the chief mechanical advantage that the plane has over the chisel is that a bearing surface is provided immediately in front of the cutter which restrains the splitting tendency of the latter, and so ensures a smoothly finished surface to the work. The closer this restraining surface is brought to the cutter, *i.e.*, the smaller the "mouth" or space between them, the cleaner will

1. Action of Cutter in a Wide-mouth Plane.

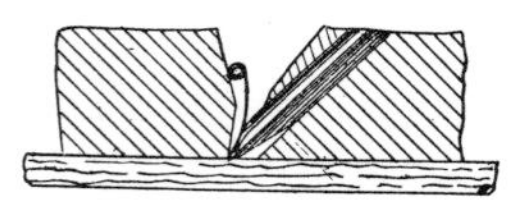

2. Action of Cutter in a Close-mouth Plane.

be the surface produced. This is shown graphically in f. 1 & 2, the first of these showing the effect of a wide mouth, where little or no restraint is offered to the splitting action of the cutter; the second, a close mouth, shows the effect of restraining the split. The *amount* of the splitting is exaggerated in the first case to make it clearer, but this is in effect what does take place when a wide-mouthed plane is used. Another property of the stock of a plane is, that it produces a surface the exact reverse of its own. For instance, a perfectly straight trying plane will produce a truly plane surface, a round-soled plane will produce a hollow, and a moulding plane, the precise opposite curves to those of its own profile.

The pitch between the cutter and the face of the stock has less influence on the surface produced than has the size of the mouth; but generally planes to be used for hard wood have their irons pitched slightly lower than those for soft wood, because the low pitch makes them work easier, and there is less tendency in hard wood to split along the grain. One of the reasons for pitching the irons high in bench and similar planes in frequent use is, that it reduces the necessity of frequent grinding to keep the heel of the iron clear of the surface.

An iron ground to a long basil at the back will enter the wood easier and also be sharpened quicker than one ground more obtusely, but at the same time it will lose its keen edge quicker, and yield more under the shocks from knots, &c., than will the thick one. In the case of bench planes and others with wide cutters this tendency to vibrate is checked by the use of cover or back irons, which are screwed tightly down on the face of the cutters as near to the edge as possible. The cover iron should be curved in its length so as to spring hard on the end of the cutter, and fit to its face with great accuracy to prevent shavings insinuating themselves between them when working, as this will cause the plane to "choke" quickly. The thickness of the shavings taken off is influenced by the distance of the cover iron from the edge of the cutter. If the exposed surface of the latter is minute, it will be unable to penetrate the wood to a greater depth than the amount of the exposed surface, whether the cutter project much or little from the stock of the plane; and the cutter being thus held very rigid, the shavings will be very thin and regular and crinkled in appearance, in consequence of the cover iron breaking them across rapidly. On the contrary, if much cutter is left exposed, coarse irregular shavings will be taken, thicker at the middle than at the edges through the yielding of the edge of the iron to the pressure unequally, and the shavings will leave the mouth in long straight lengths. Similar effects, though in a more pronounced degree, are obtained by setting the iron "coarse" or "fine" in the stock, *i.e.*, projecting the edge of the cutter much or little from the sole, especially if the conditions of the back iron and of the cutter coincide. For instance, if the cover is set "fine" (*i.e.*, very close to the edge), and the combined "iron" is also set fine in the stock, the surface produced will be the finest possible with that particular specimen; if the two are set "coarse," the shavings will be thick and the wrought surface irregular; whilst if the cover is set fine, and the combination set coarse in the stock, the plane will choke; as the shavings being taken fast and turned over at a sharp angle by the cover iron, impinge on the front of the mouth, destroying their impetus in so doing, and therefore lodge in the bottom, which they soon pack tightly. In this it is assumed that the plane is in a good condition, *i.e.*, with a narrow mouth; if the mouth be worn large the plane may not choke, but for reasons specified previously, the resulting surface will be a torn and irregular one.

Planes that are used solely for hard woods and for cutting transversely to the grain, such as Shoulder Planes, &c., have their single cutting irons turned face downwards, so that they may have a bed or support close up to the cutting edge. This arrangement has the same effect as the cover iron in the double iron type, in preventing the edge of the cutter yielding to the pressure, but it will not admit of the iron being adjusted to cut coarse shavings. Hence this arrangement is less suitable for planes working soft woods chiefly.

Saws are the most important type of wood-cutting tools in the second order, and these may be subdivided into slitting and cross-cutting kinds, not because the principle of their *action* differs, but because there are essential differences in the formation of the cutting edges, &c., due to the direction that they are applied to the material. Primarily the teeth of a saw may be considered as a series of thin chisels rigidly connected in one plane, and in the cross-cut type, approximating very closely to the action of the paring chisel; but there is an important difference between this and the first group, viz., that the chief object of the former is to produce a plain and smooth surface, that of the latter to separate the material into distinct portions. It is essential in both cases that the objects should be obtained with the least possible waste of material, and to this end plane-cutters are made as wide as possible consistent with their special uses, and saw-cutters as narrow as possible consistent with stiffness.

Saws of the first division are called **Ripping Saws**. Their teeth are spaced wide apart, with considerable depth or throat; the cutting edges are very obtuse, and with little spread or "set." These saws are used in a direction parallel to the fibres, and act by scraping or tearing a strip out equal to the thickness or *set* of the teeth, the sides of the fibres being sheared off at the same time. The shred or particle of wood removed by each tooth in a longitudinal direction is of appreciable length, and if the teeth were not widely spaced and of relatively great depth, the spaces between them would quickly get packed up with the shreds, and the teeth would cease to cut. In machine ripping saws, where the stuff is fed rapidly to the saw, a large throat or "gullet" has to be made at the root of each tooth to receive the sawdust whilst the saw is passing through the stuff.

Cross-cut Saws, attacking the wood in the direction of its greatest resistance, require their teeth to be small and spaced closely. The angle of the cutting edge is also more acute than in ripping saws. The thinner a saw-blade is the better, so that it possesses sufficient stiffness to withstand the necessary thrust without bending. If the blade is ground properly—that is, thinner at the back edge than at the teeth—little or no "set" will be required in cutting dry stuff. In cutting wet stuff, considerable "set" is sometimes necessary, as the fibres, being softened by the water, are dragged out, rather than cut, and the ragged ends spring up and bind the saw-blade so firmly that it cannot be driven through them. *Setting* is the bending of each tooth (or group of teeth) alternately to the right and left of the blade (see f. 2, p. 9), thus causing the teeth to cut a path slightly wider than the thickness of the blade, and thus allowing the latter to pass through the material with less friction. A saw with much set is more difficult to keep in line, however, than one with less, as the former drifts from side to side of the cut.

There are but few joiners' tools in the third order, the most important being the **Toothing Plane** and the **Scraper**.

Glass-paper, although a valuable adjunct to the tool list, can scarcely be termed a tool in itself. It nevertheless best exemplifies the action of all abrading tools which scratch or score the wood in various directions, breaking up the fibre into an impalpable powder, and thus producing a uniform surface.

DESCRIPTION OF TOOLS.—Having thus cursorily examined the underlying principles which govern the action of the chief types of tools, the individual varieties will now be dealt with. A brief description of their sizes, shapes, and uses accompany the illustrations. Considerations of space preclude the mention of minor or sub-varieties of the various tools, and the drawings are intended to indicate rather the especial characteristics of the type, than small differences in detail of individual makers; which was found to be the result of the use of the blocks from makers' catalogues that did duty in the first edition of this work. The author's line drawings may not be so effective pictorially, but he trusts they will be found more suitable for the purpose intended.

CUTTING TOOLS.

SAWS fall naturally first in the order of use, as the material must be "converted"—that is, cut to suitable sizes—before it can be submitted to the successive operations of planing, mortising, tenoning, moulding, &c. It may be noticed here that this order of operations, although generally followed, is varied in some machine shops, where the stuff is planed up in board form before conversion.

RIPPING SAWS.—The Rip Saw is but seldom used in these days of machinery, and it is so similar in all respects but the spacing of the teeth, which are 3 to the inch, to its younger brother, the half-rip, that it is not necessary to describe it specially. **The Half-rip Saw** (f. 1, p. 9) is used solely for cutting in the direction of the fibres in such operations as "splitting," tenon-cutting, &c. The length of blade is 28 in.; the teeth are spaced from $\frac{1}{4}$ in. apart at the point of the blade to $\frac{3}{8}$ in. apart at the heel. The best cutting angle for the teeth is 80 deg. with the line of points as shown in f. 2. The

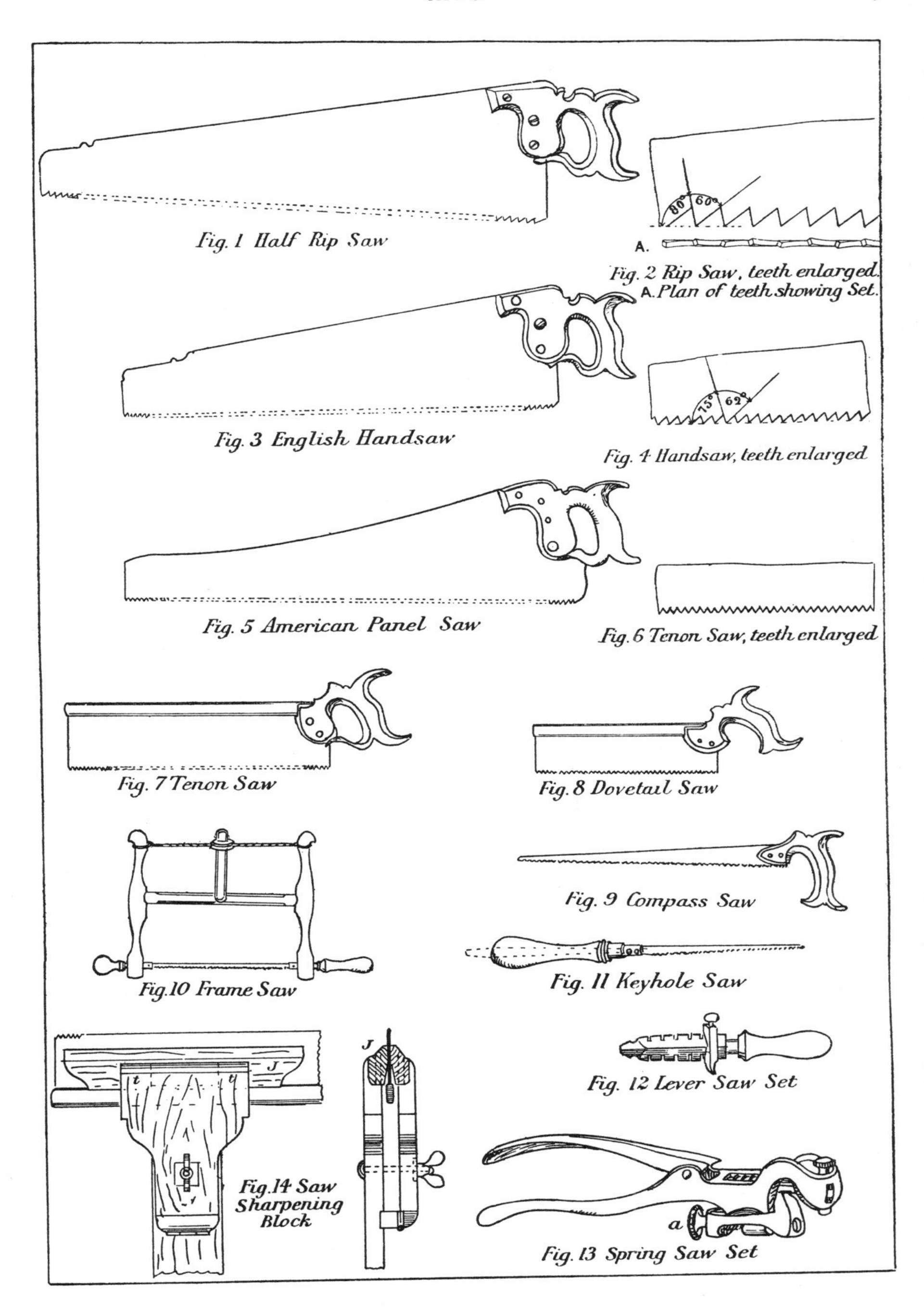

Fig. 1 Half Rip Saw

Fig. 2 Rip Saw, teeth enlarged. A. Plan of teeth showing Set.

Fig. 3 English Handsaw

Fig. 4 Handsaw, teeth enlarged

Fig. 5 American Panel Saw

Fig. 6 Tenon Saw, teeth enlarged

Fig. 7 Tenon Saw

Fig. 8 Dovetail Saw

Fig. 9 Compass Saw

Fig. 10 Frame Saw

Fig. 11 Keyhole Saw

Fig. 12 Lever Saw Set

Fig. 13 Spring Saw Set

Fig. 14 Saw Sharpening Block

angle of the back is less important so that sufficient space is provided for the saw-dust, but is usually made at 60 deg. with the front of the tooth. The angle of the front is sometimes made 90 deg. or perpendicular with the line of points. This, which is termed "giving the tooth hook," produces a rough cut, and also jars the arm of the operator in working, as the whole length of the tooth strikes the work at once. Spacing should be 4 teeth to the inch, depth of tooth $\frac{1}{4}$ in. from point to root.

CROSS-CUTTING SAWS.—The Hand Saw, f. 3, p. 9, has a stout blade from 20 in. to 26 in. long, the teeth spaced $6\frac{1}{2}$ to the inch and $\frac{1}{8}$ in. deep; angle of cutting face with line of points 75 deg. (see f. 4). This is the saw generally used for cross-cutting in conversion. For finer work on the bench a PANEL SAW is used. This is a smaller and lighter saw than the hand saw, with its teeth spaced from 8 to 10 to the inch, length of blade 16 in. to 26 in. The American pattern of these saws (f. 5) has much to commend it. The handle is thrown well forward, giving great control over the saw when used horizontally. The blade is brought under the handle, which increases the stroke and reduces the liability of "kicking" or catching of the heel in the cut at the return stroke. The "skew" or curved back by reducing the weight and friction, makes the saw much less fatiguing to work than the ordinary pattern. The blade of a saw should be thin, bright, with a high bluish-white polish, bend freely, and return after bending perfectly straight. The handle should be secured very firmly. Any saw that rattles in the handle should be rejected, as apart from the possibility of the loose blade shearing the screws through, such a one will not run true to line. The orthodox way of testing saws is to rest the back of the handle upon the thigh, grasping the blade with the fingers at each edge about 6 in. from the handle, then shaking it sharply to and fro, when any weakness in the spring of the blade or looseness in the handle will readily be detected. The shape of the handle should also be attended to, and one selected in which the hand can enter freely. Narrow thick handles cause cramp in the hand.

The Tenon Saw, f. 7, or, as it is sometimes termed, the BACK SAW, is used for cross-cutting of a finer grade than the panel saw is capable of. One of its chief uses is, the cutting of shoulders to tenons, and it would be more accurately named a shoulder-saw, as its teeth are too fine for the ripping down required for tenoning. To obtain fineness of cut, the blade has to be made so thin that it is necessary to stiffen it with a solid "back" or bar of iron or brass, which is pinched tightly on the top edge. The blade is about $3\frac{1}{2}$ in. wide, from 12 in. to 18 in. long; when over 14 in. long they are called Sash Saws. The teeth average 12 to the inch, and should be of equilateral triangle shape as shown in f. 6. These are sometimes called "peg" teeth.

The Dovetail Saw, f. 8, is a smaller type of "back" saw used for fine and light cutting either way of the grain. It is much used for dovetails in small work, hence its name. The size of the blade varies from $1\frac{1}{2}$ to $2\frac{1}{4}$ in. wide, by 7 to 10 in. long, with 15 teeth per inch. Peg teeth. Handle "open."

The Compass or Turning Saw, f. 9, has 12 teeth to the inch, shaped like rip saw teeth. A similar saw, but with longer and wider blade, is termed a TABLE SAW; both of these types have "open" handles. They are used for cutting to curved lines.

The Bow or **Frame Saw**, f. 10, p. 9, is used for cutting quicker sweeps than can be accomplished by the compass saw. It consists of a beechwood frame, with a ribbon saw secured to two revolving handles passing through its lower ends. A light stretcher across its middle keeps the arms of the frame apart, and the saw-blade is sprung tight by twisting a short bar in the loop of a cord wound round the upper end of the frame, the stretcher acting as a fulcrum, and also preventing the winding bar flying back. The handles should work rather stiffly in the arms. The better qualities are furnished with brass bushes and set screws to prevent the accidental turning of the frame. The blade is capable of complete revolution as the cut requires, but care must be taken that the whole length of the blade is on one plane, *i.e.*, not twisted, otherwise the saw will run from its course and be liable to snap. The saw-blades of these vary in length from 6 to 24 in., and in width from $\frac{1}{8}$ to $\frac{1}{2}$ in. The larger frames have wood sleeves, as in the sketch, for the lever to pass through.

The Keyhole Saw, f. 11, is the smallest saw in the joiner's kit; it is from 9 in. to 12 in. long, tapering in width from $\frac{3}{8}$ in. to $\frac{1}{8}$ in. It is fixed in the pad or handle by two set screws in the brass ferrule, which is slotted to receive the heel of the saw in such a manner that it will not pass down into the pad, but when out of use, the blade may be slipped into the handle, point first, as indicated by the dotted lines. This saw is used for cutting keyholes, and starting interior cuts for the larger saws, after a hole has been bored for its insertion.

Interchangeable Saws.—These are saw-blades of the "Keyhole" and "Compass" types, fitted to a special handle in such a manner that, whilst held as securely as in the ordinary fixed handle, they are instantly removable by the turn of a lever. These are made and supplied by Messrs Spear & Jackson Ltd., Sheffield.

PLANES.

BENCH PLANES.—The Jack Plane, f. 1, p. 12, is the first plane used in preparing stuff, its purpose being to remove irregularities left by the saw and produce a fairly smooth surface. It is also used generally for reducing scantlings quickly. It consists of a beechwood stock 17 in. long by $2\frac{3}{4}$ by 3 in., with a $2\frac{1}{4}$-in. cutting iron and similar back iron. The cutter is better parallel or gauged, as once fitted, the wedge will then always sit properly, and the size of the mouth remain the same throughout. This applies to all planes whose cutters are fixed by wedges. A $\frac{3}{4}$-in. stud or button of boxwood inserted in the nose of the plane will prevent it being disfigured by hammer marks (see f. 1 & 2). The best plane stocks are cleft or split in the natural laminæ of the wood, which ensures their remaining "true." Common qualities are cut by the saw nearly parallel to the beat of the fibres; but if these are cut across, the plane will "cast" and want continual reshooting. The stock should be so prepared that the medullary rays of the wood are perpendicular to the sole.

Single Iron Jacks are used for rough "scurfing" as a preliminary operation on extremely rough or dirty surfaces.

The Trying Plane, f. 2, has a stock 23 in. long by 3 in. by 3 in., with a $2\frac{1}{2}$-in. cutter. It follows the jack plane in reducing the wood to a truly plane surface, or in producing straight edges for joints, &c.

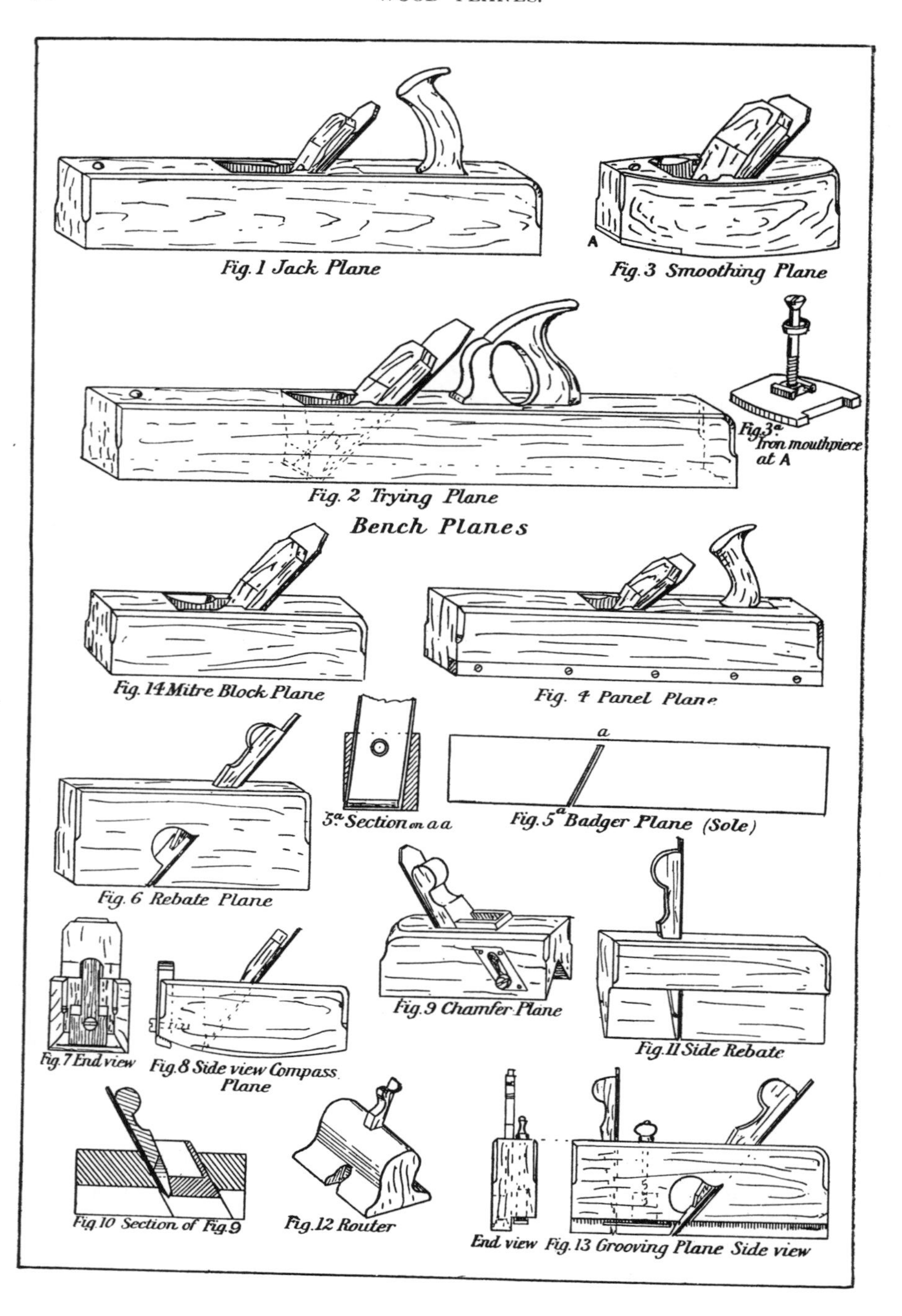

Fig. 1 Jack Plane

Fig. 3 Smoothing Plane

Fig. 2 Trying Plane

Bench Planes

Fig. 14 Mitre Block Plane

Fig. 4 Panel Plane

Fig. 5a Badger Plane (Sole)

Fig. 6 Rebate Plane

Fig. 9 Chamfer Plane

Fig. 8 Side view Compass Plane

Fig. 11 Side Rebate

Fig. 10 Section of Fig. 9

Fig. 12 Router

Fig. 13 Grooving Plane Side view

The Smoothing Plane, f. 3, is, as its name implies, used chiefly for smoothing or finishing the surface after manipulation by other planes. The standard size of stock is 8 in. by 3 in. by 2¾ in., with 2¼-in. double irons. In consequence of the frequent use of this plane its mouth wears with comparative rapidity, and to avoid the continual renewal of the mouthpiece, **Iron Fronts** as shown at f. 3*a* are sometimes used; these are attached to the stock by an iron set screw passing through the nose of the plane, and are capable of easy readjustment longitudinally as the cutter wears; but they are rather difficult to refit accurately when the sole of the plane is shot. Apart from this, they are an undoubted improvement.

METAL PLANES FOR HARDWOODS.—Smoothing Planes are shown in f. 1 & 2, p. 14. The first is the American type of malleable cast-iron skeleton plane with tote handle at rear, and hand knob at the fore end, with adjustable mouthpiece, cam-setting lever, and screw adjustment for the cutter iron.

Jack and Trying planes are also made thus. The English type of smoothing plane, with wrought-iron or cast gun-metal shell, and hardwood filling, is shown in f. 2; this has a screw lever for adjusting and holding the cutter.

The various advantages and disadvantages of the three types may be summarised as follows:—

The common Wooden Stock plane is comparatively low in price, and will stand rough usage better than either of the others, being in fact practically indestructible. It works rapidly and easily, and can be adjusted by means of the cover iron to suit hard or soft woods. On the other hand, it will not produce so highly a finished surface as a metal plane; and it requires frequent shooting, and remouthing occasionally, to keep it in good condition. The English form of metal plane will produce work of the highest class. It is of great weight and solidity, the latter quality having an important bearing on its results. It overcomes the resistance of cross grain and knots easily by its great momentum, and "tearing out" is prevented by the extremely fine mouth (see p. 6). Its disadvantages are, that it is fatiguing to work, the friction between metal and wood is greater than between wood only, and in common with all metal planes, in our moist atmosphere it is difficult to keep free from rust or verdigris, as it is made of steel or brass; and lastly, its first cost is relatively high.

The American type has for its chief recommendation cheapness and readiness of adjustment. It is easy to work in consequence of its lightness, but this quality also acts detrimentally in causing it to "chatter," which prevents the production of so high a finished surface as the English form is capable of. However, its general results are higher than those of the common wood plane. Many ingenious time-saving attachments make it a rapid worker, but it is very fragile, and will seldom survive an accidental fall from the bench. One well-known American maker has introduced a plane with a corrugated sole, with the object of reducing the "bite" of the metal. The author has not personally tried one, but the result should be good, because there is obviously less friction to overcome.

The Compass Plane, f. 7 & 8, p. 12, is a smoothing plane with a convex sole, used for cleaning up curved surfaces. It has a sliding boxwood nose-piece

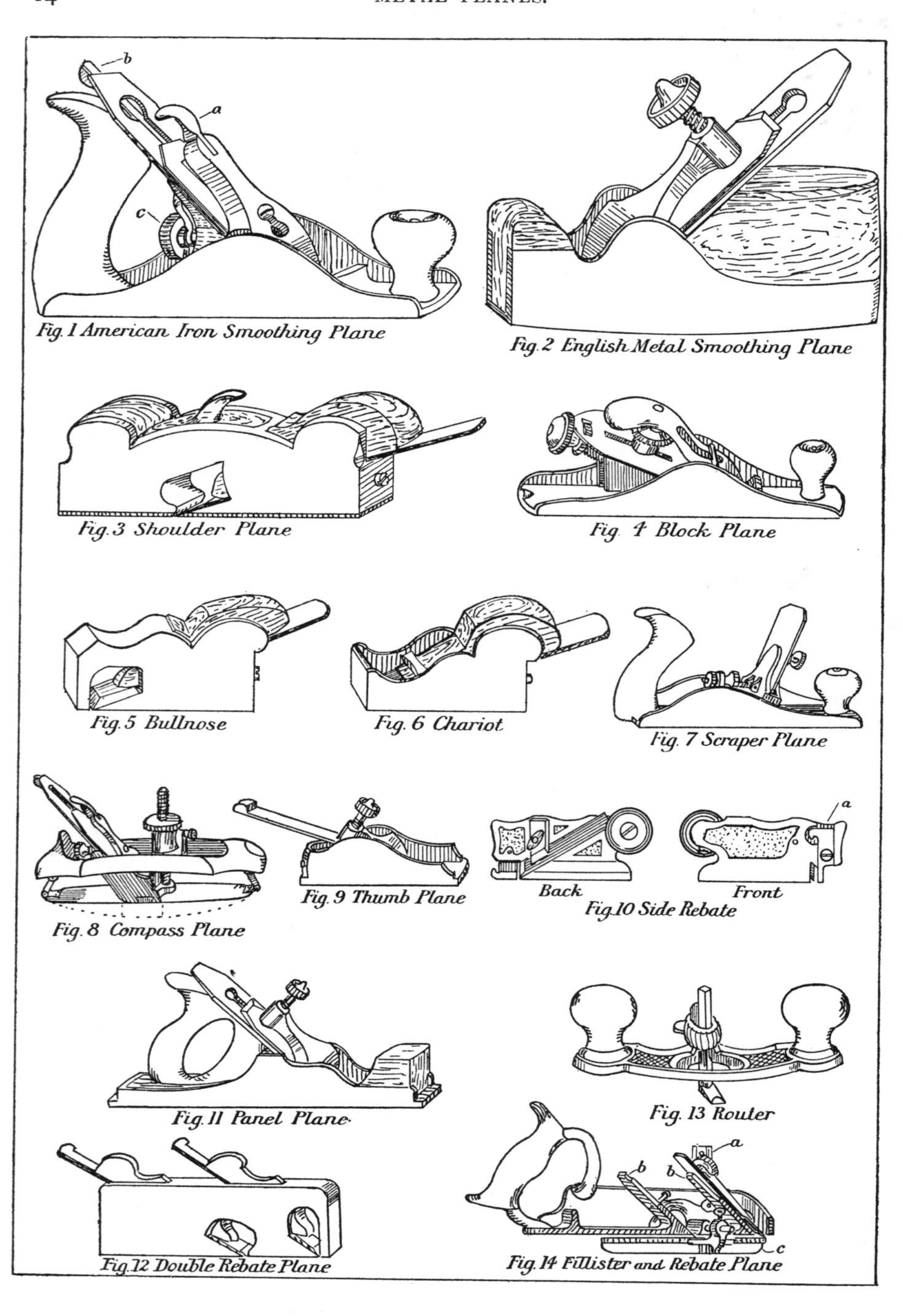

Fig. 1 American Iron Smoothing Plane
Fig. 2 English Metal Smoothing Plane
Fig. 3 Shoulder Plane
Fig. 4 Block Plane
Fig. 5 Bullnose
Fig. 6 Chariot
Fig. 7 Scraper Plane
Fig. 8 Compass Plane
Fig. 9 Thumb Plane
Fig. 10 Side Rebate
Fig. 11 Panel Plane
Fig. 12 Double Rebate Plane
Fig. 13 Router
Fig. 14 Fillister and Rebate Plane

shown hatched in the illustration, which may be moved downwards to make the sole fit the work to be planed. Fig. 8, p. 14, is the American variety, which has a malleable iron stock, carrying an adjustable cutter frame to which a spring steel face is attached. It will work either concave or convex surfaces. It has similar drawbacks to the smoothing plane mentioned previously.

The Spokeshave, f. 8 & 9, p. 17, is a variety of the compass plane, but adapted to quicker curves than the latter will enter. Fig. 8 is the older wood type of stock with double-tanged knife, now superseded by the iron stock (f. 9), with plane iron cutter and cover iron, a form much easier to sharpen than the other. They are made in several sizes with cutters from 1½ in. to 4 in. wide.

Panel Planes are made both in wood and metal, as they are required for working soft or hard wood. The former is illustrated in f. 4, p. 12, and has a removable wood slip on the right hand side (drawn in reverse), to enable the plane to clean up sunk surfaces; length 14 in., cutter 2½ in.

Metal Panel Planes, f. 11, p. 14, range in size from 9 in. with 2¼-in. irons, to 15 in. with 2½-in. irons. Longer planes in this style up to 20½ in. long with 2¾-in. cutters, are called JOINTERS.

The Badger Plane is similar in size and appearance to the wood panel plane, but it has a skew mouth, and the cutter passes through the stock at an angle with the side, as shown in the section (f. 5*a*, p. 12), thus bringing the cutter up to the extreme right hand edge of the sole (see f. 5), which enables the plane to be used for finishing sinkings, rebates, &c. Note that it is not used for making rebates, but for cleaning them off; it is both too cumbersome and too fine cutting for the first purpose, but it is a most useful tool for finishing rebates, &c., after machining. All the above are DOUBLE IRON PLANES; *i.e.*, the cutters have a "back" or cover-iron to stiffen them, as shown in f. 2 and 5*a*, p. 12. The following are all Single Iron Planes.

REBATING PLANES.—The Wood Rebate Plane, f. 6, p. 12, as its name suggests, is chiefly used for forming rebates or sinkings upon the edges of material. It has a solid beech stock, 9 in. long, 3¾ in. high, and from ⅝ in. to 2 in. wide; it is made with both square and skew mouths, the latter works the better.

A Double-iron Metal Rebate Plane is shown in f. 12, p. 14. These are intended for cleaning up hardwood rebates quickly, the front iron cutting coarse, the rear one fine; either one can be used alone when required.

These are also to be had with single irons only. The irons are used face down as in shoulder planes, for reason see page 7.

The Shoulder Plane, f. 3, is a special form of rebate plane in metal, used principally for smoothing and correcting hardwood shoulders after the saw. The casting is hollow and filled in with a hardwood core. The iron is set face down and at a low angle. The wedge projects to form a rest for the hand, and improved forms have a spur worked in the top core just over the mouth which adds to the power of the grip. Square mouths are preferable in these planes, as they are required to work both right and left hand. The sketch shows a casting with gun-metal stock, having a steel face "sweated on"; these are the best for keeping a true face.

The Bullnose, f. 5, p. 14, is another type of rebate plane in metal, its use being to finish off rebates and other narrow surfaces close up to stops or abutments. It is very essential in all planes that have their cutters face downwards that the face of the cutter be ground to a true plane, and not have to be forced into that position by the wedge. All of these planes are comparatively weak in the neck, and if wedged too tightly the sole will spring hollow and the mouth choke.

The Sash Fillister, f. 3, p. 17, and the **Side Fillister**, f. 4, are both varieties of rebate planes. They are used to form rebates or sinkings; the first on the off side, the second on the near side of the material, as may be more convenient. They are each provided with vertical and horizontal adjustments, the first by means of rising and falling stops, the second by sliding fences. They have also a tooth or cutting knife slightly in advance of the cutting iron, to sever cross grain or to cut through knots.

A Combined Fillister and Rebate Plane of American make is shown in f. 14, p. 14. This has an adjustable fence and two beds *b*, *b*, for the cutting iron, which may be used on either the front when rebating, or the rear when fillistering.

GROOVING PLANES.—The Trenching or **Grooving Plane**, f. 13, p. 12, is used for sinking trenches or grooves across the grain, as will be seen by the end view; it has a rebated sole, the cutters being in the tongue portion, which is usually made $\frac{1}{2}$ in. deep, and varies in width from $\frac{1}{4}$ in. to $1\frac{1}{8}$ in. It has a screw-stop for adjusting the depth of cut, and a double-toothed cutter for separating the fibres in front of the iron.

The Plough, f. 1 & 2, p. 17, is an adjustable grooving plane of great utility. It will sink a groove of any width between $\frac{1}{8}$ in. and $\frac{5}{8}$ in., to any depth required, up to the depth of the guide iron of about $1\frac{1}{2}$ in., and at any distance from the edge of the piece, within the length of the sliding stems. Some patterns have the front end of the guide turned up with a skate end, as indicated at *s* by dotted lines. These pass over mortises easily. There are nine irons to a set, as shown in f. 1*a*. In adjusting these, care must be taken to set the V groove in the iron accurately upon the fore end of the guide.

The Groove Router or Old Woman's Tooth, f. 12, p. 12, is used for increasing the depth and levelling the bottom of grooves formed by some other tool. It consists of a hardwood block about 5 in. long, 3 in. deep, and $3\frac{1}{2}$ in. wide (the grain running in the latter direction), with a wedge. Plough irons are used for cutters. Fig. 13, p. 14, is an American variety in metal; it is provided with two cutters only, $\frac{1}{4}$ in. and $\frac{1}{2}$ in. It is easier to adjust than the English form.

The Quirk Router, f. 10, p. 17, is a tool for sinking narrow grooves in curved surfaces, chiefly in connection with mouldings; it has three knives or cutters of different thickness, as shown enlarged in f. 11; these are adjustable in both directions, as is shown in the sketch.

Side Rebates, f. 11, p. 12, are not used, as their name would suggest, for planing the sides of rebates, but for enlarging grooves. They are made in pairs to work right and left hand. The American pattern in metal is shown in back and front views in f. 10, p. 14. These have reversible nose-pieces which enable them to be worked up to the end of a stopped groove.

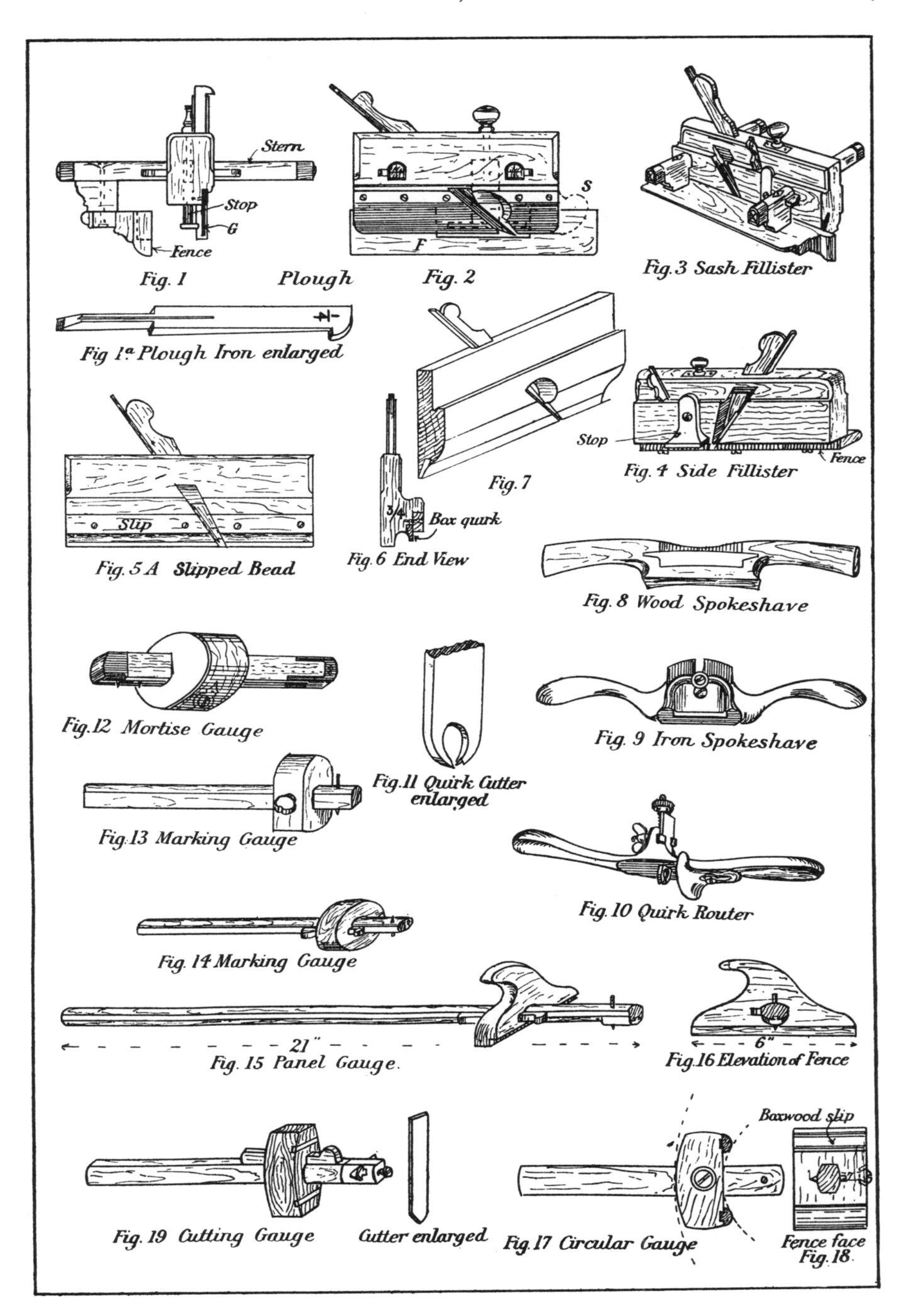
Stern
Stop
G
Fence
S
F
Fig. 1
Plough
Fig. 2
Fig. 3 Sash Fillister
Fig 1ª Plough Iron enlarged
Stop
Fence
Fig. 4 Side Fillister
Fig. 7
Slip
3/4
Box quirk
Fig. 5 A Slipped Bead
Fig. 6 End View
Fig. 8 Wood Spokeshave
Fig. 12 Mortise Gauge
Fig. 9 Iron Spokeshave
Fig. 11 Quirk Cutter enlarged
Fig. 13 Marking Gauge
Fig. 10 Quirk Router
Fig. 14 Marking Gauge
21"
Fig. 15 Panel Gauge.
6"
Fig. 16 Elevation of Fence
Boxwood slip
Fig. 19 Cutting Gauge
Cutter enlarged
Fig. 17 Circular Gauge
Fence face
Fig. 18.

Block or Thumb Planes.—These are small planes of wood or metal, chiefly used for cleaning off small surfaces where the regular smoothing plane would be too cumbrous. It has not been thought necessary to illustrate the wood varieties, which are miniature Smooth and Rebate planes, but a circular variety of the latter is shown on p. 360.

An American Metal Block Plane is shown in f. 4, p. 14. This has an adjustable mouth which enables it to be used either for hard or soft wood, and a screw adjustment for the cutter, which is fixed by a screw lever. These planes are about 6 in. in length, with 1¾-in. cutters.

The Metal Thumb Plane, English type, is shown in f. 9; it is 5 in. long, and 1 in., 1⅛ in., and 1¼ in. wide; it has a long cutter, which answers for a handle, and is secured with a gun-metal screw lever. It is, I believe, a speciality of Mr George Buck, whose address will be found in the advertisement pages of this book.

The Mitre Block Plane, f. 14, p. 12, is a short wood plane, with a relatively wide iron; its chief use is planing mitres in the screw mitre shoot, illustrated on p. 42. It is, I believe, procurable from Messrs Nurse & Co., tool makers, from whom alone it can be obtained.

The Chariot Plane, f. 6, p. 14, is a small metal smoothing plane for hard wood, 3¼ in. long, made in width from 1⅛ in. to 1½ in.

MOULDING PLANES.—Beads, f. 5 & 6, p. 17, are planes for producing a half-round moulding on the salient edge of any piece of wood used chiefly in framed openings to break the joint. They are made in sets of ten from ⅛ in. to 1 in. The skew-mouthed work best, and it is an advantage to have the smaller sizes "slipped"—that is, with the fence piece screwed on—so that it may be removed to permit the plane sticking a bead on the edge of a moulding or other sunk surface. A *Bead* always has its quirk or sinking on the inner side. A *Double-quirk Bead* has a sinking on each edge. The *Cock-bead* has no quirks, and stands above the surrounding surfaces. The planes used to stick these varieties are similar to the above.

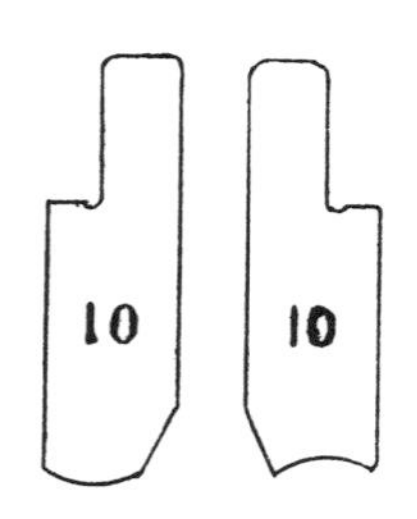

Section of Pair of Hollow and Round Planes.

Hollows and **Rounds** (see figure) are planes for producing various convex and concave surfaces in mouldings. They are made in sets of nine and eighteen pairs of opposite but similar curvature, both square and skew-mouthed. The latter work the cleaner, but are more liable to choke than the former.

The Snipe Bill, f. 7, p. 17, and **Side Snipe** may be termed complementary *Hollows* and *Rounds*. They are used to carry the curved surface below some projecting member which the ordinary plane cannot reach. The snipe bill cuts on its curved side, the side snipe on its straight side; the one continuing the moulding, the other the quirk.

Sash Planes consist of **Ovolo—Common**, f. 1, p. 19; Do. **Gothic**, f. 2; Do. **Grecian**, f. 3; **Ogee**, f. 4; **Astragal** and **Hollow**, f. 5. These are all made to suit 1½-in., 1¾-in., and 2-in. stuff. Their use is practically confined to working mouldings on window sashes. Larger sizes of the ogee plane, working mouldings 2 in. and 2½ in. wide, are called **Shop Front** and **Lamb's Tongue Planes**.

Nosing and **Scotia Planes** are used for working half-round and cavetto mouldings respectively. When both curves are united in the same stock it is termed a STAIR TREAD MOULDER.

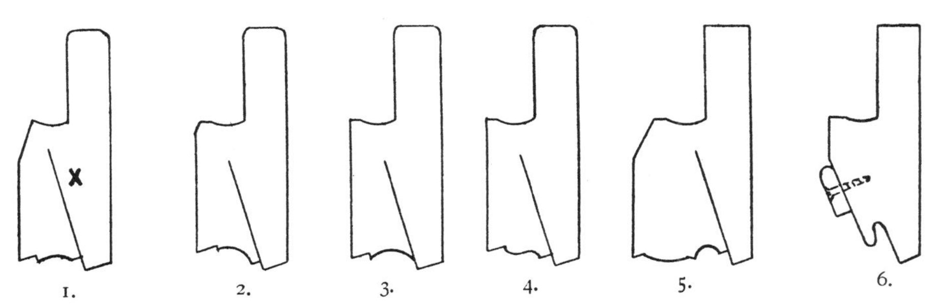

1. Section of Common Ovolo Sash Plane. 2. Section of Gothic Ovolo Sash Plane. 3. Grecian Ovolo Sash Plane. 4. Ogee or Lamb's Tongue Plane. 5. Astragal and Hollow Plane. 6. Hook Joint Plane.

The Hook Joint Plane, f. 6, is a tool used for forming the joint of that name upon the edges of casement sashes. The depth is adjustable by means of a sliding slot screwed stop, but the distance the plane works " on " cannot be varied. Smaller varieties are used in air-tight case work.

The Chamfer Plane is used for producing a regular chamfer upon the salient angle of a board, &c. There are several varieties of these planes. One of the best is shown in the sketch, f. 9, p. 12, and longitudinal section, f. 10; this has an adjustable stop, sliding in the mouth of the plane, which regulates the width of the resulting chamfer, which, however, is limited to an angle of 45 deg. with the sides. It is procurable from Messrs Nurse & Co., tool makers. The author has utilised the common rebate plane for the purpose of chamfering, by gluing slips of hardwood to the sole of the plane, as shown in f. 7. These can be arranged to produce any width and any angle chamfer.

7.

The Scraper Plane, f. 7, p. 14, is a tool very useful for giving a high finish to hard woods, an ordinary steel scraper taking the place of the plane iron, which can be adjusted to any degree of fineness, by means of the set screw shown in the sketch. The pitch of the scraper can also be altered to suit the texture of the wood scraped. This tool may also be used as a *toothing* plane, by substituting a toothing iron for the scraper.

CHISELS.

The Paring Chisel, f. 1, p. 20, is the typical tool of the chisel class. It is used for shaping and preparing the relatively long plane surfaces, more particularly in the direction of the grain of the wood, as it is invariably manipulated by steady and sustained pressure; as distinguished from the intermittent force used with other chisels, its handle is shaped to enable the hand to exercise

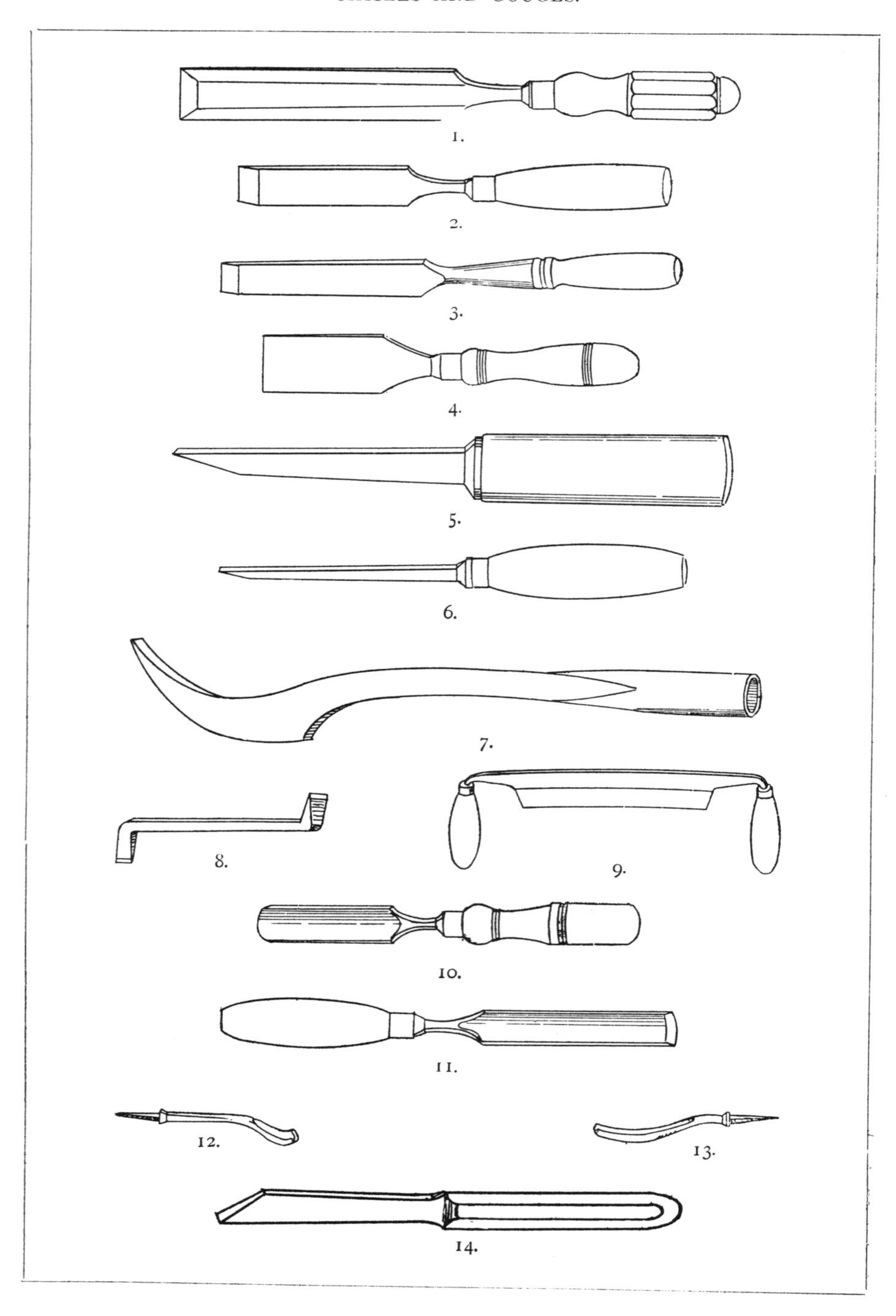
1.
2.
3.
4.
5.
6.
7.
8.
9.
10.
11.
12.
13.
14.

great control over its movements. The better forms have bevelled edges as shown, which reduces the friction when propelling the tool in a groove or trench. They are made from $\frac{1}{4}$ in. to 2 in. in width, advancing by $\frac{1}{8}$ in. to 1 in., thereafter by $\frac{1}{4}$ in.; length from 9 in. to 21 in. Sizes below $\frac{5}{8}$ in., only bevelled to order.

The Firmer Chisel, f. 2, p. 20, is, as its name implies, a chisel of firmer or stiffer substance than the paring chisel. It is a general utility tool, being used indifferently for short paring work, or for light mortising, and it is handled in a manner to be suitable either for steady pressure, or for the percussions of the mallet. It may be had with several forms of handles, but the round-swell shape shown is the best. These chisels are made in the same widths as the paring, length from 4 in. to 8 in. Fig. 3 is a socket firmer, for heavy work; the handle fits inside the steel socket instead of being secured by an iron tang driven into the handle as in other forms.

The Pocket Chisel, f. 4, is a wide, short, and very thin chisel made entirely of steel, used for cutting the ends of pocket pieces in pulley stiles of common sash frames. They advance in width by $\frac{1}{4}$ in. from $1\frac{1}{2}$ in. to $2\frac{1}{2}$ in.

The Mortise Chisel, f. 5, is made abnormally thick to prevent bending when levering the core out of mortises. As its name implies, it is solely used for producing mortises. Widths from $\frac{1}{8}$ in. to $\frac{5}{8}$ in., advancing by $\frac{1}{8}$ in.

The Sash Chisel, f. 6, is a lighter form of the same with a nearly parallel stem, used chiefly for the narrow mortises required in sash bars.

The Swan's Neck, or lock-mortise chisel, f. 7, is used for cutting across the grain at the bottom of a deep mortise. It is only made in three sizes, viz., $\frac{3}{8}$ in., $\frac{7}{16}$ in., and $\frac{5}{8}$ in., and is but little used now, such mortise being now made by means of twist bits. See p. 25.

The Drawer Lock Chisel, f. 8, is made entirely of steel. Its purpose is to cut mortises in confined spaces such as drawer openings. It has a cutting edge at each end, the edges lying in transverse directions.

The Draw Knife, f. 9, is a double-handed paring chisel of considerable width. They vary in width (or length) from 4 in. to 12 in. This tool was formerly much used for the rapid reduction of stuff to gauge, a labour that is now generally performed by sawing or planing machines.

Gouges.—These may be described as curved-faced chisels. They are used and made in similar sizes to firmer and paring chisels, the resulting face of the cut being circular instead of flat. Fig. 10 shows a firmer gouge, and f. 11 a paring or scribing gouge, so-called because it is used to scribe the moulded shoulder of framing, &c. When a gouge of either type, is curved in its length with a short bend near the cutting end, as f. 12, it is termed a Bent Gouge, and if curved throughout its length, as f. 13, a Curved Gouge. These are chiefly used by joiners in cutting the moulded surfaces of shaped work, such as handrail wreaths and the like.

The Plugging Chisel, f. 14, is a bar of low-tempered steel forged into an octagon-shaped handle at one end, and drawn out to a flat obtuse-angled point at the other. This part should be parallel on the front edge and not more than $\frac{5}{16}$ in. thick, and be slightly thinner at the back. It should also be wider at the point than at the hilt. It is used for cutting mortises in the joints of brickwork to receive wood plugs.

MARKING OR DIMENSIONING TOOLS.

GAUGES are tools for producing lines upon the surface of wood, parallel with the edge they are used upon. There are various forms and sizes, according to the kind of work they are required for.

The Mortise Gauge, f. 12, p. 17, has a stem about 6 in. long, having two steel points, one fixed near the end, and the other attached to a brass slider, adjustable by means of a screw in the end of the stem. This enables two lines to be marked at any distance apart within the range of the slider. The stock or fence slides stiffly upon the stem, and is fixed by a flush set screw, the fence determining the distance of the lines from the edge of the material. The tool is chiefly used for setting out mortises and tenons. Some gauges have *two* movable teeth, and one fixed. These are used for gauging meeting rails of sashes.

Marking or **Single Tooth Gauges** are shown in f. 13 & 14, and a **Panel Gauge** in f. 15 & 16. The latter is used for gauging panels and other wide stuff, the fence being much larger than in the common gauge. It is also

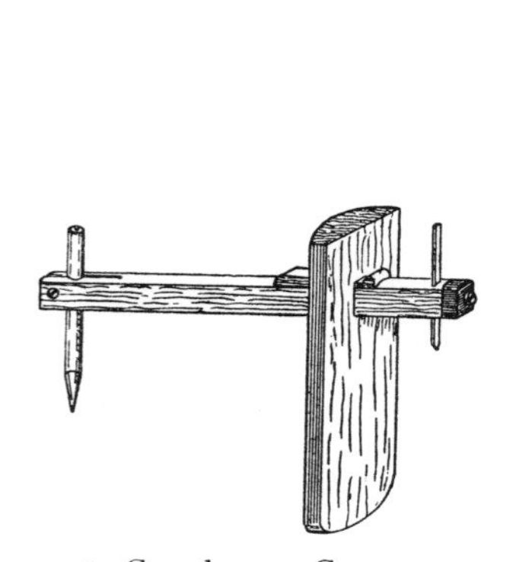

1. Grasshopper Gauge.

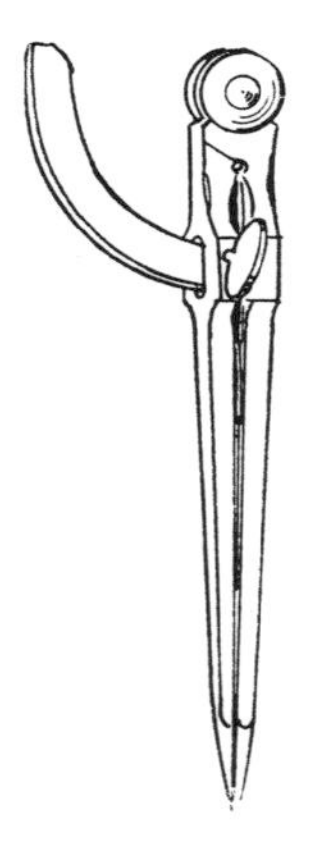

2. Compasses.

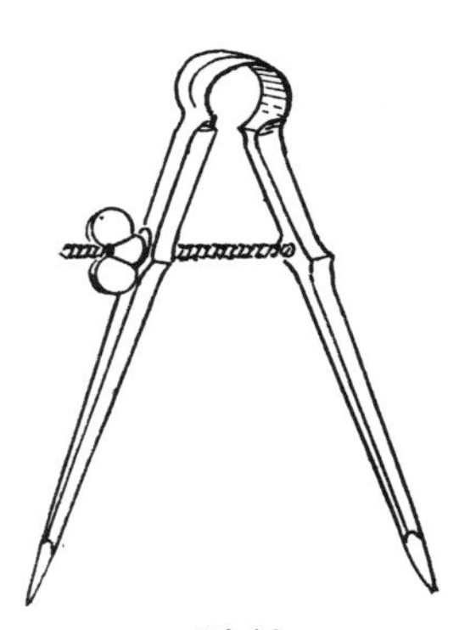

3. Dividers.

rebated on the lower edge to prevent it slipping under the pressure necessary to keep it down upon its work. The gauge point should have a hardwood slip (box for preference) under it, as shown in the drawing, to prevent the stem rubbing upon the surface of the work.

The Circular Gauge, f. 17 & 18, differs only from the straight marking gauge in having the fence shaped convex on one side, and concave on the other, to enable parallel lines to be drawn to curved surfaces, as indicated by the dotted lines.

The Cutting Gauge, f. 19, has a small adjustable knife in the place of the steel marking point of the other gauges. It is used for cutting off parallel strips of veneers and other thin stuff. The cutter, shown enlarged at A, is sharpened to a lancet point; the basil should be towards the stock.

The Grasshopper or **Handrail Gauge**, f. 1 above, is chiefly used for gauging lines upon work of double curvature, such as handrail wreaths. It has a long fence to enable it to rise over the crown of the curve, and the stem is bored and slotted to receive at one end a pencil, and at the other a steel point. These are adjustable in height or distance from the stem, and the stem is adjustable on the fence, so that markings can be made upon any shaped surface.

The Thumb or **Pencil Gauge** is illustrated and described on p. 75.

The Compasses, f. 2, p. 22, are used chiefly for gauging parallel lines to irregular surfaces, such as the scribing of skirtings to floors, &c. They are also used for the same purpose as the drawing instruments of like name, describing circles, and setting off distances with more accuracy than can be obtained by simple measurements with the rule. There are two forms—the "common," and the "wing," which has a quadrant arm and set screw. The latter only is illustrated, the first is a most untrustworthy tool. Sizes from 5 in. to 8 in. long.

The Spring Dividers, f. 3, are a lighter tool of the same description as the last, but more suitable for the bench. They are used for scribing the shoulders of mouldings, taking accurate dimensions, &c. The curved head is a highly tempered spring, tending to keep the legs open. They are closed by twisting the wing nut on the screw bar near the middle.

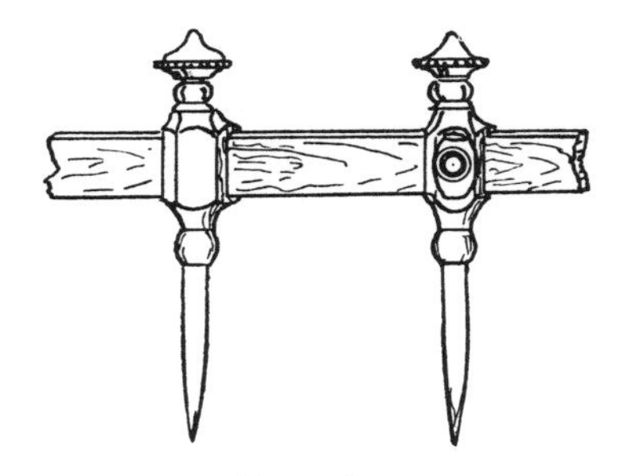

1. Beam Compasses.

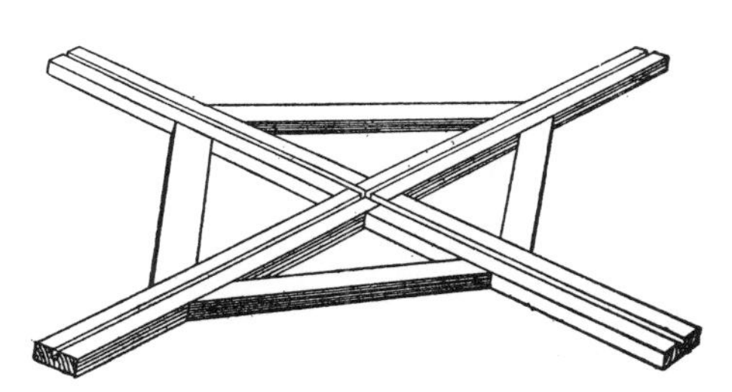

2. Trammel Frame.

The Beam Compasses, f. 1, are a pair of steel pointers fixed into brass bushes or boxes, which have slots in them through which a slender bar of hardwood is passed. The boxes have milled-edged set screws on the head, by means of which they can be secured at any part of the bar. They are used thus, as dividers, for spacing off distances too long for the spring dividers. One of the boxes has also a pencil socket at the side, and by means of this, circles can be described with the other pointer as a centre, of any radius within the length of the rod. Sometimes three legs are used on the bar in connection with the Trammel, f. 2, by means of which elliptic curves can be drawn.

The Trammel, f. 2, is a light-braced frame with two arms at right angles to each other, containing shallow grooves in which two of the pointers move, whilst the third describes the curve. The points are usually driven into a soft slip of wood, easily fitting the grooves, but even with this assistance it is difficult to draw the line evenly, and it may be better accomplished by aid of the square shown at p. 80.

The Callipers, f. 2 below, are used for ascertaining the dimensions of curved solids that cannot easily be measured with the rule.

The Marking Knife and Point, f. 1, is used for setting out dimensions when greater accuracy is desirable than can be obtained with the lead pencil, also for "cutting in," or "striking" shoulders. The severance of the fibres of the wood by the knife gives a cleaner shoulder than can be obtained from the saw alone. Care should be taken when sharpening to treat it like a chisel with a basil on one side only, and that the right hand side, when using.

1. Marking Knife.

BORING TOOLS.

The Bradawl, f. 3, is the simplest of these, consisting of a steel bar tanged at one end to fix in the handle, and ground to a double-wedge edge at the other. It should taper slightly from the cutting edge to the haft; if the taper is reversed it will split the wood in boring. It can be driven either by the hammer, or by hand pressure accompanied by a twisting motion.

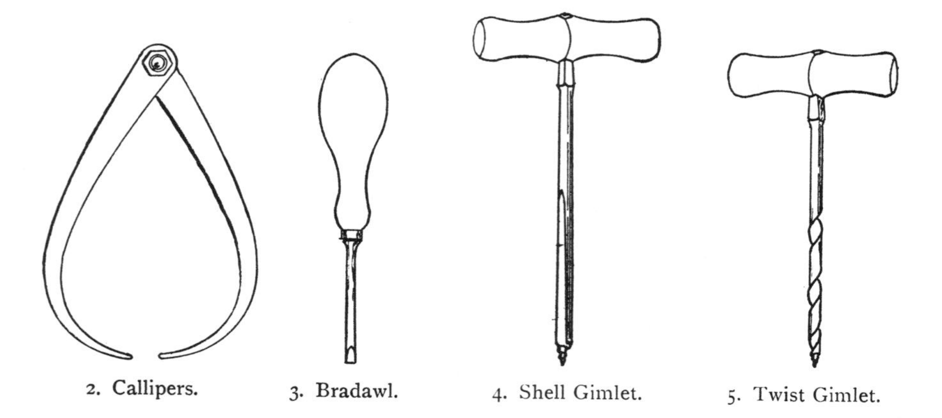

2. Callipers. 3. Bradawl. 4. Shell Gimlet. 5. Twist Gimlet.

GIMLETS are small boring tools driven by a revolving hand pressure. The **Shell Gimlet**, f. 4, has a stem ground out in part of its length to form two cutting edges at the sides, and the point is threaded to assist the tool into the wood; this is the most serviceable form, for notwithstanding that the **Twist Gimlet**, f. 5, bores quicker, it more readily splits the wood, as it is almost invariably tapered from the handle to the point, whilst the shell is tapered in the reverse direction.

Brace and Bits.—The Brace, f. 1 & 2, p. 26, is merely the handle or stock for various "bits" or cutters, &c., which are placed in it, and revolved continuously, and with high speed; at the same time great pressure can be brought on them by resting the head of the brace against the breast. Fig. 1 shows the English type, a wood stock strengthened with brass plates. Fig. 2 shows the American type, with steel stock, wood head-piece, revolving on ball bearings, as shown in the section, and having a ratchet arrangement to change the continuous to an intermittent motion—useful at times when boring in confined positions. Figs. 1 & 2 are twist gimlet bits, f. 1 being most suitable for boring in the direction of the grain, and f. 2 for boring *across* it. These are made in sets from $\frac{1}{4}$ in. to $1\frac{1}{2}$ in. diameter, and of various lengths, from the 4-in. dowel bit to the 21-in. sash bit.

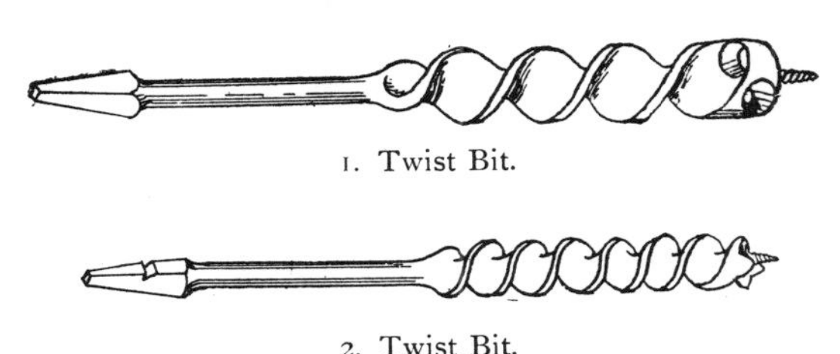

1. Twist Bit.

2. Twist Bit.

The Centre Bit, f. 3, is used for boring across grain; sizes from $\frac{1}{8}$ in. to 2 in.

The Nose Bit, f. 4, for boring across or along the grain.

The Spoon or Shell Bit, f. 5, for boring across grain only; is made for producing holes from $\frac{1}{8}$ in. diameter up to $\frac{1}{2}$ in.

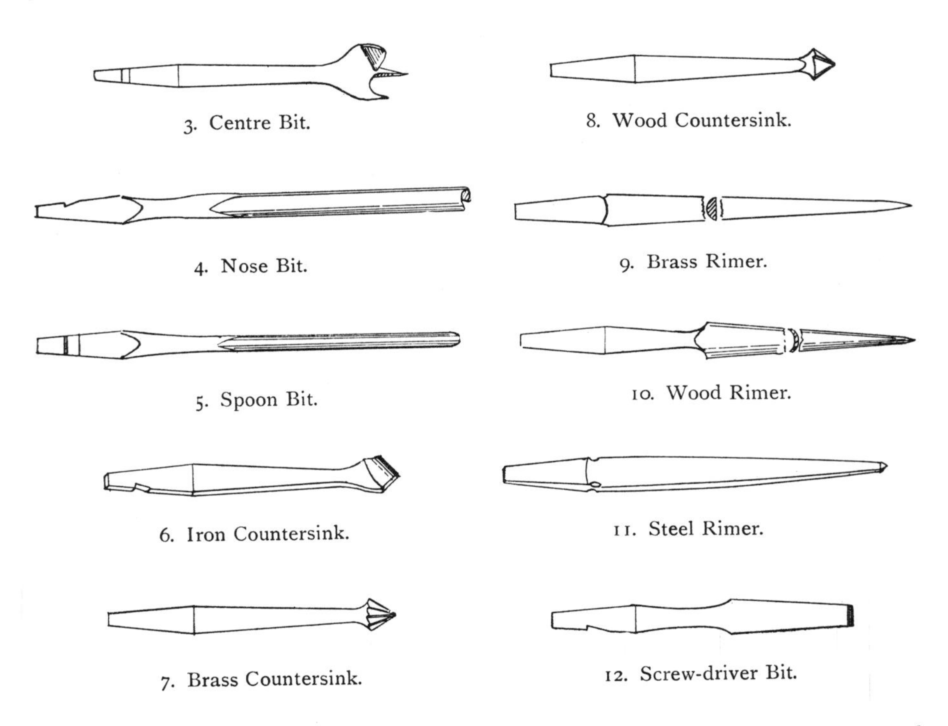

3. Centre Bit. 8. Wood Countersink.

4. Nose Bit. 9. Brass Rimer.

5. Spoon Bit. 10. Wood Rimer.

6. Iron Countersink. 11. Steel Rimer.

7. Brass Countersink. 12. Screw-driver Bit.

The Countersinks, f. 6, 7, & 8, are for dishing holes in iron, brass, and wood respectively, and are made from $\frac{3}{8}$ in. to $\frac{7}{8}$ in. diameter.

The Rimers, f. 9, 10, & 11, are for enlarging holes in similar substances.

The Screw-driver Bit, f. 12, is a common form, but Messrs Melhuish supply hand-forged bar steel screw-driver bits, 5 in. to 8 in. long, for use in the brace. These are much stronger than the common bits usually supplied with the stock.

There are sundry other appliances used in connection with the brace, such as drills for boring holes in metal, collars, or gauges for the bits (see p. 25), lengthening bars, and expansion bits, but these are not generally required.

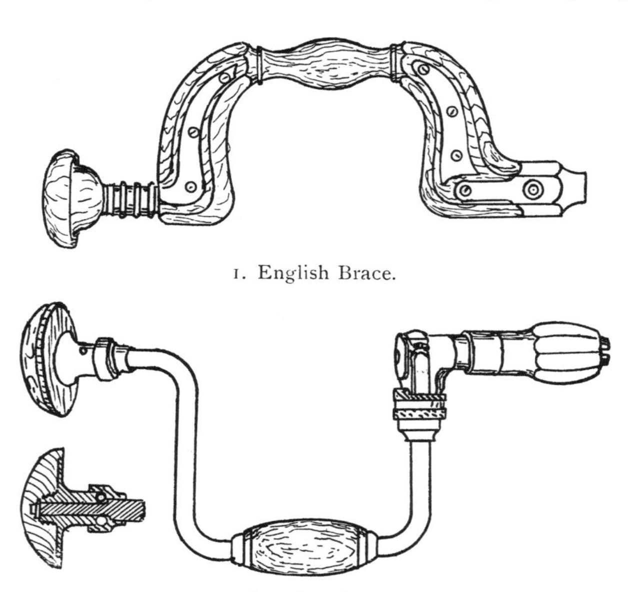

1. English Brace.

2. American Brace.

The Dowel Plate, f. 3, is a tempered steel plate with several holes drilled square through it the exact size of corresponding twist bits, its object being to produce dowels of true cylindric shape. The plate is preferably mounted in a block of hardwood about 3 in. thick, with holes bored through it slightly larger than those in the plate. These guide the pin through upright. A slight burr or lip should be made on one side of each hole in the plate to cut an air groove in the dowels, which would otherwise burst their sockets when driven. The pins should be cleft, not sawn, and roughly shaped with the chisel before driving through the plate.

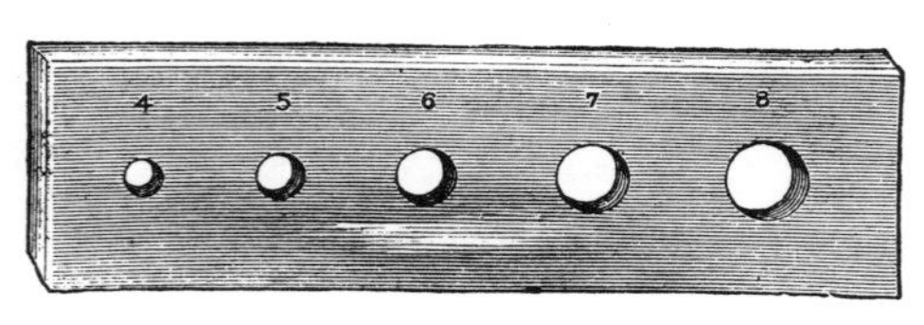

3. Dowel Plate.

MISCELLANEOUS AND TESTING TOOLS.

Screw-drivers, f. 1, 2, & 3, p. 27, are of various shapes and sizes, according to the work they are required to do. Fig. 1 shows a stumpy blade with a wide handle, the whole measuring about 5 in.; this is used for turning plane iron screws. Fig. 2 is an oval-handled spindle shank screw-driver for bench and other light work. These are called Cabinet screw-drivers by the tool dealers, and are made with blades from 3 in. to 12 in. long.

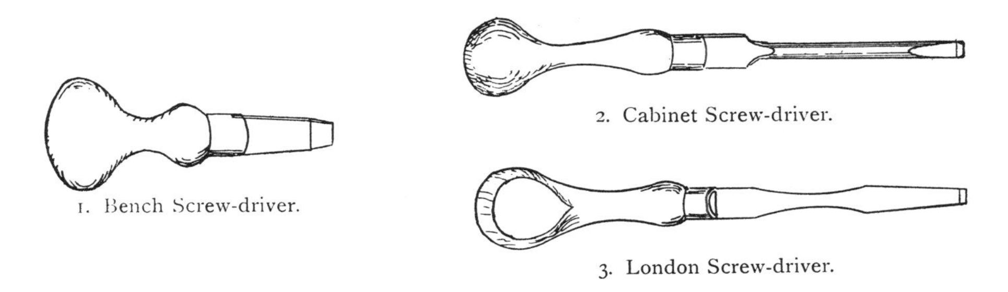

1. Bench Screw-driver.

2. Cabinet Screw-driver.

3. London Screw-driver.

Fig. 3, the flat-handle or "London" screw-driver, is a tool suited for heavier work. The larger sizes have a hole drilled through the blade for the insertion of a steel bar to increase the leverage. They are made in length of blade from 4 in. to 22 in. A screw-driver should have a long drawn-out "point." A short bevel will cause the tool to slip out of the cut of a screw.

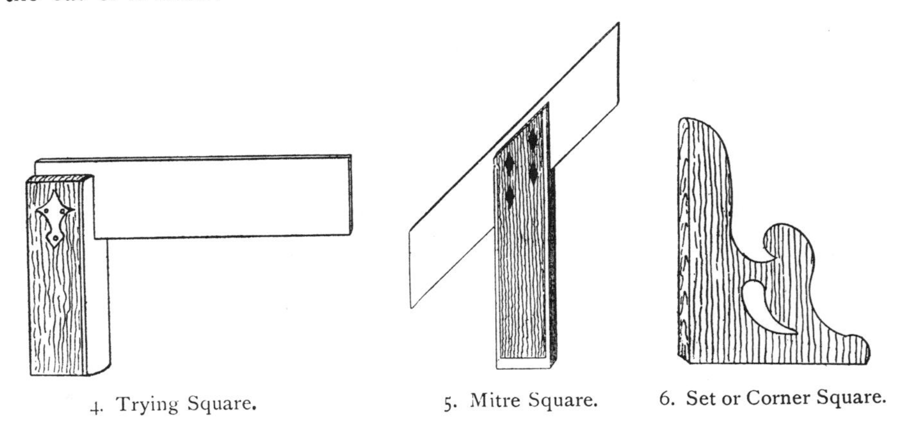

4. Trying Square.

5. Mitre Square.

6. Set or Corner Square.

The Joiner's Square, f. 4, is a tool used in the production of right angles, either in the drawing of lines or in the planing up of stuff; in the latter operation the smaller size squares are used, viz., 3, 4, or 6 in. These are termed "Trying Squares"; the larger ones with 9, 12, or 18 in. blades are simply "Squares." As the ultimate accuracy of the "setting out" of framing depends greatly on the truthfulness of the squared edges, care should be taken to select a square that is true, *i.e.*, whose blade is exactly at 90 deg. with its stock. The method of ascertaining this is to rest the stock against the straight edge of a piece of stuff, and make a fine mark at each end of the edge of the blade; then reverse the stock, and if the edge of the blade again coincides with the line, the square is true, if not, half the difference will be the amount of the error. *Note.*—Both edges should be tried.

The Mitre Square, f. 5, p. 27, has its blade set at an angle of 45 deg. with the edge of the stock. It is used similarly to the try square, but for producing lines and edges at angles of 45 deg.

The Set Square, f. 6, has two of its edges at right angles with each other, and is used for ascertaining the "squareness" of internal angles, and in fitting work together. It should be made of a piece of dry mahogany or beech about ¼ in. thick, 14 in. high, and 8 in. wide, the grain running lengthwise, and the bottom edge tongued with an ebony slip.

The ordinary **Steel Square**, consisting of an **L**-shaped piece of sheet steel, with the edges of its blades divided into inches and parts, is a carpenter's tool, but the **Universal Steel Square** shown in f. 1-4, p. 80, has various attachments and devices that make it equally useful to joiners in stair work and fixing. By its aid, bevels and lengths can be readily obtained for hip and other rafters in lanterns, roofs, and spires. The correct shoulders for diminished stiles and rails, pitch boards for stairs, angles for all the polygons and circles, and ellipses can be described with it; it can also be used to plumb and level, gauge single and double lines, gives at a glance the superficial area and cubic contents of any board or balk timber.

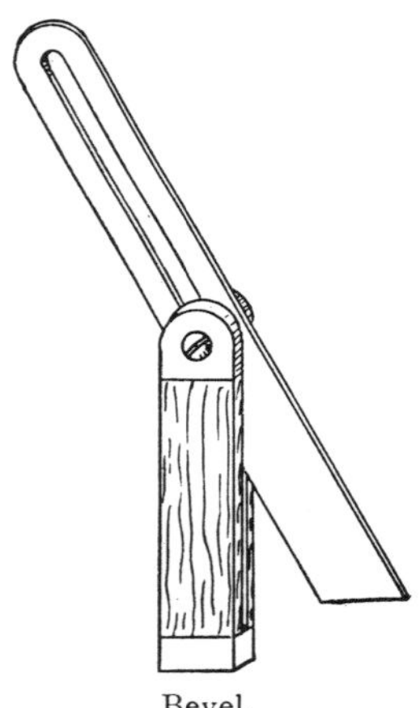

Bevel.

The Sliding Bevel (see fig.) is a tool which takes the place of the Square when any other angle than a right angle is to be obtained, the blade being adjustable at pleasure. They are made in four sizes from 6 in. to 12 in. They should be tested occasionally to see that the edges of the stock are straight throughout, and also parallel.

OILSTONES should be selected with great care, because if they are unsatisfactory no good results can be obtained even from the best tools. They should be not less than 8 in. long, and from 1½ in. to 1¾ in. wide, not more, as wider ones wear hollow quickly, and require frequent rubbing down. No especial variety can be termed the "best," because good and bad samples will be met with in all. Perhaps the safest for an inexperienced buyer to purchase is an **Arkansas**.

The Washita is also a good stone, and is cheaper. **Turkey** is invaluable when a good one is found, but the percentage of satisfactory ones is low.

Charnley Forest is a slate, and is very slow cutting, but puts a very keen edge on. It is useful as a finishing stone.

Canada is a manufactured "stone." It cuts quickly, producing a rather coarse edge, and wears away rapidly. The stone should be inserted in a wooden case, with strips of glass-paper glued to the bottom to prevent the case slipping during use.

Neat's foot or "sweet" oil is the best to use, mixed with about one-tenth the volume of paraffin oil, which keeps the stone clean and prevents the oil drying and clogging the pores. To keep the stone flat it should be rubbed down occasionally on a sheet of emery cloth glued to a piece of flat board; or a more permanent "rubber" may be formed with a piece of plate glass, using powder emery and water as an abrading material. The "cut" of a stone may be roughly tested by rubbing the edge of the thumb-nail on it. If it grips the nail and rubs it down quickly, the stone may be assumed to be a fast cutter. All four sides should be tested, as one edge will be found that will cut better than either of the others.

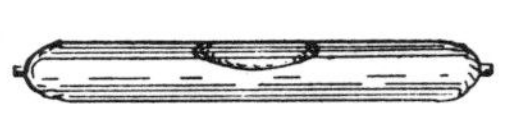
1. Spirit Tube.

The Spirit Level, f. 1 & 2, is an instrument in constant use in fixing operations. It consists of a small hermetically sealed glass tube nearly filled with spirits of wine, firmly cemented into a hardwood stock. The axial line of the tube is exactly parallel with the *under* surface of the stock. The tube is sunk flush and protected by a metal plate having a longitudinal opening crossed by a thin bar in the centre. The tube is very slightly curved in length, the convex side being placed uppermost to assist the motion of the bubble of air confined within. When it is desired to use the instrument to test a piece of work, the bottom of the stock

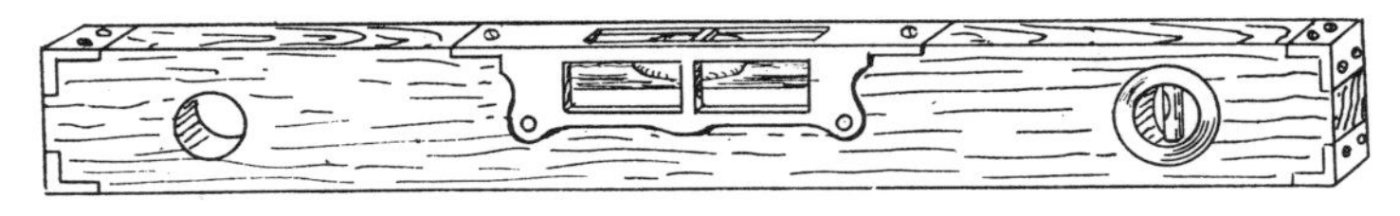
2. Spirit Level.

is laid either on the upper surface of the work or upon a parallel straight-edge resting upon it, and when it is dead level the air bubble will lie centrally under the bridge; if out of level the bubble will run to the highest end. The tube being very fragile, it is advisable to have a cardboard case into which the stock can be slipped when not in use.

The Cork Rubber is a block of cork about 4 in. by 3 in. by 2 in., rounded at the back and made quite flat on the face. It is used for applying sheets of glass-paper to work for the purpose of removing plane marks. *Note.*—The paper should not be doubled, but be cut to size of block with a margin just sufficient to enable it to be gripped tightly round the rubber, otherwise it will bulge and produce hollow rather than plane surfaces.

The Wire Rubber is formed of a short piece of coarse wire "carding" cloth fixed to a block of wood of convenient size to handle, and used for scrubbing dirty or gritty wood before planing it.

Punches are short lengths of steel rod tapered off to a blunt point. They are used for driving the heads of brads or nails below the surface, so that the hole made by the nail may be filled in with putty or composition. The point of the punch used should always be smaller than the head of the nail, so that the hole may not be unnecessarily enlarged, and squared points are better than round ones.

The Handrail Punch, f. 1, is a special form of punch used for turning the slotted nut of a handrail bolt (see f. 3, p. 72). One end is curved sharply and the end ground off to a blunt chisel edge.

The Centre Punch has its end turned to an obtuse conical point, and is used for making a slight depression in metal from which to start a drill.

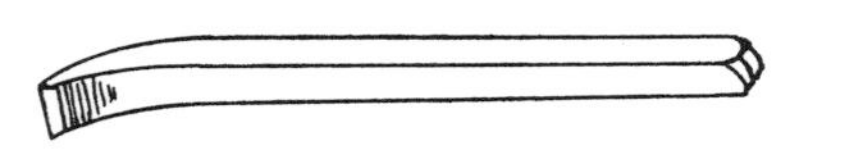

1. Handrail Punch.

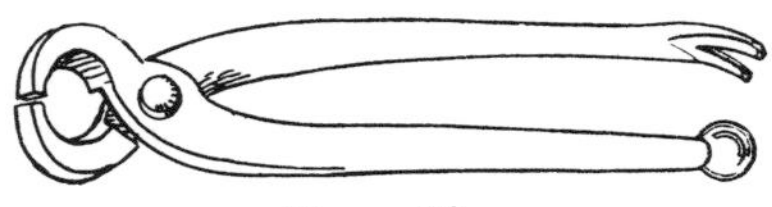

2. Tower Pincers.

The Pincers, f. 2, are known as "Tower"; another variety with more angular jaws is termed "Lancashire."

The Mallet is a percussion tool used for driving wood chisels, also light pieces of framing together. It consists of a rectangular head of beechwood from 5 in. to 8 in. long and from 3 in. to 5 in. thick, with a rectangular mortise slightly tapering in depth through which the handle passes. The handle should be of ash or other pliable wood, and comparatively slender. The shop-made article is usually too stiff in the handle, which does not "give" under the blows, and consequently transmits unpleasant shocks to the arm of the user; it should be made as indicated by the dotted lines in f. 3.

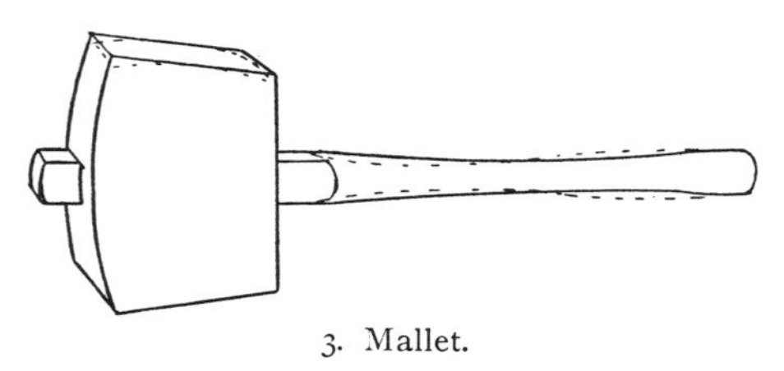

3. Mallet.

Hammers.—Fig. 4 below is the usual form of bench hammer, known alternatively as "London" or "Exeter" patterns. These have mild cast-steel heads with tempered-steel faces. They are made in twelve sizes, Nos. 4 to 6 being the most suitable for the bench. The "Warrington" Hammer, f. 1, p. 31, has a solid cast-steel head. The American or Claw Hammer, f. 2, is most useful for fixing purposes. The Framing Hammer, f. 3, is a short-handled hammer with a heavy wide-faced head, and is used in place of a mallet for driving up heavy framing.

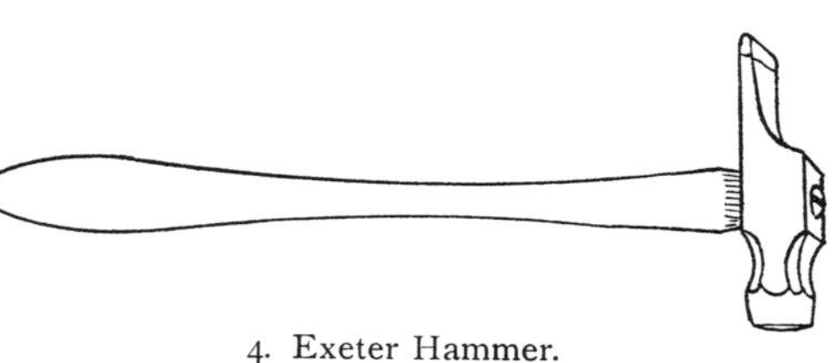

4. Exeter Hammer.

Winding Sticks, f. 4 & 5, are made of two thoroughly dry and straight-grained pieces of hardwood, such as Spanish mahogany, rosewood, or black walnut, inlaid with sighting pieces of bone or white holly. They are generally made by the joiner himself, as they are not stocked at the tool shops. Full particulars of their use, &c., are given on p. 51.

Care of Tools.—Tools should not be bundled carelessly into a basket. Such method or want of method will soon destroy the appearance and deteriorate the work of the best of tools. Chisels, gouges, and other similar cutting tools should be kept sharp, and placed in racks with edges downwards or in trays in the bench drawer. Planes when out of use for more than a few hours should have the wedges released, as the continued tension is injurious to the plane. When temporarily out of use, bench planes should have their fronts resting on a thin slip of wood screwed on the bench top to keep the cutters free from damage. On no account should planes be laid upon their sides on the bench, as apart from the danger of running the hands against the cutters, if the soles are exposed for any length of time to the action of the sun, they will cast, and if at all unseasoned, split. Bench planes, when new, should have the aperture of the mouth stopped with putty, then the mouth filled up with clear raw linseed oil, which in about twenty-four hours will have soaked into the wood, filling up the pores, and thereafter preventing the absorption of moisture. Planes should not be French polished, as this surface scratches easily, and also makes the plane difficult to hold. An occasional rub with an oily rag dipped in finely powdered bath-brick will both clean the tool and produce a mellow polish that will not scratch. Beads, moulding, planes, &c., should not be struck on their heels with the hammer, either to

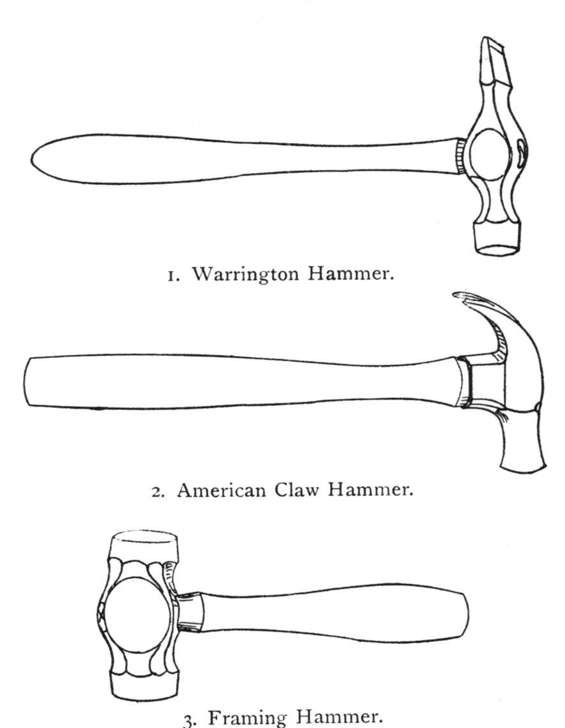

1. Warrington Hammer.

2. American Claw Hammer.

3. Framing Hammer.

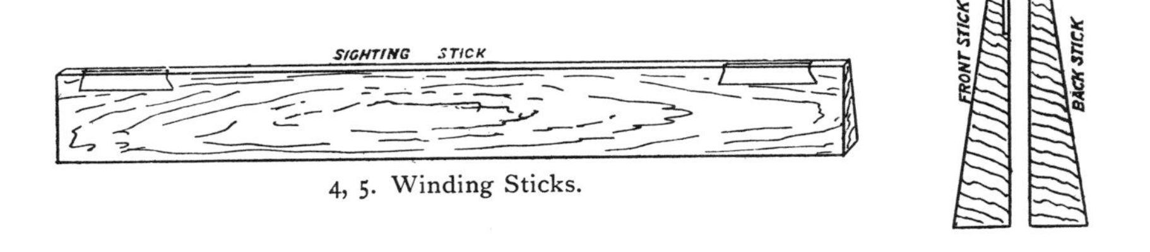

4, 5. Winding Sticks.

adjust or release the cutter. The former purpose may be accomplished, if the wedge is not tightly driven, by gripping the plane by the wedge and tang of the cutter, and striking the heel of the plane squarely on the bench top; but if the wedge is at all tight, the plane should be held at the end of the bench with the left hand, the tang of the iron resting fairly on the bench top. Then strike the shoulder of the wedge with the hammer, when it will be at once released. Metal tools should be rubbed with "sweet" oil, sperm or other non-drying oil—not linseed, which should be used only on wood tools. For metal tools that are only used occasionally, vaseline is a good preservative, but the best preservative from rust is a highly polished surface. It is the minute scratches and grinding marks that afford a hold for the moisture to set up oxidation. *To Remove Rust* from tools, scrape as much as possible off with a dull knife or back edge of a chisel. Then rub with a rag dipped in paraffin oil and fine emery powder; clean off, and rub with sweet oil. But an ounce of prevention, is worth a pound of cure; do not neglect your tools then they will not get rusty. Emery powder will clean an oilstone that has become clogged with oil, and also improve its "bite."

A New Saw Set.—The accompanying illustration shows a simple and effective little apparatus for setting saws. It consists of a cast-iron rest or holder, having guides or fences for the saw teeth to rest against, and a channel below, in which an appropriate steel anvil or setting block is wedged as shown. Two special hammers with four different faces for various sizes of tooth are supplied with the sets. The chief merit of the apparatus being that several blows may be given each tooth, which is thus "set" gradually without danger of breaking-off. The tool is made by Messrs Spear & Jackson Ltd., Sheffield.

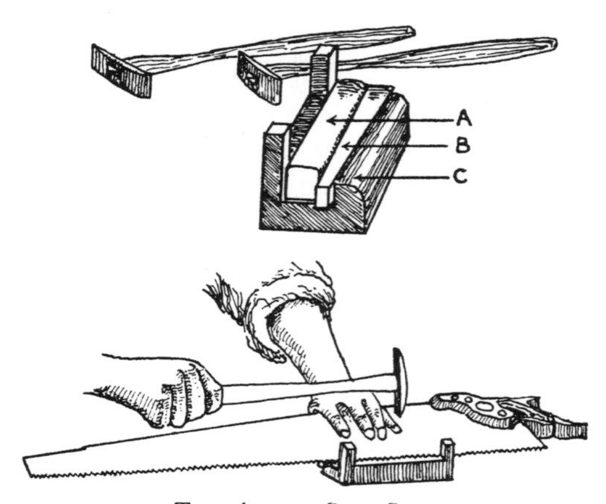

THE AETNA SAW SET.

CHAPTER III.

WORKSHOP APPLIANCES, &c.

Joiners' Benches—Merits of Single and Double compared—Standard Sizes and Construction—Gluing-up Benches—Grip Screws—Stops—Cramps, Patent, Wood, Sash, Scotch, and German—Wood and Iron Cleats—Panel, Shooting, and Sticking Boards—Mitre Blocks, Boxes, and Shoots—Construction of Screw Mitre Shoot—Thumb Clips and Cramps—Glue, Hints on Making—Glue-heating Appliances for Large and Small Workshops—Methods of Preparing Glue for various Woods—Waterproof Glue—To render Glue Joints invisible—The Workshop—Situation, Lighting, Ventilation—Loopholes and Delivery Wells—Stairs—Lighting and Ventilation—Bench Space—Storage—Effect of Machinery in Shop—Grindstones—Preserving Work.

IT is proposed in this chapter to describe sundry appliances and accessories to the modern workshop that are of general utility, reserving until the occasion arise the description of special apparatus. Of all appliances a bench is the most indispensable, and therefore it will be meet to commence with that.

THE JOINER'S BENCH is made in two forms, SINGLE and DOUBLE. The former is suitable when the workshop is lighted only at one side, the latter when lighted at both sides. Each form has something to be said in its favour, but as the great majority of benches are now made single, it would seem that this form best meets general requirements. Their respective properties may be summarised as follows:—

The Single Bench will accommodate one man only, but he can move all round his work without obstructing his neighbours. It is light to handle, economical to construct, and, most important in large shops, more of these benches can be placed on a given area of floor space, with a consequent reduction of the loss due to "idle benches," than when double benches are used, as less "walk" is required between the benches. When packed away, the single bench occupies less storage room, and is easy of transport. Lastly, it is undoubtedly favoured by the competent workman, who will always prefer to work single-handed.

Double Benches will accommodate more men on a given floor space than the single. They are useful where large pieces of framing have to be prepared, and they are perhaps more suitable for small than large workshops, but they are heavy and cumbersome to handle and expensive to make. *Details.*—The usual sizes are, SINGLE, 10 to 12 ft. long, 27 in. wide; DOUBLE, 12 to 14 ft. long, 36 in. wide, common height 28 in. A SINGLE bench is shown in f. 1, p. 35. The top is usually made of three 9-in. yellow deal boards, the front one $1\frac{1}{2}$ in. thick, the two back ones 1 in. thick, with the joints ploughed, tongued, and glued, and cleaned off flush on the top side. The legs of 3 by 4 in. deal are framed together, with mortise and tenon joints, glued and wedged. These legs are made flush with the sides, and set on the rake, the better to withstand the thrust of the planing. The sides are formed of two 9 by 1 in. boards fixed to the legs with screws, and having bevelled notched shoulders. 2 by 3 in. bearers, notched out to fit the under side of the top, connect the sides between the legs. The bottom cross-bearers should be about 9 in. above the floor, and it is advisable to cover these with a "bottom board," as shown, for the workman to keep his tools upon.

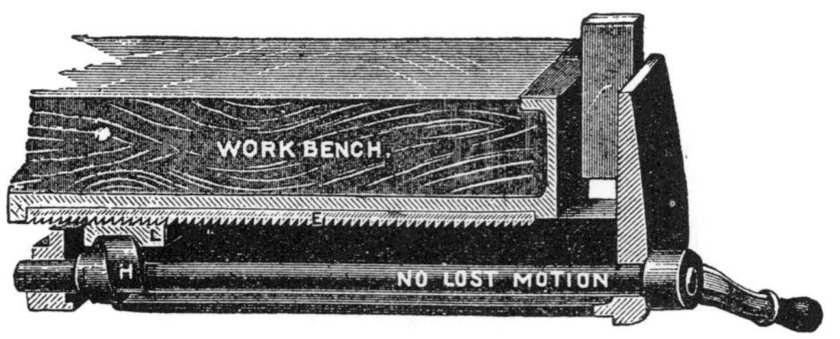

1. "Instantaneous" Bench Screw.

This provision also enables the dust and shavings to be swept from under the benches, which should be done every evening to lessen the risk of fire.

The Top should be fixed by screws in pelleted holes, and when first fixed, be allowed to overhang $\frac{1}{4}$ in. on each side to provide for subsequent shrinkage. If the old-fashioned wood screw and chock is used, the screw should not be less than $2\frac{1}{4}$ in. diameter, cut in dry beech, and be perfectly straight. The lug or slider should be at the same level as the screw, and about 12 in. behind it. It is a mistake to have them very far apart. The lug should be double tenoned to the chock and quite square with it in both directions, be made parallel and fitted easily into a case fixed to the front end of the bench, as shown in f. 2, next page, which is a section through the case. A $\frac{3}{8}$ by 2 in. mortise should be made in the top of the chock over the screw neck, and a hardwood key driven down until it enters the chase on the neck of the screw. This is to prevent the screw leaving the chock when it is unwound. All the working parts should be blackleaded. The modern "instantaneous grip" iron bench screw (see f. 1 above) is a great improvement on the old wood form, and will save its cost in a few months in economy of time. When fitting these to new benches, the upper edge of the inside cheek should be kept $\frac{1}{4}$ in. below the surface of the bench top.

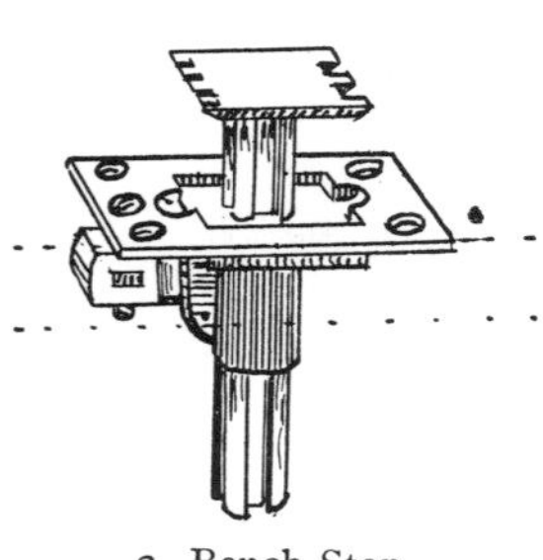

2. Bench Stop.

The Bench Stop.—Few of the various iron stops on the market meet all the requirements in a builder's workshop. Some are too light for the rough usage they will be subjected to, others have insufficient range of movement, and others get choked readily and will not act until cleared, but one recently introduced is free from many of the faults of its predecessors. It is made in malleable or wrought steel, and is on the instantaneous principle. The fixing screw

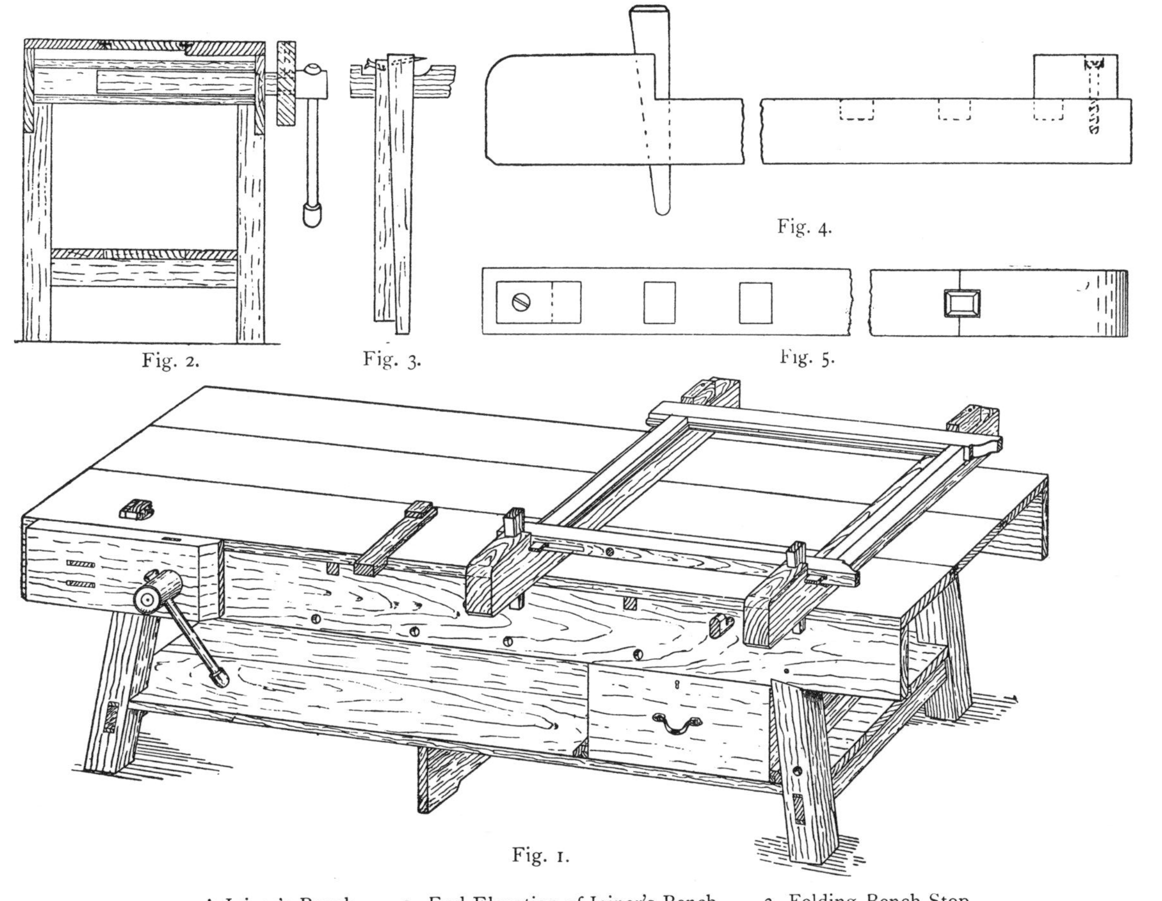

1. A Joiner's Bench. 2. End Elevation of Joiner's Bench. 3. Folding Bench Stop.
4, 5. Top and Side Views of a Wood Sash Cramp.

only requires a half-turn to release or fix the stop, when it can instantly be moved to any height by hand. It is reversible, the head-plate having two sets of teeth, fine and coarse. It sinks flush with the top when required, and the tray beneath prevents chips, &c., entering the socket. It is shown reversed in the sketch, f. 2, p. 34; the screw end should be towards the tail of the bench. This stop is procurable from several of the dealers whose addresses will be found in the advertisement pages. Wood stops, however, are much preferred by many workmen, and one of the best of these is shown in f. 3, p. 35. It consists of two pieces of hardwood 2 by $\frac{3}{4}$ in. and 12 in. long, prepared as a pair of folding wedges. At the upper end of the front one three wrought-iron nails

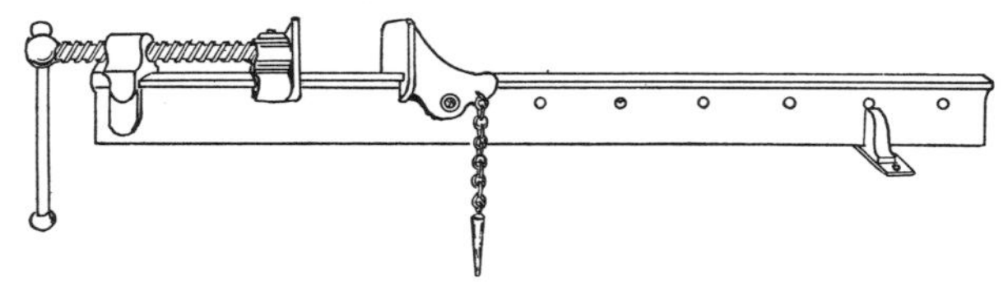

1. Bench Cramp.

are driven, and the projecting ends filed off to a chisel edge. A small screw is turned in the back piece to prevent it falling out when the front one is slackened. The front of the stop should slide up and down against the leg of the bench. All benches should be provided with a good sized lock-up drawer for the workman's tools, otherwise he can scarcely be expected to bring his more expensive tools to lie about unprotected. He will rather make shift with as few as possible, to his employer's detriment.

The Bench Hook or **Jack**, shown on the bench in f. 1, last page, is used as a rest or stop for the material when cutting, or shooting shoulders, &c. It is usually cut out of a piece of deal about 14 in. long by 3 by $1\frac{1}{2}$ in., but is sometimes made much wider, being then framed up.

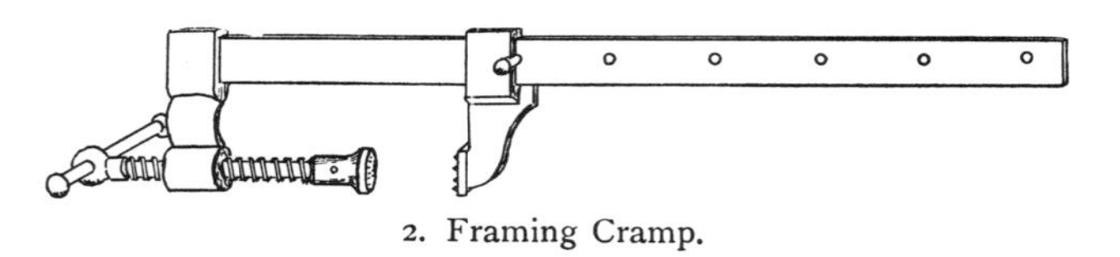

2. Framing Cramp.

Cramps are of several kinds. The **Bench** or **Sash** cramp, f. 1, is the most generally useful for light work. The bar is rolled steel of **T** section, and the shoes of malleable iron. The screw is a square-threaded, quick-action one of a length sufficient to cover two holes. For sash work, bench shoes are supplied to enable the cramp to be used with the jaws upwards. These cramps are made from 3 to 6 ft. long with 3 to 5 ft. lengthening pieces. For heavier work, the **Scotch** or **Framing Cramp,** f. 2, is more suitable. It will stand a greater strain, and the jaws have a deeper reach than those of the sash cramp, enabling it to be used on thicker stuff. These cramps are made of wrought iron throughout, and corresponding lengthening pieces can be obtained up to 7 ft.

long. For any greater length than this, wood tail-pieces are advised, as being less cumbersome. These can be made from a piece of quartering slotted for about 2 ft. to receive the cramp end, to which it should be bolted. A cleat spiked on the other end will act as a shoe.

German or **Wood Screw Cramps** are of little use, as nearly all the power applied to them is absorbed by the friction of the screw. **Wood Sash** cramps are preferred by many joiners for rapidity of wedging up, and, where a number of sashes about the same size, and especially if large, have to be glued up, they undoubtedly possess some advantages over iron cramps. They form a firm and solid bed, will remain "out of wind," they pinch up both shoulders equally, afford great facility for squaring, and the wedges are tightened and released much quicker than a screw can be wound. A small pair is shown in position on the bench in f. 1, p. 35, and f. 4 & 5 are side and top views drawn to larger scale. They are usually made from 4 to 8 ft. long, of 3 by 3 in. quartering; or, if the head is cut in the solid, of 3 by 5 in. A tapered mortise is made right through the head and bed piece to receive a long wedge, which should be of oak, about 1¼ in. thick. The outside edge of this should be square with the bed, and project beyond the head, as shown in f. 4. The shoe is formed with a lug on its

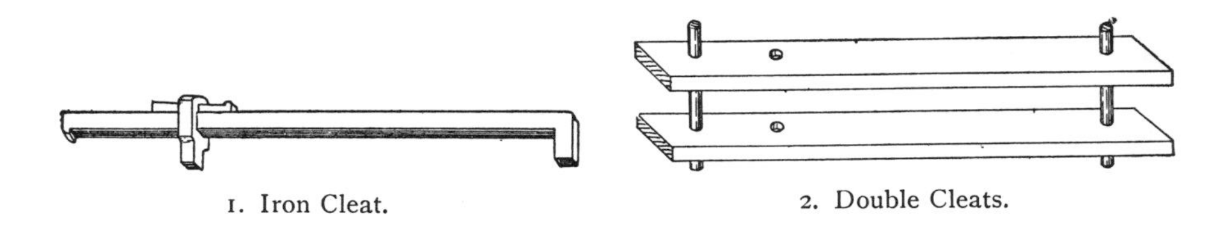

1. Iron Cleat. 2. Double Cleats.

front end to drop into the shallow mortises in the bed, and is fixed with a wood screw. The method of using them will be clear from f. 1, p. 35.

Wood Cleats for jointing purposes are now almost superseded by the **Iron Cleat,** f. 1. These are very cheaply made, easily adjusted to any width, are rigid, and practically indestructible. They consist of a ⅝ or ⅞ in. square iron bar from 20 to 38 in. long, with one end turned up at right angles, and have a sliding shoe and key wedge. When the shoe has been put on the bar, the end of the latter should be "set" up to prevent it coming off again. These cleats are very handy for many purposes in a joiner's shop, taking the place of cramps for many light jobs. They are used with a pair of wooden folding wedges. When they are made, they should be scaled, made black hot in the forge, and brushed over with boiled oil. This will prevent them rusting.

Of the wood cleats, the **Double Cleat,** f. 2, is probably the handiest for wide joints, as when closed on each side of the board it will prevent buckling; also, when many have to be used, they are much lighter to handle than iron cleats or cramps. They are made of two lengths of 1-in. stuff, 4½ in. wide, with a series of 1-in. holes bored through them. Two plugs are provided about 6 in. long. One board is passed behind the joint, the plugs placed in the nearest holes, the other board slipped on close up to the work, and the joint tightened by driving wedges under the top plug.

Box Cleats, f. 1 below, are useful for wedging up a number of thin joints, which can be placed in the cleat side by side. They are also suitable for very thick jointing, as shown in the illustration, also for fixing down veneers. When prepared for purely temporary use, housing and nailing the top and bottom into the sides will be found sufficient; but for more permanent use the joints should be bound round with hoop iron as shown in the upper part of the figure.

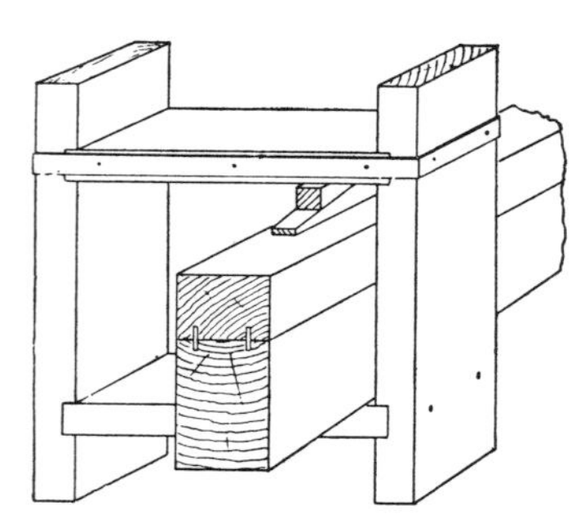

1. A Box Cleat.

The Gluing-up Bench, next page, will be found of considerable service in large establishments where many doors or large pieces of framing requiring transverse cramping are handled frequently, as its use dispenses with the clearing of the ordinary benches, the subsequent cleansing, and general interruption caused by gluing-up operations in narrow bench walks. It consists of a skeleton frame and top, firmly mortised and tenoned together with a central cramping rail, notched in flush with the cross rails, and provided with fixed and sliding shoes. A square-thread iron bench screw works in a pair of threaded plates fixed on each side of a stout head chock, for the purpose of cramping up muntings. The stiles are cramped in the usual way with iron cramps laid across the top of the frames. The two projecting cross rails are removable, and longer or shorter ones may be substituted as required, their purpose being, to hold up the stiles when knocked clear of the tenons for gluing.

Panel Boards are used for planing up panels and other thin stuff requiring a cleaner and truer surface than is provided by the ordinary bench top. They are commonly made of a short length of 1½ by 9 in. deal, with two or three screws at one end to act as stops; but as there is nothing to prevent such boards casting, they soon become useless for their purpose, and require frequent trueing up. A superior form of board is shown in f. 2, which will not cast, and is very convenient for holding the work. A frame is formed of two rails and two stiles, 3 in. wide and 1⅛ in. thick, mortised and tenoned together, and filled in with a panel of narrow battens 1¼ in. thick. These are fitted stiffly into a ⅜-in. groove, and have a little play between each other. Into the interstices thus formed a slip of wood may be inserted from which the tail of the panel may be wedged. The top side is cleaned off flush, and the ¼-in. sinking of the panel on the back will let the board lie solid on the bench. Three square wooden stops that may be knocked up or down as required, are an improvement on the screws, which are liable to chip pieces out of the stuff planed.

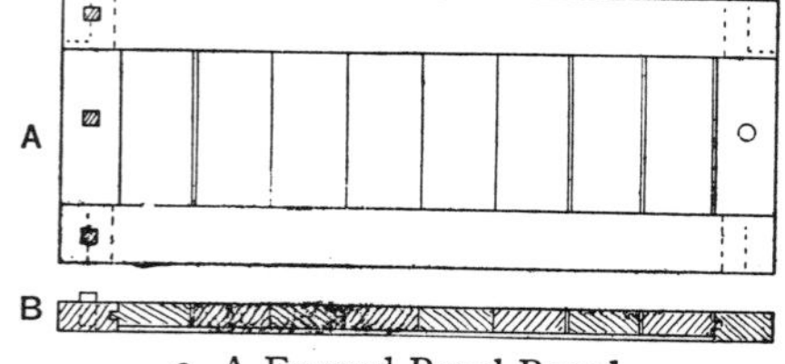

2. A Framed Panel Board.
A. Plan. B. Section.

The Shooting Board, f. 1, p. 40, is used in conjunction with the trying plane

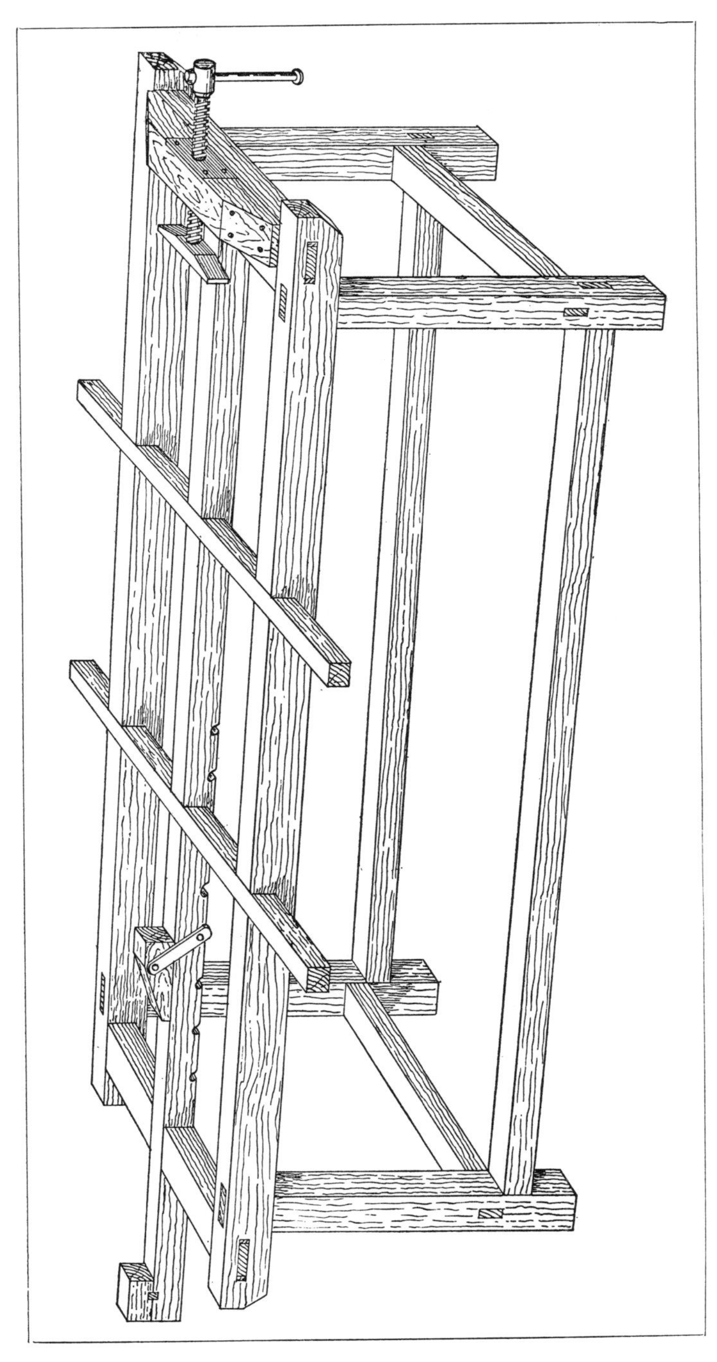

A GLUING-UP BENCH.

for the purpose of "shooting" or straightening the edges of boards that are too thin to conveniently balance the plane on, when fixed upright in the bench screw. They are made in various lengths from 3 to 6 feet. The plane bed A, $1\frac{1}{4}$ in. thick and 3 in. wide, is tongued into the top board B, which is $1\frac{3}{4}$ in. thick

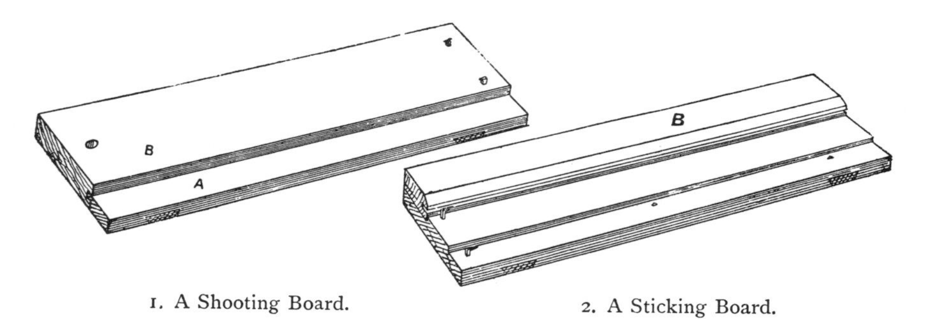

1. A Shooting Board. 2. A Sticking Board.

and 7 in. wide. These are glued together, and stiffened with hardwood dovetail keys at the back. The rebate formed by the bed-piece should not exceed $\frac{1}{2}$ in. deep to get the full use of the plane.

The Sticking Board, f. 2, will be found of service in hand shops for the purpose of rebating and moulding sash bars. It consists of a bed board of any length between 2 ft. 6 in. and 9 ft. by 9 in. wide and $1\frac{1}{2}$ in. thick, dovetail-keyed at the back to prevent warping, and with two rebates on the face, the front one to receive the sash bar whilst rebating it, and the back one to form a seat for the grooved piece B, which holds the tang of the bar whilst it is being moulded. The rebate and plough groove must correspond exactly with the rebate and tongue of the bar, and consequently a different board is required for each thickness of sash bar.

The Mitre Block, f. 3, is used as a guide to the tenon saw in cutting small mouldings, &c., to an angle of 45 deg. It is usually made out of a piece

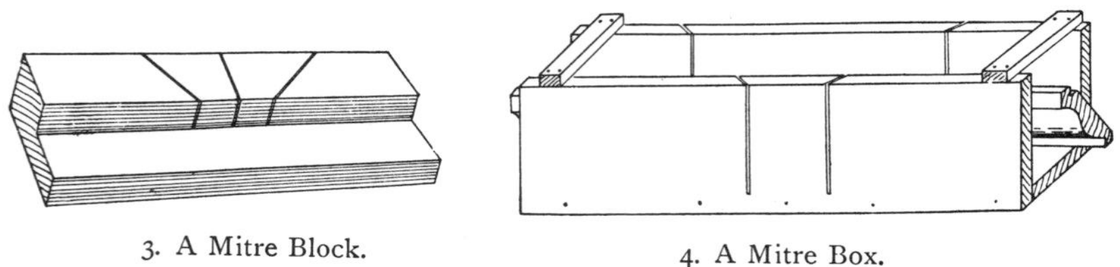

3. A Mitre Block. 4. A Mitre Box.

of oak or beech, 15 in. long, $4\frac{1}{2}$ by 2 in., rebated out $1\frac{1}{2}$ by $1\frac{1}{4}$ in. A square cut is frequently introduced between the mitre cuts, for making butt joints.

The Mitre Box, f. 4, is used for similar purposes, but for deeper mouldings than can be cut in the block. It is generally used with the panel saw. In the illustration an inclined cornice mould is arranged for cutting.

The Mitre Shoot, f. 1, p. 41, is an apparatus for planing up mitres after they have been cut in the box or block. It consists of a board rebated to form a rest for the trying plane, having two fences fixed across its top at an angle of 45 deg. with the front edge, to form rests for the moulding whilst it is being planed. It is very essential that the surface of the bed is kept parallel with that of the top board, and the method shown in the illustrations is the best way

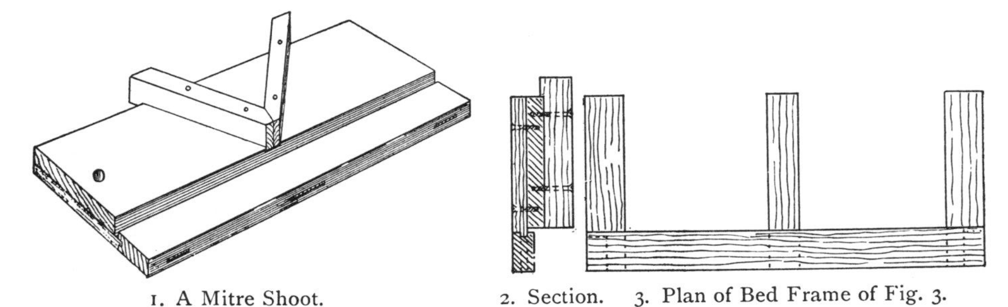

1. A Mitre Shoot. 2. Section. 3. Plan of Bed Frame of Fig. 3.

to meet this requirement. Fig. 2 is a transverse sectional elevation of the complete board, and f. 3 a plan of the bed frame. The top board may be of yellow deal, 2 ft. 6 in. long, 9 in. wide, and $1\frac{1}{4}$ in. thick, screwed to the under frame, f. 3, which is preferably made of teak, and mortised together as shown, the cross rails being thinner than the long rail, to form a sinking for the top piece, and so reduce the rebate to $\frac{1}{2}$ in. A space of about $\frac{3}{8}$ in. should be left between the edges of the top board and the plane bed to allow shavings to fall through. The fences should be of oak or deal, 2 in. square, grooved into and screwed to the top board.

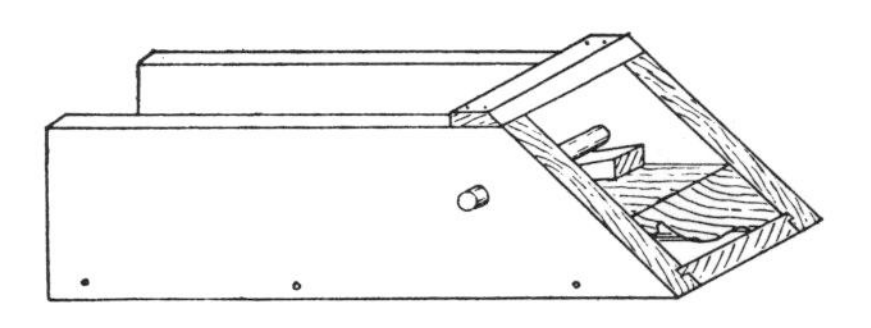

4. A Box Mitre Shoot.

The Box Mitre Shoot, f. 4, is an appliance for shooting mitres that are too deep for the trying plane to cover, when used in the bench shoot described above. It is usually made of 1-in. stuff, glued and nailed together, the side, square with the bottom, and the end cut to an angle of 45 deg. Holes are bored through the sides as shown, just above the moulding, which is wedged tightly down from a plug passing through them. The box itself is screwed to the bench whilst shooting the mitres. Another appliance for shooting mitres on wide thin stuff, such as skirtings, is—

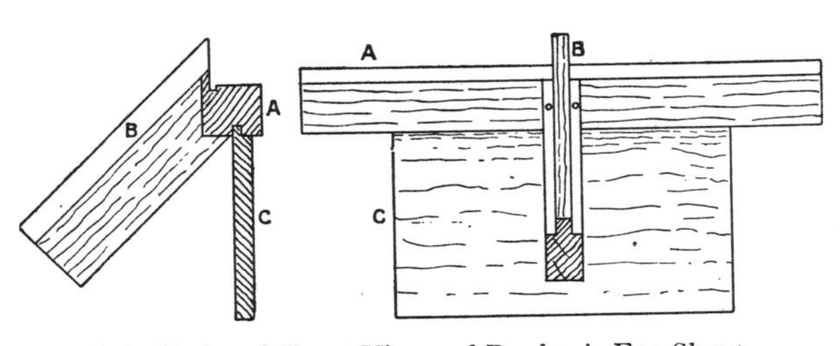

5 & 6. End and Front Views of Donkey's Ear Shoot.

The Donkey's Ear, f. 5 & 6. It is composed of a plane bed A, a fence B, and bracket C, all firmly fixed together with glue and screws. The board C is for the purpose of fixing the instrument to the bench side at a convenient height.

The Screw Mitre Shoot, f. 1, is the most perfect apparatus for shooting mitres, and is capable of doing the work of all the above-mentioned apparatus, and much more, being especially useful in manipulating very small work. A proper block plane should be used with it, care being taken not to injure the surface of the blocks. It is composed of a bed frame formed of three pieces of 1 in. by 1½ in. hardwood, 2 ft. long, with pieces 4 in. long by 1 in. square glued between them at each end, thus forming two longitudinal slots through which the lugs of the moving block pass. At one end of the bed is fixed a solid stop, 4 in. wide, with its face forming an angle of 45 deg. with the bed, and its interior end perpendicular to the same. At the other end of the bed a 3-in. beech block is fixed containing a hand screw. Between these two fixed points the movable block works, actuated by the hand screw. This block is exactly the same size and shape as the end fixed block, and it is prevented from tilting by a cross piece fixed to it under the bed. Both blocks are built up of a number of pieces of hardwood, such as Spanish mahogany, with the grain crossed, but all showing end grain on the mitre face, as this direction offers the greatest resistance to cutting by the plane when in use. Fig. 2 is an enlarged detail of the bolt and plate which connects the block to the hand screw. The bolt is threaded with a coarse screw at one end to fix in the wood screw, and a fine thread at the other to receive the locking nuts. A washer is welded to the bolt at the end of the thread, and the lock nuts are arranged so that the bolt turns easily in the hole in the fixing plate.

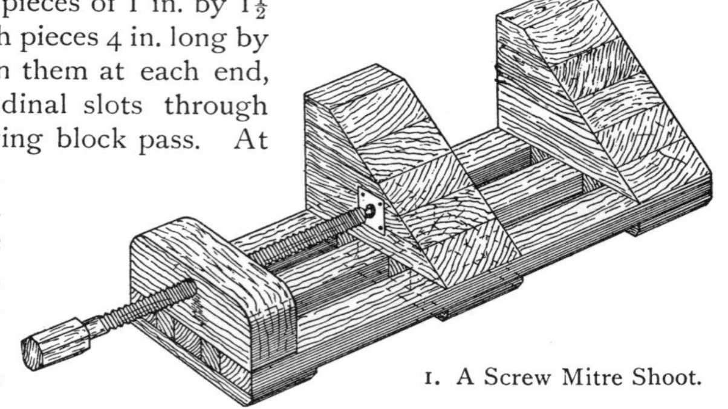

1. A Screw Mitre Shoot.

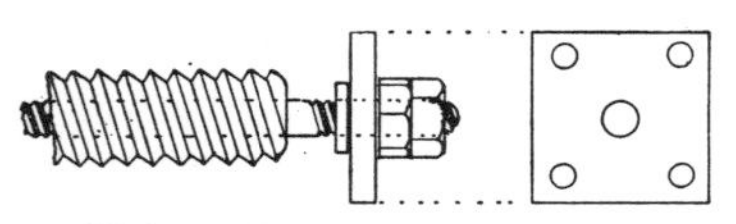

2. Enlarged Details of Connecting Screw and Plate of Shoot.

Clips or **Thumb Cramps,** f. 1 & 2, p. 43, are handy bench appliances for securing work whilst setting out, holding templets, &c. To prevent the roughened feet of the cramp bruising work, hardwood shoes should be fitted tightly on the ends, and a piece of zinc bushed over the face of the button on the end of the screw, which will prevent the button riding out of the central line of the screw, and so bending it, which it will otherwise do, and render the cramp useless.

Gluing Appliances.—Glue is such an important factor in the construction of modern joiners' work that the method of preparing it deserves some consideration. Glue is, briefly, the gelatinous extract of bones, hides, and horns of animals and fish. It has a great affinity for water, and will absorb it from the atmosphere, however old it may be, becoming viscid if sufficient moisture is present.

Fish glue is the strongest, and that known as Scotch and Russian is of this kind. When in the dry state it should be hard, tough, clear of spots, and translucent, and when placed in water should swell and dissolve slowly. Glue is prepared for use by dissolving it in water, in a water bath, as dry heating will spoil it. The slower it is dissolved and the hotter made, the better. Glue that is frequently remelted loses its strength. The best way to make it in shops where large quantities are required, is to melt it in bulk in a large cauldron with a closely fitting cover that will not permit the escape of the vapour, which carries off the adhesive constituents with it; and ladle it out as required into kettles that may be heated either in hot water or steam chests. This reheating is better done with steam than with hot water, as the tanks are seldom emptied and cleansed, and the water consequently soon becomes foul and greasy, and if such water is added to the glue to thin it, the glue is rendered practically useless, the grease forming a kind of soap with the glue, and preventing its drying. All vessels that contain glue should be tinned inside, as many kinds of glue contain free acid, which is introduced in the manufacture to clear them. This combines either with iron or copper, discolouring the glue and causing it to stain many woods, such as oak and mahogany. Inner glue vessels should be pierced with

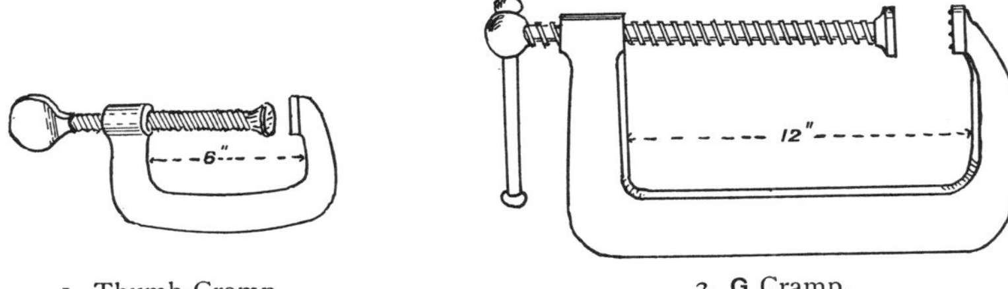

1. Thumb Cramp.

2. G Cramp.

a ring of small holes near the top, which will prevent, by the escape of steam, the wasteful caking of the glue on the sides of the pot, and a stout wire should run across the top of the pot to wipe the brushes upon. Glue may be rendered water-resisting by the introduction, when melted, of certain substances, which, however, all slightly decrease its adhesive qualities, and act differently on various woods. For use with other woods than mahogany and black walnut, a tea-spoonful of powdered bichromate of potass to half a gallon of glue will answer. For resinous woods and others than oak and mahogany, a tablespoonful of quick-lime to a quart of glue is a good compound. A small quantity of powdered chalk mixed with glue increases its strength and hastens its drying. Glue may be coloured to match oak, by the addition of yellow ochre; mahogany, litharge of lead; pitch pine, peaflour; basswood, gamboge; and dark glue in joints, &c., may be bleached white by rubbing lightly with a solution of oxalic acid (*a poison*).

THE WORKSHOP.—This chapter will be closed with a few remarks upon the general arrangements of a joiner's shop. Circumstances of locality, require-ments, and available space will of course largely determine the position and size of the shop, but it is hoped that the following suggestions, the fruits of a varied experience, may be of some service to those contemplating the erection of workshops.

SITUATION.—Obviously, where it can be obtained, the ground floor is the best position, especially if adjacent to the machine shop. Such a position will economise time in the transport of material from one shop to the other, and reduce the possibility of damage during such transit. If not opening one into the other direct, a covered way should connect them. Where this position cannot be obtained, a floor immediately above the mill is the next best, and here a large trap or opening in the floor should be provided for the ready passage of work to the mill. If this trap can be arranged against a side wall opposite the ends of the benches it will be the most convenient. A pulley and fall should be fixed over the centre of the opening, and a door should be hung on the wall side with counterbalance weights. A long narrow opening will be found more serviceable than a short wide one. No machinery should be nearer than 4 ft. from this opening.

A LOOPHOLE or door in an exterior wall will be required to pass out finished work to the carts, &c., and it should be furnished with a strong spur and fall or a jib crane. A hinged sill board, 18 in. wide, and the length between the jambs hung with strap hinges and supported by end chains, is a useful addition to the loophole, giving the men engaged in lowering more command over the load. A greenheart or iron roller should be fitted into the outer edge of the sill.

THE STAIRS should be wide, not less than 3 ft., and of easy going, as they will be constantly used for transferring light loads to the mill, &c. No winders should be used, and landings should be avoided if possible. The well should be lined with an easily removable bulkhead.

LIGHTING.—Side lights are better than top lights, and they should commence just above the bench level, and run as high as possible. The sashes are better pivoted horizontally than hung or sliding, as the first mentioned can be kept open in wet weather. If the frame is high enough to introduce a transom, the lower lights can be fixed and the fanlights hung as hoppers.

ARTIFICIAL LIGHT.—Gas is more suitable as a light than electricity, though other considerations are in favour of the latter. Gas brackets should be of the folding elbow type, fixed about 2 ft. above the benches, with one light for each bench. If T pendants are used, they should have a universal joint and pulleys to raise them out of the way in the daytime. If electricity is used, the swinging incandescent glow lamp with metal reflector will be found the more suitable, as stationary arc lights are very unsatisfactory, causing intense shadows to be thrown on the work on its off side, necessitating continual rearrangement of the work to obtain requisite light, with consequent loss of time.

VENTILATION.—Nothing is better than a louvre or lantern ventilation in the roof, and if the shop is ceiled, a shaft should be made in the centre with a louvre at the top. If such provision is made for the escape of the warm vitiated air, the windows and doors may be depended upon to keep up the supply of fresh air; and this is a matter worthy of consideration, because men working in a moist overheated atmosphere suffer from depression and lassitude, and a decreased output is the result, whilst the dampness also affects injuriously the material. If, for instance, stuff that has been stored for a considerable time in drying rooms or sheds is brought into a badly ventilated workshop filled with the exhalation of say forty or fifty men, directly it is planed it will begin to absorb the moisture and swell; then, however good the construction may be, in work framed from such stuff, every shoulder will be open a few months after it has been fixed in a dry situation. It is a *sine quâ non* for turning out first-class joinery that the air of the workshop should be *dry*. A high temperature is not required except in the veneering room, but it should not be allowed to fall much below 50 deg. Fahr., otherwise all glued work will turn out unsatisfactory.

BENCHES.—The most convenient arrangement of benches is to have a row along each side of the shop with their fore ends towards and about 18 in. from the windows, and with a run down the middle of the shop between the tail ends that can be utilised for fitting up large work and temporarily stacking material till required. If this space is wide enough, a row of standards should be fixed along the centre, with movable transoms between them at about 7 ft. 6 in. high to rest framing, &c., against. The space between the benches depends very much upon requirements, but it will not be found economical to work with less than 20 in.; 2 ft. is a good general average for a bench walk. It is not advisable to lay down machinery in the joiners' shop if first-class work is desired. The vibration set up in the floors by quick-running machines will prevent anything like accurate fitting on the part of the workman, and the inevitable dust is fatal to a high finish. If a grindstone runs in the shop, it should be match-boarded round to prevent the grit escaping. Large covering sheets should be provided to cover up partially finished work at night and week-ends, and all finished work should be primed or polished as soon as possible to prevent the access of damp.

CHAPTER IV.

WORKSHOP PRACTICE.

Method of Sharpening and Setting Saws, Making a Sharpening Block—Sawing—Ripping, Cross-cutting, Tenoning, Shouldering—Devices for Cutting Straight—Planing—Sharpening and Setting Irons—Fitting Back Irons—Roughing—Taking Out of Wind—Shooting—Jointing—Dealing with Crooked Stuff—Gluing Up—Use of Dogs—Making Ploughed Joints—Cross and Feather Tonguing—How to Use the Plough—Trenching—Housing—Rebating—Sticking Mouldings—Beads—Hook Joints—Dowelling—Mortising—Stub Mortises—How to Use the Mortise Machine—Use of the Shoulder Plane—Shoulder Tonguing—Franking—Mitreing—Use of Templets, Shoots, &c.—Mitreing Small Breaks—Scribing Mouldings—Mitre Scribing—Moulds that cannot be Scribed—Scribing Irregular Surfaces—Spiling to fit Irregular Curves—Dovetailing—Angle of Pins—Spacing—Size of Dovetails—Drawer Dovetailing—A Dovetail Square—A Dovetail Marker—Secret Dovetailing—Cabinetmakers' Methods—Cutting Dovetails in the Band Saw Machine—Boring with Bradawl, Brace and Bits—Gauge Collars—Fox-wedging—Draw-boring—Pinning—Nailing—Bradding—Screwing—Description of Joiners' Nails, &c.—Glass-papering, Preparation of Various Woods for—Burnishing, Stopping—Pelleting—Removing Bruises and Stains from Wood—Setting Out Work, what to do and what to avoid—Allowances—Setting Out Slips—Setting Out for Machine Work—Peculiarities of Machinists—Cleaning Off—Painted Work—Polished Work—How to Sharpen and Use the Scraper—Protection of Finished Work, &c.

THIS chapter is devoted to a series of brief, but it is hoped sufficiently clear explanations of the chief mechanical processes of the workshop that collectively comprise Joinery. This gathering of the various elementary operations into one chapter, rather than scattering them throughout the whole book, will, it is thought, commend itself to the reader, as the many subjects to be dealt with will thus not be overloaded with explanations and repetitions otherwise necessary. Only the more general operations are here dealt with, as special methods and contrivances will be explained as the occasion arises. So much depends, in the production of sound, clean-fitting, and well-finished joinery, upon the correct handling and manipulation of tools, that no apology is needed for the introduction of this elementary chapter; and as it is mainly intended for the young apprentice, who will more readily comprehend a pictorial illustration than a written description, photographs of the tools as they should be held when applied to the work are inserted, and the beginner is advised to practise himself in the positions and methods of holding the tools therein shown.

Setting and Sharpening Saws.—The object of "set" upon saws has been explained on p. 8, and the appearance of the edge when properly set is shown at A, f. 2, p. 9. Saws are first "set," then sharpened. The best method of setting saws is to hold the blade upon a steel block, having its edge slightly chamfered or bevelled, the amount of the desired "set" and then striking the teeth alternately with the point of a setting or "pene" hammer, but this method requires some considerable skill to ensure uniformity, and several tools have been invented to ensure this. Most of these act by leverage, bending the teeth over gradually the required amount, but again, unless great care is used, the blade of the saw will be buckled or strained. A common form of **Lever Set** is shown in f. 12, p. 9; it has a movable fence that can be fixed with a thumb-screw, and a series of notches for different size teeth. The fence is set the right distance from a notch, which is then slipped over alternate teeth, and the handle swayed down until the fence touches the blade. All the teeth on one side are set first, then the other side is similarly treated.

The Spring Grip Set.—Fig. 13, p. 9, is an improvement upon this, as the fence may be set with great accuracy by means of the milled screw *a*, and no matter how hard the levers are gripped, only the desired amount of "set" is given. When the saw has been set, it is ready for sharpening, *i.e.*, the filing of the teeth to a cutting edge; for this purpose the saw is placed in a **Sharpening Block** as f. 14, p. 9, which shows front and end views of the block. This consists of a long leg-piece for fixing in the bench-screw, to which a flap is hinged, both leg and flap being fixed at the top, by mortise-and-tenon joint to parallel jaw pieces J, usually of hardwood. A screw bolt with winged nut passes through both cheeks, for the purpose of drawing them together when the saw is placed in position, with its edge about an inch above the jaws.

A tenon saw is shown in the block, in the sketch. The method of filing varies with the saw, but always one side is finished at a time, as the pitch of the file is better preserved by this method. Coarse-teeth saws, such as the half-rip, the hand, and the compass saws, require the file to be held level, and the point directed very slightly towards the heel of the saw, *i.e.*, about 5 deg. out of square with the blade. With panel, tenon, and bow saws the file should incline about 5 deg. out of level, the point being highest, and point about 10 deg. out of square across; the dovetail saw 15 deg. across, and level. Parallel triangular files are the best, and the correct size should always be used, or the teeth will not be cut uniform. The best method of sharpening, in the writer's opinion, is to first run an oilstone along the teeth, which will bring them to one level (or a file if very uneven), and also show a bright point on each tooth. Next file each tooth, two steady strokes throughout; remembering that it is the front edge of each tooth that has to be sharpened; the back of the next tooth does no cutting, and is only filed to keep the proper shape. Then examine the saw, and if any teeth still show a bright tip, give these a stroke or two more, until the tip disappears, the actual points should be practically invisible. If any teeth show excessive setting when "sighted" from the handle, a rub on the oilstone will take it off—do not try to bend it back.

SAWING, ripping, or "splitting down." The photograph on next page shows the position the workman takes in the operation of ripping down a board to a line. The cut is commenced by resting the saw near its heel upon the line, with the thumb of the left hand lying by the side of the blade as a guide, then draw the saw backwards (holding it at an angle of about 45 deg.) for about a foot. This makes a slight kerf in the end of the board. Next push the saw forward the same distance. Repeat the strokes, gradually lengthening them until the full length of the blade is employed. For the first few strokes only sufficient force to send the saw forward should be used, otherwise it will "kick" and jump the cut, to the detriment of the left hand of the operator if he is not alert. When the blade is buried its whole depth in the cut, more force may be applied by throwing the weight of the body with the forward movement of the arm, but even then the saw should not be forced too fast, or it will be liable to buckle—that is, the blade will become permanently bent. To ensure the saw running true and cutting square, assuming that it is properly set and sharpened —without this, good results cannot be obtained — the operator must lean over the cut so that his head is directly above the saw, when a glance down the blade with one eye closed, will show if the saw is out of wind with the line.

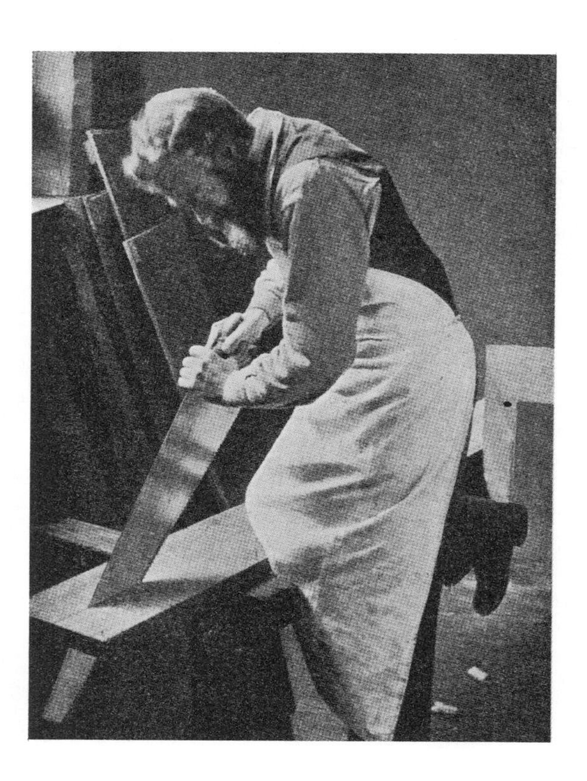

Position for Ripping Down.

In **Cross-cutting** with the hand or panel saw, the commencement of the cut is the same as in ripping. The saw must be drawn backwards for the first stroke. Less force is used than in ripping, and the strokes are quicker. Care must be taken not to draw the blade out of the cut, or the return stroke may buckle the saw. In cross-cutting hardwoods the saw should be used at a low angle with the surface. In some instances it will require to be held quite horizontal to prevent the board splitting.

Tenon Saw.—In cutting shoulders with this saw it should be held as shown in photo on p. 49, the line in this case being first cut in with a knife or chisel, and the saw worked close up to the left side of the cut, commencing with the drawing cut, and working the saw comparatively slow, but steadily and continuously, until the cut is finished. The handle should be grasped easily, with just sufficient firmness to control the saw, but a very tight grasp should be avoided, as the strain on the muscles will cause the hand to tremble, and the saw will roll, producing a bad cut. The chief difficulty with beginners using this saw is to keep the blade upright and the cut square. This may be overcome by holding the left thumb against the blade, as shown in the illustration, and letting the saw work easily and freely in the cut. It will be found an assistance in guiding the saw to the line in wide shoulders, &c., to hold a strip of wood about $\frac{3}{8}$ in. thick, with its edge shot straight and square, exactly to the left side of the knife cut (note shoulders are always cut with the tenons on the right hand), and pressing the saw against this as it is worked. This device should only be used until the novice is able to keep the saw upright, or in very wide shoulders, as, if constantly relied upon, he will not obtain a complete mastery over the saw.

Cutting Shoulders.

PLANE IRONS.—Releasing the Iron.—To remove the irons of a jack or trying plane, grasp the plane in the left hand with the fingers on the face and the thumb in the throat, pressing on the back iron, strike the nose or front of the plane a few sharp taps with a hammer, squarely, so as not to bruise the plane. The wedge should not be driven harder than sufficient to hold the iron firmly, or the plane will be curved hollow on the face. When the wedge is released, drop the hammer and remove wedge and iron with the right hand. The left thumb will have prevented the iron slipping down and injuring the mouth. Unscrew the cap iron by holding the top end in the left hand, resting the fore end on the bench, and turn with a wide screw-driver. Do not unscrew the iron whilst holding it in the palm of the hand, in case the screw-driver should slip. Smoothing plane irons are released by striking the heel of the plane upon the bench.

To Sharpen the Iron.—Place a few drops of sweet oil on the oilstone, and grasping the iron firmly in the right hand, with the palm downwards, apply it to the stone at an angle about halfway between that of the grinding basil and the pitch of the iron when in the plane. Rub it to and fro nearly the length of the stone, pressing the edge firmly down with the fingers of the left hand. Endeavour to keep the top end of the iron moving in a line parallel with the face of stone, which will produce a flat bevel. An undulating motion must be avoided, as this will produce a "round edge," necessitating frequent regrinding. After rubbing the back for a couple of minutes, turn it over with the face held quite flat on the stone, and give it a slight rub to remove the wire edge. Take care not to put any bevel upon this side. Wipe the oil off with shavings and buff the edge by drawing it across the palm of the left hand in backward strokes, turning the left hand over on each side of the iron. Examine the edge; when held to the light it should be invisible, or it may be tested by gently applying the ball of the thumb in a sliding motion across the edge, when if sharp, it will be felt to grip the skin.

Jack Plane Irons are sharpened with a slightly convex edge, colloquially the "corners taken off." **Trying Plane Irons**, straight, up to within $\frac{1}{8}$ in. of each side, from which points the edge is slightly rounded. **Smoothing Plane Irons**, practically straight, with just a trifle rubbed off at each corner. The object of this is to prevent the plane leaving marks where the edges of each shaving finish.

The Set of the Iron is the amount of cutter face exposed below the edge of the back iron, and the iron is said to be set "coarse" or "fine" according to the amount exposed. The set regulates the thickness of the shavings removed, and is varied according to the nature and kind of wood to be planed. If this is soft or mild, about $\frac{1}{8}$ in. is suitable for Jacks, $\frac{1}{16}$ in. for Trys, and a full $\frac{1}{32}$ in. for Smooths; but if the wood is hard or cross-grained, about one-half of each of these settings will be required in each case.

To Set the Iron, enter the screw of the back iron in the slot, and bring the back iron up to within $\frac{1}{4}$ in. of the edge. Then tighten the screw. Finish the setting by driving the back iron up to its place, striking on the set screw. The **set** of the iron may be termed the fine adjustment, and the **Setting of the Plane** the coarse adjustment in regulating the thickness of shavings; the latter is accomplished by resting the heel of the plane on the bench, inserting the iron in the mouth, cover upwards. Hold it in position with the thumb, as in releasing, then enter the wedge, and tap home lightly. Turn the plane backwards until the fore edge is level with the eye, when the projection of the cutter can be seen. This should be regular across, scarcely discernible in the smoothing plane, about $\frac{1}{32}$ in. in the Jack plane, and between the two in the case of the trying plane.

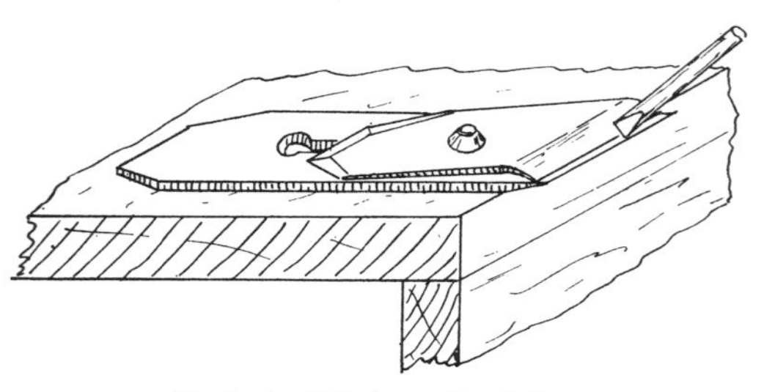

Method of Fitting a Back Iron.

To Fit the Back Iron.—This iron should be kept with a fine, but not a cutting edge, which must be made to fit the face of the cutter accurately, for if it does not, the plane will quickly "choke" in working, in consequence of shavings driving between the two irons, and it is often astonishing, to the inexperienced worker, how minute an interstice between the irons will cause the plane to choke. Having sharpened the back iron to a fine edge on the oilstone, screw it fairly tight to the cutter, leaving about $\frac{1}{16}$ of the latter exposed, as shown above. Next holding the pair very firmly on the bench, draw a hard bradawl, or the broken-off end of a saw-file, along the edge of the back iron as shown ; this will strip off a fine shaving of metal, and the operation should be repeated until it is impossible to see the slightest trace of light between the irons when the edge is held in a strong light, and the joint at the side, which is looked through, is shaded by the hand.

Planes choke sometimes, through the wedge not fitting accurately, for the great force with which the shavings enter the mouth of a plane, will drive these apparently fragile substances into a solid mass into the slightest interstice in the throat of the blade. As it is difficult to see, whether the wedge fits accurately throughout its length, try it by vibration. Drive it up on the iron tightly, then holding the plane by the sole, balanced in the hand, tap the face of the wedge with a hammer ; if it is loose anywhere, the vibration set up will be felt, or it may be detected by the sound, which is sharp instead of dull. Assuming the wedge ill-fitting, slacken it, and try to detect the place where it *binds*, by pushing it home with the hand, and moving it from side to side, ease off the place which is found to be tightest, either with plane or a wood file, and repeat the operation until it is equally tight everywhere. If the iron does not bed squarely, pack the low part with brown paper glued in. Do not ease the plane, as the next iron used may be out of truth in another direction.

Facing Up, or "taking out of wind," is the producing of a practically true surface upon a piece of wood so that every part lies in the same plane. It is the preliminary operation upon nearly all stuff that has to be dealt with by the joiner. First remove the rough surface with the Jack plane, working the stuff straight across, testing it by turning the plane upon one of its lower edges, when the light passing between the board and the plane will show whether the former is hollow or round. Next ascertain whether the surface is twisted by applying the **Winding Sticks.** These are two parallel strips of hardwood about 16 in. long, 1¾ in. wide, ⅜ in. thick at the lower edges, and ⅛ in. at their upper. Their object is to multiply any inaccuracy in a surface to which they are applied, so that it may be readily discernible. Arrange them near the ends of the stuff and parallel to each other. Then standing about a foot away from the one that is least lighted, lower the head until the eye is level with the top edge of the near strip, when it will be readily seen if the edge of the distant one lies parallel with it. If this is the case, the stuff is "out of wind"; but if not, notice which end of the distant strip is the higher. Then plane off that corner and the one also diagonally opposite, half the estimated quantity required off each. Gradually work towards the middle of the board, where, as the twist changes its direction, nothing will be removed. Finally, when the sticks have been brought into the same plane, proceed to straighten the surface in the length. With the Trying plane held firmly with both hands, commence at the near side, and endeavour to take a shaving the whole length of the piece, repeating the stroke in successive widths of the plane across the stuff. At first perhaps, only short shavings will be removed, which will, however, gradually get longer as the inequalities are reduced, until the differences will be so minute that the iron will bite throughout the whole length. If the surface planed is more than twice the length of the plane, a straight-edge may be used to test its truth longitudinally.

Shooting and **Squaring the Edge** is the succeeding operation to the above, and consists of producing a perfectly straight surface, at right angles to the face that has been taken out of wind. Stuff when so treated is termed "faced, shot, and squared up." After the rough has been removed by the Jack plane, the Trying plane is applied in the manner shown in the photograph below. The left hand grips the stock just in front of the cutter, with the fingers lying on the sole, and acting as a guide to prevent the plane running off the edge. The shooting should commence at the highest point, working down to the lower—that is, if the edge is "round," commence in the middle; if "hollow," at the ends. Lengths of 5 ft. and upwards should be tested with a straight-edge. Shorter than these will be worked accurately by the plane, provided that its face is true and that it is kept firmly down to the work. Apply the try square (p. 27) occasionally to test for squareness. Having faced and shot the piece, it can be gauged to a parallel width and thickness, the excess being planed off to the gauge marks.

Method of Shooting Joints.

Jointing.—Long joints are prepared by fixing the boards to be jointed separately in the bench screw, and shooting and squaring the edges as above described, then testing them by placing one above the other, and applying a straight-edge across the face to ascertain their truth. The necessary corrections can usually be made by taking a shaving off the front or back of the edge of the board fixed in the screw. A long joint should be shot slightly hollow in length to counteract the effect of cramping up, and also to ensure that the ends of the joint are tight up, as these are the most likely points for the air to enter and break the joint. If the boards to be jointed are in winding, and this is often the case with thin stuff, apply the testing edge near the middle of the length, and shoot the joints straight, disregarding any inaccuracy at the ends, as the cleats will pull them up to their place. When the edges are all shot and marked as in f. 2, p. 54, cleats, f. 1 & 2, p. 37, are set to the proper width and wedges prepared. If three boards have to be joined, place the middle one in the screw, and rest the edge of the top board against it, as shown in f. 1; then proceed to apply the glue—which should be very hot, and of such consistency as to run freely from the brush but not to break into drops—to the edges quickly; then turn the boards together, and standing at one end, with an assistant at the other, grasp the ends of the upper board near the joint, and rub to and fro steadily, in strokes of about 2 ft., until the bulk of the glue is rubbed out, and the joint begins to "drag." Then the pair should be lifted carefully from the screw so as not to break the joint. If this does happen, the joint should be immediately re-rubbed, and deposited upon two pieces of stuff, as shown in f. 1. The third board is then secured in the bench screw, and its edge liberally glued, when the first two are carefully lifted upon it and rubbed as before. The cleats are next applied and wedged up, one near each end, and the others at intervals of about 3 ft. on each side of the joint alternately, if single cleats or cramps are used. In wedging the cleats up, use the wedges in pairs, and drive the one resting on the edge of the board. If a screw cramp is used to tighten up the joint, either set its face close to the work, or put packing pieces between, whichever is more convenient.

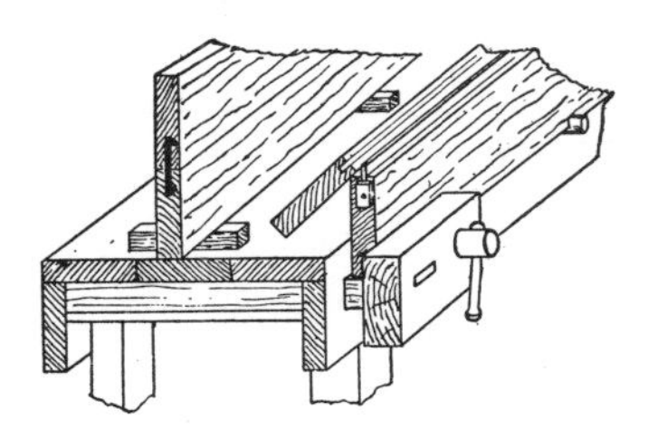

1. Method of Jointing.

2. A Joiner's Dog.

In all glueing operations remember to "hasten slowly." Have everything ready to hand so that no delay occurs. Apply the glue quickly and freely, then get the joint together before the glue has time to chill. Rub steadily until the glue begins to drag, and endeavour to stop rubbing when the boards are flush at the required end. If the joint "sets" before the board is in position, drive it up to its place with a hammer. Any attempt to jerk it up by hand will probably break the joint.

Joiners' Dogs, f. 2, may be driven in the ends of the joints to steady them whilst moving, or to hold short ones whilst drying; but they should not be left in more than an hour or so, because, if the boards shrink, the dogs not yielding will cause the joint to open.

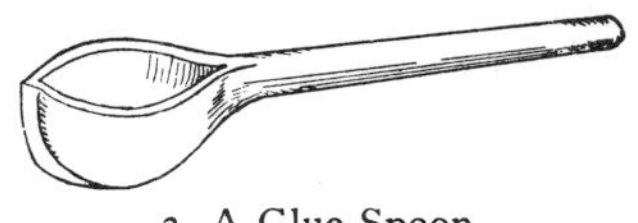

3. A Glue Spoon.

Ploughed and Tongued Joints require, after the edges are shot, a $\frac{1}{8}$ by $\frac{1}{2}$ in. groove run in the centre of each edge, and cross tongues fitted on. These should slide easily, but not be loose enough to rattle. Their ends must be cut square, and a small stop bradded at each end of the groove in which they are placed to prevent them running out during the rubbing (see f. 1 above, & 3, p. 137). To glue these joints, having fixed the second board as shown in f. 1 above with the top one resting on it, remove the tongues and lay them along the board or the bench in regular order, then nearly fill the lower groove with glue by aid of the **Glue Spoon,** f. 3, commencing at the fore end and working backwards. Insert the tongues in the same order, knocking them well down with a mallet. Apply hot glue to the edges and both sides of the projecting tongues with the glue brush, and proceed as described in the former paragraph.

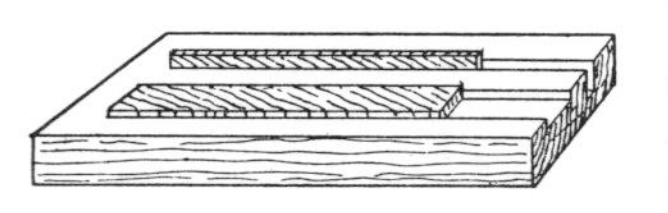

1. A Tonguing Board.

Cross Tongues, f. 4, p. 137, are prepared by running a cutting gauge, f. 19, p. 17, set to $\frac{7}{8}$ in. across the squared end of a wide thin pine board, and reducing them to the required thickness by planing them in a **Tonguing Board,** f. 1, which is a short piece of deal having two stopped grooves in it to receive the tongues, one for edging, the other for thicknessing.

Feather Tongues, f. 5, p. 137, are prepared by cutting the board off to a long bevel, by which means longer tongues can be obtained out of a narrow board in the same time that cross ones can be cut, but the ends of these require squaring, and they are liable to swell and force off the joint.

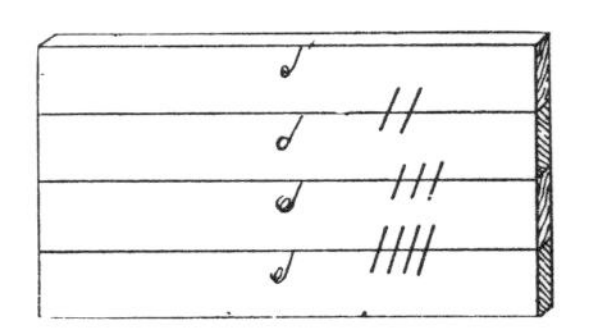

2. Method of Marking Joints.

Short Joints in thin stuff are prepared in a shooting board, f. 1, p. 40. The boards are numbered as they are intended to be placed (see f. 2). The first board is laid face down and its edge shot with the trying plane, as shown in the photo below; the corresponding edge of the second board is shot with its face upwards, and the opposite edge with the face down, and so on, the edges of each joint being planed in reversed positions. This is to counteract any want of truth in the shooting board or side of the plane, and to ensure that when the boards are placed together they shall be straight across the face. The plane iron should be set square for this operation. The wood is held down firmly with the left hand, whilst the plane is worked with the right grasping the stock, just behind the wedge.

3. Making Joints with Shooting Board.

PLOUGHING, as the formation of grooves in the direction of the grain is termed, is accomplished as follows:—We will assume that a board is required to be ploughed on the edge to produce a groove $\frac{1}{8}$ in. wide and $\frac{1}{2}$ in. deep as used in a tongued joint. Fix the board on the bench screw face out, set the plough iron projecting $\frac{1}{16}$ in., and with the groove in its back resting on the V edge of the guide plate. Arrange the wood fence exactly parallel to the skate or guide plate, and at such distance from it as will bring the cutter in the centre of the edge of the board. This is done by slacking the side wedges with a hammer, and striking the stems on either end with a mallet as required, then tighten the wedges. Set the stop with the thumb-screw $\frac{1}{2}$ in. deep, measuring from the edge of the cutter. The method of holding and working the plough is shown in the photo below. The forward stem is grasped with the left hand on the near side, and the fence kept pressed closely up to the work. The rear stem is grasped with the right hand on the off side, and the bulk of the driving of the tool is done with this hand. Care must be exercised that no downward pressure is made on the right hand, otherwise a rolling motion will be produced, and the groove will not be perpendicular. The ploughing should be commenced at the fore end and work backwards. Short quick strokes produce the best work.

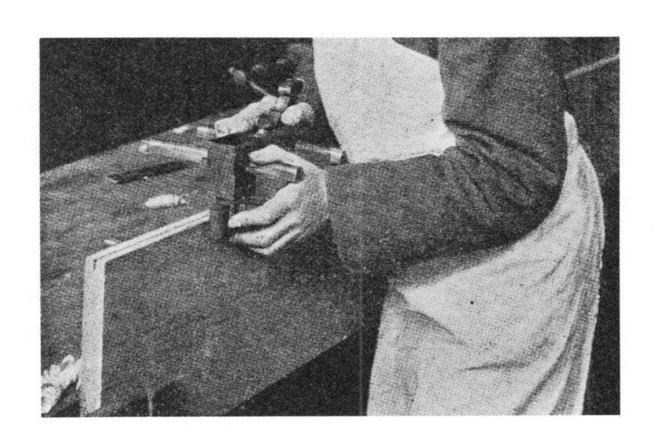

Method of Ploughing.

Trenching or **Cross-grooving** is executed with the trenching plane, f. 13, p. 12. A line is drawn across the stuff to correspond with the right-hand side of the desired groove, and a strip of wood to act as a guide for the plane is fixed on this line. If the groove is on an unseen surface the strip may be bradded on, but if this method of fixing is not permissible, it may be secured by clips or handscrews. This plane should be worked from the near to the off side, keeping the fore end of the plane well down, so that the cutting knives, which are just in front of the iron, shall sever the fibres in advance of the latter. The set stop, if arranged the required distance from the iron, will gauge the depth of groove correctly. If the groove has to be stopped, it must be commenced immediately behind the stop by mortising out sufficient to receive the length of the plane in front of the cutter, when the remainder can be finished with the plane. **Cross-grooving** can also be done without the aid of the plane, as follows:—Set out the size groove required with the marking knife, f. 1, p. 24, and if it has to be stopped, mortise out about 2 in. behind the stop. Then run in a tenon saw to the lines on each side to the required depth, and remove the core roughly with a chisel. Next set a plough iron in the **Old Woman's Tooth,** f. 12, p. 12, to the exact depth of the groove, and work it backwards and forwards until it will no longer bite.

Rebating or **Rabbiting.**—Rebates are produced by first marking their dimensions with a marking gauge, then either fixing a temporary fence to one of the lines to act as a guide for the rebate plane, as shown below or by ploughing a groove to one of the lines, and chipping away the remaining wood with a firmer chisel, and finishing off to a regular surface with the rebate plane. Wide rebates, such as the one shown in the door jamb, f. 1 below, may take two or more grooves to break up the width of the core, and these are finished off with the badger plane or panel ditto, f. 5, p. 12. Narrow and special rebates, such as those on the backs of sashes, are worked with the **Side Fillister**, f. 4, p. 17, and the **Sash Fillister**, f. 3, p. 17. The latter is set and used in a similar manner to the plough, but care must be taken to set the cutter precisely in line with the marking tooth.

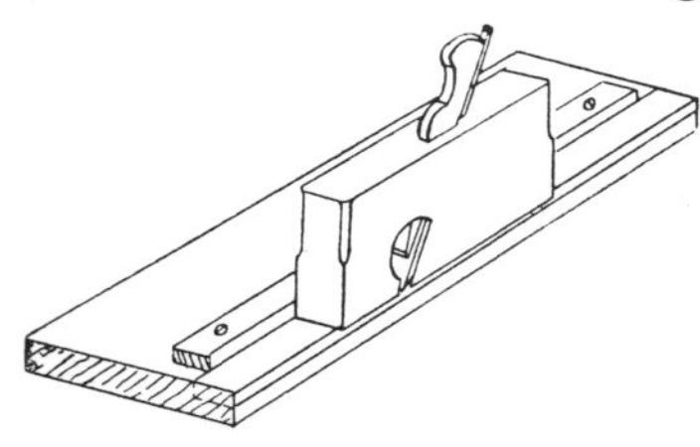

1. Method of Forming Rebates.

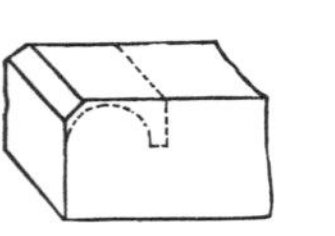

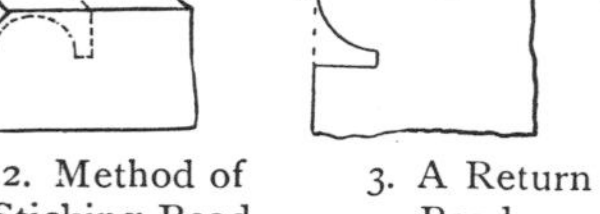

2. Method of Sticking Bead.

3. A Return Bead.

Moulding or **Sticking**, as it is usually termed, consists in working upon the salient angles of framing, boards, &c., sundry curves and combinations of facets known as mouldings. Many of these are produced by special planes of reverse contour, such as are shown in f. 1 to 5, p. 19, whilst others have to be formed with various hollows and rounds aided by the plough and rebate plane.

Beading is produced with the bead plane, f. 5, p. 17. The salient angle of the edge to be stuck should be taken off with the jack plane, as shown in f. 2 above, the cut commenced at the fore end, the plane held upright and worked in long strokes. It should not be worked down to the stop, but only low enough to just round off the surface. In certain cases, such as door panels, the bead may require working below the surface, but generally it should be flush. A return bead, f. 3, is produced by first sticking a bead on one surface, then turning the work round and sticking the same bead on the adjacent surface. Before, however, working the second bead, the cutting iron should be knocked in slightly, so that it will not cut a second time on the outside, otherwise the section will be flattened at the point marked X in f. 3. In sticking the various sash mouldings shown in f. 1 to 5, p. 19, the planes are not held upright as the bead plane is, but lying over towards the workman until the line drawn on the fore end of the plane and marked X in f. 1, p. 19, is vertical, or in line with the face side of the stuff to be worked, and the plane is worked down until the stop or rebate on the off side rests upon the edge of the work. These planes are generally made in pairs, one slightly larger than the other. The No. 1 or smaller one is used to remove the bulk of the core; the No. 2 to finish the moulding accurately and clean, thus dispensing with glass-paper.

In sticking the **Hook Joint** for sashes, the plane shown in f. 6, p. 19, is used. The edge of the stile is first worked to the required bevel, which varies with the width of the sash, then a rebate is taken out to a similar bevel and down to the top of the hook, then worked with the plane which is held over so that its fence lies fair against the bevelled edge. The movable stop is arranged to regulate the depth of sticking, which is usually half the thickness. The pairing stile is treated in a similar manner.

Dowelling is a method of connecting boards, &c., by the insertion of short wood pins, termed dowels, in the joints. It is used in positions where tongued joints, which are much stronger, would be unsightly. The pins should be made of some tough hardwood such as oak or beech, cleft in the direction of the grain; roughly trimmed to a round section, then pointed and driven through the dowel plate, f. 3, p. 26. The dowel should not exceed in diameter one-third the thickness of the piece containing it, and in length eight diameters, otherwise it will be liable to break in driving. Dowels are usually driven perpendicular to the surface in both directions, but if entered at any other angle, all of those in the same joint must be parallel to each other. To mark a joint for dowelling, lay the two pieces together back to back as shown in f. 1, and square lines over the edges about 12 in. apart; run a gauge mark from the face side along the middle of each edge, and bore the holes at the intersections with a twist bit. Should the bit run out of square, enter the dowels in the holes, and with a straight-edge held over their axis, mark their direction on the face of the board. Withdraw the pins and place the other board in position, producing the lines just marked on to this board, when the second lot of holes can be bored in the corrected direction by fixing a straight-edge to the line and keeping the bit parallel with it in both directions. Dowelled joints must be shot quite straight and be pulled up with cramps, as obviously they cannot be rubbed.

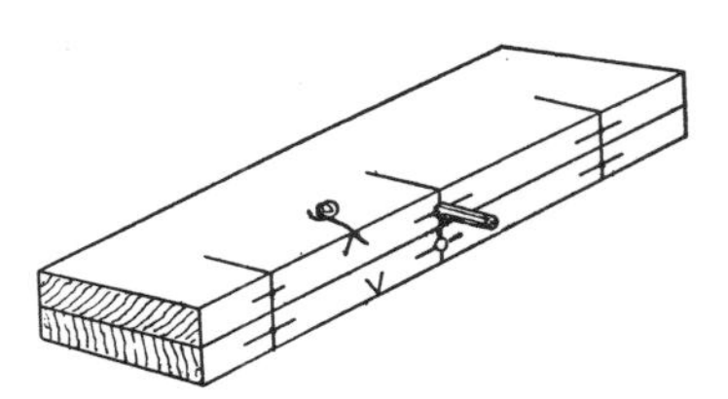

1. Method of Setting Out Dowels.

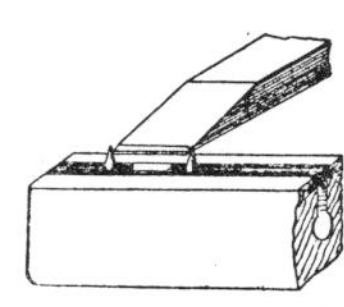

2. Method of Setting Mortise Gauge.

Mortising.—Mortises up to $\frac{5}{8}$ in. wide are made with mortise chisels, f. 5, p. 20; over that width with firmer chisels. In setting the gauge for a mortise, the chisel should drop just within the points of the gauge teeth, as shown in f. 2. In mortises that are made through the stuff, care must be taken to gauge both edges from one side. Commence the mortise in the middle of its length, and on the back or wedging side, and take small cuts, working towards the near end, and removing the core as you go; then turn the chisel round and work back to the off end. Cut half-way through from each side. If the work is small, it may be fixed on the bench screw, the workman standing close beside the work and just behind the mortise, so that by looking up the edge of the chisel he can see if it is out of wind with the gauge line. Larger work is fixed on the mortising stool or horse, f. 3, by wedging it between the horns. In this case the workman sits astride the work to keep it steady. After the mortise has been cut, the core must be punched through with a core-driver. This is a piece of hardwood about a foot long, and of such size as to pass through the mortise easily. Clean off the inside of the mortise (if a "shelf" has been made, *i.e.*, if the chisel has not been driven through so as to meet in the centre) with the float, f. 1, p. 58. This is a very coarse cut file, with safe or smooth back and edges, about 12 in. long and $1\frac{1}{8}$ in. wide. It is frequently made by the workman himself by

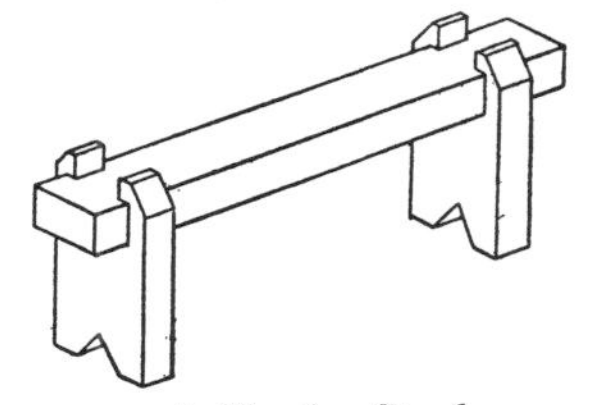

3. A Mortise Stool.

grinding down an old flat file, softening it in the fire, and filing up coarse teeth as shown, then rehardening it.

In Stub Mortises, *i.e.*, those which do not pass through the stuff, a convenient method for gauging the depth of cut is shown in f. 2 below. A piece of white paper is glued on each side of the stem of the chisel with its bottom edge to the required depth. In using a **Hand Mortising Machine**, set the work on a strip of deal, so that the chisel shall not be driven or fall on to the iron bed. Arrange the height of the bed or standard, whichever moves, so that at the finish of the down stroke the chisel will be nearly through the work. Set the chisel with its face towards the reversing handle and at right angles to the fence. Fix the work with the screw fence against the mortise, tighten the screw only sufficiently to hold the work down, and place the face side towards the fixed fence. Work the back edge first. The wedging, in soft woods may be formed with the back cut of the chisel, but for hard woods the work should be pitched up on a wedge piece until the wedging line is vertical, then the cut can be made with the face of chisel. After all the mortises have been made, remove the chisel and substitute a core-driver, lowering the standard or raising the bed until the core-driver will pass clear through the mortise. Drive the core out from the face edge, as the wedging affords more clearance, and prevents the edges being broken out. All mortises in hard wood should have one or more holes bored through them before mortising, to afford clearance for the chisel, which will otherwise be drawn out of the socket by the great friction. A grease mop should also be frequently applied to the sides of the chisel.

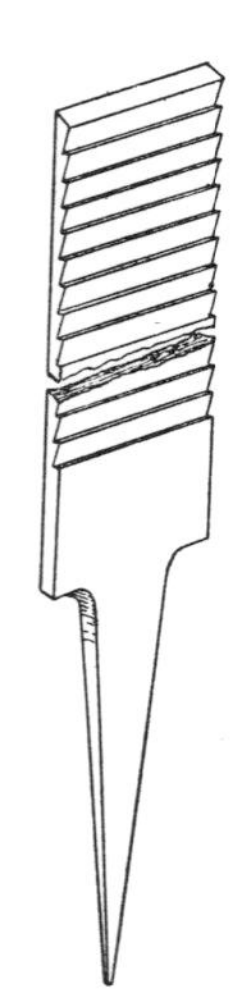

1. A Float.

Tenoning.—In cutting tenons by hand it is advisable to cut out the haunching between and beyond the tenons before ripping them down, as shown at *a a*, f. 7, p. 143, as this much lightens the labour and enables the inner edges to be gauged, which ensures cutting them to a more regular thickness. When commencing the cut, fix the rail in the bench screw at an angle as shown in the photo opposite, so that the gauge lines on both edge and end can be seen. Cut one side down to the shoulder line; then turn the rail round with its other edge in same position and saw down that side in like manner; then fix the rail upright and finish the cuts square through down to the shoulder, being careful not to cut beyond it. In cutting double tenons, bore a hole the size of the space between them, immediately above the haunch, before ripping down, as shown in f. 2, p. 59. Remember that in preparing framed work by hand the mortises and tenons must be cut first; then any ploughing, rebating, moulding, &c., worked; and finally the shoulders cut, the latter always being the last operation before fitting, as the cheeks of the tenons are required to act as guides for the plough, &c. For the same reason, rebating and grooving must precede moulding.

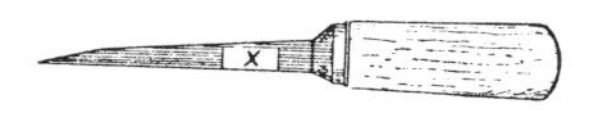

2. Method of Gauging Depth of a Mortise.

Shouldering.—The shoulders of deal framing are better left as cut by the saw, unless some considerable fault requires correction, as the rough surface produced by the saw causes great friction and assists the glue in holding the shoulder up, and trifling inaccuracies in soft wood will be overcome by the crushing in of the fibres in cramping up. When, however, a clean shoulder is desired for varnished or polished work, and in the case of all hardwoods which will not yield perceptibly to lateral pressure, the shoulders must be shot with the shoulder plane, f. 3, p. 14, and the operation is well shown in the photo next page. The iron must be set fine, and a small slip of wood held close against the off edge of the shoulder to prevent the latter breaking out as the plane passes over it. When the edges of the rail are moulded, the plane should be worked from each edge towards the middle of the shoulder. Square shoulders should not be undercut as shown on the upper side of tenon in f. 2, p. 60, for such work is simply "scamping," and although appearing close on gluing up, will show an open joint on cleaning off; the correct method is shown on the under side. A bevelled joint, such as occurs in a chamfered rail, of course requires an undercut shoulder, but this is accurately fitted wood to wood. It is, however, not a joint to be recommended even when so constructed, as it soon becomes faulty through the wood shrinking.

1. Cutting Tenons.

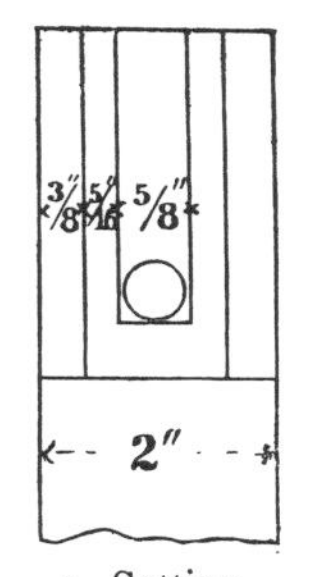

2. Setting Out of Double Tenons.

Shoulder Tonguing is employed in mortise and tenon joints when the shoulders exceed $\frac{1}{2}$ in. in depth; $\frac{1}{8}$-in. grooves $\frac{3}{8}$ in. deep are worked in the shoulders of rail and stiles. If the framing has moulded edges, the grooves must be stopped within the moulding line. Oak cross tongues are fitted tightly in the grooves and glued in the rails, and then treated as tenons. When the grooving is done by machinery, the framing must be dry wedged up and cleaned off flush at the shoulders before grooving, as the machine will work from the face of the stuff. When the grooving is done by hand, the necessary lines should be gauged from the faces of the mortise and tenon, not from the face of the work, and in this method the preliminary flushing off is not required. To execute the grooving in the rail shoulders, prepare two slips of wood, one exactly equal in thickness to the space between the tenon face and the inside groove line, the other equal to the distance between the tenon and the outside of the groove, less the thickness of the saw blade to be used in cutting. Then apply them to the shoulder, as shown in f. 3 below, and run the saw down upright beside them. Remove the core with an eighth of an inch chisel. To mark the stile, the tenon should be entered in the mortise a short distance, and the thinner slip laid on it, and a mark made along its side on the stile. Then add to the slip a short piece of the tongue stuff, and make the second mark. Mortise the core out with a small chisel, and regulate the depth with the old woman's tooth (see f. 12, p. 12).

1. Shooting Shoulders.

Franking is a method of forming joints in framed work by which the member containing the mortise has its shoulder stopped on one or both sides of the mortise by a projection or spur that is sunk into the shoulder of the rail at the root of the tenon, as shown in f. 4 below, the object being to strengthen the stile by the avoidance of haunching, and also to prevent the possibility of water gaining access to the mortise through the shoulder. Fig. 1, next page, shows the application of franking to a sash bar. The cut bar is first mitred, then the lower part of the mitre is cut away nearly up to the sight lines, thus allowing the mortised bar to be continued

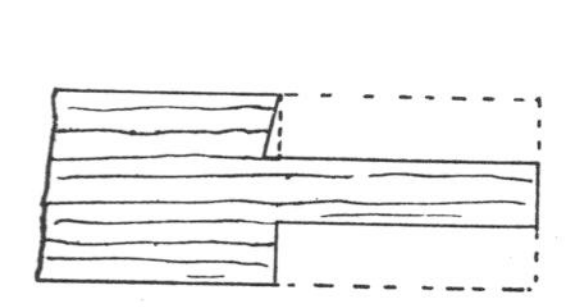

2. Correct and Incorrect Way of Cutting Shoulder.

3. Method of Shoulder Tonguing.

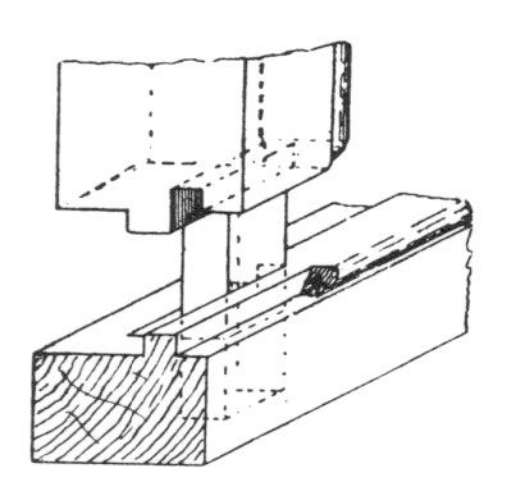

4. A Franked Shoulder.

nearly solid on each side of the mortise. The triangular part within the dotted lines of the section shows the uncut portion of the bar. In connection with this method of fitting it is usual to halve the tenons of the cut bar, as shown, to increase their length.

MITREING is the formation of a joint that bisects the internal and external angles of the pieces joined (see f. 2, A and B). A right or true mitre forms an angle of 45 deg. with the sides of the pieces, but any other angle of intersection is also termed a mitre. Those that are curved as in C, f. 2, are

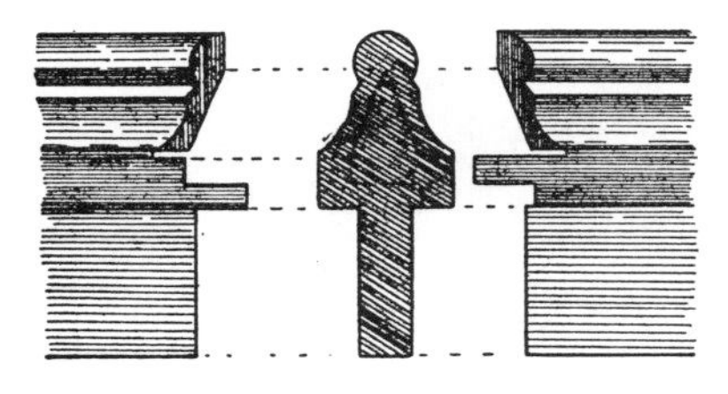

1. A Franked Sash Bar.

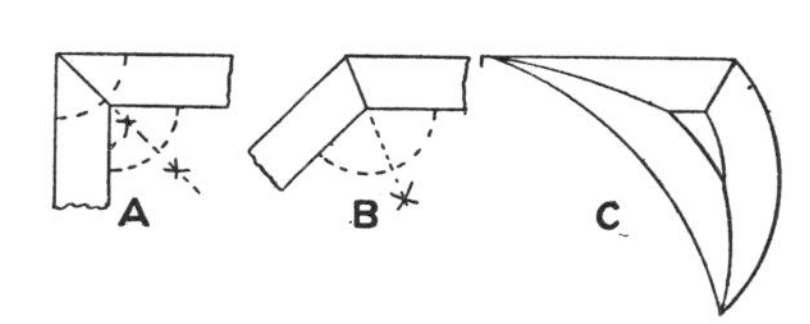

2. Diagram of Mitres.
A. Square Mitre. B. Obtuse Mitre.
C. Circular Mitres.

distinguished as circular mitres. The object of using this joint, which is constructively one of the weakest used in joinery, is that moulded surfaces that have to be changed in direction shall not be stopped abruptly nor continued in unsuitable curves. The mitred surfaces are prepared in various ways according to circumstances. Mouldings that are planted in frames are first cut off rather full to length in the Mitre Block, f. 3, p. 40, then placed in the Mitre Shoot, f. 1, p. 41, and planed down to their exact dimensions. Bolection Mouldings—that is, those rebated at the back—when being planed in the mitre shoot, should not be allowed to rest upon their backs, but be bedded on a strip placed in the rebate, as shown in f. 3. The strip is planed to equal the depth of rebate between the surfaces of panel and framing. If this precaution is not taken, the mitres will be slack inside when sprung into position. Cornice or any similar moulding that does not lie flat on its back, must be placed in the Mitre Box for cutting, in exactly the same relative position to the horizontal that it will occupy when fixed; then the saw will produce a mitre that will be vertical when the moulding is in position (see f. 4, p. 40). The mitre may be shot either by fixing it in a box shoot, f. 4, p. 41, so that the plane of cut lies in the same plane as the end of the box, which is done by resting the bottom edge of moulding upon the side of the box, then planing down with a block plane or a trying plane until the surface is flat; or the moulding may be fixed in the bench screw and the mitre planed with a smoothing plane, and tested with the mitre square, f. 1, p.

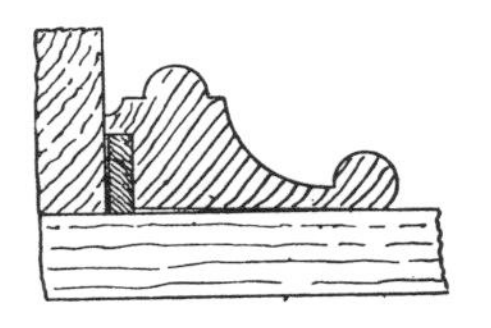

3. Method of Shooting Mitre of a Bolection Moulding.

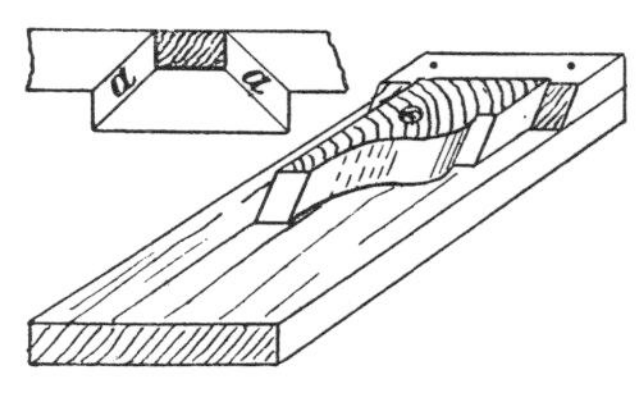

4. Method of Shooting Small Breaks.

28, which is held to the *face* of the moulding, the stock being kept parallel to the edges.

Mitreing Breaks.—In mitreing a deep moulding, such as a plinth or cornice around a pilaster, small breaks, *i.e.*, return pieces, occur on the edges as at *a a*, f. 4, p. 61. These pieces being much less in length than they are in width and depth, are difficult to hold whilst planing. Fig. 4, p. 61, shows one method by which they can be planed without injury to the edges. A strip of hardwood, a trifle thinner than the break to be planed, with its edge bevelled to 45 deg., is fixed to a panel board. A hole is bored through the centre of the break, and countersunk on each side so that a screw can be inserted and turned below flush. This secures the piece, which can be planed and tested with a mitre square, then reversed and finished. Very thin pieces may be held sufficiently by chalking the panel board.

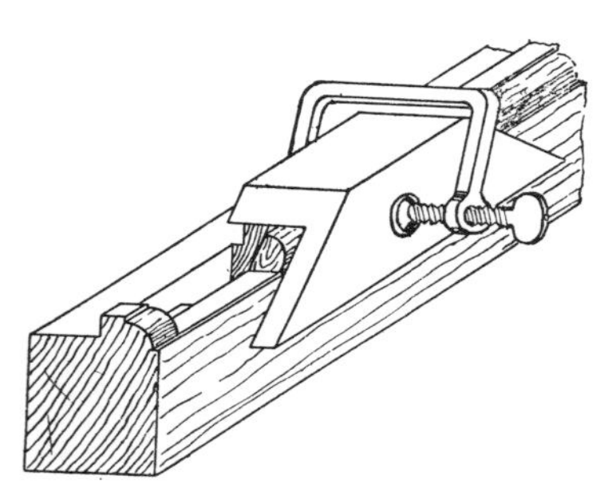

1. Method of Using Mitre Templet.

Mouldings stuck in the solid, on framing are mitred by aid of the **Mitre Templet,** f. 1, which is a short rectangular block of hardwood rebated out square on the inside and having its ends cut to angles of 45 deg. each. It is applied as shown on the sketch, with its square edge to the sight line of the mitre, and held in position by the left hand of the operator, or when both hands are required for cutting, by a screw clip or handscrew; and the mitre made with a paring chisel, which is pressed firmly on the templet with the right hand, whilst it is thrust forward by the right shoulder. Fig. 2 is a reverse or **Panel Templet** used for cutting the internal mitres on bead flush panels.

A Combination Panel and Mitre Templet is shown in f. 3. These are made in steel or brass, and although not so easily damaged as the wood variety, their range of depth is very limited.

A Square Templet is shown in f. 4. This is used for marking and cutting butt shoulders.

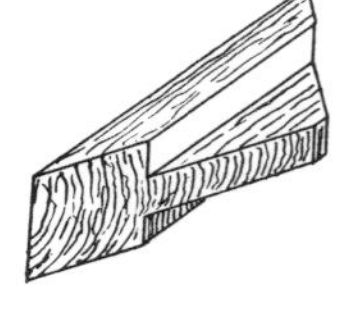

2. A Panel Mitre Templet.

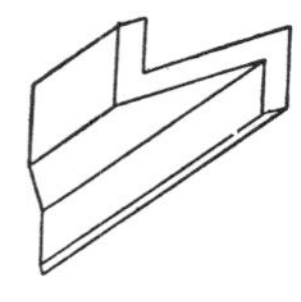

3. Brass Combination Templet.

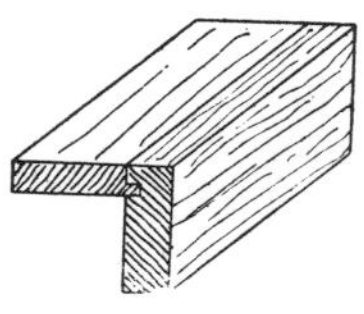

4. A Square End Templet.

When mouldings have to be mitred around frames that are not at right angles, the mitre lines may be found readily by cutting in the moulding in the clear, as shown at *b b*, f. 1 next page, and applying the ends of the adjacent piece or a waste short end of the moulding to the front edge, and marking its width as shown by the shaded parts, then a line drawn from this point to the back extremity will be the mitre. The process repeated with the end pieces will give the mitre upon these.

Scribing is a term applied to two distinct methods employed in making joints—(1) That in which the end of a piece of moulding is cut to fit the profile of another intersecting it, the finished joint having the appearance of a mitre; (2) the fitting of the edge of a board or frame to an irregular surface. Scribing as applied to a moulded frame is executed as follows:—The piece of framing containing the mortise has its moulding cut away to form a square shoulder, as shown in f. 2, but sufficient moulding is left on to run about $\frac{1}{4}$ in. below the line of intersection of the rail moulding. The moulding on the rail is first mitred with the templet, as shown in f. 1, p. 62, then the mitred portion is cut away with scribing or inside gouges and chisels, back to the line of mitre and down to $\frac{1}{4}$ in. below the sticking line, as shown at A, f. 2. Sash bars and other narrow rails are scribed right through, as shown in f. 3, as are also the shoulders of sash rails when machine-worked. In scribing sash bars, a scribing templet made to fit the moulding, and with its end cut to a reverse of the mould, is sometimes used; but the outline of the scribe can better be obtained by first mitreing the moulding on each side of the bar with the side templet, f. 1, p. 62, and cutting away the mitre portion with a gouge; and also by this method there is less danger of splitting out the under face of the bar. The **Scribing Block,** f. 4, should be used to steady the bar whilst cutting. This is made of a piece of soft deal or pine, shaped to fit one side of the bar, with the moulding seat cut across the end grain, so that the gouge as it passes through the scribe may bury itself in the end, thus preventing its edge becoming dull and the off side of the bar from breaking out.

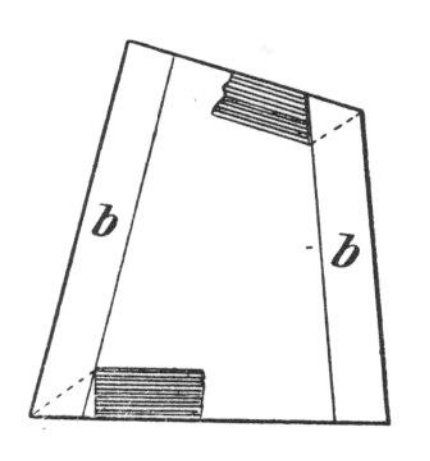

1. Method of Mitreing Moulding around a Panel.

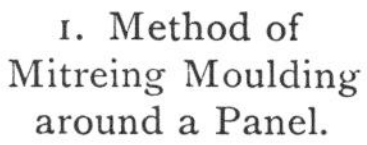

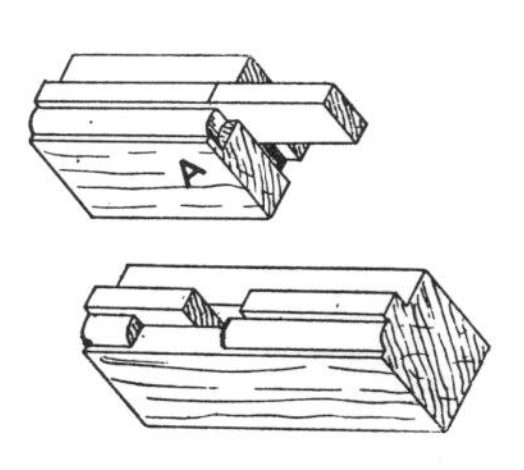

2. A Scribed Shoulder.

3. A Scribed Joint in Sash Bars.

4. A Scribing Block.

Undercut Mouldings, or those having any inner member deeper than the outer one in the direction that the frame would go together, cannot be scribed. These must be mitred. Mouldings also that have some part of their contour almost vertical are difficult to scribe, or rather to preserve the scribed portion intact during fitting, and such mouldings are usually part scribed and part mitred, as shown in f. 5 & 6. The rail moulding is first mitred right through, then the flat part of the mould is scribed away, leaving the outer part as a mitre, and a corresponding piece is

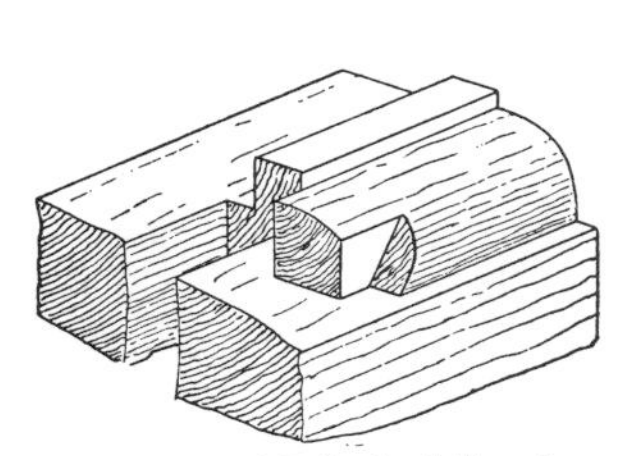

5. Part of Stile in Mitred and Scribed Shoulder.

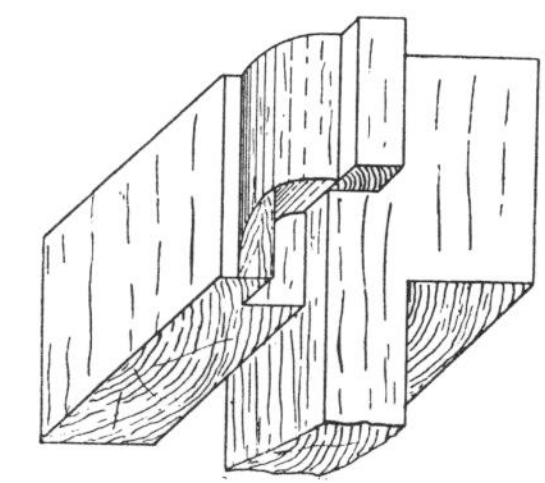

6. Part of Rail in a Mitre and Scribed Shoulder.

mitred on the stile portion, the remaining part of this moulding running on under the scribed part. Where it is possible to use the scribed joint it is always to be preferred to the simple mitre, as any shrinkage of either member of the frame in the latter case produces an unsightly gap, whilst shrinkage does not affect the appearance of the scribed joint. It may be noted here, that in the case of polished work, the portion of the moulding running under the scribe should be polished before putting together, so that, should shrinkage occur, the bare wood will not be exposed.

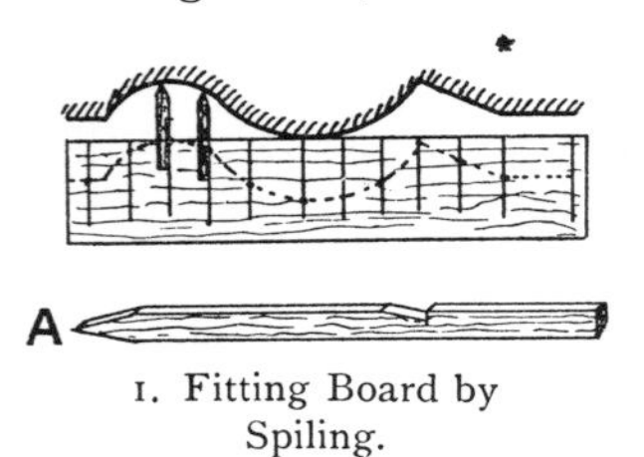

1. Fitting Board by Spiling.

Scribing of the second kind, *i.e.*, the fitting of an edge to an irregular surface, such as a skirting to a floor, is managed thus. The board to be fitted is placed in position with its upper edge level, the lower end being packed up for the purpose with a wedge or block. Then a pair of compasses, f. 2, p. 22, are spread, equal to the greatest distance of the lower edge from the floor, and drawn along the face of the stuff in a perpendicular direction, with one leg following all the irregularities of the floor, the other making a corresponding scribe on the face of the stuff. All the wood below this line is removed with saw, draw-knife, or chisel, as may be most convenient, slightly undercutting the edge, which will then fit accurately to the floor, and the top edge will be level in length.

Spiling is another method of fitting to an irregular surface allied to the above, but used when the depressions are too great to be spanned by a pair of compasses. This is illustrated in f. 1, and consists in marking off on the piece to be fitted a number of points at equal distances from the curved surface which are obtained with a rule or light rod termed a spile, cut off to a convenient length for handling, with a notch in one side for the purpose of holding a pencil to mark the points. The spile is applied at various points in a direction perpendicular to the edge of the board to be fitted, or in other words in parallel lines. This may be facilitated by drawing on the surface, before offering up, a number of lines with a square, and keeping the spile to these lines when marking off the points. The distance of the marking notch from the point of the spile A, f. 1, is made equal to the distance of the edge of the board from the greatest depression in the curve. The various points being obtained, they are joined up into a continuous line either by freehand or by bending round a thin strip to act as a guide for the pencil.

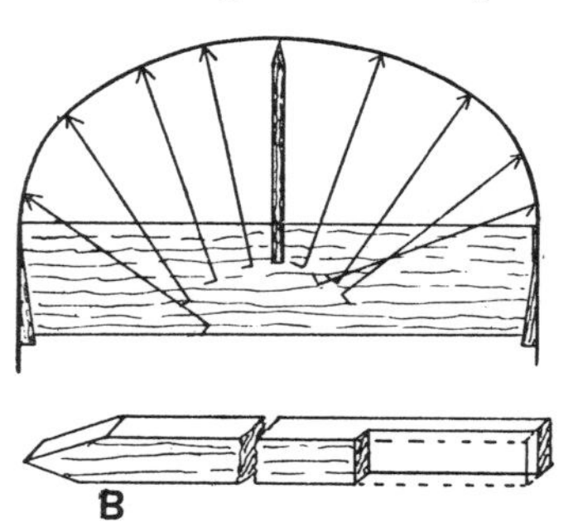

2. Taking Dimensions by Spiling.

A variation of the above, used when the framing to be fitted is too large to be conveniently placed in position for marking direct, is shown in f. 2, which illustrates the method of fitting the head of a frame to an elliptic opening. A thin board or templet is fixed by means of wedges in a convenient position for marking, with its face in the same position that the frame will occupy, and its upper edge lying in or parallel to the springing line. Then the spile cut to the shape shown at B, f. 2, p. 64, one end pointed and the other square, is applied in various positions, as the exigencies of the curve may determine, and a pencil line drawn on the templet down the right-hand side and the square end of the spile, as shown in the sketch; the arrow lines representing the edge of the spile. When a number of these positions have been marked, and they may be as varied and as numerous as possible, the templet is removed to the face of the frame, and its upper edge made to coincide with the springing, or lie in the same relative position it occupied in marking, and the spile again laid in its original positions, as indicated by the marks on the templet, and pencil marks made at its pointed end, which being joined up, will outline the required curve. For convenience of marking, the end of the spile may be rebated out, as shown by the dotted lines, so that it will fit down over the templet and rest flat on the work. The taking of spilings is a convenient method of fitting surfaces around columns or into polygonal openings.

DOVETAILING.—There are two methods practised in preparing dovetails by hand. In the one the pins are cut first, the core cleaned out from between them, and the sockets marked by scribing round the pins with a marker, as shown in the sketch, f. 3, p. 67. In the other, the sockets are first cut, then without removing the core the pins are marked by running the saw in the same cuts upon the end of the piece to contain the pins. Each method has its advantages and disadvantages. The first method gives the opportunity of correcting any fault that may occur in cutting the pins, and therefore is much favoured by the inexperienced dovetailer. It is also necessary in lap and secret dovetailing, where neither pins nor sockets come through; and it is very suitable for small work where little cleaning off is required. This method, however, demands great attention in the marking to ensure correctness, and it is very slow in comparison with the second method, by which a large number of sockets can be cut at the same time, thus dispensing with many markings, and ensuring symmetry and regularity, which adds greatly to the finished appearance of the work. The method of marking the pins is easy and accurate, the one drawback being that an appreciable quantity of stuff has to be cleaned off at the finish to remove the saw marks. This, however, can be met by using thicker stuff, or setting the work out rather full to size. The first method is the one more generally practised by cabinetmakers, the second by joiners, for obvious reasons, the work of the former being usually small and consisting of few pieces alike, that of the latter heavier and in greater number of repeats.

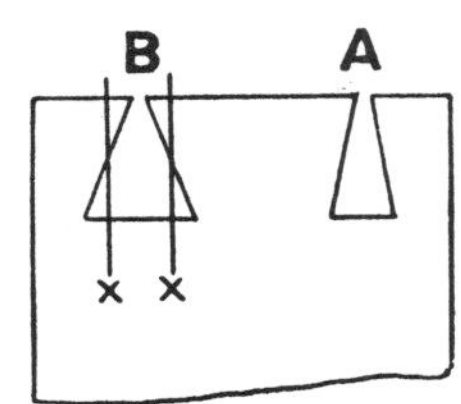

Correct and Incorrect Angles for Dovetails.

Angle of Dovetails.—There is a tendency among the inexperienced to make the angle of the sides too acute. Such dovetails, though appearing strong, are really weaker than if cut more nearly parallel. Thus, as shown above, the dovetail at A is much stronger and of better appearance than the one at B, although it is considerably smaller, because in the first case the fibres of the wood are less cut across than in the second, and in driving the dovetail B together, the portions of the pin and socket respectively on each side of the lines × × are liable to break off. All that is required to make a dovetail joint effective is that the pin should be slightly larger inside than outside, and even this has been proved unnecessary by the experience of machine-made corner locking, in which the friction of the accurately fitting interlocking pins is so great that once driven together they cannot be separated without breaking. The slight bevel on the pins of the dovetail, however, whilst giving it the appearance of greater strength, also compensates for any slight inaccuracy in the cutting. A very useful angle suitable for all dovetails is 80 deg. with the end, or 10 deg. out of square, and a handy tool for marking them with is the **Dovetail Bevel**, f. 1. This is a piece of mahogany $2\frac{1}{2}$ in. long, $1\frac{1}{2}$ in. wide, and $\frac{1}{4}$ in. thick, rebated out $\frac{1}{16}$ in. on each side to within $\frac{3}{4}$ in. of the end, with shoulders cut to an angle of 80 deg.

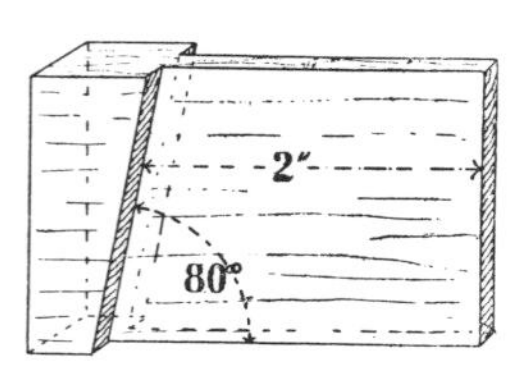

1. A Dovetail Bevel.

Size of Pins.—The strongest arrangement in dovetailing is to make the pins in one piece, equal in size to the spaces between the sockets on the other, as shown in f. 3 & 10, p. 141. This form is used for cisterns and other heavy work, where strength is of more importance than appearance. The spaces should not, however, exceed 3 in. in width, as the effect of shrinkage becomes appreciable in that width. To space the pins equally, having squared over the thickness of the end upon one of the sides, draw a pencil line in the centre of the thickness as at *a a*, f. 10, p. 141. Determine the number of pins to be used, say two, and two sockets. This will give four equal divisions, calculating the two outside half sockets as one. Assuming the side 9 in. wide, divide this by 4, giving $2\frac{1}{4}$ in. as the width of pin. Now set off on line *a a* from the edge half a space, viz., $1\frac{1}{8}$ in., then three spaces $2\frac{1}{4}$ in. wide, finishing with the remaining $1\frac{1}{8}$ in. Draw the bevel lines through these points with the dovetail bevel, f. 1, and the pins and sockets on either side will be equal. For lighter work, and where the appearance is a consideration, the pins are made much smaller than the spaces, generally in the ratio of 1 to 4, but a maximum space should be $1\frac{3}{4}$ in.

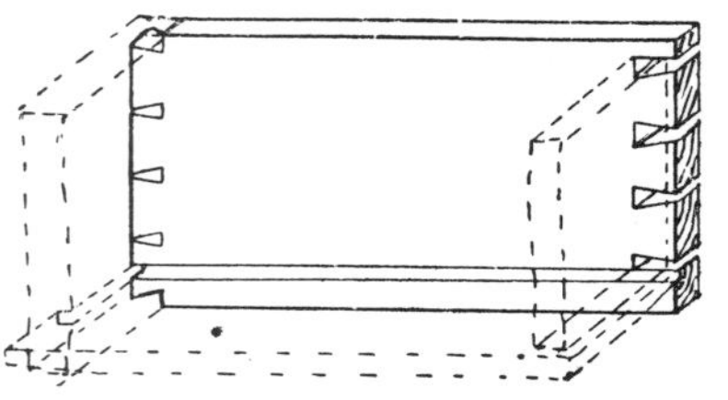
2. Inside View of a Drawer Side.

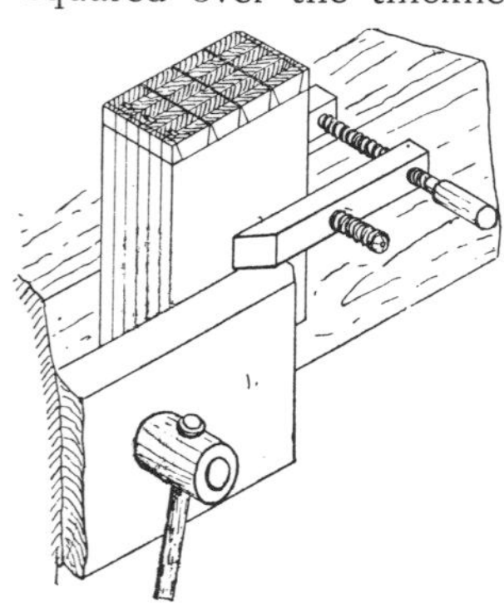
3. Method of Cutting Sockets in Drawer Sides.

In **Drawer Construction** the pins upon the front are kept one quarter of the thickness back, so that no end grain may be visible in front, and the thickness of the pin where it joins the lap is usually equal to the thickness of the saw blade used in cutting the sockets. The pins in the back are generally made stouter and spaced wider for the sake of economy. A half pin should be cut on the bottom edge, which is kept flush with the upper side of plough groove, and a similar pin upon the top edge, when it is kept below the sides, as must be done in any close-fitting case, to let the air escape from behind the drawer. This arrangement is shown in f. 2, p. 66, which represents one side of a drawer viewed from inside, with the position of the back, bottom, and front indicated by dotted lines. When a number of drawers of similar size have to be made, the sides are placed altogether in pairs, secured with hand-screws and squared off to neat length, then placed in the bench screw, as shown in f. 3 opposite, the sockets spaced and all cut through at once, cutting gauges being set to the depth of sockets, and run all round as shown. The gauge that is used to mark the front end of the sides should also be used to mark the lap on the front, and must be run on the end from inside. The pins are marked by setting a Front up in the bench screw, with its grooved side towards the bench, resting a side upon it with the end accurately to the gauge line of lap, and kept at the right height by a small slip inserted in the two plough grooves, then steadily drawing the front end of the dovetail saw through each cut in the side, which will mark the exact size and position of the pins upon the end. This operation is shown in the photo above. The pins should be cut slightly larger than marked. The saw should be run down just on the outside of the saw kerf. The core can be cut away with the bow saw, and the shoulders and back finished quite square with chisels. It will be found very convenient to keep two special $\frac{1}{2}$-in. chisels for dovetailing, with their edges ground in pairs to an angle of 80 deg. for clearing out the corners, and f. 2 illustrates a handy little square for testing the back and shoulders. It is made either of brass or hardwood $\frac{1}{8}$ in. thick, with the edges *a a* parallel, and *b* at right angles with them. When the core of the sockets is cut out the inside edges should be slightly chamfered off to prevent them getting broken when the parts are knocked together.

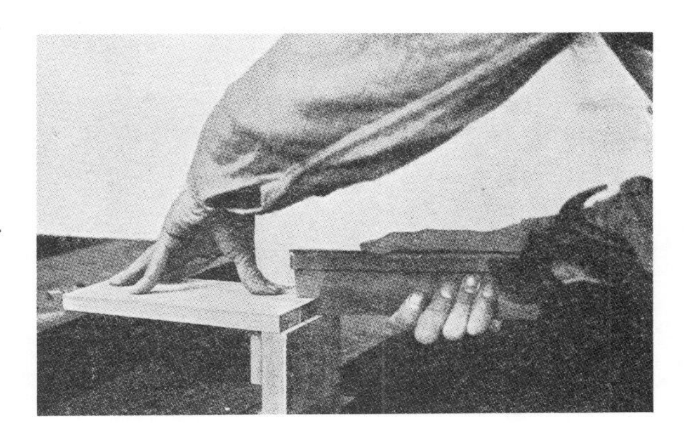

1. Method of Marking Pins of Dovetails in a Drawer Front.

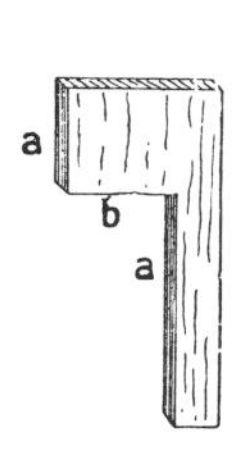

2. Dovetail Square.

3. Method of Marking Sockets in Secret Dovetail.

Mitre or **Secret Dovetailing** is used in joining the salient angles of plinths, cornices, desks, &c., in high-class work, and as mentioned before, it is necessary in this form of dovetailing to cut the pins first. The method of marking the sockets is clearly shown in the sketch, f. 3, p. 67. The sockets in this case must be cut within the lines so marked. After the marking is done, the ends of the pins are cut off ¼ in. back, and a mitre formed on the lap. A convenient way of shooting the mitre is shown in f. 1, where a piece of stuff B equal in width to the dovetail piece is shot to a bevel of 45 deg., and secured to the face of the work with thumb cramps. This acts as a rest for the shoulder plane, as shown, and enables the mitre to be shot with ease and accuracy.

Boring.—In boring holes with a Bradawl by hand pressure the tool should be revolved as it is pressed forward; but in driving with the hammer this is less

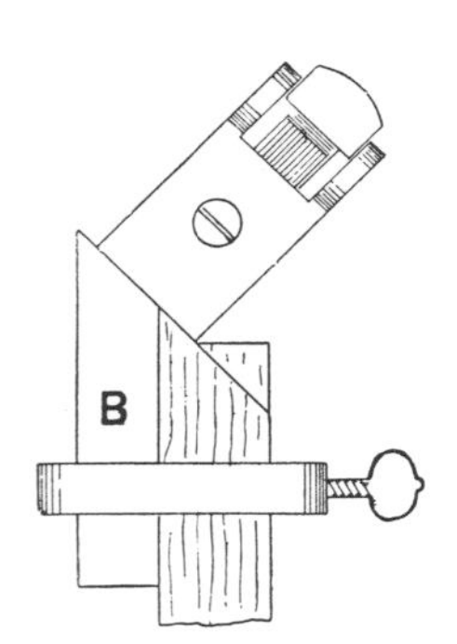

1. Method of Shooting the Mitre.

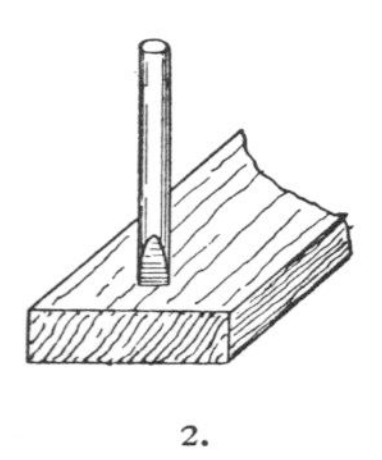

2.

3. Method of Upright Boring.

needful, but it should be turned to and fro before it is withdrawn. It must always be entered with the edge across the grain, as shown in f. 2, to prevent splitting. Gimlets should not be used for boring into thin wood nor near edges; as they are made tapering, and thus have a tendency to split the wood. This may to some extent be counteracted by securing the stuff in a bench or hand-screw, but generally a Bit is to be preferred. In boring with Brace and Bits, suitable bits should always be selected for the work. Shell or spoon bits, f. 5, p. 25, are suitable for boring across the grain, but not with it, in which direction they will run out of truth through the core not clearing. Centre bits will only cut across the grain. Nose bits, f. 4, p. 25, are primarily intended for boring in the direction of the grain, but they can be made to bore across if the hole is commenced with a bradawl. Twist bits, f. 2, will bore in any direction, and will make a much cleaner and truer hole than any other kind. They require considerable care both in use and storage. The hole should be commenced exactly in the direction it is intended to go, as any attempt to alter the direction during the boring will result in a bent or broken bit, both equally useless. The method of holding the brace when boring upright is shown in the photo opposite. The left hand holding the stock should be steadied by resting the forehead on it, and the direction of the bit should be guided by taking it out of wind with a straight-edge fastened on the directing line. Holes may be conveniently and accurately bored in a horizontal direction by resting the stem of the bit upon a block equal in thickness to the distance that the bit is from the surface of the support. This is shown in the photo above. When it is requisite to bore a hole any certain depth, a collar or sleeve of wood, f. 2, is secured to the bit by two small screws, leaving the required amount uncovered. Each size bit requires a separate collar, but they are readily made by boring holes in each end of a cylinder of wood, one to fit the shank, the other the stem of the bit.

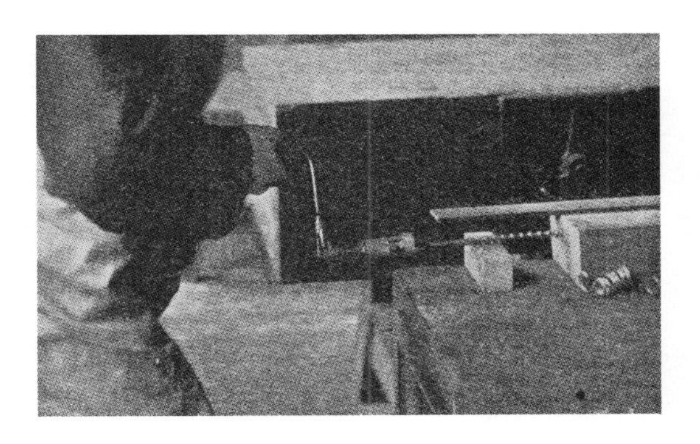

1. Method of Horizontal Boring.

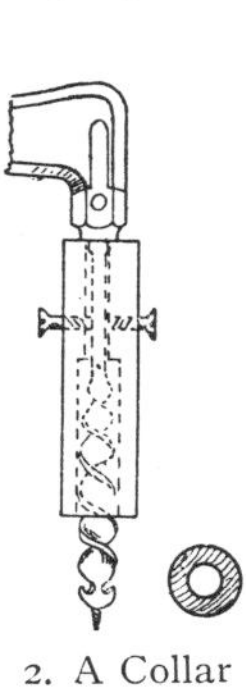

2. A Collar for Bits.

Fox-wedging, or foxtail-wedging as it is sometimes called, consists of a stub mortise and tenon, made to fit accurately in depth and thickness, the ends of the mortises being undercut to receive the spreading tenon, which is forced out as it is driven home by a number of small wedges inserted in the end, as shown in f. 4, p. 150. This effectually prevents the withdrawal of the tenon, which should present the appearance of f. 5, p. 150, if ripped down the centre. **To Make this Joint**, the mortise should be bored out to a uniform depth, the tenon cut $\frac{1}{8}$ in. short of the depth, and spaces for the wedges as shown, the longer ones on the outside diminishing gradually towards the centre where the splitting must be slight, or it will extend beyond the shoulder. The wedges should be of hardwood, straight-grained and tough, well fitted, with slight taper, and planed quite smooth. After fitting the rail accurately without the wedges, and it should fit very tight in the length of the mortise to prevent splitting, fix a hand-screw close to the shoulder as additional protection. Well glue the wedges and tenon,

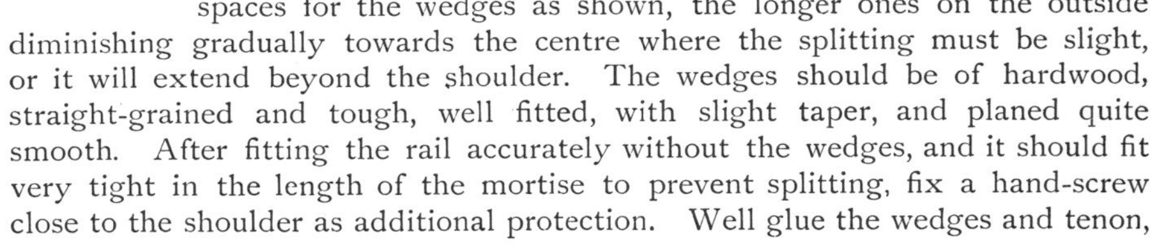

and enter the wedges in the cuts, glue the mortise, and cramp up steadily but continuously. If the operation is suspended but for a moment, the wedges will set and break, spoiling the joint. The cramp should be left on the joint until the glue is hard.

Drawboring is a method of tightening up the shoulders of framing in mortise and tenon joints, when from the size it is not convenient to use cramps. It is done with the aid of a tool called a drawbore pin, which is made in two forms. The one shown in f. 1 is known as a joiner's pin. It is a highly polished low-tempered steel bar, tapered in length, and fixed in a round handle. The other form, chiefly used by carpenters, as shown in f. 2, is made of malleable iron, somewhat stouter, and with a head formed in the solid containing a hole used for the insertion of a lever to twist the pin out with. The method of using the pin is as follows:—Drive in the tenon until the shoulder is close up, then mark the outside edge of the stile upon it. The tenon having been left purposely about 1½ in. longer, withdraw the tenon and bore square through with a ⅜-in. or ½-in. bit, the centre of the hole being on the line and in the middle of the tenon. When the tenon is again inserted, the hole will project half-way; if the pin is now driven in as shown, the shoulder will be brought up tight, f. 1.

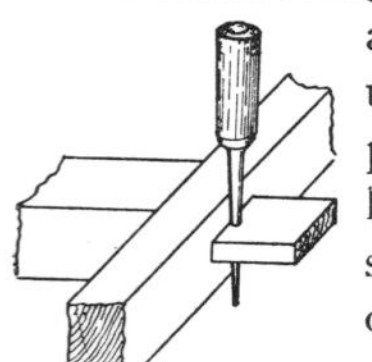

1. Method of Drawboring.

2. A Drawbore Pin.

Drawbore Pinning is a method of fastening a stub tenon by means of a hardwood pin driven through the cheeks of the mortise. It is chiefly confined at the present time to rough outside work, or imitation mediæval joinery, as it is neither so effective or of so good an appearance as a glued and wedged joint. The hole is first bored through the mortise sides, then the tenon is inserted, and the centre of the hole marked upon it by twisting the bit used for the boring backwards. Then the hole is bored through the tenon $\frac{1}{16}$ in. nearer the shoulder, and when the pin is driven in it

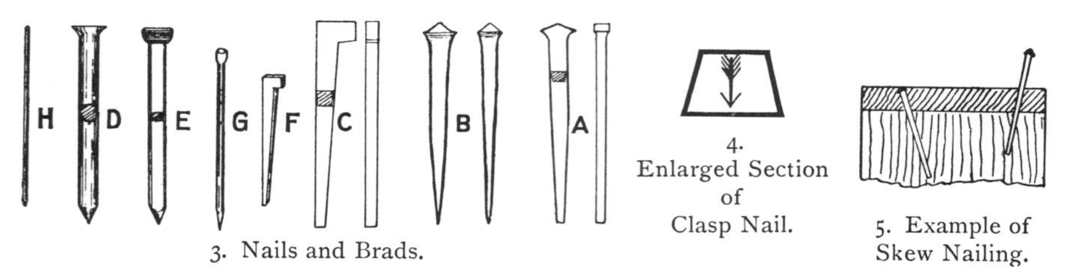

3. Nails and Brads.

4. Enlarged Section of Clasp Nail.

5. Example of Skew Nailing.

continually exerts a strain, tending to keep the shoulder up. The pin should be cleft, and is better left with a number of sharp arrises than shaped in a dowel plate. When used in painted work, it is usual to countersink the ends of the pins below the surface and stop them flush with putty, but the subsequent shrinkage of the wood invariably betrays the presence of the pins.

Nailing, Bradding, and Screwing.—The various metal fastenings used in Joinery comprise nails, cut or clasp, wrought ditto, French and American wire ditto, floor brads, joiners' brads cut and wrought, panel pins, needle points, f. 3, wood screws, f. 2, p. 72, handrail bolts, and handrail screws, f. 3, p. 72, and fig., p. 73. Clasp nails, A, f. 3, p. 70, are the most generally used kind. They will drive in all soft woods and many hard varieties without splitting, if the wider sides which are parallel are placed in the direction of the grain. When driven near the end of a board they should have holes bored for them. These nails and also cut brads are slightly wedging in section, as shown in exaggerated degree in f. 4, p. 70, and in driving will drift towards the wider side, as indicated by the arrow. This drifting must be counteracted, when it is desired for the nail to go in straight, by inclining it slightly towards the narrower side. It is, however, often an advantage to have the point drift away from the surface, and the nail should be inserted with its wider side towards the side that it is desired the point should go. When driving nails into end grain they should be skewed towards each other, as in f. 5. There is less danger of splitting in thus crossing several rows of fibres, and the nails hold much better. Nails are usually driven ⅛ in. below the surface with a punch, and the holes stopped with putty for painted work, or coloured stopping for polished woods.

Wrought Nails, B, f. 3, are much tougher than clasp, are more suited for driving in hardwoods, or in positions where their points can come through and be clinched. They require holes to be made for them, as they taper on both sides. Both the above-mentioned kinds are made in sizes from 1 in. to 4 in., advancing by ½ in. Larger sizes are termed spikes. They were formerly sold by the hundred—3d. for 1¼ in., 4d. for 1½ in., 6d. for 2 in., and so on, and are sometimes still spoken of as sixpenny and tenpenny nails, &c. They are now all sold by weight, but brads are still sold by the thousand.

Wire Nails are of parallel section throughout—the French, circular as at D, f. 3; and the American, elliptic as at E. These may be driven in any direction without splitting, and require no boring. The former hold best, but the latter make the smaller holes. They are made in sizes from 1 in. to 5 in.

Floor Brads are shown at C, f. 3, and are, as their name indicates, used chiefly for fixing floor boards. Sizes from 1½ in. to 3 in.

Brads, F, f. 3, require similar treatment to cut nails. The projecting head is always driven in the direction of the grain, not across it. Brads are concealed in hard wood, by carefully lifting a chip along the grain with a bradawl as in sketch, and gluing it down again after driving in the brad. Brass brads are made for use in oak and other acidiferous woods, sizes increasing by eighths of an inch from ¼ in. to 2 in.

Method of Concealing Brads.

Panel Pins, G, f. 3, are specially made wire brads; they do not hold so well as the cut, but make a smaller hole. They are difficult to drive without bending. The finer sizes require boring, and the points should be dipped in grease. The heads must be cup-shaped, not flat.

Needle Points, H, f. 3, are fine polished steel pins similar to sewing needles without eyes. They are made in six degrees of fineness, and are chiefly used in fixing small mouldings or thin mountings. They are driven into the wood and snapped off flush, being very hard and brittle. They require great care in driving. The points should be dipped in grease before driving. **Wood Screws**, f. 1 & 2, are made in iron and brass, in lengths ranging from $\frac{1}{4}$ in. to 6 in., and in diameter from $\frac{1}{16}$ in. to $\frac{5}{8}$ in. They are flat-headed, and are intended to be sunk flush with or below the surface. Projecting or **Round-headed Screws** are shown at A, next page. These are used for fixing metal to wood. All screws should have holes bored to the depth of their shank, of a size to take the shank easily. The thread will cut its own way in all except very hard wood, where this will also require to be bored for, with a smaller awl. Care must be taken in turning in screws to keep the screw-driver upright, or it will slip and burr the edges of the cut, producing a very unworkmanlike appearance. Holes for screws should be countersunk with either conical or rose bits, f. 7, p. 25, unless the head is required to be sunk, when a centre-bit hole to the required depth is first bored, and then the hole for screw within that, as shown in f. 1. After the screw is inserted, the hole is plugged with a pellet of wood similar to the work and in the same direction of grain, the operation being termed **Pelleting**. The pellets are usually turned in sticks, as shown in f. 1, A being the elevation and B the plan. The latter indicates the way the grain runs.

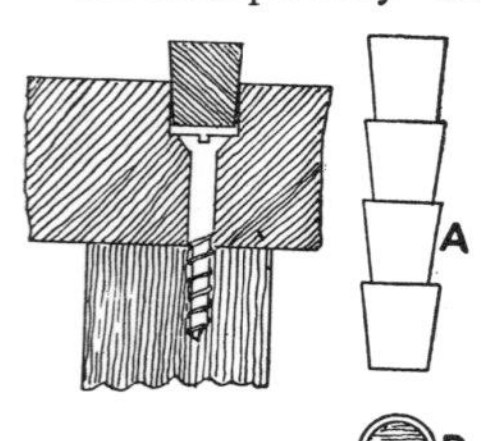

1. Pelleting.
A. Stick of Pellets.
B. Section.

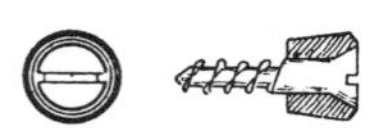

2. Brass Screw Cup.

Brass Cups, f. 2, are used to receive the heads of screws in beads, mouldings, &c., that have to be removed occasionally. The cup should be fitted in first by boring a centre-bit hole rather smaller than the top of the cup, and slightly less in depth. A drop of glue should be placed round the hole and the cup driven in with a smart blow of the hammer; a piece of steel or hard wood should be interposed to prevent driving it below the surface. Then the hole for the screw can be bored. Iron screws should always be greased before insertion, to prevent rusting, and to facilitate their subsequent removal, especially when in oak, which rapidly corrodes iron; for permanent work, brass screws should be used in this wood. A rusty screw can sometimes be started, when other means fail, by heating the head with a red-hot iron bar.

Fig. 3 is a **Handrail Bolt**. These fastenings, originally intended for securing the butt joints in handrails, are now used generally for fastening any butt joint that requires pressure to draw and keep it up. The bolt is screwed at each end, and has a square nut C at one end, and a round or "live" nut A at the other. This nut is driven round by the handrail punch, f. 1, p. 30, which enters the slots shown on the edge of the nut. A washer B is interposed between the live nut and the wood to prevent the former burying itself when turned.

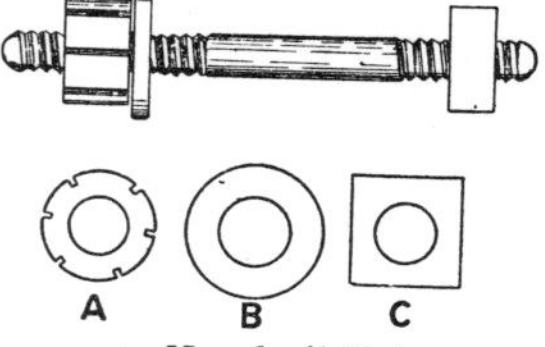

3. Handrail Bolt.
A. Live Nut. B. Washer.
C. Dead Nut.

The **Handrail Screw**, next page, is a tapered screw threaded at each end, and is used for holding up small joints in handrails, &c. The screw is turned into one piece until half the shank is embedded, then the other part of the joint is turned on to the projecting end and twisted until tight. This screw requires a little manipulation to get it exactly in the right position to ensure the correspondence of the sides of the moulded rail.

Glass-papering.—Glass-paper is used to bring work to a uniform surface by removing plane marks and other slight irregularities, and is not intended or should be allowed to take the place of the plane in reducing surfaces to a common level. Such a proceeding is wasteful and unworkmanlike—good workmen are very sparing of glass-paper, especially of the coarser grades. Glass-paper is made in seven grades, numbered O or flour, 1, 1½, F2, M2, S2, and 2½. For cleaning off deal or other soft wood that has to be painted, F2 (fine 2) and 1½ are the most suitable, and should be rubbed across the grain diagonally. A piece of cork about 3 by 5 by 2 in. thick makes the best rubber, and the paper should be cut into quarters, used in one thickness, and wrapped close to the rubber. Pine or deal that is to be left in the white or varnished should be rubbed lengthways of the grain, not across. The shoulders may be flushed off with a few circular rubs, but finished straight; 1½ and 1 are the papers for this work. Hardwoods that have to be polished are prepared with Nos. 1½ and 1, or for a high finish with "flour" paper. The first paper should be rubbed with the grain; the second passed all over in small circles, thoroughly cutting up the first made scorings; the final rubbing being with the grain, and in the case of framed work finishing accurately at the shoulders. If this procedure is followed, the surface will be dead flat, and not show scratches when polished. The coarser grade paper is used for rubbing down mouldings. Rubbers should be made the exact reverse of the moulding or some part of it, and be made of soft pine, preferably cork-faced. Hard rubbers destroy the paper rapidly. The paper should not overhang the ends of the rubber, and should be bent round closely to it, otherwise the sharp arrises of the moulding will be destroyed. Rubbing should be done quickly but not heavily. The latter will generate heat, soften the glue attaching the glass to the paper, causing it to come to the surface, clogging the wood, and in some cases tearing strips of fibre out. In cleaning up resinous woods, the surface and the paper should be rubbed with chalk to prevent clogging. Glass-paper should not be used damp, as the glue is soft, and causes the dust to cake on the surface. Woods of a woolly texture, such as Honduras mahogany, sequoia, black walnut, &c., are more rapidly cleaned up by **damping** their surfaces after the first rubbing. Hot water should be rubbed on with a sponge, only just sufficient to colour the wood. This will "raise the grain," *i.e.*, make all the short ragged ends of the fibre that have been pressed down into the pores curl up, when, after they are dry, they can be easily cut down with the second paper. It is next to useless damping down before papering, as then only the larger fibres left by the cutter will be lifted, and if too much water is applied the fibres will be rendered soft and limp and will not rise out of the pores. Some woods are of so woolly a texture that a good surface cannot be obtained without first binding the grain—that is, after the first rubbing, going over the surface with a brush dipped in size or very weak glue, which must be left to get thoroughly dry, and then dusted with powdered

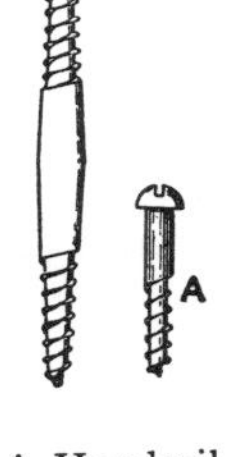

A Handrail Screw and a Round-headed Screw.

whiting before rubbing up. A high finish is given to mouldings worked on very hard wood, such as ebony or Spanish mahogany, by burnishing them after papering with a rubber made of the same wood, oiled and rubbed rapidly.

Stopping and **Removing Defects.**—Various cements are used for filling nail holes and other slight imperfections on the surface of work. Those used for soft woods are usually composed of glue and whiting, or other colouring matter, such as yellow ochre. They should be used hot, heaped up in the hole, as they contract in the drying. Stopping for hard wood has usually wax or resin as a base. Plain yellow beeswax is a good stopping for oak or teak; the wax melted over the gas and mixed with Indian red or alkanet root for mahogany; brown umber for walnut; and yellow ochre for pitch pine. All holes to be stopped with cement should be undercut to key in the cement. Large holes must be stopped with inlays of similar wood. In painted work these may be any shape. Where the defect is small enough to be bored out, pelleting is the most economical. In polished work, lozenge-shaped or saw-tooth inlays are best, as shown in figure. The inlay should be cut first with the edges under square. Mark the outline with a marking point, and mortise out carefully to the line, glue both and drive in. Let the glue dry hard before cleaning off.

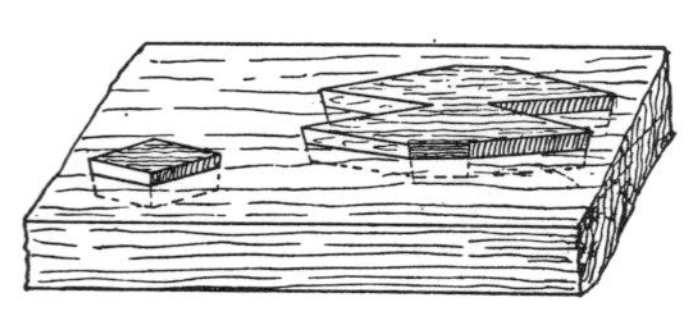

Stopping with Wood Inlays.

Bruises may be removed by wetting the depressed part with hot water, or if very deep, by laying a wet piece of flannel on the bruise and ironing it with a hot iron.

Iron and **Glue Stains** may be removed from coloured woods with dilute oxalic acid (*poison*).

SETTING OUT.—In setting out work from rods that have been prepared by the foreman, as is the usual practice in large workshops, it is not wise, unless positive instructions to that effect have been received, to work to the shoulder lines shown thereon, as these are seldom laid off with sufficient exactness for the purpose. They are merely indications of what is required. The exact amount of the sticking of planes or machine-cutters must be ascertained and allowed for. The lines struck on the work should be the "sight lines," as shown on the rod—that is, the inside edge lines of the stiles and rails in framing, and the outside face lines in fittings, cases, cupboards, &c. All allowances for mouldings, rebates, housings, &c., should be made from these lines, and in the subsequent fitting the shoulders, &c., should be eased back if required, until the sight lines coincide with the surfaces they represent. This will ensure the finished work being the right size. Allowances for sticking &c., are best set off with spring dividers or special marking slips. These latter are short prisms of hard wood planed to definite sizes, which should be marked on them. They are much safer to use than trusting to rule measurements. It is a further advantage to have the slips in various coloured woods for their easier identification.

In setting out for hand work every mortise and tenon shoulder must be squared over, and in soft woods the shoulder lines should be neat length, as they are generally put together from the saw; but in hard woods the shoulders should be left slightly "full" to allow for shooting. In preparing work for machinery, the method of marking and amount of allowance must depend upon the methods pursued in the particular mill, and close observation should be made by the workman of the peculiarities of various machinists; some of these will work accurately and even blindly to lines, altering their machine to suit any variation in the piece; others will work on the "safe side"—that is, leave lines in, shoulders full, grooves tight, &c. Others again, with more confidence than consideration, will set their machines to one piece, and run through all that is required to that size or shape, paying no heed to lines on the stuff. Each of these idiosyncrasies require special treatment, but generally it is better to set out sight lines accurately, shoulder lines full, grooves and housings bare. Gauging stickings and rebates with marking gauges is not advisable, unless in cross-grained stuff, and if they are numerous and varying, they should be run on with a pencil gauge, f. 1. For exceptions to this rule see Chapter VII. under "Rebating."

Mortises should all be set out and gauged as in hand work, but tenons and shoulders, if more than one alike are required, need not be marked on

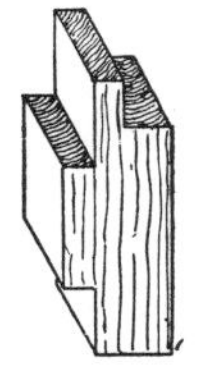

1. Pencil Gauge.

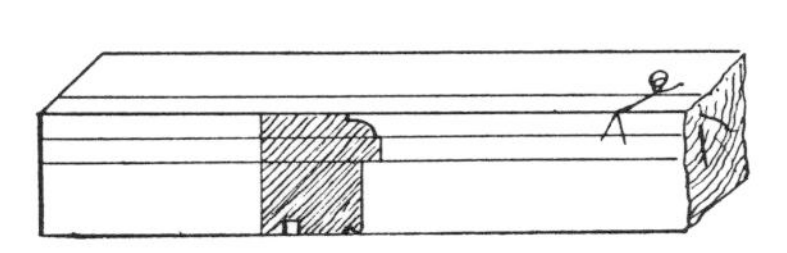

2. Jamb Set Out for Machining.

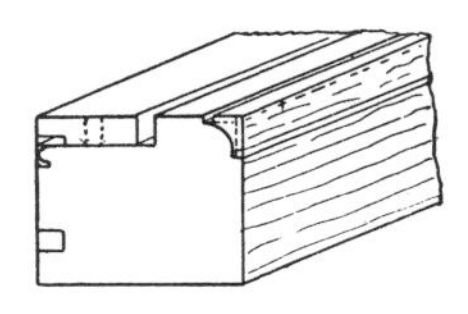

3. Jamb Set Out for Hand-working.

more than one piece, as the tenoning machine, which forms the shoulders also, when once set, cannot well vary.

Plough Grooves should not be gauged on both sides, but cross grooves and housings are better knife-marked in. Generally all shoulders and cross grooves should be knife-marked in, and all mortises pencil-marked.

True Sections should be drawn on the side of the piece, as shown in f. 2, which represents a portion of a door jamb lined ready for machining. If more than one alike is required, one section is sufficient, with a written reference to the number wanted. Sections of mouldings are best placed on the ends of the stuff. In setting out frames, &c., that start from the floor, the floor line should always be drawn on with the knife, so that it shall not be lost in the working. The presence of this line saves much time when fixing the work.

Fig. 3 illustrates the different treatment required for work to be prepared by hand. The same jamb is represented as in f. 2. Here, however, the section is drawn upon the end, and the lines for the mouldings, rebates, and grooves gauged upon the sides as shown. Small plough grooves would be run in to commence the moulding, and rebates taken out, as indicated by dotted lines.

In Setting out Dovetails for Machine Work, when a dovetail machine is used, only one set of pins and one set of sockets will need to be marked, as the machine, once set, will make any number of repeats; but where the dovetails are cut on the band saw, all the pieces containing similar sockets, such as drawer sides, should be fastened together in pairs in one large block, either by bradding one to the other, or by gluing two strips of paper across each side near the ends, and then gluing all together by these strips, which will hold them quite steady whilst cutting, and they may be readily separated with a chisel afterwards. The two outside sides should be marked in with a templet as in f. 1, and the pins marked by the corresponding templet, f. 2. It will be noticed that these templets, which are carefully fitted together, are made in "reverse." The reason of this is that the sockets will be marked the thickness of the line

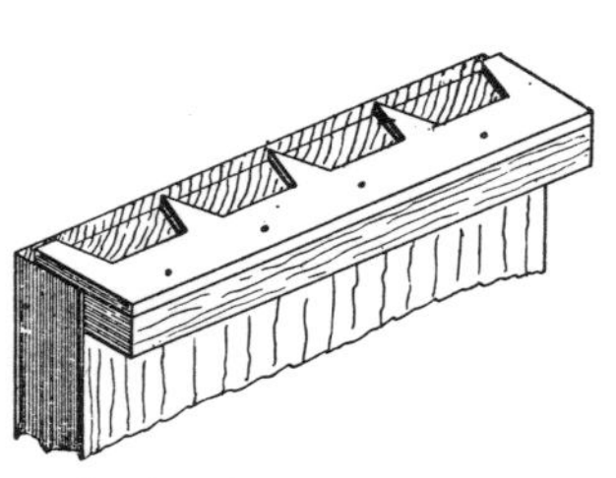

1. Setting Out Dovetail Pins for Machine Cutting.

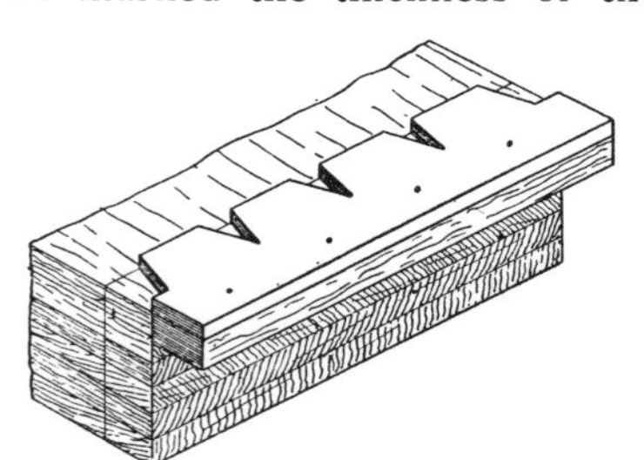

2. Setting Out Dovetail Sockets for Machine Cutting.

smaller, and the pins an equal amount larger than the templets, which will allow for contingencies. The pins on the fronts of drawers are run in by tilting the table of the band saw to the pitch of the dovetails, and fixing a wedge-shaped fence across in front of the saw to give the piece the correct angle with the line of blade, as shown in the sketch, f. 3. The core of the fronts can be removed by the "elephant" or recessing machine.

Cleaning Off is the production of a finished surface, flat, straight, or regularly curved, and free from torn up or stripped fibres, and as applied to framed work, consists in reducing the inequalities at the shoulders and elsewhere to a dead level with the trying plane, then going all over the surface with the smoothing plane to remove the "tears" occasioned by the coarser set iron of the former. This operation is succeeded in the case of work to be painted by "papering up" with glass-paper across the grain, as described under the head of glass-papering. In the case of work to be polished, **Scraping** precedes the papering up. This is the removal from the surface of a series of extremely fine shavings by means of a small thin sheet of steel called a scraper. The scraper removes all the marks left by the smoothing plane, and produces a very highly finished surface.

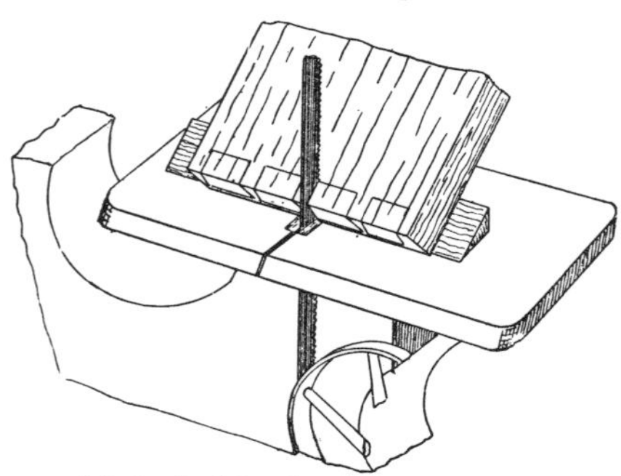

3. Method of Cutting Lap Dovetails on Band Saw.

In Cleaning Off a door or other similar framing, commence by levelling down the muntings to the rails, then the rails to the stiles, then try up the stiles straight through, work the plane diagonally across the shoulders, which will avoid the danger of breaking out the rails. Test all the parts with a straight-edge, and try for winding by laying the straight-edge from corner to corner. If the work is in wind, it will be hollow on one diagonal and round on the other. Remove the high places.

To Smooth Up, set the smoothing plane iron as fine as it will possibly bite, go over muntings, rails, and stiles in the same order as in trying off, but plane in the direction of the grain, finishing the rails exactly at the shoulders. The shoulders should be tested with the back of a foot-square to see that they have not been made hollow in the smoothing. Lift the plane off when it reaches the end of a rail, and be careful not to cut across the grain of the rail. In smoothing the stile, such cross planing, though not visible in the "white," will be much in evidence after polishing. After all marks have been removed by the smoothing plane, the work must be scraped.

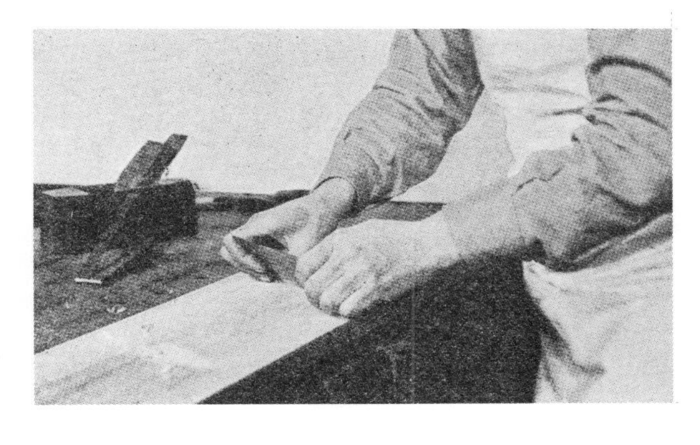

1. Method of Using Scraper.

In Using the Scraper, grasp it firmly with the fingers of each hand around the ends, the thumbs behind and the fingers in front, as shown in the photo above. It must not be bent, or hollows will be produced. Tilt the scraper away from the body at an angle of about 45 deg., and press steadily but lightly, passing the tool in short strokes about a foot in length over the whole surface until every mark and irregularity has been removed. This state of the work can best be ascertained by drawing the hand slowly over its face palm downwards. If the scraper is thin and liable to bend, it may be stiffened by inserting it in a holder or pad, f. 2. This is made of a piece of pine with a saw-cut on one edge to hold the blade. The pad is also useful to hold the scraper whilst sharpening it.

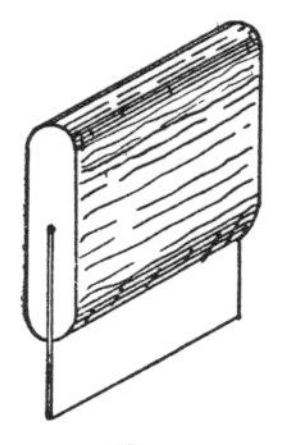

2. A Scraper Pad.

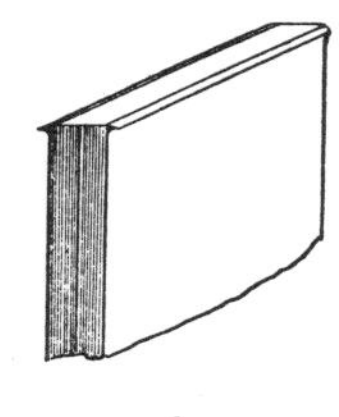

3. Edge of a Scraper Magnified.

To Sharpen the Scraper.—At the first time of using, the edge must be filed perfectly straight and square; then all the file marks must be obliterated by rubbing carefully on an oilstone until a polished straight edge is obtained. In this operation the scraper must be held perpendicular to the stone. Then lay the scraper on its side close to the edge of the bench, give it several sharp rubs with the steel sharpener held quite flat on each side, to remove any burr left by the stone. Next, holding the scraper upright on the bench, and grasped firmly in the left hand, draw the steel in one hard, steady, but rapid stroke along the edge from end to end. The sharpener must be held slightly under a right angle with the face, the handle being inclined towards the body, so that a burr is produced on the near edge of the scraper, which is then turned over and the operation repeated on the other edge. Fig. 3, p. 77, which is a section of a scraper, shows the burr or sharp cutting edge (much exaggerated). The edge is best produced by one sharp rub; a second stroke will frequently destroy the burr produced by the first, as it is difficult to keep the hand exactly at the same angle.

Protection of Work during Transit.—All open frames—that is, those having only three sides, such as door frames, jamb linings, &c.—should be braced diagonally whilst on the bench, to prevent them getting racked out of square when handled. The brace must form a triangle with the head and one side, and may be cut in the rebate or nailed on the edge, as most convenient; and they should also have a rough stretcher fixed near the bottom to preserve their parallelism. Fitted abutments and joints should have deal packing pieces bradded to them; and fitted sashes, short packings nailed at each end of the stiles. Sashes are generally fitted into the frames on the bench, then removed, marked on the edges with a chisel for easy identification, and then handed to the glazier. After glazing, they are weighed to ascertain the necessary counterbalancing weights required; then sent to the building and stored in a safe place until the plastering is finished, when they are hung in the frames. Doors and framed work are sent out from the shop with the horns left on as a protection to the rails. The top horns of sash frame linings are usually left on for fixings. In work that has to be fitted to walls or floors a sufficient extra width should be left on for scribing thereto.

Polished work, such as framing, counter tops, &c., should have pieces of coarse felt interposed between them, to prevent scratching, and be enclosed in rough crates, or have short packing pieces screwed on the top and bottom ends.

CHAPTER V.

THE STEEL SQUARE IN JOINERY.

The "American" Square: its uses in Carpentry—The "British" Square: improvements by the Author described—Methods of obtaining Roof Pitches and Bevels—Describing Polygons of from five to ten sides—Uses in fixing as a Plumb Rule, as a Level, as a Mortise or Panel Gauge—Marking the Shoulder of a Diminished Stile Door—Obtaining the Mitre Lines of Double Splayed Linings—Bevels for Hoppers and Trays with Splayed Sides, Mitre and Butt Joints—Obtaining true shape of Skylights in a Hipped Roof—Setting out a Triangular Louvre Frame; Bevels for Louvres, marking the Housings—Setting out Stairs—Step Ladders—Describing Ellipses with the Square—Describing Circles.

THE ordinary or "American" Steel Square is generally considered more adapted to Carpenter's than to Joiner's use, and in the companion volume to this book its uses in roofing and other framing has been fully dealt with; but the several improvements made in this tool by the author have considerably enlarged its scope of usefulness, so that many operations in Joinery that have hitherto required the preparation of special drawings or moulds to obtain the necessary bevels, &c., may be more readily done by its aid and a few simple calculations. Some of these improvements are patented, and the tool is known as "Ellis' Universal," otherwise "The British Steel Square." It can be obtained of the maker, Mr G. Buck, 242 Tottenham Court Road, London, or through any large tool dealer. This square is provided with an adjustable steel bar or fence, by which it can be converted into triangles of various dimensions or degrees, and it is the sundry properties of the right-angled triangle that are utilised to obtain dimensions or produce required bevels. These principles, having been fully explained in "Modern Practical Carpentry," need not be repeated here: space only permits of brief references to a few of the operations that this square may be applied to. The more important side of the improved square is shown in f. 1, p. 80—the edges divided into inches and parts. At the middle of the length of the blade is a semicircle divided into degrees, which may be utilised for obtaining given angles. Certain lines are engraved upon both arms radiating from the centre of the arc—these are the common roof pitches to which the fence, shown in f. 2 & 3, is adjusted for obtaining lengths and cuts of rafters and braces: other pitches are obtained from the index arc. Upon the left hand arc lines representing the sides of various polygons are engraved. A bevel set to the edge of the square and the indicated line will give two sides of that polygon, and the operation repeated for the required number of sides will complete the

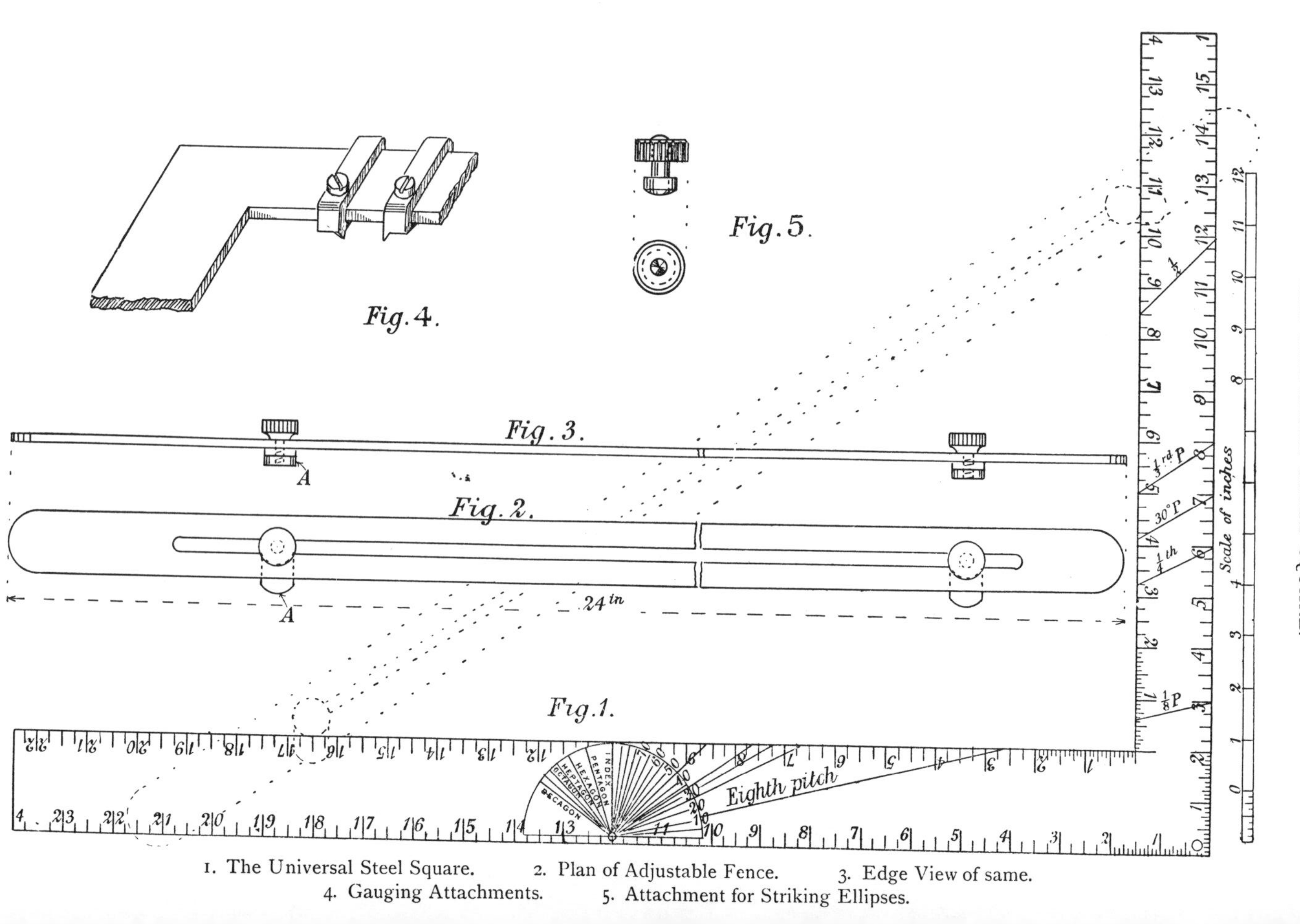

1. The Universal Steel Square. 2. Plan of Adjustable Fence. 3. Edge View of same. 4. Gauging Attachments. 5. Attachment for Striking Ellipses.

figure. A plummet hole (not shown in the drawing) is made in the blade, and by aid of a small plumb-bob work can be levelled or plumbed when fixing it. Fig. 4 shows the gauging attachment, by means of which single or double lines can be accurately gauged from $\frac{1}{16}$ in. apart up to 22 in. Fig. 5 is the pin attached to the fence for the purpose of describing circles and ellipses, as explained below.

To Obtain the Bevels for a Diminished Stile Door, see f. 1 & 2, Plate I.—Set the adjustable fence, by means of the thumb-screws, to the width of the rail upon the tongue or narrow arm (less any sticking) and to the amount of diminish upon the blade or wide arm (plus any sticking)—these dimensions in the example are 10 in. and $2\frac{3}{8}$ in. respectively. Then hold the fence to the edge of rail, and the blade will give the shoulder cut. By applying the Square to the sight lines on face of the stile, as shown in f. 2, the tongue of the square will give the corresponding shoulder for the stile. If the back edge of the stile is worked from, apply the square as shown in the dotted lines.

To Obtain the Bevels for Splayed Linings, see f. 3 & 4.—Let *a b*, f. 3, represent the face of a splayed jamb-lining in plan, and *b′ b″*, f. 4, the elevation of the front edge; *c c′* is the soffit to which it is mitred. It is required to find the bevel for the mitre joint *b″* 4. This is shown on the developed side, which is merely drawn to make plain the application of the square, and is not necessary in practice (this holds good of most of the drawings here given, which are not necessary when the reader has mastered the principle of the tool).

Rule.—Set the fence so that, measuring from the heel *x*, the width of the lining $a\ b=9$ in. is shown on the blade, and its run, or the amount it is "out of square," $c\ a=4$ in. is shown on the tongue. Then apply the fence to the inside edge of the board, and the tongue will give the correct bevel as shown.

To Obtain Bevels for Hopper or Tray Sides, see f. 5 & 6.—The first is a part plan and the second a section of a tray with inclined sides. The methods of obtaining the bevels for these geometrically are shown on p. 225, but, as will be seen, it can be accomplished much quicker with the Square. First, to obtain the splay of the sides, draw the run *a b*, f. 6, or, in the absence of a drawing, determine the amount of pitch. Set the fence to this amount on the tongue, as shown on the right hand of f. 5, and set the width of the side *a c* on the blade; then, with the fence held to the bottom edge, the tongue will give the cut for the side joint. The method of obtaining the mitre cut for the edge, as shown on the left hand, is similar. Take the actual thickness of the side, *d e*, f. 6, upon the blade, and the run, *d f*, on the tongue, as shown in the plan—the tongue gives the cut. It will be noticed that the square is drawn to double the scale of the rest of the figure. Any dimensions may also be doubled, without affecting the angles, for convenience of setting out.

For a **Butt Joint**, as at F, take thickness of edge *c g*, f. 6, on blade, and run of side at level of lower edge *h i* on tongue (or a larger triangle of similar angles may be obtained by producing the edge *c g* to cut the ground line; then take length of this produced edge on the blade and the run *a b*, f. 6, on the tongue); with the fence applied to face side, the tongue will give the butt joint.

Hipped Skylights or Lantern Roofs.—See f. 7, in which the dotted triangle represents the plan of the end light, and the grained parts the true shape of stile and rail. The drawing is made to show the application of the square, but the only things necessary to obtain the bevels are the width of the roof, 6 ft. 10 in., and its rise, 3 ft. 3 in. With these, the length of the hip, or the stile of the light can be found. First take half the width of roof, 3 ft. 5 in., on the blade, and the rise, 3 ft. 3 in., on the tongue; measure the diagonal joining these figures across the heel of the Square, which will be found to be 4 ft. 9½ in., the *run* of the hip. (*Note.*—Inches on the square represent feet, and the $\frac{1}{12}$th divisions inches.) Next take the run, 4 ft. 9½ in., on the blade, and the rise, 3 ft. 3 in., on the tongue; then the diagonal will be found to measure 5 ft. 9½ in., the nett length of the outside edge of the light on its joint lines. Having obtained these dimensions, set the fence to read 4 ft. 9½ in., the run of hip, on the blade, and 3 ft. 5 in., half the width of the light, on the tongue, and apply the square as shown in the drawing—the tongue will give the shoulder cut. To find the edge or joint bevel; set the fence to the length of the edge, 5 ft. 9½ in., on the blade, and the rise of the light, 3 ft. 3 in., on the tongue. Cut the horn of the stile, square from the face, then rest the tongue on the face, and the fence will give the joint bevel. This is exactly the same procedure as finding the "backing" of a hip.

To Obtain the Bevels in a Louvre Frame.—See f. 8 & 9, which are elevation and central sections respectively of a triangular ventilator, such as is used in stables, &c. To find the end cuts for the boards, take 12 in. on blade, and run of the frame in the width of a board, as *a b*, f. 8, on the tongue, and tongue gives the end cut. To find the bevel of the housings, take 12 in. on the blade, and the thickness of frame on the tongue, hold fence to opposite side of frame, as shown in dotted lines, and the tongue gives the trench bevel. The edge is developed to show this plainer.

Setting out Stairs.—For Cut Strings.—Set the fence to the going or width of the step upon the blade, and the rise or height of the step upon the tongue, and mark from top edge of string. For Close Strings.—Run the nosing line down at the required distance from top edge, as shown in f. 10, place the square as shown to the required figures—in this case, 10 in. on blade and 6½ in. on tongue—set fence to edge of board, and line out with wedge templets held to edge of the square.

Step Ladders.—Fig. 11 shows the method of setting, being the same as for cut strings. The mortises may be set out from inside of blade by noting the figure that cuts the face edge when lying on the pattern step.

Striking Ellipses for sash and door rails, centering, &c. This is done with the square, as shown in f. 12. Draw two lines upon the stuff at right angles to each other, as A A and B B. Mark off on these the span and rise of the desired curve, then arrange the square, as shown to the right, driving in two brads or screws to keep it in position. Set the adjustable fence with its two thumb-screws, so that measuring either from one end of the slot or from the notch in the end of bar to the first and second pins respectively, their distances therefrom shall equal the rise and half-span (semi-minor and semi-major axes of the ellipse); then, placing a pencil at the end of the slot (or in the notch at end), move the bar along the edges of the square, and one-quarter of the ellipse

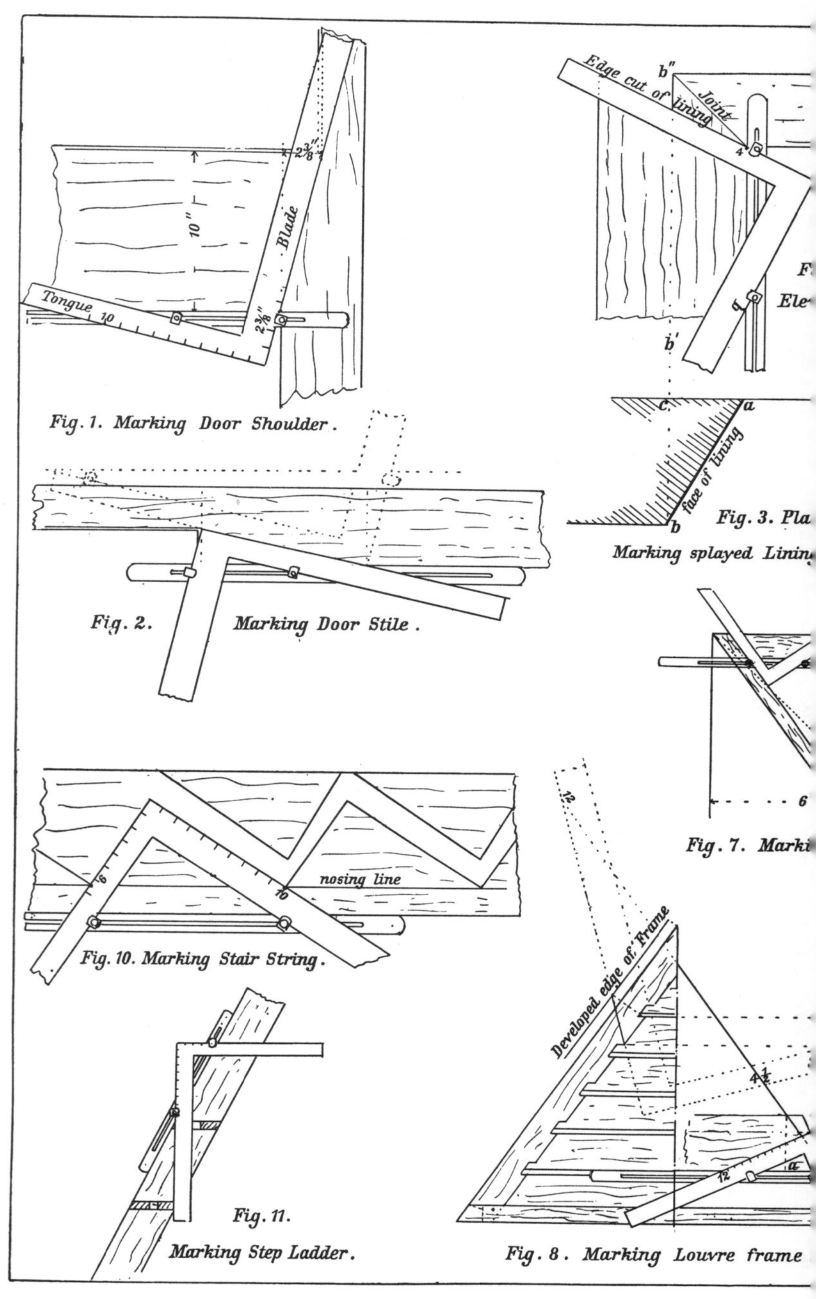
Tongue
10
Blade
10"
2⅜"
Fig. 1. Marking Door Shoulder.
Fig. 2. Marking Door Stile.
Edge cut of lining
b"
Joint
4
q
b'
c
a
face of lining
b
Fig. 3. Pla
Marking splayed Linin
Fig. 7. Mark
nosing line
6
10
Fig. 10. Marking Stair String.
Fig. 11.
Marking Step Ladder.
12
Developed edge of Frame
4½
a
Fig. 8. Marking Louvre frame

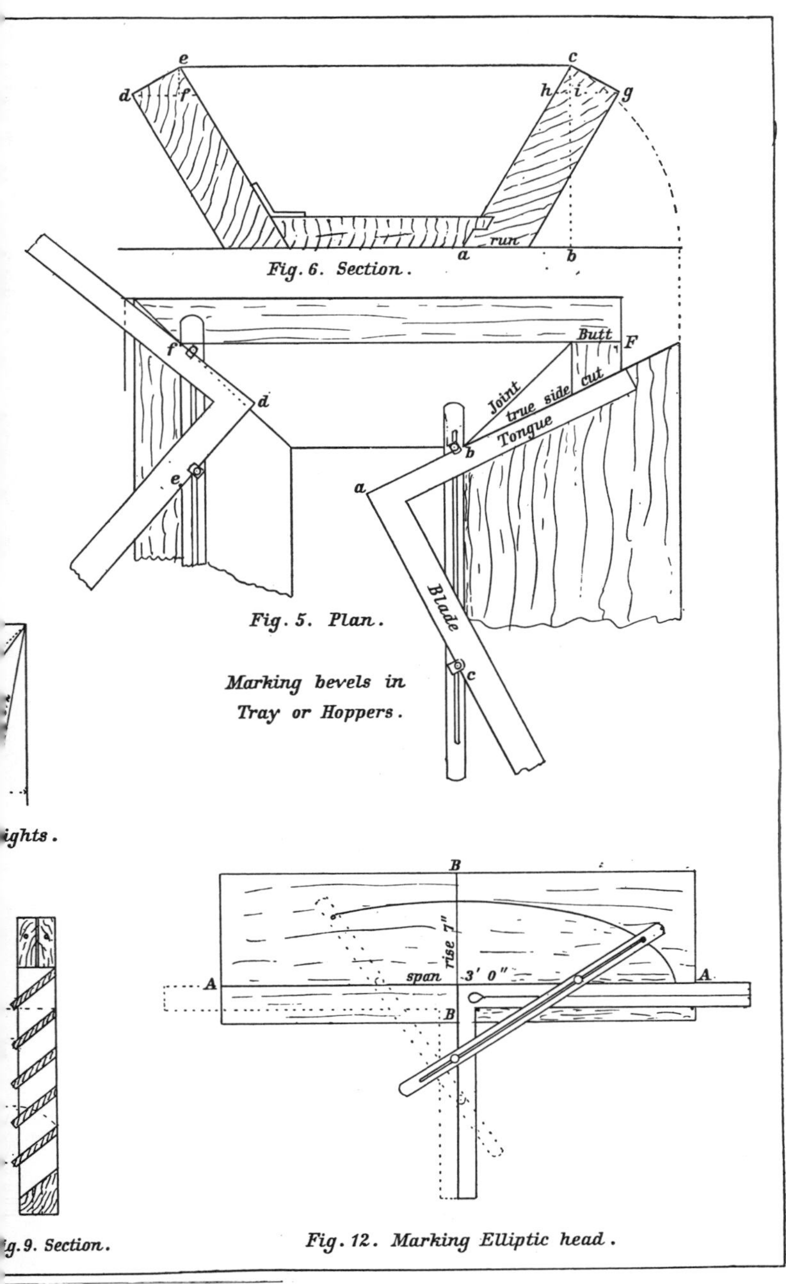

Fig. 6. Section.

Fig. 5. Plan.

Marking bevels in Tray or Hoppers.

ights.

ig. 9. Section.

Fig. 12. Marking Elliptic head.

[To face page 82.

will be drawn. To complete, turn the square over as indicated by the dotted lines, and repeat the action.

Circles of any size up to 4 ft. diameter can be described by driving a bradawl at one end of the slot in the fence to act as a centre, and arranging one of the buttons to the required radius. Place a pencil in its notch, and sweep around the curve.

CHAPTER VI.

JOINERY MACHINES.

Introduction—Scope and Limits of Remarks—Advantages of Skilled Operators—Most Useful Machines to Small Builders—Sawing Machinery: Circular Saw Benches, Classification of, according to Saws, according to Feeds, Description of Feeds, Comparison of Rising Table and Rising Spindle Benches—Description of Various Makers' Specialities—A Plain Circular Saw Bench—A Rising Table Saw Bench—Automatic Feed Ripping Bench—Cross-cutting Benches: Classification—The Joiner's Cross-cut Bench—A Combination Ripping and Cross-cutting Bench—Best Part of Saw to cut with—The Three-Spindle Saw Bench: its Uses—American Type of Tilting Table Saw Benches—Band Sawing Machines: Description of Various Types, Advantages of, Comparison with Circular and Frame Saws—A Jobbing Band Saw: Extemporised Fence, Requirements of Pulleys, Saw Blades, Sizes, Guides, &c. —An American Band Saw—A Band Saw with Swinging Frame—A Combination Circular and Band Saw Bench—Planing Machines—The Surface Planer: Dangers of, Speed of, Cutters' Capabilities—Portable Electric Planer—The Panel Planer—Over and Under Planer: Advantages of—Hand and Roller Feed Thicknessing Machine: Capacity and Speed—Spindle Machines: Description and Uses of, Method of Fixing Cutters, Varieties of Cutters, Cutting Action, Fences, Advantages of Double Spindles—Single Spindle Moulder—Double Spindle Moulder: Speed, Capabilities—A Spindle Machine with Tenoning Attachment—The Routing Machine —An Improved Elephant—Shaping and Recessing Machines of Various Makers—Tenoning Machines: Classification, a Typical Machine—A Joiner's Tenoner—Automatic Double-End Tenoner—Mortising Machines—A Power-driven Reciprocating Mortiser—The Continuous or Chain Mortiser: Advantages of this Type—A Hand Mortiser—A Combined Hand or Power Mortiser—A Combined Mortising, Boring, and Moulding Machine—Automatic Dovetailing Machine—A Saw Sharpening Machine.

INTRODUCTION.

THE increasing use of machinery in the preparation of joinery since the publication of the first edition of this book, renders necessary a fuller treatment of the subject than was at that time deemed needful in a work the chief aim of which was to give information upon the methods of handwork.

To deal, however, with *all* the machines that are employed in woodworking operations in the builders' shops was impossible in the space at the author's disposal, therefore the machines dealt with are those only that may be advantageously employed by the joiner, in the more mechanical and laborious operations connected with the craft, or in specially difficult work where mechanical accuracy is considered essential.

It may here be interpolated that in the author's opinion it is a mistake on the part of many manufacturers to recommend their machines for their supposed advantage in displacing skilled labour. His own observations, during employment in various large shops using machinery, has led him to conclude that not only will a competent joiner-machinist turn out better work than the average labourer, but he will, by his knowledge of the requirements of the specific work in hand, often utilise a machine in a manner, and to an end, never dreamt of by the untechnical operator. It should be remembered that a machine, however well constructed, cannot *think*.

The author has been requested to point out the best kind of machine for various purposes. After considerable investigation he concludes that there is no "best." If it were possible to combine all the several excellences of the various makes in one machine, even then it would not be invariably the "best" in all circumstances.

Combination machines, which are the kind more particularly dealt with herein, since they are the most likely to be the first set up in a small shop, are generally out of place in a large machine shop with ample space and driving power. Again, many types of combination machines have accessories designed for local requirements that would rarely be used in other places. Obviously to purchase these would be a waste of money to a builder in a neighbourhood where they were not required, although indispensable in the first-mentioned case, and so on. In the author's opinion the five most generally useful machines in a shop of limited accommodation and requirements are, a circular saw bench, a band saw machine, an overhand planer, a spindle moulder, and a mortise machine. In many cases the planer may be dispensed with, as a great deal of small planing work can be done upon the spindle moulder.

This chapter consists of a brief description of sundry machines and their capabilities, with a few general remarks on the characteristics of the various types; the method of using them is dealt with in the succeeding chapter.

The photographs of machines that illustrate the descriptions have been obtained specially for this work, with the object of giving the reader information of the latest patterns of machines; the names of the respective makers are appended, so that those interested in any particular example may be enabled to make further inquiries. The specialities of a few American firms are introduced for the benefit of readers in the U.S.A. Whilst all the firms that are quoted are makers of considerable repute, it must not be inferred that any not mentioned are necessarily inferior; want of space, rather than of material, has prevented reference to many others.

SAWING MACHINERY.

Circular Saw Benches are of three distinct classes, severally designed that they may best meet particular requirements. The first class, used for ripping or cutting in the direction of the grain, are in their larger forms confined to the conversion of balk and log timber into flitches, &c., and are termed breaking-down benches. Formerly these were known as "mills," but this designation, leading to confusion with the building in which the machine was housed—the saw-mill—is now abandoned. A smaller variety, termed re-sawing benches, are used chiefly for ripping stuff already sawn into market forms, or flitched in the breaking-down machines, into smaller scantlings. The second class, called cross-cutting or dimensioning benches, are, as the name implies, used for cutting across the grain, also for the conversion of joinery stuff to required dimensions, for its ready manipulation upon other machines; in other words, the first class of machine does the work formerly done with pit saws and half-rip saws, and the second the work done similarly with the hand and panel saws. The third class are termed combination benches; these are designed to some extent to fulfil the duties of the first two, and in addition, are capable of performing several other operations such as boring, grooving, mortising, &c. **Saw Benches** are also differentiated by the method of "feeding," *i.e.*, the way in which the timber is brought up to the saw for cutting; they are thus classed as Hand-Feed and Power-Feed. In the former the timber is simply fed up to the saw by the muscular exertion of the sawyer. In the writer's opinion this is a method that cannot be economically employed for cutting stuff of greater depth than 8 in., except in the case of such soft timbers as American pine and whitewood. For deeper cutting than this it is advisable to use some form of power-feed. The simplest of these is the Rope Feed, in which the timber is brought up to the saw at a regular and continuous speed by means of a rope fastened to its end by a hook, and wound upon a drum at the back of the bench, which should be driven by a belt from the saw spindle, so that the speed is proportionate to the speed of the saw. For still heavier cutting, such as balk timbers and hardwoods generally, Continuous or Roller Feeds are employed, whilst for the conversion of round timber, rack-feed benches are the best. Roller-feeds consist of (with sundry minor variations according to make) a series of small anti-friction or "dead" rollers working in the fence behind the saw, and one or more fluted "live" rollers bearing against the outside of the timber; these latter are made to revolve by endless chain gearing or bevel wheels on the connecting spindles driven from pulleys on the countershafts, and they drive the timber forward continuously at any desired speed.

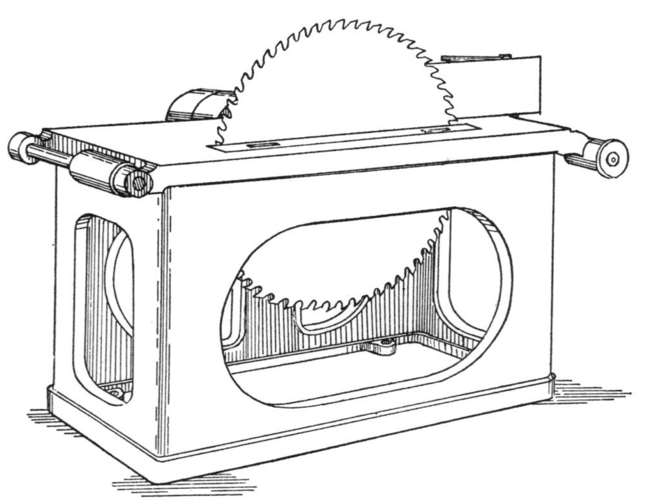

A Plain Circular Saw Bench.

Rack Benches have a travelling table in place of the usual fixed top, which moves on rollers and carries the timber past the saw. Underneath the table is a rack or series of teeth, which engage with a revolving toothed wheel or pinion, driven from the countershaft. The gearing is so arranged that the feed forward is comparatively slow, to suit the speed of the saw in use, whilst the return motion of the table is much accelerated, usually about three times faster than its forward motion. There are several other methods of feeding, such as the Automatic Slide Bench for short stuff; the Endless Chain, with lifting links or dogs to catch the timber, &c., but these are too special in their uses, to come within the purview of these notes. The larger benches have "fixed spindles," that is, the saw is mounted upon an arbor, or spindle, that revolves in its bearings, but is otherwise immovable; the difference in depths of cut required, being obtained by the use of saws of different diameters. In the smaller benches, differences in depth of cut are obtained by either the rising and lowering of the spindle, or of the top of the bench, in addition to the use of saws of various diameters. There is considerable difference of opinion among users as to which is the better method of adjustment, movement of the spindle or of the table. The disadvantage of the former is, that the length of the driving belt is altered when the spindle carrying the pulley is raised or lowered. When the machine is working at "short centres," *i.e.*, with the countershaft near to the bench, the belt must be long enough for the highest position, and it then frequently slips when in the lower; this is not such a drawback where a long belt is used, as the sag of the belt compensates for its looseness. The disadvantage of the rising table is, that when lifted to its greatest altitude, it is too high for the average man to work at with convenience. This drawback, however, is more readily overcome than that of the loose belt, by the use of a footboard, or raised platform placed around the bench.

A Plain Circular Saw Bench, as made by Messrs Reynolds & Co., is shown in the sketch opposite; it may be obtained with either a "fixed" or a rising spindle. The dimensions of the table top are 4 ft. 2 in. by 2 ft., and saws up to 30 inches diameter can be used, the maximum depth of cut being 11 inches. The frame is cast in one piece, and the rear end of the table is fitted with a bearing roller for carrying off heavy timber. The fence is capable of a fine parallel adjustment to the saw, by means of the hand-screw S, and a coarse adjustment by hand; it may also be tilted at any angle up to 45 deg., or turned completely off the table when cross-cutting is to be done. This latter movement will be better understood by reference to the plan of a similar bench shown on p. 110.

A Rising Table Saw Bench as made by Messrs Sagar & Co., is shown in the two views, f. 1 & 2, next page. These are made in three sizes from 4 ft. by 2 ft. to 5 ft. by 2 ft. 8 in., the latter taking saws up to 38 in. diameter. The table is attached at each end to wide slides working in dovetail grooves, which are adjusted by bevel wheels actuated by the large hand wheel A. The makers claim great rigidity and accuracy of movement of the top by this method. Two fences are provided, the ordinary one shown at F, f. 1, having the usual quick adjustment by lever *b*, and fine adjustment by finger wheel and screw C. It will angle up to 45 deg., slide longitudinally to suit the saw in use, and turn right off the bench upon the sleeve and spindle shown at the end. The special fence F, f. 2, is of extra length, and is fitted at the top with a dovetail slide to which a tenoning apparatus can be attached, as shown in dotted lines. The tenons are cut by a pair of small saws, mounted on the spindle with spacing collars between them. A cross-cutting slide is also provided, which runs in the groove shown at E, f. 1, and upon an adjustable bracket not shown in the sketch. This is used for shouldering, squaring up panels, &c.

1. Rising Table Saw Bench.

An Automatic Feed Ripping Saw Bench, with rising and falling table, as made by the H. B. Smith Company, New Jersey, U.S.A., is shown in the photograph, f. 1, Plate II. The method of adjusting the height of the table is novel; movement is given to the table by means of the hand wheel seen at the front of bench, which actuates, by means of worm gearing, heavy steel links at each of the four corners. Two vertical studs at the ends of the table, sliding in stout guides, ensure perpendicular movement. The automatic feed is accomplished by means of a saw disc or spur wheel, mounted on a radial arm in line with the saw; thus the marks made by the spurs in the timber, fall in the saw cut. A discharge roll with spreader is mounted at the back of the saw, and this may be corrugated or plain, according to the state of the timber manipulated. There are also anti-friction rolls in the table top. The whole of the automatic feed

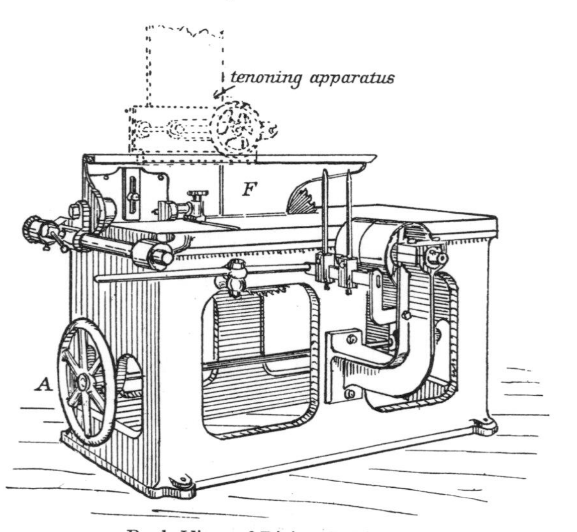

2. Back View of Rising Table Bench.

Combination Saw Benches.

1. Automatic Feed Rising Table Saw Bench.

2. A Three-Spindle Saw Bench.

[*To face page* 90.

frame may be thrown upright upon a pivot at the rear end of the bench, having the top clear for hand feeding. The adjustable **T** fence travels on a railroad at the front of the bench, and is instantly clamped in any position by a lever. Its distance from the saw is indicated on a gauge under the rail.

CROSS-CUT SAW BENCHES are of three kinds—(1) those in which the saw spindle moves either horizontally or radially, carrying the revolving saw through the timber, which remains stationary; (2) those in which the timber is carried across a fixed saw upon a travelling table actuated by hand; (3) those in which the timber is fed automatically to the saw upon a reciprocating table. There is a fourth kind of cross-cutting saw known as the **Pendulum Saw**, which, however, can scarcely be termed a "bench," as the arms upon which the saw is mounted swing from the roof of the shop, and the revolving saw is carried radially across the timber, which rests upon trestles or any convenient support.

A Joiner's Cross-Cut Bench in class 2, which is made by Messrs Thos. Robinson & Son Ltd., is shown below. The bench is a cored casting 5 ft. by 2 ft. 6 in.; the saw runs transversely to the front in self-oiling bearings. The wood to be cut is placed on the travelling table to the right hand of the saw, and is moved across the latter by hand. The table travels on a pair of rollers running on **V** tracks. On the left side of the saw is an adjustable and removable arm having a sliding stop or gauge which can be set to the required length of the timber from the saw.

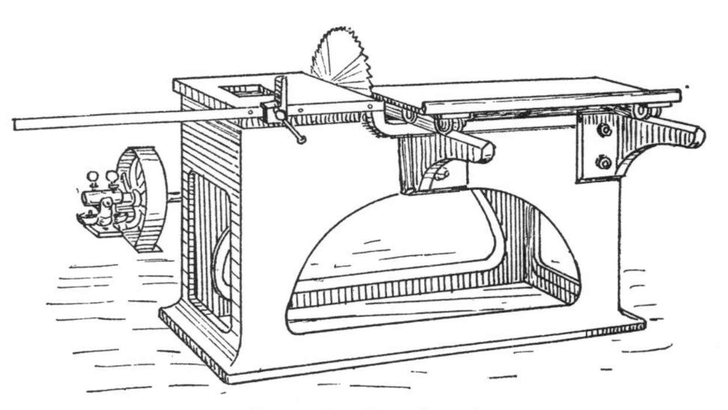

A Cross-Cut Saw Bench.

The saw is 20 in. in diameter, and will cross-cut timber of any size up to 8 in. thick and 24 in. wide. If a rip saw is used instead of a cross-cut, many odd jobs can be managed on this bench, such as taper cutting, slotting joists, cutting wedges, &c. Some of these are further referred to under "Methods of Using," in the next chapter.

A Combined Ripping and Cross-Cutting Bench, made by Messrs Kiessling & Co., is shown on next page. This has an adjustable spindle actuated by the hand wheel A; the table top has two parallel grooves running parallel with the saw, in which are fitted accurately two slides attached to a light fence B. This enables short stuff to be carried across the saw at any angle, so that mitreing, square cutting, shouldering, &c., can be done. By the use of small saws, adjusted to barely pass through the thickness of the wood, very fine and clean cuts can be obtained. It should be known that a circular saw cuts best and cleanest at its upper edge, directly above the spindle; this is due to the angle at which the teeth strike the wood; therefore to obtain clean surfaces, the wood when possible, should always be cut with the upper part of the saw. The special guard shown at the back of the saw acts also as a riving knife to prevent the stuff binding the saw when deep cutting is done.

The Universal Three-Spindle Saw Bench shown in the photograph, f. 2, Plate II., is a machine of great utility, recently patented by Messrs A. Ransome & Co. It is suitable for ripping, cross and mitre cutting, grooving, rebating, beading, chamfering, sticking sashes, and similar small work, and last, but not least, by special arrangement of the table and fence, compound angles up to 45 deg. either way may be cut with great accuracy. Three spindles, carrying respectively a ripping and a cross-cutting saw and a cutter block, are fitted to a triangular frame which revolves upon a central shaft, the arms of which are so arranged that the driving belt, by an ingenious adaptation of the equilateral triangle principle, is always at the same

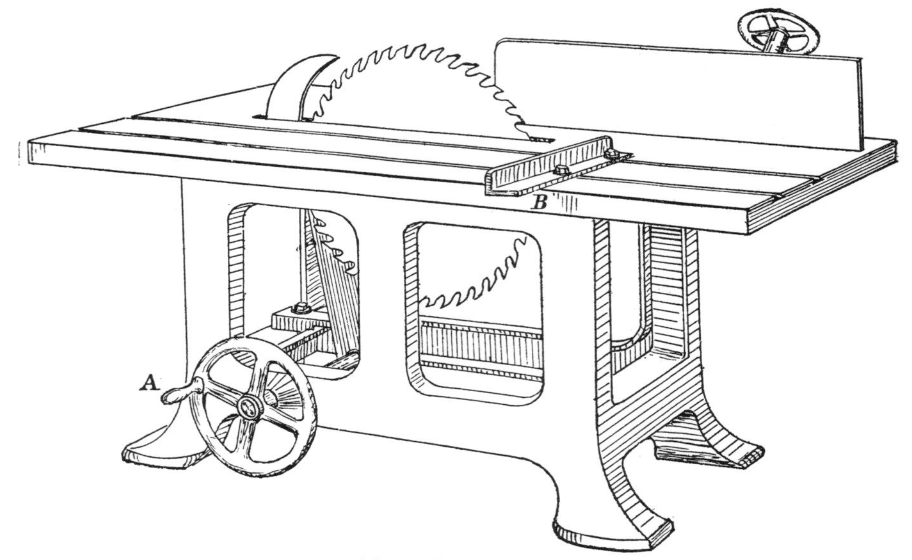

A Rising Spindle Saw Bench.

tension, whichever spindle may be uppermost and in use. Either of the respective spindles can be instantaneously brought into operation by turning the hand wheel shown at the back of the machine, and locked in position by means of the lever in front. The hand wheel seen at the front of the machine will tilt the table to any angle between the horizontal and 45 deg., and it is locked in position by a lever at each end. A wide filling in, or finger plate, is fitted to the table, and the makers recommend the removal of this, and the filling of the space with a piece of hardwood, through which the saw is allowed to cut its own path, when any accurate and clean cross-cutting is required, this method dispensing with the "packing" of the saw. The table has a wide groove sunk near its front edge, in which travels the connecting arm of the fence, which, as will be clear from the photograph, is capable of adjustment at any portion of

PLATE III.

AMERICAN UNIVERSAL SAW BENCHES.

1. The H. B. Smith Universal Saw Bench.

2. Cutting Compound Angles on the Colburn Saw Table.

[*To face page* 91.

PLATE IV.

AMERICAN UNIVERSAL SAW BENCHES.

1. The Colburn Tilting Table.

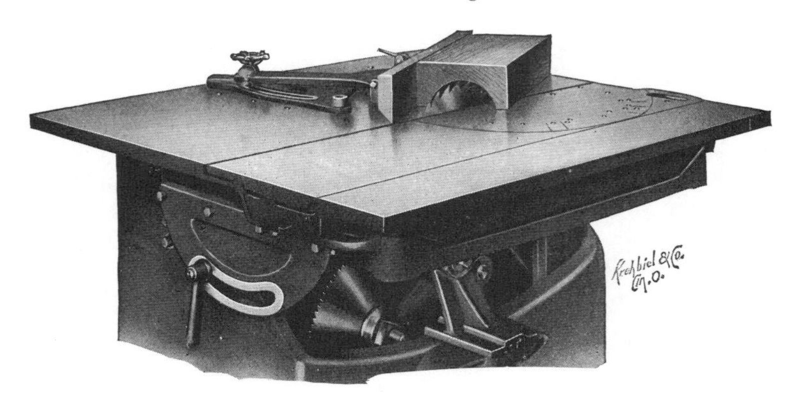

2. Hollowing Core Boxes on the Colburn Saw Table.

[*To face page* 91.

the table or at any distance on either side of the saw, and at either vertical or horizontal angles up to 45 deg. The makers do not mention in their description, but it is quite obvious, that with a little manipulation such operations as tenon cutting at various angles, moulding of sash brackets, ploughing architrave mitres, and sundry other odd jobs, can be readily done with this machine.

The American Universal Saw Bench, illustrated by the photograph, f. 1, Plate III., is made by the H. B. Smith Machine Company, of Smithville, New Jersey, U.S.A. It is a machine with several features in common with the one previously described. The table will tilt to an angle of 45 deg. for bevel cutting or grooving, and it is adjustable vertically, by hand wheel, to suit the required projection of the saw. It is also provided with cross-cutting and slitting fences; the former are attached to slides, which work in dovetail grooves on each side of the saw, and are capable of horizontal adjustment at any angle. The slitting fence works on a railway at the front of the table, and can be fixed at any distance either before, or behind, the saw. A 12-in. slitting saw is supplied with the machine, but cross-cutting and grooving saws can be supplied if desired. At the back of the bench a rising and falling table is attached for boring purposes, the saw arbor being drilled out at the end, to receive augers or boring bits. The table slides also horizontally to carry the work up to the bits, and the fence may be placed at any angle to the bit if angular boring is required.

A Universal Saw Table, of somewhat different construction from the above-mentioned benches is the "Colburn" shown in the photographs, Plates III. and IV. It is made by the Colburn Machine Tool Company, of Franklin, Pa., U.S.A. The table in this machine will tilt for angle cutting, as indicated in f. 1, Plate IV.; it has also a wide sliding table to the left of the saw, which travels quickly and easily on dustproof roller bearings in scraped trackways. This table is used for cross and diagonal cutting, also for cutting compound angles, as shown in f. 2, Plate III. A pivoting fence is supplied for this purpose, and a scale of angles is engraved on the table, with holes and taper pins for setting the fence, so that angles of any required number of degrees may be cut with accuracy. When the bench is used for ripping, the sliding table is locked, and the splitting fence shown on the right of the saw in f. 1 can be fixed at any angle or position on the table. One of the particular uses of the rip saw is shown in f. 2, Plate IV., where the fence is set diagonally to the saw for the purpose of cutting out a round core box, as required by pattern-makers, or for half-round guttering and the like. The semicircle is struck on one end of the piece, which is then run over the saw to make a succession of parallel cuts; the saw is gradually raised or lowered, as required, by means of the hand wheel P, and a worm gearing actuating the saw arbor. There are two saw spindles, and a dado head or grooving tool, up to 2 in. wide, may be mounted on one of these.

THE BAND SAWING MACHINE consists of an endless band or "ribbon" saw, revolving around the circumference of two (or sometimes three) large wheels, lying in the same plane, but at some little distance apart; the wood to be cut, being fed up to the blade in the portion travelling between two wheels; only one wheel is driven by power, the second (and the third when used) is moved by the saw itself as it travels around, the purpose of this wheel being to keep the saw in tension, and in the required path. The saws are arranged to run both vertically and horizontally, but the latter method is chiefly confined to very large machines used in the flitching of log and balk timber, and is outside the scope of this work. The vertical machines are of two kinds; the heavier form, often called a band-mill to distinguish it from the lighter jobbing or variety saw, is used for the resawing of deals, and planks, and boards, and it is provided with mechanical fences and feeds to regulate the thickness of the board cut off, and the speed of cutting. The band saw is, speaking generally, the fastest and most economical saw in use. It can be fed faster than a circular saw, cutting the same depth, will waste less than half the quantity of wood in saw kerfs, and requires much less power to drive, whilst the cutting is much cleaner and more regular. As compared with the frame saw, the speed is from five to fifteen times faster than the latter, if allowance is made for setting and changing the frame. The power required is about one-eighth that required to drive a log frame, and the quality of the output is nearly equal.

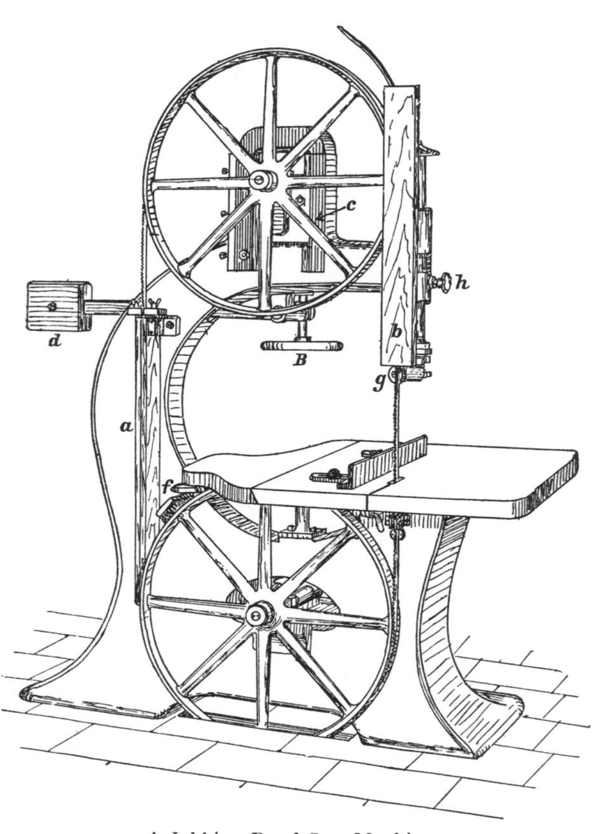

A Jobbing Band Saw Machine.

A Variety or Jobbing Band Saw machine is shown in the sketch on this page. This has an open frame of box section, described by engineers as a "cored casting." The lower or driving wheel has fixed bearings. The upper or "loose" wheel has three adjustments—(1) the block carrying the spindle and bearings can be quickly raised or lowered in the carriage way *c*; by turning the hand wheel B this adjusts the top pulley when mounting or removing the saw; (2) the bearing block is hinged at the top, and by twisting one of the screw bolts shown, can be thrown out of the vertical line, thus

BAND SAW MACHINES.

1. An American Tilting Table Machine.

2. A Tilting Frame Machine.

[To face page

adjusting the pulley to counteract any tendency of the saw to run off the edge; (3) the bearing block is attached to the lever shown just above wheel B; this lever is pivoted in the frame, and has a sliding weight *d* at its outer end, which can be adjusted to throw the saw into any desired tension. This also automatically compensates for any contraction or extension of the saw when in use, and goes far to reduce breakages, which are chiefly due to the crystallisation that follows unequal contraction on cooling. The pulley should not be less than 30 in. diameter, as smaller pulleys cause the saw to break. The top pulley should be as light as possible, and the bottom one heavy, to act as a fly wheel and steady the running. The front table is mounted on a pivot, and may be canted at an angle with the saw for bevel cutting (an example of this is given on p. 76). The back table is fixed; the provision for fences varies with the makers; some tables have parallel dovetail grooves cut in the face, in which slide lock-nuts, for fixing the fence to; in others, as shown in the sketch, two or more holes are drilled, and the tail of the fence is slotted, to slide upon a screw bolt. In many cases no provision at all for a fence is made, as many persons do not utilise the band saw for anything but curved work. An extemporised fence for such machines is described on p. 114.

Band saw blades vary in width from ¼ in. to 8 in.; the latter size, as may be surmised, is only used in the larger machines; for joiners' use the smaller sizes from ¼ in. to 1½ in. are the more useful. A hole of 1 in. diameter can be cut with the ¼-in. saw, and a 1½-in. blade will cut a curve of 20-in. radius without straining. The saw blade is supported by anti-friction guides both above and below the table, the one above, being adjustable to suit the material operated on. There are several varieties of these guides; one of the best is shown at *g* in the sketch. This is a hardened steel wheel mounted on ball bearings; it is brought up close to the back of the saw, and revolves with the latter, which travels across one side of its face. The height is adjusted by the screw wheel *h; a* is a wooden safety guide for the protection of the return blade; *b* is a wooden shield for the protection of the operator, with a wing at top designed to throw the saw back on the pulley, when it breaks; as a matter of fact, there is perhaps less danger to the operator in the use of a band saw than in any other power-driven machine. The author has, on two or three occasions, had the experience of a saw breaking and falling over his shoulders without doing the slightest damage, for the saw ceases running instantly when it breaks. A brush for clearing the pulley of sawdust is shown at *f;* this would be in a more suitable position at the near side of the wheel.

The Band Saw shown in f. 1, Plate V., is made by the Silver Manufacturing Company, of Salem, Ohio, U.S.A. It has an easily adjustable tilting table that can be fixed rigid instantly, in any position, by the turn of a hand wheel. It has all the usual appliances; the tension of the saw is provided by a spring. Rubber bands are cemented to both wheels, which are of cast iron, and are made in three sizes, viz., 26-in., 32-in, and 36-in. diameter. Wood rim wheels are provided if preferred. Messrs M'Dowall & Sons make a band saw machine, in which the **C** frame, carrying the saw, is mounted on a pivot at the base, and moves by rack and pinion gearing on a quadrant arm, to any angle between 90 deg. and 45 deg, the table remaining horizontal; this greatly facilitates the manipulation of heavy timber that requires cutting at an angle. This machine is illustrated in f. 2, Plate V.

A Combined Circular and Band Saw Bench, suitable for small and medium size shops, is shown in the photo below. This is made by Messrs John Pickles & Son. It consists of a band saw and circular saw mounted on the one frame, but running independently of each other. The circular saw

A Circular Saw and Band Saw Combined Bench.

has a rising and falling spindle mounted in phosphor bronze bearings, carried by a bracket, which is raised or lowered by a hand wheel. The end of the saw spindle is fitted to receive boring bits, with which slot mortising, boring, housing, &c., can be managed, also cutter blocks to carry planing (small), moulding, tonguing, grooving, &c., irons. The front of the bench is fitted with a sliding plate by means of which cross-cutting to length, and to any angle, can be done, and the fence is fitted to receive a tenoning slide. The band saw table is of good size, and has the usual tilting adjustment.

PLANING MACHINES.

The Surface Planer, or Overhand Planer (see sketch), is one of the simplest constructed machines in use, withal it is probably the most dangerous to inexperienced operators, as the high speed with which it is necessary to drive the cutters renders them absolutely invisible when working, and, as the work has to be fed by hand, the danger when approaching the cutter gap is obvious. The machine illustrated, which is made by Messrs Kiessling, is constructed with three guides or fences to keep the stuff in position, which considerably reduces the risk of accident. The machine consists of the usual cored casting, having two planed tables, each moving independently of the other, upon angle carriages

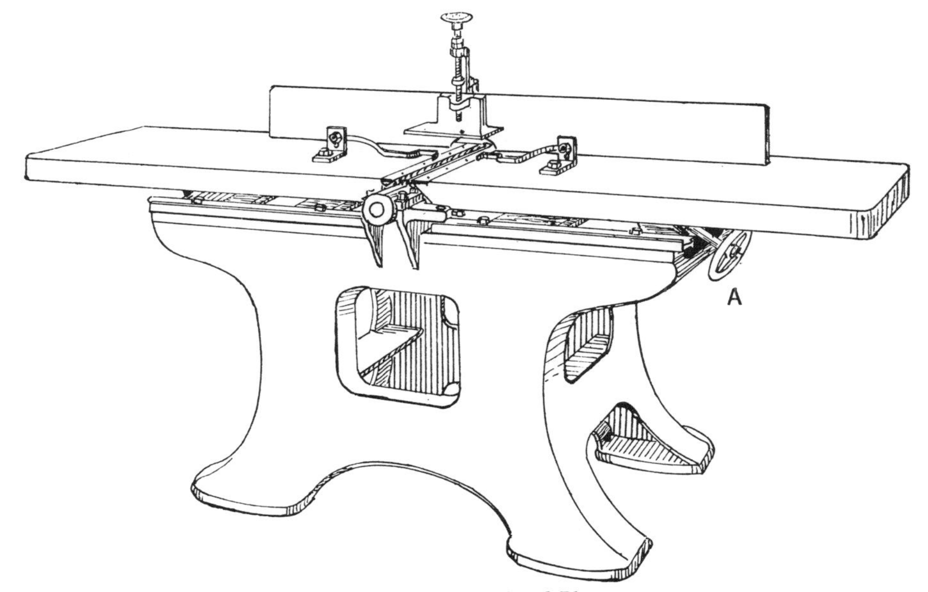

The Surface or Overhand Planer.

actuated by hand wheels, one of which is shown at A; as these are turned, so the tables are raised or lowered, and move also backwards or forwards as desired, to open or close the cutting gap. In the top of the frame and immediately below the table is mounted, in long bearings, a spindle carrying a square cutter block, to which the knives or plane irons are attached; these are revolved at high speed between 4,000 and 6,000 ft. per minute at the cutting edge, or from 700 to 900 revolutions of the countershaft per minute. These machines are made in several sizes. The smallest will plane stuff up to 12 in. wide, and the largest up to 36 in. A skilful operator can plane stuff out of winding, make joints ready for gluing up, rebate, chamfer, and bevel with the ordinary cutters, and, by the use of special cutters, can groove, tongue, bead, flute, stick mouldings, &c. Some hints on the management of this machine will be found at p. 117.

A Portable Electric-Motor Surface Planer is shown in f. 1, Plate VI. This is made by Messrs John M'Dowall & Sons, and is an extremely useful machine on a large building, where it can be rolled from one part to the other, wherever planing is required. It would also be very handy to a small contractor to send out to various jobs, as, being entirely self-contained and carrying its own driving gear, it could be used in any place where electric current is available. The switch board and fuses are shown open in the photograph, but in use these are all enclosed in a neat iron case. Electric power is in many respects superior to steam for driving wood-working machines, as the motor is continuous in its action, not intermittent, as are reciprocating or oscillating engines, with obvious advantage to the smoothness of the running

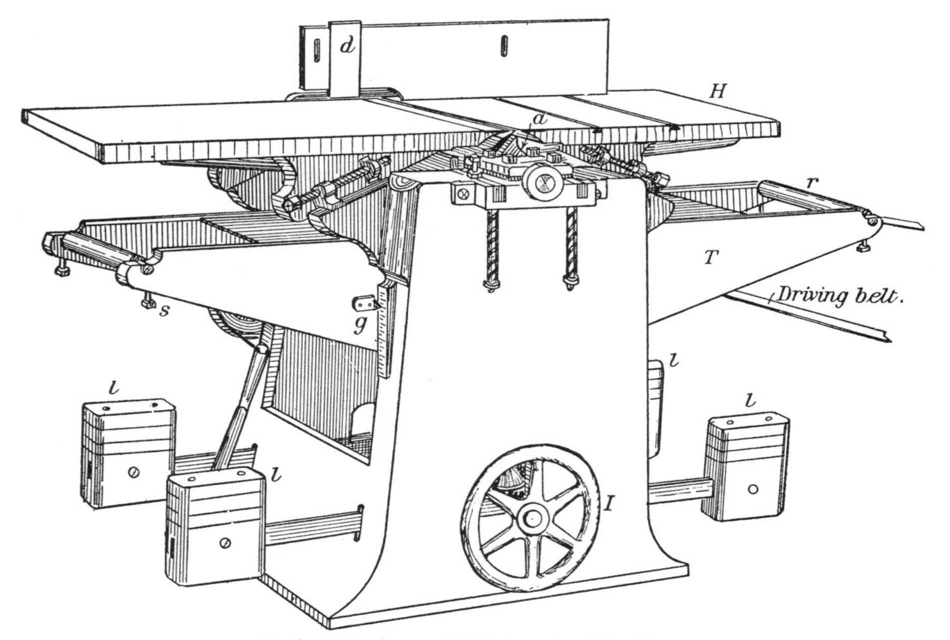

Surface Planing and Thicknessing Machine.

of the machines. This particular machine has the usual rising and falling table mounted on slides, with adjustable angling fence, and can be used for surfacing, jointing, rebating, chamfering, bevelling, moulding, grooving, &c.

The Panel Planer or Under-cut Machine is somewhat similar to the surface planer, but in this type the table, which is all in one piece, rises and falls *beneath* the cutter block ; thus the stuff, having been first passed through the machine and surfaced, can be turned over, and the table readjusted to reduce the stuff to any required thickness. The feed in this case is by rollers, driven from the countershaft.

Planing and Thicknessing Machines are otherwise described by some makers, Over and Under Planers, and familiarly in the machine shops as "buzzers,"

PLANING MACHINES.

1. Portable Electric Planer.

2. Hand and Roller Feed Planer and Thicknesser.

[*To face page* 96.

from the noise they make when working. This machine is a combination of the two last described, doing all the work that can be done upon either, and of course taking up only the space required for one, and requiring only the same driving power as either one machine. There is one serious drawback, however, to these machines when working the thicknessing table. When the stuff has passed under the end of the upper table it is out of control of the operator, and, if wet or cross grained, it is liable to slip diagonally across the table, and either get smashed up under the cutter, or jam the machine and cause a stoppage, sometimes breaking the knives. The pieces may, of course, be wedged in between the edges of the table, but this interferes with the speed, and is therefore generally neglected. The machine shown in the sketch on opposite page is made by Messrs M'Dowall, and is fairly typical of its class. These makers regulate the pressure of the feed rollers by means of weighted levers, as shown at *l*, *l*, *l*, *l*, a method that gives better results than stiff springs when planing thin or irregular boards. Stuff in this machine is taken out of wind by feeding over the hand table H, when it can be immediately returned under the cutters, without alteration of the latter, upon the thicknessing table T, which is adjusted to the required gauge by the hand wheel I, a mechanical index *g* showing the exact thickness at which the machine is set without the need of measurement; this, however, is apt to get misleading if care is not taken when re-setting the cutters to pass a piece of stuff through the machine, and adjust the irons until the gauge registers correctly. Another advantage of this machine is that all the gearing is at the off-side, so that there is nothing to distract the attention of the operator when passing stuff through it. The fence on the top table slides in dovetail grooves which keep it perfectly rigid; it has a drop slide *d* to prevent thin stuff slipping under the edge of the back fence, and it can be tilted to an angle of 45 deg. for bevelling, chamfering, &c. The cutters are arranged obliquely upon the block to take a shearing cut, which reduces the resistance, and is less likely to tear up the grain; it also drives the stuff towards the fence. Some of these machines are provided with cutter blocks on vertical spindles at the rear of the thicknessing table which plane the edges of boards, &c., thus bringing stuff to a width and thickness at one operation.

A Hand and Roller Feed Thicknessing Machine, shown in Plate VI., is made by Messrs J. Sagar & Co. Here again the frame is a solid casting, *i.e.*, it is not bolted together in parts, a method which invites vibration, and consequent bad results in planing. The cutter block is of the square taper form, giving a shearing cut with straight cutters, and is cast solid on the spindle, which is afterwards engine-turned and squared to ensure perfect balance. The gearing, as the photo shows, is all at the back of the machine, out of the way of the operator. In addition to the list of operations given under the heading "Surface Planer," this machine is capable of planing taper, such as is required by pattern-makers, plane and mould skirting, or match-lining, at one operation, and its planing capacity is from $\frac{1}{16}$ in. thick up to $1\frac{1}{2}$ in., at speeds from 9 ft. to 40 ft. per minute.

SPINDLE MOULDING, &c., MACHINES.

SPINDLE MACHINE is the generic term applied to a class of machine of great utility in a joinery establishment. In these planing, moulding, and other cutters are mounted upon the extremity of a vertical spindle, revolving at a high speed, and protruding from the middle of a small table, which, being accessible all round, offers facilities for the manipulation, &c., of curved or irregular work such as no other machine affords. With the addition of suitable fences straight work of moderate dimensions can be managed on these machines, but their chief purpose is the execution of work of single and double curvature. The spindle in these machines is made to rise and fall by means of the hand wheel and screw *h*, f. 1, p. 99, for the purpose of adjusting the cutters to the height of the work, and the cutters may be driven in either direction, left or right, as best suits the work in hand, by having extra pulleys upon the countershaft driven from the main shaft by "open" and crossed belts. When these machines are used for planing, the cutters—ordinary machine plane irons of small size—are fixed to square CUTTER BLOCKS (f. 1 above) by means of dovetail nuts and screw bolts; these blocks slip upon the end of the spindle resting upon a square shoulder, and are fixed by lock nuts screwed on the end of the spindle. For moulding and rebating purposes thinner "irons" are often used, and these are fixed in slotted nuts or collars, as shown in f. 2. In the case of machines that can be driven at very high speeds, an altogether different kind of "cutter" is used, which is not a cutter at all in the sense that a chisel is, but rather a scraper. These cutters are made of softened steel, which is, after grinding to the required shape, tempered, and then the edge turned, precisely as is a joiner's scraper. They are mounted by fixing them in a slot or keyway cut through the axis of the spindle head, in which they are secured by a long screw bolt, as shown in f. 1, p. 123. The machine fitted with this type of cutter was originally called a "French spindle," but now most machines are made, so that either type of cutter can be used in them: the spindle head is drilled out to receive "loose" heads, which may be either "French" spindle heads, or slot-nut spindle heads. These loose heads are secured variously by different makers; some taper slightly in length, and are held by their tightness of fit; they are released by driving a wedge in a slot below the end; others are fixed by a small bolt through the side of the

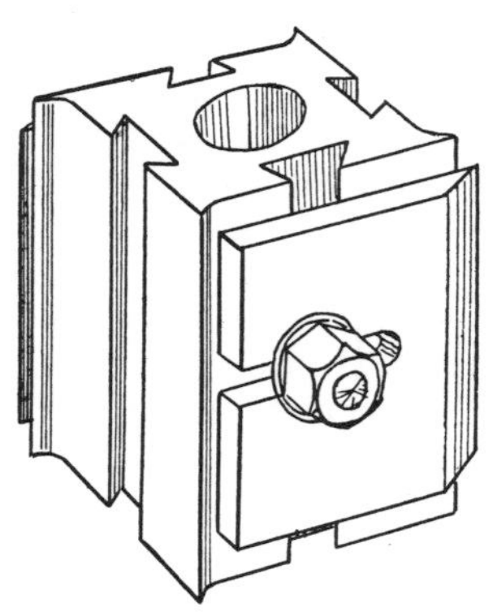
1. Cutter Block for Spindle Machine.

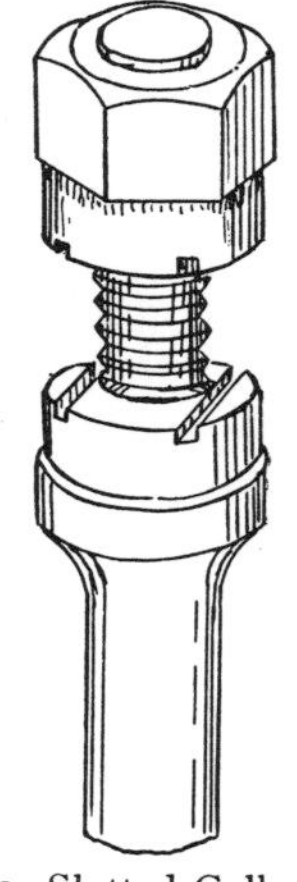
2. Slotted-Collar Spindle Head.

3. A Solid Moulder Cutter.

socket, others by a cotter pin. Another form of cutter is shown in f. 3, p. 98. This is a solid steel head turned to a required profile, then portions cut away radially, the sharp edges of the remaining portions providing a series of cutters that are exactly alike; they are useful when large quantities of a given section are required. These cutters may be driven in either direction, and the profile is not altered in sharpening. Various fences are supplied with these machines; their use is described in the next chapter. It is often necessary in moulding circular work to reverse the direction of the cutters as the grain alters. All spindle machines are fitted to do this without any re-setting, but the drawback with a single spindle machine is, that it must be stopped, and opposite hand cutters fixed when the direction is changed. This means a serious loss of time when much irregular work has to be done, and to obviate it larger machines are made with *two* spindles, which revolve independently, and in opposite directions. These, of course, still require "pairs" of cutters (unless the above-mentioned "solid" form is used), but no time is lost in setting up and changing the irons; the operator has to merely step from one spindle to the other, as he wishes to change the direction of the cutting.

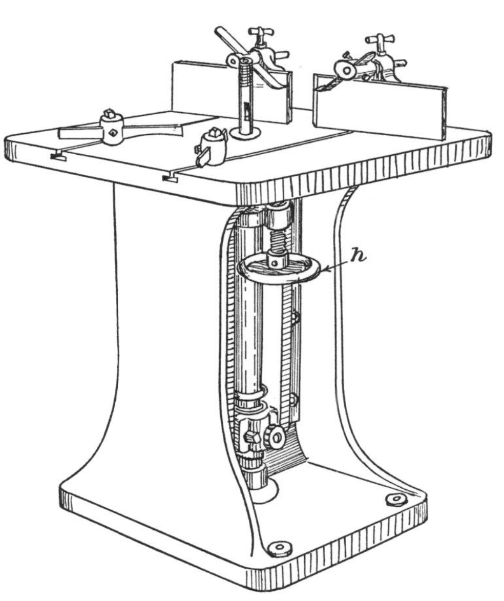

1. A Single Spindle Moulder.

2. A Double Spindle Moulder.

The Single Spindle Moulder shown in f. 1 is Messrs Reynolds & Co.'s pattern. The size of the table is 2 ft. 6 in. square, and it will plane or mould up to $4\frac{1}{2}$ in. deep. It is fitted with the various fences and guides shown, and in addition, square cutter blocks, and filling-in rings for the spindle mouth.

A Double Spindle Moulder of Messrs John M'Dowall & Sons' make is shown in the sketch, f. 2. The table is 4 ft. 3 in. by 3 ft.; the spindles have a vertical movement of 5 in., which makes them especially useful in moulding handrail wreaths, and the pulleys make 1,000 revolutions per minute. Straight and curved planing, ditto moulding, rebating, chamfering, grooving, and tonguing, &c., can be done on these machines, also by the aid of sundry attachments, boring, tenoning, dovetailing, &c.

A Single Spindle Moulder with Tenoning Apparatus attached is shown in the photograph, f. 1, Plate VII., which is self-explanatory. This is made by Messrs W. B. Haigh & Co., and the machine is so constructed that a dovetailing apparatus can be readily substituted for the tenoning mechanism when desired, or either of them readily removed, when the machine is ready for the usual work of its class.

The Router or Elephant Machine is one in which the revolving spindle is carried in a slide at the end of an overhanging arm *above* the table, and, by means of compound slides, the work operated upon is carried laterally or transversely under the cutters, which may be boring bits for trenching or grooving, or shaped cutters for recessed mouldings. Messrs J. Sagar & Co. have just placed on the market a greatly improved "elephant" machine, to which they give the compendious title of "*The Improved Double Spindle Moulding, Shaping, Trenching, and Recessing Machine.*" This is shown in the photograph, f. 2, Plate VII. The chief novelties in its construction are—(1) it is furnished with *two* spindles, concentrically mounted, above and below the table, working separately and independently; (2) the pedestal supporting the table revolves a quarter turn in either direction. This, with the transverse slide of the table upon its bearings, gives an almost universal motion to the work fixed upon it, so that, in addition to the capabilities of the single spindle moulder, the upper spindle will house and cut grooves in every direction, throat and weather sills, mould the edges of sunk panels, Gothic tracery, or any shaped work that is difficult to manage on the bottom spindle. One of the great advantages of this machine is the wide space between the frame and the overhanging spindle—3 ft., which allows very wide stuff to be manipulated. The spindles may be run in either direction to suit the grain. On the floor are scattered various accessories supplied with the machine. Another form of these machines recently introduced is the **Shaping and Recessing Machine**, f. 1 & 2, on p. 101. This carries two separate spindles, one running within the table frame, as in the ordinary spindle machine, the other working above the table as in the ordinary elephant machine. The chief improvement is, that instead of having to move the work about upon the table to suit the direction of the cutting, the top spindle is so arranged that it can be moved either vertically, horizontally, laterally, transversely, or angularly across the table, thus providing for every possible direction of cut, whilst the work remains securely fixed to the bench. The methods by which these movements are made vary slightly in different makes, but they are all in principle as shown in Messrs Reynolds' type in f. 2. Here the overhand spindle works in a vertical carriage *g*, which is mounted on a long transverse slide *e*, that is pivoted to the carrying arm A; the spindle is brought down to the work quickly by pulling a hand lever (not shown), and fed down slowly, for boring, &c., by the hand wheel *f* attached to its carriage. The lateral transverse is accomplished by turning either wheel *a*, *a*, attached to the endless screw. The lower spindle *d* runs independently, and its head being removable, the whole extent of the table

VERTICAL SPINDLE MACHINES.

1. Spindle Moulder with Tenoning Apparatus.

2. The Elephant Moulder and Router.

[*To face page* 100.

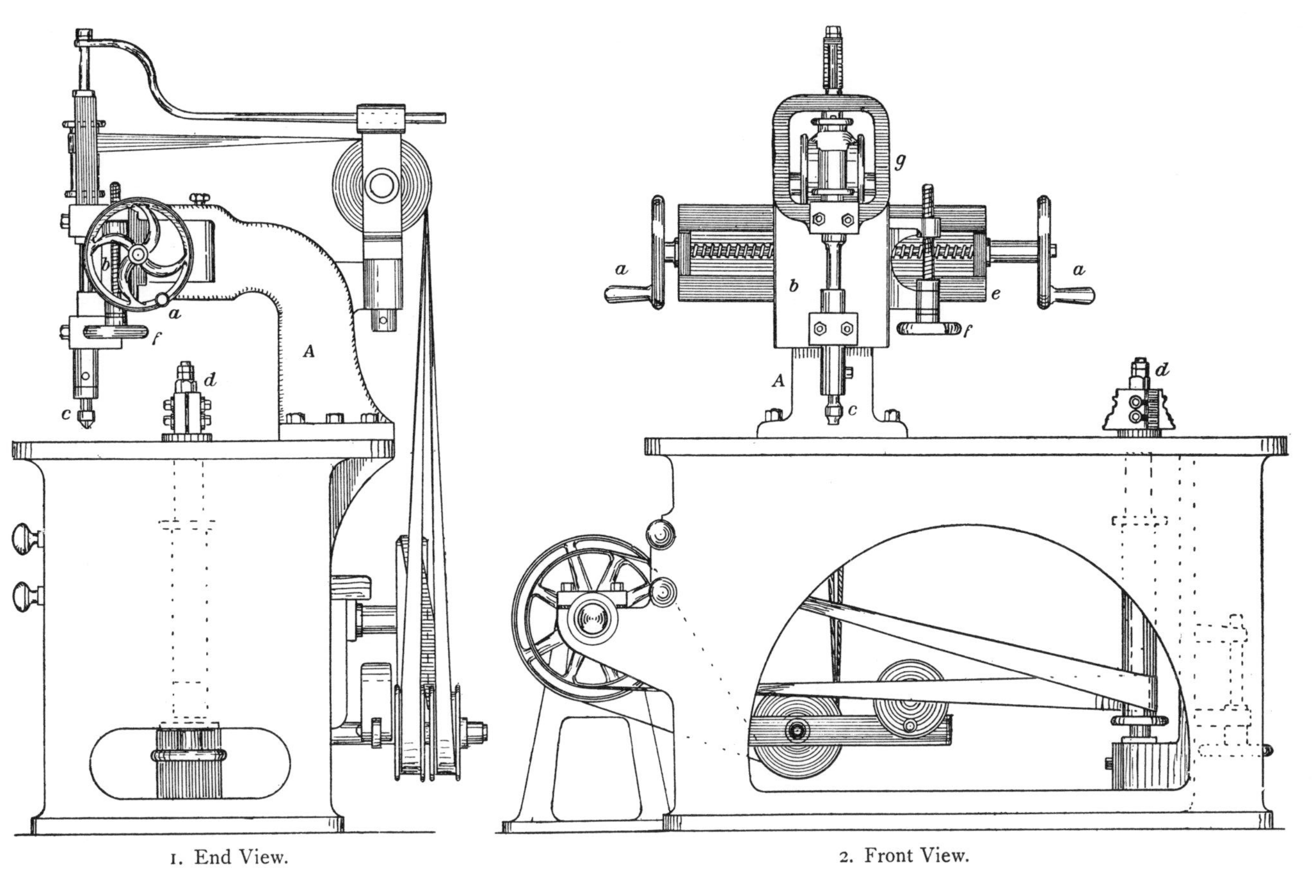

1. End View.

2. Front View.

DOUBLE SPINDLE MOULDER AND RECESSING MACHINE.

can be made available for receiving large work. Messrs M'Dowall make a machine with an additional radial movement of the carrying arm A, which is of advantage in large sweep work, also for throwing the upper spindle clear of the bench when it is desired to use the lower spindle. In Messrs T. Robinson & Son's make the spindle carriage is swivelled, so that the *bit* may work either vertical, horizontal, or at any angle between these directions. This firm also supplies various attachments by which square and ordinary turning can be done, such as moulding balusters, table legs, brackets for chimney pieces, columns, &c., also dovetailing, tenoning, scribing end grain grooving, &c.

JOINT-PRODUCING MACHINES.

Tenoning Machines are of two classes: first, those in which the cutters attack the wood, chisel fashion, cutting across the grain, either vertically or horizontally, the shoulders being sometimes made with small circular or reciprocating saws. In the second class the adzes or cutters are revolved on an axis parallel with the grain of the wood, which is fed between them upon a travelling table. The former method is more suitable for heavy work, such as piling, mill-framing, &c., and need not be considered here, the second being almost universally employed in joinery mills and shops.

A Tenoning Machine of simple construction is shown in f. 1 & 2 opposite, which are drawn rather to show the principles of construction of this class of machine than as a pattern now followed closely by any maker. The machine consists of a heavy cast-iron frame, carrying two horizontal steel spindles having square cutter blocks fixed at their inner ends, and wide pulleys of small diameter at their outer ends. The cutter blocks are made conical, and the adzes are bevelled in the opposite direction to bring the cutting edges parallel with the axis of the spindle; the result of this dual sloping is, that one side of the edge of the cutter strikes the wood in advance of the opposite side, thus making a shearing cut which prevents splitting, and causes less vibration in the overhanging spindles than parallel cutters would. A disc is attached to the end of the cutter block, in which lancet knives are fixed for cutting the shoulders across. These are shown at *l* in f. 1; *a* and *b* are the cutter block pulleys; the cutter spindles are mounted in a sliding carriage, and are capable of vertical and horizontal adjustments independently to suit the requirements of the tenon, whose thickness is gauged by the distance apart at which the two cutter irons are set; *c* is a loose tension pulley used to compensate for the differences in length of the belt, as the two fast pulleys are moved nearer together. The wood to be operated upon is fixed to the table by a hand lever *d*, which may be locked in position by the clip bar *e*, thus leaving both hands of the operator at liberty. When one tenon is cut, the sliding stop *s* is set to the required length of the rail between the shoulders, and, the rails being reversed, are passed through a second time, as shown in f. 2. The drawing shows the machine as set for cutting double tenons, the haunching, or space between the tenons, being formed by a horizontal drunken saw working on a vertical spindle just in advance of the cutters. A chamfer scriber is shown beneath this, which is adjusted by a spacing collar.

TENONING MACHINE.

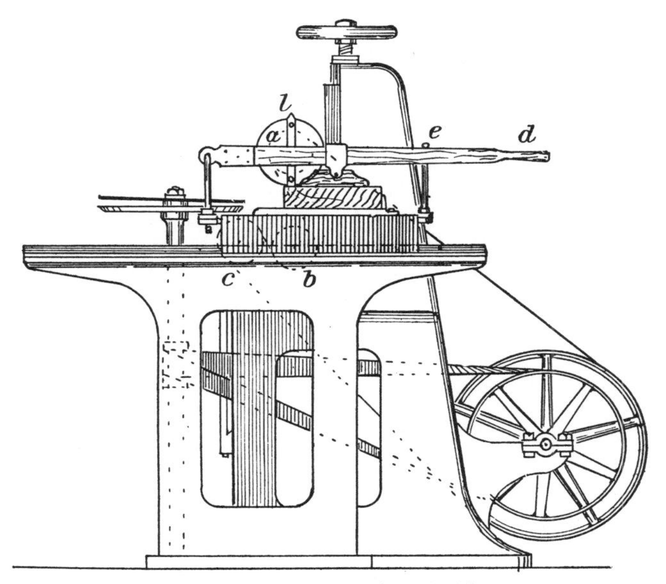

1. End View of Tenoning Machine.

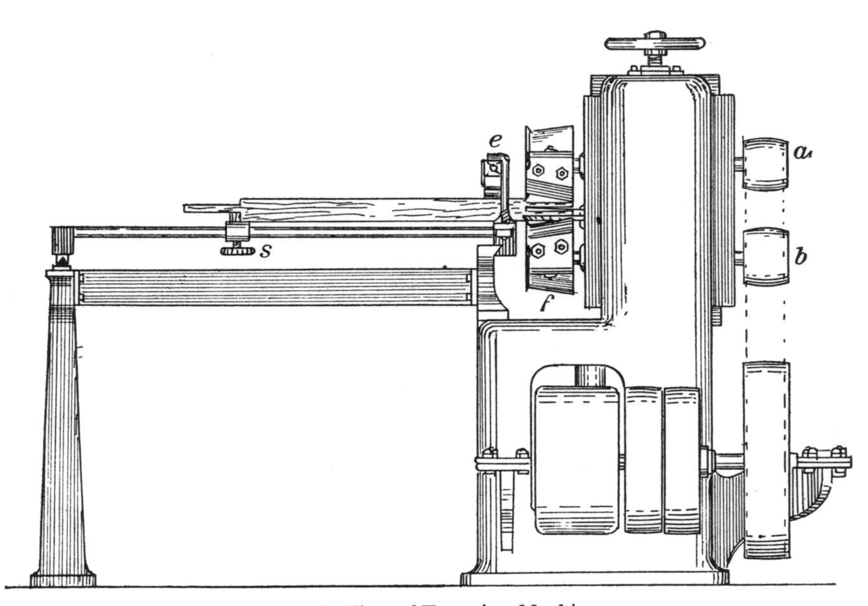

2. Back View of Tenoning Machine.

The photograph, Plate VIII., shows the latest type of **Tenoning Machine,** produced by Messrs Robinson & Son Ltd. It is very solidly constructed, withal the parts are excellently balanced and move with perfect freedom. The travelling table works on V guides and ball bearings, and it has a quick action screw clamp for fixing the work, in addition to the usual hand clamp. The length, or shoulder gauge, works on two parallel tubular guides, which keep it square and rigid when set. This machine will cut tenons, single or double, up to 5 in. long, on stuff up to 14 in. by 16 in. An expanding cutter is supplied for the top spindle; with this trenching or cross-grooving can be done, and if a small circular saw is substituted for the cutter block, panels and other thin stuff can be cross-cut on the table. An expanding cutter fixed on the lower spindle will haunch rails, and the vertical spindle shown at the back of the carriage will scribe single or double shoulders, core between tenons, or any similar end grain ploughing or moulding.

The Automatic Double End Tenoning Machine, or four cutter, shown in the photograph, Plate IX., contains Messrs Ransome & Co.'s latest improvements in large tenoning machines. This machine, once set, which is done with a pattern rail, will cut with rapidity any number of tenons with great exactitude between the shoulders of any length from 6 in. to 6 ft. 6 in. in the clear, with tenons up to 6 in. long, on timber up to 8 in. thick. The stuff is fed up to the knives by means of endless chains having studs or dogs at intervals of 2 ft., the rails, &c., being dropped between these as they rise over the ends of the carriage, guarded by the semicircular boxes seen at the front of the machine. The machine may also be used for squaring off to length large timber when a quantity of uniform length is required. All the cutter spindles have compound movements, and the right hand headstock slides laterally on the frame, being moved as required by rack and pinion gearing.

Power Mortising Machines.—The photograph, f. 1, Plate X., is a graphic illustration of Messrs A. Ransome & Co.'s improved type of high speed plunger chisel mortising machine. The main standard is a hollow box casting, made all in one piece, with a wide base to be bolted to the foundation, which, if on the ground, should be of solid concrete, and if on the upper floor, should have a brick pier beneath it; so secured, the machine may be driven at a speed of 500 revolutions per minute, without objectionable vibration. The chisel, which has a stroke of 5 in., is held in a stout steel spindle secured to a gun-metal block which slides on the planed faces of the standard within two V guides, which are adjustable for wear. The chisel is automatically reversed, and locked by a spring stop, engaging with projections upon the grooved wheel shown at the top of the spindle. This is revolved by a leather band running on a grooved pulley on the driving shaft. The chisel is instantaneously released by the insertion of the key, seen hanging by a chain, in the drift way in the spindle. The table is furnished with a long sliding fence, having a flange at the top edge, which prevents the wood rising by the upward motion of the chisel; this fence will take wood up to 11 in. in depth. The table will incline in either direction to an angle of 10 deg. for wedging purposes. The table is adjusted for height by the hand wheel and screw directly below, and in line with the chisel, which then transmits the thrust to the base of the machine. The work is brought up to meet the chisel by

A JOINER'S TENONING MACHINE.

[To face page 104

AUTOMATIC DOUBLE END TENONER.

[*To face page* 104.

1. Power Mortising Machine.

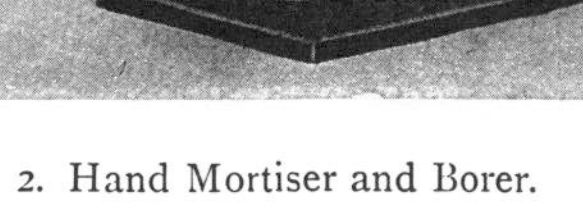

2. Hand Mortiser and Borer.

3. Boring, Mortising, and Moulding Machine.

[To face page 105.

Mortise Machine.

Plate XI.

1. The "Chain" Mortise Machine.

2. The Machine fitted with Hollow Boring Chisel for Small Mortises.

pressure of the foot upon the balanced lever table seen at the bottom of the machine, thus obtaining a graduated stroke. The machine has also a boring spindle at the side of and in line with the chisel spindle, which enables the operator to pass the wood under either tool as required without transverse adjustment, thus saving much time. The spindle has a range of 12 in., and the hand lever which is used to bring the bit down to the work is counterbalanced to automatically withdraw it. The mitre wheels actuating the spindle are enclosed in a cast box to protect them from dust and damage.

A Hand Mortising and Boring Machine is shown in the photograph, f. 2, Plate X. This is made by Messrs Pickles & Son. Pulleys are fitted to the boring attachment, so that, if desired, the bits can be driven by power. The frame is a very strong casting, all in one piece, with extra wide base. The table has the usual transverse and longitudinal motions, and a vertical adjustment by hand wheel and screw. It is provided with a deep fence. The chisel spindle is actuated by a cranked and forked lever, to which it is attached by a pair of connecting rods that keep the thrust central and make the working smooth and easy. The machine will take timber up to 12 in. by 6 in., and is provided with a set of chisels and bits.

A Vertical Boring, Mortising, and Moulding Machine recently introduced by Messrs W. B. Haigh & Co., is shown in the photograph, Plate X., f. 3. This is a very solidly constructed machine, taking up but little floor space, and it may be driven either by electric motor or steam power. The table has a compound traverse, and its carriage has a vertical movement on the face of the standard, the lateral movement of the table is by means of rack and pinion and large hand wheel, for the quick action necessary in slot mortising, which is done with an auger bit. Various cutters can be inserted on the end of the spindle, and the machine to a limited extent will take the place of a recessing machine and a single spindle moulder, although its chief claim is to be a rapid borer and mortiser.

The Epicyclic Chain Mortise Machine, shown in the photograph, Plate XI., is patented and made by Messrs W. B. Haigh & Co. With this machine the mortise is made by means of an endless chain, which revolves around two small sprocket wheels carried by the steel bar shown in the photograph, f. 1; each link of the chain is formed into a cutting edge at one end, which may be readily sharpened by a special grinding apparatus supplied. The chief advantage of this type of machine is that hard, or soft, or knotty wood may be mortised with equal facility, no preliminary boring being required. Mortises may be made through timber up to 12 in. deep, or stopped at any required depth. Shallow grooves can also be made with the cutters, and there is absolutely no danger of bursting the sides of the mortises, however thin, as in the "plunger" type. The machine is power-driven, but the chain is brought down to its work by hand lever, thus graduating the cutting as required. As no shorter mortise than $1\frac{1}{2}$ in. (the width of the chain carrier) can be made by means of the chain, when small mortising has to be done, as in sash bars, a hollow square chisel containing a boring bit is substituted for the sprocket, as shown in the second photograph.

The Automatic Dovetailing Machine shown in the photograph, Plate

PLATE XII.

A DOVETAILING MACHINE.

[*To face page* 106.

THE COMPLETE JOINER.

[To face page 106.

XII., is made by Messrs Sagar & Co., and is especially useful for the finer class of dovetailing, chiefly required in joiner and cabinet shops. This machine will cut the pins and sockets of a drawer front and side at one operation; the pieces are secured in their relative positions, as shown in the photograph, and traversed across the head of a revolving spindle carrying a suitable conical shaped cutter. Drawer sides, and other work which is to be dovetailed "through," are secured in the long slide seen at the top of the machine by means of screw clamps, and the slide is moved up to the cutter by automatic mechanism, set to the required "pitch" or spacing of the pins. Three standard size cutters are supplied with the machine, as shown in sketch above, but any others desired can be supplied. The work of the machine is very clean and accurate, and the speed about thirty-four dovetails per minute. Its use as an *automatic* machine is confined to dovetailing with uniform spaces between the pins. When irregular spacing is required, the work must be set out in the usual way and fed by hand.

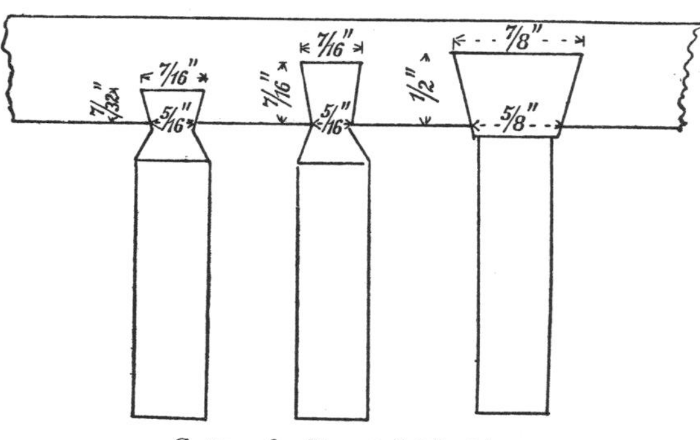

Cutters for Dovetail Machine.

COMBINATION MACHINES.

General Joiners.—The photograph, Plate XIII., shows the latest form of this combination machine as made by Messrs A. Ransome & Co. As the following list of operations that may be performed upon it will indicate, it now well deserves the name the patentees give it, viz., "The Complete Joiner." It will take saws up to 30 in. diameter and deep cut 12-in. planks, or cross-cut up to 6 in. thick; plane, thickness, and bead, at one operation, boards up to 12 in. wide by 4 in. thick; joint edges, groove and tongue, or rebate, up to 4 in. wide; stick straight mouldings, single or double, *i.e.*, worked on all four sides up to 12 in. wide; work circular mouldings or rebates up to 3 in. wide; groove with drunken saw up to 1½ in. wide; cut tenons (with cutter irons) up to 4½ in. long, with equal or unequal shoulders; slot mortise from ⅛ in. to 2 in. wide, and any length; bore holes from ⅛ in. to 2 in. diameter, either with or across the grain. The saw spindle being separate and independent from the planing spindle, two men can work at the machine at one time without in any way interfering with each other. The saw table is made to rise and fall, by the hand wheel to be seen at the bottom of the photograph, and the fence may be used square or bevelled, or may be turned over the end of the table to leave the latter clear when cross-cutting. There is a vertical spindle at the end of the saw table for working

PLATE XIV.

THE VARIETY WOODWORKER (AS A SURFACE PLANER).

[*To face page* 107.

PLATE XV.

THE VARIETY WOODWORKER—

1. As a Moulding Machine. 2. As a Tenoning Machine.

[*To face page* 107.

irregular mouldings, and two other adjustable spindles at the other side of the machine for edging and grooving or bringing stuff to a width. The stuff is fed to the planers by a pair of fluted rollers, and these have a variable feed motion to suit hard or soft wood. The tenoning apparatus works on a carriage way that can be removed if required. The boring and mortising table is shown on the floor in front of the machine.

The Variety Woodworker, of Messrs Thos. Robinson & Son Ltd., illustrated by the photographs on Plates XIV. & XV., is another high-class combination machine of great utility in small establishments, and, for the multitude of operations that may be *profitably* carried out upon it, is remarkably simple in construction and operation. The photographs speak for themselves as to what can be done upon it. Plate XIV. shows the machine in its simplest form with all the supplementary apparatus removed. In this condition it is ready for planing or surfacing up to 12 in. wide ; or by bringing the fence forward to the required width of a rebate, the latter operation may be accomplished without alteration of the cutter block. If the fence is adjusted to an angle, which can instantly be done by turning the small wheel shown at the front end, chamfering may also be done, and, with the addition of suitable cutters, which can be attached to the cutter block without removing the plane knives, ploughing, beading, matching, &c., can be done. In the second photograph, the machine is shown prepared for sticking mouldings, the table lowered beneath the overhanging cutter block, and a light power-feed roller attached. This will feed satisfactorily stuff up to 4 in. by 2 in. Chip, and front guards, and a pair of pressure springs are supplied, as shown. By substituting a circular saw for the cutter block, and raising the table as required by the hand wheel and screw seen underneath, an efficient ripping or cross-cutting bench is formed, and a sliding fence can be fitted in the groove, seen in the front edge of the table. With this adjunct, which takes but a few seconds to fix, mitreing or dimensioning, squaring off panels or drawer stuff and the like, can be done with hollow ground saws that leave the cut as smooth as if planed. In the third photo the machine is shown cutting tenons, an ordinary tenoning cutter block replacing the moulder shown in the previous view. The stuff is cramped, as shown, to a sliding table with a fence at the rear end, against which the fair edge of the rail is placed, and to ensure accurate shoulders, when one side of the tenon has been cut by running under the cutter, the table should be brought up above it, and the stuff passed *over* the cutter. Of course, when a number of similar rails have to be tenoned all would be passed under the cutter, before raising the table for the second cutting. A further attachment to the front side of the machine, not shown in the photographs, enables slot mortising, boring, and housing to be done.

An Automatic Circular Saw Sharpening Machine.—This is an exceedingly useful tool in small and medium size Works. It is absolutely automatic ; once set it may be left unattended until finished. The cutting is done by corundum wheels ; and every tooth is perfectly topped, backed, and gulleted as desired, in one, or several operations. As the saw is moved around against the revolving wheel, a uniform diameter is produced ; and every tooth is made different in shape, or an exact duplicate of its fellow, as desired. The machine is made by Messrs Henry M'Ewen & Co., of Hertford Road, Downham Road, London, N.

CHAPTER VII.

MACHINE SHOP PRACTICE AND METHODS OF USING MACHINES.

Introductory Remarks—Sawing Machinery—Circular Saw Benches—Forms of Saw Teeth, Precautions to ensure Satisfactory Work—Method of Trueing a Saw—Packing, its Objects—Method of Correcting Faulty Running—Fences, their Uses, Descriptions of Various Kinds—Methods of Dealing with Crooked or Warped Timber—Ripping, Saws to Use—Deep Cutting, Adjustments of Fence, Manipulation of Timber—Bevel Cutting, Uses of Compound Fence, Special Mouthpiece—Rebating with Saw, Sizes—Grooving, with Thick Saws, with Drunken Saw, Method of Adjustment, Patent "Drunk Saw"—Band Saw Machines, Treatment of Saws, Correcting Faulty Cut—Cutting Tenons—Handrail Wreaths, Preparation of Block—Circle-on-Circle Work, Lining-out, Cutting, Preparation of Moulds—Ellipse on Circle to Cut-Planing Machines—Overhand Planer, Requirements of Cutters, Importance of Balance of Parts—Method of Producing Plane Surfaces with Revolving Cutters, Cause of "Corduroy Surface"—Setting the Machine, a Dangerous Practice—Taking Stuff out of Winding—Jointing—Rebating with the Planer, Stopped Rebates—Chamfering—Grooving, Tonguing—Beading—Moulding, Fitting False Top to Machine—Sticking Straight Mouldings—Curved Mouldings, Use of Templet—Spindle Moulders, Planing on, Setting the Fences, Speed—Rebating and Moulding—Thicknessing—Double-Faced Mouldings—Panel Raising—Curved Work—The French Spindle, Advantages of—Wood Curved in Plan—Sticking Circular Sashes—Treatment of Cross-Grained Wood—Use of Templets—Work Curved in Elevation, how to Mould—Construction of Fences, Use of Saddle—Work of Double Curvature—Method of Moulding a Handrail Wreath—Cutting Tenons, Obtaining Shape of Moulding Cutters, &c.—Using the Hand Mortise, Setting the Chisel, Cutting the Wedging, Driving the Core, Treatment of Hardwood.

THESE somewhat disjointed notes are intended to be of service to the inexperienced machinist and the joiner who may desire to utilise machines in the most profitable and skilful manner. Want of space precludes a more complete treatment of the subject, but it is hoped that the operations selected will be sufficiently typical to indicate further applications of the methods suggested by them.

SAWING MACHINES.

CIRCULAR SAW BENCHES.—The saw should suit its work, *i.e.*, a cross-cut saw must not be used for ripping, and *vice versa*. In the figure on next page is shown some typical teeth for circular saws. A is a good form for soft woods, B a good form for hard woods, C a bad form, having insufficient gullet or space for sawdust; also the cutting face is nearly radial, therefore the whole tooth strikes at once, making the feed hard and the surface rough; D and E are cross-cutting teeth for hard and soft woods respectively. See that the teeth are sharp, all of equal length, and suitably set; keep the saw truly circular, so that *all* the teeth do a share of the work. To ensure this, either rub the saw down with a piece of whetstone whilst running slowly on the spindle, applying the stone at a slight angle across the teeth of the saw; or make a radial gauge to test the truth of the teeth when filing them with any thin slip of wood mounted on the sharpening mandrel.

This, if not exactly the same diameter as the saw spindle, should be made up to it with wood collars, which can be made by boring a hole to fit the mandrel, then from the same centre describe a circle equal in diameter to the saw spindle, cut this out at the band saw, and the collar or sleeve is ready. Fit the radius gauge on this, sweep it round the teeth, and, having found the lowest, bore a small hole in the rod, touching the top of the said tooth, and drive in a peg, then file down the rest of the teeth until the peg will just clear them as it is swept around.

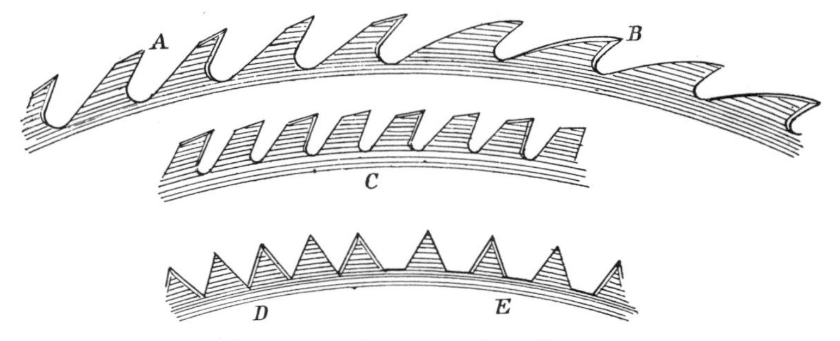

FORMS OF CIRCULAR SAW TEETH.

A & B. Ripping Teeth. C. Ditto, bad form. D & E. Cross-cutting Teeth.

Packing the Saw.—Circular saws require "packing" to make them run steady. With small saws this is usually done by adjusting wood pegs at opposite sides of the saw through holes in the frame provided for the purpose; with larger saws the packing is done with oiled cotton waste, spun yarn, oakum, or sometimes linen rag. The chief object of the packing is to provide a smooth, homogeneous filling between the saw blade and the edges of the gap in the table, to prevent its vibration, but beyond this it is further utilised to heat the saw by its friction where required, and so correct irregular running due to unequal tensions in the blade. The larger saws are made, by judicious hammering, rim bound, *i.e.*, the metal near the edge of the saw is in a state of greater denseness than that near the hub; this is done to counteract the centrifugal stress when running; if the saw is heated near its centre this part expands, and the more dense rim offering great resistance to the expansion of the plate, it bulges, or buckles sideways. The packing is therefore adjusted to prevent this by driving it in tightly near the rim, and gradually easing off the pressure towards the centre; thus the rim, more heated than at the centre, expands in greater ratio, and so equalises the tension. Occasions arise when it is necessary to reverse the process, difficult to explain in writing, but soon learned by watching the behaviour of a saw under different treatments of the packing. Fig. 1, p. 112, shows in part plan and section the "packing box" of a circular saw bench; *f* is the finger plate, closing the aperture for the insertion of the saw; *a, a* are pieces of wood bolted at the bottom of the saw gap; *m* is a wood mouthpiece at the feed end of the gap; *p, p* is a yarn packing. The back of the saw is not packed, but unless the gap is edged with wood strips it is best to put two similar pieces to *a, a*, at the back of the saw to prevent the teeth striking the iron edges.

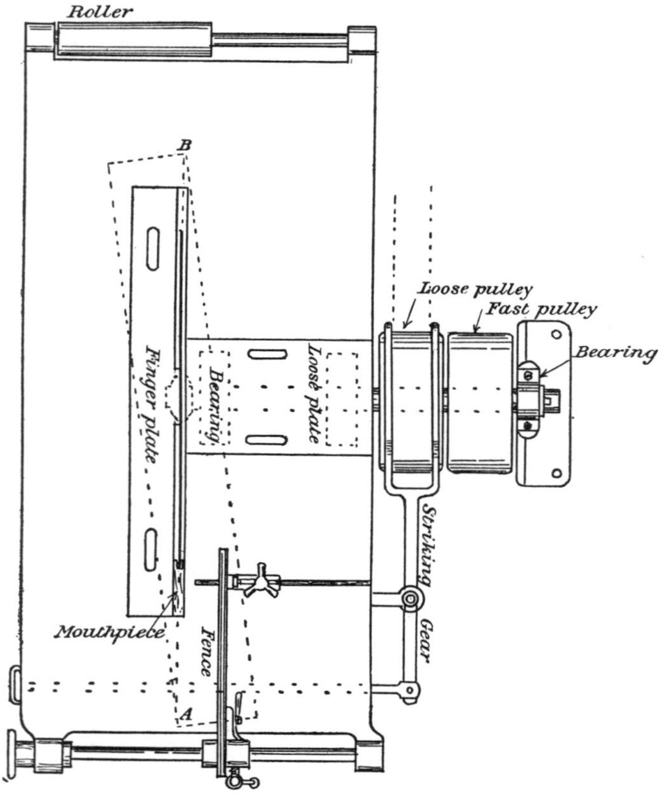

CIRCULAR SAW BENCH TOP,
Showing Method of Cutting Diagonally.

FENCES are used to ensure the feeding of the stuff parallel with the saw blade in both directions, and are only serviceable when boards or other stuff of uniform width or thickness throughout is required. When, for instance, a board is to be cut diagonally, as shown in the dotted lines in the plan of a bench top alongside, the fence is dispensed with, and the board is fed to a line drawn upon it. If, however, a quantity of similar pieces had to be cut, it would be worth while making a running piece, the reverse of the piece cut off, to be attached to the boards by two cross battens, with spikes in them to bring the inside edge in line with the saw, when the pieces could be run against the fence in the usual manner.

A Plain Fence is one capable of adjustment only to and from the saw, and is usually fixed by means of screw bolts in slotted brackets, as in f. 1, p. 112. **Compound Fences** are pivoted at the top edge to a frame, and can be set at varying angles with the bench top; they can also be moved horizontally to suit the diameter of the saw with or without alteration of their distance from it. Usually they have both a coarse and fine adjustment, and can be tilted off the table without disconnecting them, as shown in the photo, p. 94. As this type slides on a spindle at one end, and in a dovetail groove at the other, no trouble is experienced in adjusting them; in fact, if the saw is mounted correctly on its spindle no adjustment is necessary, save setting at the required distance from the blade; but with plain fences a straight-edge should be laid across the face of the saw, and the front end of the fence first set at the required distance, then the rear end is arranged exactly the same distance from the straight-edge. The "lead" that is sometimes advised in cutting crooked deals, *i.e.*, making the opening a little wider at the front end of the fence than at the back, is, in the opinion of the writer, better dispensed with. If a deal is very crooked, it is better to remove the fence altogether, and work to a gauge line on the edge, which may be chalked to render it more readily visible, or, better still, cut it with the band saw, when a $\frac{3}{4}$-in. iron rod may be used as a fence. It is sometimes advisable to fix a wood bed-piece or "face" to the fence, for instance, when stuff that has cast hollow is to be cut. If this is fed with the hollow side to a narrow fence, the top edge falls over. If the round side is placed next the fence it will wobble, but if two strips or ribs are nailed at top and bottom edges of the face piece, the deal will travel steadily against these. In the adjoining figure is shown how such a "face piece" may be fixed to the iron fence.

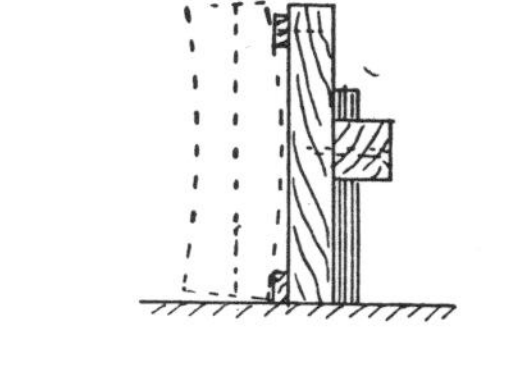

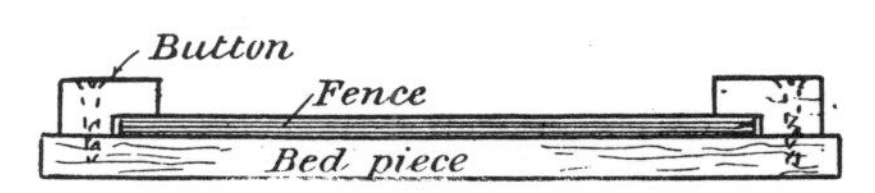

ADJUSTMENT TO FENCE.
1. Section. 2. Plan.

In ripping down, the smallest saw that will go through the stuff should always be used, except in the case of hand-feeding deep stuff, then the larger the saw the easier the feed. In the first case, the advantage lies with the increased speed, which gives a cleaner cut; in the second, saws cutting with the top teeth force the timber backwards, whilst those cutting with the teeth near the axial line merely force it on to the bench.

Deep Cutting.—The saw having been properly packed, the next thing is to arrange the fence parallel with the saw at the required distance from it. This should be "full" to allow for the set of the saw. The best way is to make a trial cut in a waste piece of stuff, and set accordingly. If one of the up-to-date benches is used, as shown on pp. 86 & 88, no difficulty will be experienced with the parallelism of the fence, but fences that are simply secured by bolts through slotted angle brackets at the back require careful adjustment, and the fence should be tested with a straight-edge laid against the face of the saw. Keep the back end of fence just beyond the front edge of the saw, as shown on opposite page. If a long heavy plank is to be cut, the sawyer should take the rear end, keeping himself in line with the saw that he may be able to watch the cut, and raising his end slightly so that the plank shall clear the end of the table, and bed firmly at the cutting point, he presses forward steadily and regularly without unduly forcing the pace. The labourer, after starting the cut by pressing the fore end hard to the fence, moves to the middle of the plank, where he helps to hold up the weight until the cut is half-way through, when he proceeds to the leading-off end and pulls out steadily, first however, driving a wedge in the cut to prevent the teeth scoring on the rise, unless the bench is provided with a riving knife, as shown in f. 1, p. 90, when the wedge is unnecessary.

The hand should be kept away from the inside plank when nearing the end of the cut, and a stick with a pointed spike in the end, or a notch, as shown in f. 2, used to push the wood past the saw.

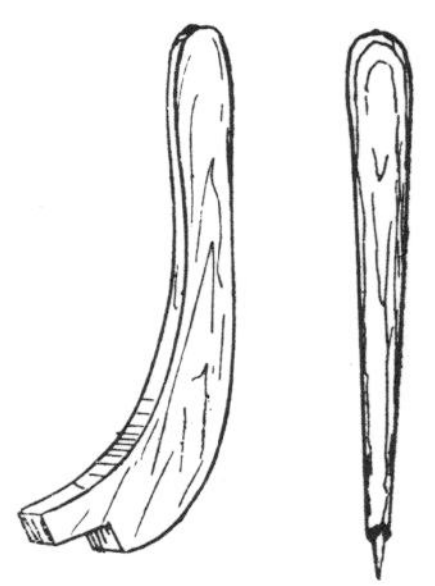
2. Notch Sticks.

1. Bevel Cutting with a Circular Saw.

Bevel Cutting, such as is required for arris rails, window sills, weather boardings, reducing scantlings for moulding, &c., can be readily accomplished by aid of the compound fence. Nothing further is necessary than to set it at such an angle that when the stuff is resting against it, the cutting line is perpendicular to the bench, which is easily found by marking a line upon the end of the stuff, and testing it with a set square (one of these should be in the sundries drawer of all machines). If a plain fence is used, remove the usual mouthpiece from the saw gap, and replace it by one having a **V** groove cut in its top edge, as shown at *m* in f. 1 adjoining. This is merely to keep the lower edge of the stuff in line with the saw. To ensure it travelling upright bring the fence up to the inner edge of the stuff, as shown in full and dotted lines in the section above, f. 1, which give two extreme positions. Hard wood should be used for the mouth-block. It can be cut to shape with the band saw.

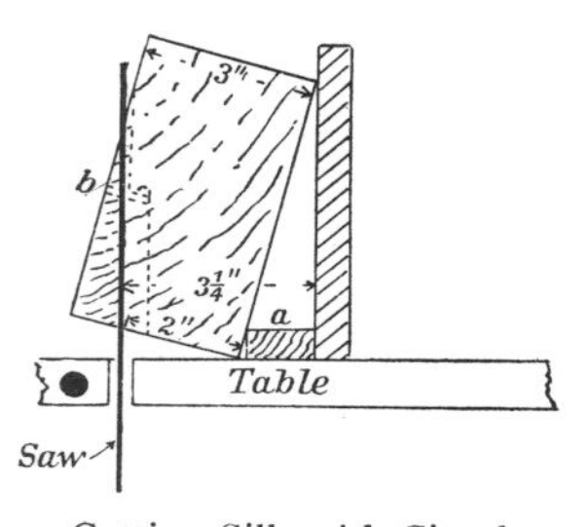

3. Cutting Sills with Circular Saw.

Window sills are frequently cut by aid of a slip or batten placed on the table against the fence, as shown at *a*, f. 3. This should be notched out to clip over the end of the fence to prevent it slipping forward when working. When the sill is to be double weathered, the saw is first run in at the point marked *b*, then run up to meet this cut on the dotted line.

Rebating with the Saw.—Use a small saw, about 10 in. or 12 in. diameter, and either lower the spindle or raise the table, according to the type of bench in use, until the top of the teeth stand above the bench $\frac{1}{32}$ in. less than the required depth of cut. Fix the fence temporarily at an equal distance from the *outside* of saw that the back of the rebate is from the face of the stuff, and, having marked the rebate on the end, bring the piece up to the saw to see that it is set exactly; if so, tighten the nuts, and run through all one way first; then reset the saw and fence to the second depth in like manner, and run these through. It is usually better to make the deeper cut first.

Grooving with the Saw.—If a special thick saw is used, nothing further is necessary than to select one of the right thickness, and to fix it to the required depth, adjusting the fence as necessary. But grooving is more often done with a thin saw set obliquely on the spindle. This is termed a **Drunken Saw**. Its action is shown in the section adjacent. Taking a vertical section, the teeth at the upper part cut one side of the groove, and those directly opposite cut the other side. This is indicated by the dotted line, the remaining teeth cut away the "core," moving gradually across from one side to the other as they come around. As, however, the cutting diameter of the saw is less in the inclined direction than on the transverse diameter, the path of the teeth is elliptic; and they produce a groove, hollow at the bottom, which must be afterwards squared out with a router, unless there is sufficient to make it worth while to stone down the saw to an elliptic shape. Of course, in grooves that are not seen, as for panels, the routing is unnecessary. The pitching of the saw is accomplished either by means of three set screws in the collar, or by wedges inserted between the collar and the saw, or, as shown above, by a pair of special collars, wedge shape in section. These are made in sets, to produce grooves of definite widths with saws of a certain diameter. As the latter wear in use, the pitch is increased by packing the collar with brown paper until the reduction is sufficient to suit another pair of collars.

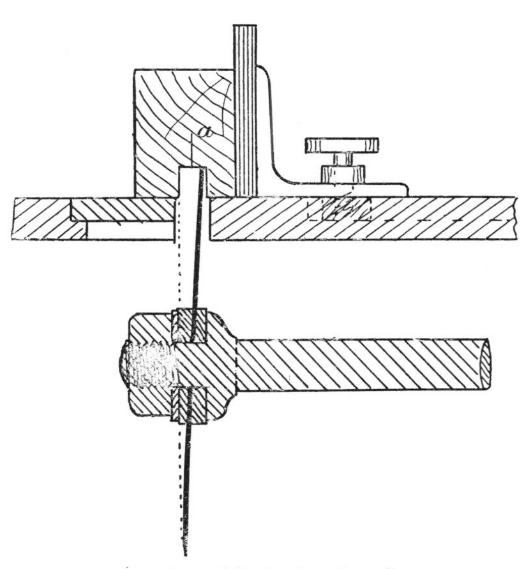

Grooving with the Drunken Saw.

The stuff should not be fed rapidly, otherwise some portions will be left uncut, and the edges of the groove consequently be left ragged.

"Miller's Patent Drunk Saw" has a ball-and-socket attachment, which makes it instantly adjustable to cut grooves from $\frac{1}{8}$ in. to 4 in. wide, up to 5 in. deep. The saws supplied have both ripping and cross-cutting teeth on the same blade. The cross-cutting teeth are on the highest and lowest points, therefore strike the stuff first and make a clean cut to the edge of the groove, whilst the ripping teeth remove the core between. "Oval" saws are also supplied to form round-bottomed grooves. They are manufactured by the Carron Company, Glasgow.

THE BAND SAW.—An essential preliminary for obtaining good results with the band saw, is always to change the saw when it gets dull. When a saw is not sharp it necessarily cuts slow, and the feed is involuntarily increased. The result is, the saw turns aside, avoids the hard places, and the cut is untrue and irregular. Another important point is the correct adjustment of the top guide, which should be brought up exactly touching the back of the saw whilst the latter is at rest, also be brought down as near to the work as possible. Then the saws should suit the work, wide bands for straight cutting, narrow bands for small curves and fine moulding. In cutting curves, if the saw runs from the line, it is better either to run the stuff back and commence the cut at the other end, or, when this is not possible, to cut into the faulty place from the outside, and start the cut afresh from that point. To attempt to force the saw into the right path will, if it does not break the saw, merely cause it to follow the line at top, whilst pursuing the original path below. The curve should be watched slightly ahead of the saw and fed up in a sweep. Fences are not used in curved cutting, but are needed in straight work of any length over a foot or so. **Tenons** on door or sash rails, &c., may be easily cut with the band saw if a temporary fence is made with a piece of wood shaped like a set square, and about $1\frac{1}{2}$ in. thick. This should be held with the left hand against the rail exactly opposite the saw, and the rail can then be fed past it with the right hand, running the saw down the gauge lines. **Handrail Wreaths** of double curvature may be "bevelled" with the band saw by preparing a bed-piece or block, as shown in the sketch above. This block should have its lower end cut quite square with its sides, and its upper end cut to a plane containing the tangents, as described at pp. 339 and 346; then the wreath piece, first lined out and roughly cut to the face mould, is fixed upon the inclined surface of the block by means of a screw at each end, as

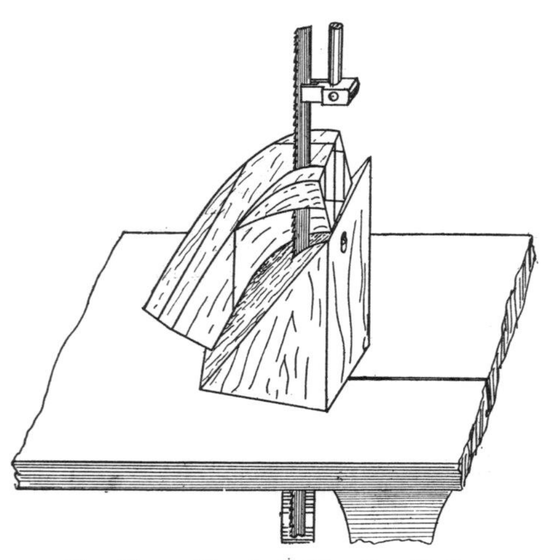

Bevelling a Handrail Wreath with the Band Saw.

shown, when the lines on the joints will be perpendicular to the saw table, and the wreath and its carrier may be passed around the saw, as shown. When the outside cut has been made, the lower edge may be squared off from it by keeping the wreath bearing solidly upon the table, on its convex side, as it is moved around to the line drawn in the manner shown in f. 3, p. 357. This cannot be done with the top edge because there is seldom thickness enough to get the top line marked upon the side of the piece.

Circle on Circle Work, such as door, sash, and frame heads, may be cut out with accuracy and despatch with the band saw if manipulated as follows. The methods of preparing the necessary moulds are detailed at p. 277, and to facilitate this description it is assumed that they are prepared. Let it be required to cut out the head of a solid frame, as shown in plan and elevation in f. 1 & 2, p. 277, the head to be in two pieces with joints at the crown and springings. Having cut out the stuff to the rectilineal figures shown in plan

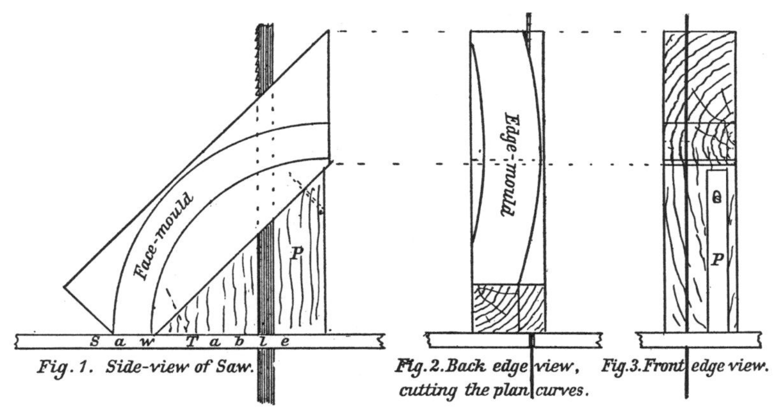

Fig. 1. Side-view of Saw. *Fig. 2. Back edge view, cutting the plan curves.* *Fig. 3. Front edge view.*

Cutting Circle on Circle Work on Band Saw.

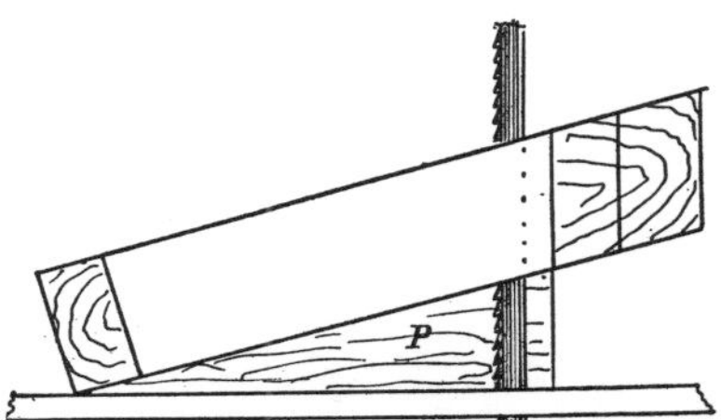

Fig. 4. Cutting the elevation curves.

and elevation, and prepared the moulds, plane the face side and back edge. Line out the face mould on the face of the stuff as shown in f. 1, and the plan mould on the back edge as shown in f. 2. The piece should be cut off to the joint lines *c-b′* and *c-s*, f. 1, p. 277. This, if cut square to the face, will be long enough to make the proper bevel joint later; the method of obtaining this bevel for the edge cut is explained at p. 277. When lining on, from the moulds, adjust the ends of the face mould upon the two joints, and the end of the edge mould accurately at the top end, keeping its bottom end at the same distance from the face that it is in the plan. Next prepare two bed-pieces out of 2-in. stuff, as shown at P, f. 1, 3, & 4. These are for the purpose of tilting the head-piece up whilst cutting, to the same positions that it is shown in the plan and elevation, p. 277. The shapes of the bed-pieces are indicated at f. 1 & 2 by the dotted triangles S *f b* and C C *c* respectively. Screw and brad the bed-piece on, as shown in f. 1, p. 115, so that the brads are not in the path of the saw, and cut the plan curves first. Do not run the cuts completely through at the lower end, as the two waste pieces are required to retain their position whilst making the transverse cuts. Having cut the plan curves, remove the first bed-piece, and attach the other one, as shown in f. 4, to cut around the elevation curves. The saw may be run completely through in this case, but the two waste pieces must be tacked on again, so that the other two cuts may be finished, then the half-head will be of its exact size and shape, ready for planing, &c.

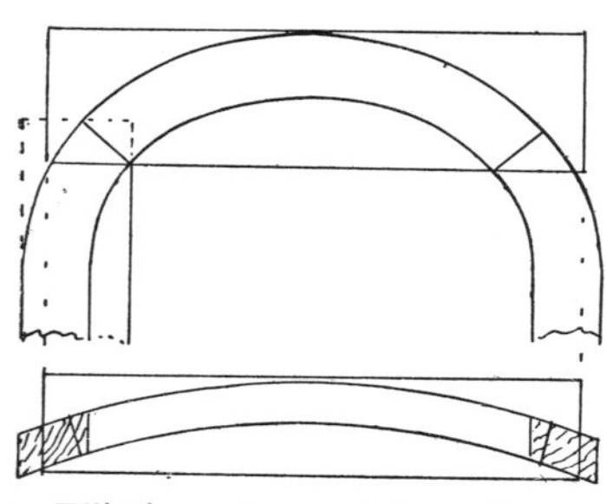
1. Elliptic on Segment Door Frame.

An Elliptic on Circle Door rail is shown in f. 1 on this page, and this would be jointed as shown, the rectangles showing the dimensions of the piece required. The central piece of the top rail, after lining out on the stuff exactly as drawn in the plan and elevation, can be cut out by the band saw in the manner shown in the sketch, f. 2, the plan cuts first, then turned over upon the side *a*, and the elevation cuts made, the direction of the two cuts being at right angles to

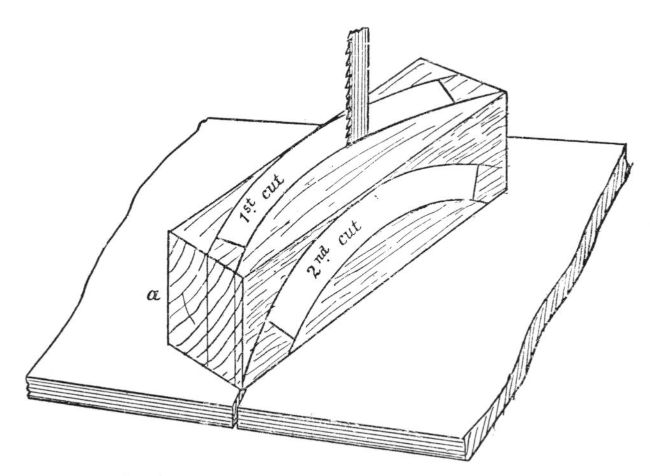

2. Cutting out Double Curved Work on Band Saw.

each other. As described above, the first cuts are not taken completely through at once, as the two waste pieces are required, one for a bed, and the other for a mould ; but after the second cuts are made the first ones are finished.

PLANING MACHINES.

Overhand Planer. — The cutter iron in all planing machines must obviously, if a truly plane surface is to be produced, be absolutely straight upon the edge, and the two irons (two are almost universally employed) must balance, not only in weight, but in their position upon the block; the effect of even a minute variation of either of these things is at once apparent in the work. The tremendous centrifugal forces generated by the swiftly rotating cutters set up a vibrating motion in the spindle, which is communicated to the cutters, and

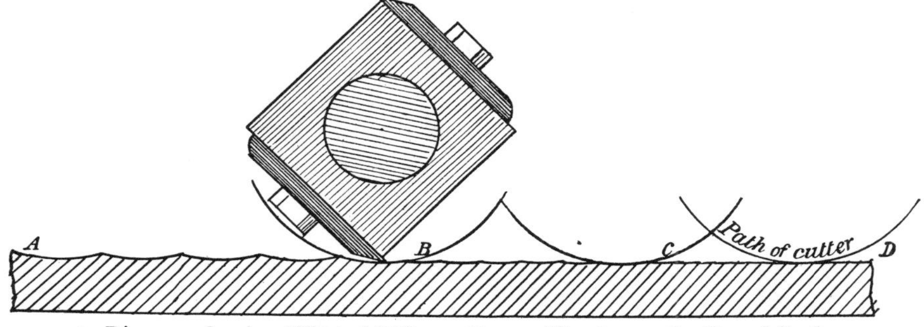

1. Diagram showing Effect of Different Rates of Feed upon the Planed Surface in a Planing Machine.

produces undulations on the surface of the work. Fig. 1 represents the method in which an approximately plane surface is produced by revolving cutters. A series of hollows is formed by the sweeping cut of the iron, with small ridges between them. The nearer these ridges are together, the nearer the machine-planed surface approaches, to the surface produced by the reciprocating motion of the hand plane, or the more perfect flat produced by passing the wood over the fixed knife of a "lightning planer." The slower the wood is fed over the cutter, the more revolutions are made upon it within a given area, and consequently the nearer the ridges are together, the finer the surface produced, as shown in the diagram, f. 1, where A to B indicates the effect produced on a board by rapid feeding, B to C the surface produced by a slower feed with the same speed of cutter, and C to D the practically straight surface resulting from a still slower feed. It should be noted that these surfaces have been actually produced in the drawing by a series of circular arcs similar to the paths of the cutter. The effect of a slight difference in the "balance" of the cutters is shown (exaggerated) in f. 2, where the full line indicates the path of one cutter, and the dotted line the path of the other; the perceptible result of this in any piece of stuff that is badly machined is the "corduroy" appearance of its surface. The deeper undulations that are sometimes observed at regular intervals are caused

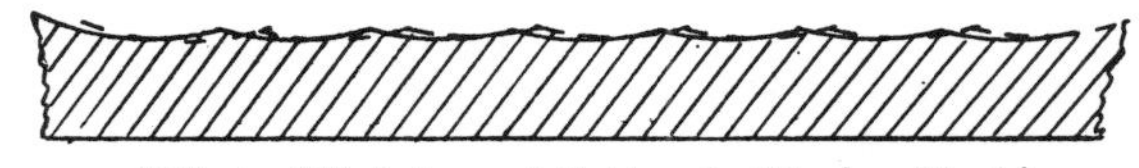

2. Effect of Unbalanced Cutters in Planing Machine.

by the vibration of the cutter, set up by a lap in the driving belt as it passes over the pulley at each revolution.

Surfacing.—Fix the cutters firmly, bed each on strips of cardboard, brown paper, or pieces of thin leather, and tighten each nut equally. Revolve the cutter blocks until the knife edge is perpendicular to the table, then screw the back table forward, or up, until a piece of straight stuff laid across the cutter edge exactly touches it; test it throughout; the cutters may require slight adjustment on the block. Next move *down* the front table just so much as the thickness of cut is intended to be; this can only be determined by experience with the tool and the particular wood that is to be planed. The dimensions of a cut cannot be usefully stated in figures. Set the smallest possible difference and try on a waste piece, then gradually augment the difference until a desirable surface is produced with a maximum depth of cut. Remember never to place the hand over the cutter even if a thick piece of wood intervenes; it is a dangerous habit to acquire, and sooner or later an accident is sure to occur if it is done.

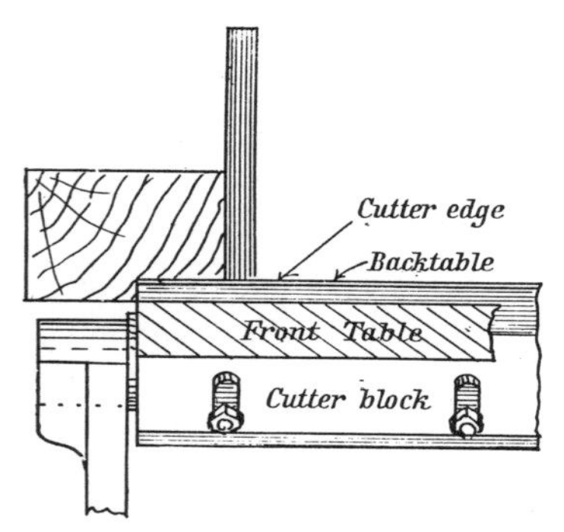

1. Rebating with an Ordinary Planer.

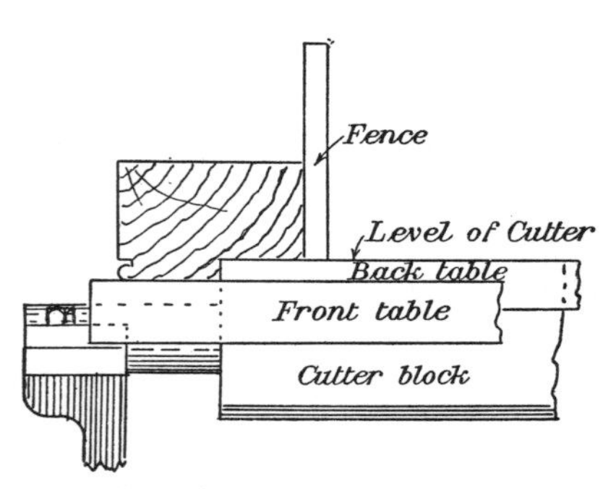

2. Rebating on a Rebated Planer.

When small stuff is planed, use a holding down fence, as shown in f. 1, p. 95. With wide, thin boards, use the knife guard, p. 130, and pressure springs; those that will keep the work down in advance of the cutter are the best, as the stuff is lifted in the up stroke, and drawn towards the table by the cutter which is leaving the cut.

To Take Out of Winding pass the stuff over the cutter as often as necessary to bring it to a true surface, keeping its high corners at equal distances above the table. If very much twisted, a small packing piece should be kept under the high end farthest from the cutter, whilst taking the first cuts, to keep the piece steady.

To Plane Hollow Stuff, or to **Joint Edges.**—If very irregular, pass each end over the cutter once or twice, for a foot or so, according to the length of the piece, then pass right through. If many glue joints have to be made, it is best to "rough" them all through first, that is, give them a coarse cut and fast feed, then run them through again with a lighter cut and slower feed.

To Rebate, lower the front table to the depth of the required rebate if it is not more than ½ in. (deeper rebates than this should be done in two operations), then adjust the fence at a distance from the edge of table equal to width of rebate, as shown in the section of a planing machine, f. 1 opposite. Some overhand planers have a special back table for rebating purposes with a recess on the edge about ½ in. deep. The advantage of this is, as shown in the section opposite, f. 2, that the whole of the piece of stuff that is rebated is supported by the table, the adjustments in this case being for rebates the same depth as recess, *i.e.*, ½ in. Set the cutter flush with edge of recess and top of back table, then sink front table to level of recess in back table. Set the fence the required width of rebate from edge of recess, when the state of things will be as shown in f. 2, the dotted line indicating the cutter block. If the required rebate is to be *less* in depth than the table recess, set as above, but place a slip of wood in the recess equal in thickness to the difference between the actual rebate in the table and the required rebate; for instance, a ¼-in. rebate would require a ¼-in. slip on the table. In some of Messrs M'Dowall's planers the tables slide laterally; this produces the same effect as a recessed table, with the additional advantage that the "recess" may be made of any width. In rebating with the planer, the face of the rebate should be gauged, or the cutter edge will leave it ragged.

Stopped Rebates, that is, rebates that do not go through from end to end of the piece. These are required in such things as casement frame jambs, circular head door frames, &c., and they are a ticklish job to handle, for should not great caution be used in dropping the work upon the cutter, it will be thrown back violently. The exact manipulation depends upon the part of the stuff that the "stop" is at; if the rebate is stopped at both ends, and is near the middle of a long piece, rest the far end on the back table, and gradually lower the end in hand upon the cutter (having first arranged it so that the cutter is just in front of the stop); hold firmly against the thrust back of the cutters; if it is a heavy piece it will be advisable to have some one also to press down close by the cutter. If the stop is near the feeding end, the best method is to chop away about 3 in. with a chisel, nearly down to the gauge line, and holding, or having the piece held up, level, drop slowly on to the cutter, and at once push ahead when it bites. In both these cases the back table is lowered level with the front one. If the stop is at the rear end no difficulty in feeding occurs, but as the cut is hidden by the stuff, a plain mark must be made on the back of the piece exactly above the stop; when this is reached either lift off, or, better still, stop the machine. In this case the back table is kept level with the cutter.

Plain Chamfering.—Adjust the fence at an angle, so that when one side of the stuff is resting on it, the chamfer face is parallel with the table, as shown here. Lower the front table as much below the cutter, as the face of the chamfer is perpendicularly from the salient angle. This is most usually found by trial, by drawing the chamfer on the end of the stuff and moving it up to the cutter, when the knife is at its highest point above the table, raising or lowering the front table accordingly, then adjusting the back

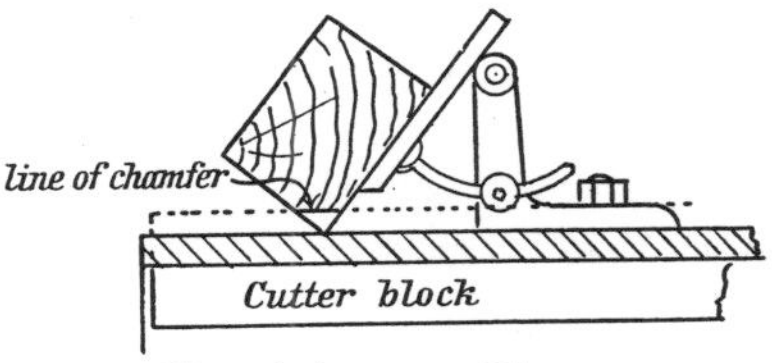

Chamfering on a Planer.

table to the same level as the cutter, so that the chamfered surface leads off on it. The dotted lines indicate the back table, also top edge of cutter.

Stop Chamfering is managed similarly to stop rebating, which see.

Grooving, Tonguing, Beading, and Moulding.—Either of these operations can be done on an overhand planer by attaching appropriate cutters to the block. For an odd job in between the usual planing, the cutters may be attached in pairs to the two vacant sides of the planer block, which is almost invariably provided with dovetail grooves for that purpose. If, however, a large quantity has to be done, it will be more economical to remove the plane

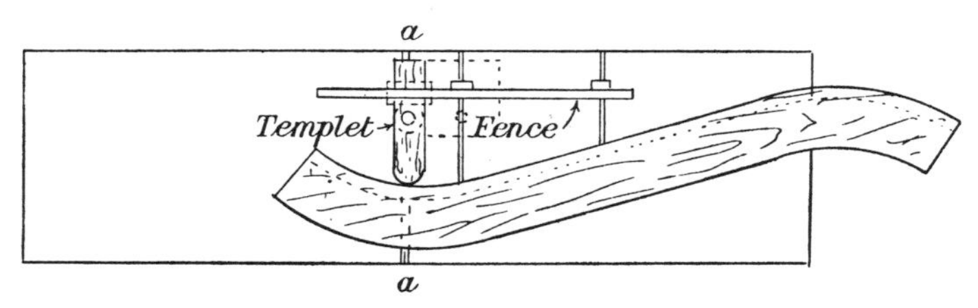

1. Moulding Curved Edge on a Planer.

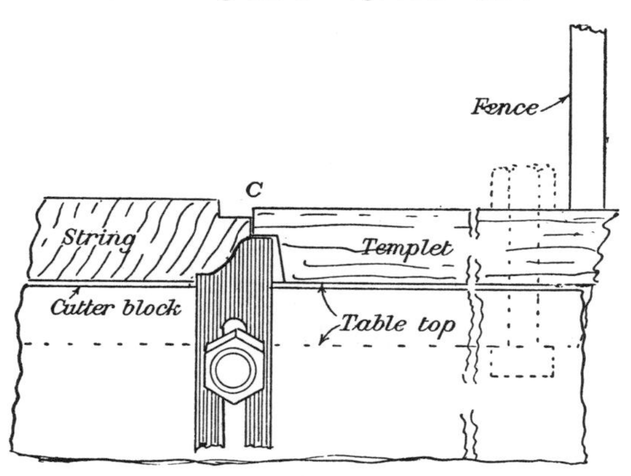

2. Enlarged Section on *a—a*, Fig. 1.

METHOD OF MOULDING CURVED EDGES ON A PLANER.

cutters. If a "solid" spindle, *i.e.*, one in which the square cutter block is cast in one piece with the spindle; or if of the "sleeve" type, remove the long cutter block, and substitute a short block to take the moulding knife. This can be adjusted at any convenient part of the spindle by means of loose collars. The gap in the table will be considerably increased with these cutters on, as their projection is necessarily greater than in the case of plane cutters, and it is sometimes wise, especially if the stuff to be worked is short, to provide a false top to the machine. This may be an inch board planed to thickness and secured to the table by nuts and bolts, if there are any holes provided, if not, by wood buttons screwed to the overhanging edges. Having secured the board, start the machine and gently lower the table, when the cutter will penetrate the false top, making a gap for its path, and leaving all else covered up. Some machines are provided with "bridges," that is, loose plates of metal with pins on their edges which fit into holes in the edge of the table, and so fill in the gaps.

Straight Mouldings of comparatively small section, may be safely worked on the Surface Planer, by the use of the holding-down fence, shown in f. 1, p. 95, but large mouldings should be worked upon an undercutter with suitable feed rollers.

Light Curved Mouldings, such as those on stair strings, circular architraves, and the like, can be worked on the overhand planer by means of a circular templet fixed to the table, as shown in f. 1 & 2 opposite. The templet should have a round end of sufficiently small radius, to prevent any part but its extreme end

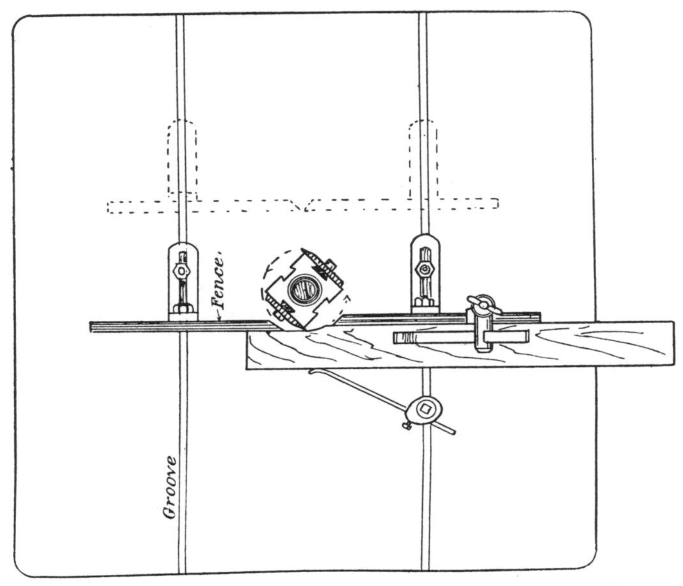

Method of Surface Planing—Plan of Table of a Spindle Moulder.

touching the curve to be worked, and this point must be placed exactly over the top of the cutter. The templet must be thick enough to reach a "square" upon the work above the moulding, then the cutter will revolve under it, in a path that may be cut by itself, leaving sufficient wood above to act as a guide to the stuff, as shown at *c*, f. 2. The templet may be secured in any convenient way, either by widening out the sides, as indicated by dotted lines in f. 1, until they cover a hole or groove in the top, or it may be secured by the fence and guide springs.

IRREGULAR MOULDING MACHINES.

SINGLE SPINDLE MACHINE.—Straight Work.—Surface planing or squaring: Raise the spindle to the full extent, slip on the square cutter block similar to f. 1, p. 98, with a pair of wide irons, set to cut in the direction shown by the arrows in plan on p. 121, that is, against the feed, then lower the spindle until the lower edges of cutters are just below the table, and place a suitable filling-in collar around it. Next adjust the fences *in front* of the cutter block, *i.e.*, they must cut through the fences in a similar manner to a horizontal cutter, through the table of a planing machine. Place the fences as near to the revolving cutters as possible, so that they do not actually touch any part. The back or leading off fence should be exactly in line with the edge of the cutter when at its greatest projection. This will be clear on inspection of the drawing. The front or feeding fence should be slightly within or behind the path of the cutters, the exact amount must be found by trial, but this distance determines the thickness of chip

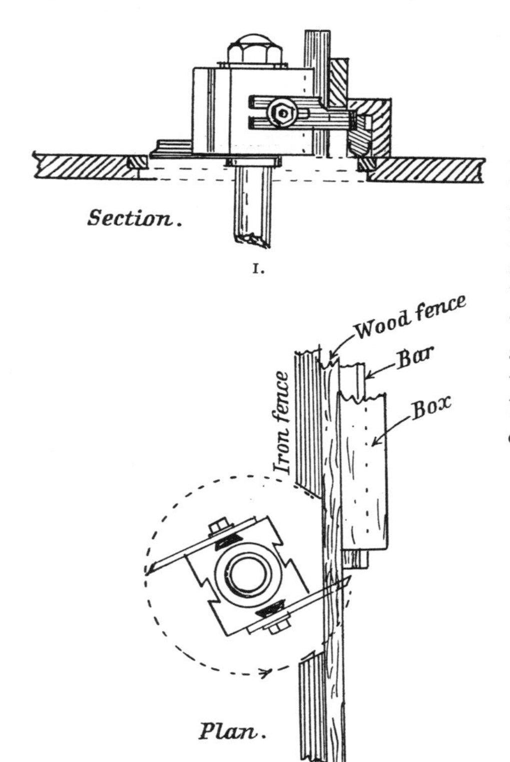

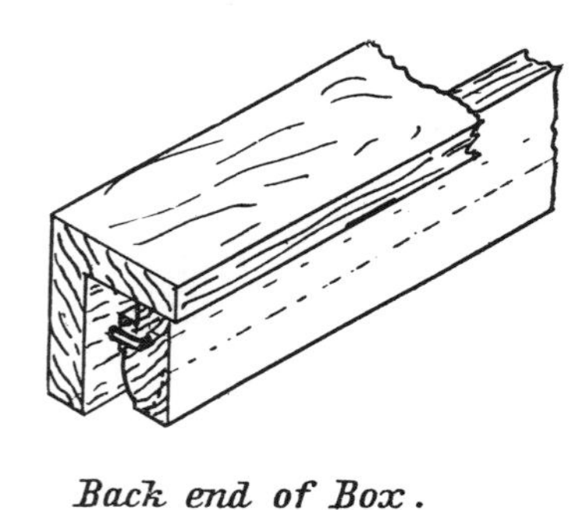

Method of Sticking Sash Bars on a Spindle Moulder.

taken, which must depend upon the state and kind of wood treated. If the fence is of the "fixed" type, that is, not capable of being set to any angle, nothing further will be required than to fix the pressure springs as shown, just tightly enough to keep the work up to the fence, and down to the table; but if it is a variable fence both portions must be tested for squareness. The work may be fed through fast or slow, just as the finish is required to be passable or high. Hard wood must always be fed slower than soft wood, and knots should be fed very slowly or pieces may be knocked out of the cutters.

Rebating, Moulding, &c., on the face: Proceed as described for planing, but use suitably shaped cutters, instead of plane irons, fixed on the cutter block. See f. 1 above.

Thicknessing.—In this operation the two fences should be brought quite close together and in line, and be placed the exact distance behind the cutters that the stuff is required to finish in thickness, as shown in dotted lines in f. 1, p. 121, the stuff passing *between* the cutters and the fence. Of course the stuff must be faced up, as described previously, before thicknessing.

Sash Bars and other thin, double-sided mouldings are done in a "box" or saddle, as shown in f. 1, 2, 3, opposite. The first side is worked in the usual way, running it through against the fence, the moulding iron fixed on one side of the block and the rebating iron on the other; but when the other side has to be worked, there is danger of the bar tilting over, unless some device is used to prevent it; one of the easiest ways is as shown. Two pieces of wood are nailed together at right angles to each other, of such size that the edges are exactly flush with the surface of the bar, or other work to be moulded.

At the rear end a tenter hook, or small screw is driven, so that it will fall upon the unworked square, as shown in f. 3; this prevents the piece flying back. Some machinists have a small spike in the upper side of the box to prevent the bar tilting; others brad a slip in the angle to fit into the first rebate, for the same purpose. A wood fence should be fixed to the iron fence (which must be kept back clear of the cutters) having a hole cut in it just sufficient to let the cutters work through, and the box with its contents is passed along in front of this.

Raised Panels may be worked with the French Spindle cutter, as shown in sketch below. The piece above the cutter is merely a filling-in piece. The tail of the cutter is made the same shape as the front edge to provide the necessary "balancing," but having a shade less projection to avoid touching the cut surface. These panels should be squared off accurately to size, then all four sides passed over the cutter as shown; the result will be perfect mitres and a smooth surface, good enough for painting or varnishing, but requiring fine glass-papering if intended to be polished. These instructions for working over the cutters are contrary to the practice of some machinists, but it will be found safer for the operator, and produce better work on panels that are cast than by working *under* the cutter. It is advisable to pass the work through twice, the first time with the cutter set back a trifle, then the finishing cut to the proper depth will leave a very smooth surface.

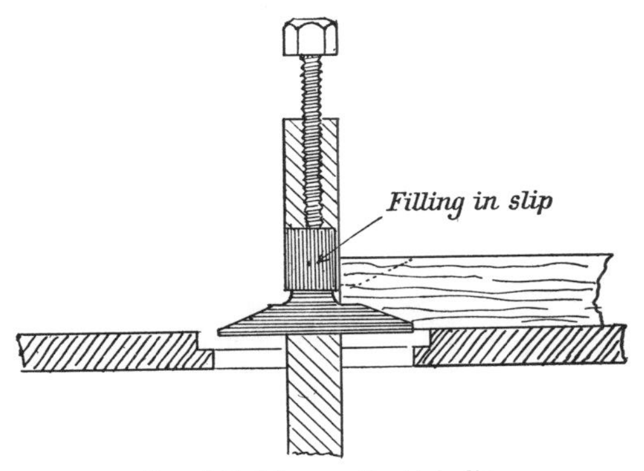

Panel Raising on the Spindle.

Curved Work.—If this is done on a single spindle machine, the moulding cutters must be made in pairs, that is, of opposite hands, because the direction of the grain changes owing to the irregular path through the wood that the curve takes, and the direction of the spindle has to be reversed that the cutter may not be driven in a direction to tear up the grain, and as bevelled cutters only cut in one direction, two are required. To avoid the trouble of changing, most single spindles now are provided with loose heads in which very thin cutters can be used, either the slot collar, f. 2, p. 98, or the French spindle, p. 123; the latter is a scraper, pure and simple, and although it does excellent work on hard woods, it is comparatively slow in action, and its use is limited to moulds of small projection. The cutters used in the slot collars are about $\frac{1}{4}$ in. thick, a medium between the thin French scraper and the thick bevelled cutter knives; they are distinguished as "flat steel cutters," and are made with the profile or cutting edge square, *i.e.*, filed at right angles to the surface; they may be revolved in either direction, cutting equally well either way. Only one cutter is required, but a balancing blank must be placed in the opposite slot shown in f. 2, p. 99, so that the collar may bear equally at each side of the spindle.

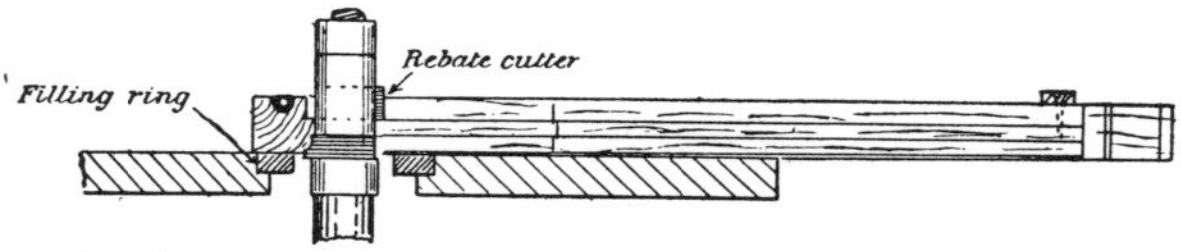

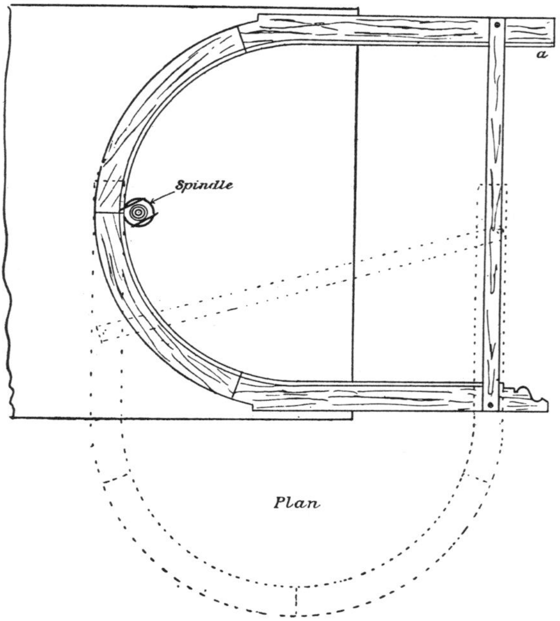

MOULDING AND REBATING A CIRCULAR SASH ON THE SPINDLE MACHINE.

1. Section. 2. Plan.

Work Curved in Plan.—This is the simplest to manage on the spindle; it comprises such things as circular-headed sashes, shaped stair treads, bath and W.C. tops, &c. Figs. 1 & 2 above show how a circular-headed sash is managed. Use flat cutters, one to cut the moulding, the other the rebate: place concentric collars above and below, as shown in f. 1 & 2, and project the cutters the exact amount the moulding, &c., has to be stuck down; thus the "square" of the sash will work against the spindle collars as a fence. The sash should be fitted up in the square, all the joints being dry; in some shops the meeting rail is fitted in first, but this is not advisable if first class work is desired, as there is always a possibility of the machine sticking slightly more or less than has been allowed. As shown in the drawing, the radial joints are made with handrail bolts; also the mortises for the meeting rail, but in place of the latter a rough stretcher is screwed on the back, to keep the stiles in position. Commence by removing one screw and pushing the stretcher aside, as shown in dotted lines, feed up the stile to the cutter from the end (the horn is left square as at *a*). As soon as the mortise is passed, stop the spindle and screw on the stretcher, then start again and continue until the other end is reached, removing the stretcher as before. The work should be eased off when the joints are reached, so that it may be done as it were, in two cuts, which will prevent the cross grain tearing out. In exceptionally cross-grained wood it may be necessary to stop and reverse the spindle, as it begins to cut upwards, and great care must be used to take a light finishing cut right across the joint. When a number of **Curved Mouldings** or any flat, shaped work is required all to one pattern, it is more economical to make a deal templet to the required curve, and fix this by bradding or screwing to the back of the work, which has been cut to shape in the band saw. It ought to be lined out with the same templet, but sawn full, *i.e.*, about $\frac{1}{16}$ in. left on for planing. The stuff is then pressed against the cutters until the templet rests against the spindle, when the ensuing surface, whether square or moulded, will be everywhere parallel with the edge of the templet. Figs. 1 & 2 show a console in the rough, with the shaped templet attached. If only one is required it would be quicker to plane the edge to shape first, and then mould, in which case the template would be unnecessary.

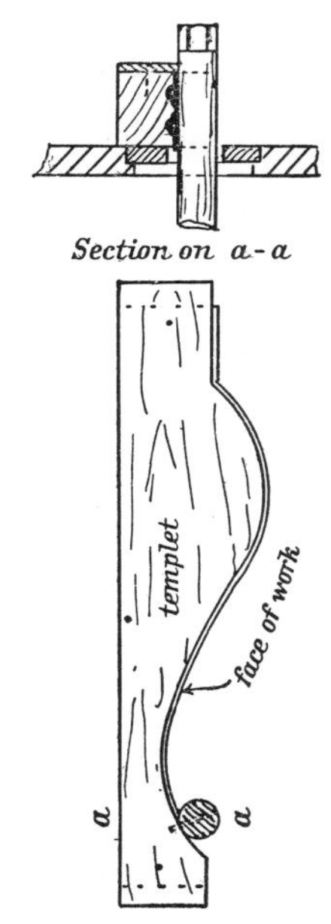

MOULDING CURVED WORK ON THE SPINDLE.

1. Section. 2. Plan.

It may be pointed out that the spindle will not work out at an internal angle, as at the head of this console, this part would have to be finished by hand. Many machines are supplied with an adjustable ring fence similar to f. 3. This is placed over the spindle, and the work is pressed against it, the cutters projecting beyond the fence the distance it is desired to stick the moulding down. It is more suitable for use with the cutter block, than with the French spindle.

3. A Ring Fence.

Work Curved in Elevation, such as architraves, overdoors, rails of soffits, &c., have to be applied to the spindle in a vertical direction, and they require a saddle templet as shown in f. 1 & 2, next page, shaped on one edge to fit the inside curve of the piece to be worked, or slightly quicker, and straight upon the other. The lower edge must generally be secured to a base board, that in turn can be secured to the table top, either by screw bolts through convenient holes, or by wedging pieces of wood in the dovetail grooves, and turning screws into these. A long spindle head should be used, which will bring the cutter up to the crown of the curve, and the highest part of the templet must be opposite the cutter; these heads are usually supported in a top bearing, formed by an L bracket as shown. In the case of elliptic or compound curves, the saddle should be made to fit the smallest or quickest part of the curve; it will then, of course, not fit accurately the larger parts, but with a little care in holding the piece firmly, it can be passed through steadily enough to obtain a true section. If the piece is much wider than it is thick, as shown in the illustration, short side fences will be required to prevent it tilting; these can be fixed upon the face of brackets as shown at *a, a*, the leading-off piece should be cut roughly to the section of the mould, so that the work will bed firmly on it after passing the cutter. In some extreme cases it may be necessary to unscrew the table from the machine, and fix the saddle on the pedestal, as the

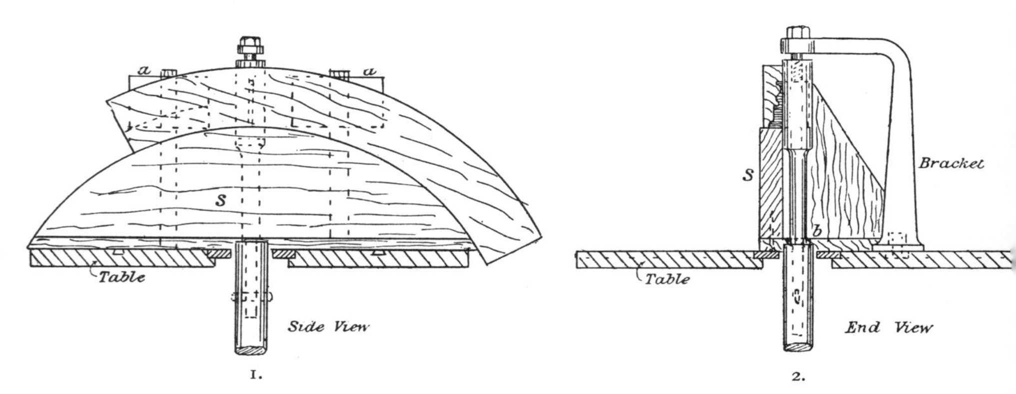

Working a Curved Architrave on a Spindle Machine.

curve may pass within the width of the table. In most cases where the edges of the moulded piece are concentric as shown in f. 1, the mould may be worked without the saddle by turning the cutter upside down and working the moulding from its back edge. When, however, the back edge is straight, this cannot be done. A parallel or concentric piece is shown on the saddle, because it was desired to show the spindle head and fences above it.

Moulding Handrail Wreaths and other work of double curvature is done upon a dome-shaped block, which has a hole bored through its centre for the purpose of dropping it over the spindle, as shown on next page. This block, termed in the mill a "dumpling," is made of sufficient height to enable the lower end of the work to clear the table, when the upper end is applied to the cutter. It has a small flat made upon its top for the work to rest upon, and it is secured to a base board with screws, and the latter is bolted to the table as shown in the section, which illustrates the method of starting the moulding of a Handrail wreath upon a French Spindle. Three irons would have to be used for working this moulding, the bottom half of the lower bead should be left for the last operation, and the rail turned upon its side to complete. The rail must be turned upon its back, to work the top mould, because the stuff should always be applied upon the top of the cutter to prevent accidents. The piece must be held down upon the blocks very firmly, and in such a position that the cutter is radial to its curve in one direction, and parallel to its "springing line" in the other.

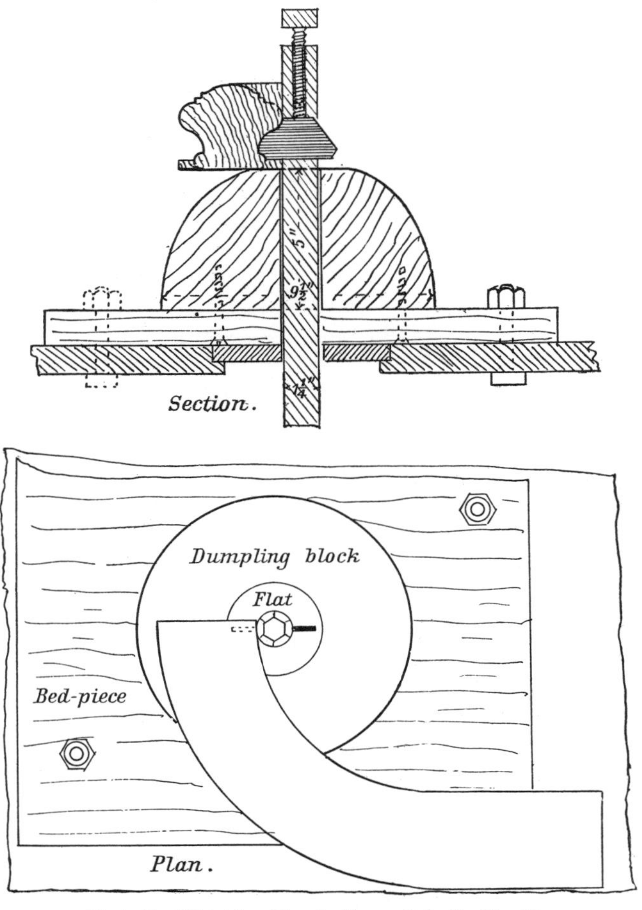

Method of Sticking Handrails on Spindle Moulder.

Tenon Cutting may be done upon the spindle by means of two small circular saws, with a spacing collar between them, as shown on next page. Set the straight fence close up to the teeth of the saw on each side, with a wooden fence fixed on the face, having a saw cut in it, through which the saw protrudes; this can be made by the saw itself by pushing the fence towards it, when set to the right height. Cut the stuff which is to be tenoned, all off to an equal length from the shoulders, and the saw must project from the fence exactly the same distance. Adjust holding-down springs on the fence, to just permit the stuff to pass under, then slide it past the saw. A block shaped like a set square, may also be used behind the rail, to prevent it "kicking." Messrs Robinson & Son fit a special sliding table to one of their spindle machines, to which the rails may be clamped and travelled past the spindle, which in this case carries two narrow cutter irons.

Obtaining Shape of Moulding Cutters.—When the cutter is mounted in the axis of the revolving spindle, as in the case of a French spindle, all that has to be done is to file or grind it to a true reverse of the required section; but if it is to be fixed to a square cutter block, then the edge will strike the material at an angle other than a right angle, dependent upon the size of the block, and the cutter must be shaped accordingly. This will be understood upon inspection of sketch opposite which is the elevation and plan of a moulding block and iron. A represents a section of the required moulding on the line *a-a* at right angles the axis of the spindle. The extreme point of the moulding 8 is projected into the plan, and the large circle represents the path of the cutter at that point. Obviously if the iron was mounted on line *a-a*, it would require to be a reverse of the moulding, but if it is moved out parallel from this line to the face of the block, its members must be shorter, according to their distance from the centre of the spindle. To find the true shape, drop projectors from the various points in the moulding, as 1, 2, 3, 4, 5, 6, 7, 8, to cut the line *a-a* on the centre of the spindle; describe concentric arcs from these points, intersecting the face of the cutter in points 1′, 2′, 3′, 4′, 5′, 6′, 7′, 8′, project these points into the elevation as shown by the dotted lines, and draw horizontal projectors from the original points to meet them, and the profile drawn through their points of intersection will be the true shape of the iron. In practice it is not necessary to make an elaborate drawing showing all the parts; all that is necessary is to draw a line *a-a*, representing the axis of the block, and a second line parallel thereto at a distance equal to half the thickness of the block. Draw the given section above this, in its relative position, and the path of the widest part of the moulding gives the key for the parallel arcs representing other parts of the moulding.

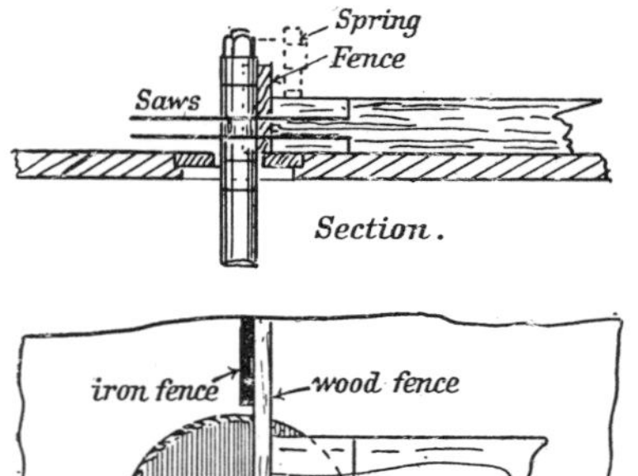

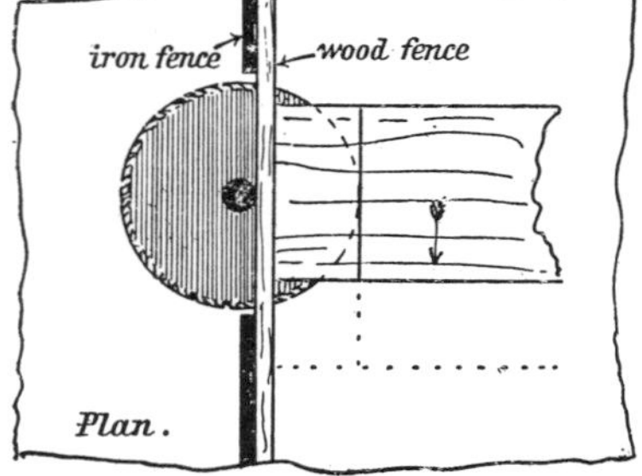

Cutting Tenons on a Spindle Machine.

MORTISING: With a Hand Machine.—The machine is prepared by fixing the appropriate chisel (which should also be used to set the mortise gauge by) in the holder, placing a bed-piece of wood about 1 in. thick, and equal in width to the thickness of the stuff to be mortised, upon the iron bed of the machine. Next place the stile in the machine, with one end under the chisel, and adjust the height, so that the chisel will go about two-thirds through. See that the face of the chisel is square with the fence, then place the piece directly under the chisel by turning the bed wheel, adjusting it laterally with the small wheel. The "face" of the stuff should be placed towards the fence, and the back edge up. Stab in the chisel near one end of the mortise, and work towards the other end with successive strokes, each deeper than the other, until the chisel takes its full stroke; do not take big jumps when cutting, short, quick strokes close together make the cleanest work. When the wedging line is reached, reverse the chisel and cut the wedging with the back of the chisel, run the bed forward, and finish the other end of the mortise similarly. Always when reversing the chisel try it for accuracy with the gauge lines before driving in, as it frequently happens, especially if the chisel has been strained, that it will not reverse correctly. If the error is serious it may be corrected by introducing a shaving between the head of the chisel and the socket. Release and refix the next mortise under the chisel, and repeat the aforesaid operation until all are done, then turn the piece over lengthwise and cut through from the face edge, taking care to finish exactly to the sight lines of the mortise, and hold the large wheel steady when making the last stroke, to prevent the bed running forward, and so making the end of mortise standing, which will prevent the tenon coming up to the sight line. Having passed all the stuff through the machine, remove the chisel with the key-wedge provided, and insert the core-driver in spindle. Take away wood bed-piece, place the mortise over the hole in the bed-plate, wedging side down, raise the bed until the core driver will just reach the plate, adjust the mortise directly under it, and immediately opposite the grip fence, which will prevent the sides bursting out should the core jam; having tightened the fence

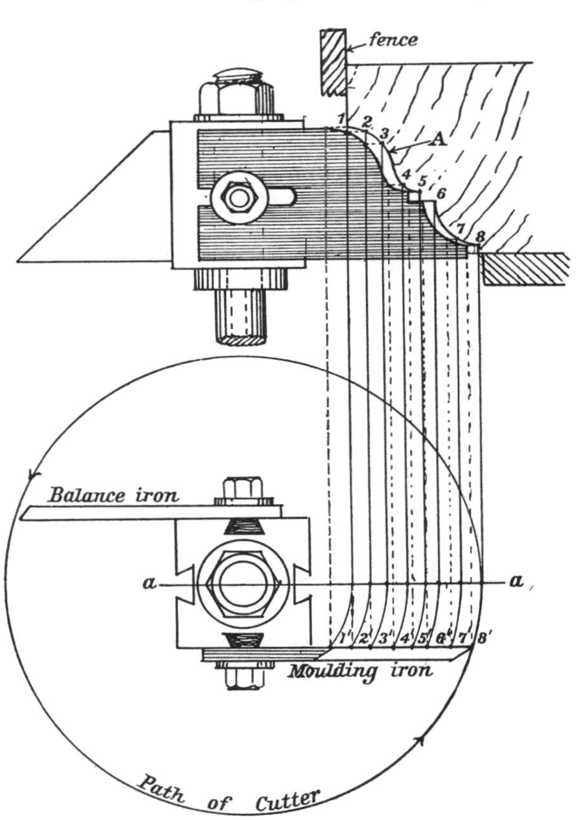

Method of Obtaining Shape of Moulding Cutters.
1. Section. 2. Plan.

drive through. If hard wood is to be mortised, one or more holes should be bored through the mortise with a twist bit to ease the chisel, otherwise the chisel may jam and be drawn out of the socket. Use the grease pad freely. When the wedging is over 2 in. deep, it is advisable to make a taper bed-piece on which to rest the stuff for wedging; this brings the back of the wedging perpendicular, and it can then be cut with the face of the chisel half-way through the depth of the stuff.

A Hand Guard for planing machines.—The "Boomerang" guard shown below is supplied by Messrs Sagar; it can be fitted to any planer, and works automatically, a spiral spring near the pivot keeping it pressed tightly up to the stuff passing through the machine. Thus the cutter is always covered, as shown, so that even should the operator's hand slip it could not reach the knife.

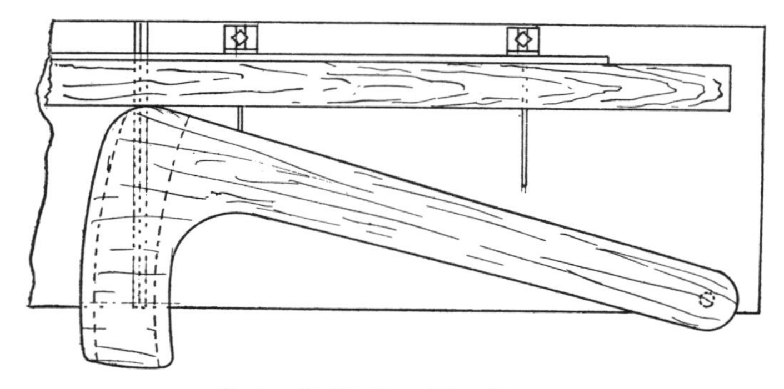

Spring Knife Guard for Planer.

CHAPTER VIII.

PREPARATION OF WORK FOR MACHINING.

Introduction to and Scope of the Chapter—Duties of the "Cutter-out," Methods of Conversion, Need of Care in Selection of Timber, Attention to Peculiarities of Machinists, How Stuff is Selected to Prevent Warping, &c., When Hand Saws must be used, Selection of Stuff for Rails, Panels, &c., Allowances for Cutting and Planing, Method of Lining-Out—Order of Machining—Preparing for Mortise Machine, When to Gauge, Marking Stub Mortises—Preparing for Tenoning Machine, Square Work, Setting Out Bevelled Shoulders, Splayed Shoulders—Methods with Hardwoods, Precautions against Indifferent Workmen—Scribing, How Indicated, Scribing Sash Bars—Preparation for Grooving—Plough Grooves, How Treated—Housing in the Recessing Machine, How Marked; Dovetail Grooving, Stair Strings—Rebating, Various Methods of Marking—Boring—Moulding—Single Faced Solid, How Marked; Double Faced Solid, How to Treat, Running Mouldings, Preparing Section.

A FEW general remarks upon the methods of preparing work for machining, as practised in large works where a separate staff is employed in the machine shops, may be useful, in addition to those incidentally given in Chapter VII. It will readily be understood that minor details may vary in particular cases.

Dimensioning or cutting to rough sizes is usually done by a man whose time is devoted entirely to this work, known as the "Cutter-Out." His duties are to take off the quantities from the rods, which are entered upon paper lists or small material boards, as described at p. 395 (this work is sometimes done by the "Setter-Out"); then select suitable stuff from the yard or mill and line it out, so that when sawn and planed it will bear the given dimensions of the list. Obviously he should be a man able to read drawings readily, and have a practical acquaintance with timber, its qualities and defects, to avoid wasteful conversion, and also have intimate knowledge of the stock in the yard, &c. He must also be acquainted with the methods of the various machinists he has to supply, and the requirements of these will differ widely, as their knowledge and skill. In converting hardwoods especially, care must be taken to select stuff that will match in the grain and colour, to use stuff about equally seasoned; for instance, in a mahogany door, the stiles should not be cut out of stuff that has been stacked two years and the rails or panels from stuff that has been only a few months in the yard; the difference in the shrinkage that goes on for some time after working up will spoil the appearance, however well finished on the bench. Planks and boards must be sorted also, according to the length of the pieces they have to be converted into. For long lengths, such as stiles or rails of long framing bars, &c., planks should be selected that have formed part of large trees; this can be generally decided by the size of the "beat" or figure. Pieces cut from such planks will "stand," that is, not be liable to cast hollow or round when cut, which will occur when a long piece is cut out of a small tree plank, which contains wood, at one end, of a much older growth than that of the other. Many woods also will cast freely, if the fibres are not severed all across the face at once. Stuff of this kind should not be run through the cross-cut saw, but cut across with a hand saw, held quite flat. Cross-grained stuff should not be used for rails, as the tenons would be weak, and probably break off in the fitting. The appropriate place for cross-grained "figury" wood is in panels. Long lengths should be lined out with the chalk line, lining the waney edges also, if any. The planks should also be inspected at the ends for nails, stones, &c., which will dull the saws if not removed. Avoid having knots upon the edges of stuff that has to be rebated or moulded; better to waste stuff getting rid of them, than time afterwards filling in. The usual allowance for saw cut is $\frac{1}{8}$ in., and $\frac{1}{8}$ in. for each edge, and $\frac{1}{16}$ in. for each side planed, but this latter allowance must be regulated according to the condition of the stuff. Preserve all short ends over a foot long; they will work in for short rails, muntings, &c. Line the stuff out, so that it may be cross-cut through, or, when it must be mixed, place the short pieces on the outside, then run these through lengthwise, to be afterwards separated on the cross-cut bench.

The General Order of Machining where a considerable quantity of stuff has to be dealt with is preliminary trying-up, and edging, and thicknessing; then setting-out, either in the mill—this is usually confined to common work of easy duplication—or in the joiner's shop, when difference of details are numerous; then mortising, tenoning or dovetailing, grooving or rebating, and finally moulding.

Preparation for Mortise Machine.—Set out length and position of each mortise, just as described for hand work (see p. 57), but the wedging lines may be omitted. Thickness or "gauge" lines may be omitted in common deal work if fairly good machinists are employed, but should always be used in hardwood or any double-moulded work, for it is better to correct a faulty mortise when the gauge lines are showing than it is to ease down the mouldings to make them intersect. Stopped or stub mortises, should have their depth shown by a line on the *back* side of the stuff, as this is the visible side when in the machine. *Note.*—For machine work it is best to use a coloured pencil for face marks so that the face side may be readily identified. When other gauge or sight lines have to be made on the work it is advisable to distinguish the mortises by two crossed lines drawn roughly from opposite corners of each mortise.

Preparation for the Tenoning Machine: Common Work with Square Shoulders.—Draw the shoulder lines in pencil on one piece of each length required. If the shoulders are of unequal length attention should be drawn to it by writing on the pattern piece, also how many are required to pattern. If the shoulders are bevelled across the width, one of each should be marked accurately on the face; the machine can be set to cut bevel shoulders as readily as square. If splayed or undercut shoulders, as in chamfered work, the splay may be drawn on the edge, or better still, let the machinist obtain a piece of the chamfered work that has to be fitted, and set his machine to that; splayed shoulders are cut with a special "scribing" cutter. In the latter arrangement set the pattern rail out with a square line to the longest point of the shoulder.

For Superior Work in hardwood it is advisable to knife cut in all shoulder lines, more especially when there are internal divisions, just as in hand work; first, because if the stuff is at all cross-grained or of woolly texture, a sufficiently clean shoulder for polished work cannot be obtained from the machine, and if the shoulder plane is used there is danger of splitting out at the end unless the wood has been first knife cut; second, machinists sometimes cut the first shoulder rather slack, and correct the total length by leaving the other one full. This, of course, throws the internal spacing of bars or muntings out, and if there are no knife lines to indicate what has been done, the fitter up is put to considerable trouble and loss of time to get the bars, &c., straight. If the knife line is visible at one end and missing at the other, he knows what has happened, and has a chance to correct it before knocking together. Machinists who do this sort of thing not unnaturally object to the knife lines for obvious reasons.

Scribing is not generally indicated upon sash stuff and the like, because it may be assumed that the necessary operation is known to the machinist, and all that has to be done is to adjust the scribing cutter to the required depth, and as the stuff leaves the tenon cutters it passes across the scriber, which hollows the shoulder right through; but special scribing such as double-moulded door frames, &c., which is often done on a spindle machine or the elephant, should have each shoulder that is to be scribed marked plainly with some understood mark; the author always used a capital S in blue pencil. If the scribe has to be stopped, the depth should be indicated on the face of one shoulder. Sash bars should be clamped about half a dozen together, on the tenoning machine, and scribed at one operation before moulding.

Preparation for Grooving.—Ploughing: Common Work.—Pencil a section on one piece showing the size of the groove required, as shown in f. 2, p. 75. If the plough groove occurs on some but not all of the pieces in a job, a pencil line should be run down the fair side of the groove where required, and specify size of groove in figures thus, /$\frac{3}{4}$. If the groove is to be stopped, mark this plainly on the offside. Stuff is always put through the machine with either the face or the face edge towards the fence, therefore marks on that side are not visible. In superior work the fair side of the groove should be marked with a cutting gauge, the tooth set "fine."

Cross Grooving or Housing, or, as it is sometimes termed by machinists, "Recessing," is usually done with the overhand spindle. If the latest form of machine is in use with compound travelling tables, such things as cupboard sides, drawer cases, shelving, &c., only require setting out of the grooves upon the edge (these should be marked a trifle smaller than the thinnest board that is to go into them). As the right sized cutter will be employed, and as the tables traverse with exactitude, squareness is insured; but when the type of machine requires the work to be moved by hand under the bit, it is best to pencil the grooves right across the face, showing the stops, if any.

Parallel Dovetailed Grooves are treated in similar manner, but tapered dovetail grooves should be marked in on both sides, as the bit must be run through twice, *i.e.*, a conical bit of a size to cut the smaller end of the groove is first run through, then the stuff is moved to bring the other side of the groove under the bit, and passed under again; of course, if one side of the groove is required square, as in shelving, a square bit is substituted for the second cut. Stair strings must be set out in the same way as in hand work.

Preparation for Rebating.—Straightforward work, such as sash stuff, door frames, jamb linings, &c., will be sufficiently indicated by an accurately drawn section upon one piece, with depth of rebate figured in, as $\frac{1}{4}$ in. on sash stiles, $\frac{3}{16}$ in. on bars, and $\frac{1}{2}$ in. on door jambs, &c. If beading is to follow the rebate, the setting-out slip used by the joiner as allowance for the shoulder should be sent to the machinist, so that he may adjust the rebate to suit the bead iron, that the combined "stickings" shall equal the slip. The reverse of this operation is in use in some shops. The machinist sets up his machine and rebates and beads a short length, which is then sent to the joiner's shop as a guide for setting out purposes. More complicated rebating, such as rebates on French casement jambs, where the rebate is inside, below the transom, and outside above it, with of course, the under sides of head and transom rebated differently, require, to avoid mistakes, that the rebate should be lined down in pencil for soft wood, or gauged for hard wood; and it is often advisable to chalk the end of a jamb, &c., and pencil in the section of rebate where it is required; three minutes over the operation may save an hour in replacing a spoiled piece of work. It must be remembered that the machinist seldom sees either the "rod" or the drawings.

Preparation for Boring.—When a series of holes are required, the centres only should be marked on a line or lines parallel with one of the outer edges, and one hole outlined, or the diameter figured in; if they are to be stopped, the depth also should be figured in. Should the line-of-centre not be parallel with the edge of the stuff, either the fence of the machine must be set at an angle or a taper piece interposed to throw the line parallel with the straight fence.

Preparation for Moulding.—*Sticking in the Solid.*—One section only should be drawn for each shape, and the number of repeats specified. If the mould is stopped, mark the stop accurately, also put a warning mark on the outside face. Double-moulded work should not be stuck until the framing has been knocked together, and flushed off, otherwise the back moulds will not intersect. *Planted or Loose Moulding.*—This only requires a section, which may be taken with tracing paper, and glued to a piece of stuff with the quantity written on.

CHAPTER IX.

JOINTS.

Classification of Joints—Description and Uses of—Glued Joints—Dry Joints—Angle Joints—Framing Joints—Shutting Joints—Miscellaneous Joints, comprising Square, Groove and Tongue, Dowelled, Rebated, Bevel Tongued, Matched, Tongue and Bead, Tongued Mitre, Stopped Mitre, Lipped Mitre and Dovetail, Box Rebate, Rebated Mitre, Double Tongue Mitre, Circular Tongued, Feather Slip, Slip Dovetail, Dovetail Key, Common Dovetail, Lap Dovetail, Secret Mitre Dovetail, Dovetail Housing — Description of Tenons — Definition of Single and Double Tenons—Pair of Double Tenons—Triple Tenons—Stub Tenon—Stump Tenon—Proportion of Tenons—Barefaced Tenon—Dovetail Tenon—Oblique Tenons—False Tenons—Rule Joint—Watertight Joint—Single and Double Lap Joints—Hook Joints—Joints for Curved and Swing Doors, how to Obtain—Hammer Head Tenon and Key Clamps—Slot Screwing—Buttoning—Counter Top Joint—Fox-wedging—Forked Tenons.

ALL the usual forms of joints in use by skilled workmen have been grouped together for illustration and description in this chapter, with references to their most suitable application in practice, thus avoiding the necessity of burdening succeeding chapters with elementary details. Many fancy and elaborate joints to be met with in books on the subject have been omitted here in consequence of their unsuitability for practical work.

JOINTS may be arranged for convenience of reference into six groups :—

1. GLUED JOINTS, or those used for uniting the edges of boards, &c., to produce wide surfaces.

2. DRY JOINTS.—Those used for connecting the edges of boards, with provision for alteration in their dimensions.

3. ANGLE JOINTS.—Those used in connecting the ends or edges of boards and frames at various angles.

4. FRAMING JOINTS.—Those used specially in the construction of framed work.

5. SHUTTING JOINTS.—Those used in the fitting together of moving parts of framed work.

6, MISCELLANEOUS.—Various joints and connections in use for special purposes, and not coming under either of the above descriptions.

GLUED JOINTS, p. 137.

The Square Joint, f. 1, is the simplest form of joint it is possible to make, and is more suitable for thin than thick boards.

Ploughed and Tongued Joints, either double or single, f. 2 & 3, are the most common method of jointing boards. They are much stronger than the square joint in consequence of the increased surface to be glued afforded by the sides of the tongue, and they have the additional advantage that in the event of the joint breaking the passage of light and air will be intercepted by the tongue. The double tongues, f. 3, are used in thick stuff.

Tongues in glued joints should be either **Cross**, f. 4, or **Feather**, f. 5, and not more than ⅛ in. thick, as thicker tongues would weaken rather than strengthen the joint, by reducing the area of the glued surface. Cross tongues can be used much thinner than feather tongues, as they are stronger, the latter being cut diagonally across the grain, as shown in f. 5.

Dowelled Joints, f. 6, are formed by inserting cleft oak pins or dowels in the edges of a square joint about 12 in. apart. They are not suitable for stuff less than ⅞ in. thick, and are only used in situations where tonguing would be objectionable, as for instance in work with moulded ends. They are generally more suitable for cabinet work than joinery.

DRY JOINTS, p. 137.

Matched and **Flooring Joints** of several kinds are shown in f. 7-15.

Ploughed and Tongued Joints, f. 7 & 8. The first has a wood tongue cut the *length* way of the grain, and fitting easily in the grooves, its purpose being simply to prevent dust and draught passing through the joint when the boards shrink, and not, as in a glued joint, to add strength to the joint. In the second example the tongue is of hoop iron, which is used because a thinner groove is required to accommodate it, and the groove can consequently be kept nearer the under surface, thus giving a greater depth of wearing surface.

Filleted Joint, f. 9. The tongue here called a fillet, because of its extra thickness, which is not less than ⅜ in., is placed in rebates formed in the under edges of the boards. It is only used in flooring upwards of 1½ in. thick, and has the advantage over ploughed and tongued that a single board can be removed without disturbing the remainder.

Rebated Joint, f. 10. This is a variation of the filleted, in which the fillet is formed in the solid. By its use one row of floor brads is hidden, as shown in dotted lines. It is an old form of joint not now much used, having been superseded by the machine-wrought matched joints, in which the fixings are all hidden.

The Bevel Tongue and Rebate Joint, f. 11, is fixed at one edge only, as shown, the other being left free to swell and shrink, but kept from rising by the lip of the neighbouring board. The joint is effective but expensive, as ⅝ in. is lost in covering space on each board. It is chiefly used in solid parquet and polished floors.

The Bevelled Tongue and Groove Joint, f. 12, is used in foreign prepared

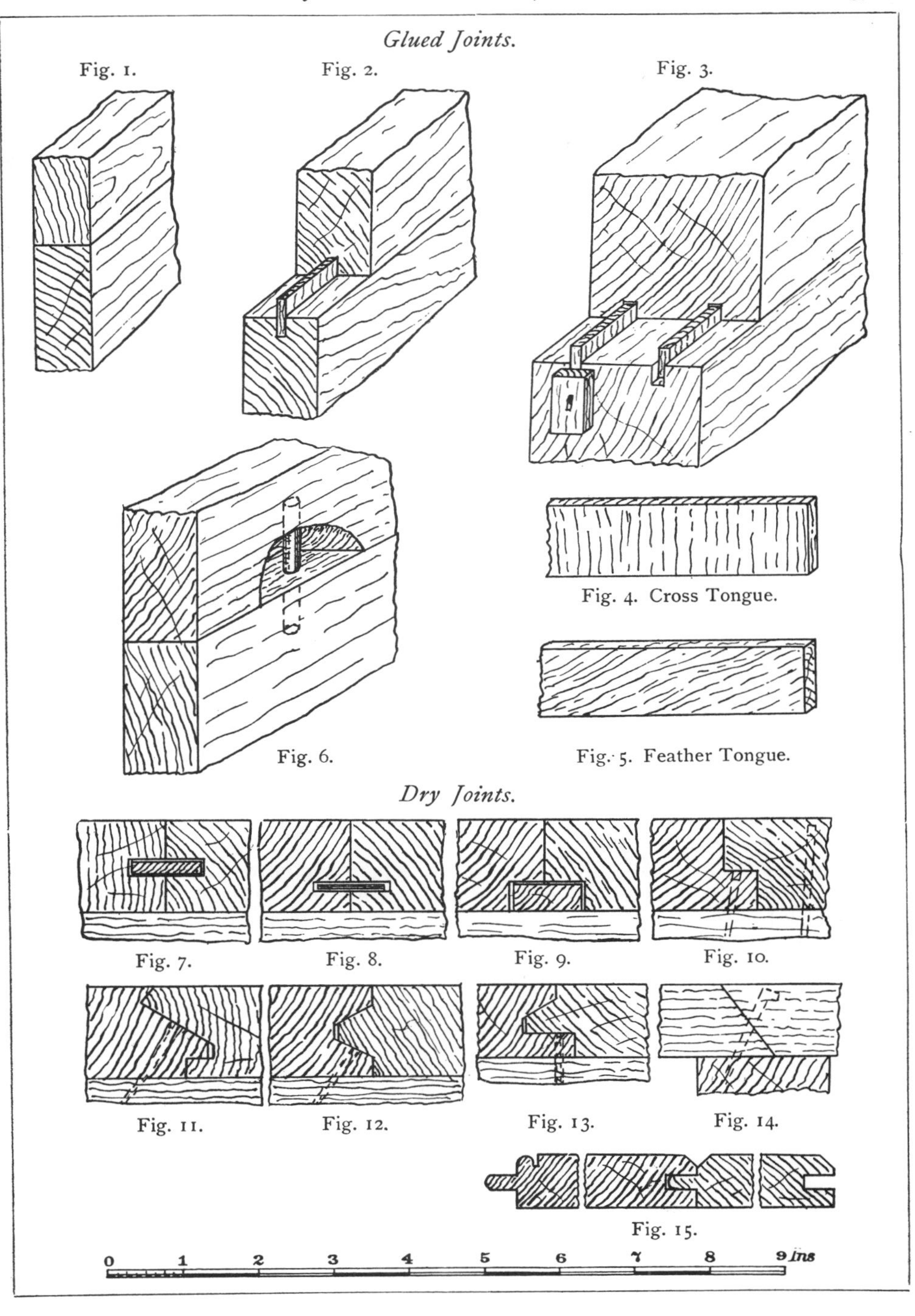
Glued Joints.
Fig. 1.
Fig. 2.
Fig. 3.
Fig. 4. Cross Tongue.
Fig. 6.
Fig. 5. Feather Tongue.
Dry Joints.
Fig. 7.
Fig. 8.
Fig. 9.
Fig. 10.
Fig. 11.
Fig. 12.
Fig. 13.
Fig. 14.
Fig. 15.
0 1 2 3 4 5 6 7 8 9 ins

flooring. The joint can be secret nailed, but is usually double nailed in the ordinary way.

The Tongue and Lip Joint, f. 13, is used in secret fixing, and is slightly more economical than f. 11.

A Splayed Heading Joint as used in flooring is shown in f. 14.

Matched Joints as used in matchboarding are shown in f. 15. The difference between a matched joint and a tongued joint is, that in the former the tongue is worked on the solid, and in the latter is formed by a separate slip inserted in plough grooves.

ANGLE JOINTS, p. 139.

The Butt or **Square Joint**, f. 1, must be secured by nailing.

The Groove and Tongue Joint, f. 2, is suitable for either edge or end joints. When used as an edge joint it is usual to bead one of the edges to hide the joint.

The Rebated Joint, f. 3, is used only for edge-jointing, and is secured by nailing.

The Return Bead or **Staff Bead Joint**, f. 4, is used chiefly to secure the salient angles of framing.

The Double Tongued and Beaded Joint, f. 5, is used for similar purposes to the former, but in a higher class of work, no other fastening but glue being necessary.

The Plain Mitre, f. 6, is a joint chiefly employed in the salient angles of skirtings, &c.

The Tongued Mitre, f. 7, is similar to the above, but has a cross tongue inserted at right angles to the joint. This form is chiefly employed when the mitre is used for end grain jointing. When the internal angle is not seen, it may be strengthened by gluing in angle blocks, as indicated by the dotted line.

The Stopped Mitre, f. 8, is used, when one piece is thicker than the other, and both sides can be seen.

The Lipped Mitre, f. 9, is one of the best of the mitre joints, easily made and very strong; it must, however, be secured with nails or brads.

The Lipped and Tongued Mitre, f. 10, is an elaboration of the last. Used chiefly for polished work where nailing is unsuitable. This joint is secured with glue only.

The Lipped Mitre and Dovetail, f. 11, is an excellent joint for uniting narrow portions of fittings having the grain running in different directions, as indicated by the sectioning.

The Box Rebate, f. 12, is used instead of a mitre where the sharp angle is not desired. The pieces are secured by nailing or occasionally by keying, as at X, Fig. 18.

The Rebated Mitre Joint, f. 13, is used in connecting frames at other than right angles. One of the edges may be beaded or **V** jointed, to hide the joint.

The Double Tongue Mitre Joint, f. 14, is used in high-class joinery. It is an easy joint to make either by hand or machine, and is very efficient. It is secured with glue.

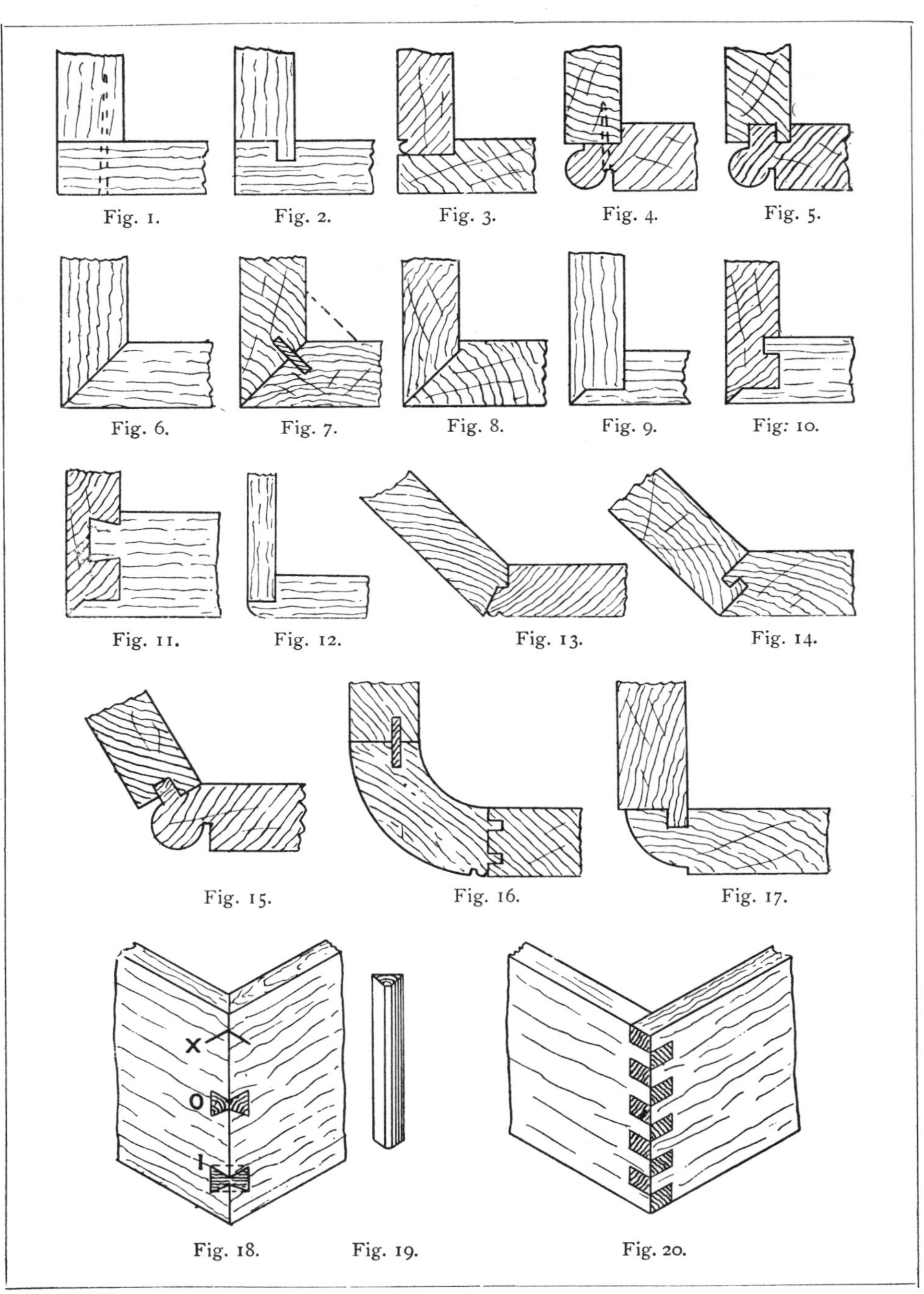

Fig. 1. Fig. 2. Fig. 3. Fig. 4. Fig. 5.

Fig. 6. Fig. 7. Fig. 8. Fig. 9. Fig. 10.

Fig. 11. Fig. 12. Fig. 13. Fig. 14.

Fig. 15. Fig. 16. Fig. 17.

Fig. 18. Fig. 19. Fig. 20.

ANGLE JOINTS.

The Double Tongue and Beaded Joint as varied to suit an obtuse angle is shown in f. 15, p. 139.

Single Tongued and **Double Matched Joints** as used in connecting circular corners to return frames are both shown in f. 16. The double matching is used when the joint is to be left dry for easy separation subsequently; the tongues are made to fit tight, and are secured by screwing from the interior edge of the stile.

The False Rule Joint, f. 17, is an adaptation of the groove and tongue joint, f. 2, an ovolo moulding being worked on the edge to hide the joint. Chiefly used in the construction of window and door furnishings where required to match a rule joint.

Keys.—Fig. 18 shows two methods of securing a plain mitre joint by means of inserted slips. The upper one marked X is called a **Slip-Key**, or a slip feather, and consists of a thin slip of hardwood glued into a saw kerf made across the mitred angle. The one marked O is called a **Slip-Dovetail**, a dovetail key, f. 19, is inserted in a dovetail socket cut straight through the mitre, as shown by the dotted lines at I, and having, when cleaned off, the appearance of a dovetail on each face. The shaded part at I indicates the socket.

Corner Locking, f. 20, is a machine-made joint, lately introduced from America. It is used instead of dovetailing, as it is much stronger than the latter, although not so neat in appearance.

DOVETAIL JOINTS, p. 141.

There are three types of dovetailing, each with variations, mainly confined to the size and spacing of the pins; these variations are sometimes used as descriptions of the types. The first kind, called through or **Common Dovetails**, show the end grain both of the socket and pin pieces. The second kind, called **Lap Dovetails**, show the end grain of the pin piece only. The third kind, called **Mitre** or **Secret Dovetails**, show the end grain of neither piece, both pins and sockets being entirely hidden in the thickness of the joint.

COMMON DOVETAILS, f. 1 & 2, are single dovetails, and show the methods of framing a bracket or bearer. The object of the first is to prevent the horizontal piece being lifted; that of the second, to prevent the vertical piece being drawn out.

Cistern Dovetail, f. 3, is so termed because this form is generally used in the construction of wood cisterns. Fluid pressure being equal in all directions, the dovetails are spaced so that an equal amount is removed from each side. The method is fully described in Chapter IV., p. 65.

Packing Case or **Box Dovetails**, f. 4, are a variation of the above, the spaces varying from twice to three times the width of a pin.

Fig. 8 shows the application of the common dovetail to splayed work, such as trays and hoppers. The pins must be parallel with the edges.

LAP DOVETAILS.—The Drawer Front Dovetail, f. 5, is used in the construction of drawers and similar fittings where one side only of the article is seen.

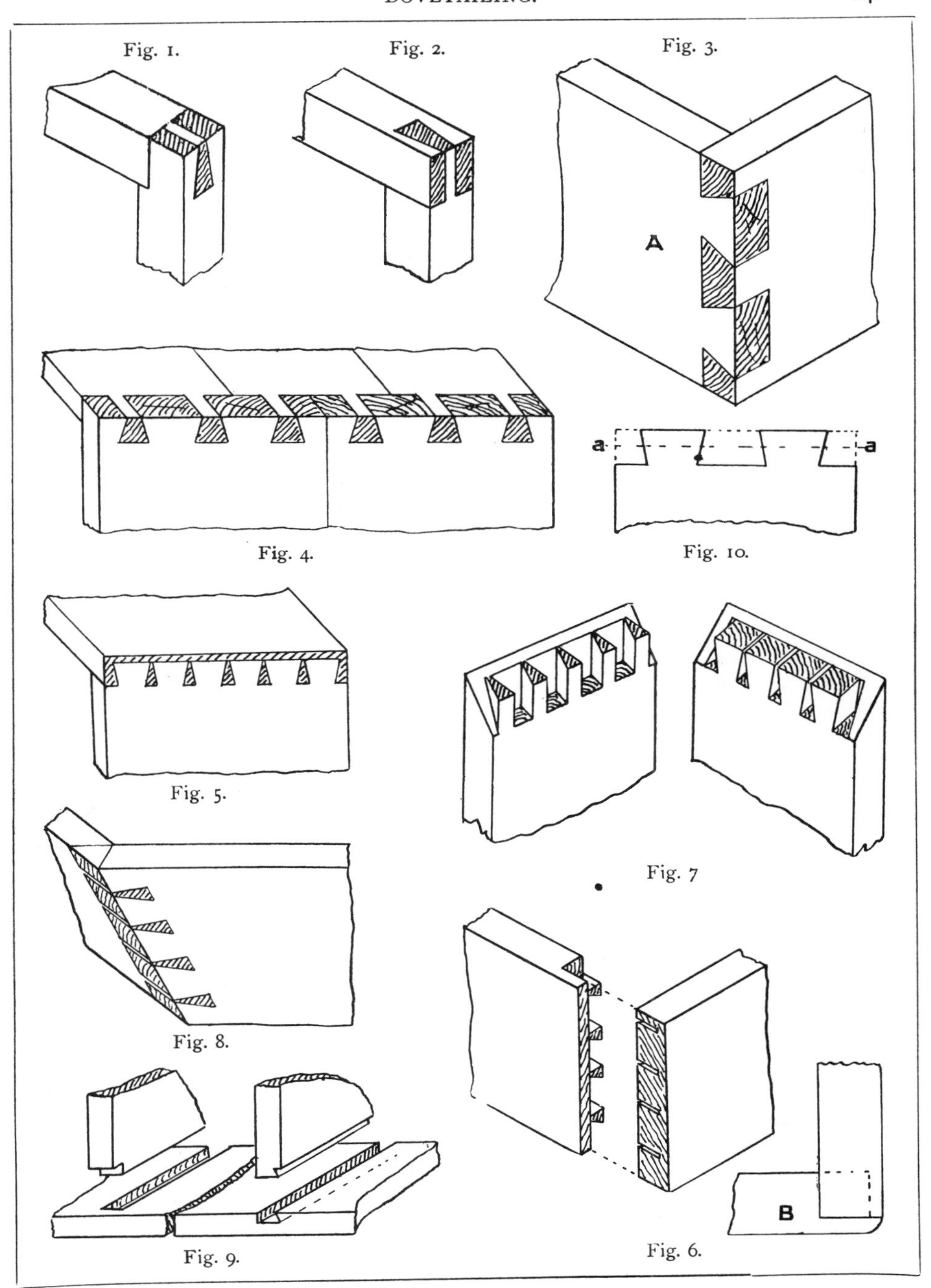

DOVETAIL JOINTS.

The Stopped Lap Dovetail, f. 6, p. 141, is used in the construction of boxes, trays, &c. The lap is frequently rounded as shown at B, p. 141.

Mitre Dovetails are shown in Fig. 7. These are used for connecting salient angles of plinths, &c., in good class work.

HOUSING.—Two methods are shown in f. 9; that on the left is known as a **Plain Stopped Housing**, that on the right as a **Straight Dovetail Housing**. A variation of the latter, where one end of the dovetail is made wider than the other, so that it shall tighten as it goes in, is termed a **Tapered Dovetail Housing**. All three methods are employed in the construction of movable fittings.

FRAMING JOINTS, p. 143.

The joints used to connect the sundry members of framed work which lie in the same plane, are all variations of the mortise and tenon joint, which consists in its simplest form (see **The Plain Tenon**, f. 1) of a parallel-sided projection formed on the end of one piece, and a corresponding cavity or slot in the other to receive it, the two being secured together by wedges or pins, aided usually by glue.

A Haunched Tenon, f. 2, is one reduced in width so that there shall be sufficient wood beyond the mortise to resist the pressure of wedging. These are used on the outside members of a frame. The short stump left in cutting down the tenon is called the haunch; when a special stub mortise is made to receive this, it is called the haunching. Usually, however, the groove formed to receive the panel in the frame acts as the haunching. The abutments at the root of the tenon are the shoulders, and the pieces removed from the sides to form the tenon are the cheeks. The sides of the mortise are also called cheeks. Tenons are also haunched to reduce them to suitable proportions, as in f. 3, which shows a **Pair of Single Tenons** haunched suitably for the bottom rail of a door.

The Double Tenon, f. 4, is employed in very thick framing; and a **Pair of Double Tenons**, f. 5, is used in the middle rails of doors that are intended to receive mortise locks. The corresponding mortises and haunching in the stile are shown in f. 6.

The Best Proportions for framing tenons are, one-third of the thickness of the piece mortised, and one-half the width available for tenon in the rail, no single tenon to exceed in width six times its thickness. Thus a 4½-in. top rail, grooved ½ in. deep for the panel, would have a tenon 2 in. wide; and a 9-in. bottom rail would be divided as in f. 7. With 1½-in. haunch at the bottom, and ½-in. groove at the top, taken from 9 in., 7 in. would be available for tenon. Half of this, viz., 3½ in., is too wide for any tenon less than ⅝ in. thick, and if placed in the middle of the rail, would not be in the best position to prevent it casting, so it is divided into two tenons 1¾ in. wide as shown in f. 7, which represents the end of the rail, with tenon, gauge, and shoulder lines set out ready for cutting. Double tenons should be looked upon as a necessary evil, and be sparingly employed, as two thin tenons are not equal in strength to one thick one of corresponding substance. When using them in a lock rail, the

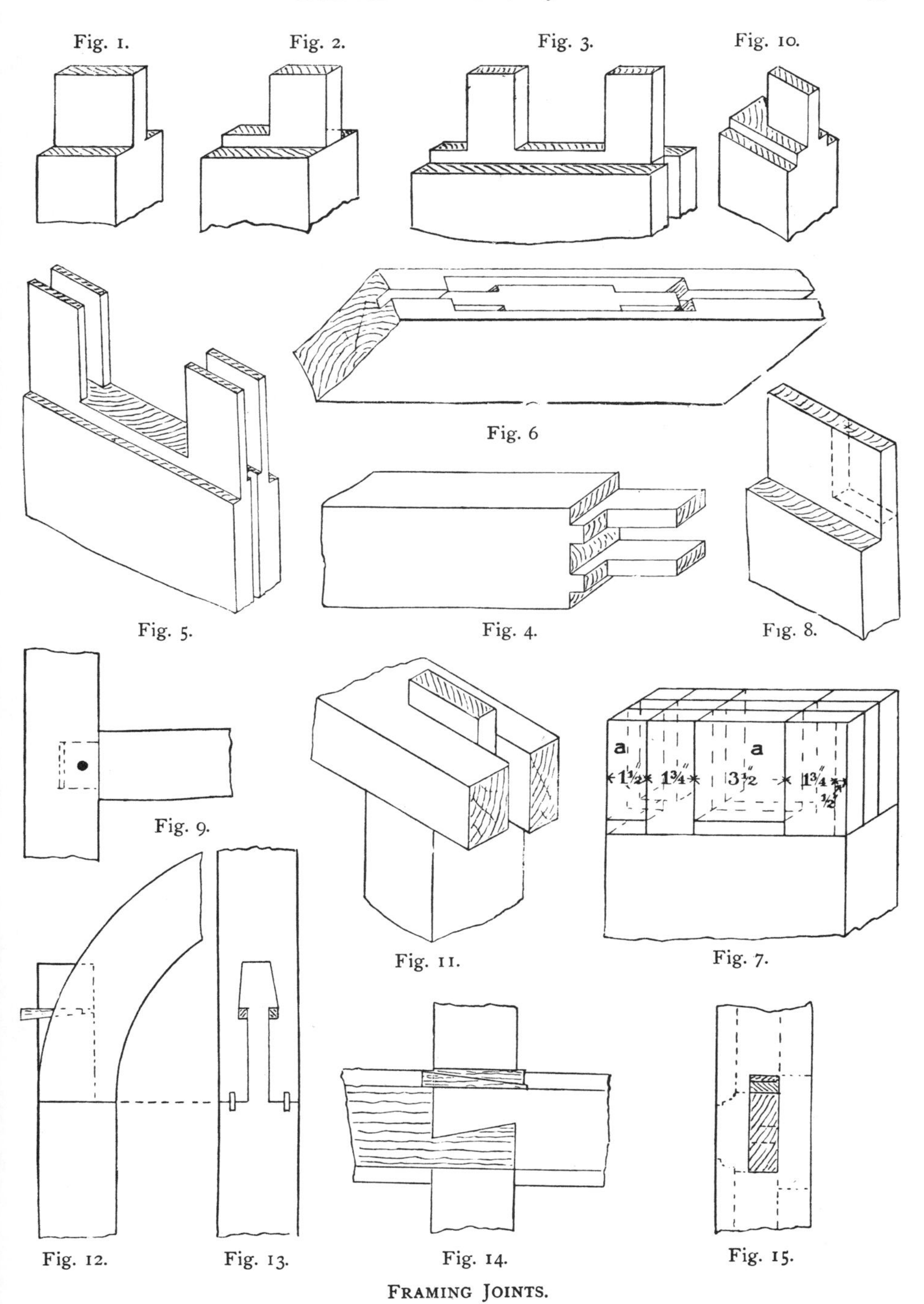

FRAMING JOINTS.

inner face of the tenons should coincide with the position of the lock, and be a distance apart equal to its thickness, usually ⅝ in.; their thickness nearly equal to one-half of the remaining cheek, as shown in f. 2, p. 59, this arrangement giving a slightly greater thickness to the mortise cheeks, to allow for "cleaning off." **The Haunching** should be as shown in f. 5, for, to frank between the tenons, as is sometimes advised, is a waste of time, the lock mortise subsequently cutting all of the franked part away.

A Barefaced Tenon, f. 8, is employed when the rail is thinner than the stile, and it is desirable to keep the mortise near the middle.

The Stub Tenon and **Mortise**, f. 9, is one that is not taken completely through the stuff. They are fastened either by pinning as shown, or by fox-wedging, f. 5, p. 150.

The Stump Tenon, f. 10.—The stump is the short thick tenon at the root of the main tenon. It is employed when, for constructive reasons only, a thin tenon can be used to keep the shoulder up, the stouter stump taking the cross strain; it differs from a haunch in that it is thicker than the superior tenon.

The Slot Mortise, f. 11, is always used in conjunction with a plain tenon; it is employed in inferior work only; the joint is secured by a pin or screw.

The Hammer-head Tenon, f. 12 & 13, is used chiefly in circular-headed frames to connect the head to the jambs. When the shoulders are more than ¾ in. wide, they should be tongued, as shown in f. 13.

The Dovetail Tenon, f. 14 & 15, is used to connect two pieces end to end such as transoms in wide framing. The mortise is placed in the rebate, and made wider than the tenon, to provide a passage for the overlapping part. After the tenons are inserted, a pair of folding wedges are driven in over them, filling in the clearance, and fixing the tenons firmly together.

Oblique and **Circular** Framing Joints are illustrated in f. 1 to 5, opposite page. The method shown in f. 1 can be used when the angle between the two members of the frame is not very acute. When the inclination is great, the method shown in f. 2 must be employed; inserted tenons of hardwood are fixed in the rail as shown, square with the shoulder, and wedged to the stile in the ordinary way. Figs. 3, 4, 5, are respectively a plan, elevation, and section of a circular rail and stile with inserted tenons, the grain of the rail being too much cut across to allow of a solid tenon being used. *a* and *b* are the upper and lower keys. Fig. 6 is an angle or **Box Tenon** as used for the corner posts of lanterns and similar frames.

SHUTTING JOINTS, p. 147.

The Bead and Rebate Joint, f. 7, is used on the meeting edges of pairs of doors. **The Bevelled Rebate**, f. 3 & 8, is used on the edges of casements and circular doors. **The Rule Joint**, f. 9, is used for tables and shutter flaps. **The Cock Bead and Rebate**, f. 10, is a **Watertight Joint**, used for the hanging stiles of casements in solid frames. **The Single Lap** and **Double Lap Joints**, f. 11 & 12, are suitable for casements with two or more leaves folding upon each other. **The Hook Joint**, f. 13, is used for the meeting stiles of a pair of casements.

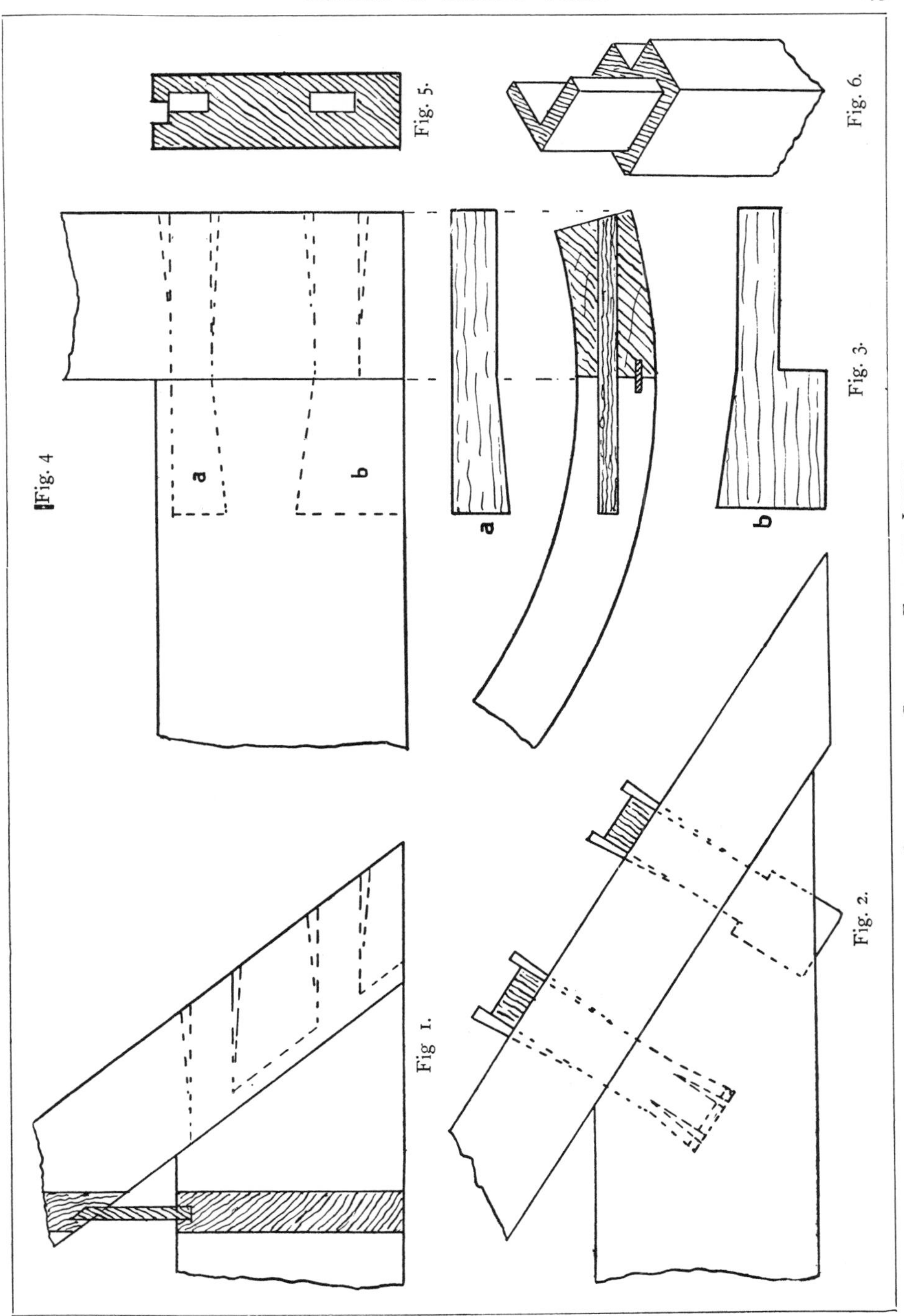

OBLIQUE AND CIRCULAR FRAMING JOINTS.

BEVEL JOINTS.—Figs. 1 to 4, p. 147, show the method of ascertaining the necessary bevel to give the edges of doors and rebates of frames, to enable the former to open freely and fit the latter accurately when closed. The principle involved is the same in each example, and is based on the fact that a "plane, tangent to a circle, is outside that circle," and therefore clear of it during its movement, the plane in this instance being the surface of the rebate, and the circle, the path of the door. A line drawn at right angles to the radius line of a circle, is tangent to the circle at the point where the radius line cuts the curve. In f. 1 draw a line *c b* from the centre of motion to the bottom of the proposed rebate, and draw *a b* square to *c b*. *a b* will then be the joint line, affording the necessary clearance, as *b d* is the path of the point *b*. In the case of a **Pair of Doors**, f. 2, the line *b c* is drawn from the hinge to the bottom of rebate, and *a b* made square with it. In like manner *e c* will give the diagonal from which to obtain the inner half of the rebate, the point *e* is discovered by moving a set square with its right angle upon the inside line of the door and its side against the point *c* until sufficient distance from point *b* has been reached to form the rebate, usually $\frac{3}{8}$ in. The lettering is similar in the other examples, which will render them clear.

Circular Doors, f. 3 & 4, represent both the joints for a single door and a pair of doors, the one opening inwards, the others outward. The relative thickness in each case is exaggerated, to show the necessary bevelling clearer.

Pivoted or Swing Door Joints are shown in f. 5 & 6. To ascertain the shape of the back edge of the hanging stile, draw a portion of the plan of the door in its two extreme positions—that is, closed and wide open—as in the diagram. The centre of the square formed by the intersecting lines will be the position of the centre of the pivot upon which the door moves. With this point as centre, and point *a* as radius, describe the arc *a b*, which will be the shape of the door edge. It is usual to make the joint on the jamb to a slightly quicker curve, so that the door in moving round shall not rub on the back, and yet show a close joint on each side when closed. This precaution is necessary, because the parts are usually polished, and should they touch a squeaking noise is caused.

To obtain the curve, move forward the compasses to point *c*, and taking point *a* as before for radius, describe the arc on the jamb. The joint for the front edge is found by describing an arc from the pivot as a centre, with a radius equal to the width of the door, as shown by the dotted line.

Fig. 6 is another form of pivot joint used in screen and other dwarf swing doors, which are equal in thickness to the frame in which they work. The joint has the same appearance when the door is open as when shut. The dotted lines show the position of the door when open. The edge of the door is made three-quarters of a circle, and the adjacent stile is curved to the same sweep with a very slight clearance.

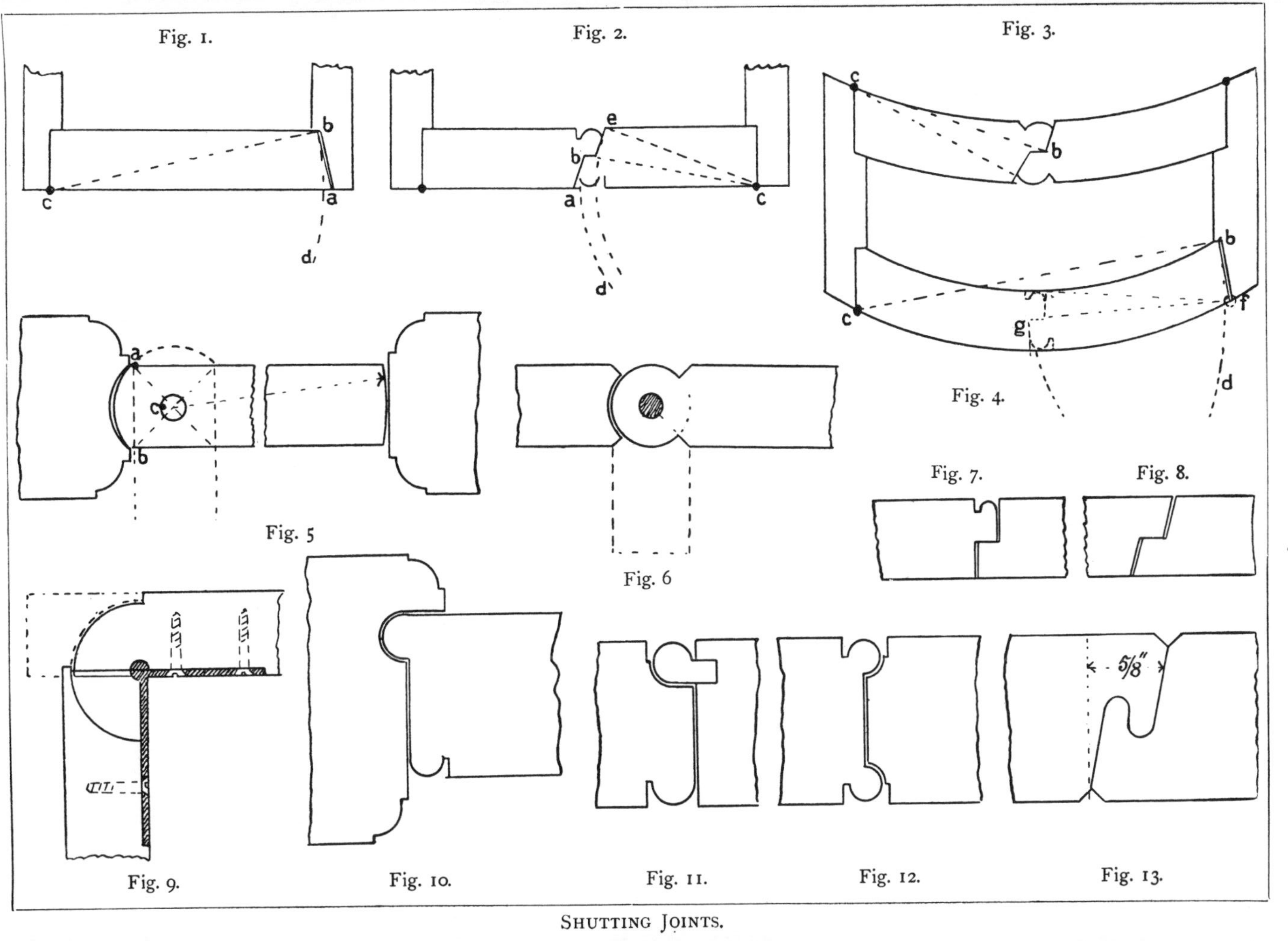

SHUTTING JOINTS.

MISCELLANEOUS JOINTS, pp. 149 & 150.

The Triple Tenon, f. 1, is used in the construction of solid casement frames of considerable thickness.

The Dovetail Key, f. 2, is used to prevent wide panels, &c., warping. The key, usually of hardwood, is left projecting at the back, as shown in the section, and is tapered in length to facilitate the fitting in. It should only be driven sufficiently tight to hold in place, so that the panel may swell or shrink with impunity. Generally the larger end is placed upwards when the key is vertical. In cases where horizontal panels in dado framing are made in long pieces, they are hung to the keys which are fixed at the top ends, the smaller end being then upwards.

The Double Dovetail Key, f. 3 & 4, is used for the purpose of connecting two pieces lengthwise. The key, which should be of hardwood, is secured by glue, or screws, according to the position of the joint.

The Hammer-headed Key Joint, f. 5 & 6, is used for connecting portions of the heads of solid circular frames. Another method used for the same purpose is shown in f. 7, p. 150. In this case a handrail bolt, f. 3, p. 72, is inserted in the centre of the thickness, as shown in the section, f. 8, the nuts are dropped into mortises made in the back of the frame, and movement of the joint is prevented by two cross tongues.

Fig. 7, p. 149, shows two methods of **Clamping** a wide board to prevent its warping. (A **Clamp** is a piece of wood framed to another piece in the same plane, but in transverse direction of grain.) The upper part in the illustration is **Mitre Clamped**, the lower **Plain Clamped.** The latter method is only used in inferior work, and in such case the tenons are only stubbed in and are sometimes omitted altogether, and a solid tongue and groove is substituted. When the edge of the piece is moulded, the clamp or mitre must start from the back of the moulding.

Slot Screwing, f. 8, shows a method employed to secure a board, with freedom of movement laterally, when one side is not seen, as in the case of drawer bottoms, stair treads, &c.

Secret Fixing is a variation of slot screwing used in good class work, when it is not permissible to show the fixing on either side. It is used in securing mouldings to panels, architraves to grounds, skirtings to backings, and polished work generally. A series of stout screws are turned into the backing, in vertical or horizontal lines as may be more convenient, their heads projecting uniformly about $\frac{5}{16}$ in., as shown at A, f. 9. Corresponding holes are bored in the back of the piece to be fixed, and slots made wide enough to take the shank of the screw as at C. These slots are made in the direction the piece is to be driven on, and are undercut, as shown at B. When fitted, the back is glued, the piece entered on the screw-heads, and carefully driven down into position.

Buttoning.—Fig. 10 shows another method of fixing a wide board so that it may swell or shrink without splitting. A ledge or batten running transversely to the panel is rebated or grooved on its edges, and a rebated block of wood B, called a **Button**, is fixed at intervals on each side with screws. The lower edge of the button should be tapered towards the rebate, so that the

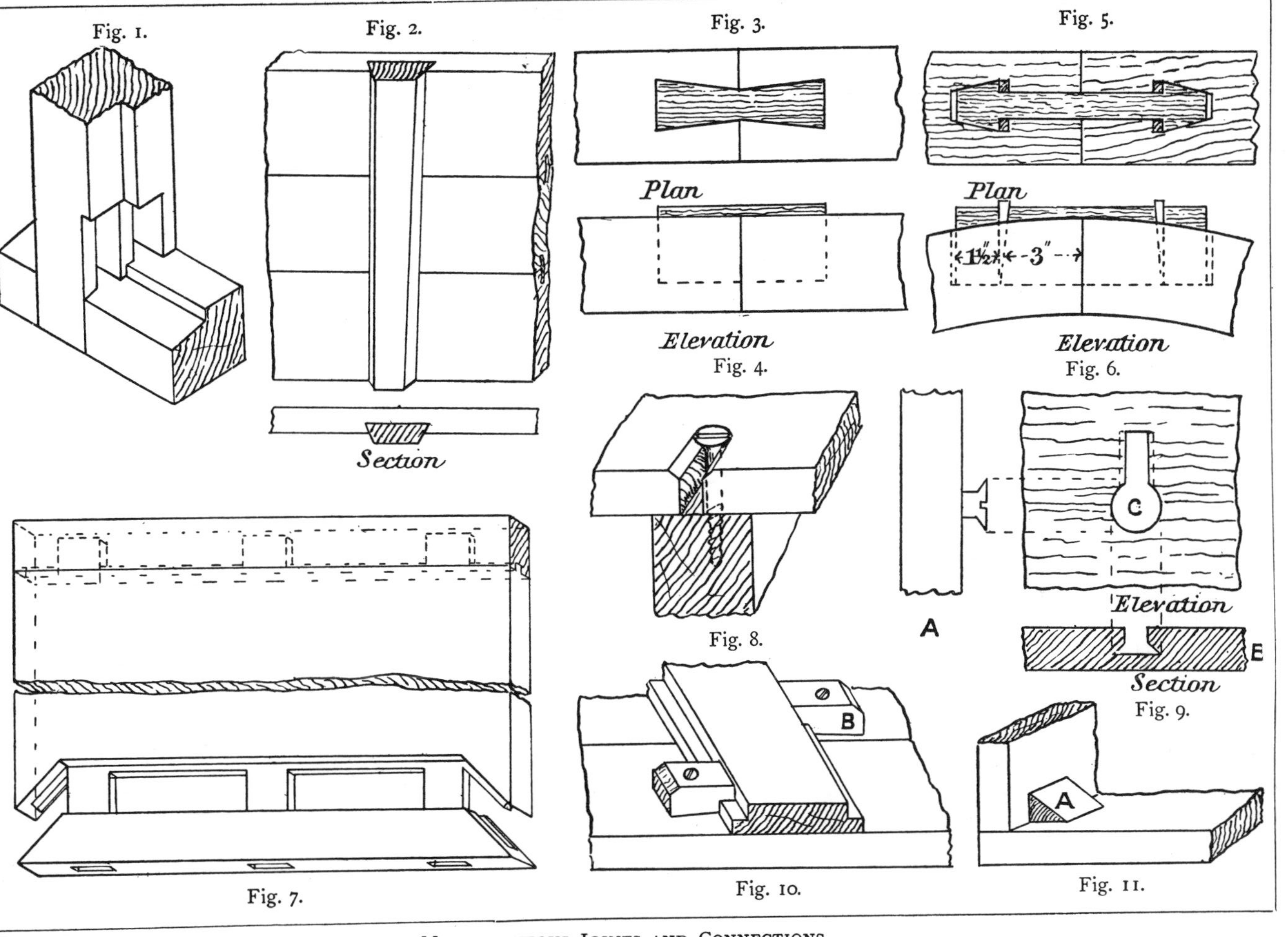

MISCELLANEOUS JOINTS AND CONNECTIONS.

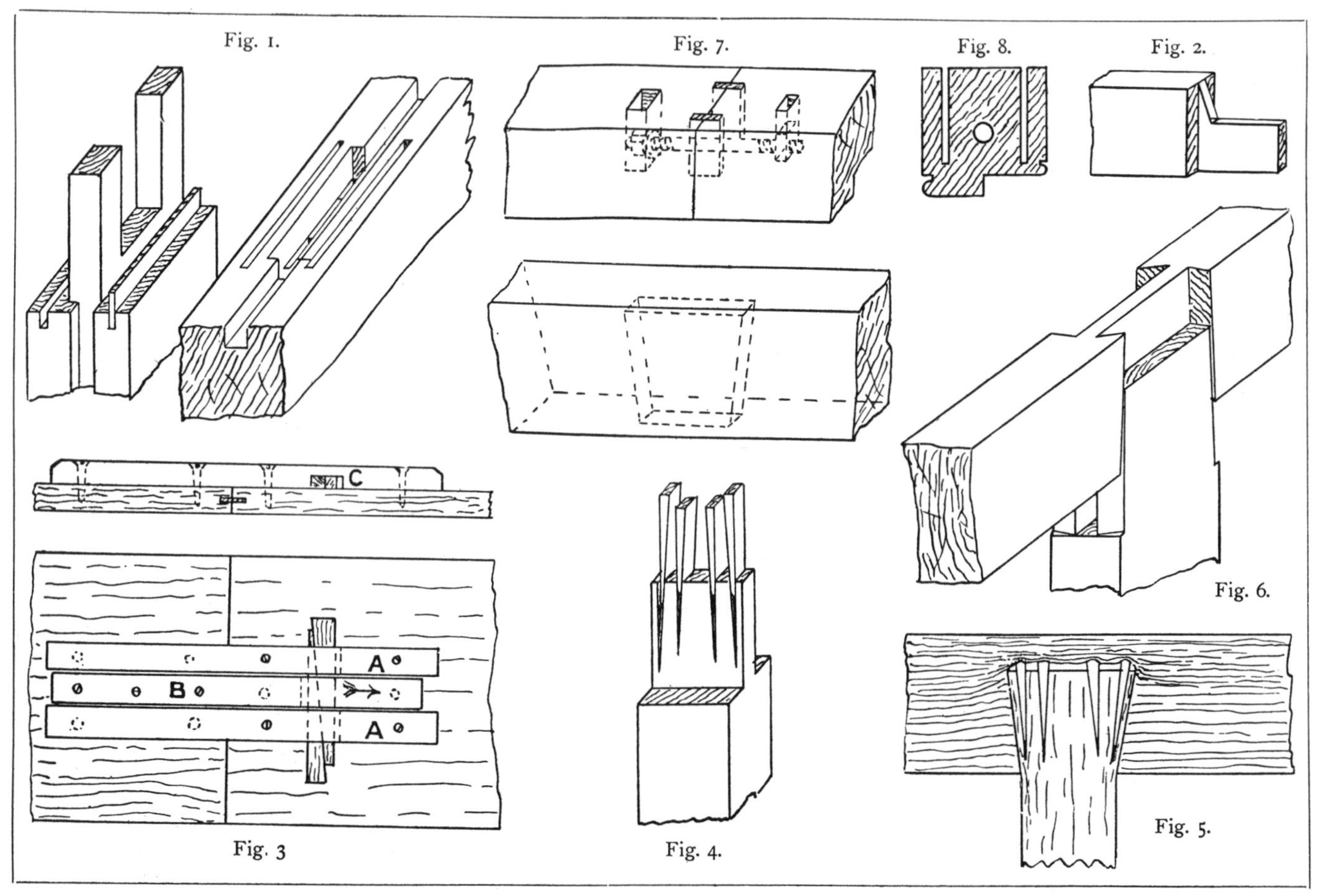

MISCELLANEOUS JOINTS.

tongue on the latter may bind tightly. Table and counter tops are secured by a variation of this method.

A Blocked Joint is shown in f. 11, p. 149, A being an **Angle Block** glued into the internal angle to strengthen the butt joint, a method in frequent use in all kinds of fittings.

The Shoulder Tongued Joint, f. 1, p. 150, is a method of strengthening shoulders in thick framing, used in preference to double tenons.

The Table or **Taper Haunching**, f. 2, is used on the top rails of dwarf doors and similar situations, where it is desired to show a square shoulder; also used in table rails, to avoid reducing the strength of the leg.

Counter-wedging, f. 3, is a method of securing the butt joints in wide thin surfaces, such as counter tops, stair strings, &c. Two strips of deal A A are screwed to the back or under surface of one of the pieces overhanging the end, and parallel to each other. A similar strip B is secured to the other piece between them, a slot or keyway, as shown at C in the elevation, having been previously cut through the under sides of the three. The slot in the piece B must be slightly nearer the joint than those in the pieces A A. Then when the joint is brought into position, and a pair of folding wedges driven through the slots, the piece B will be forced forward in the direction of the arrow, and draw the joint up close. Afterwards the loose ends of the pieces are screwed down, as shown by the dotted screw-heads.

Fox-wedging or **Fox Tenon** is a form of mortise and tenon joint used in superior hardwood fittings when it is not desired to show the end of the tenon through the edge of the stile. It is a joint requiring great accuracy in fitting. The methods employed are fully described on p. 69, Chapter IV. Fig. 4 shows the appearance of the joint when ready for putting together, and f. 5 its appearance after cramping up, if it were cut down the middle.

Forked Tenons.—Fig. 6 shows a method of connecting a munting to a rail, when it is desired that the munting should run through, and the rail also be continuous; this is called a Forked Tenon, and is used in wide doors, counter-fronts, pilasters, &c. The shoulders of the rail are slightly undercut, to keep the ends of the fork down, and the latter are slightly tapered, so that they may tighten up to the shoulders as they are driven in. This taper is much exaggerated in the sketch to make it plainer.

A Tease Tenon is somewhat similar to f. 10, p. 143, and consists of an ordinary tenon reduced in its outer half to a square section. Its purpose is the connection of two rails which are at right angles to each other to a tranverse post or rail, the lower or root end of the tenon passing through the first rail full width, and the part passing through the outer rail being reduced in width until its section is square.

A Notched or **Halved Tenon** is shown in connection with f. 1, p. 61. It consists in reducing the opposite sides of two tenons equally, so that combined, they will only require the same size mortise that either tenon would otherwise require.

CHAPTER X.

DOORS AND PANELLING.

Classification—Essentials of Design and Construction—Description and Examples of Ledged-and-Braced Doors, Requirements, Construction—Panelled Doors, Types of Panelling, Methods of Fixing—Sunk and Bolection Mouldings, securing the Mitres to prevent Panels Warping—Standard Sizes for Doors—Materials, Arrangement of Figure in Panels and Framing, &c.—Folding Doors, how Hung—Double-Margin Doors—An Eighteenth-Century Door, Modern Construction—Gluing Up—Precautions with Tenons — Entrance Doors — Diminished Stile Doors — Vestibule Doors—Swing Doors—Methods of Cutting Diminished Stiles, Setting-out Rails and Stiles, Use of the Shoulder Square, Precautions, in Second Seasoning—Jib Doors, Construction of Joints—A Dwarf Screen Door—Joint in the Capping—A Concealed Door in a Dado, Construction of the Movable Panel—Gothic Door, Tracery Panels, Fixing the Lights, Describing Four-centred-head—A Sliding Fire-Resisting Door—A Five-Panel Interior Door with Framed Linings—A Stop Chamfered Door—Interior Door with Glazed Panels—Method of Setting Out and Preparing a Panelled Door by Hand—Machine-made Veneered Doors—A Hospital Door.

DOORS are named in accordance with their modes of construction, position, style, or the general arrangement of their parts, also by the method in which they are hung as, LEDGED, LEDGED-FRAMED-AND-BRACED, PANELLED OR FRAMED, ENTRANCE, VESTIBULE, SCREEN, SASH, DIMINISHED STILE, DOUBLE MARGIN, GOTHIC, DWARF, FOLDING, SWING, JIB, WAREHOUSE HUNG, &c. The essentials in the design and construction of doors are, for the first, that they shall have a due proportion to the building or place they have to occupy and be suitably ornamented; for the second, that their surfaces shall remain true and their parts be so arranged and connected that their shape will be unalterable by the strains of usage or the effects of weather. The various examples illustrated will indicate the points to be considered in designing doors for sundry situations, and the methods of construction herein described will supply the necessary information to meet the constructive requirements.

Ledged and **Ledged-Framed-and-Braced Doors** are shown in f. 1 and 2, p. 153, and f. 1, 2, 3, p. 154. These doors are suitable for positions where one or both sides are exposed to the weather. Little or no attempt is made to ornament them—economy of cost, strength, and utility being the chief requirements of this class of door, which are fitted to coach-houses, W.C.'s, and outhouses generally.

The plain **Ledged Door**, f. 1 & 2, opposite, is composed of battens A, from $\frac{3}{4}$ in. to $1\frac{1}{4}$ in. thick, ploughed and tongued in the joints with straight tongues which should be painted before insertion, nailed to three ledges B, from 1 in. to $1\frac{1}{4}$ in. thick, usually with wrought nails long enough to come through and be clinched on the back side. The ends of the ledges are better fixed with screws, and their top edges as well as those of the braces C, should be bevelled to throw off the water, as shown in the detail, f. 3. The lower edges may be throated, or bevelled under, as shown. The braces should be placed so that their lower ends are at the hanging side, for if in the opposite direction, they will be useless to prevent the door racking. Their ends should be notched into the ledges about 1 in. deep and 1½ in. from the ends, with the abutment square to the pitch of the brace. Narrow doors are sometimes made without braces, but they seldom keep "square." These doors are hung with wrought-iron strap hinges called cross garnets, which should be fixed on or opposite the ledges, whether placed on the face or back of the door.

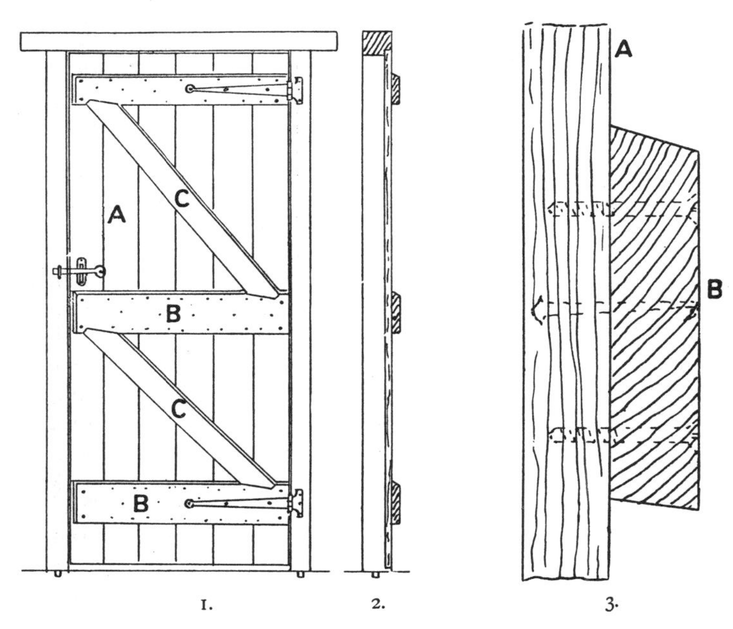

1. A Braced Ledged Door. 2. Section. 3. Enlarged Detail of Ledge.

The Framed-Ledged-and-Braced Door, f. 1 to 3, next page, differs from the former in that, the battens and ledges are enclosed on three sides by a frame, of a thickness equal to the combined thickness of the battens and ledges, so that it is flush on each side with them. The boards are tongued into the frame at the top and sides, and the ledges are framed into the stiles with barefaced tenons (see f. 8, p. 143). The braces should not be taken into the angle formed by the stile and rail, but be kept back from the shoulder about 1 in., as shown. If the brace is placed in the corner, the strain thrown on it has a tendency to force off the shoulder, unless the door is very narrow, when the brace will be nearly upright. These doors, as in fact all framed work exposed to damp, should be put together with a quick drying paint instead of glue in the joints, because ordinary glue has such an affinity for water that it will soften in damp situations, releasing its hold, and also be the means of setting up dry-rot in the timber. The battens in these doors should be made $\frac{1}{16}$ in. slack for each foot of width to allow for subsequent expansion, or otherwise the shoulder will be forced off. The frame-

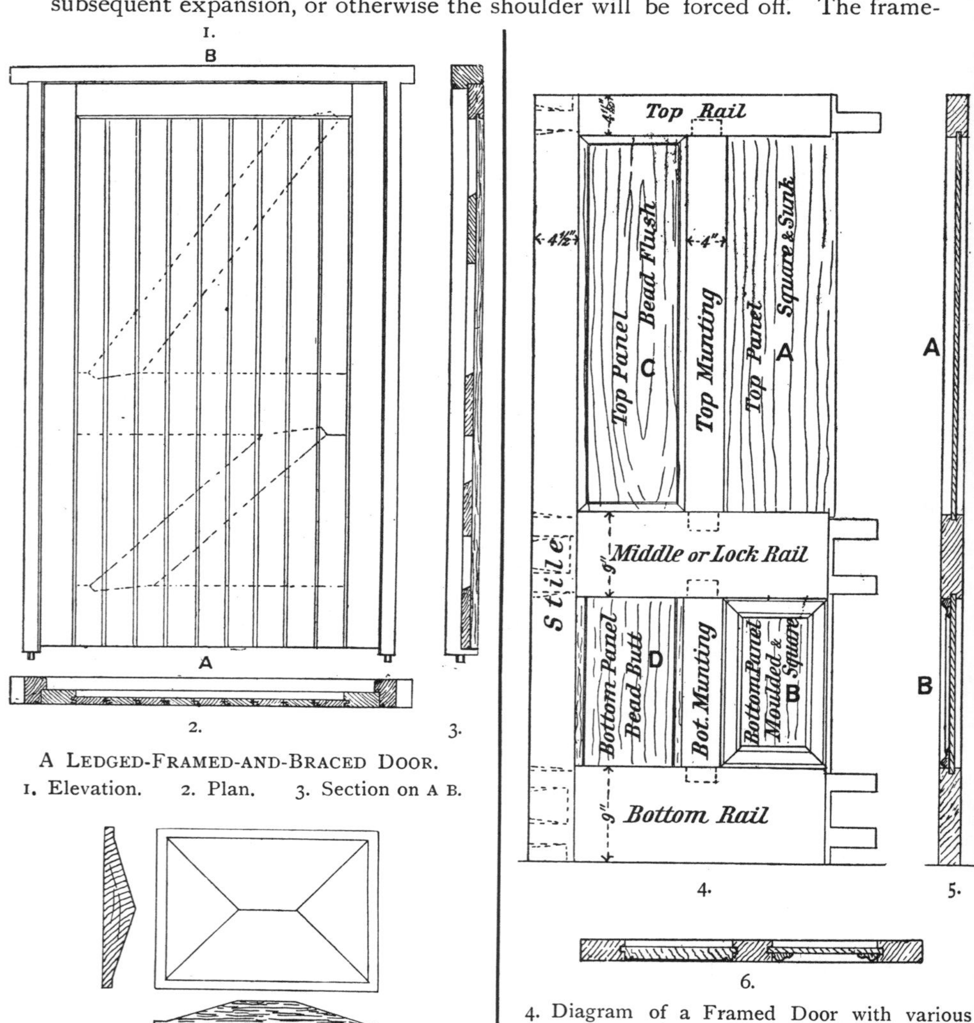

A Ledged-Framed-and-Braced Door.
1. Elevation. 2. Plan. 3. Section on A B.

7. Plan and Sections of a Chamfered Raised Panel.

4. Diagram of a Framed Door with various Panels.
5. Vertical Section at A.
6. Horizontal Section below Lock Rail.

work of these doors is first made and wedged up, then the battens folded in and driven up into the top rail and nailed to the ledges, after which the braces are cut tightly in, and nailed to the battens in turn, and the whole cleaned off together. In large gates of this description it is usual to stub tenon the braces into the rails, in which case they must be inserted first and wedged up with the framing.

FRAMED or PANELLED DOORS are of several kinds, distinguished by the number or treatment of their panels, or by the arrangement of the mouldings: in reference to the number of panels they are termed two, three, &c., panel doors; in reference to the mouldings as under:—

Square and Sunk.—When a thin panel is used without mouldings, as shown at A in the elevation diagram of a Framed or four-panel door, f. 5, p. 154.

Moulded and Square.—When one side of the panel is moulded and the other plain, as at B.

Bead Flush.—When one side of the panel is flush, or nearly so, with the frame, and has a bead worked round the edges to break the joints, as at C.

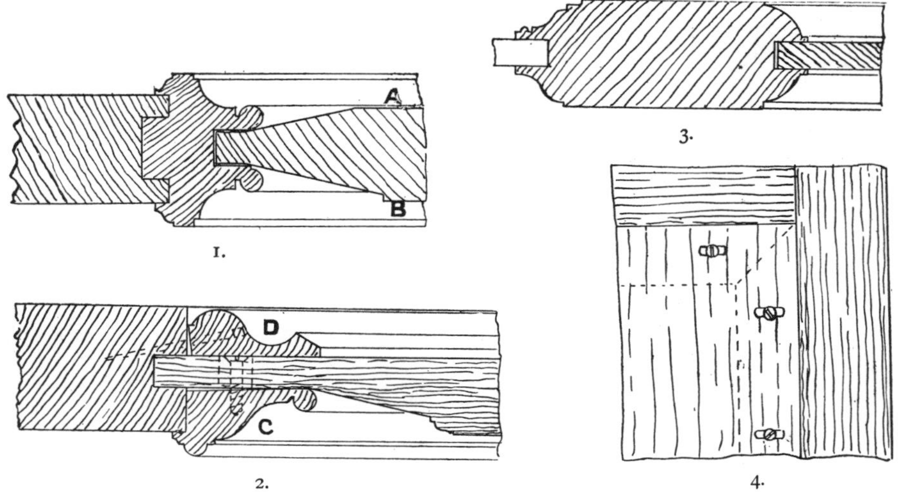

1. Double Bolection Moulding. 2. Raised, Chamfered and Moulded, and Sunk and Moulded Panel. 3. Solid or Stuck Mouldings. 4. Panel Slotted for Screws.

Bead Butt.—When the bead is worked only on the sides of the panel, as at D, the end butting against the rails.

RAISED PANELS.—When the centre part of the panel is thicker than the margin. There are four varieties of raised panels:—

1. **The Chamfered.**—In this the panel is chamfered down equally all round, from the centre to the edge when square, or from a central ridge if rectangular, as shown in f. 7, p. 154.

2. **Raised and Flat or Chamfered and Fielded.**—When a chamfer is worked all round the edge, leaving a flat in the centre, as at A, f. 1, above.

3. **Raised, Sunk, and Chamfered**, as at B, f. 1.—When the chamfer starts from a marginal sinking below the face. This example would also be described as "double," or twice raised, &c.

4. **Raised, Moulded, Chamfered, and Seated**, as at C, f. 2.—When the edge of the sinking is moulded, and a flat or seat is provided for a frame moulding.

Stop Chamfered.—When the edges of the framing are chamfered and stopped near the shoulders, as shown in f. 1 & 2, p. 175.

Bolection Moulded.—When the panel moulding stands above, and is rebated over the edges of the framing, as shown at f. 4, p. 174, and f. 2, p. 155, C.

Double Bolection Moulded. — When the moulding on each side of the door is made in one solid piece, grooved to receive the panel, and is itself grooved and tongued into the framing. This variety is shown at f. 1, p. 155.

Constructive Memoranda.—The outside vertical members of doors (in common with all framed work) are called stiles. The one the hinges are fixed to, is called the hanging stile, the one containing the lock, the striking stile. In a pair of doors, the two coming together are called the meeting

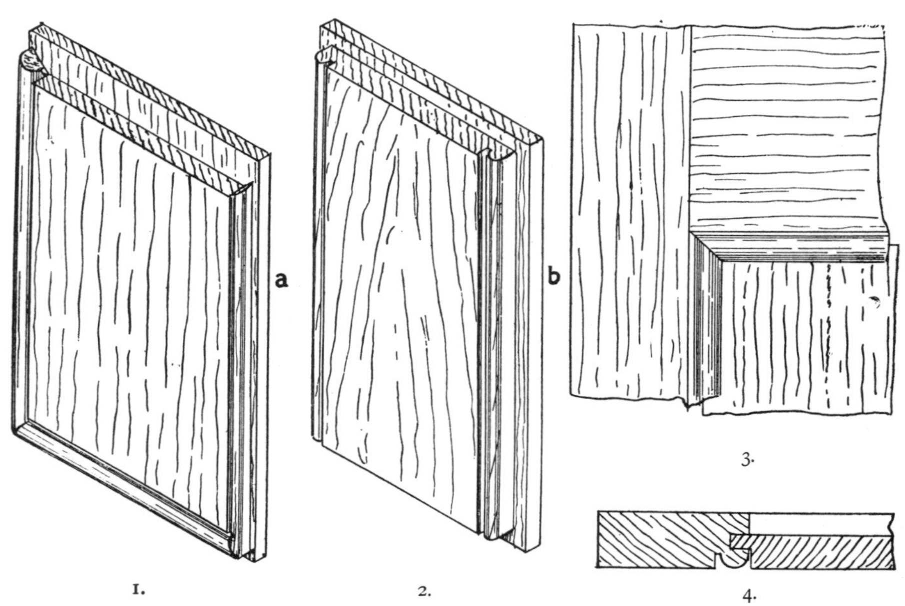

1, 2. (*a*) Bead Flush and (*b*) Bead Butt Panels. 3. Elevation, 4. Section } Solid Bead Flush Panelling.

stiles. The inside vertical members are the mountings, or more commonly muntings. The horizontal members are rails; respectively, top, frieze, middle or lock, and bottom. The panels are named similarly. When the grain of these runs horizontally, they are said to be "laying" panels; when vertically, "upright." Doors are called Solid-moulded, when the moulding is wrought or "stuck" in the substance of the framing itself, as is shown at f. 3, p. 155; and Planted, when the moulding is worked separately and bradded around the frame, as shown at B, f. 5, and B, f. 5, p. 154. These are also called sunk mouldings.

Bead Flush Panels are commonly made as shown at *a*, f. 1 above, but such panels will, unless made of thoroughly seasoned stuff, inevitably split when drying. The correct way to obtain the effect of bead flush panelling is to work the beads upon the edges of the framing, as shown in f. 3 & 4 opposite.

Bead Butt Panels are better kept about $\frac{1}{32}$ in. below the framing, as a truly flush surface is difficult to prepare through the yielding of the panel, and, when produced, will seldom last, as the shrinkage of panel and frame is unequal. Planted-in "sunk" mouldings should be fixed to the framing, not to the panels, as shown in f. 2, p. 155, D; for if fixed to the panels, when the latter shrinks the moulding will be drawn away from the frame, leaving an unsightly gap. The back edge of the moulding should be bevelled under as shown, so that when bradded in, the front edge will keep close down to the panel. As it is not permissible to brad polished mouldings, except in the case of inferior work, these are usually glued to the frame, and their back edges should of course be square. The panel should be polished before the moulding is planted in, so that in case of shrinkage a white margin will not be shown. When, however, the moulding is wide and thin, it is unavoidable that it be fixed to the panel to keep its front edge down, and to overcome the difficulty of shrinkage the method shown in f. 1 is resorted to. The stiles in vertical, and the rails in laying, panels are prepared with a second shallow groove as shown, and the moulding, made with an extra wide quirk, is cut in and pressed back into the groove. It is glued to the panel, and moves with it without showing a gap. The cross pieces across the ends of the panel have their quirks shot off to the proper width, and are sprung in after the side pieces are in place, as shown at *b*. These are dowelled but are not glued, or where there is a moulding at the back, the face side may be slot screwed.

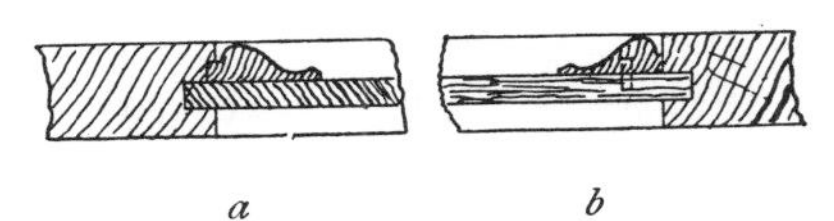

1. Method of Fixing a Thin Moulding. *a*. Along Panel. *b*. Across Panel.

Bolection Mouldings are intended to be fixed to the panels, which is rendered necessary by their great width and thickness, and they are rebated over the edge of the framing to prevent the interstices produced by the shrinkage showing. From $\frac{1}{8}$ to $\frac{3}{16}$ in. is sufficient for this purpose, and more should not be given, or the edge will be liable to curl off. To prevent the panel being split by the fastenings when it shrinks, the moulding is fixed by "slot screwing," as shown in f. 2, p. 155; these slots are cut across the grain; see the back view, f. 4, p. 155. The moulding should also be screwed together at the mitres, and the latter may be grooved and tongued with advantage, as shown in f. 2, 3, and 4 adjoining, and dropped into the panel as a complete frame, where it is fixed as described, the heads of the screws being covered by the interior moulding, which, if a sunk one, is glued to the frame, and if a bolection, dowelled to the panel, or, as is sometimes done in inferior work, bradded, and the holes filled up with coloured stopping. It is not good con-

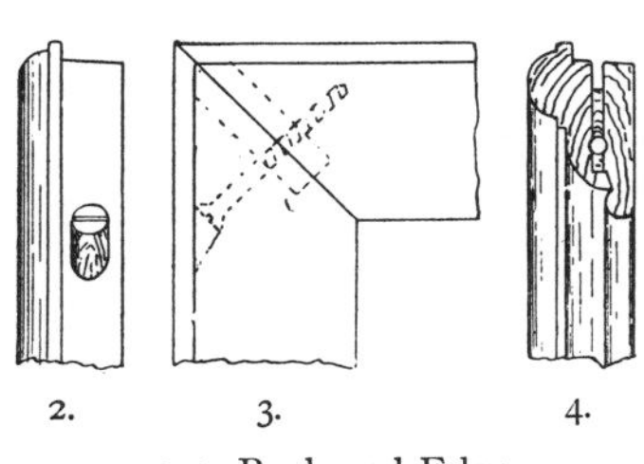

2-4. Back and Edge Elevations of a Moulding Mitre Tongued and Screwed.

struction to glue the moulding to the panels in any case, as the alteration of size in the latter, due to the state of the atmosphere, is very liable to cause them to split, if fixed immovably, or when swelling, to disarrange the mitres. This is less likely to happen if the mouldings are bradded in, if the brads are not placed too thickly, as they yield slightly to the pressure. Framing is usually grooved ½ in. deep for the panels, and the latter given ⅛ in. play sideways, but fitting close lengthways of the grain. This is sufficient for panels up to 2 ft. wide. Over that width the grooves should be ⅝ in. deep, and the panel enter ½ in. at each side. Ordinary dry stuff will eventually shrink about ⅛ in. to the foot, and will swell equally if exposed to damp. When wide panels are used they will warp less if glued up in several pieces, as the pull of the fibres is lessened by the cutting, and the effect of the warping is diminished in the same ratio as their width. Much can be done to ensure the permanent flatness of panels by paying attention to the way the boards have been cut from the tree. The direction of the annual rings on the ends will indicate this, and the various pieces should have their similar sides placed together. What is meant by this will be rendered plain by an examination of the diagram below. When a panel is glued up with the hollow or heart sides of the rings all on one face as at A, and the board warps, it will cast in one continuous curve, as shown in the unshaded diagram, whilst if glued up with the heart sides reversed alternately, as shown at B, it will assume the serpentine shape shown in the unshaded diagram. Boards cut radially or with the annual rings perpendicular to the surfaces, as at C, will swell and shrink less than the others, and will not warp perceptibly.

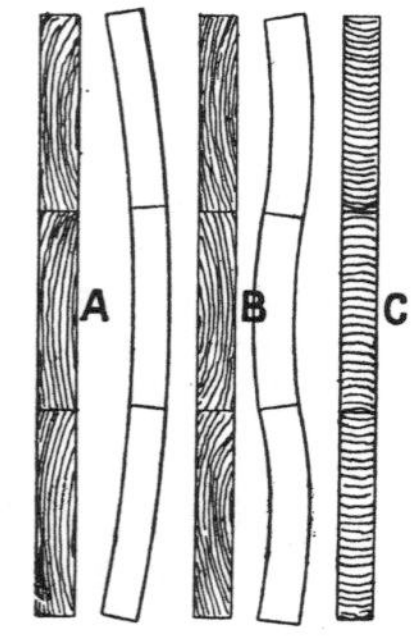

Diagram illustrating the Effect of Position on the Parts of a Panel.

Proportions.—The size of doors depends so much upon the scale and design of the buildings they occupy, that no definite data can be given, within reasonable limits, for important doors; but it may be pointed out that very large doors not only tend to dwarf a building or a room, but they also take up a great deal of space in opening, and the difficulty of preserving their accurate fitting increases in direct ratio with the size. The following may be taken as an indication of the more usual dimensions given to ordinary good class dwelling-house doors:—Entrance Doors from 7 ft. to 8 ft. 6 in. high by from 3 ft. to 4 ft. 6 in. wide, 2 to 2½ in. thick. Reception Rooms, 7 ft. by 3 ft. 3 in. by 2 in. Bedrooms, 6 ft. 8 in. by 2 ft. 8 in. by 1½ in. Details of interior doors: stiles and top rails, in common work, out of 4½ in., muntings and frieze rails 4 in., middle and bottom rails 9 in. Superior doors vary much, but generally stiles and rails are somewhat wider than the above, muntings and frieze rails narrower. Height of lock rail usually 2 ft. 8 in. to its centre. This is a convenient height for the handle, which is generally placed in the middle of depth of rail. When an entrance door is approached from a step, the middle rail is kept about 6 in. lower, to bring the height of the handle convenient.

MATERIALS.—Common Doors, both internal and external, are made of "yellow deal" or Baltic pine throughout. A better class of interior doors have yellow deal frames and American pine panels. The latter wood should not be used for external work, as it is far too soft, and decays rapidly if exposed to dampness. Superior Internal Doors are made throughout of Honduras mahogany, black walnut, and oak; also of pine and baywood, veneered with Spanish mahogany. External Doors of oak, teak, walnut, and pitch pine.

In constructing doors of any of the above-mentioned figured woods, great attention must be paid to the arrangement of the members, so as to balance the figure, and this may also well be studied in the conversion of the plank. For instance, two stiles, each having pronounced figure at one end, and the other end plain, should have the figured ends placed at the bottom. This gives the effect of solidity, whilst the reverse would make the door look top heavy. Similarly the upper rails should be plain, the lower figured. It must be understood the above only applies when the wood is a mixed lot. When the wood is handsomely figured throughout, the point of most importance is the effect of its position upon the figure, and this is so great that in some of the light-coloured woods, wainscot and baywood for instance, a piece that in one position will appear richly figured will in another show quite plain and dull. The best way to judge the effect is to prop the pieces up in the approximate positions they will occupy when finished, facing a top light. Then when standing a few feet off, the play of light on the fibres will be observed. Deep-coloured woods, such as teak and Spanish mahogany, may have their figure brought out by slightly oiling them, which will facilitate their arrangement. Panels also require balancing, the more heavily marked ones being placed below plainer ones, and symmetrically arranged, either in pairs, or figured in the centre and plain outside. In all cases where the figure is coarse, taking a truncated elliptic shape, the base or wider part should be kept downwards, as shown at A in the accompanying sketch. The panel at B is upside down, from an artistic point of view. This arrangement is known in the workshop as "placing the butts down," although as a matter of fact the width of the figure is not due to its being towards the butt end of the tree, but merely to the accidental position the surface of the board occupies with relation to the annual rings, which are more or less waved in length, due to crooked growth, and the board passing through them in a plane, their edges crop out on the surface in irregular elliptic shape.

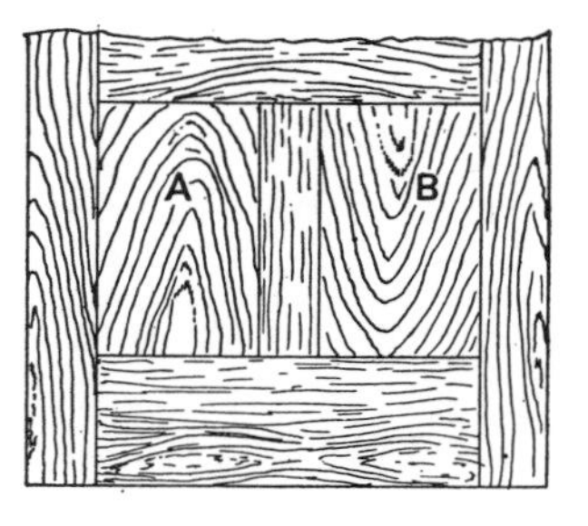

Method of Arranging Panels, Rails, &c., according to Figure.

Wide entrances are frequently fitted with doors hung in two leaves, and rebated together in the middle, as shown in f. 2, p. 165. This is done to economise space in the lobby, or to reduce the strain on the frame and hinges. These are called **Double Doors**, also Pairs of Doors, and sometimes incorrectly Folding Doors, because the latter are necessarily hung together, and fold one upon the face of the other, as shown in f. 1, 2, 3, & 4, p. 161. These doors are used to divide one large room into two smaller ones, and should be arranged, as shown in f. 2, for one leaf to open against the other, the first remaining bolted, or for the two together to fall back against the wall. The enlarged details, f. 4, show how the leaves are rebated together and hung. The butts for the wall stiles are necessarily wider than the others, to allow of the doors opening over the architrave. The centre of the hinge must project half the distance from the wall that the highest part of the architrave is, so that the leaf will lie closely on its surface, as indicated by the dotted lines in the plan (see also f. 4).

Double Margined Doors are wide single doors so framed as to appear pairs of doors. They are used in openings too wide proportionally for a single door, but where half the opening would be rather small for convenient passage. Figs. 1 to 5, p. 162, show elevation and sections of a **Double Margined Entrance Door**, typical of the style in vogue in the latter part of the eighteenth and early part of the nineteenth centuries.

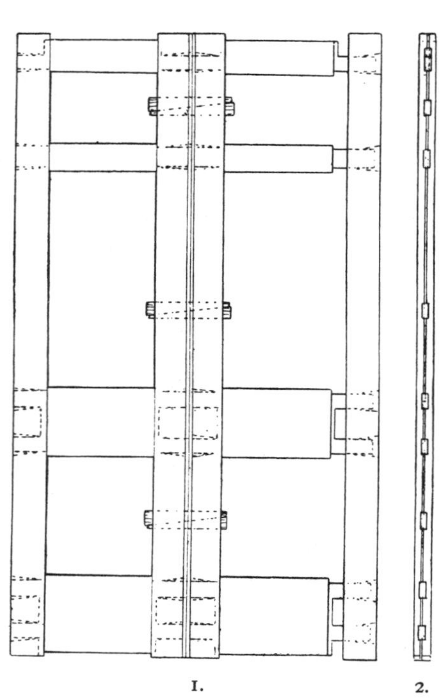

1. Method of framing-up a Double Margin Door.
2. Joint Edge of Meeting Stile.

These doors are made in two ways. In the earlier method, the middle dividing stile, which in this case is constructively a munting, is made in one piece, and forked over the top and bottom rails, which are continuous, as shown in f. 6, p. 150. The intermediate rails are stub-tenoned to the middle, and through tenoned and wedged to the outside stiles, but unless the stub tenons are fox-wedged, the shoulders are very liable to start, for which reason the method of construction now to be described is generally preferred. The door is composed of two separate pieces of framing each complete with two stiles and a set of rails that are tenoned through and wedged up. The two portions are then united by a ploughed and tongued and glued joint, which is hidden by a sunk bead in the centre, as shown in the detail, f. 1 adjacent, and the parts keyed together with three pairs of hardwood folding wedges. The door is sometimes further strengthened by having flat iron bars sunk and screwed into the top and bottom edges. The actual process of putting the door together is as follows:—After the various rails and panels have been duly fitted and marked, each leaf is taken separately and the stiles knocked on. The one intended for the meeting stile having been glued, is cramped up and wedged. Then the meeting stiles are shot to a width, grooved, jointed, and rebated for the beads, as shown in the detail, f. 4, p. 162. The ends of the tenons and wedges should be cut back $\frac{1}{8}$ in. to prevent them breaking the joint when the stile shrinks. The mortises for the keys will have been made when the mortises for the rails were

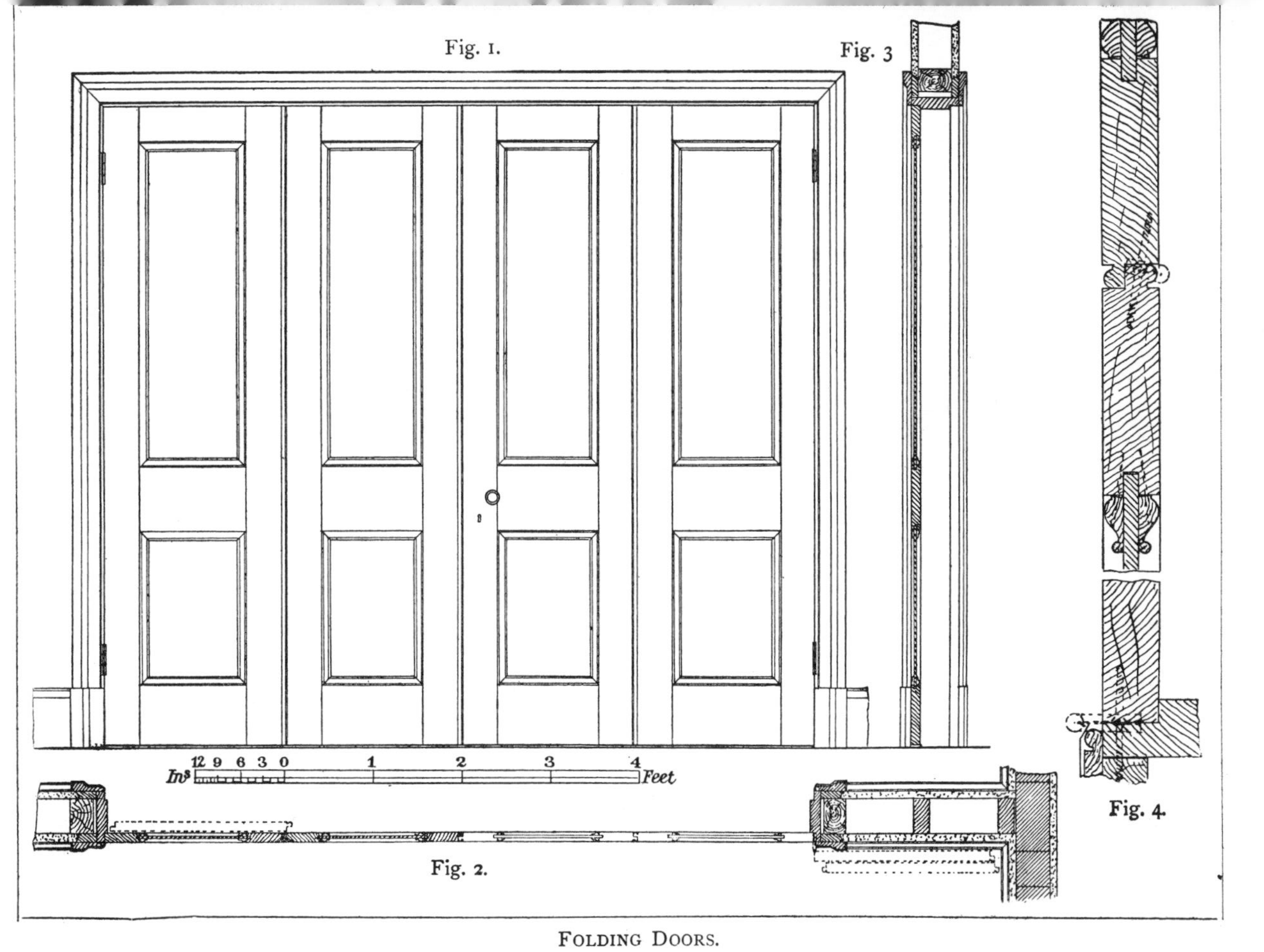

FOLDING DOORS.

1. Elevation. 2. Plan. 3. Section. 4. Details of Joints, &c

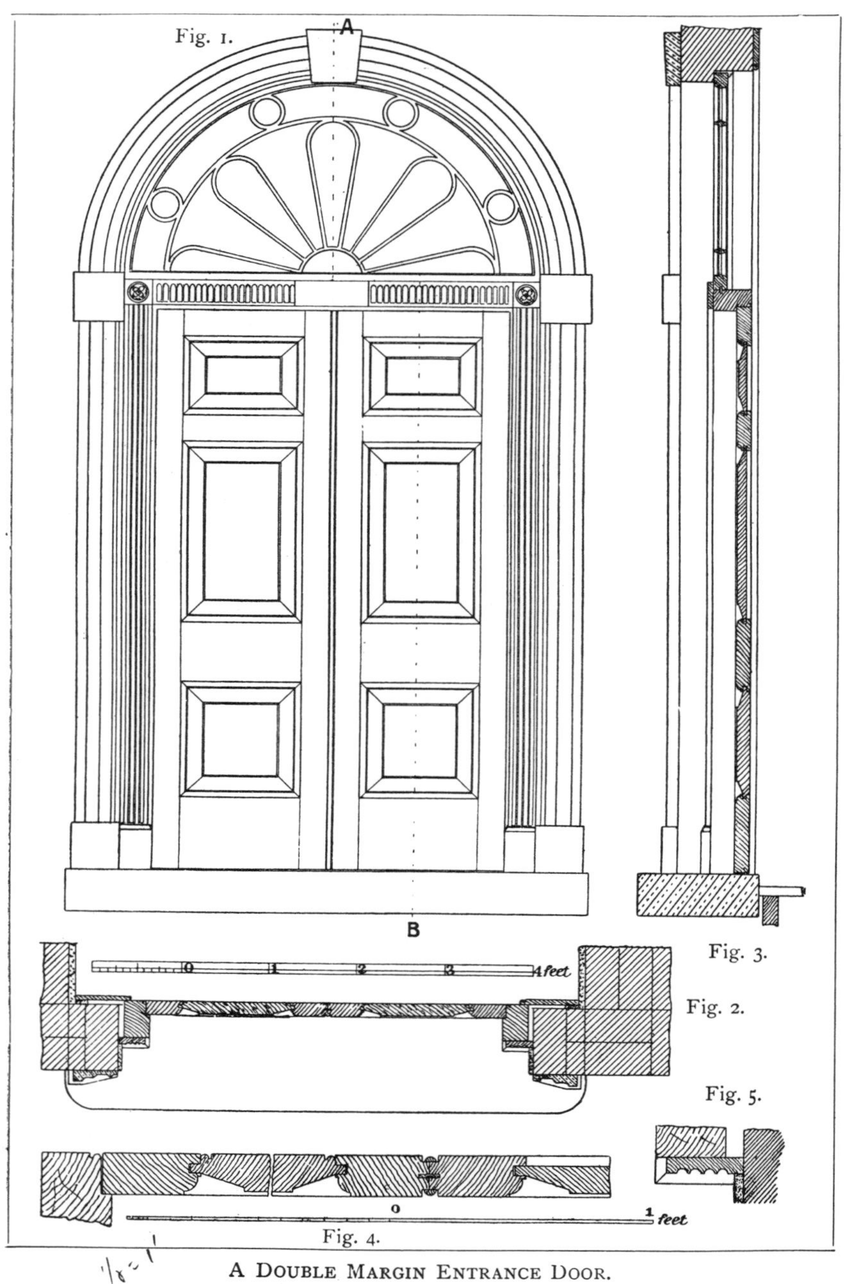

A DOUBLE MARGIN ENTRANCE DOOR.

(*Late Eighteenth-Century Style.*)

1. Elevation. 2. Plan. 3. Section. 4, 5. Details.

done, and cross tongues are next glued in, the joint rubbed, the two stiles pinched together with handscrews, and the oak keys, well glued, driven in. At this stage the frames are stood aside to dry, after which the projecting ends of the keys are cut off, the panels inserted, and the two outside stiles glued and wedged in the usual manner. After the door is cleaned off, the grooves to receive the beads are brought to their exact size with side rabbit and router planes. Should iron bars be used, they are inserted in grooves made after the door has been shot to size. The bars should be about ½ in. shorter than the width of the door, so that their ends may not be visible when the door is brought to size.

Diminished Stile Doors are usually glazed in the upper part, the stiles above the middle rail being reduced in width to increase the lighting area. Hence they are sometimes termed sash doors, though strictly, sash doors are those in which a separate sash is inserted in a rebate in the upper part of the door, the remaining portion of the thickness being utilised for the reception of a shutter which lies flush with the outside of the door. Two examples of diminished stile doors are given on pp. 164 & 165, the first being a modern shop or **swing door**, the latter a typical eighteenth-century **vestibule** or screen door, used to enclose the entrance hall. These are in pairs, rebated together at the meeting stiles, and hung to a solid rebated frame with transom and fanlight above. They are frequently made of mahogany. The bars, finishing ⅝ in. thick, are scribed and dowelled at the intersections. A ¾-in. ovolo moulding is worked in the solid around the lower panels.

Shop Doors, of which f. 1, p. 164, is a type, are generally made in mahogany not less than 1¾ in. thick, hung in substantial frames with pivots, the lower one attached by a brass shoe S, to a spring enclosed in a metal box sunk in the floor. This shoe lies when the spring is at rest, along the opening, as shown in the plan, f. 2. When the door is pushed in either direction the spring is brought into action, and tends to throw the shoe carrying the door back into its usual closed position. The older forms of these spring hinges cause the door to swing too violently, and pass the jamb several times before coming to rest. Newer forms have additional springs or pistons which check the swinging and cause the door to close steadily.

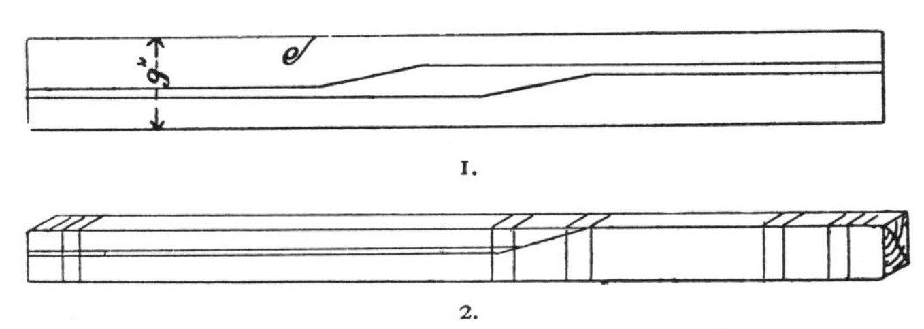

1. Method of Cutting Out Diminished Stiles.
2. Setting Out of a Diminished Stile.

The Diminished Stiles are sometimes cut out in pairs from a board or plank, as shown in f. 1 above. When this is done, the back or outside edge is shot straight and the setting out made thereon, the two portions being gauged to width also from the back; but the method more suitable for machine working is shown in

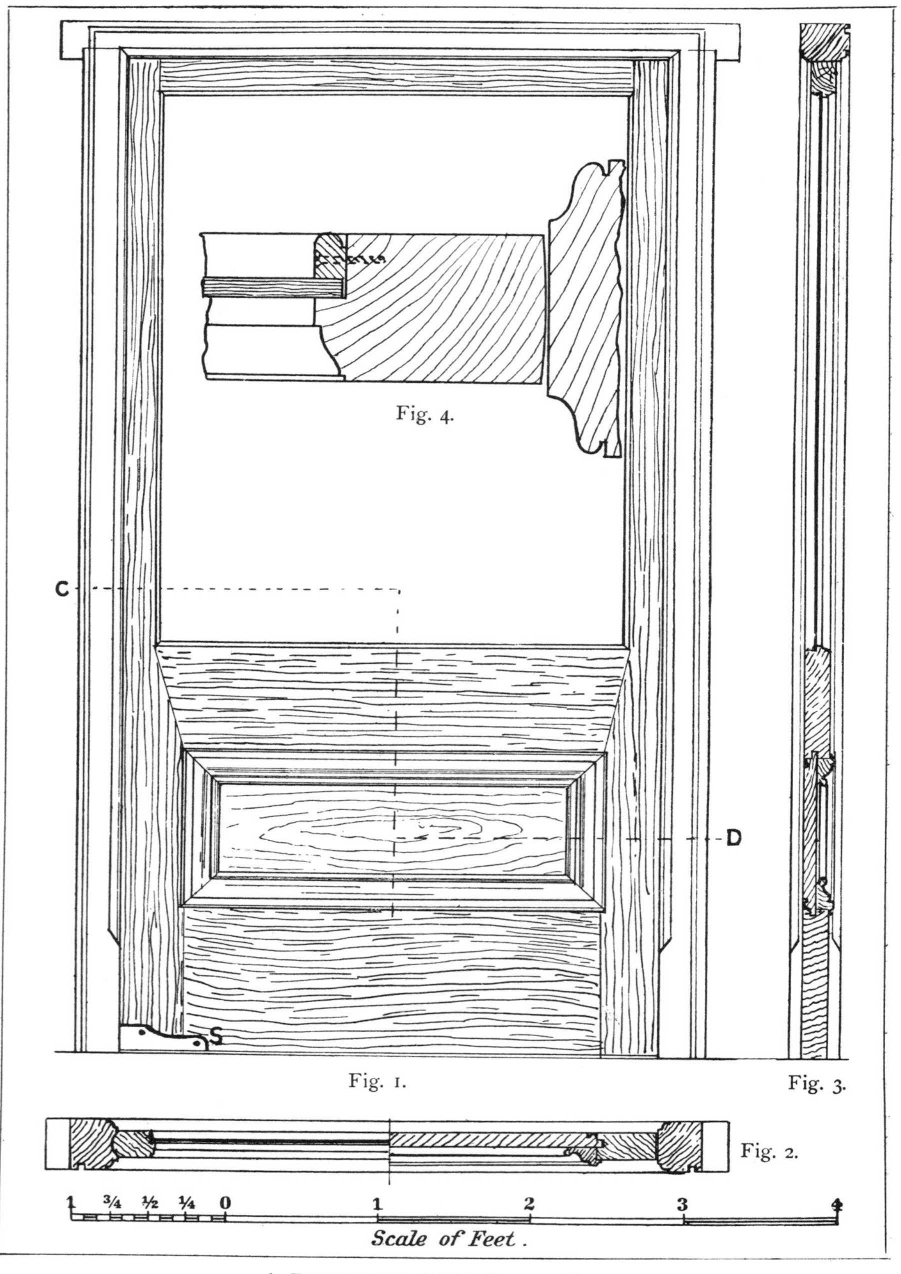

A DIMINISHED STILE SWING DOOR.
1. Elevation. 2. Section on C D. 3. Vertical Section.
4. Detail of Upper Part of Stiles.

A Pair of Vestibule Doors. (*Eighteenth-Century Style.*)
1. Elevation. 2. Plan.

f. 2, p. 163. Here the stile is cut out parallel to the full width, the face edge shot in the usual way, and the setting out made upon that, the diminish being gauged from inside. In this method the mortises are made before the diminished part is cut out, to render that operation easier for the machinist. He should not, however, mortise right up to the sight lines on the diminished part, because should the chisel be at all out of upright, when the waste is cut away the mortise will be found beyond the sight line, which will be a serious defect.

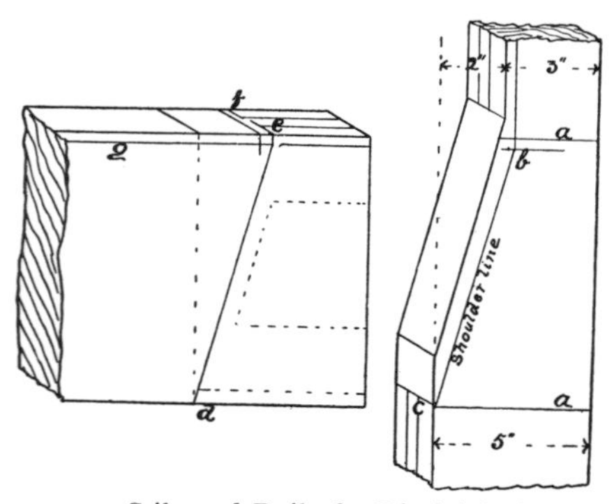

1. Stile and Rail of a Diminished Stile Door.
Method of Obtaining Shoulder Lines.

Setting Out the Bevelled Shoulders of the stile and rail.—Taking the stile first, having, as described in the first method, gauged and faced up the inside edges, and set out the width of the rails and mortises on the back edge, square over on each side the sight lines of the middle rail as at *a*, *a*, f. 1. Then draw a second line representing the depth of sticking of the rail on the face side, and the rebate on the back side, as at *b*. Next run gauge lines down on each side of the diminished part as working lines of rebate and sticking, and from the points of intersections of the stickings and rebates respectively, draw in the shoulder lines to the sight line of the lower edge of the rail at *c*. The only difference to be made when the stile is prepared by the second method is that the sticking and rebate gauges would be run from the original face edge instead of the actual diminished edge, and as before stated, the sight lines would be marked on the face edge instead of the back, and squared down to the intersections. See also p. 81.

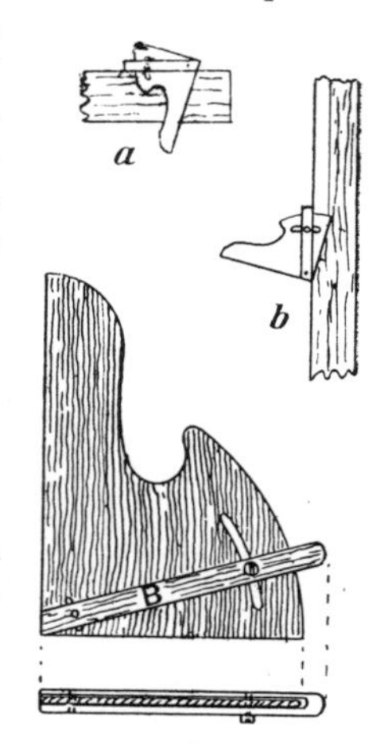

2.
An Adjustable Square for Bevel Shoulders.

To Set Out the Rail.—Mark on the bottom edge the "width" or sight lines of the widest part of the stiles. Square this point on to the upper edge as shown by dotted line in the sketch, f. 1, and set off therefrom the amount of the diminish on the stile; as shown by the dotted line on the stile in the example this is 2 in. This line, knifed in on the edge, is the "sight line" of the upper part of stile. Again set off beyond this line the amount of the sticking and rebate shown in the sketch by the lines *e* and *f*. Next run the sticking and rebate gauges on front and back sides, as shown at *g*, and square down the lines *e* and *f* to meet them. Then draw the shoulder lines from the intersections to the point *d* on each side. Having thus found the shoulder lines upon one rail, bevels may be set to them and used to mark any number. A contrivance sometimes used when a large number of similar shoulders have to be set out is shown in f. 2.

This is known as a **Shoulder Square.** It consists of an ordinary set square provided with a movable fence or bar B, which is slotted to pass on both sides of the square, and is pivoted near the right angle. A set screw near the outer end of the bar passes through a concentric slot, and fixes the fence in any desired position. The pivot works tightly in a small slot to allow the lower edge of the bar to enter the right angle, and the outer edge of the square is also made a concentric curve to permit the easy passage of the end of the bar. The theory upon which the action of the tool is based is, that the angle between the parallel bar B and either of the edges of the square is the complement of the remaining angle, the two combined forming a right angle, which is the desired angle between the edges of a rail and stile. Its application is shown in the upper part of the figure; *a* being the rail, *b* the stile. Either a rail or stile is first set out as described above, the edge of the square set to the shoulder line, and the bar brought up to the face of the work and fixed with the set screw when it is ready for application to the other piece. The moulding upon a diminished stile should not be mitred but continued on to the shoulder, and the rail scribed over it, to prevent an open joint occurring should the rail shrink. When doors, after knocking together, are stored for a second seasoning, a slight difference will have to be made in the setting out of the shoulders of the middle rail. The wider part of the stile will shrink more than the narrow part, and consequently if the shoulders are set out accurately at first as described above, when they are refitted the shoulders will be found short at the lower ends. To prevent this, allow about $\frac{1}{32}$ in. extra on the lower part of the shoulder at each end of the rail.

Jib Doors are those used to mask the presence of an opening in a wall which would otherwise destroy the symmetry of the apartment by breaking the continuity of the skirtings, dado, &c. They are not primarily constructed for the purpose of concealment, but as stated, to preserve an unbroken surface to the wall, &c., and the necessary joints and hinges of the moving parts are made as unobtrusive as possible. The side which is in the room is flush panelled, and the door hung with pivots so that it lies exactly in the plane of the wall. It is fitted closely into its frame, which is also flush; and the various finishings are continued right across its surface, so that at a short distance it is practically invisible.

Fig. 1 next page shows a portion of a wall containing a jib door, and f. 2 is a vertical section through the opening (if the opening is to be veiled also upon the other side of the wall, a similar door would be constructed in that face). Fig. 3 is an enlarged section of the bottom rail and skirting. Fig. 4, ditto of the top rail, showing the pivot and picture rail used to mask its presence. Fig. 5, plan of the shutting stile and part of jamb lining. Fig. 6, plan of the hanging stile. The dotted lines show the position of the door, &c., when open. The portions of the skirtings, dado rail, &c., that are fixed to the door, run beyond its edge at the hanging side, the joints of the various members forming parts of concentric circles struck from the centre of the pivot; these are shown by the three circular arcs in f. 6. To find the position of the joints, draw the plan of each member of the moulding as shown, and from the centre of pivot draw a line at an angle of 45 deg. The intersection of this line with the various members will give the radii required to describe the arcs that form the path of the moving pieces, as shown by full lines in the drawing. These doors

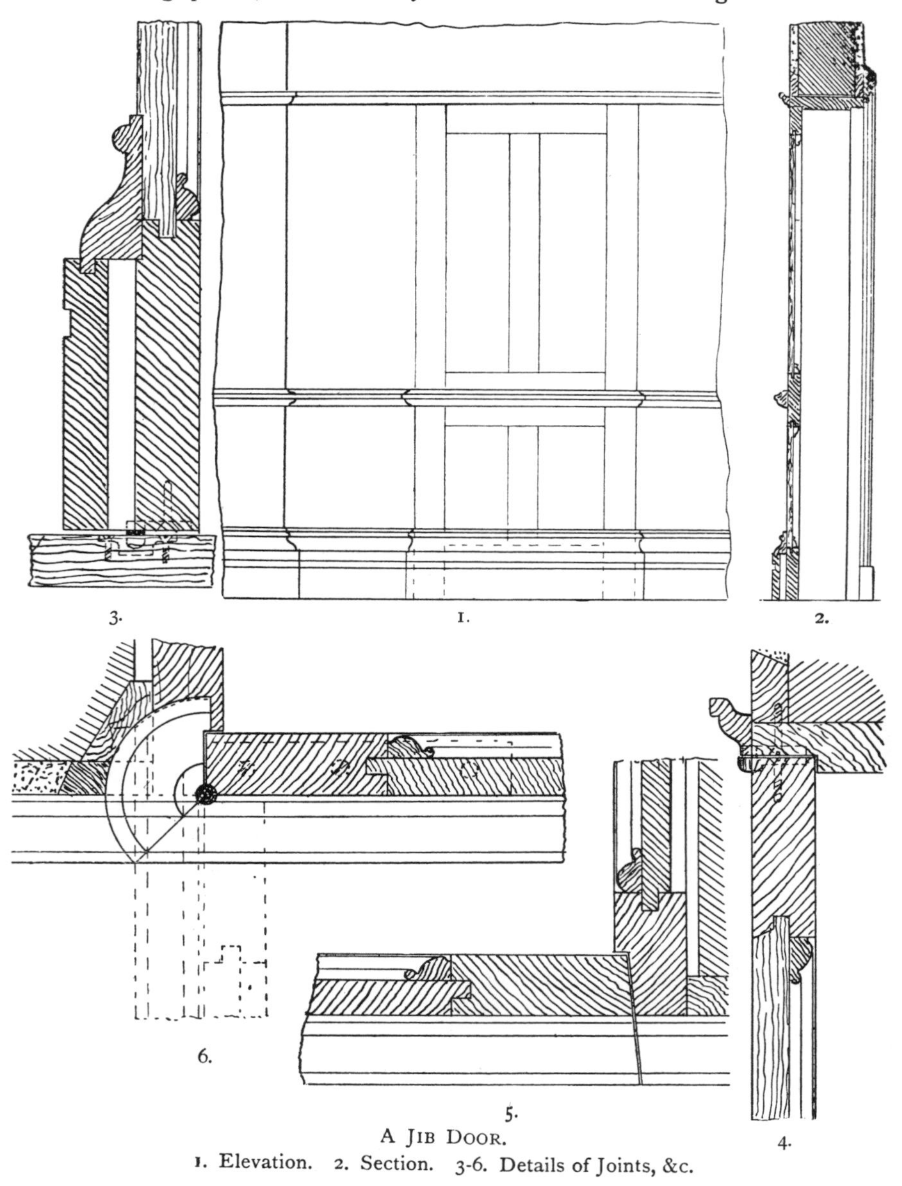

A JIB DOOR.

1. Elevation. 2. Section. 3-6. Details of Joints, &c.

require to be made of thoroughly dry straight-grained stuff, to prevent casting, and the ends of the panels that are not covered should be undercut, as shown at f. 4. The surface is sometimes covered with thin canvas glued on and then

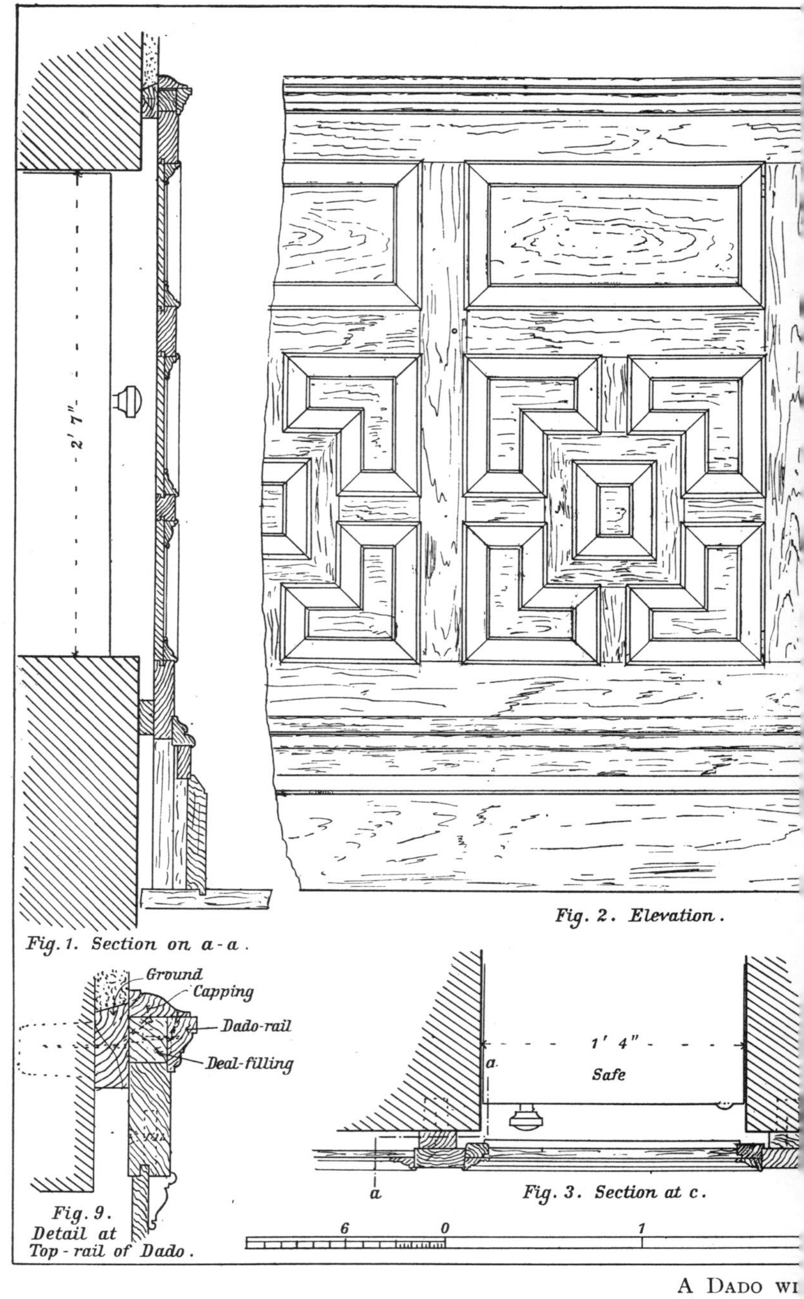

2' 7"
Fig. 1. Section on a-a.
Fig. 2. Elevation.
Ground
Capping
Dado-rail
Deal-filling
Fig. 9.
Detail at
Top-rail of Dado.
1' 4"
a
Safe
a
Fig. 3. Section at c.
6
0
1

A Dado wi

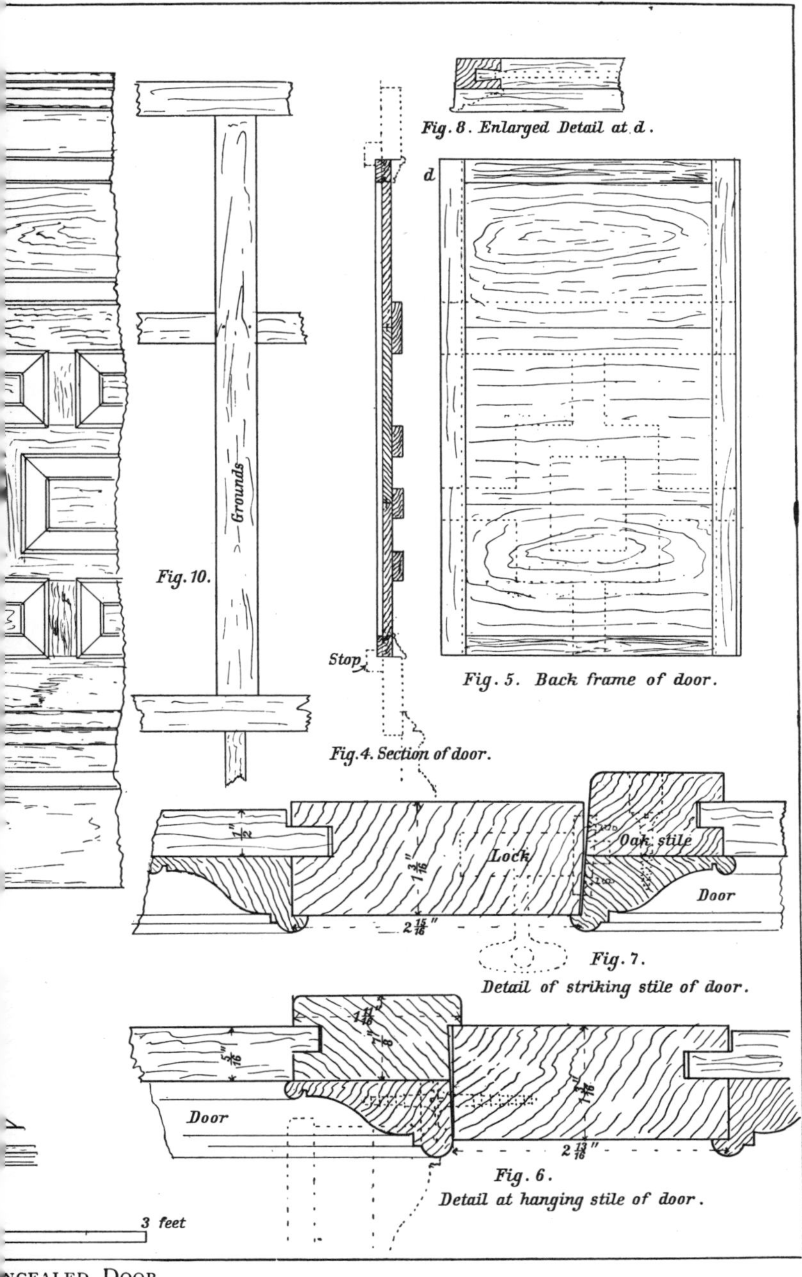

NCEALED DOOR.

[To face page 169.

painted. When the doorway is in frequent use the frame is sometimes made of **T** iron.

A Concealed Door in a Dado, giving access to a hidden recess in a wall formed to receive a safe or the like, is shown on Plate XVI. Fig. 1 is a vertical section through the dado, which is constructed in the usual manner of panelling, the muntings stub mortised and tenoned to the rails, and screwed from the back,

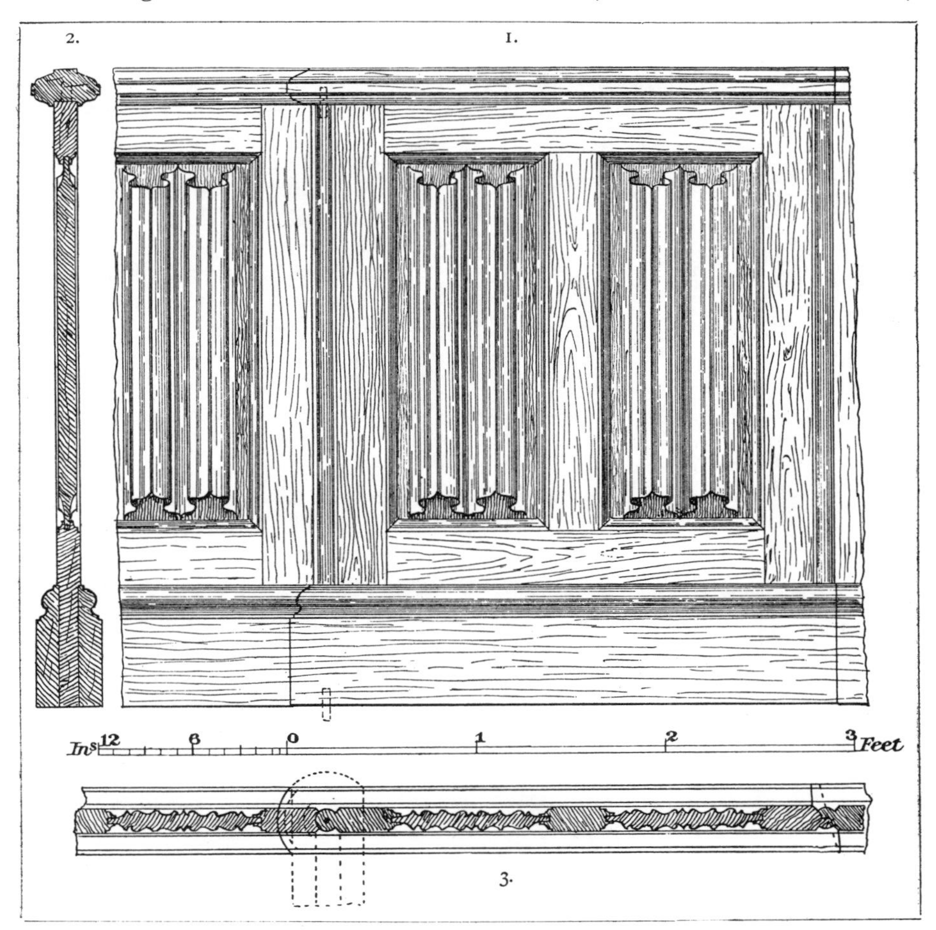

A DWARF SCREEN DOOR.
1. Elevation. 2. Section. 3. Plan.

as shown in the detail, f. 9, the end stiles tenoned and wedged, and the panels grooved in. For constructional purposes of the secret door, the whole of the panels are designed "laying," but if the lower ones are preferred "standing" the door panels could be veneered. Fig. 2 is an elevation of part of the dado containing the door; the joints of this and also the hinges are indicated more plainly than they would appear in reality, given the careful workmanship required for such work. Fig. 4 is a section of the door, and f. 5 an elevation, with the face rails and mouldings removed to show the construction clearly. The position to be occupied by the rails is indicated by dotted lines. Whatever the material of the dado, the back framing of the door should be either of European oak or beech, and straight-grained pieces should be selected to avoid warping. The construction of the door is clearly shown in the enlarged details, f. 6, 7, & 8. The frame is dovetailed together, and the panel glued up in one piece, and grooved in $\frac{5}{16}$ in., the joints arranged to fall under the rails; these are to be glued, and screwed to the panel; some allowance at top and bottom should be made for swelling, none for shrinkage, as the mouldings hide that. The moulding is designed with an astragal on its outer edge for the purpose of concealing the hinges, which are ordinary brass butts. These should be procured first, and the moulding struck to suit the knuckle, which is embedded in the member as indicated by dotted lines in f. 6. The edge of the moulding at the hanging side forms the joint; on the striking side it is rebated on the face to cover the joint. The lock, which may be of the railway carriage door type, with a round straight key as making the least obtrusive keyhole, should be sunk in the striking munting, and not in the door itself; this will prevent the bolt making any mark on the surface of the framing. If preferred, the lock may be dispensed with, and a snap-bolt with push button used instead, such button to form part of the carved mouldings. The dado is fixed to framed grounds fastened to plugs in the wall, as indicated in f. 9, where also the fixings of the various members are shown. It is perhaps needless to point out that the several nails and screws,

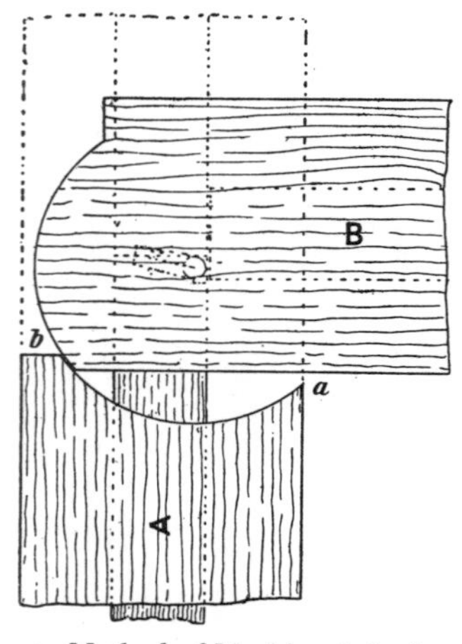

1. Method of Marking Joint in Capping of Screen Doors.

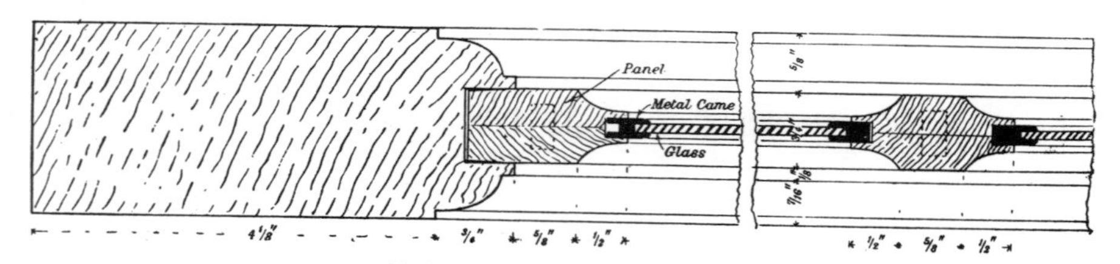

2. Enlarged Detail of Gothic Door.

though shown in the one drawing, do not all come in the same section. An elevation of part of the grounds is shown in f. 10; the skirting is blocked out about every 4 ft. with $1\frac{1}{2}$-in. blockings, also fixed to plugs in the wall.

A Dwarf Door such as is used in chancel screens or pews is shown in elevation in f. 1, p. 169, section in f. 2, and plan in f. 3. The panels are moulded and carved in the style in vogue in the sixteenth century, which is termed "linen-fold"; the material is almost invariably oak. These doors are usually hung with pivots fixed in the floor and capping. The joint, known as a sunk astragal, is fully described in the chapter on joints. The plinth and capping are cut in a similar manner to that described for a jib door. Another method of hanging these doors is shown in f. 1 opposite. In this case a pair of ordinary butts are sunk flush into a bead on the frame, and the joint in the capping is part of a circle which is struck from the centre of the hinges. A small clearance should be given at *a* to prevent the capping on the door bruising the joint when open. The movement of the door is checked by a stop fixed to the floor. To determine the radius of the joint sweep, draw the door and capping as shown in its two extreme positions, and the intersection of the capping lines on the hinge side, plus the clearance, will give the necessary radius. The stop *b* may be made in any position suitable for the section of the capping, but should cover the end of the moving part when open, to preserve the joint from damage. In the diagram the shaded part A represents the fixed capping, B the moving part wide open, and the dotted lines the same

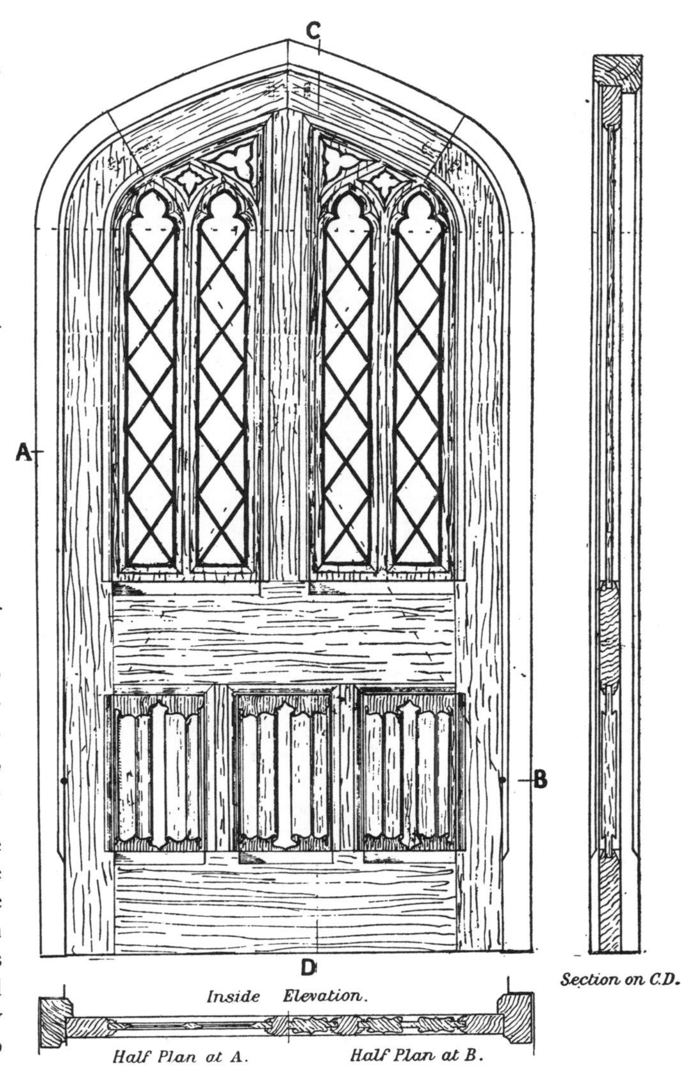

A Gothic Style Door.

parts closed. Any door less than 5 ft. in height is termed a dwarf door. A small door framed within a larger one is called a wicket.

A Gothic Door of the Tudor period style is shown on previous page. The upper panels are of pierced tracery, carved with a flat hollow, as shown in the enlarged detail, f. 2, p. 170, and filled in with leaded lozenge-shaped lights. The lower panels are moulded in relief, with wave-like flutes, the ends of which are carved in scrolls forming the so-called " linen-fold " panels. The moulding around the edges of the door framing is a sunk ovolo, struck in the solid and stopped in line with the rails. The rails are chamfered on the top edges; these also are stopped at the muntings to provide butt joints. The lower edges are moulded to correspond with the stiles, and the shoulders are sunk and mitred, or scribed. In the early Gothic work the joiner followed the mason's methods of construction, making the mitre in the solid, but doubtless soon found that his material did not behave in a similar manner to stone, its shrinkage destroying the intersections of the members with the solid mitres, and later the mouldings were stopped and the shoulders butted without mitres. As skill in joinery increased, scribing was introduced, which gave the effect of a mitre without its drawback of an open joint, when the material shrank. Both of the latter methods are illustrated in this door. Details of the construction of the upper panels are given in f. 2, p. 170. The tracery panels, if prepared by hand, would be better made in two thicknesses, to be glued together after the metal work had been fitted in, small dowels to be inserted as indicated by the dotted lines, to ensure adhesion of the two leaves. If made in a machine shop, the panels would be more readily prepared in the " solid " and grooved out on the spindle, the lower end of each opening grooved right through, to permit the passage of the light, after which a core would be glued in to keep it in position. The dotted lines on the elevation indicate the radii of the segmental arcs forming the doorhead. It is upon these lines that the joints should occur; four centres are used, two on the springing line and two below. To find their position, divide the width of the door at the springing into four equal parts, and upon the portion of the springing between the two outer division points, construct an equilateral triangle, produce the sides to intersect the outer vertical edges of the door. These points will be the centres for the upper segments, the aforesaid angles of the triangle giving the centres for the small arcs.

A Pair of Sliding Doors are shown in f. 1 & 2 opposite.—These are so constructed as to be fire-resisting. They are made of English oak, with a solid framing $3\frac{1}{2}$ in. thick, double panels $1\frac{1}{4}$ in. thick, the middle rails and muntings, in two thicknesses, enclosing a space 1 in. in depth, which is filled up with slag wool. This material, which is light in weight and fireproof, is interlaced through a steel wire netting, which is nailed to the inside of one set of panelling and the top rail before putting together, the object of the netting being to keep the protective covering of wool in place should one side of the door get burned through. The doors run on wheels let in the solid rail at the bottom and in an iron channel in the floor which is tightly packed with the slag wool to prevent fire creeping under. The top of the door is also supported by wheels carried by straps bolted to the face, and running on an angle bar cemented into a chase in the wall. Another iron bar is inserted in the wall over the opening, and pro-

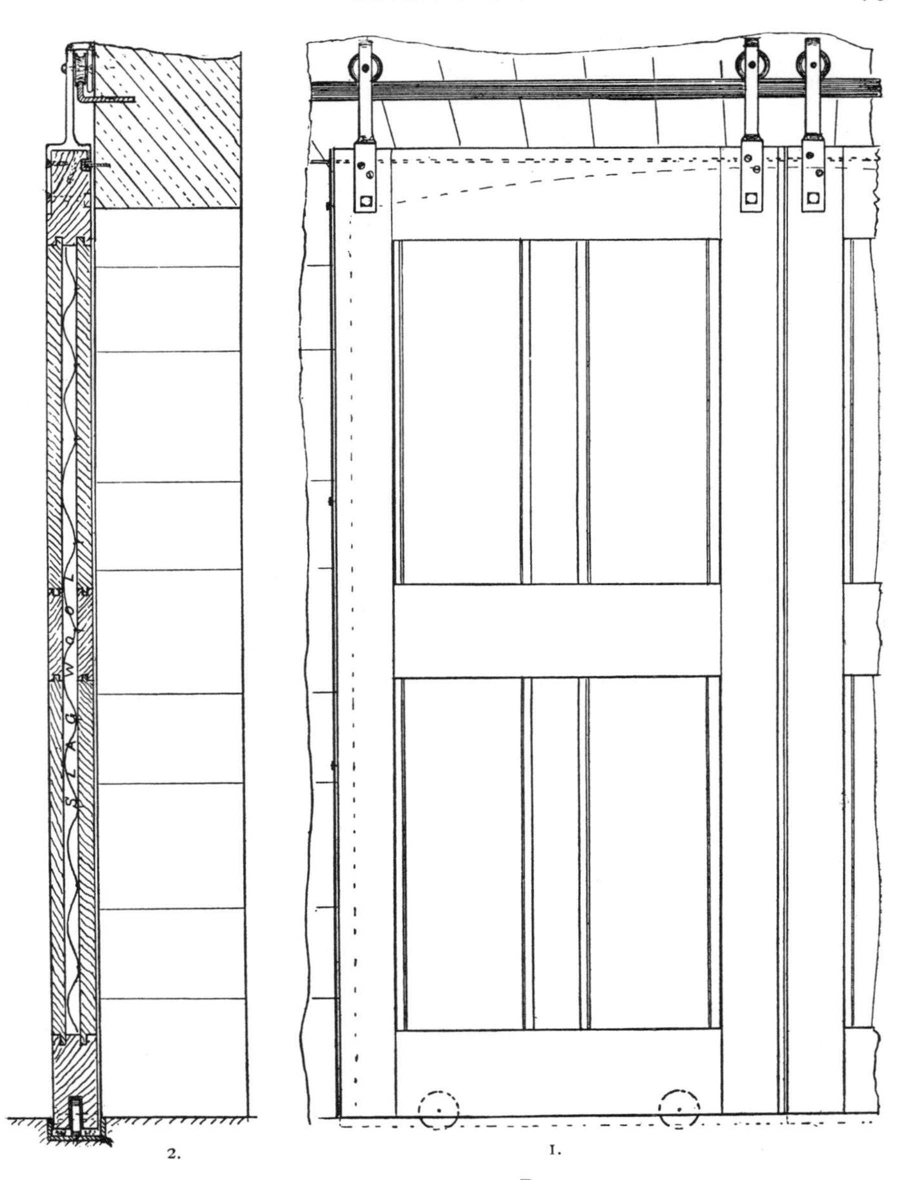

SLIDING FIREPROOF DOORS.
1. Elevation. 2. Section.

jects ¾ in. into a groove worked across the face of the doors, and similar iron bars, sliding laterally on studs on the back edges of the doors, fit into corresponding chases in the wall when they are closed, and thus effectually cut off all communication between the two rooms.

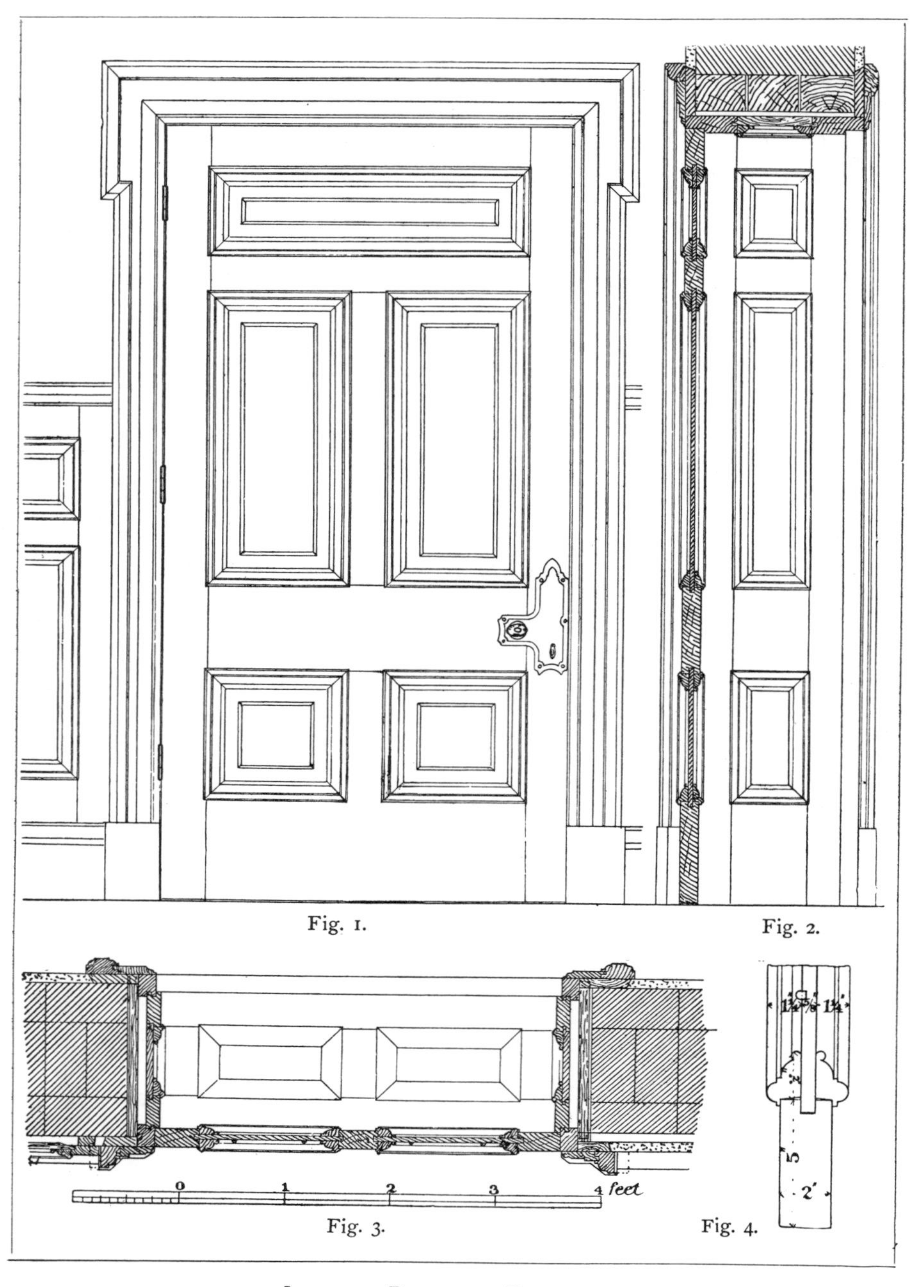

INTERIOR DOOR AND FINISHINGS.

1. Elevation. 2. Section. 3. Plan. 4. Detail of Stile.

A Superior Interior Door of modern design is shown in f. 1 to 4 opposite. Fig. 1 is the elevation, f. 2 is a vertical section through the opening, showing the jamb linings, and f. 3 is the plan showing in outline the framed soffit lining. Fig. 4 is an enlarged section of one stile and part of panel, &c.

A Stop Chamfered six-panelled interior door is shown in elevation in f. 1 below, and f. 2 is an enlarged detail section. Figs. 1 to 3 on next page are elevation and sections of a modern glazed or **Sash Door**, suitable for closing the end of a passage or lighting an interior room.

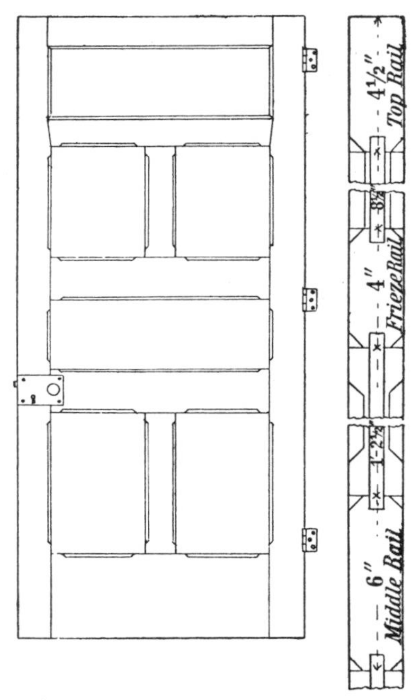

Six-panelled Gothic Style Door.
1. Elevation. 2. Detail Section.

Method of Setting Out and Preparing a Panelled Door by Hand.—Assuming the rod set out as shown in f. 4, next page, and the stuff planed to size and out of winding, proceed to lay a stile on the rod face downwards, and strike up the sight lines of the rails marked with an X in the figure upon its edge. It should be noted that a good workman always uses a pencil for marking mortise, and a knife for shoulder, lines. If the knife is used for the mortises, the marks will show on the face after the work is cleaned off. Next turn the stile up on edge, and set off ½ in. within each line for the plough groove, and space out the mortises as described at p. 144, Chapter VIII., squaring them over to the back edge, and marking the wedging; ¼ in. is sufficient in hard wood, and ⅜ in. in soft wood. Pair the other stile with the one marked, and fix them together with hand-screws; square the lines across the second stile, next turn a stile down, and lay the muntings on it in their approximate position, and knife-cut in the sight lines of the rails slightly full. These must be squared over on each face for the shoulders.

To Mark the Rails.—Lay the middle one on the width rod, and square up the sight lines of the stiles and munting, the former with the knife, the latter with pencil. Set in ½ in. on each side of the munting sight lines for the ploughing, and between these will be the mortise. Lay the top and bottom rails above and below the middle, and square all three over. Square over on each side the shoulder lines, mark the width of the tenons, and cut out the haunchings. Next set a mortise gauge to a suitable chisel, and in the centre of the thickness of the stuff, and gauge the mortises and tenons all from the face sides. These may now be cut as described in Chapter IV., p. 59. Then grooves are ploughed for the panels ½ in. deep in the desired position, which may vary according to the thickness of the panel or the depth required for the moulding. In any case the groove must either fall within the lines of the

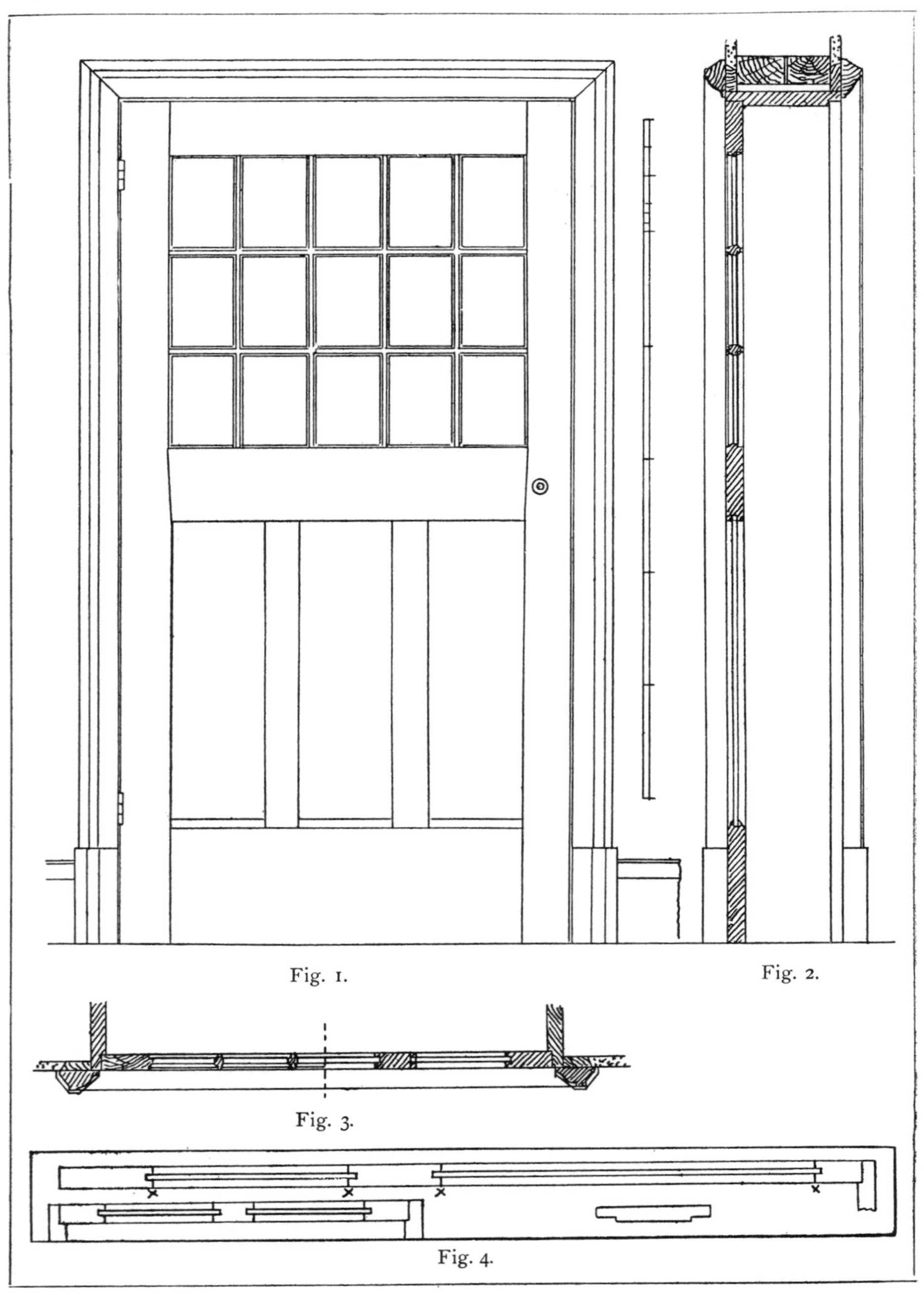

A MODERN SASH DOOR.

1. Elevation. 2. Section. 3. Half Plans above and below Lock Rail.
4. Height and Width Rod of a Door.

mortises or be stopped at them, for if they fall outside and are run through, one or both sides of the haunching will be cut away, rendering the haunch useless. After the ploughing is finished, the shoulders may be cut, and the fitting proceeded with. Both the edges of the stiles and the shoulders of the rails and muntings should be first tested for "square" and corrected if required. Each rail should be fitted in its appropriate mortise, and when driven up, tested for "upright" with a straight-edge, as shown in f. 1. The straight-edge should be held firmly to the face of stile, just clear of the shoulder, as shown, when it can be seen if the rail lies in the same plane. If it does not, either the tenon or shoulder wants easing on the "hard" side. After all fitting is finished separately, knock all together to see if the frame is out of winding. Commence with the muntings; insert these in the rails, and turn the latter down on the floor on their ends, and drive on a stile, then turn the frame over, and knock on the other stile. The frame may then be laid on the winding blocks and cramped up, when any faults in the fitting will be seen, and can be marked for correction when again taken to pieces to insert the panels. These latter are cut off to size, nett length, and $\frac{1}{8}$ in. narrow (some joiners cut them $\frac{1}{16}$ in. small all round, but this is not *necessary* unless the panel is a raised one, which requires to be worked equally from its edges all round), and are faced up true or at least regular across with the trying plane, then turned over and mulleted to thickness. This operation is illustrated in f. 2. A mullet is a block of wood ploughed with the identical iron used for the frame-grooving, and is used to gauge the thickness. After the mulleting, the panels are smoothed up and glass-papered. The frame is next knocked entirely apart, so that all the mortises and tenons can be glued, the former with the mortise stick, f. 3, which is simply a thin piece of wood shaped as shown, with several saw kerfs in it to hold the glue, which is thus transferred to the sides of the mortises. The muntings should be first cramped up; then the rails, commencing with the middle one. The cramp, if only one is used, should be placed between the tenons; but two are better, one at each edge of the rail. Drive these wedges equally, and well home. They should be cut so as to pinch inside harder than out, so that when the stile shrinks it may be able to move from the outside towards the shoulders. Cramp and wedge the bottom rail next, and the top one last. Drive the outside wedges slightly harder than the inside, to cause the rails to pinch the muntings tightly. The ends of the wedges are better cut off whilst the glue is soft. Stand the door aside for the glue to dry, about four hours in summer and twelve in winter will be required for this. Then the shoulder can be cleaned off, and the door finished as described at p. 76.

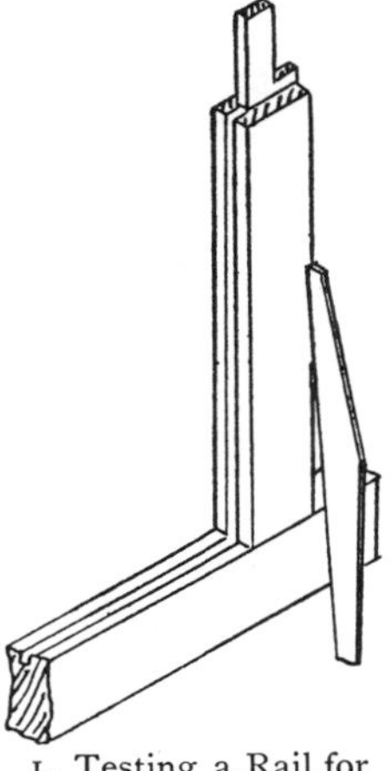
1. Testing a Rail for Truth.

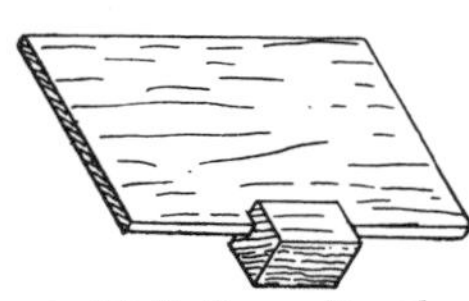
2. Mulleting a Panel.

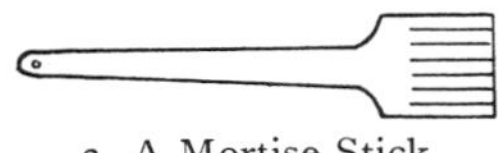
3. A Mortise Stick.

Machine-made Veneered Doors.—Doors constructed as shown in the sections, f. 1 & 2, below, are termed compound doors. The first method is patented, and the process is worked by the Gilmour Door Company, of Canada and London. The object of the process is not economy of cost; which is about the same as that of solid hardwood work of similar design, but the method is said to render impossible the warping of stiles and rails, or the shrinkage of the parts causing open joints or splitting of the panels; it also ensures the door always remaining true, as by the method of building up the members, "casting" is eliminated. The stiles, rails, and muntings of the door or other framework are prepared by gluing together strips of soft pine about 1 in. in thickness, and of a width equal to the desired thickness of the framing, less the clear thickness of the veneers (bare $\frac{1}{2}$ in.). The outside strip on the stiles is of hardwood, to correspond with the face veneer, and is $\frac{7}{8}$ in. thick. When glued up, these pieces are run through a thicknessing machine, then multiple-grooved as shown, and duplicate grooves made in the back of the veneer, which is then glued and run through a rolling machine; this drives out the superfluous glue and so interlocks the parts that it is an impossibility for the veneer to peel off. The parts are then treated exactly as are the parts of a solid door, mortised and tenoned, or dowelled together, as may be preferred. The panels are made of five-ply veneer, two plies on each side of a pine core, in transverse directions of the grain. Another method of constructing veneered doors is shown in f. 2. This is largely practised in the United States. The section is made through the middle rail of a glazed door or screen, and incidentally shows an excellent way of fixing the mouldings and panels. The frame is built up as in the last case, of strips of clean dry pine, glued together, then faced on each side with hardwood heavy veneer. Instead of fixing the panels into plough grooves in the frame,

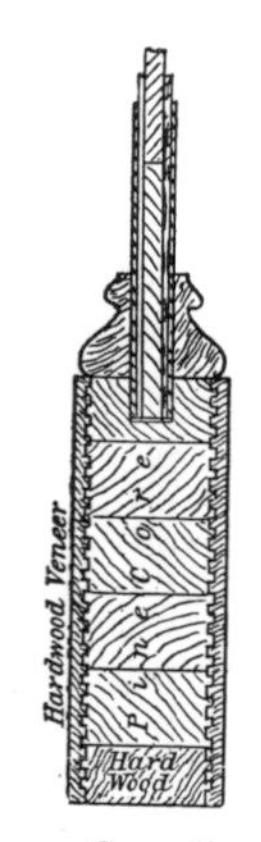

1. Canadian Machine-made Door.

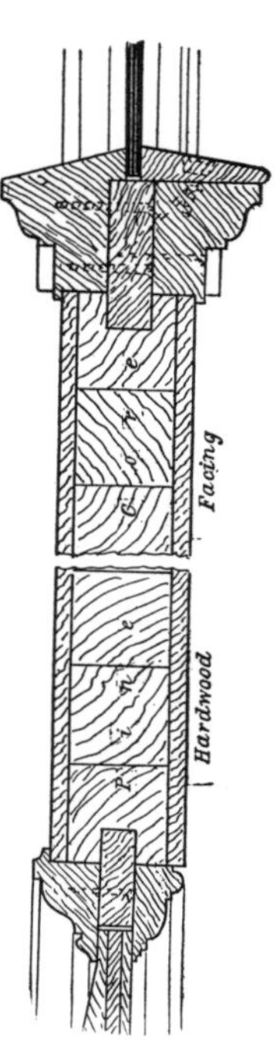

2. American Method of Constructing a Veneered Door.

3. Veneered Hospital Door.

hardwood strips or splines, equal in thickness to the panel or its tongue, are fitted tight into the grooves, and the mouldings, made half-an-inch wider than the spline, are glued or screwed to the latter, thus allowing either the framing or the panel, to shrink without disturbing the mitres.

A Slab or Hospital Door, built on the compound principle, is shown in f. 3. The core of the framework is built up with narrow strips, as described in the preceding method, but in this case the core of the panel is made of equal thickness to the frame, and is glued up in somewhat wider strips, the ends having ½-in. tongues worked on them, which are sunk into grooves in the stiles. The panel is fitted tightly between the top and bottom rails, and the whole is glued together, forming a solid slab, upon which two-ply veneers are laid in transverse directions. The face veneers are generally made up of 6-in. strips, as shown, running the lengthway of the door. No mouldings or other ornaments are used in these doors, the object being to avoid any lodgment or receptacle for dust or germs.

CHAPTER XI.

DOOR FRAMES, LININGS, AND FINISHINGS.

An Eighteenth-Century Door and Frame, Details — Construction of Solid Frames — Methods of Fixing—Vestibule Frames—Segment and Circular-headed Frames—Formation of the Heads—Built Up—Solid—Position of Transoms—Construction of Frames for a Warehouse-hung Door—Loophole Frames—Jamb Linings—Definitions —Construction of Plain, Framed, Double Framed, and Skeleton Linings—Methods of Fixing Painted and Polished Finishings—Grounds—Rough, Framed, Skeleton, and Moulded—How they should be fixed in relation to the Openings—Method of Wedging Up—Architraves—Single and Double Faced—Framed Architraves—Treatment of Heavy Architraves—Framed Mitres—Secret Fixing of Architraves, Operation described—Foot Blocks—Alternative Methods of Fixing.

DOOR FRAMES fitted into external walls, or other positions more or less exposed to the weather, such as in vestibules, shop and warehouse entries, and the like, are usually made in the SOLID, that is, the various members in each case consist of a single piece of timber of considerable substance, having the necessary rebates, mouldings, &c., worked therein; whilst the frames fitted to interior openings are usually composed of several comparatively thin members built up to form a casing or LINING to the opening. There are, however, exceptions to this custom, as the following example shows. The outer vertical members of a solid frame are called posts or jambs; interior ones, mullions; the horizontal members sill, transom, and head. The jambs are framed between the head and sill, chiefly that the ends or horns of the latter may run beyond the frame, and so provide fixings that can be built into the wall; also because the shoulders of the post form a better abutment for carrying any load that may be thrown on the head, than the edge of a tenon would if the rails were framed between the posts. Transoms are cut between the jambs, and also between the mullions when these are used.

An Early Eighteenth-Century Doorway, still in existence in the Grosvenor Road, Westminster, is shown in the photograph facing p. 181, and several large-scale details of the door and linings are given on p. 181. The hood above the door, and the frieze below the head, are both richly and boldly carved, as are also the two massive consoles or modillions carrying the corona. The hood projects 2 ft. 9 in. from the wall. The opening, from sill to soffit, is 9 ft. 9 in. high, and 4 ft. wide between the jambs. The height to the transom is 7 ft. 2 in. A pair of $2\frac{1}{4}$-in. doors fill the opening; these have 7-in. stiles, 7-in. top rails, $9\frac{3}{8}$-in. frieze rails, 10-in. lock rails, and 8-in. bottom rails. A cavetto moulding is stuck in the solid on the inner edge, forming, with the $1\frac{3}{8}$-in. Greek ovolo

PLATE XVII.

RENAISSANCE DOORWAY (EIGHTEENTH CENTURY), GROSVENOR ROAD, WESTMINSTER.

[*To face page* 181.

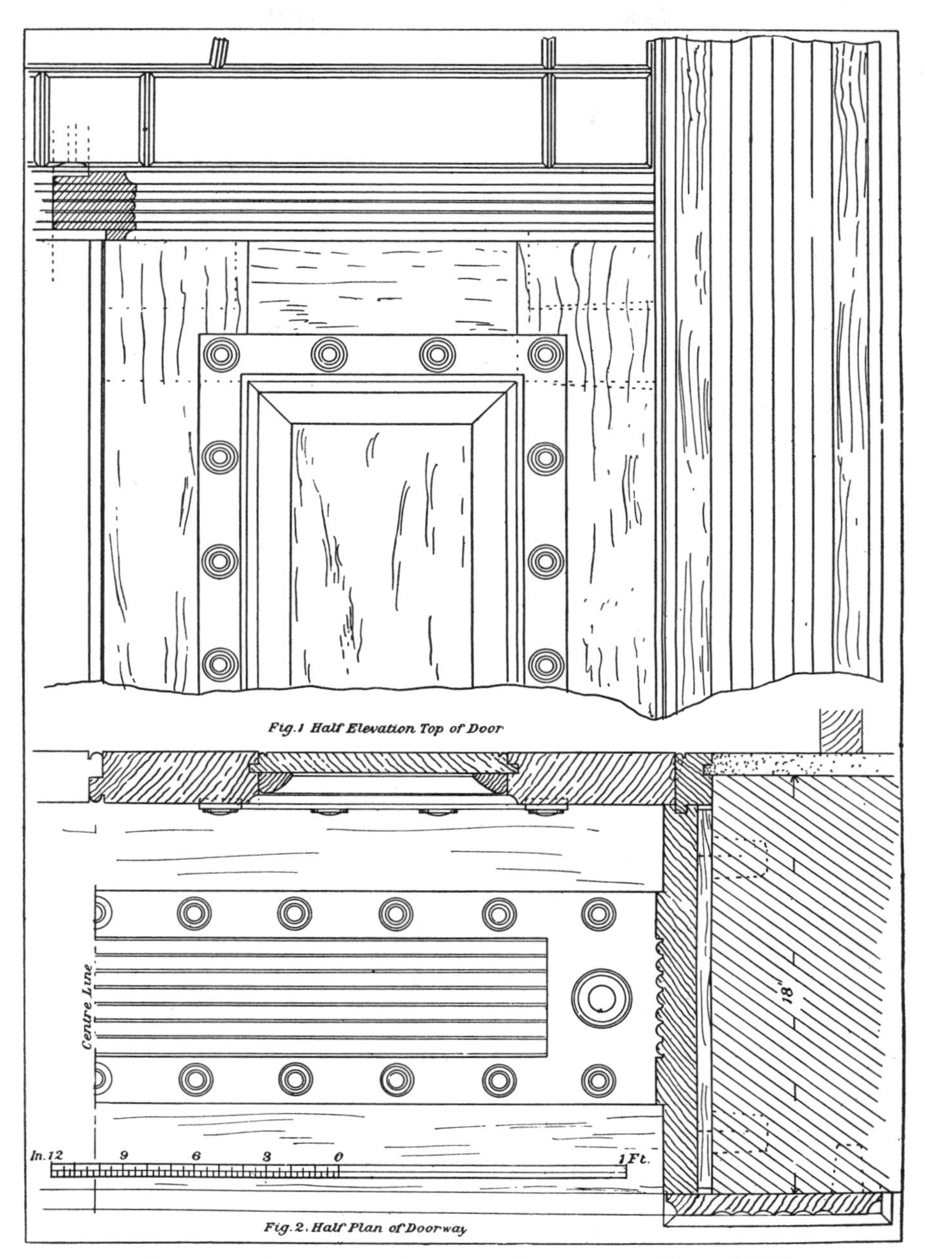

ENLARGED DETAILS OF DOORWAY—PHOTOGRAPH PLATE XVII.

planted on the panels, a bold and effective moulding. The panels are 1 in. bead-flush at the back. An imitation iron studded band (relic of a less secure age) is sunk in the framework around the panels, as shown in the details, f. 1, p. 181. The jamb and soffit linings, which are shown in the half plan, f. 2, have studs or pateras corresponding with the door. The central portions are raised and reeded. These linings, being of considerable width, are probably jointed at the rebates, but it is impossible to ascertain this, as the whole of the woodwork is heavily coated with paint. Two 9-in. fluted pilasters, planted on the face of the wall, and running from the step to the under side of the consoles, give a very substantial appearance to the linings. The fanlight is a thin glazed frame sunk into rebates upon the inside of the soffit and transom, as shown in section, f. 1, p. 181. Originally this doubtless contained a lantern in the middle opening, but this is now plain glazed.

A Small Solid Frame is shown in interior elevation at f. 1 above, and plan f. 2. On the left-hand side the plaster and linings are removed to show the method of fixing. In this case the head is built into the wall and the arch turned directly upon it. In a better class of work a lintel would be provided to carry the core of the relieving arch, and the head of the frame would be fixed

1.

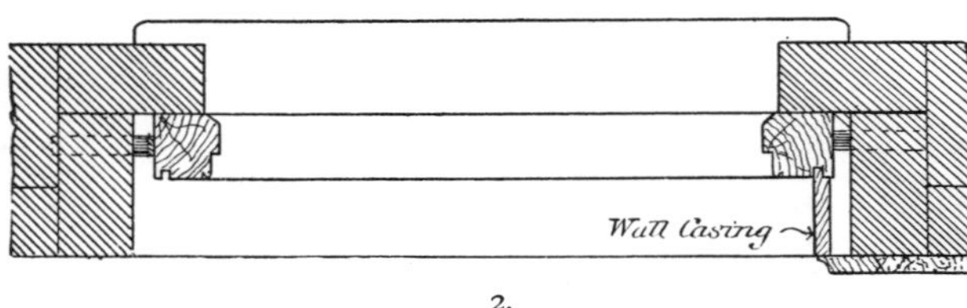

2.

A Solid Door Frame.
1. Elevation. 2. Plan. 3. Plug in Foot of Jamb.
4. Iron Shoe for Foot of Jamb.

by wedging and nailing to the lintel. The foot of the frame, when an oak sill is not framed in, is secured to the stone sill by iron plugs that are driven in the ends of the jambs, and fitted into mortises in the stone, as shown in f. 3.

In damp situations, or where the frame is fixed in a concrete floor without a sill, an iron **Shoe**, cast to the same section as the jamb, is fitted on to the end of the latter, and bedded in the concrete; this is shown in f. 4. Additional fixings are sometimes required when the jamb is not straight, or the frame is large, and also in stone walls when it is not convenient to build the head in. One method of doing this is shown in the drawing. It consists of a piece of bar iron with one end turned up at right angles, and having a hole punched in the end. This is secured to the back of the jamb by either a screw or clout nail at a convenient height, and built into a joint of the wall. The frames are also fixed after the opening is formed, by driving folding wedges between the back of the jambs and the brick-work, and nailing them through the edge of the frame. An air space should be left around the frame to prevent dry-rot, which is frequently set up when frames are fitted closely to green brickwork.

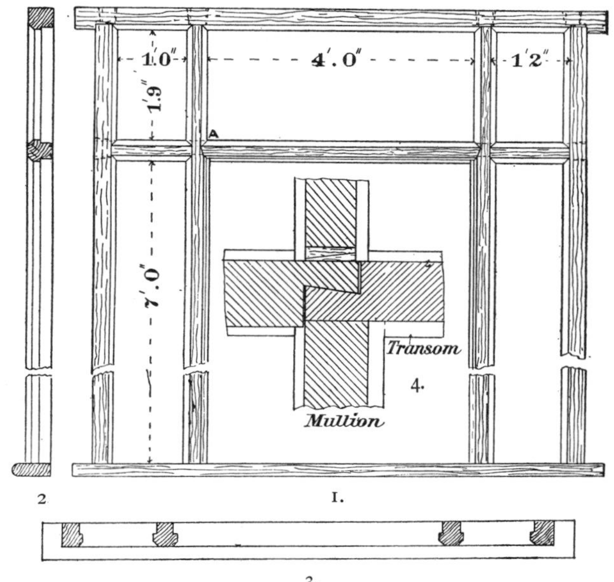

A Solid Vestibule Frame.
1. Elevation. 2. Section. 3. Plan.
4. Detail of Joint A.

A Vestibule Frame is shown in f. 1, 2, & 3 on this page. This is fitted with mullions and transom to receive fixed fanlights and framed sidelights to match the door. Fig. 4 is an enlarged section of the joint at A. This joint is fully described in Chapter VIII. Handrail bolts are sometimes used instead of this form of joint for the purpose of connecting the two portions of the transom together.

A Segment-headed Frame is shown in f. 1, next page, and enlarged details of the joints in f. 2, 3, & 4. The heads of these frames are cut out of the solid when the rise will permit of their being cut from deals of ordinary width. When this cannot be done, they are made in two lengths, jointed at the crown, and fastened with a handrail bolt. The horns are taken out level at the springing line, and the back is made roughly parallel to the shoulders for convenience in fixing. When the frame is 4 in. and upwards in thickness, double tenons should be used, as shown in f. 3; and if the position in which the frame is to be fixed does not admit of the horns being left on, the mortises should be haunched back, as shown in f. 4, although the horns would not be cut off until the frame was fixed, as they would be required for the purpose of cramping up the frame.

1. A Segment-headed Frame.

Tenon Rebate Mortice Horn

2. 3. 4.
2. Detail at Shoulder. 3. End of Jamb.
4. End of Head in Plan.

A Semi-headed Frame may have its head cut out in two or three lengths and bolted together, but is frequently built up as shown in f. 5, 6, & 7 below. The jambs and transom are worked solid, but the head is formed in two thicknesses glued and screwed together, one layer being in two lengths, the other in three, so as to break joint. This is both a strong and economical way of forming a head, because the grain is less cut across than it would be, in a head cut out of one thickness, and the labour of rebating is also dispensed with, the inner ring being kept back ½ in. to form the rebate. The head is fastened to the jambs by hammer-head tenons and shoulder tongues, as shown in f. 6, and double tenons are used for the transom to avoid cutting the root of the head tenons. The transom is better kept about 3 in. below the springing as shown, to ensure a strong joint. But if the exigencies of the design necessitate its being placed at the springing, then the jamb should be carried above the springing, and a portion of the curve worked upon it, because if the two joints are made together, the connection will be very weak.

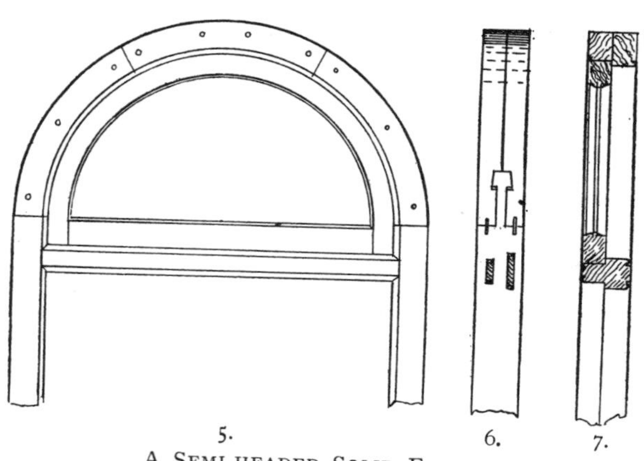

5. 6. 7.
A Semi-headed Solid Frame.
5. Elevation. 6. Edge View. 7. Section.

The construction of the frame for a **Flush-hung**, or, as it is sometimes termed, a "**Warehouse-hung**" **Door** is shown opposite. The door in this case is hung to its outside edge instead of to

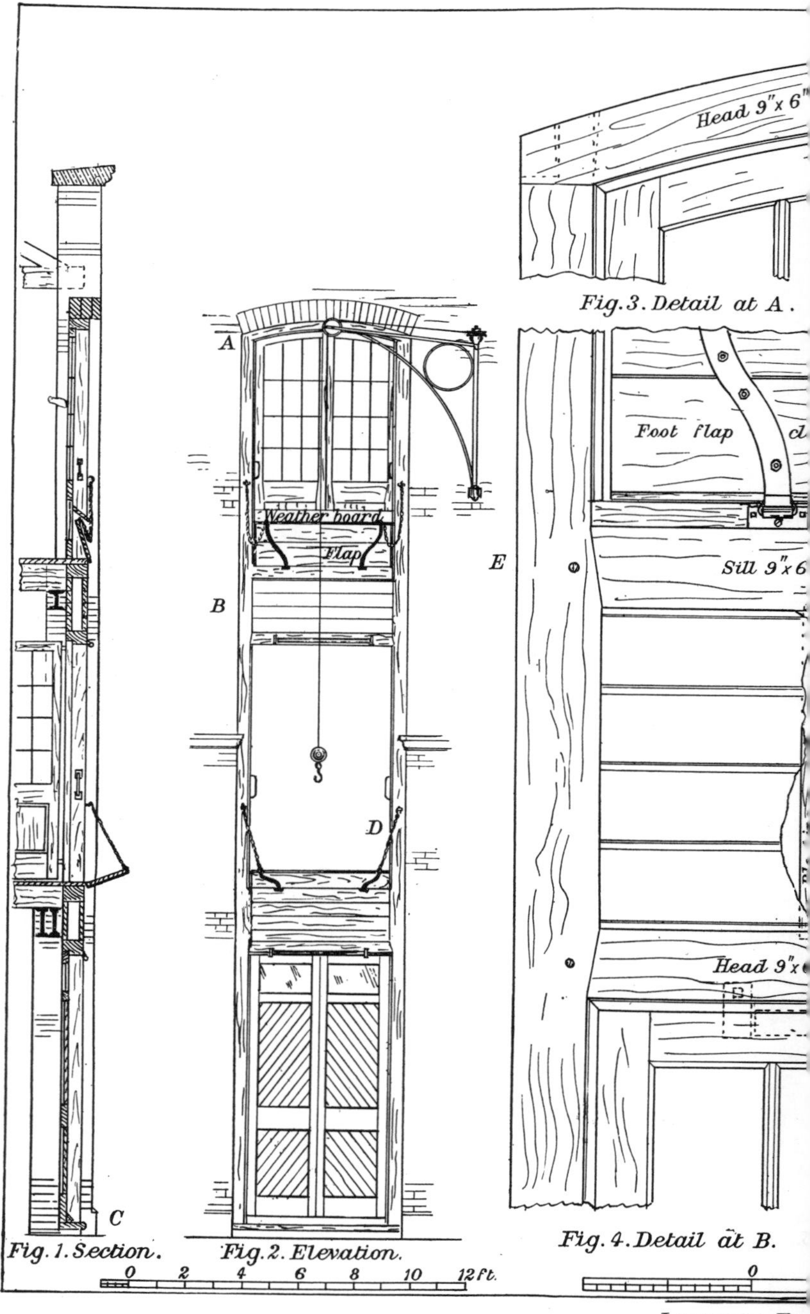

Fig. 1. Section. Fig. 2. Elevation. Fig. 3. Detail at A. Fig. 4. Detail at B.

LOOPHOLE FR

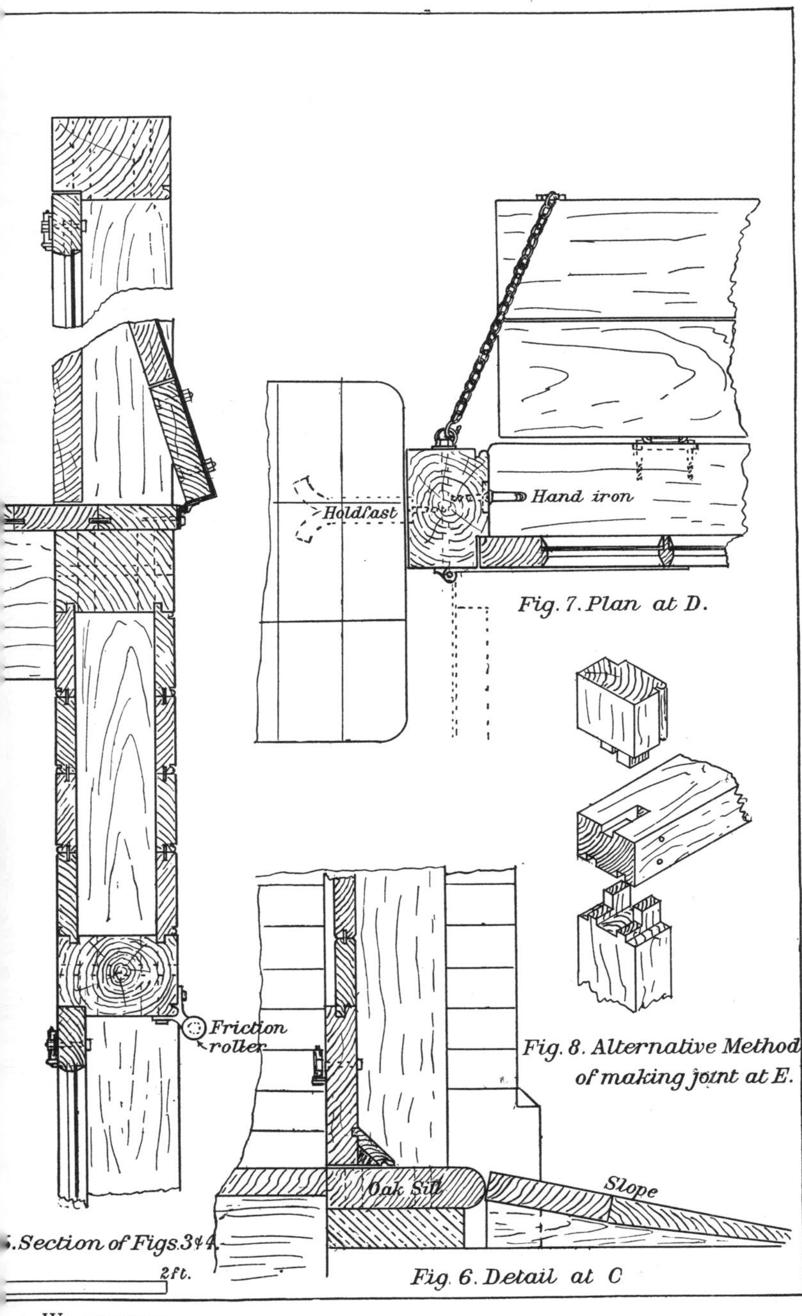

A WAREHOUSE.

[*To face page* 185.

the inside, as is usual, so that when opened it shall show an unbroken surface flush with the jamb. Its position when open is indicated by the dotted lines. The hanging jamb is made with an extra deep rebate equal to the thickness of the door, plus $\frac{1}{8}$ in. for clearance, formed by grooving and tonguing two pieces

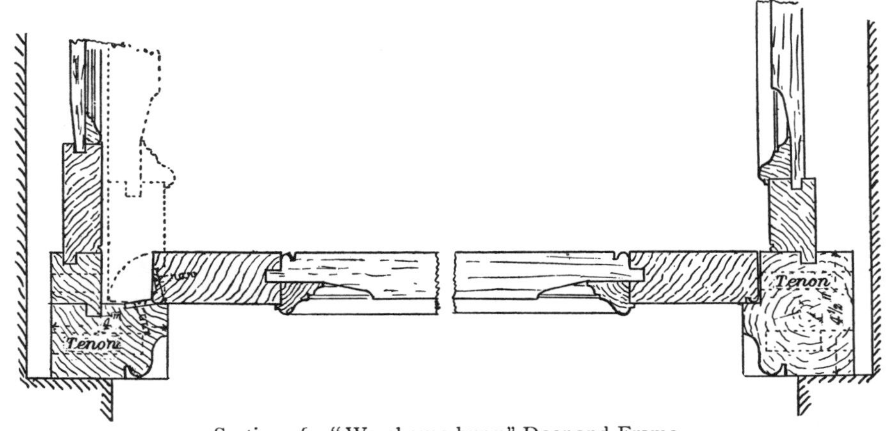

Section of a "Warehouse-hung" Door and Frame.

together as shown. By this construction, apart from its economy, the jamb will be less liable to cast than if rebated out of the solid. The opposite or striking jamb is rebated in the ordinary way. The lobby framing facing the door when it is open is made to correspond with it, that the construction lines may carry across the soffit symmetrically. These doors are chiefly used in offices, banks, &c.

Loophole Frames.—A three-floor loophole frame for a warehouse is shown in Plate XVIII. These frames being subject to much rough usage in the loading and unloading of goods, require to be of substantial dimensions and construction. The jambs are generally made continuous throughout the entire height of the frame, and the intermediate sills and heads are framed between them with double mortise and tenon joints, drawbore pinned. The upper head and the base sill are framed in the way common to all solid door frames, which is clearly shown in the details, f. 3 & 6. The space between the intermediate sills and heads is filled in with a double panel, formed of $1\frac{1}{2}$-in. boards grooved and tongued into the framing; $1\frac{1}{2}$-in. blocking pieces are housed in between the heads and sills, as shown in f. 4, to stiffen the panelling. The heads are arranged to come level with the floor joists, and the flooring is carried across the top. Usually the outside board to which the foot-flap is hung is of oak or teak, the frame itself being of Dantzig fir or pitch pine. A $\frac{7}{8}$-in. bead is run around the outside angle of the frame to remove the sharp arris, and the rebate for the door is either $\frac{5}{8}$ in. or $\frac{3}{4}$ in. deep; the doors are from 2 in. to $2\frac{1}{2}$ in. thick. Usually the upper doors are glazed, and in the case of lofty doors in the ground floor the top panels of these are also glazed. The foot-flaps, which are of elm or oak 2 in. thick, and from 15 in. to 24 in. wide, are hung with wrought-iron strap hinges, so that the top surface is flush with the floor. The outer edges are supported by chains hooked to the sides of the frame, and the flap is kept from 4 in. to 6 in. out of level, so that the persons using it can tell by the feel when they are on the flap. Friction, or guard rollers of iron, are fixed to each head below the crane to prevent the chain chafing away the woodwork. These frames are usually set up, and built into the opening clear between the jambs. At various points as the wall rises, fanged hold-fasts are fixed to the back of jambs and built into the wall as shown in f. 7, the joint between the frame and the wall being pointed in cement. It is sometimes not possible to obtain timber long enough to make the jambs continuous, and they may then be scarfed, or two separate frames made, and connected as shown in f. 8. All the joints are painted with a mixture of red and white lead in linseed oil. The doors are hung, as shown in f. 7, with the pivot of the hinge well back on the frame to cause the doors to swing back flush with the jambs, as shown in dotted lines (this is the real "warehouse-hung" door, see p. 185). The foot-flaps fold up against the doors, and are often covered with a weather board, as shown in f. 1 & 2, to prevent the rain getting down behind them and running in under the doors.

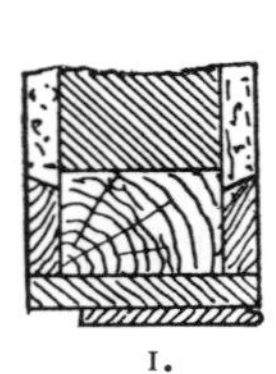

1.

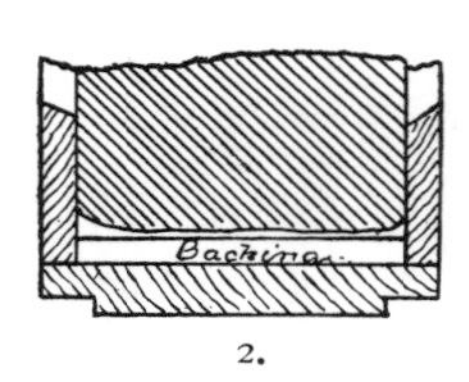

2.

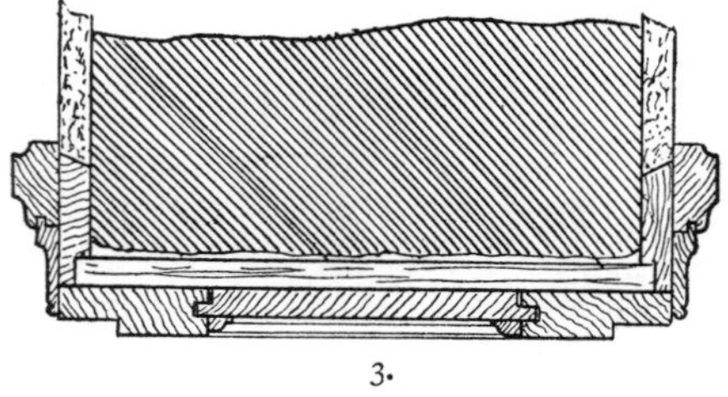

3.

1. A Single Rebated Plain Lining. 2. A Double Rebated Plain Lining.
3. A Double Rebated Framed Lining.

JAMB LININGS.—There are four varieties of these interior door casings, viz., Plain, Framed, Double Framed, and Skeleton Framed, these names defining the method of construction. Other sub-names are also used denoting the nature of the ornamentation, as in doors, with which they agree in the general arrangement of their parts. The term **plain** is applied to any wall lining however it may be treated, if it is made of one flat board or surface.

A "Set" of Linings comprise a pair of Jambs and a Head or Soffit. The flat surfaces, against the edge of which the door rests when closed, are called Stops; and in common work these are merely nailed upon the surface of the main lining, and are kept back from the edge sufficiently far to form a rebate for the door, as is shown in f. 1 above. In better work the rebate is formed in the solid, as shown in f. 2, the lining in such case being thicker. Not less than 1½ in. stuff should be used for any lining to which a door has to be hung, as the rebate takes ½ in. out of the thickness, leaving only 1 in. for screw-hold for the hinges. This is often supplemented by hinge blocks, glued at the back of the lining in a suitable position for the hinges, as shown on one side of the isometric sketch of a set of **Plain Jamb Linings**, f. 1 opposite. When the lining has a rebate on both edges, as in f. 2 & 3 above, it is said to be "double rebated." Plain linings are not suitable for walls thicker than 14 in., in consequence of the amount of shrinkage, which disarranges the finishings; and even in 14-in. work they are better framed.

Framed Linings are shown in f. 3, p. 186, and f. 2 below. These are formed of panelled frames from $1\frac{1}{2}$ to 2 in. thick, and can be made of any width to line any opening upwards of 9 in. thick. The rebates, as will be seen, are worked in the solid upon the edges of the stiles. The frames are put together with mortise and tenon joint, usually stubbed and screwed from the back, because if tenoned

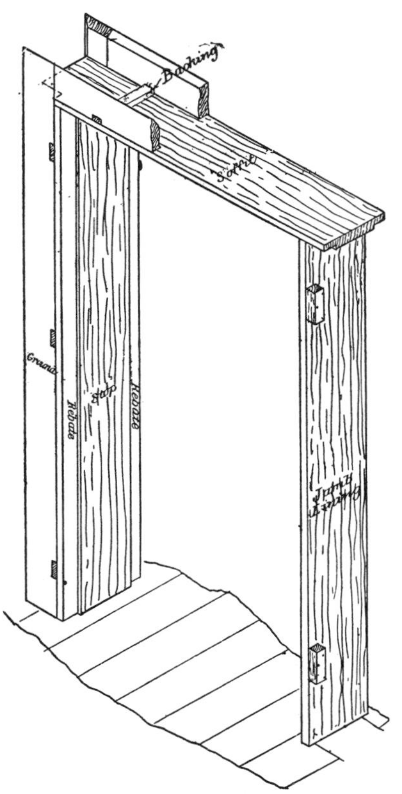

1. Isometric Sketch of a Double Rebated Set of Plain Linings with Double Set of Grounds.

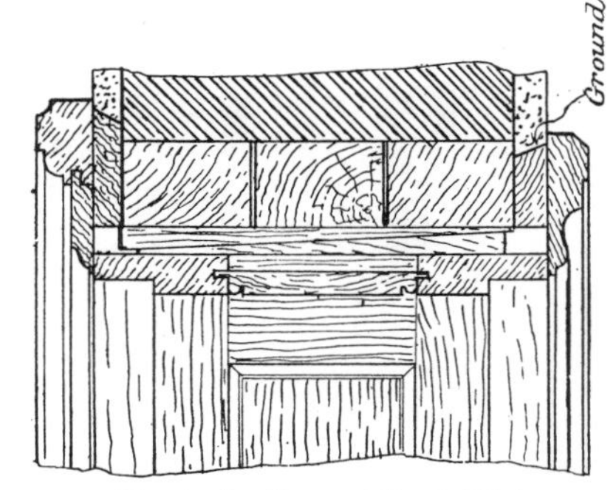

2. Head of a Framed Jamb Lining.

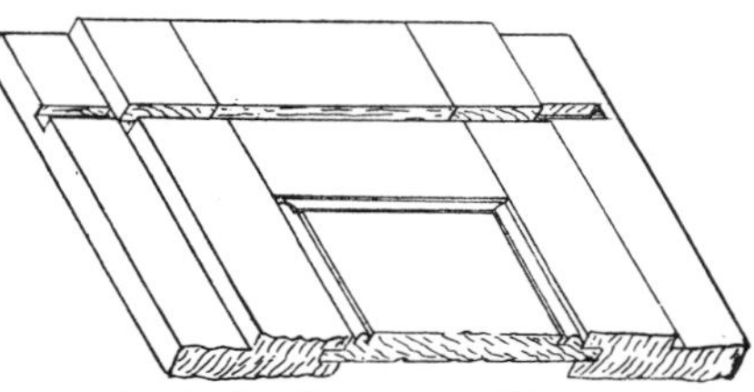

3. End of Soffit showing Housings.

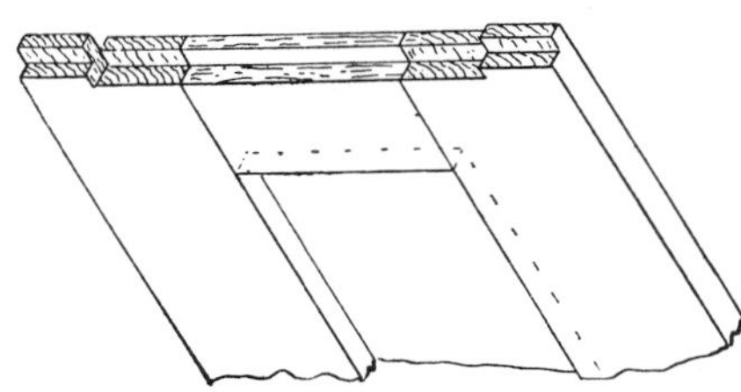

4. End of a Jamb Lining showing Tongues.

through and wedged, when the stile shrinks the ends of the tenons protrude, forcing off the architraves. The method of fixing the soffit to the jambs is shown in f. 3 & 4 on this page. A $\frac{1}{2}$-in. groove is made right across the stop, and a corresponding tongue worked on the end of the jamb. The two are fixed together by nailing. When the rebates are more than $1\frac{1}{2}$ in. wide, these are also grooved and tongued, as shown. The grooves are stopped, in polished work, as shown on the further side of f. 3 above, but are cut right through, for painted work.

A Double Framed Lining is one in which the part forming the stop is framed and panelled, and the rebate on the edges formed with separate pieces of square stuff, connected to it by double grooving, as shown in f. 1 & 2 below. This makes a first class set of fittings, as the stop frame, is tenoned through and wedged (the mortises, of course, are kept below the rebate), and the rebate frame, is made substantial enough to carry any door. The tongues should be fitted hand tight, but not glued, the object being to fix the rebate frame immovably to form a good foundation for the architraves, and to carry the

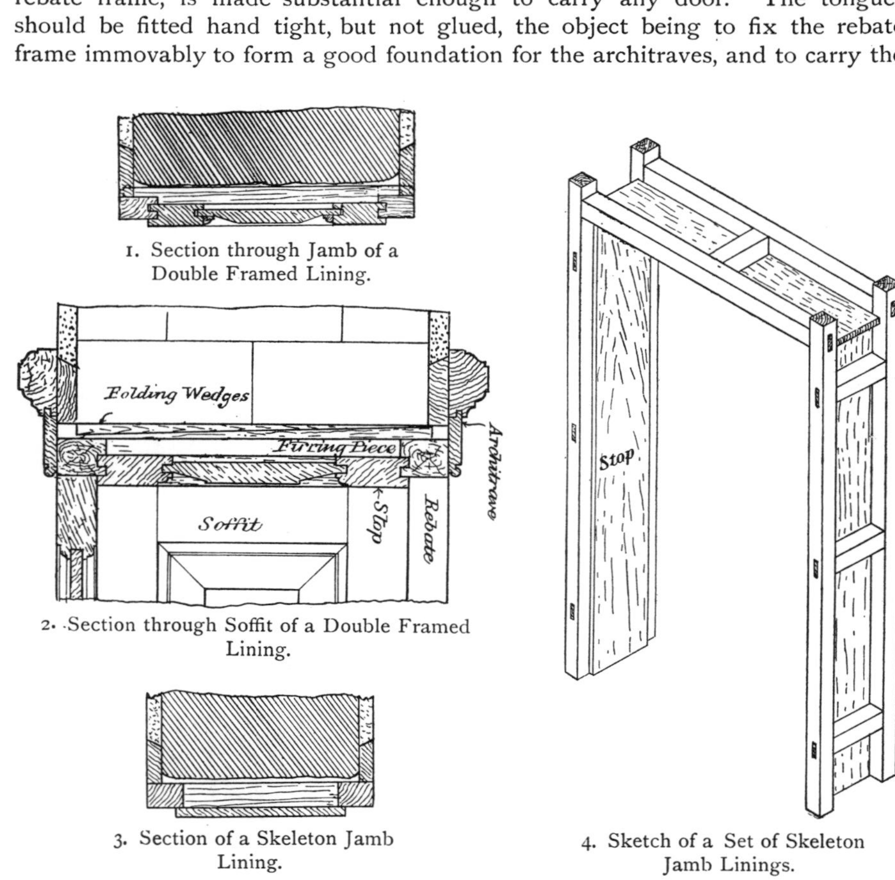

1. Section through Jamb of a Double Framed Lining.

2. Section through Soffit of a Double Framed Lining.

3. Section of a Skeleton Jamb Lining.

4. Sketch of a Set of Skeleton Jamb Linings.

door; whilst the bulk of the framing is free to swell and shrink without disarranging anything. There should be $\frac{1}{8}$-in. clearance between the edges of the grooves and the sides of the panels, when they are rebated in, as shown in f. 3, p. 186, and f. 2 above, to allow for swelling, and in addition in all good class work, the backs of the linings are painted to protect them from dampness.

Skeleton Jamb Linings, f. 3 & 4 above, are so-called because the part forming the rebate, is an open or skeleton frame; the stop, a plain $\frac{1}{2}$-in. board, is simply nailed upon its surface, with the soffit running through, and the jambs butted beneath it. This forms a cheap and fairly good lining, but if used in walls thicker than 9 in. the stop is liable to be split, through the shrinkage; as it must necessarily be fixed at both edges. The linings are fixed to the grounds in various ways, according to their quality and the method of fixing the latter. In polished work of the best class, secret fixing is employed, as will be described later. In ordinary painted work the linings are nailed direct to the grounds, when the latter are fixed as they should be, as shown in f. 3, p. 186, or f. 1 opposite, and in cases similar to f. 2 opposite, to Firring pieces.

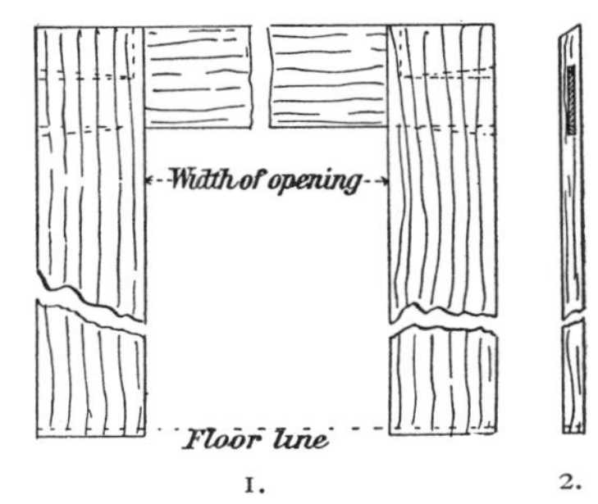

Face and Edge Views of a Set of Grounds.

GROUNDS, f. 1 & 2, are light frames forming a boundary to all openings in plastered walls, their purpose being to act, as the name implies, as a groundwork for the lining, architraves, &c., and also as a margin and gauge for the thickness of the plaster. They are either bevelled or grooved on the back edges to form a key for the plaster, and should be framed up perfectly square, with their faces and inside edges true, straight, and square, to ensure good fitting in the finishings. They are fixed by nailing to plugs or wood bricks in the walls, and should be secure, plumb, and out of winding. They are usually prepared out of 1-in. stuff, their width depending upon the width of the architraves, which should overhang them about $\frac{3}{8}$ in. (as shown in f. 2 opposite). More cover than this is not advisable, or the fixing for the outside of the architrave will be lost, and, for the same reason, much bevel should not be given to the back edge. A $\frac{1}{4}$ in. will provide quite sufficient key for the plaster, and also, as the grounds have to be bevelled before gluing up, a very thin edge is liable to be crushed in cramping them up. The sets of grounds on each side of an opening are connected across the jamb, by 1 by 2 in. slips called Backing pieces, dovetailed, and nailed to the edges. These are shown in f. 2, p. 186, f. 1, p. 187, and f. 1 opposite; their purpose is to tie the grounds together, and also act as additional fixing points for the linings; their presence ensures an air space behind the linings, which is necessary if the work is to remain sound any length of time. Grounds are, however, very often fixed flush with the brickwork, as shown in f. 2 opposite, which does not permit of the use of backings; it is a very inferior method, the linings either touch the walls, or if made smaller, have to be fixed with folding wedges and firring pieces, which do not afford so firm a fixing. When the architrave is of such width that the sides of the ground are required more than 5 in. wide, it is not advisable to make them in one piece, because from their position they are very liable to cast, and so spoil the appearance of the finishings. In such cases, the grounds should be formed as skeleton frames with rails and stiles, where required, from 3 to 5 in. wide; see f. 1 next page.

Architrave Grounds or Moulded Grounds are those moulded or beaded on their inside edges, as shown in f. 2 opposite. They are secured to plugs and to the edges of the linings; which in such cases must be fixed first. The moulded ground is more frequently used in window and shutter openings than in door openings. All joinery finishings are fixed to grounds, but these are seldom framed, except in the cases of door, window, and shutter openings, all other work being fixed to rough grounds, which will be illustrated in their appropriate place.

Framed Grounds are usually wedged up and sent out from the shop in pairs, as shown at f. 3 & 4; the latter is an edge view to larger scale.

1. Portion of Set of Skeleton Grounds.

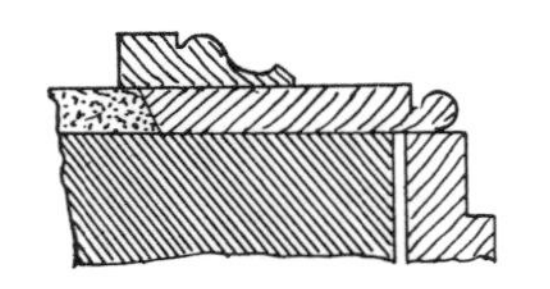

2. A Moulded Ground.

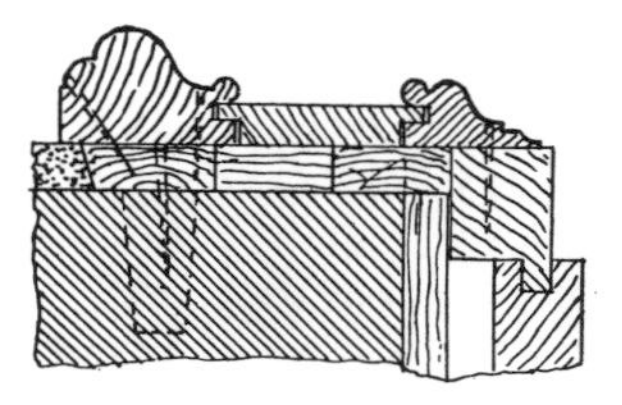

5. Framed Architrave and Grounds.

3. 4.
Method of Arranging "Grounds" for Gluing Up.

Two sets are lightly nailed face to face, with heads reversed, in which position they can be knocked apart and glued, then cramped, squared, and wedged up with facility.

ARCHITRAVES are the moulded borders or frames to window or door openings. When square at the back, as in f. 2, they are termed **Single-faced**; when moulded at the back, as in f. 2, p. 187, **Double-faced**. No architrave should be made wider than 6 in. in one piece, as there is great danger of its splitting when shrinking, as it is necessarily fixed at both edges; but when this size is exceeded, the moulding should be made up of two or more members, grooved and tongued together at some convenient point, as shown in f. 2, p. 187, and f. 5 above, the latter being an example of a very wide so-called **Framed Architrave**; then, if fixed at both edges, it will be free to shrink in the middle.

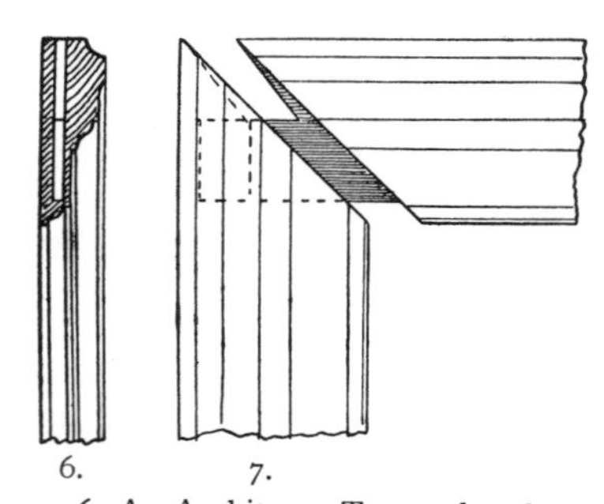

6. 7.
6. An Architrave Tenoned and Mitred. 7. Face of Mitre.

Heavy Architraves are ploughed and cross tongued in the mitres, and in some cases framed or stub mortised and tenoned, as shown in f. 6 & 7 above. These are glued and fixed with screws turned in from the back. Small handrail bolts are also used to draw up the mitres in thick mouldings.

Sets of architraves so treated are framed up on the bench, provided with stretchers at the bottom end, and braced to prevent racking during transit, and are fixed complete. Smaller ones are mitred and fixed piece by piece. In the best class of polished work, secret fixings are used for securing the architrave to the grounds. The method is shown in f. 1, 2, & 3. A number of stout screws are turned into the grounds with their heads projecting uniformly about $\frac{1}{4}$ in. Corresponding holes and bevelled slots are made in the back of the moulding, which is dropped on the screws, and carefully driven down in the direction of the arrow. The architrave should be driven on to the screws dry in the first place; two workmen being employed in the operation, one upon a scaffold, carefully driving down the architrave equally on each side; the other pressing the moulding tightly towards the grounds, and striking a piece of wood with a heavy hammer over each screw, as his fellow strikes the top. After being duly fitted, the architrave may be knocked up again, and the front edges only, lightly glued before replacing.

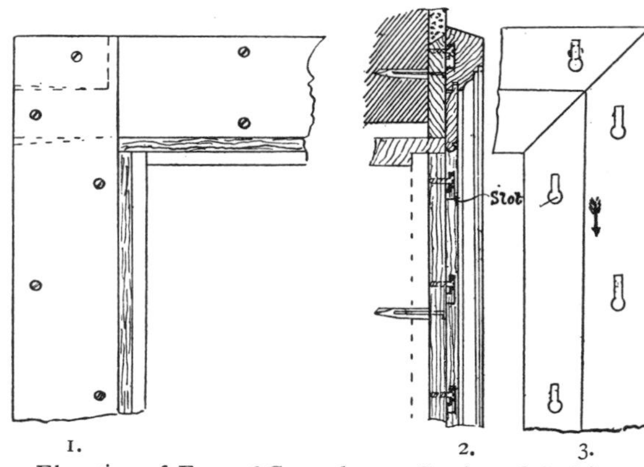

1. Elevation of Face of Grounds. 2. Section of Architrave. 3. Elevation of Back of Architrave.

4. 5.

4. Method of Fixing Plinth to Architrave. 5. A Better Method of same.

Foot or **Plinth Blocks** are used at the bottom of all architraves in good work, and are secured in two ways (see f. 4 & 5). The latter is the better way, as there is less difficulty in fitting the shoulder, and also less danger of splitting the plinth, when driving the dovetail tenon in; this is glued and secured with a screw turned in the back. A handscrew may be used to grip the sides of the block whilst driving the latter on, to prevent its splitting.

CHAPTER XII.

WINDOWS, FITTINGS AND FINISHINGS.

Classification—Materials of Construction—Description of Various Kinds of Windows—Detailed Description of a Double-hung Cased Sash Frame—Constructive Memoranda, Fixing Guard Beads, Ventilating Pieces, Why Sills are Weathered and Throated, Precautions with Sills, Pocket Pieces, comparison of Methods of Cutting: Axle-Pulleys, Position of—Details of Setting Out and Constructing a Sash Frame—Use of Sill Square—Sash Details, Standard Sizes of Parts, How to Deal with the Bars, Joints in Meeting Rails—Sash Lines, how Fixed—Double Windows, construction of Sash and Casement Types—Segmental Window Frames, with Square-headed Sash, with Parallel-headed Sash, Tenoning the Rails—Methods of Fixing Frames Flush with Outside Wall—A Circular-headed Sash Frame, construction of the Sashes by Hand and Machine—Frames, Circular in Plan—Venetian Frames, Single and Double Boxed, Solid Mullion, methods of Hanging the Sashes—A Borrowed Light—French Casements, Details of Construction, Patent Water Bar, Fittings for Casements—New Weather-tight Joints for Sash and Casement Windows—Special Water Bars and Joints for Casement Sills—Bay Windows, A Solid Frame with Folding Shutters, A Cased Frame with Circular Boxing Shutters, construction of the Roof—Centre-hung Casements, method of Hanging, How to Cut the Beads—A Bull's Eye Frame—Hopper and Hospital Lights—Yorkshire Lights—Dormers—Framed Skylight—Construction of Lanterns, Forming Shoulders of Angle Posts, Obtaining Shape of Top Lights, To Find Length and Backing of Hips, To Obtain Section of Hip, Ridge, and Bars in a Moulded Roof, To Obtain the Mitre Cuts for Hips, &c.

WINDOW FRAMES may be broadly separated into two classes, SOLID and CASED, according to the method adopted in their construction. Those in which the lights or sashes are hung to counterbalancing weights, having to be provided with a box or receptacle for the weights, are built up with a number of thin linings. These are termed cased or boxed frames, whilst those in which the lights are hung, either by hinges or pivots, are made solid throughout. "Cased" frames are occasionally made with solid heads, and solid sills are used invariably in both classes. Frames of both types for ordinary purposes are made of yellow deal, *i.e.*, Baltic pine, and the sills of oak, which should be one of the European varieties, the American kinds are too soft and unenduring for the purpose. For superior work, oak, teak, or pitch pine is used instead of deal for the heads, jambs, and linings, and teak for the sills. The latter wood is superior to all others in its resistance to weather and freedom from casting. The sashes are generally made of deal, although oak and mahogany are sometimes used in first-class work; the latter wood is of good appearance, but will not weather well, unless kept painted upon the outside. Each of the above-mentioned classes comprise several varieties, approximating in their general construction, but with differences in detail, which have obtained them special names. For instance :—

CASED FRAMES include boxed **Sash Frames**, f. 1, 2, & 3 overleaf, with two sashes hung with weights and sliding vertically. These are further distinguished as "single" or "double" hung, as one or both sashes slide. They are also known by the number of lights the bars divide the sash into, the one in question being a "four-light" window. A cross bar in each sash would make it an "eight-light" window. The shape of the heads also serve to distinguish these frames, such as Circular, Segmental, Gothic, Elliptic, &c. TWO-LIGHT or MULLIONED FRAMES contain two pairs of sashes in the same plane. DOUBLE WINDOWS consist of two or more sashes or casements lying one behind the other in the same frame, p. 200. VENETIAN FRAMES, p. 206, contain three pairs of sashes lying in the same plane. BAY WINDOWS are composed of three frames, either cased or solid, two being placed at an angle, usually 45 deg., with the centre one, as shown in Plate XXI. A BOW WINDOW is a curved bay.

SOLID FRAMES include **Casement Frames**, p. 209—when the lights are hung with hinges they are termed casements; when these run to the floor, so that they may be used as doors, they are known as French casements; SOLID BAY FRAMES, Plate XX.; PIVOT FRAMES, p. 215, also called centre hung sashes, when the pivots are placed horizontally, as shown. Similar frames, circular in elevation, are called BULL'S-EYE FRAMES, p. 217; HOPPER or HOSPITAL LIGHTS, p. 217; YORKSHIRE or SLIDING LIGHTS, p. 217; DORMERS, vertical roof lights, p. 217; SKYLIGHTS, inclined roof lights, p. 219; and LANTERNS or raised skylights, p. 221.

A Double-Hung Cased Sash Frame of good construction is shown in elevation in f. 1, next page, vertical section f. 2, and horizontal section f. 3; and f. 1 & 2, p. 195, are enlarged details of the same. The names by which the various parts of a sash frame are known are indicated upon the illustration, as follows:—H, the head; OL, outside lining; IL, inside lining; P, the pulley stile; C, sill; V, ventilating piece; G, guard or staff bead; J, parting beads; K, parting slip; N, pocket piece; BL, back lining; U, grooves for jamb lining; T, top rail of sash; M, the meeting rails; B, the bottom rail; S, stiles. An inspection of the details will show that the various members are all grooved and tongued together. This method is now universal, except in the very commonest description of work, as the additional cost of grooving and rebating is very trifling when machinery is employed; but for the sake of completeness, the construction of a cased frame without tongues has been shown, f. 3, 4, & 5, p. 195. The parts are merely nailed together, and it is a very inferior method. Fully dimensioned details of the finished sizes required for the different thickness of sashes, &c., are shown on pp. 195, 199, & 200.

Constructive Memoranda.—In the best class of work the guard beads are rebated into the pulley stiles, &c., as shown in f. 1 & 2, p. 195, and are fixed with cups and screws. This ensures that the sash will always travel freely, as the beads once properly fitted, will not bind on the sash when replaced, as is too often the case when they are simply bradded in square. Whether rebated or not, the bead should always be wider than the thickness of the lining, so that the

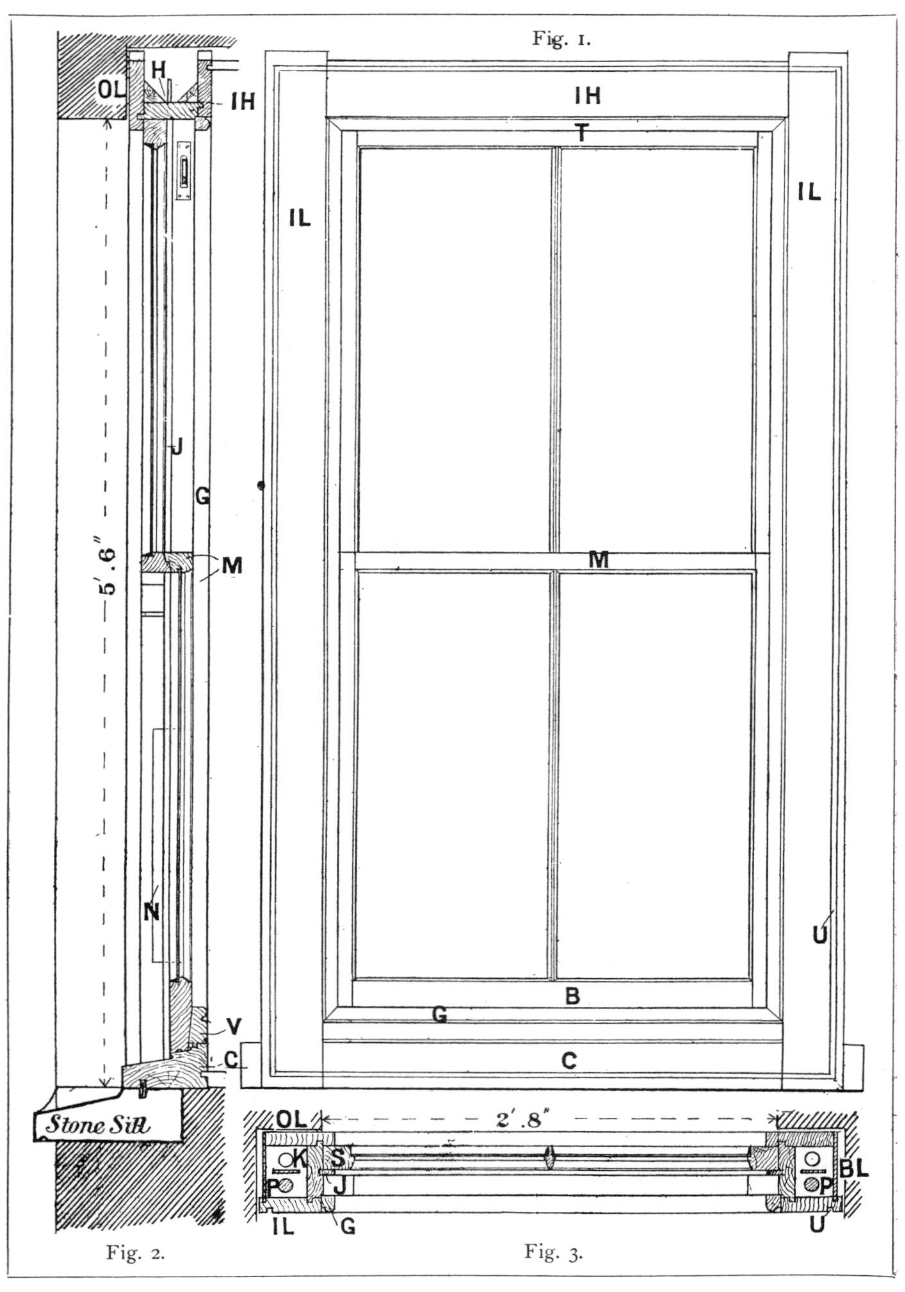

A CASED SASH FRAME.
1. Inside Elevation 2. Vertical Section. 3. Horizontal Section.

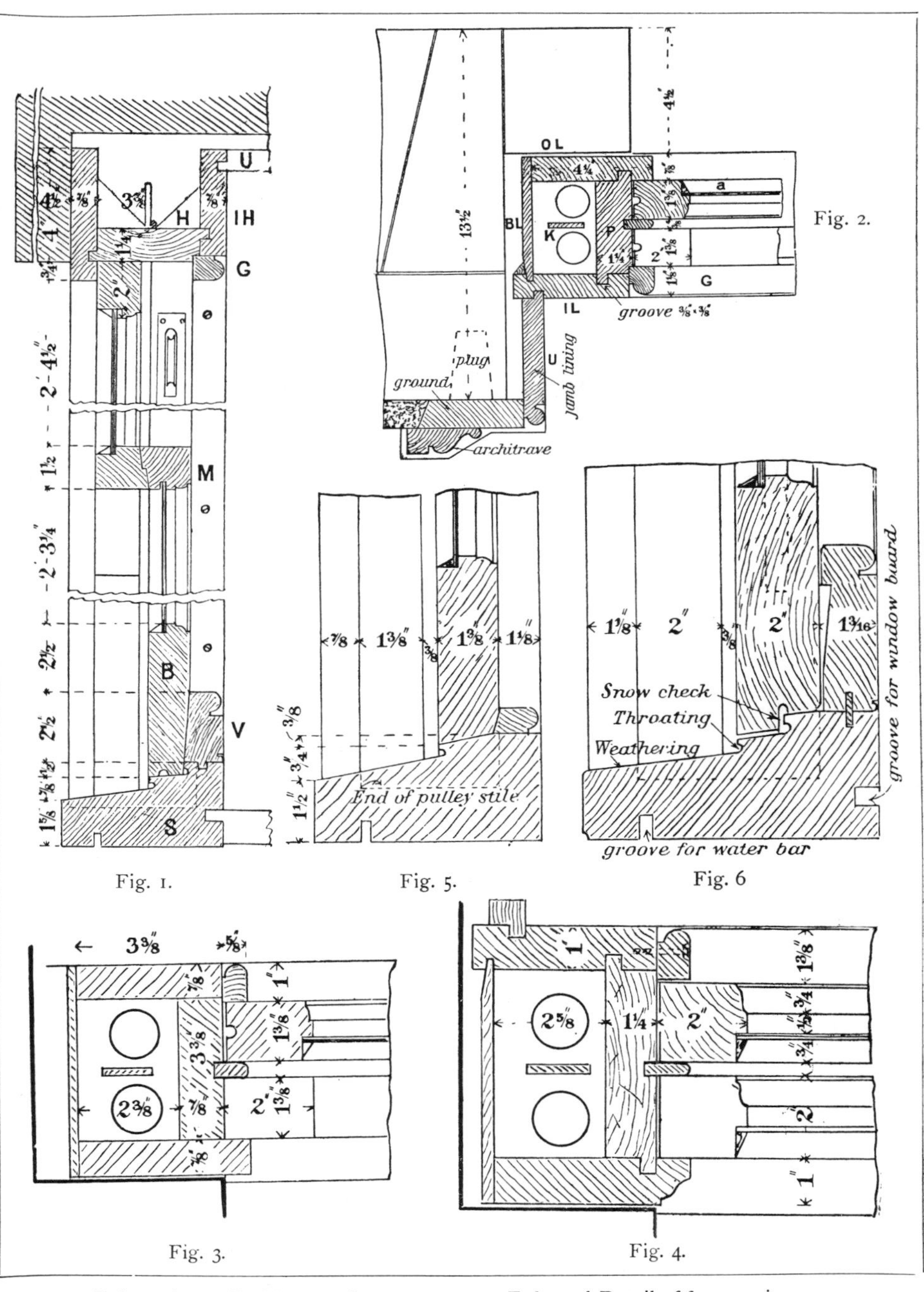

1. Enlarged Detail of f. 2, previous page.   2. Enlarged Detail of f. 3, previous page.
3. Section of Casing of a Common Sash Frame with $1\frac{1}{2}$-in. Sashes.
4. Section of Casing of a Superior Sash Frame with 2-in. Sashes.
5. Part Vertical Section of Frame, f. 3.   6. Part Vertical Section of Frame, f. 4.

joint of the latter may be covered, as shown in f. 3 & 4, previous page. The sill bead is advantageously made $\frac{1}{8}$ in. wider than the side ones, and bevelled as in f. 5, p. 195, so that the sash when down is pressed tightly between it and the parting beads, to prevent rattling; and prepared thus, the edges of the sash and bead are also less liable to be knocked off. When ventilating slips are used, these also should be bevelled for similar reasons, as shown in f. 1, p. 195, although some joiners object to this on the ground that it causes draught when the sash is slightly open. By these the section shown in f. 6, previous page, is preferred, and in this case the arrises of the bead and sash should be taken off as shown. The purpose of the **Ventilating Piece** is that the sash may be opened a few inches, admitting air between the meeting rails, but avoiding a direct draught at the bottom.

Sills are weathered—that is, made to slope towards the outside—to throw off the rain-water readily, and **throated**—that is, grooved in the edge of the rebate to

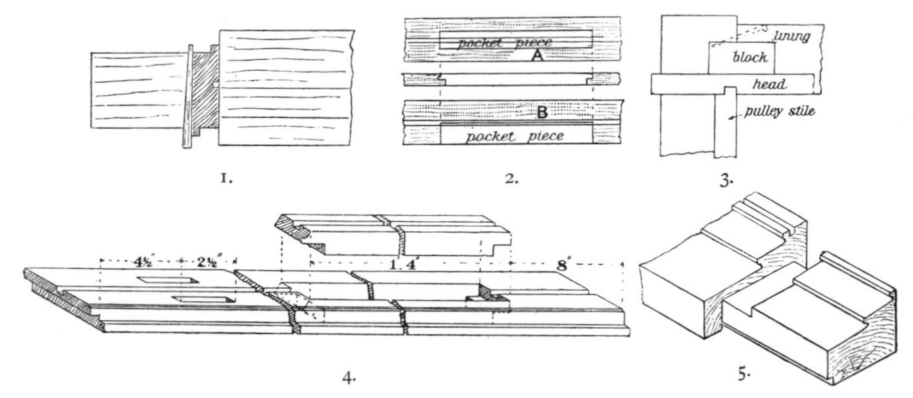

1. Method of Fixing Pulley Stile in Sill. 2. Two Methods of Cutting Pockets.
3. Joint in Head of Sash Frame. 4. View of a Pulley Stile and Pocket Piece.
5. End of Oak Sill showing Sinkings.

break the flow and prevent the water being drawn between the sash and sill by capillary attraction. To be effective the Throating (see f. 6, p. 195) should not be less than $\frac{1}{4}$ in. wide, but its depth is of less importance. The sill shown in f. 2, p. 195, is termed "twice" or "double" weathered, rebated, and throated. The sash is also shown throated in the rebate; this is sometimes termed a "snow-check." The outside rebate of the rail should not quite touch the sill, thus the space between the throats gets choked up with snow, which prevents more drifting through. The sinking in the ventilating piece, is to reduce the friction between it and the sash; it also acts as an additional check against water getting in. Sills should be grooved at the bottom to receive a water bar of 1 in. by $\frac{3}{16}$ in. galvanised iron, bedded with white lead in a similar groove in the stone sill. The groove should be kept as near the outside edge as possible, to preserve the bulk of the sill from rotting; the best position, is in line with the outside linings, which when sunk across the sill will form one side of the groove. A strip of sheet lead is sometimes used instead of an iron bar, and if of sufficient substance is more lasting than the latter. The sill should be made $\frac{1}{8}$ in. wider than the rest of the frame on the outside, to compensate for its greater shrinkage; if finished off flush, it will eventually shrink below the linings, exposing their unpainted edges to the weather. Pulley stiles are housed, wedged, and nailed into the sill, as shown by the dotted lines in f. 1 & 5, p. 195, and in plan in f. 1 opposite. This sinking, which should be made as small as possible, consistent with providing for sufficient substance in the wedge to drive without breaking; is taken down until it is $\frac{1}{4}$ in. below the weathering at the outer edge, as shown in the sketch, f. 5 opposite; more than this is unnecessary. The upper end of the stile is housed and nailed, or preferably grooved and tongued into the head, as shown in the section, f. 3. A pocket hole is cut in the lower part of each pulley stile, for the insertion of the weights. This is covered by the **Pocket Piece**, made to fit accurately, as shown in sketch, f. 4 opposite. It is rebated at each end about $\frac{1}{2}$ in., the top end being undercut, so that it will not check the descent of the sash should it accidentally project. The size of the hole varies; it is usually $\frac{1}{4}$ in. wider than the diameter of the weight, and $2\frac{1}{2}$ in. shorter in the clear than the latter. Pockets in inferior work are cut through from the inside edge of the stile, and the piece taken out is, by gluing pieces behind the end rebates, utilised for the pocket piece; a saw cut is run down the parting groove, and both edges are thus covered by the beads, which is its only advantage. The ends are cut with the special chisel shown in f. 4, p. 21 *ante*, and the rebate is formed by splitting through the edge, but the method is not to be recommended, as in addition to seriously weakening the pulley stile, the inside lining cannot be nailed throughout the length of the pocket. In the former method, when the frames are to be painted, a small bead or chamfer should be run on the outside edge of the pocket piece, to cause a break in the joint, so that the paint may not be damaged by the subsequent removal of the pocket piece. Fig. 2 opposite illustrates both methods, A being the centre cut, and B the side cut method, the section being the same in each.

The Axle Pulleys are sunk in flush with the face of pulley stile, the top of the plate being kept from $1\frac{1}{2}$ to 3 in. down from the head, according to length of sashes. The mortises are made of such size that the cases will just go in without rattling, clearances being cut for the bushes at the time of insertion. The mortises must be made exactly in the centre of the path of each sash, and are best set out equally on each side of the parting-bead groove.

Setting Out and Making a Sash Frame.—Considerable care must be taken in setting out the pulley stiles, as upon their accuracy depends the proper working of the sashes. The width of the pulley stile regulates the thickness of the frame, and is itself regulated by the thickness of the sashes, which are generally prepared in reputed sizes, viz., so-called $1\frac{1}{2}$-in. sashes are made from one cut stuff, which finishes, with care, $1\frac{3}{8}$ in. bare; reputed $1\frac{3}{4}$-in. sashes finish $1\frac{5}{8}$ in. bare; and 2-in. finish $1\frac{7}{8}$ in. bare. Hardwood frames should have the paths set out nett to the above *real* dimensions, and painted frames full, to the same. For instance, taking the stile in f. 2, p. 195, commence gauging from the inside edge, run a gauge line down on the face $\frac{3}{8}$ in. on, for the tongue; then another $\frac{1}{4}$ in. further on for the guard bead rebate; then one $1\frac{3}{8}$ in. full from the last; then $\frac{3}{8}$ in. nett for the parting bead; then another $1\frac{3}{8}$ in. full, but this line is gauged upon the *back* side; then a further $\frac{3}{8}$ in., which is the full width of the stile. If all the gauging is thus done from the face edge, any difference that may exist in the linings, &c., will not affect the working of the sashes. The gauging for the depth of the rebates should also be made from the face of the stile. Note, in a machine shop one stile only need be gauged, however many there may be required of the same setting. In working by hand, all similar grooves and rebates should be finished throughout, before the tool is altered (see also p. 74).

The length of the pulley stile is set out from the housing in the sill (which should be marked upon the rod as described previously) to the under side of the head, plus $\frac{3}{8}$ in. for a tongue. The width of the frame between the pulley stiles is set out upon the sill, and this line is squared right round for the shoulders of the linings, a useful tool for the purpose of marking the top being the **Sill Square**, shown in plan and section in f. 1. This consists of a piece of deal a trifle wider than the sill and about 10 in. long and 2 in. thick, rebated to fit the sill, with its ends made quite square with the front edge, which has a narrow fence screwed on, as shown. The end of the sill, when cut for the linings, presents the appearance of f. 5, p. 195, and the outside lining is notched out as at *a*, f. 2 below, in line with the groove, which allows it to shrink without exposing an open joint to the weather. It may be remarked that in the case of a frame rebated for a bead, as in f. 3, p. 194, the inside shoulders of the sill would require to be the depth of this rebate longer than the sight line.

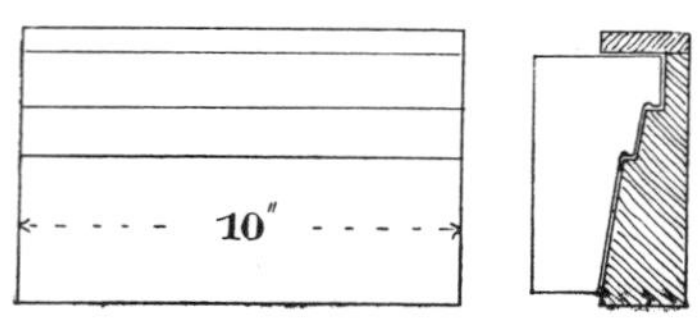

1. Plan and Section of a Sill Square.

In **Assembling** the parts of the frame, the pulley stiles are first wedged and nailed in the sill, out of winding with each other. Next the head nailed on, and the skeleton frame laid on the bench, inside uppermost, in the position shown in the sketch, f. 1 next page, the sill being screwed to a piece of quartering fixed near the end of the bench. It is then squared by testing its lengths diagonally with a rod, and fixed by a nail driven temporarily through the head. The inside linings are next fitted and nailed on (if painted, through the face; if polished, skew-bradded through the edge), the head lining being kept flush at the back with a skew nail, as shown in f. 3, p. 196; the inside is then planed over and the guard beads mitred in. These should be stop mitres, as shown in f. 2, next page, the head and sill beads running through, and the side ones notched over them. The frame is next turned over, re-squared and re-fixed, and the parting beads cut in; these should go in hand-tight. The outside linings are then fitted, the lower end cut first, as above described, to fit the sill, then tapped on the tongues; one end of the head lining squared off and entered on the tongue, so that its inner edge may be scribed on the side-piece, which is then taken out and mitred, as shown at *b*, f. 2 above. Before nailing these on, the lower end must be smoothed up, as the sill projects in the way of the plane; the head lining is

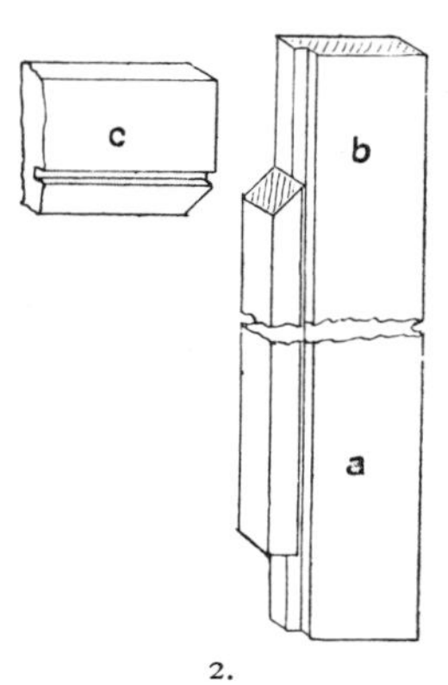

2. Detail of Sash Lining.

next marked and cut tightly between the shoulders, mitred back to the groove, as at *c*, f. 2, previous page, and nailed in ; the parting slips inserted, and the back

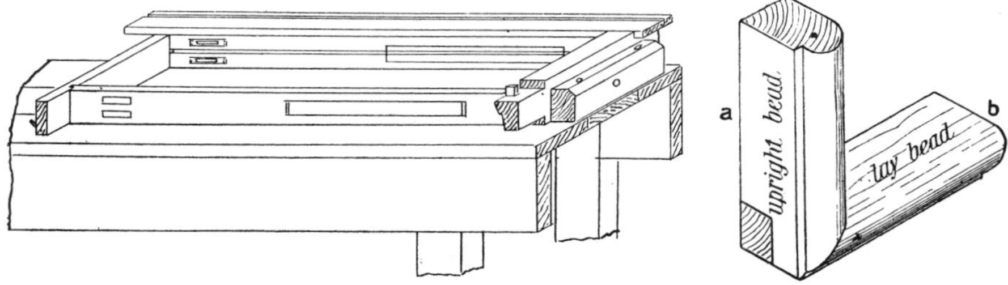

1. Method of "Putting Together" a Sash Frame. 2. Stop Mitre in Bead.

lining nailed on. A narrow margin is then planed up all round, the head blocked, and the frame is ready for fitting the sashes.

Various details of sashes are given in f. 4, 5, & 6, below. The stock sizes of moulding planes are, irrespective of section, for $1\frac{1}{2}$-in. sashes, $\frac{1}{2}$ in. on, $\frac{3}{16}$ in. down : size of the square, and of mortise chisel to be used, $\frac{5}{16}$ in.; $1\frac{3}{4}$-in. sashes, $\frac{5}{8}$ in. on, $\frac{3}{16}$ in. down, square $\frac{3}{8}$ in.; 2-in. sashes, $\frac{3}{4}$ in. on, $\frac{1}{4}$ in. down, and $\frac{1}{2}$ in. square; depth of rebate in all, $\frac{1}{4}$ in. Bars, $\frac{5}{8}$ in.; depth of rebates, $\frac{3}{16}$ in.; stouter bars, $\frac{1}{4}$ in. deep. Stiles and top rails are usually 2 in. wide; meeting rails, $1\frac{1}{4}$ to $1\frac{1}{2}$ in.; bottom rails two and a half times the width of top rails.

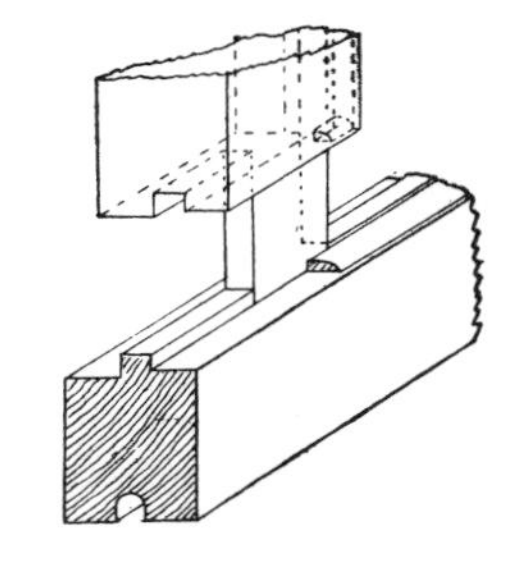

3. Method of Haunching Sashes.

Top and bottom rails are haunched in the manner shown in sketch, f. 3. The top meeting rail is mortised its full width when a bracket is used, as shown in f. 4 & 5.

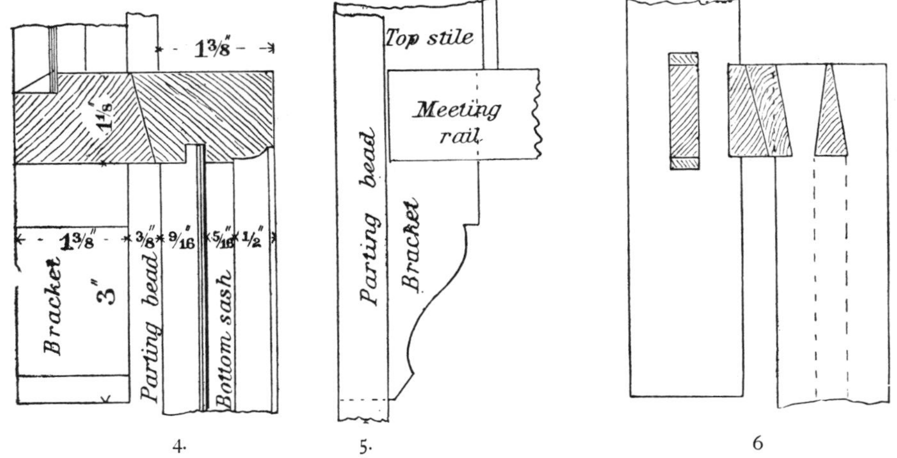

4. Section of Bevelled Meeting Rails. 5. Side Elevation of f. 4 showing Bracket.
6. Outside Edges of Sashes, f. 4.

The bottom meeting rail is now generally dovetailed, as shown in f. 6, p. 199—formerly they were slot tenoned and screwed; the dovetail occupies the same position as the tenon, and the overhanging portion of the rail should be housed across the stile in a dovetail sinking. Meeting rails are necessarily $\frac{3}{8}$ in. thicker than the stiles, &c., to cover the space between the two sashes occasioned by the parting beads; and the oversailing portion is either plain bevelled as in f. 6, p. 199, bevel rebated as in f. 1 & 3, or double tongued or hook jointed as in f. 2. When the joint is bevelled, it is not advisable at first to work the bevel quite down to the level of the stile at its lower edge; but leave it standing $\frac{1}{32}$ in. for fitting, which may then be done with the rebate plane, without damaging the stiles. Fig. 3 shows the rebate in the end of the meeting rail to accommodate the parting bead.

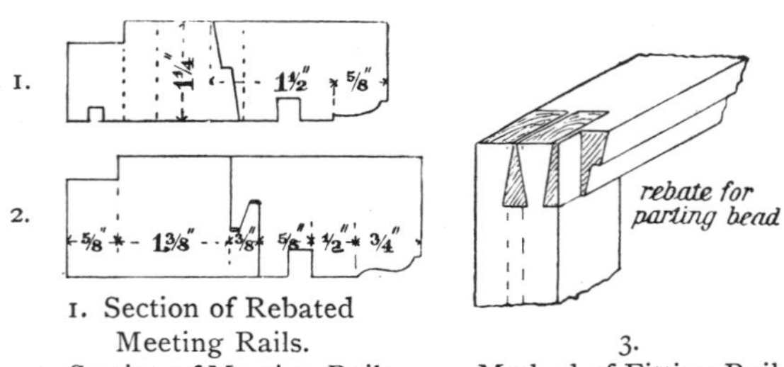

1. Section of Rebated Meeting Rails.
2. Section of Meeting Rails with a Hook Joint.

3. Method of Fitting Rail over Parting Bead.

Brackets, shown in elevation, section, and edge view in f. 4, 5, 6, previous page, are usually formed on the stiles of top sashes, and occasionally upon bottom ditto, when they are large, to strengthen the joint of the meeting rails. When cross bars are used, the upright bar should run through, and the horizontal bar be cut, otherwise the lifting of the sash will be liable to break the joint in the centre. It is advisable in making sashes to take all mortises for the bars right through, so that they may be wedged and straightened if crooked. Drive the wedge on the hollow side first. When a bar is to be cut into several short lengths, stick and rebate it in one length, scribe also, and do not cut off until ready to put together, when each shoulder should be marked so as to "follow on."

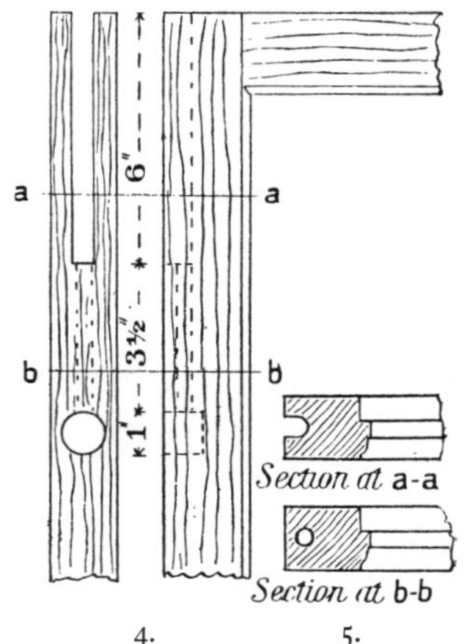

4. Side and Edge of Sash showing Groove, &c., for Line.
5. Sections of Stiles.

Sash Lines in common work are fixed in a plough groove in the back of the stile by means of clout-headed nails, but in good work they are passed through a hole in the bottom of the groove and knotted, a hole being bored lower down to receive the knot, as shown in the elevations and sections of a sash stile, f. 4 & 5.

Double Windows are used chiefly in places where there is much traffic, to exclude noise and dust. They may consist of two pairs of sashes, one in front of the other in the same frame, as shown in the section, f. 5 & 6, opposite, or one pair of sashes with a single or a pair of casements in front, as shown in the section,

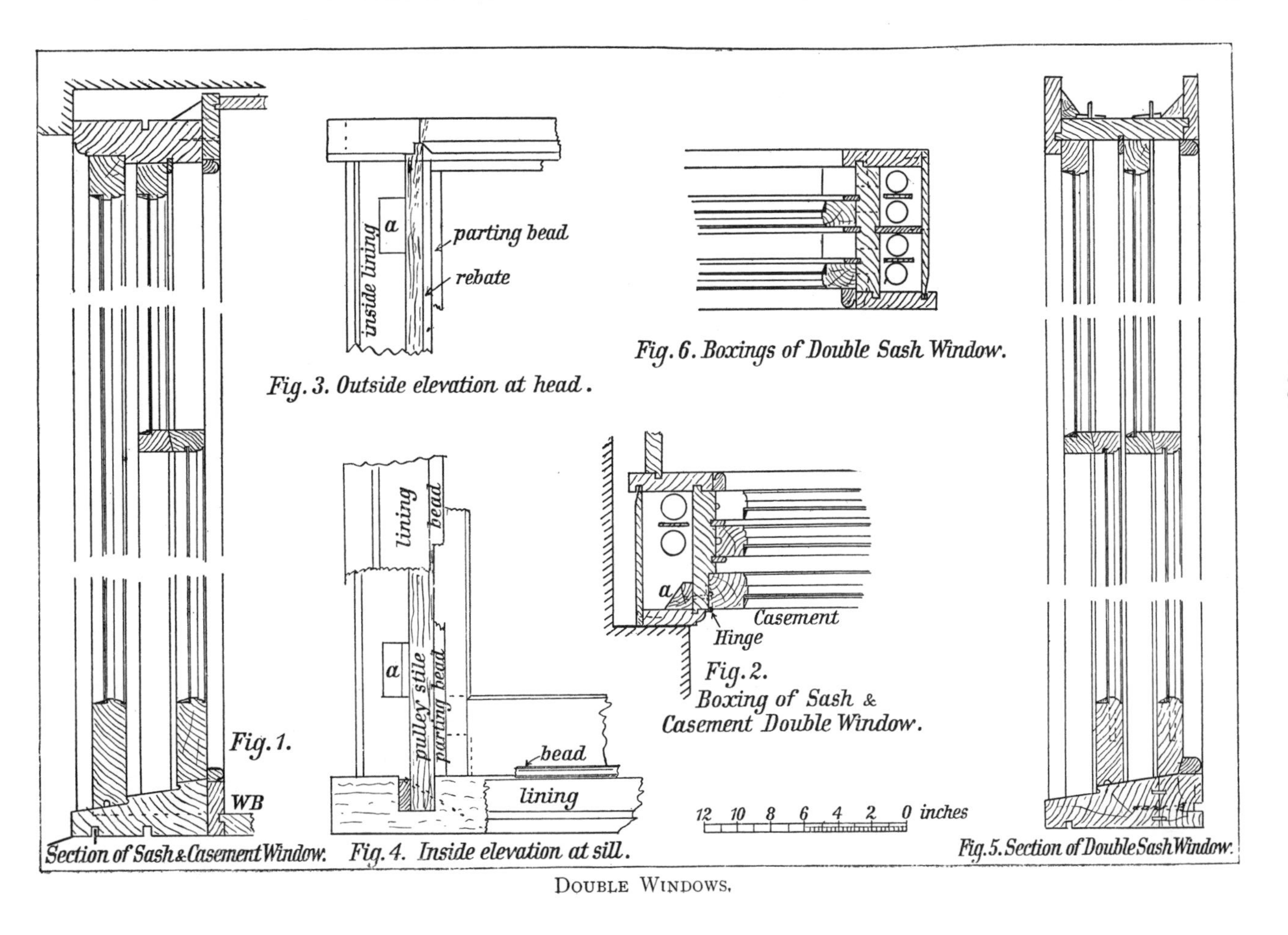

DOUBLE WINDOWS.

f. 1 & 2, p. 201. The chief constructional difficulty met with in these frames is due to the great width of the sill and head, which are liable to cast, causing the lights to jam. This is especially the case with the sash casement window type, f. 1, and to prevent it the inside linings may be planted on, thus reducing the width of the solid members. The boxing linings are in all cases planted on in the usual manner, as shown in f. 2. A wide plough groove made on the backs of sill and head, as shown in f. 1, will also assist in keeping them true. Where two pairs of sashes are used, as in f. 5, the sill and head are necessarily still wider, and the former may be jointed and screwed together as shown. By judiciously arranging the grain so that the circumferential shrinkage of each piece will

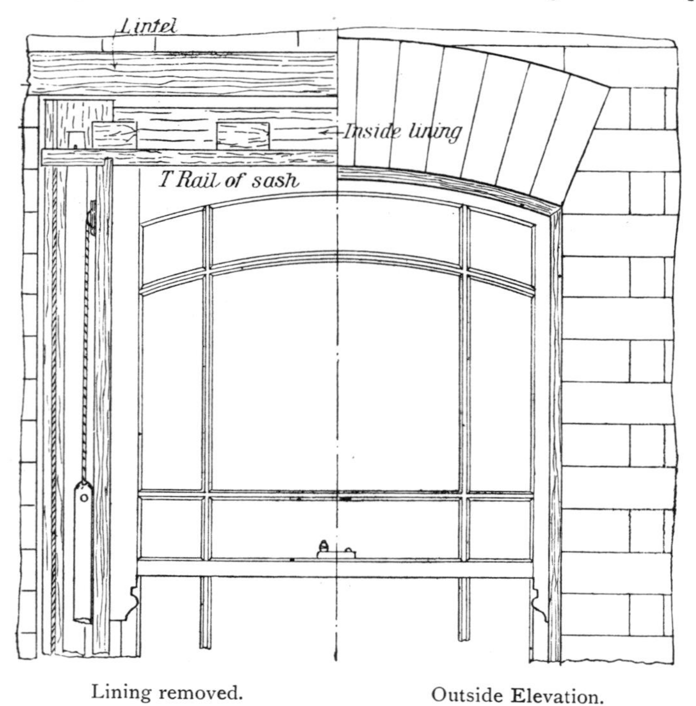

Lining removed. Outside Elevation.

A Segment-headed Sash Frame.

counteract that of the other (this is fully explained on p. 418), the sill may be kept perfectly true; the head, being thin, and well nailed to the pulley stiles, will not be able to cast much. Fig. 3 is an outside elevation at head of the frame shown in f. 1, with the outside lining removed. The block shown at *a* in the several drawings is placed opposite the position of the hinge of the casement, and is glued on the stile before fixing the lining. Fig. 4 is an inside elevation of the lower end of frame, with various parts broken away to show the construction. Fig. 6 is a section through one of the boxings of the frame shown in f. 5; the dotted lines indicate the position of the pocket pieces.

A Segment-head Sash in a **Square Headed Frame** is shown above.

The frame in this case, with the exception of the outside lining, which follows the sweep of the sash head, is made similarly to the ordinary window frames described at pp. 193 & 196, and has the same appearance on the inside. The outside lining and the top rail of the sash, however, are much wider, the under edges being shaped to correspond with the sweep of the arch, giving it the appearance shown in the half outside elevation. In the left-hand half of the drawing the outside lining is supposed to be removed, showing the construction. In making the sash, the tenon of the top rail is kept up a short distance from the inside edge, as shown in the detail alongside, to keep it clear of the short grain. Bars placed near the sides of the lights, as shown in sketch opposite, are termed marginal bars.

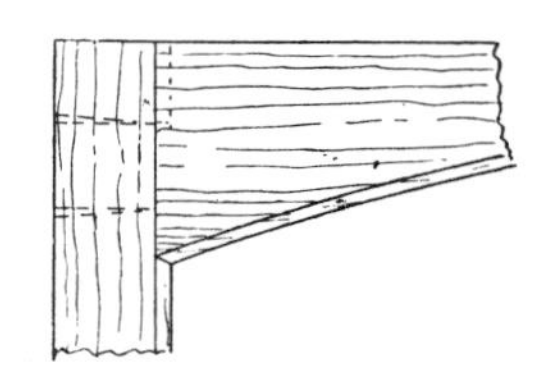

Method of Tenoning Rail in Segment Sash.

A Segment-Headed Sash Frame fixed at the outside of the opening is shown on next page. In this case the head of the frame is shaped to follow the sweep of the arch, and the sash rail is parallel ; the head is built up in the solid, as described in detail, in the next case. The outside head lining is shaped to fit the opening, but the inside lining is trimmed straight to fit under the lintel ; apart from this, there is nothing calling for explanation. In f. 4 & 5 are shown enlarged details of alternative methods of fixing window frames at the outside of openings. The common method is shown in f. 4 where the opening is made square through, and the frame is set back about 3 in. from the face of the wall. The frame is fixed by wedges driven between the brickwork and the ends of the head and sill ; it is also wedged down from the lintel, as shown in f. 1. An outside architrave or trim, is scribed to fit the brickwork, and nailed upon the face of the linings. A better method is shown in f. 5, where a half brick reveal is formed in the outside of the opening and the frame made to fit it more closely, the joint between the frame and brickwork being pointed with cement, or a cement fillet run in a groove in the architrave ; this method effectually prevents any rain getting in when the woodwork shrinks. The frame is fixed by nailing it to plugs in the wall.

Semi-headed, or, as sometimes termed, **Circular-headed**, frames are constructed in two ways ; in the one, the head is formed by bending a veneer around a cylinder and backing it up with cross pieces to the required thickness ; this method being more applicable to hardwood frames or those of double curvature, is fully described in the chapter upon Curved Work. The other and commoner method is shown in f. 1 & 2, p. 205. The head of the frame is built up in three thicknesses, the segments of each ring breaking joint, the whole being glued and screwed together. The two pieces forming the path for the sashes, are about 2 in. thick and of the requisite width to suit the thickness of the sashes, and are divided by a $\frac{3}{8}$-in. piece which is eventually worked to the section of and continues the parting bead round. Tongues are not necessary on the head because of the extra thickness of the latter, and the linings are glued on as well as blocked. After the head is made up, it is cut off some 3 or 4 in. above the springing line, as shown, to form a stop for the sashes. The exact height depends upon the sweep. The pulley stile must be carried up high

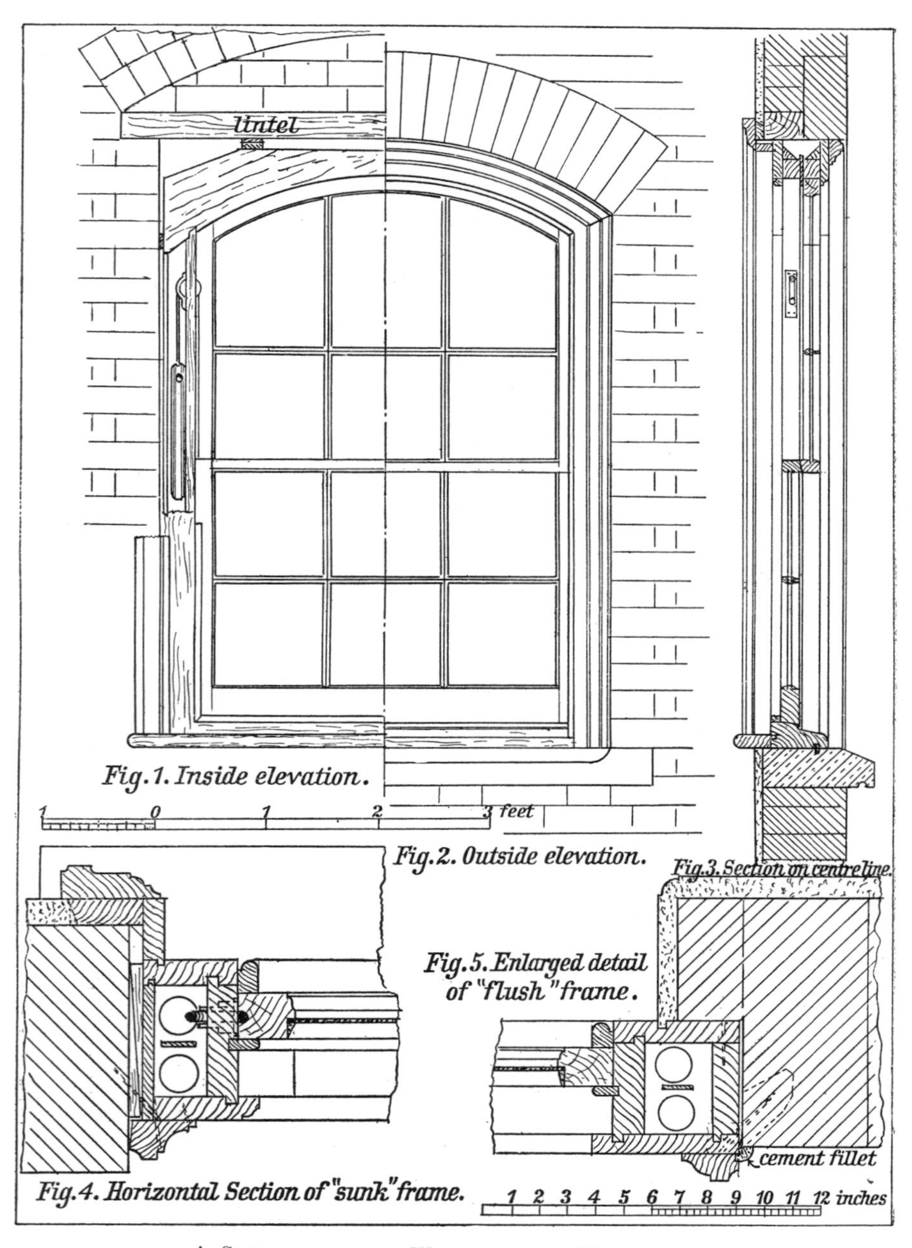

A Segment-headed Window fixed Externally.

enough to obtain a projection of at least $\frac{3}{8}$ in., and is shouldered out and carried up behind the head, to which it is fixed with glue and screws. Care must be taken in making these joints to have the vertical one exactly square with the springing, otherwise the stiles will be bent and cause the sashes to run badly. The object of the stop upon the head, is to prevent the bottom sash jamming if run up into the head, and also the glass of the top sash breaking, through the sash striking at the crown. The linings are jointed at the springing with a stub tenon or grooved and cross tongue, see f. 3. The pulleys in these frames must be kept just below the springing, as shown in f. 2. The top sash stiles are carried above the springing and shaped out to fit the frame, as shown in the enlarged detail, f. 4. The joint between the stile and head of the sash is made

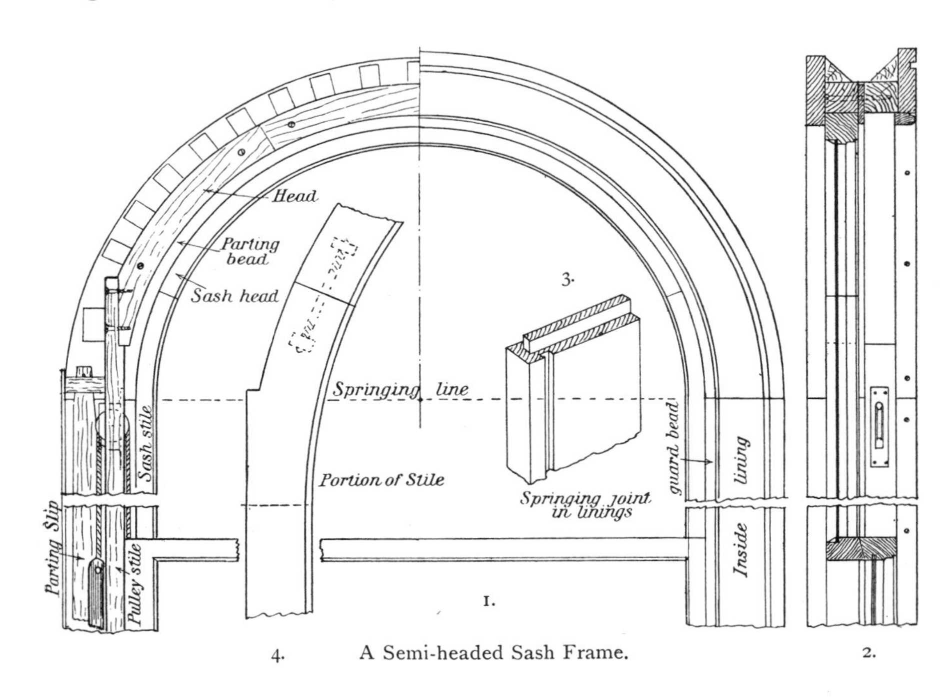

A Semi-headed Sash Frame.

radial, and fixed with a handrail bolt, a cross tongue being inserted in the rebate side, and the head is jointed similarly at the crown. The method of construction where machinery is employed is detailed at p. 125, but in a hand shop the several pieces would be rebated and stuck separately and fitted together similarly to a square sash, save that when gluing up, one half of the head or top rail would first be fixed to each stile, then the ends of the meeting rail inserted in the mortises, the nut entered on the point of the handrail bolt in the top joint, then the various joints well glued and brought together as quickly as possible. A stretcher cut off to the length between the sight lines on the meeting rail should be inserted parallel with the latter when cramping up, to prevent the bending of the stiles which would start the radial joints. It is perhaps unnecessary to remark that it is an imperative to "square" this sash, as it is a square-headed one, and that the squaring points are the intersections of the flat on the edge of the stiles, with the inside of the meeting rail and with the springing line. The afore-mentioned stretcher placed at the springing affords convenient squaring points at the top. It is advisable when working the circular parts by hand to cut them off ½ in. longer than is required at the joints to set out the moulding and rebate by templet on the joints, and, after working them as accurately as possible, to cut off to the proper joint line and fit each separately, and clean off the moulding with the files and glass-paper. Any attempt to do this with planes will probably result in spoiled joints, through the cross grain splitting out at the ends.

Sash Frames Circular in Plan should have their pulley stiles parallel on the face, not radial, because in the latter case the sashes would have to be inserted from outside, a very awkward arrangement. The inside edges of the sashes, however, should radiate, or practically be made square to the portion of curve within the width of the stile, see f. 3, Plate XXXI.

Venetian Frames are those divided by mullions into three or more openings. There are three varieties of these, known respectively as **Solid Mullion, Boxed Mullion**, and **Double Boxed Mullion Frames.** In the first of these only the centre pair of sashes is hung, the side ones, which are usually smaller, being fixed, and solid mullions are used to divide them. Fig. 1, next page, is the elevation, f. 2 vertical section, and f. 3 horizontal section of a frame of this description, and enlarged details are given in f. 4 to 9. The cords for hanging the centre pair of sashes are taken through the top of the mullions, over pulleys placed as shown in f. 4 & 7, into the side boxings. The cords are concealed from view in the side openings by the rail of the top sash and a cover piece or wide guard bead, which are both grooved for the passage of the cords, as shown in the sketch, f. 8. Purpose made axle pulleys are sometimes used for these frames, but generally the ordinary ones are utilised by the top part of the plate being cut off, so that the wheel may be brought close up to the head of the frame, and a small portion of the plate should be left above the wheel for insertion in the head, as there are no other means of securing the top end. Heads are generally made solid in these frames, and the mullions are housed into them, similar to the pulley stiles, and the mullions should also be tenoned and wedged to the head and sill. They are kept flush upon the inside, and are double tongued into the outside lining, as shown in the isometric sketch, f. 9. The linings should be stub tenoned into the sill, and the latter allowed to run through continuously; it considerably weakens the sill to cut the linings through it, and when they shrink an unsightly gap is left on each side. When the head is a cased one, the head linings should run through between the two outer linings, and the mullion linings be cut between the head and sill.

A Double Boxed Venetian Frame is shown on plan at C, f. 1, and elevation in f. 2, p. 208. The frame is constructed in this manner when all the sashes have to be hung, and the opening is large, and is divided by wide stone or other wide piers, affording plenty of room for two sets of weights. **A Single Boxed Venetian** frame is shown at D in same illustration, and is used when the piers are small or entirely absent, so that it becomes necessary to use a narrow

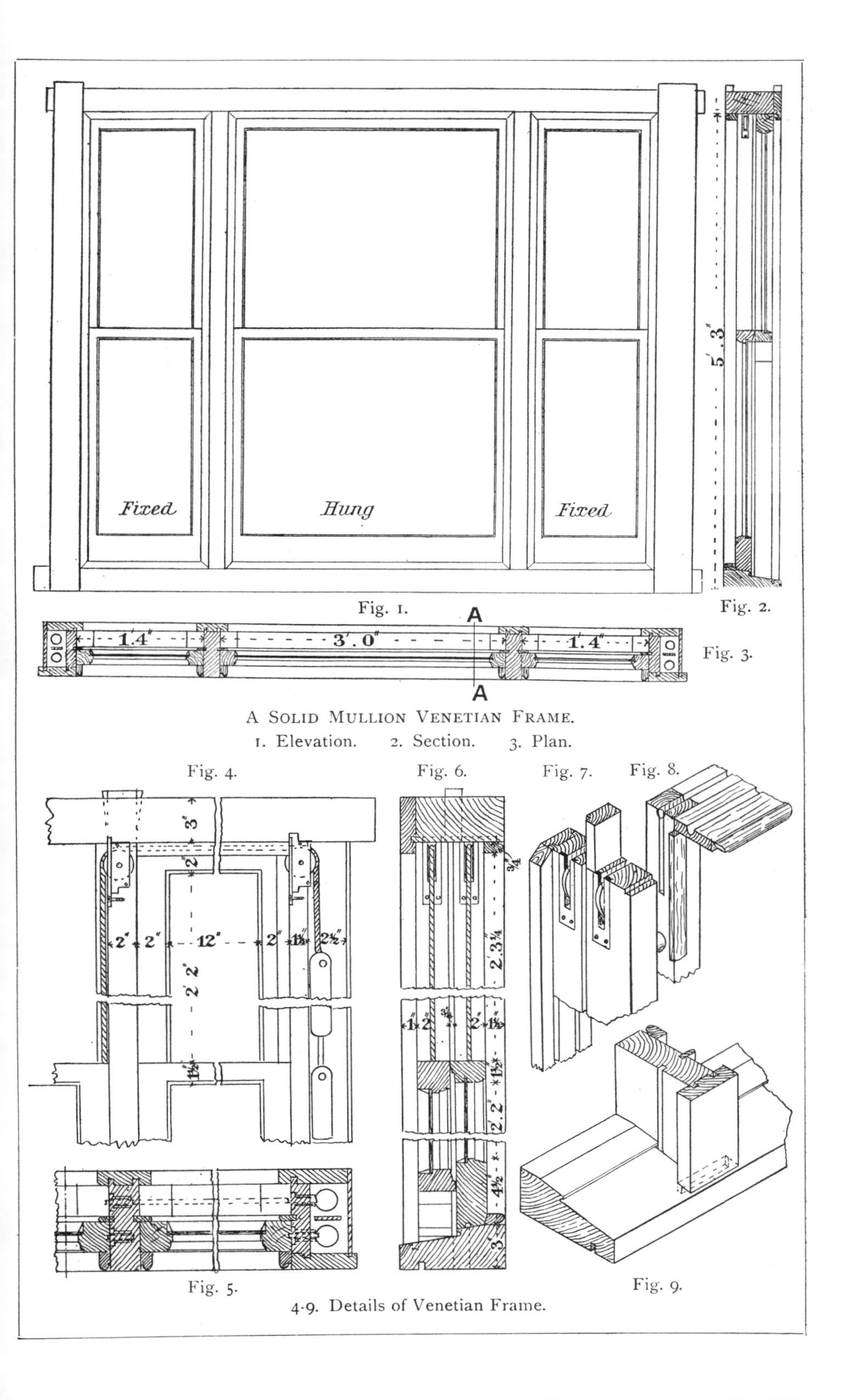

A Solid Mullion Venetian Frame.

1. Elevation. 2. Section. 3. Plan.

4-9. Details of Venetian Frame.

mullion. In these cases square leaden weights take the place of the usual round iron ones, and each of these weights supports one side of two sashes,

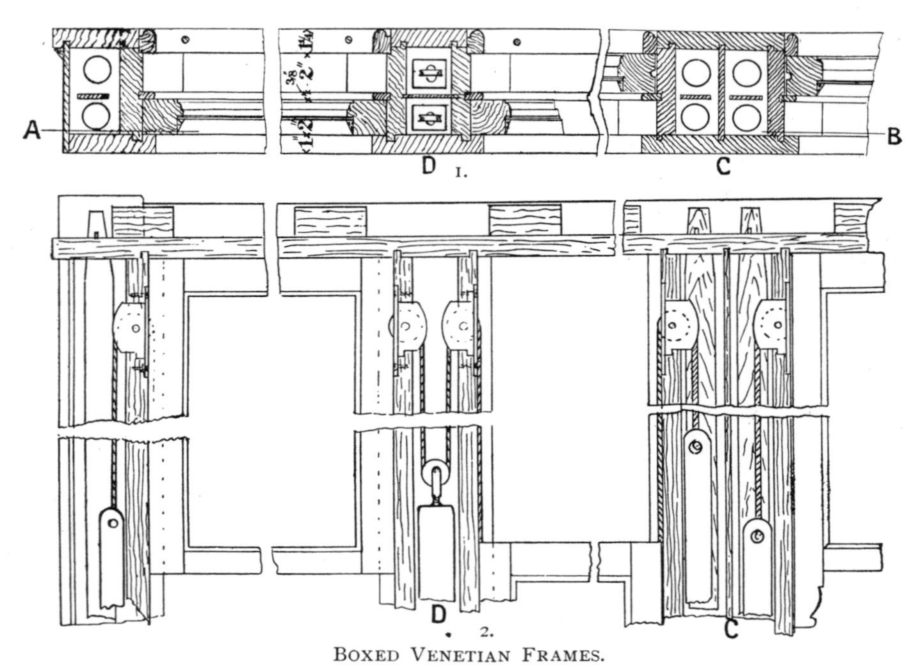

BOXED VENETIAN FRAMES.
1. Plan. D Single Boxed. C Double Boxed. 2. Longitudinal Sections on A B.

the cord passing from the sash in one opening over the pulley in that stile down to a pulley fixed in the end of the weight, and up over the opposite pulley to the corresponding sash in the next opening. The method is clearly shown in f. 2, which is a longitudinal section of the upper part of the frame. A difficulty sometimes occurs in this arrangement when the side lights are much smaller than the centre ones, the weight necessary to raise the larger being too heavy for the smaller. This may be overcome by inserting a small cork ball in the edge of each stile of the smaller sash. The hole to contain it should be dished at the bottom, so that the balls press somewhat tightly against the pulley stile but move freely in the hole. The friction produced will prevent the sash running up too easily.

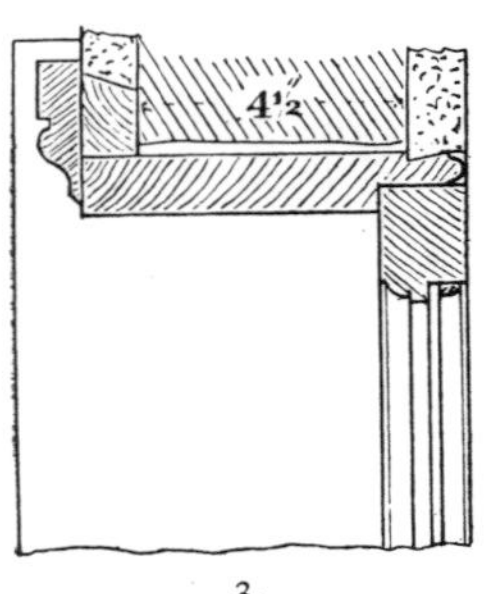

3.
Section of a Borrowed Light.

A Borrowed Light is one used in a corridor or other interior wall to transmit light obtained from an outside window. Fig. 3 above is a section through one side of an opening containing a borrowed light.

SOLID FRAMES.—**A French Casement Frame,** with the middle pair of casements opening inwards, the sidelights fixed, and the fanlights hung at the bottom and also opening inwards, is shown in broken elevation in f. 1, ditto vertical section, f. 2, horizontal section, f. 3, and enlarged details in the drawings on next page. The outside vertical members of the frame are called jambs, J; the inside ditto, mullions, M; the top horizontal member, the head, H; the division member, a transom, T; the bottom one, the sill, S. The members of the casements and fanlights bear the same names as in sashes. In these frames all rebates, mouldings, &c., are worked in the solid, and the joints made with mortise and tenon and put together with a round quick drying paint.

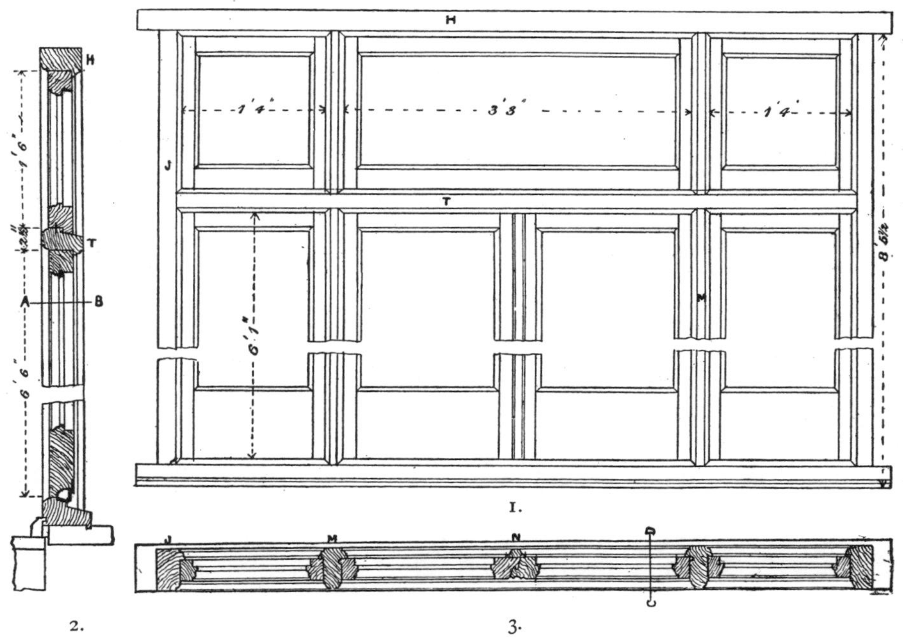

A FRENCH CASEMENT WINDOW.
1. Elevation. 2. Section on C D. 3. Plan.

When the casements open inwards there is much difficulty in making them weathertight, as of course the pressure of the wind tends to force them open. Numerous devices have been tried for this purpose, and those illustrated herewith are among the most satisfactory, but good workmanship is needed to make them effective. The hanging stile of the casement has a ½-in. cock-bead worked upon the outside edge, which fits into a corresponding groove in the mullion (see f. 2, next page); the latter should be stopped at the sill level to prevent the access of water to the wall by way of the mortise. The meeting stiles are fitted with a special joint called the hook joint, as shown in the detail, f. 1; double tenons must always be used on these meeting stiles, as the hook would cut through a central tenon, and the piece on the outside projecting lip being short grain, would fall out. The joint at the sill is often made with the patent water bar shown in f. 4 below. This consists of a $1\frac{1}{2}$-in. brass bar extending across the whole width of the opening, having its ends sunk slightly into the mullions. It is attached to a rebated casting, by several small hinges, the casting being fitted into a sinking, in and screwed to, the sill, as shown in the sketch, f. 1 opposite. When the casement is open the hinged bar lies down nearly flush with the sill, as seen on the left of the sketch; and as the casement closes a projecting spur screwed to its lower edge, and shown at S in the sketch, passes under the bar through the slot *a*, in the casting, lifts it up, and when the door has passed over it, presses it firmly against the guard bar C, screwed on the outside of the casement; the bar is then in the position shown in the right of the sketch, the casement being removed to show the position clearer. The bar being sunk into the mullions at each end, must be fixed to the sill when fitting together and wedged up with the frame. Fig. 3 on this page shows the method of preparing the transom for a fanlight hung to the head, and f. 4 the variation when it is hung at the bottom. When the casements open outwards the problem of keeping them weathertight is much simplified. Two methods in common use are illustrated in f. 3 opposite, and f. 2, Plate XX. In the first of these security depends upon the throatings and the moulded weather board *b*, which throws the water clear of the sill. Should any blow in, it will be caught in the groove *a*, and run out through the weep hole. In f. 2, Plate XX., a wrought-iron bar with its upper edge turned over, forming

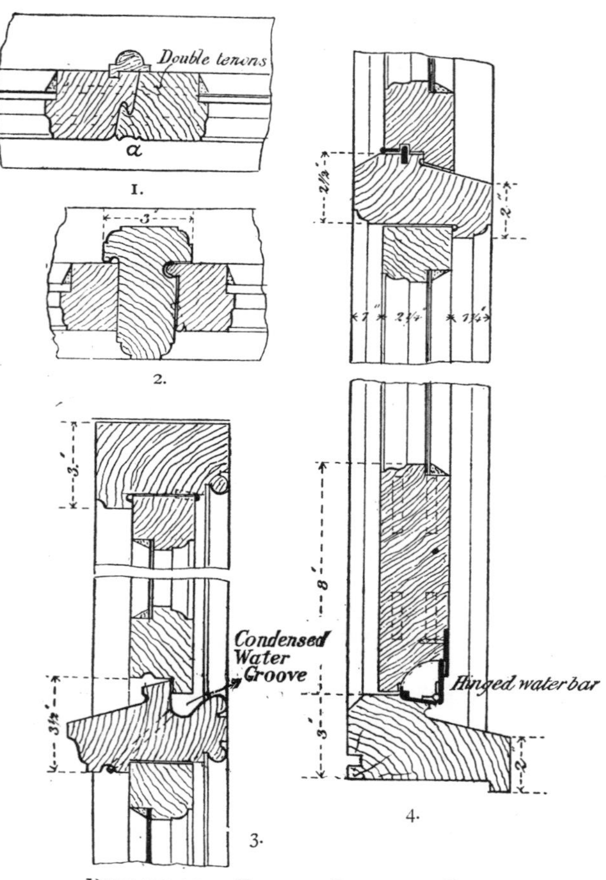

DETAILS OF A FRENCH CASEMENT FRAME.

1. Section of Meeting Stiles. 2. Section of Hanging Mullions. 3. Section of Head and Transom. 4. Section of Sill and Alternative Joints in Transom.

a throat, stops the entry of the water. This is shown to larger scale in f. 4, Plate XIX.

French Casements are usually fastened with an espagnolette bolt extending the whole height of the sash, and fastening it at top, bottom, and centre by one turn of the handle. These bolts work in a brass tube about $\frac{3}{4}$ in. in diameter, and the hollow shown at *a*, f. 1 opposite, is provided to receive them. Another and in several ways a better form of joint for casements than the hook joint is shown in f. 2 below. This is known as the Janus bolt joint, and consists of a $\frac{5}{8}$ by $\frac{1}{4}$ in. copper bar fitted into a brass channelled casing which is sunk flush in one of the rebated edges. A corresponding grooved casing is sunk in to the edge of the other casement, and when the door is closed and the lever turned, the bolt

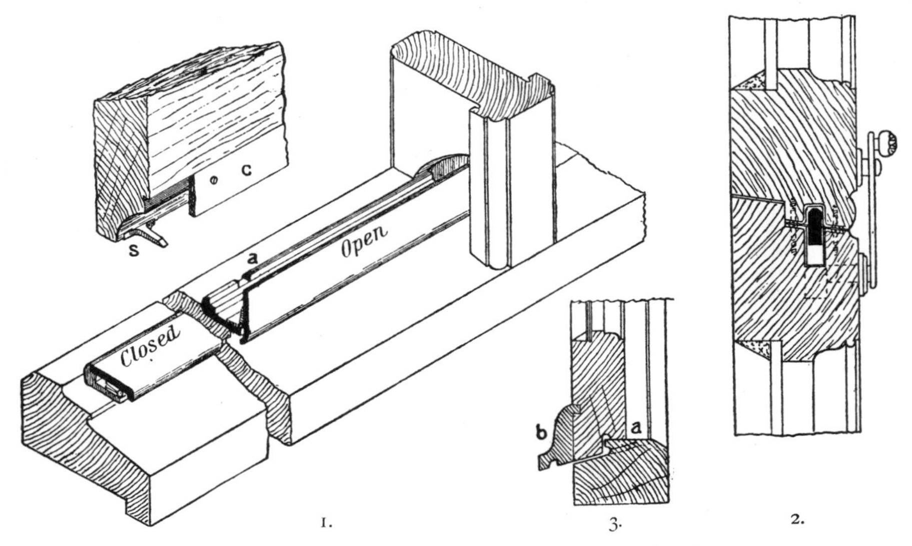

1. Sketch of Adams' Patent Water Bar for Inward Opening Casements.
2. Section of Casement Fitted with Janus Bolt.
3. Method of Making Joint at Sill in Outward Opening Casements.

is made to advance into the groove, thus providing a metal tongue throughout the length of the joint, the bolt, which being divided and notched at its middle, is simultaneously projected into sockets at each end, so fastening the doors at the top and bottom. One of the chief advantages of this joint is, that should the doors be in winding, or the stiles crooked, it makes no difference to the effectiveness of the bolt.

Nos. 1 to 8, next page, illustrate several improvements in the construction of **Cased Sash Frames** and **French Casement Frames**,* designed to render such frames wind and water tight, by means of checks easily and economically

---

For models illustrating these methods, the author was awarded the Bronze Medal of the Joiners' Company, at the Exhibition of Works in Wood, held at the Carpenters' Hall, London, in 1901.

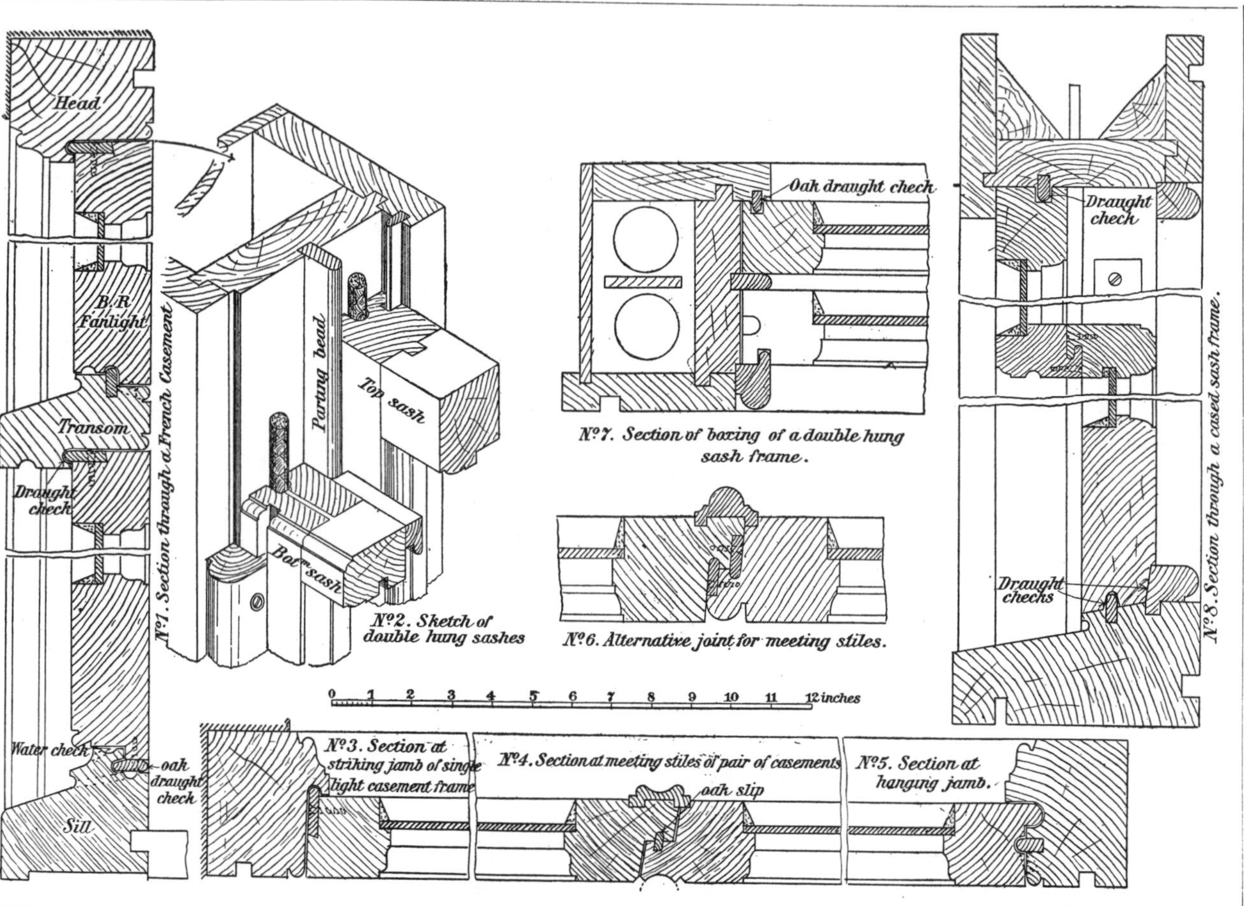

NOS. 1-8. DETAILS OF IMPROVED METHODS OF CONSTRUCTING WEATHERPROOF SASH AND CASEMENT WINDOWS.

applied, and forming part of the constructions. No. 8 is a broken vertical section of a double hung sash frame ; a draught check is provided in the head of the top sash by a $\frac{1}{4}$-in. projecting bead fixed in the head of the frame, and fitting into a groove in the head of the sash. A similar bead is fixed in the sill, where it will also effectually prevent any water getting through, and these beads should be inserted out of the line of the tenons in the sashes. The sill guard bead is bevel rebated into the sash, which causes the latter to be forced tight against the parting beads when shut. The meeting rails are made flush with the stiles on both sides, and hardwood rebated slips are glued and screwed to them, as shown, after the sashes are cleaned off and fitted. These may of course be made in the solid, but though more expensive to make, are in no way more serviceable. No. 7 is a horizontal section through one side of the same frame, showing the weather checks for the stiles of both sashes. The oak slip shown in the outside lining need only run down as far as the meeting rails, but the inside bead must be tongued throughout. When the sashes are fitted too thin for the frame, this method will prevent any rattling or draught. The isometric sketch No. 2 shows perhaps more clearly how the various parts are arranged. No. 1 is a vertical section through an inward opening French casement frame, and the weather checks are provided by a series of oak beads glued and screwed to the lights as shown. The projecting bead fixed to the bottom of the sash is bevelled on the upper side, and is made to fit tightly in the groove when the light is bolted up. The bead is bevelled to throw off any water that may be driven under by the wind, which is caught in the groove beneath, and discharged through the outlet shown by dotted lines. This should be in the middle, and the groove be given a dip from each end. The first groove on the edge of the sill will, however, catch most of the water driven in, the lower one being merely precautionary. The sashes should be fitted in accurately before the rebates are made for the various slips, which should thereafter be fitted tightly in the grooves and fixed to the sashes, due allowance being made for paint in the joints. No. 3 is a section through the striking jamb of a single light frame, and will not need any further explanation. No. 4 is a section through the meeting stiles of a pair of casements. These are first plain rebated together, then corresponding grooves are worked in the faces of the rebates, and an oak slip glued in the one in the inside rebate. An ordinary cover stop is fixed on the outside of the left sash, but this is double tongued in the sash to prevent it curling off and exposing the joint. No. 5 is a section through the hanging jamb, and shows a second cock bead inserted in the joint, which provides an efficient draught check, the outer one occasionally failing through shrinkage or too much play being allowed when hingeing. The butts in this case should be sunk flush in the edge bead. In the cases shown, the lights all lie flush with the inside of the frame, but in the case of these being sunk below the face, the only variation necessary would be to stick the joint bead upon the lights instead of around the frame. In No. 6 an alternative method of jointing the meeting stiles is shown, which is more suitable for a thicker sash. A bead is screwed on the edge of each light after they are rebated, thus forming a particularly strong hook joint with much less trouble than the usual method.

Various Weather Bars for Casements are shown on Plate XIX. Figs. 1, 2, & 3 is the "Perfect Simplex Weather Bar," patented by Messrs Elliotts, of Newbury; it consists of two cast channel bars of either steel, zinc, or brass, one fixed to the sill and one to the casement, as shown in the drawings. They are only suited for inward opening casements, but are very effective for these. Fig. 1 shows the bars fitted to a casement window that is hung flush inside. Fig. 2 shows the bars sunk into the sill of a French casement; in this case a drip pipe is fitted to the sill to convey any water that may drift in, outside. Fig. 3 shows the method of applying the bars to an existing casement without any special preparation, save the moulded weather board screwed to the casement. The bar on the sill requires to be bedded in red lead in this case. Figs. 4 to 7 show various patented weather bars made by Messrs James Hill & Co., of London. These are made in brass, zinc, or copper, as required. Figs. 4, 5, & 6 are for outward opening casements; the latter is particularly effective. Fig. 7, termed "The Registered Bar," is claimed to be the simplest, strongest, and easiest to fix, of any on the market. The sill is usually prepared as shown in full line, but the author considers that the section to the dotted line *a* would be an improvement. Figs. 8 & 9 show a very simple and effective arrangement of ordinary flat bars, designed by the author for outward and inward opening casements respectively. The foot bar on the sills prevents wear, as well as providing a water check.

A Bay Window with solid frame and casement lights is shown on Plate XX. Two methods of fitting such a window, with folding shutters, are given in the plan. In the half plan at C, the shutters fold into a boxing projecting into the room, and at D they fold back upon the face of the wall, which is splayed to receive them. The sills of the frame are mitred at the angles, the joint cross tongued and fixed with a handrail bolt, which should be painted with red lead before insertion. The joints in the head are halved together, the mullions stub tenoned and fixed with coach screws. The jambs are tenoned and wedged into the head and sill. The transom tenoned into the jambs and mullions, and secured with bolts. The mullions may be worked in one piece as shown at D, or built up as at C, and tongued and screwed together.

A Cased Frame Bay Window is shown in the half plan, half inside elevation, and central vertical section, Nos. 1, 2, and 3, Plate XXI. This window is composed of three ordinary sash frames, the sills connected at the angles either by halving and screwing, or by mitreing them and fastening the joint with a handrail bolt. The heads are tied together with a short piece of 1-in. stuff screwed across the top of the joints, and the joints in the linings are covered by the mullions of the blind frame B. The latter, made 2 in. wide, forms an enclosure for venetian blinds. Boxings are formed in the elbows between the sash frames and the interior face of the wall, the front of the opening being finished off with a moulded ground and architrave. These form receptacles for the folding shutters, which are curved in plan, and when opened out, convert the octagonal bay into a segmental niche. The window back and the seat beneath, are also curved to parallel sweeps. The window board also follows the sweep, and is rebated to receive the shutters, a shaped bead being fixed on the soffit to form a stop at the top. Nos. 4, 5, and 6 on the same plate illustrate another method of finishing a bay window. In this case the frame is solid, and is fitted with outward

PLATE XIX.

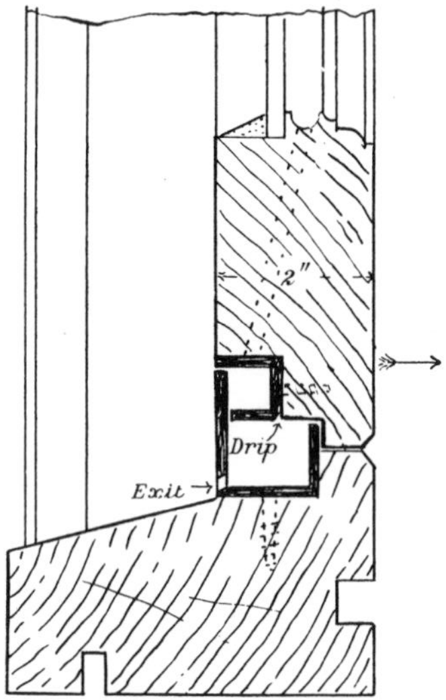

Fig. 1. Casement Window.

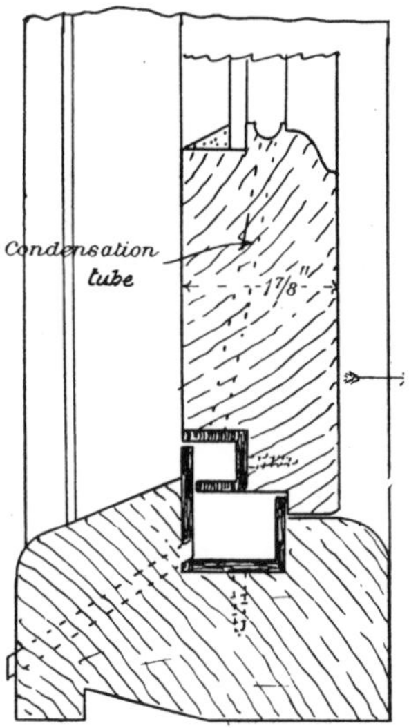

Fig. 2. Casement Door.

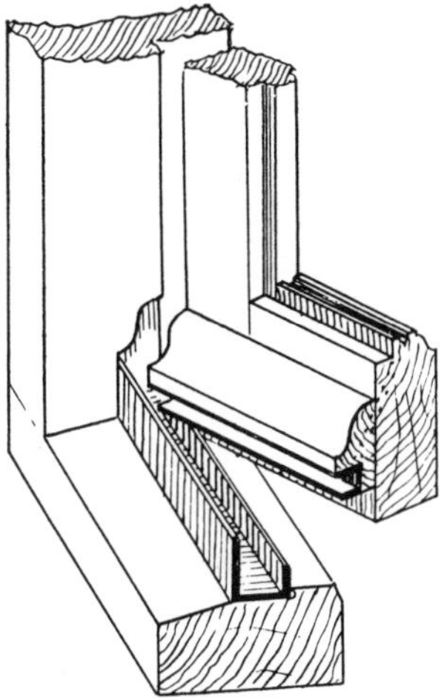

Fig. 3. Fitted to Old Casement.

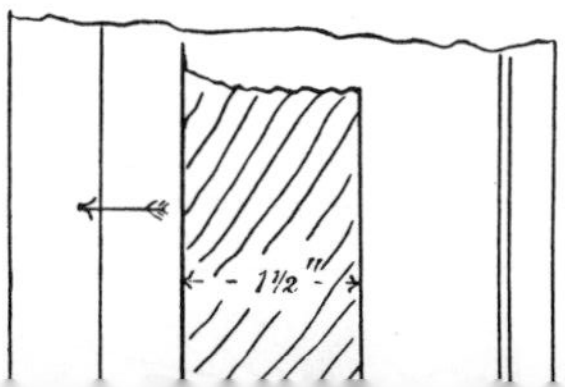

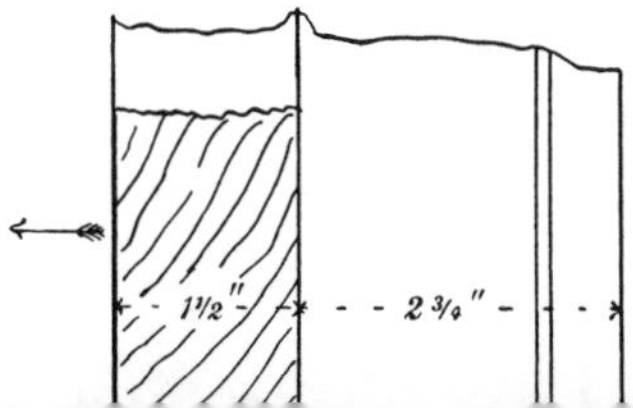

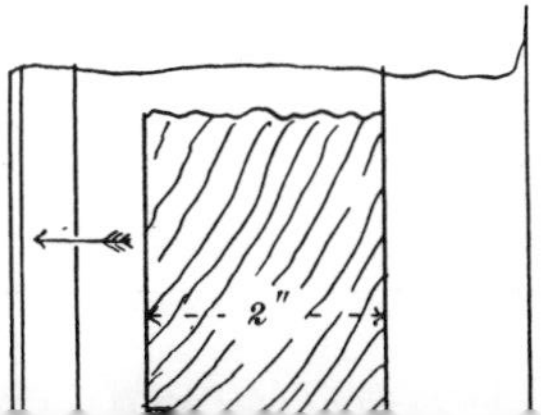

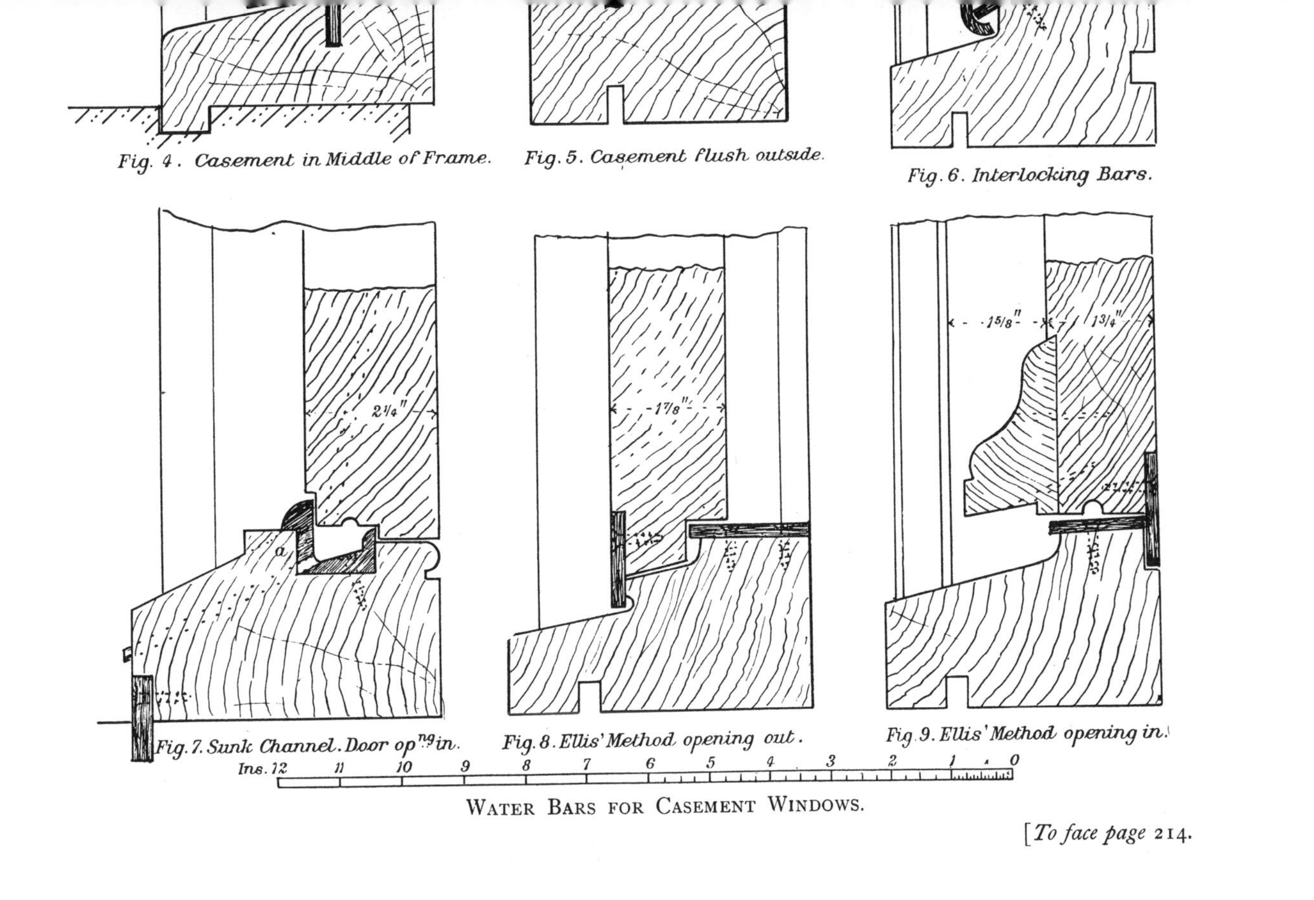

WATER BARS FOR CASEMENT WINDOWS.

[To face page 214.

Plate XXI.

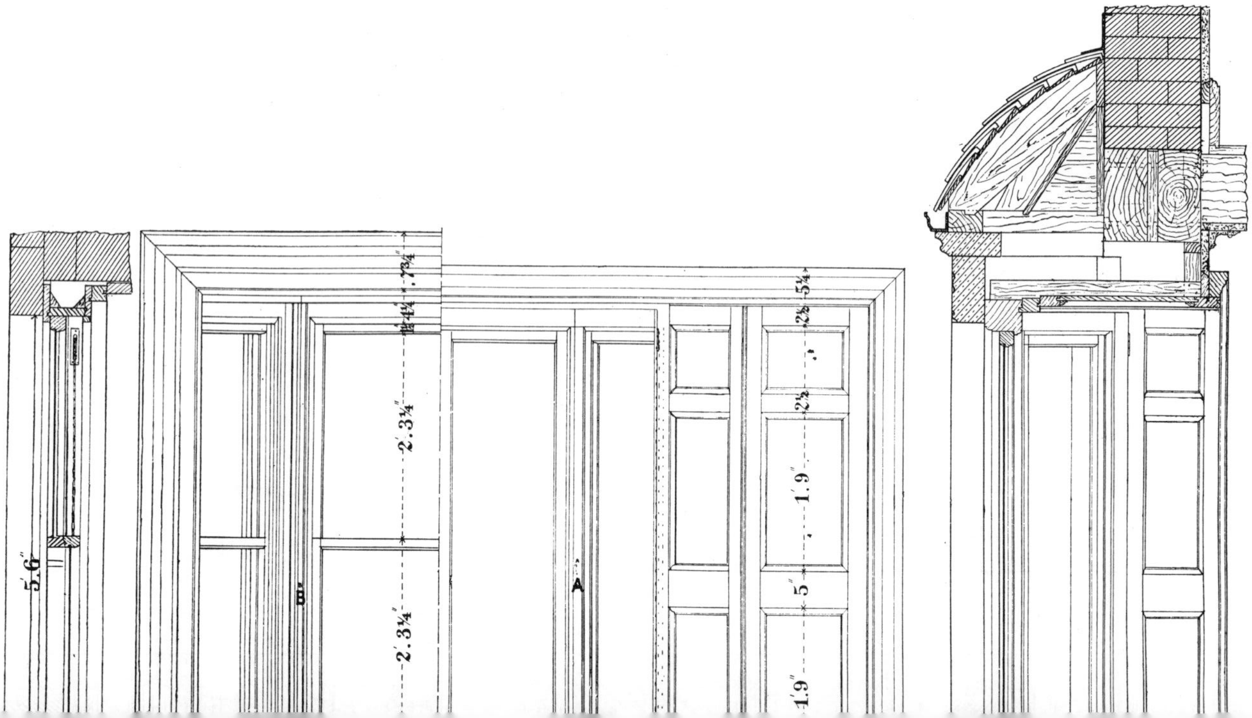

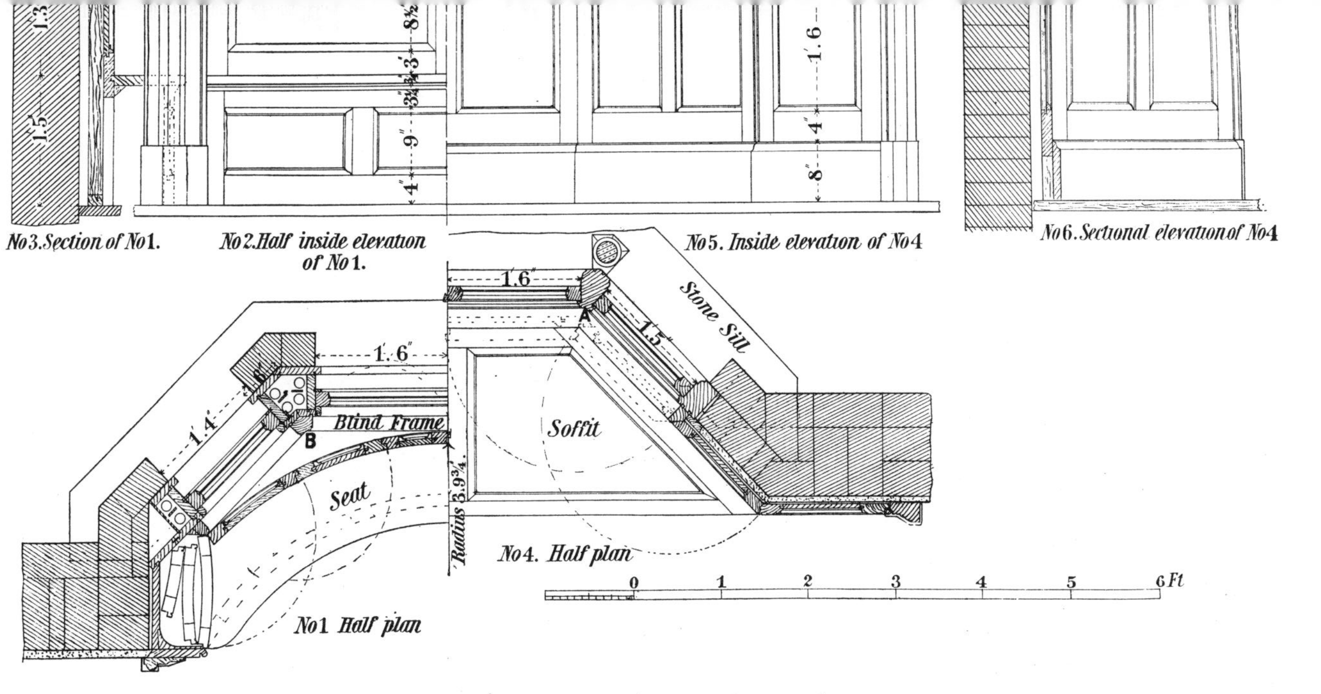

Designs for Bay Windows with Folding Shutters.

Nos. 1 to 3. Elevation and Sections of a Cased Frame. Nos. 4 to 6. Elevation and Sections of a Solid Frame.

[*To face page* 215.

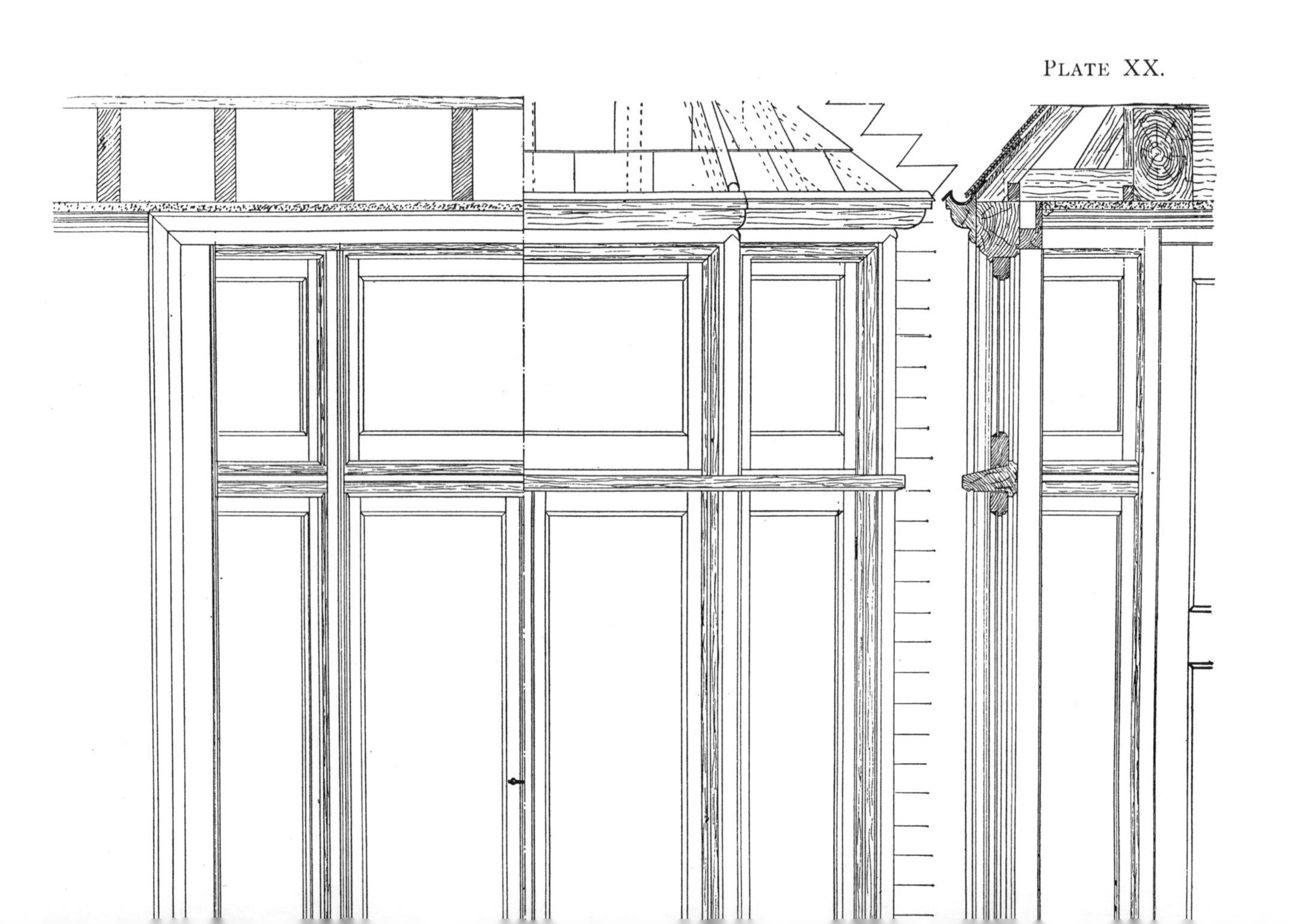

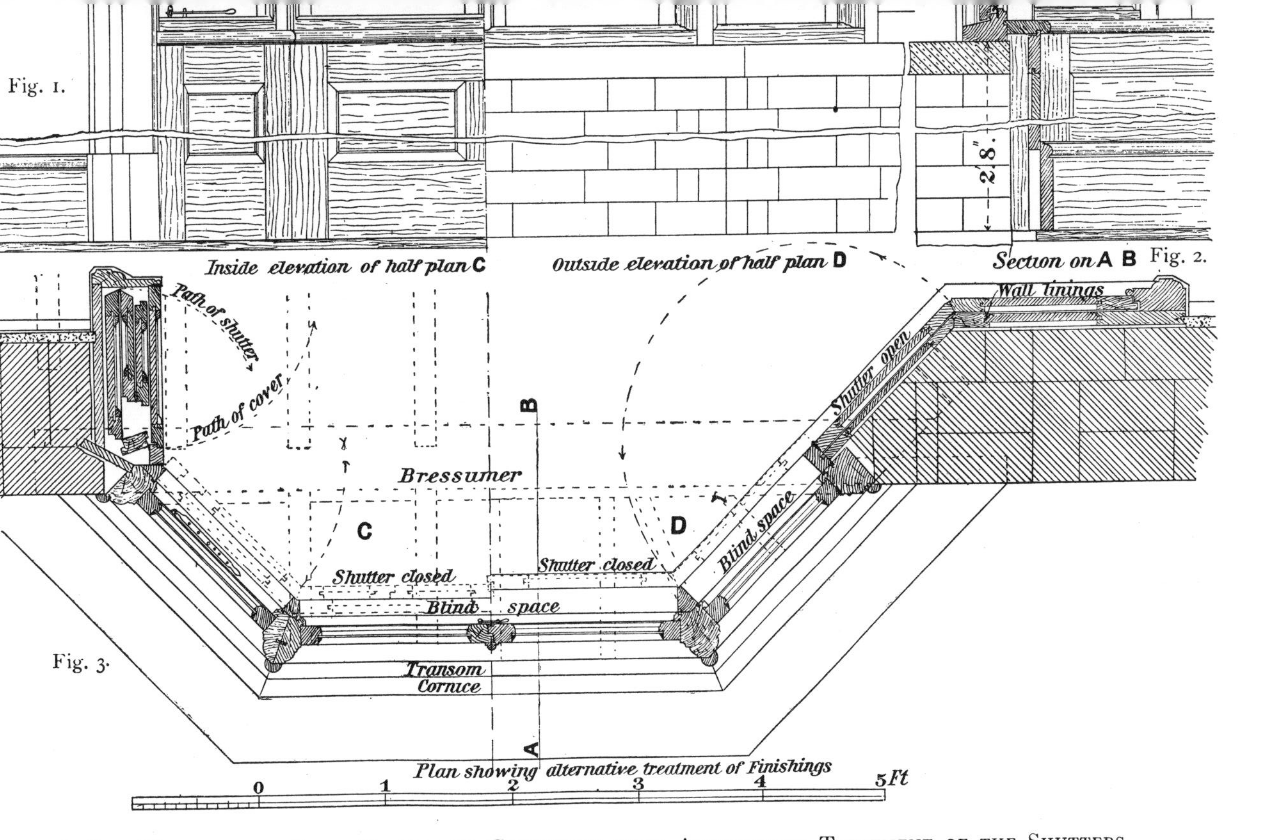

A Bay Window with Solid Frame and Casements, with Alternative Treatment of the Shutters.

[*To face page* 214.

opening casement lights. A blind frame is provided, and the shutters fold on to the face of the jamb and wall, the outer edges passing behind the rebated edges of the architrave; the latter is continued down to the floor, and elbow linings, to correspond with the shutters, are fitted beneath the window board. These are fitted to the window back in the manner indicated by the dotted lines in the plan No. 4. The section No. 6 shows the treatment of the roof of the bay, which is segmental in section and covered with shaped pan-tiles. The ribs,

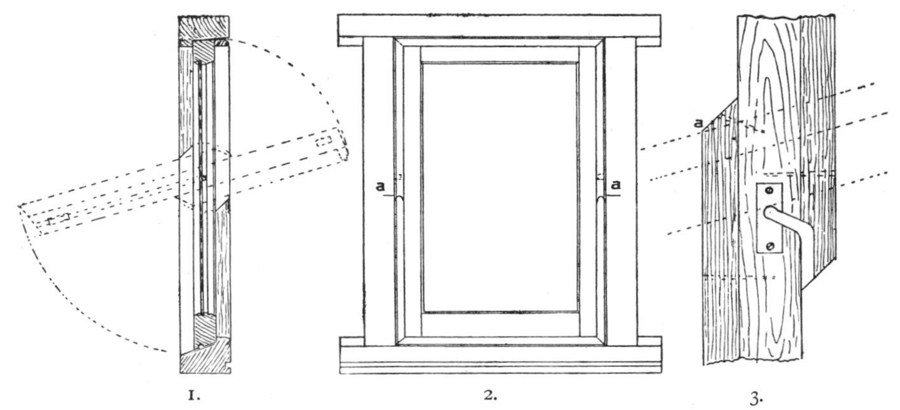

1 & 2. Section and Elevation of a Pivot Hung Sash. 3. Method of Finding Cuts in Bead.

which are elliptic at the hips, are notched into a wall plate resting on the stone cornice, and are nailed at the top into a shaped rib fixed on the face of the wall; the ribs are covered with weather boarding, which affords a good fixing for the tiles. The wall is carried by a breastsummer formed of two 12 in. by 6 in. balks bolted together, with spacing fillets between, and the soffit is carried by three brackets fixed to the breastsummer and the head of the window frame.

A Centre-Hung Sash in a solid frame is shown in elevation in f. 2 and section f. 1. These are used chiefly in warehouses, lanterns, and other inaccessible positions, the lights being opened and closed either by cords and pulleys or by metal gearing. For small lights the frames are usually made out of $4\frac{1}{2}$-in. stuff by 2 or 3 in. wide. Small lights are pivoted horizontally, large ones vertically. The pivot should be fixed to the frame, not the sash, from $\frac{1}{2}$ in. to $1\frac{1}{2}$ in. above the centre, according to the weight of the bottom rail. The lower part of the sash should exceed in weight the upper part, just sufficient to keep it closed; its action may be easily ascertained by inserting two bradawls in the stiles, and balancing them on the fingers. The sash is inserted and removed from the frame, either by means of plough grooves in the edges of the stiles, as shown by the dotted lines in f. 1 above, or by cutting a notch through the face of the stile for the passage of the pin, which is concealed when in use by

4. View of Slotted Bead in a Pivoted Sash.

the guard beads (see f. 3 & 4, p. 215). This latter is the better method, as it does not reduce the strength of the sash, as does the former, by cutting away the wedging. The stop beads at the sides are cut in two, one part being fixed to the frame, the other to the sash. Their joints can be at any angle greater than that made by a line tangent to the sweep at the point of intersection *a*, f. 3, but for the purpose of using the Mitre Block, they are generally made at an angle of 45 deg. A curved joint has no advantage over a straight one, except in being more expensive.

To hang the sash.—Insert the pivots in the frame quite level, but do not screw them. Then with the try square resting on the *top* of the pins, square lines

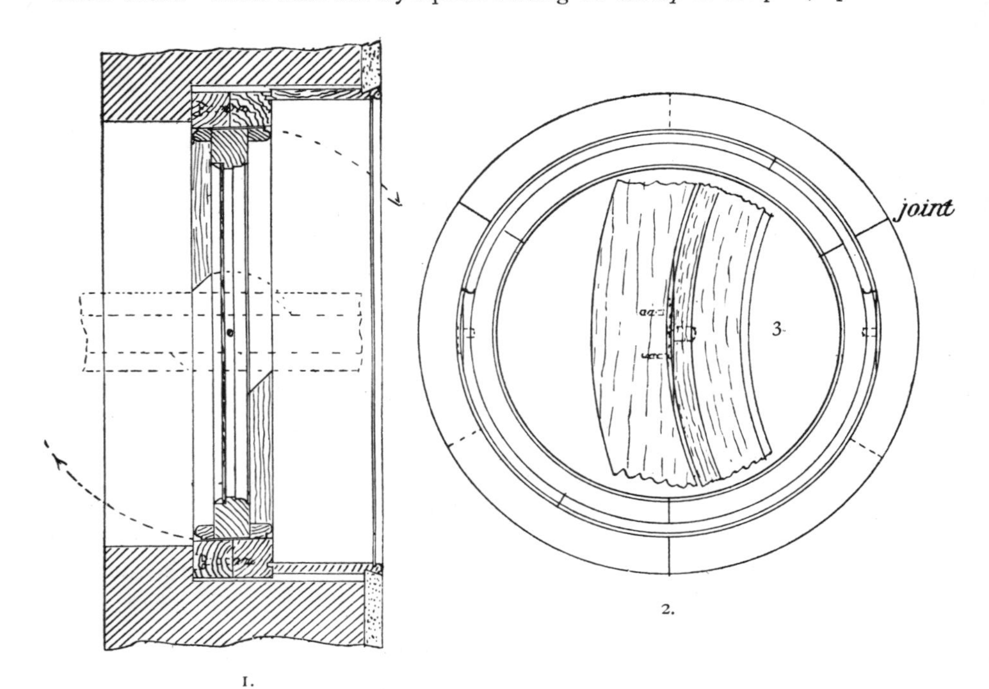

A BULL'S-EYE FRAME.
1. Section. 2. Elevation. 3. Detail of Pivot Joint.

across the jambs. Then remove the pivots and insert the sash, which should be fitted rather tightly at first, and square the lines on to the sash. Return these on the edges, and keep the edge of the hole in the socket plate to the line, and the plate itself, in the middle of the thickness. After the socket is sunk in, and the notches cut, test the sash and correct the joints, which should be a bare $\frac{1}{8}$ in. clear all round.

To find the position to cut the beads.—After fitting them round, remove them and open the sash to the desired angle, which should be less than a right angle, so that the water may be thrown off. Lay the beads upon the sash upside down for convenience of marking, and draw a line along their edge upon the jambs at the point where the line meets the faces of the frame; square over lines as at *a*, *a*, f. 2, p. 215; the position is shown in f. 1, the outer dotted lines indicating the beads. Next replace the beads and transfer the marks to them, cutting them off in the mitre block (remember that the mark is the longest point of the mitre). The upper portions outside and the lower inside are fixed to the frame, and these are shaded in the drawings. The remainder of the beads are fixed to the sash. The above describes the method when the sash is grooved. Where the beads are slotted, a variation must be made with reference to the top cut (see f. 3). In this case the sash must be drawn out and rested upon the pin, then the bead laid on it and marked as before, the intersection *a* giving the mitre point. The beads must be slotted as shown in the sketch, f. 4, p. 215, before nailing to the sash.

A Bull's-eye Frame with a pivoted sash is shown in f. 1 & 2 opposite, and enlarged detail of the joint in f. 3. This frame is built up in two thicknesses, glued and screwed together, each ring being in three pieces breaking joint. The beads may be steamed and bent round, or worked on the edge of a board that has been cut to the sweep, and cut off in two lengths. The sash is made in three pieces, with butt radial joints bolted together. To enable the sash to open, a plane surface must be provided at the centre, equal to the thickness of the sash and beads, as shown by the dotted lines in f. 1. Having fitted the sash in, and the beads around each side, brad them temporarily to the sash, lay a straight-edge across it parallel with the centre, and square up with the set square a line at each side equal in length to the thickness (see detail, f. 3), then cut the pieces so marked off, with a fine saw, both beads and sash, and glue them to the centre of the frame, then fix the pivots to these pieces and proceed as in a square frame.

Hopper Frames or **Hospital Lights**, as they are also called, in consequence of their frequent use in these institutions, are constructed usually as shown in f. 1, 2, 3, next page. The frame is an ordinary square rebated solid one, filled in with several fanlights, which are rebated together, and pivoted at their lower corners into the frame, side pivots being screwed on the face of the lights (see detail, f. 4). The lowest light is hinged to the sill, beads are fixed on the jamb linings at a convenient angle, and the lights when open, rest on these. The lights are usually connected by geared rods, so that all can be opened at once. In another variety the frames are fitted with a single pane to open into a metal hopper which is glazed at the sides, the pane being fixed in a metal frame which is hung to the bottom of the hopper.

Yorkshire Lights, f. 1 to 4, p. 219, are solid frames containing a light that slides horizontally on a metal bar. The half of the frame not filled by the sash is glazed, the jamb and mullion and part of the head and sill being rebated for the purpose. A bar is also frequently inserted, as shown in the elevation, f. 1. Fig. 2 is a section to a larger scale, and f. 3 a plan. The bottom rail of the sash should be lipped over the stile, as shown in f. 4, to prevent the latter splitting out in working. Brass shoes, f. 5, are inserted in the bottom rail to prevent wear, and to cause the sash to run easier.

Dormer Windows are those used to light rooms formed in the roofs of buildings. They are invariably vertical, which position differentiates them from

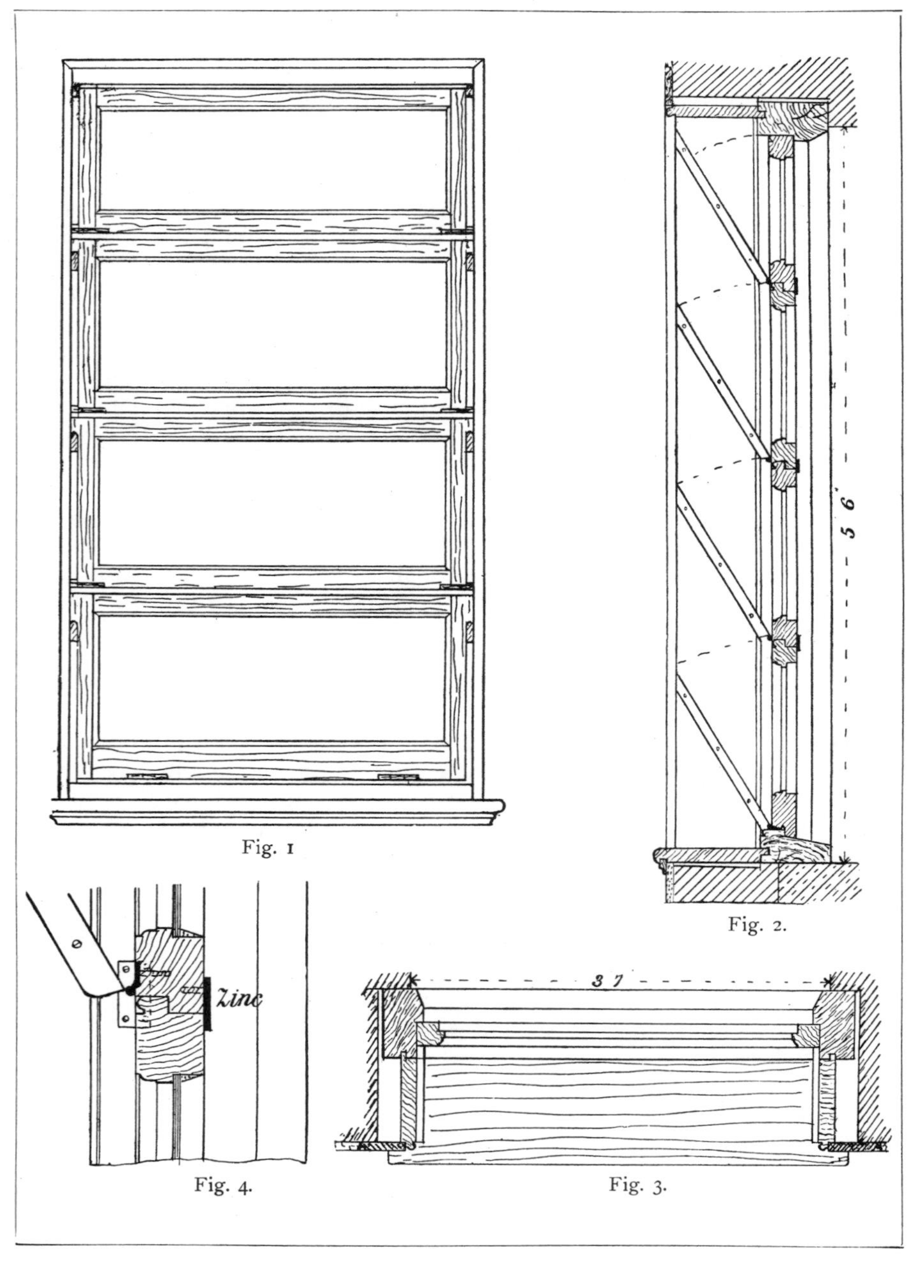

A HOPPER FRAME OR HOSPITAL LIGHT.

1. Elevation. 2. Vertical Section. 3. Horizontal Section. 4. Detail of Pivot.

skylights, that always more or less follow the inclination of the roof in which they are placed. Dormers are variously fitted with cased sash frames, sliding sashes, or casement frames. Figs. 1, 2, & 3, next page, illustrate the construction of the latter kind, having a gable or pediment head, a solid frame, and a pair of outward opening casements. The sides of the dormer are boarded on the outside and covered with lead, the inside of the studding being lathed and plastered, as shown in the detail, f. 3. Fig. 4 is an isometric sketch of the inside of a dormer fitted with a cased sash frame. The roof is a flat one, but made sloping towards the front, and the outside is covered with sheet lead, the roof having a bottle nose drip around the edges. The flooring and plastering are omitted from the drawing, the better to show the construction of the dormer. The two trimming rafters at the sides of the opening are ½ in. stouter than the remainder. Two short trimmers are framed between these, the one to carry the roof and the

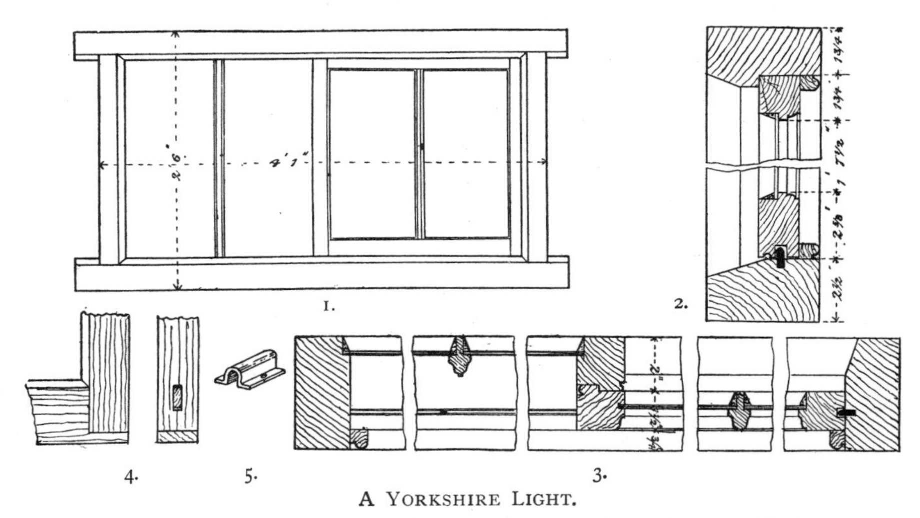

A Yorkshire Light.

1. Elevation. 2. Vertical Section. 3. Plan. 4. Joint at Bottom of Sash. 5. Metal Shoe.

other the sill of the frame, also the ends of the trimmed rafters. Two studs are carried up from the floor at the front of the opening to carry the outer ends of the plates or capping pieces, and the sash frame is fixed against these. Short studs are cut between the rafters and the capping, being stub tenoned to the latter; the ceiling joists are diminished in depth towards the front to give a fall to the roof, and are nailed to the top of the plates. Below the window, and in a continuous line across the rafters, are nailed vertical studs called ashlering; these are eventually lathed and plastered. Fillets are nailed on the sides of the trimming rafters between the studs to form a fixing for the laths, as shown in f. 2 & 4, next page.

A Small Framed Skylight is shown on p. 221. The curb or lining to the opening may be dovetailed at the corners, or tongued and nailed, and runs above the roof about 6 in. to form a gutter. The light, if hung, must

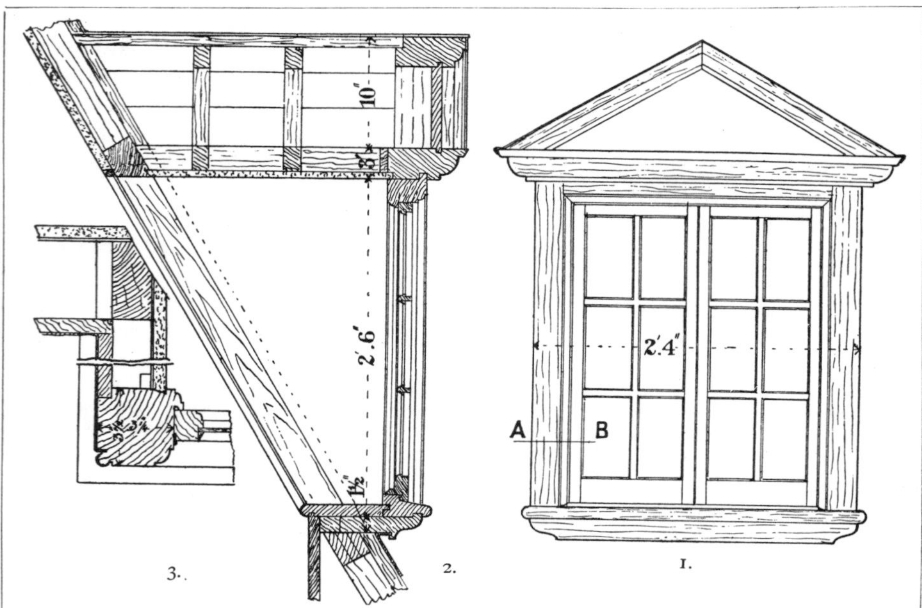

1-3. A Dormer Window, with Section and Detail.

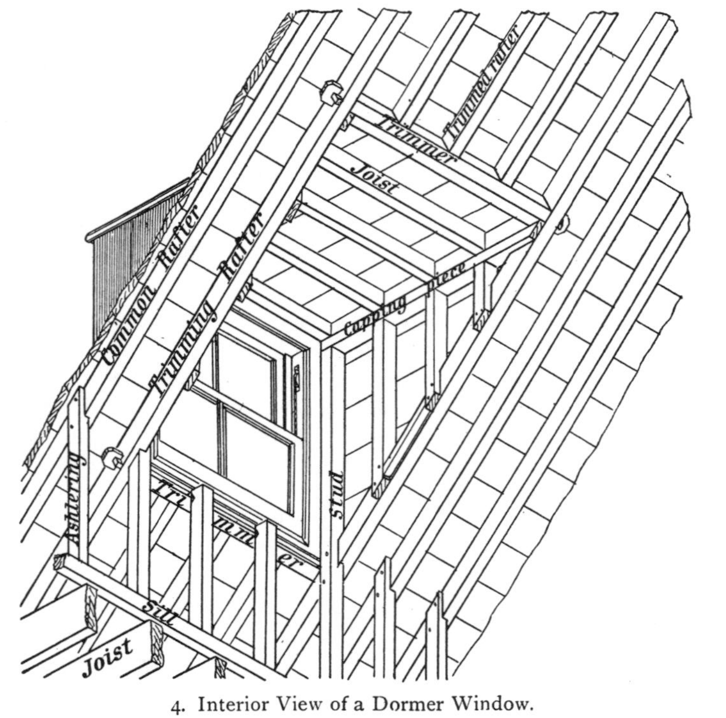

4. Interior View of a Dormer Window.

be hinged at the top, and should overhang all round, the lower side being throated. The lead flashing is sometimes brought up over the edge of the curb, and dressed into a small gutter to intercept any water that may drift through the joint. The bottom rails of skylights are always kept below the level of the glass, so that the water may escape easily. Top rails should be ploughed, as the putty would be liable to fall out of rebates. A dishing must be made under the tail of each pane of glass, as shown in the detail, f. 1, by dotted lines, and in plan on next page, to allow of the escape of condensed vapour upon the under surface of the glass. The tail of the bar should finish just below that of the glass. It is not a good method to carry the glass right across the rail, as it is liable to get broken by the casting of the latter, induced by the heat of the sun through the glass.

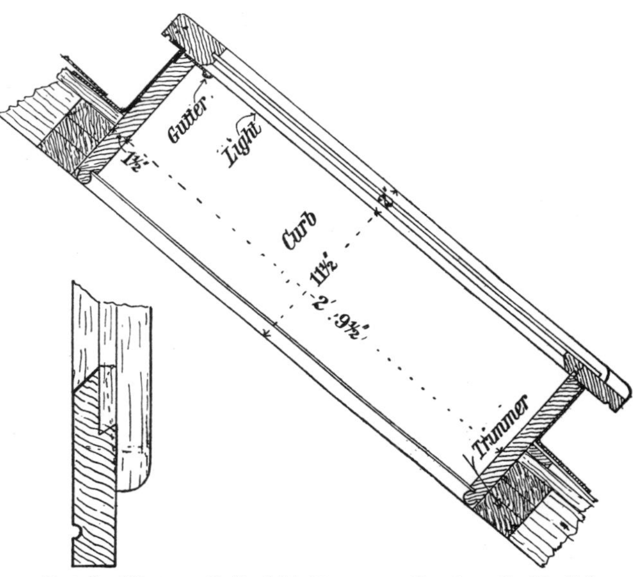

1. Detail of Bottom Rail of Light. 2. Section of a Skylight.

A Lantern Light, suitable for a billiard-room, is shown on pp. 224 & 225. Fig. 1, p. 224, is a sectional elevation, f. 2 a part plan. Figs. 3, 4, & 5, p. 224, and 1, 2, 3, 8, 9, p. 225, are enlarged details, and f. 4, 5, 6, & 7 are geometrical diagrams connected with the same. The lantern stands upon a rough-framed curb, resting on the trimming joists of the lead flat. The side lights open outwards, and are hung at the top. The corner posts and mullions are rebated in the solid. The head, has the bead forming the rebate planted on, as it is not required to withstand any shocks. The roof (skylight strictly) is formed with moulded hips and ridge filled in with framed lights. The curb and trimmers are covered with a framed and moulded lining, which finishes flush with the ceiling. An enlarged detail of a corner post, mullion, and adjoining parts is given in f. 3, p. 224, showing alternative methods of construction, both of which are good. Figs. 4 & 5 show alternative methods of framing the angle of the sills. In the former two sides of the sill are mortised, tenoned, and wedged together; in the latter, mitred, tongued, and bolted. The post is stub tenoned and secured with a coach screw, or when the lantern is not very high, by a screw bolt taken right through sill, post, and head. In making the joint at the foot of the corner post, it should be observed that the shoulders cannot be cut with the saw, as they form an internal angle, the inside of each shoulder rising higher than the outside. They are usually cut square off to the longest point, and cut back to the proper shoulder with the chisel. The head or plate is halved together at the corners, as shown in f. 9, p. 225. The true shape and size of the skylights is not shown either in the elevation or plan because the real length of an inclined line is not seen unless it is drawn upon a plane parallel with itself. Now the hip lines in question are not parallel with either the vertical or horizontal planes—in other words, with either the elevation or plan; and before their length can be ascertained, they must be revolved until they are parallel with one or other of the planes; either will do. The plan, has here been selected for illustration, and to avoid confusion with the constructive lines, a separate plan diagram has been drawn in f. 7, p. 225, one-half illustrating the method of finding the true shape of the lights, the other half the method of finding the length and backing of the hips. Let *a*, *b*, and *c* represent the extremities of one of the hip lights in plan. The length of the bottom edge between *a* and *c* is a real length, because it is parallel with the ground, and if the length of the other two sides is found, the true shape of the light can be drawn. Draw a line from the apex *b* perpendicular to *b a*. Make it equal in length to the height of *b* above *a*, as ascertained by reference to the section, f. 1, p. 224, where the method of finding the height is indicated. Erect a perpendicular from the lowest point of the back of the light as *a a'*. Intersect this by a horizontal line drawn from the highest point of the back of light *b*, and *a a'* is the vertical height. Join *a* to *b'*, the height as just found, and this line will be the true length of the side *a b*, and as both sides are alike, of *c b* also. With *a b'* as radius, and *a* and *c* as centres, describe arcs intersecting in *e*. Join this point to *a* and *c*, and the triangle *a e c* will be the true shape and size of the light on its top side. Lines drawn parallel to these lines, at a distance equal to the width of the stiles, will give the shoulders, and the bar lines continued from their plan will locate the mortises in the stiles.

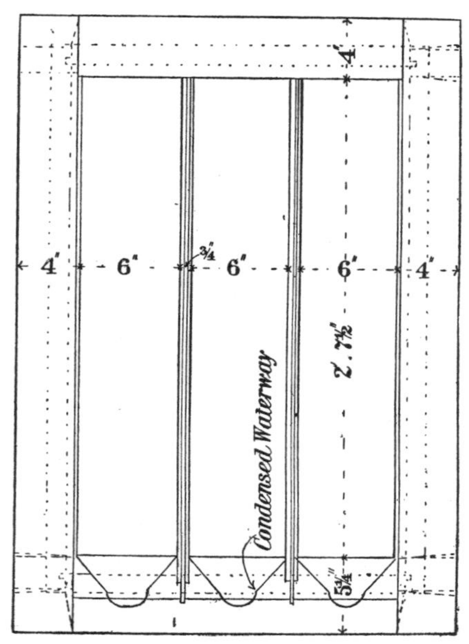

Plan of a Skylight.

To Ascertain the Size of the Side Lights.—Square out lines from the two highest extremities (only one is here given), as at *i*, f. 7, p. 225. From the two lower extremities cut off on these lines lengths equal to *a b'*, as previously found, then join these points, and the shape will be found as indicated by dotted lines. These dimensions will of course be set out full-sized, and a little extra should be allowed in the width of the stiles for fitting, cramping, &c.

To Find the Length and Backing of the Hip.—The backing, it may be explained, is the bevel given to the top edge of the hip, so that it may lie in both planes of the roof surface. Let *o f*, f. 7, represent the seat of the hip; erect *o o′* perpendicular to this, and make it equal to the height of the top end of the hip over the bottom end; in this case it will be the same as the rise of the light in f. 1, next page, all being flush on top (it must be noted that the scale of the drawings is different). Join *o′* to *f*, and the true length of the hip at the centre of its thickness is seen. Next, at any point *n* in the line *o′ f*, draw a line perpendicular to it, cutting the plan of the hip in K; through this point draw a line at right angles to the plan of hip, cutting the lower edge of the light (which in this case represents the projection of the hip) in points *g* and *h*. With *k* as a centre and *k n* as radius, describe an arc cutting the plan in *l*, join this point to *g* and *h*, and the triangle so formed contains the dihedral angle between the two planes of the roof, and a Bevel set to one side, as shown in f. 7, will give the backing bevel. The hips and ridge being moulded, and not lying in the same plane, will require different sections of mouldings to enable them to be mitred properly at the joints. The ridge is usually the determining member, and assuming that section given, as A, f, 5.

To Find the Section of the Bed Mould *b*, f. 1, p. 225, divide the profile of the Ridge mould A into any number of parts, as shown, and project the points to a horizontal line at the top, numbering them for convenience of reference 1, 2, 3, 4, 5. Draw lines also from the same divisions, the rebate, &c., parallel with the pitch, in this case 30 degrees; set off from a perpendicular line *o o* the required thickness of the bed mould; draw a horizontal line at the top of this, and transfer the points 1, 2, 3, 4, 5 to it in reverse order; drop perpendiculars from these to intersect the corresponding lines drawn from the points on the moulding, and the intersections will be points through which to trace the curve. If a Bar was required, its section would be found similarly by drawing a line *x x* at right angles to the pitch, and setting off the horizontal projections on each side of this, as shown in dotted lines. In like manner the section of a jack-bar can be obtained by setting off the horizontal projections of the hip moulding, as shown in f. 6, parallel on each side of the line x^2 x^2, and drawing the profile of the moulding through the intersections with the corresponding inclined lines 1, 2, 3, 4, 5.

To Obtain the Section of the Hip B, f. 6, draw its determined thickness, in the present case equal to the ridge, also a line representing its bevel or backing, as found in f. 7. Draw a line as x^2 x^2 at right angles to the backing pitch, and make it equal in length to *x x*, f. 5. Through the end of this line and parallel with the top edge draw a line which will determine the depth of the hip. From the points 1′, 2′, 3′, 4′, 5′, obtained from the line *x x*, f. 5, draw lines parallel with the back edge, and intersect them by perpendiculars dropped from the level line *a b*, containing the corresponding horizontal divisions. Draw the curve through the intersections and repeat on the opposite side. Fig. 4 shows a method of obtaining the true bevel for the joint between the ends of the hip and ridge. The thin lines are an enlarged reproduction of the joints shown in the plan below, and which are obtained by drawing the joint lines from the intersections outside to the intersection of the centre lines.

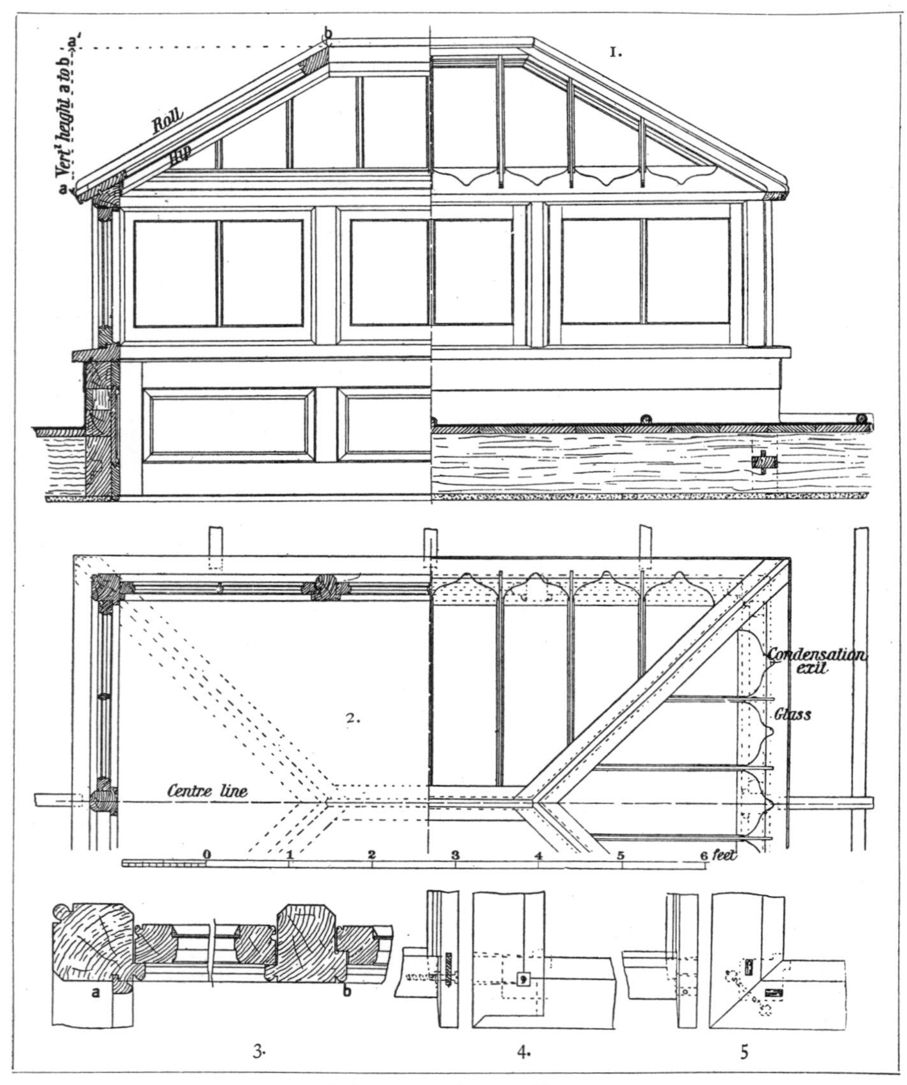

A Lantern Light, No. 1.

1. Half Inside and Half Outside Elevations. 2. Half Horizontal Section and Half Plan of Roof. 3-5. Enlarged Details.

Part of the plan is drawn full size as shown, also a line representing the hip as in elevation (not its true pitch, remember). Join the extremities of the joint of one of the hips, as *a b*, from where this line cuts the centre line of that hip at 1; erect

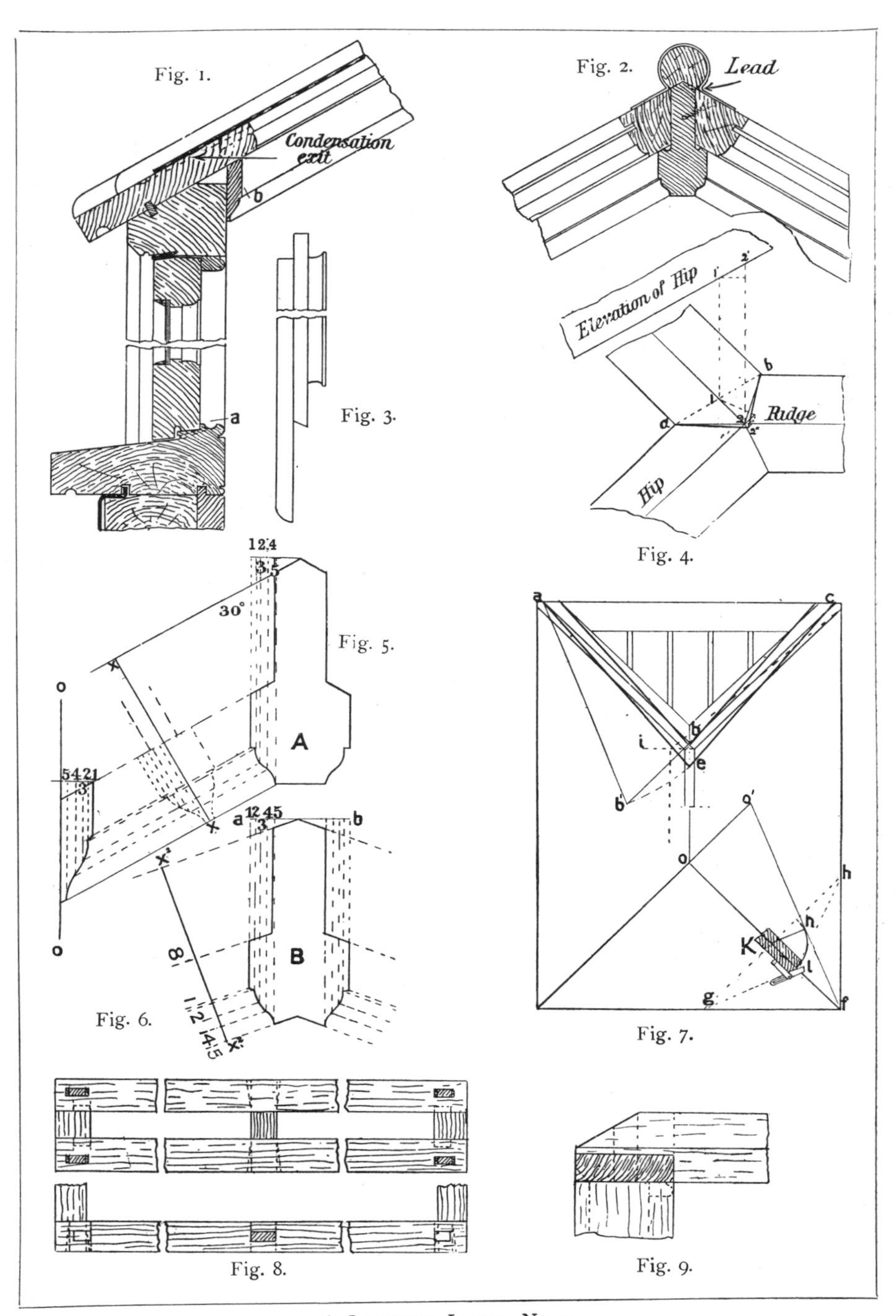

A Lantern Light, No. 2.

1, 2, 3, 8, 9. Constructional Details. 7. Part Plan and Elevation of Curb.
4, 5, 6. Diagrams for obtaining Bevels and Sections.

a perpendicular cutting the elevation in point 1′. Draw a perpendicular also from the end of the hip at point 2 in plan to 2′ in the elevation, which will show how much point 2 is higher than point 1. Draw a line perpendicular to the line 1 2 in plan, and make it equal to the height that 2′ is above 1′ in the elevation. Join the end of this line to point 1 as shown by the dotted line, and throw this line down to the centre line of the hip produced, as shown by the dotted arc at point 2″. Draw lines from this point to *a* and *b*, which will be the true bevels for the joint to be applied before the hip is backed. These true joints for the hip are shown by strong lines in the diagram. The thin lines will be correct for the ridge joints; the down or side cuts of the ridge will be square from its top edge; and the down or side cuts of the hips are shown at the angle *o′* in the elevation of the hip *o o′ f*, f. 7.

When it is desired to obtain the joint bevel for the hip, to be applied after the latter is "backed," the bevelled edge of the hip must first be developed as described for f. 4, p. 288.

For other methods of obtaining the above-mentioned bevels see p. 82.

CHAPTER XIII.

SHUTTERS, BLINDS, AND FINISHINGS.

The Construction of Interior and Exterior Shutters and Blinds for Houses and Shops—Folding, Sliding, Lifting, Revolving Spring Coils, &c.—Formation of Boxings for Shutters and Blinds—Difference between Jamb and Elbow Linings—Boxing Shutters for Bay Windows—Shutters for Square and Splayed Openings—Soffits—Window Backs and Elbow Linings—Wood, Iron, and Steel Laths for Coils—Methods of Connection—Fixing Coils and Blinds, Precautions to take—Lifting Shop Shutters, Methods of Fastening—Parliament Hinges—Venetian Shutters.

INTERIOR SHUTTERS include FOLDING, SLIDING, BALANCED, and ROLLING. EXTERIOR comprise HANGING, LIFTING, SPRING, and VENETIAN SHUTTERS.

FOLDING, or as they are frequently called, **BOXING SHUTTERS**, because they fold into a boxing or recess formed between the window frame and the walls; are composed of a number of narrow leaves framed or plain, as their size may determine, rebated and hinged to each other and to the window frame. They should be of such size that when opened out they will cover the entire light space of the sash frame and a margin of a $\frac{1}{4}$ in. in addition. Care must be taken to make them parallel, or they will not swing clear at the ends; and as a further precaution, they should not be carried right from soffit to window board, but have clearance pieces interposed at their ends about $\frac{3}{8}$ in. thick. The outer leaf, which is always framed, is termed a shutter; the others are termed flaps. It is not advisable to make the shutters less than $1\frac{1}{4}$ in. thick, and flaps over 8 in. wide should be framed; those less than 8 in. may be solid, but should be mitre clamped to prevent warping. In a superior class of work the boxings are provided with cover flaps which conceal the shutters when folded, and fill the void when they are opened out. The sizes and arrangements of the framing are determined by the general finishings of the apartment, but it is usual to make the stiles of the front shutters range with those of the soffit and elbow linings. When venetian or other blinds are used inside, provision is made for them by constructing a block frame from $2\frac{1}{2}$ to 3 in. thick inside the window frame, and the shutters are hung to this. (See Plate XXII.)

The leaves are hung to each other with wide hinges called back flaps, that screw on the backs of the leaves, there not being sufficient surface on the edges for butt hinges. In setting out the depth of boxings, at least $\frac{1}{8}$ in. should be allowed between each shutter to provide room for the fittings; the shutters are fastened by a flat iron bar hung on a pivot plate fixed on the inner left-hand leaf, and having a projecting stud at its other end which fits into a slotted plate, and it is kept in this position by a cam or button. Long shutters are made in two lengths, the joint coming opposite the meeting rails of the sashes; these are sometimes rebated together at the ends.

A Window fitted with Boxing Shutters is shown in sectional elevation on Plate XXII. It has a cover flap, and spaces for a blind and a curtain; one-half of the elevation shows the shutters opened out and the front of the finishings removed, revealing the construction of the boxings, &c. The plan, f. 2, is divided similarly, one-half showing the shutters folded back, with portions of soffit, cornice, &c.; the other half gives the plan and sections of the lower parts; the dotted line showing the window back.

Fig. 3 is a vertical section, and f. 4 an enlarged section through the boxings. The framed pilaster covering the boxing is cut, at the level of the window board, and hung to the stud A′, this being necessary for the cover to clear the shutters when open (see dotted lines on opposite half). The cover flap closes into rebates at the top and bottom, as shown in section in f. 1. The window back is carried behind the elbows, and grooved to receive the latter. The rails of the soffit must be wide enough to cover the boxings, and should have the boxing back tongued into it, as shown in the section, f. 1. When the linings of an opening run from soffit to the floor uninterruptedly, they are called **jamb linings**; but when they commence, as in the present instance, under the window board, they are termed **elbow linings**, the corresponding framing under the window, being the **window back**. Other arrangements of boxing shutters are shown in f. 1 & 2 opposite, also in f. 3, Plate XX., in the preceding chapter.

Sliding Shutters are used instead of folding shutters in thin walls, and consist of thin panelled frames running between guides or rails fixed on the soffit and window board, which are made wider than usual for that purpose. When open, they lie upon the face of the wall adjacent to the opening. They are so seldom used that they do not call for illustration.

BALANCED OR LIFTING SHUTTERS are shown in section in f. 3 opposite, and plan in f. 4. They consist of thin panelled frames, the full width of, and each half the height of the window, hung with weights in a cased frame in a similar manner to a pair of sashes. The frame extends the whole height of the window, and is carried down behind the floor joists to the set off. The window board is hinged to the front panelling or shutter back, so that when the former is lifted, the shutters can pass down behind the back out of sight. Cover flaps hung to the outer linings on each side, close over the face of the pulley stiles and hide the cords. When it is desired to close the shutters, the cover flaps are opened out flat, as shown to the left of plan. The window board can then be lifted, and the shutters run up, a pair of flush rings being inserted in the top edge of each for that purpose. The window board is next shut down, the inside shutter brought down upon it, the outside one pushed tight up to the head, and the meeting rails, which overlap an inch, fastened with a thumb-screw. The bottom rail of the upper shutter is made an inch wider than the other, for the purpose of showing an equal margin when overlapped. Square lead weights have generally to be used for these shutters in consequence of their comparative heaviness.

ROLLING OR SPRING SHUTTERS are made in iron, steel, and painted or polished woods, and though somewhat monotonous in appearance, are, in conse-

Breastsummer
C
Blind Frame
Blind Space
Curtain Box
Soffit Lining
Cover Flap

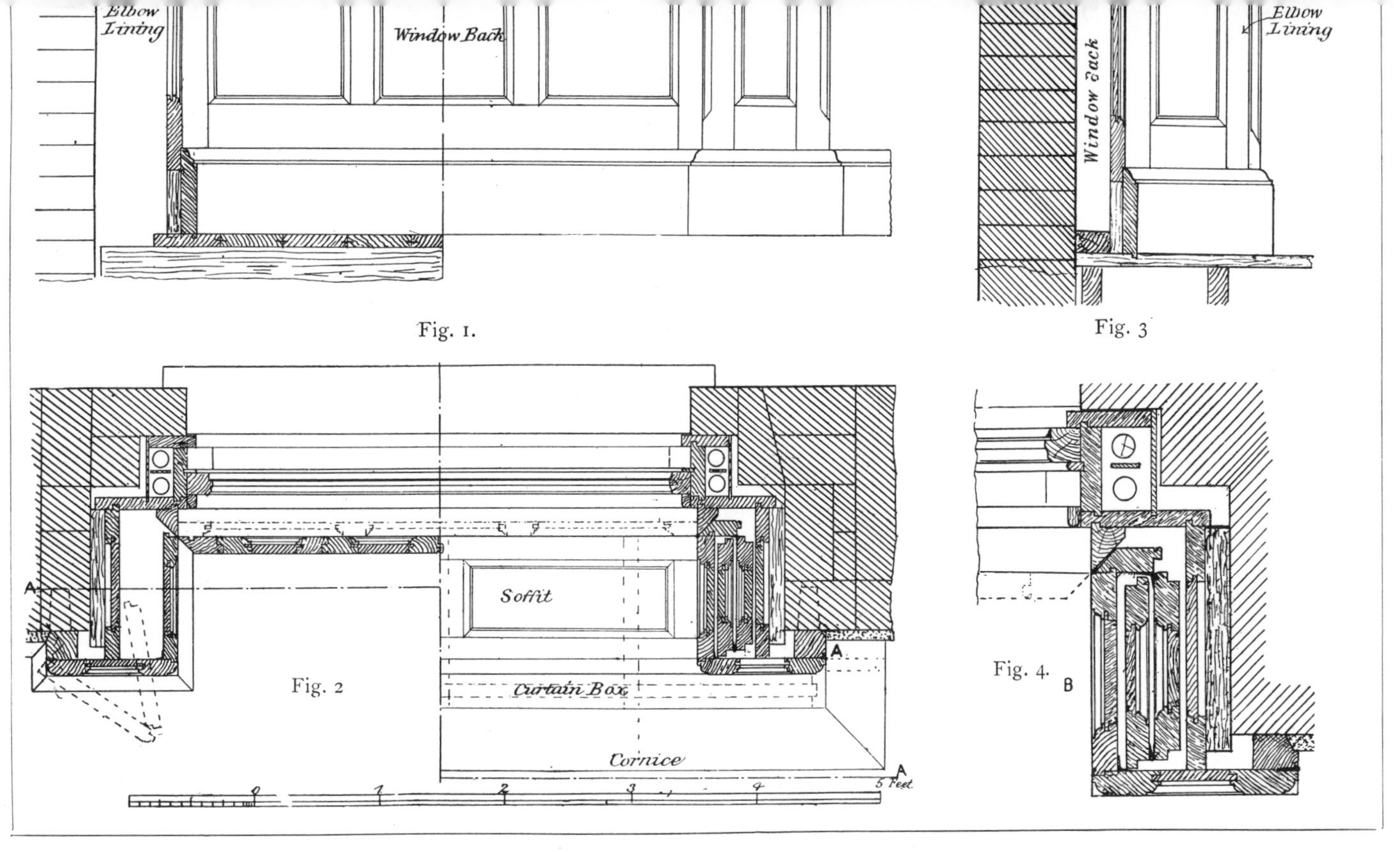

WINDOWS FITTED WITH BOXING SHUTTERS.

1. Sectional Elevation on A A. 2. Plan. 3. Section. 4. Enlarged Detail of Fig. 2.

[*To face page* 228.

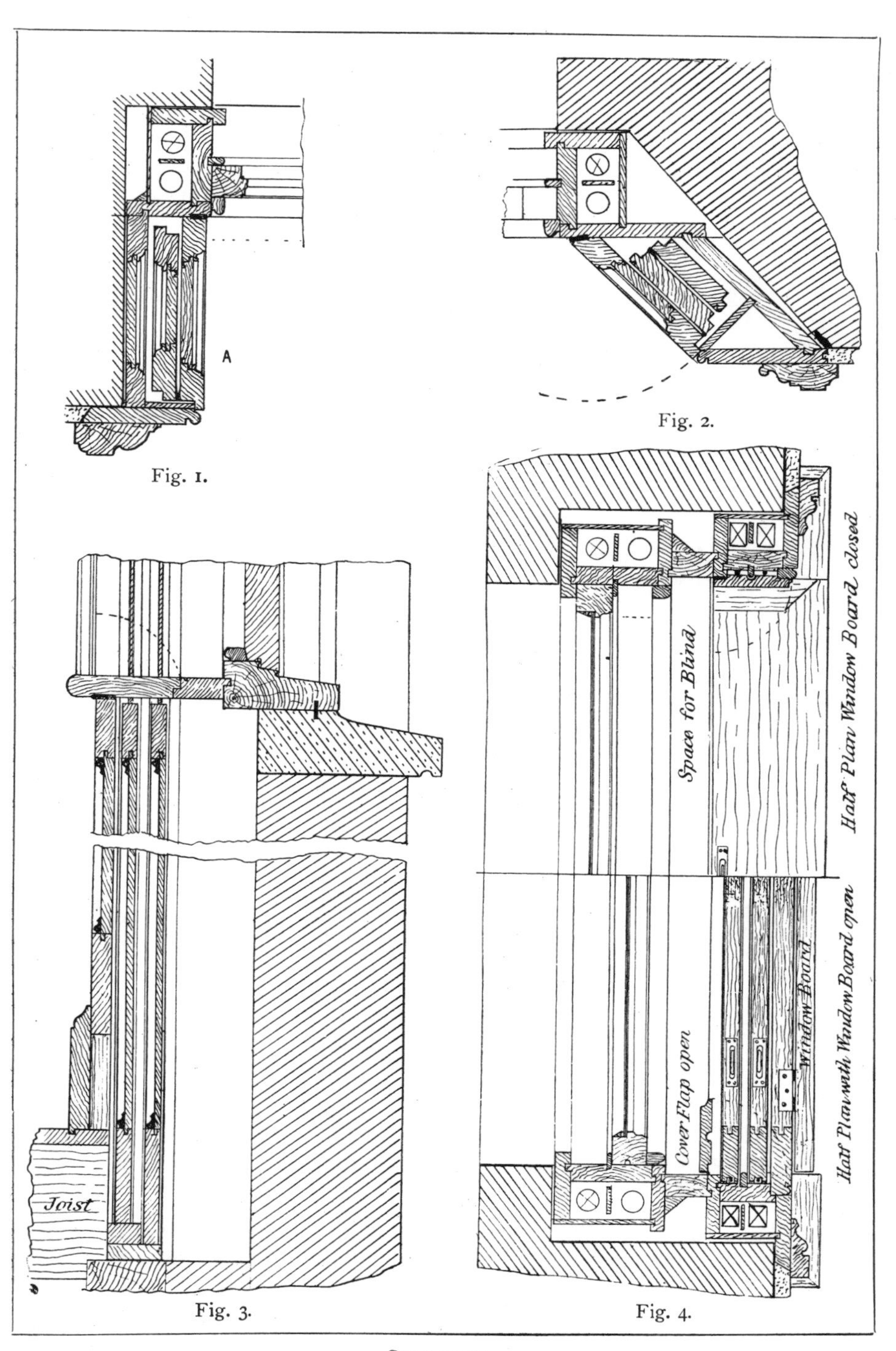

SHUTTERS.

1. Plan of Boxing Shutters in Square Opening. 2. Plan of Boxing Shutters in Splayed Opening. 3. Section of Lifting Shutters. 4. Plan of Lifting Shutters.

quence of their convenience in opening and closing, and in the case of the metal kinds, the additional security against fire and burglary, fast superseding all other kinds both for internal and external use, especially in shops and public buildings. They consist, in the case of the wood varieties, of a series of laths of plano-convex, double convex, oval, or ogee section, fastened together by thin steel or copper

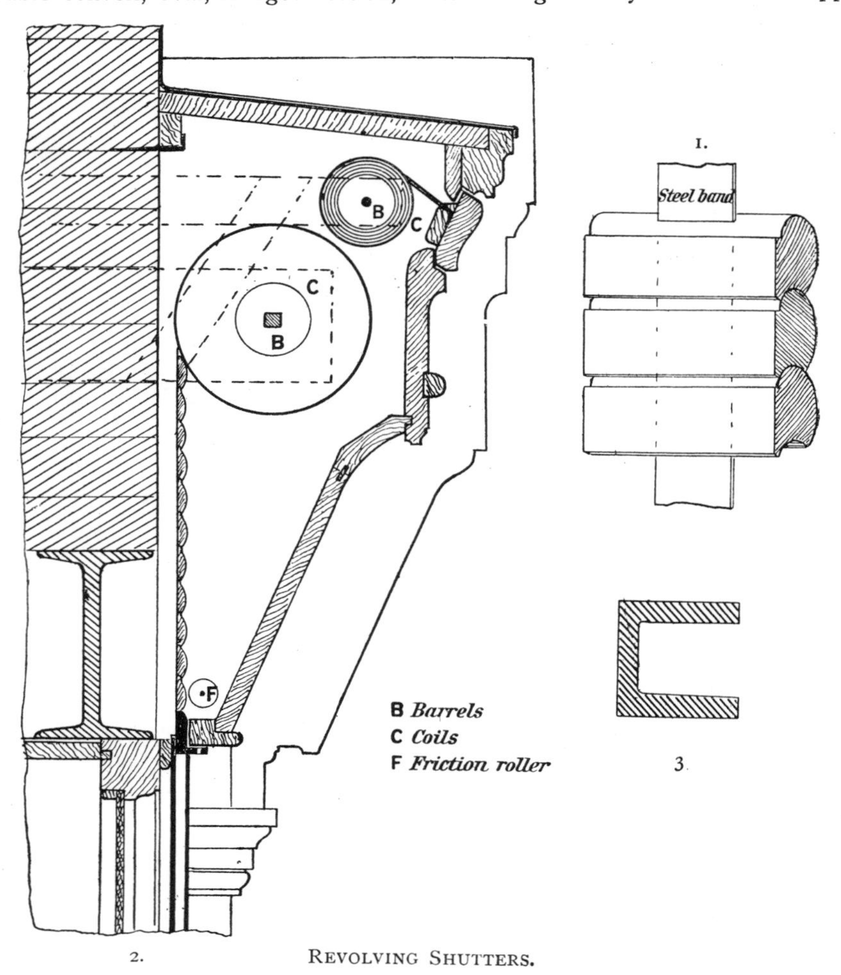

REVOLVING SHUTTERS.

1. Detail of Shutter Connections. 2. Section of Spring Shutter in Shop Front.
3. Channel Iron for Shutters.

bands passing through mortises in the centre of the thickness, as shown in f. 1, and secured at their upper ends to the spring container. These metal bands are supplemented by several waterproof bands of flax webbing glued to the back sides of the laths. The upper edge of each lath is rounded, and fits into a corresponding hollow in the lower edge of the one above it; this peculiar overlapping joint, whilst preventing the passage of light, &c., between the laths, enables the shutter to be coiled around the barrel. The spring barrel, is usually made of tinned iron plate, and encases a stout spiral spring wound round an iron mandrel with squared ends, to which a key is fitted for winding the spring up. The ends of the mandrel project from the case and are fixed in bracket plates at each end of the shutter—in the case of shop fronts, in a box or recess behind the fascia; one end of the spring is secured to the barrel, the other to the mandrel; the shutter is secured to the barrel by the metal bands mentioned above, and the normal condition of things is, that when the shutter is coiled up the spring is unwound. The pulling down of the shutter winds up the spring, and the tension is so arranged that it does not quite overcome the weight of the shutter and the friction, when the shutter is down, so that the latter must be assisted up with a long arm. Great care should be taken in fixing these shutters to arrange the barrel perfectly level and parallel with the front, and to securely fix the brackets. These may be bolted to the girder or breastsummer, as shown in f. 5, Plate XXIV., and f. 3, Plate XXV., or screwed to fixings plugged in the wall, as shown by dotted lines in f. 2 opposite. Where possible a wood groove should be formed on each side of the openings for the shutters to work in, but iron channels, f. 3, are frequently used. These are cemented into a chase in the wall or pilaster. Fig. 2 opposite is a section through a shop fascia, showing shutter and blind barrels fixed to the face of the wall, where no provision has been made beneath the girder.

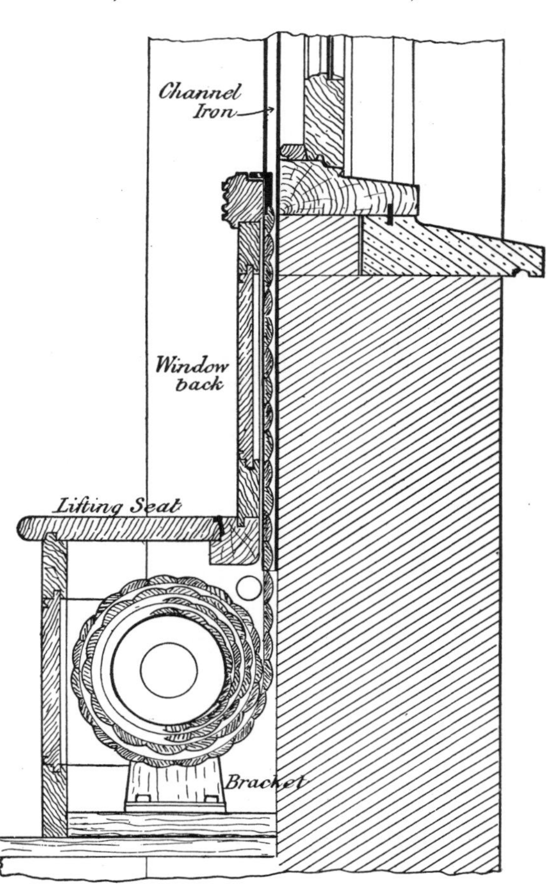

Section of an Inside Revolving Spring Shutter.

An Inside Revolving Shutter is shown above. The barrel is fixed in a seat formed beneath the window, the top of which is hinged, to gain access to the coil. The minimum space required for a coil, for a shutter about 6 ft. high, is 10½ in. A friction roller F should be fixed close to the back rail to prevent chafing as the coil unwinds; the shutter is lifted by means of a flush ring in the L iron bar at the top, and this ring engages with a tilting hook in the soffit to keep the shutter up. The metal varieties of these shutters are wound up by the aid of balance weights or bevel wheel gearing.

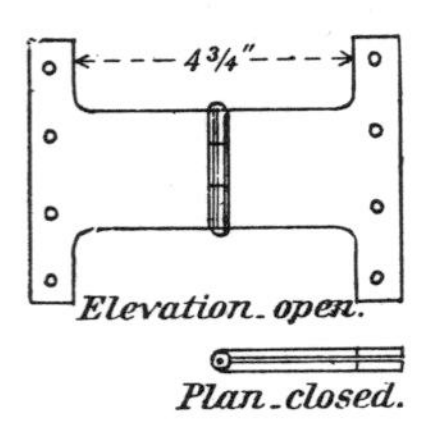

1. A Parliament Hinge.

The Hanging or **Wall Shutter** for dwelling-houses, now nearly obsolete, is simply a thin panelled frame composed of a pair of stiles and three rails, with two bead butt panels, hung to the outside lining of the window frame in such a manner that when opened back it will lie close to the face of the wall, where it is fastened with a turn-buckle. The H or "parliament" hinge with which they are hung is shown in f. 1.

Outside Lifting or **Shop Shutters** are shown in elevation in f. 2, and an enlarged section in f. 3. These are made in sets to cover the shop front. They range in width from 11 to 18 in. The higher they are, the narrower they are made, to reduce the weight. They are usually made out of 1¼-in. stuff for the frames,

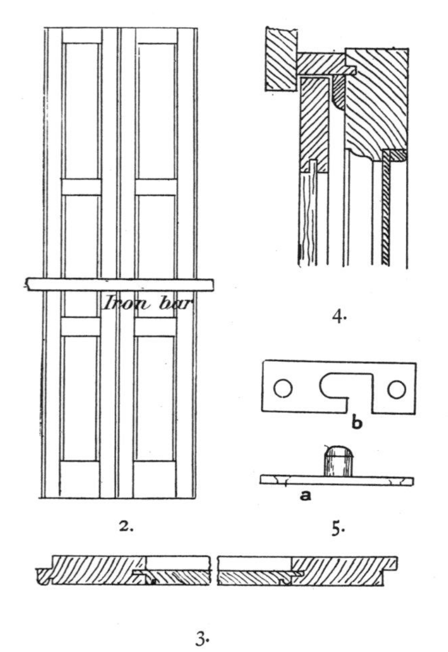

6. 7.

OUTSIDE LIFTING SHUTTERS.

2. Elevation of Two Shutters. 3. Section Enlarged. 4. Method of Fixing at Head. 5. Fixing Studs for Bottom.

6. Elevation of a Venetian Shutter.
7. Section of a Venetian Shutter.

with ⅝ in. for the panels, bead butt or bead flush on the outside; the edges are rebated together and beaded, as shown in the section, f. 3. Wrought-iron angle shoes are fitted to the corners to prevent wear. The shutters are fixed at the top end, as shown in f. 4. A channel is formed between the guard bead on the shop front, and the lower edge of the fascia board, into which the end of the shutter is slipped, the lower end resting on the stall-board, where it is secured in two ways. First by a wrought-iron bar which runs across the face of the shutters, one end entering a slotted plate in the pilaster, the other fixed by a long screw bolt passing through the story post. In the second method, a small iron stud *a*, f. 5, is fixed in the stallboard in the position occupied by the centre of each shutter, and a slot plate *b*, f. 5, is sunk in the bottom edge of the shutters with the notch inside. The shutter is entered upon the pin and pushed tight up to its neighbour, and the last one is secured by a screw bolt passing through the bottom rail of the front and fixed inside with a butterfly nut.

The Venetian Shutter or sun blind, f. 6 & 7 opposite, consists of two rectangular frames about 3 in. deep, formed of ¾-in. stuff dovetailed together at the angles, and filled in with ½-in. louvre boards housed ¼-in. deep in the sides, the louvres arranged so that the lower edge of one just covers the upper edge of the one beneath. The guide frame should be of oak or teak, and fixed to the face of the wall by plugs and screws or wall hooks. A ¼-in. play should be given to the shutters sideways, and a pair of brass pivot rollers inserted in the bottom of each to enable them to move easily. Sometimes the louvres are pivoted at their ends into the sides of the frame, and are connected together by a wire running from top to bottom of the inside face; this enables them to be adjusted at various angles to suit the light. In this case the top and bottom rails of the sliding frame must be made thicker to stiffen the joints at the angles.

CHAPTER XIV.

SHOP FRONTS AND SHOP FITTINGS.

Characteristics of Modern Shop Fronts—Technical Terms—Requirements for Various Trades — Linendrapers—Milliners — Tailors — Butchers—Fishmongers—Grocers—Confectioners — Chemists — Greengrocers — Cheesemongers — Jewellers — Public-Houses, &c.—Heights of Stallboards for the Various Trades—Suitable Woods to Use for Different Trades and Situations—How to Ventilate—To avoid "Fogging"—Double Windows—To prevent Fascias splitting—Method of Fixing Fronts—Glazing to prevent breakage—Heights of Counters—Underfittings for Various Trades—Method of Fixing Spring Shutters and Blinds—Details in the Conversion of a House into a Shop—Provisions of the London Building Act in reference to Shop Fronts.

THE chief characteristic of the modern shop front is the absence of vertical division bars; the whole front is usually filled with one plate of glass stretching from boundary wall to the doorway, the containing frame being kept as small and unobtrusive as possible. This paucity of woodwork renders it almost impossible to design shop fronts with any pronounced individuality in their treatment, the supposed requirements of the shopkeeper overriding all considerations of architectural effect. The result is the monotonous repetitions of big sashes mounted on narrow strips of framing and surmounted by heavy name-boards which line the thoroughfares; broken here and there by the ornate public-house front, with its superfluity of meretricious detail. Before noting the requirements of the various trades which should govern the arrangement of the front within the limits mentioned, it may be well to explain the technical names applied to the several parts comprising a shop front.

The term SHOP FRONT embraces the whole of the wood or metal work between the pavement and breastsummer in height, and between the party walls of the building in width. The **Sash** is the frame containing the glass. The **Stallboard** is the stout sill upon which the sash sits; this piece of timber is so called because it was formerly the custom, when open fronts were used, to project it some distance into the street, forming a stall for the display of the wares; it was sometimes a fixture on brackets, but more frequently hinged, so that it might be folded down at night. The custom still lingers to some extent among fishmongers and butchers, but generally, since the introduction of closed fronts, the stall itself is placed inside, where it becomes the window-board or **Show-board**. The framework below the stallboard is called the **Stallboard Framing** when panelled, and **Stallboard Light** when glazed. The board above the sash is called the **Fascia-board**, and the moulding at the top is the **Cornice**. The panelling over the doorway is the **Soffit**. The narrow soffit under the fascia

SHOP FRONT AT NO. 5 QUEEN VICTORIA STREET, LONDON, E.C.

George Walton, Architect.]

[*To face page* 234.

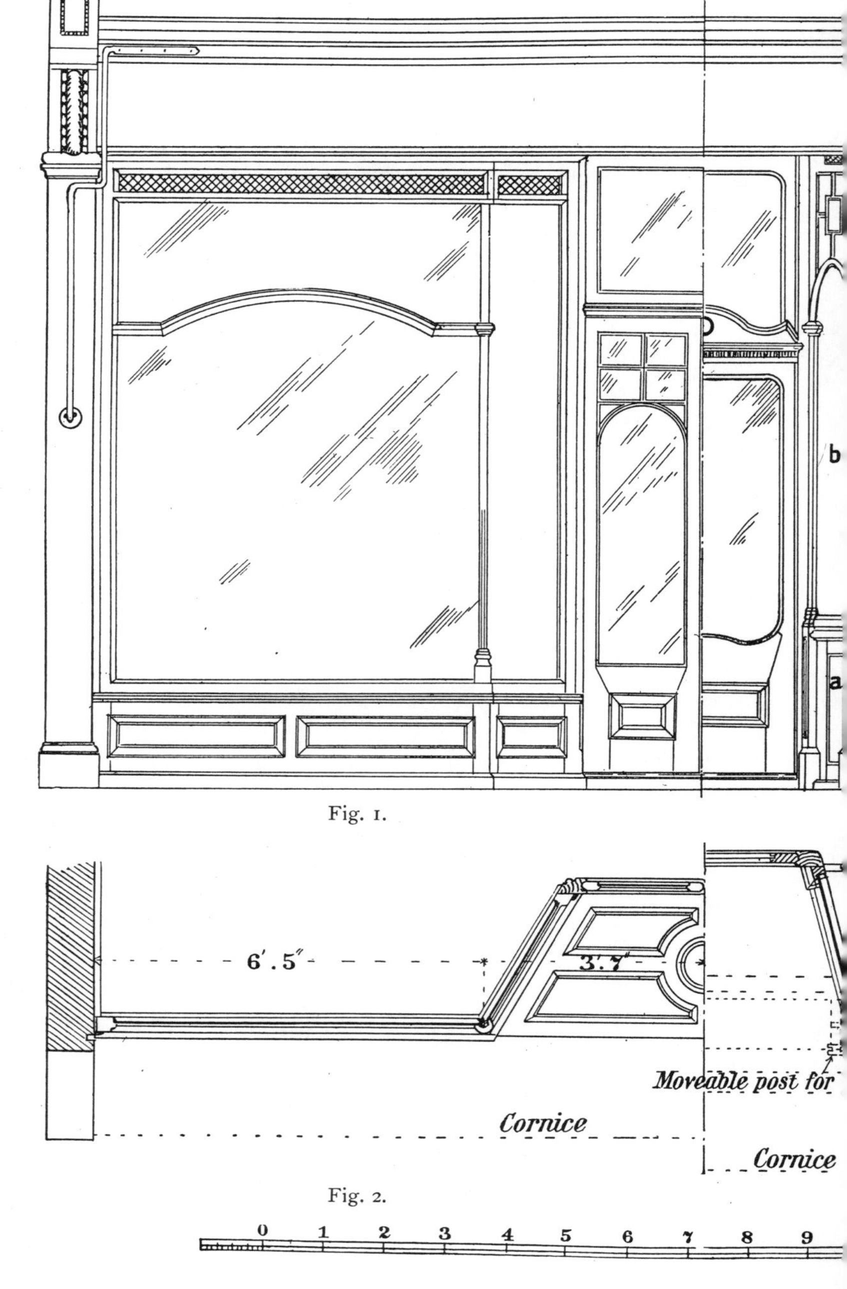

Fig. 1.

Fig. 2.

1, 2. Part Elevation and Plan of a Modern Shop Front. 3, 4, 5. P

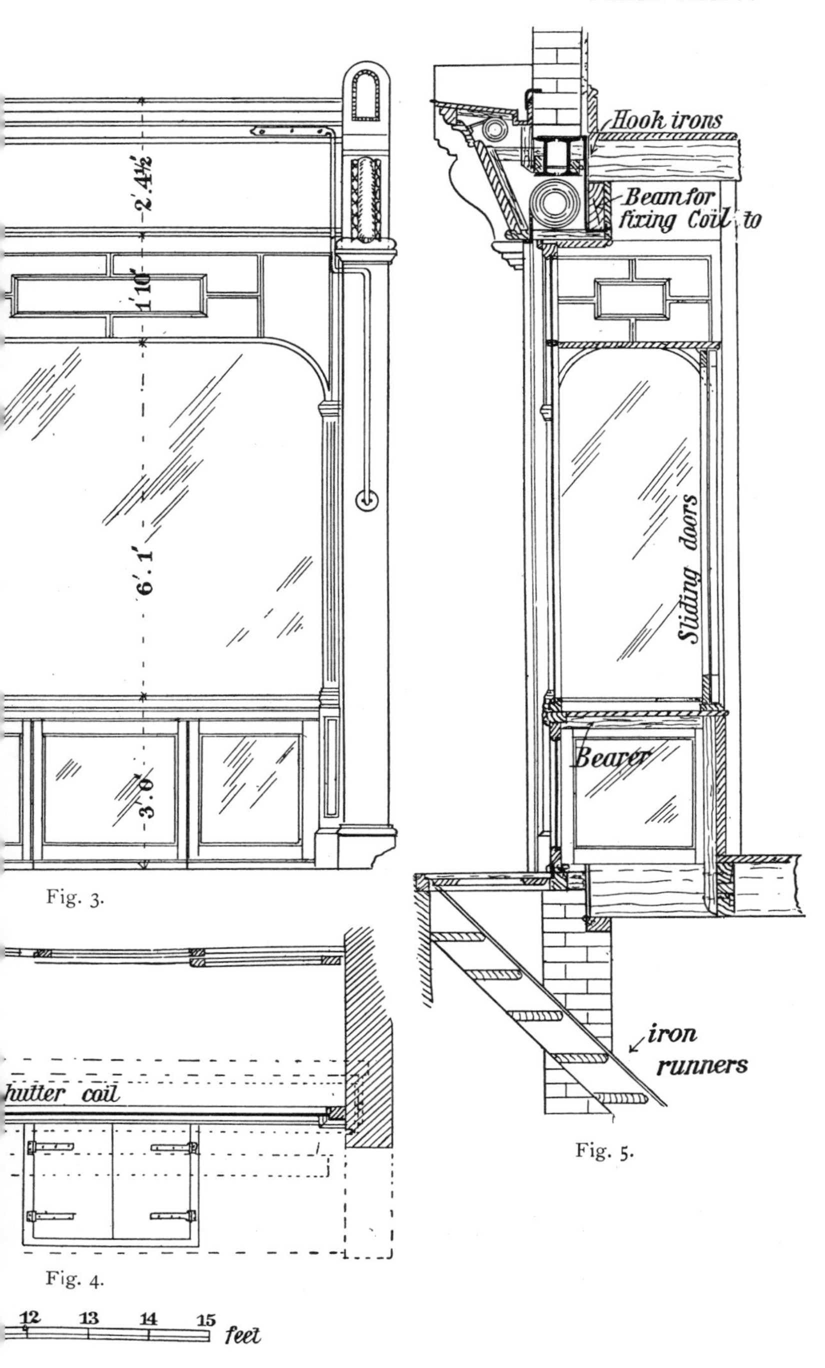

Fig. 3.

Fig. 4.

Fig. 5.

VATION, PLAN, AND SECTION OF A SHOP FRONT, WITH LIGHTED BASEMENT.

[*To face page* 235.

is the **Planceer**. The frames or square columns dividing adjacent fronts are **Pilasters**, and the moulded stone projection above them are **Trusses**.

Traders dealing in bulky goods, such as tailors, bootmakers, outfitters, &c., require their goods to be viewed from above and from the front, also to have ample height to display them in. To meet these conditions the stallboards are kept low down, frequently they are only a few inches above the pavement, and the windows are generally enclosed with glazed sliding lights. The fronts may with advantage be made massive, and of richly coloured wood, to contrast with the sombreness of the contents. Linendrapers and milliners prefer polished brass fronts, as harmonising with the brilliant colours of their wares. Traders dealing in provisions or other perishable goods, and those whose stocks include

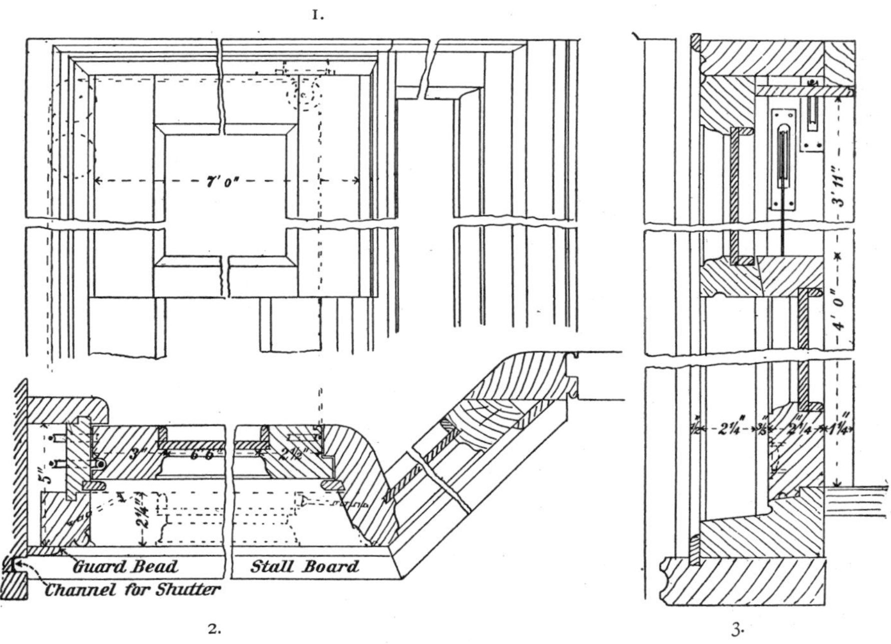

Details of a Shop Front with a Lifting Sash.

numberless small articles, such, for instance, as butchers, fishmongers, grocers, and confectioners, require that their goods shall be brought near to the level of the eyes, and the usual heights for stallboards in these trades are from 2 ft. 6 in. to 3 ft. Most of these trades do not require the upper parts of the window for show, and the front may be broken up with bars, or otherwise ornamented in this part, as shown in the designs in f. 3, Plate XXIV. Butchers, fishmongers, greengrocers, and cheesemongers require their windows to open, as their stocks deteriorate quickly if confined; this requirement, and the difficulty of preventing the goods touching and soiling the sashes and adjacent parts, render it necessary that these shall be of the plainest description, free from small or intricate mouldings that would be difficult to keep clean. In f. 1 to 3 above is shown the construction of a front suitable for many of these trades. The window on one side of the doorway is divided into two sashes. The top outside one is fixed, and the bottom one, which moves up inside it, is hung to balance weights by wire cords that pass over pulleys in a boxing formed on the wall side. Both weights are placed on this side, so that the angle bar may be reduced to as small dimensions as possible. The cord from the right-hand side is carried over a pulley screwed to the head of the frame behind the top sash, and the boxing is carried down below the stallboard and through the floor if necessary, to obtain sufficient depth for the weights to drop; if the latter is inconvenient, a sufficient drop may be obtained without it, by fixing a pulley in the head of the weight and bringing the cord up through it to a hook in the head; the weight will then only need to drop half the distance that the sash rises. Both cords may also be attached to one weight if it is sufficiently heavy to counterbalance the sash. Butchers' and fishmongers' shops should have slate or marble show slabs carried on stout framed brackets; the former level, the latter sloped towards the street, with gutters dished in around the front and ends.

Teak, Oak, Black Walnut, and Mahogany, the latter polished natural or "bright," also ebonised, are the chief woods used for shop fronts. The first mentioned is more suitable for large and massive fronts with boldly struck mouldings, as used for banks, large stores, offices, &c. It is the most durable and also the most expensive of the four. Oak is not suitable for fronts facing south, as it soon bleaches in the sun to a pale grey; wax polish is more effective on oak than French polish. Black walnut does not make an effective front, and is mainly used because of its cheapness. Mahogany is the more generally used wood, and if thoroughly well hand-polished (not brush-polished), will retain its brilliancy for years. Ebonised fronts are used for those shops whose contents are brilliant or rich coloured, such as jewellers', silversmiths', silk mercers', chemists', &c. In these fronts the very smallest possible amount of wood is used, in many instances the half-round bars being less than $\frac{1}{2}$ in. in thickness. A full-size section of a bar suitable for a jeweller's window is shown above.

Section of a Bar suitable for a Jeweller's Shop.

Ventilation should be provided for by making fanlights to open, and framing in ventilators in the top of the sashes, either as small gratings in the top rail, as shown at f. 3, Plate XXIV., or as pierced metal panels between the two rails, as shown in the design, f. 1. If this is not done, the windows will "steam" in the cold weather through condensation upon the cold glass. Where this fault occurs, and for some reason efficient ventilation cannot be provided, the only prevention is to have a light frame containing a second plate of glass about 1 in. inside the front; this one will intercept the moisture, and being warm, will not fog or "steam." Many shops are now provided with "blue flame" gas burners in the showboards, which are kept burning during damp weather to prevent steaming. These act by warming the inside of the glass of the front, also by drying up the moisture in the air.

Two Designs for Shop Fronts are shown in f. 1 & 3, Plate XXIV., and various details are given in f. 1 to 6 below. The front, f. 3, Plate XXIV., has a glazed stallboard light to the basement; the centre sash is hung at the side; a pair of flaps give access to the basement. The inside rail of the flap frame is removable after the light is opened, lifting out of notches in the end rails for the purpose of giving a clear way to the ladder beneath, which also is fitted with rails for a travelling hoist. **Fascia-boards** should be fixed so that they may swell and shrink without danger; this is best provided for by turning screws in the face

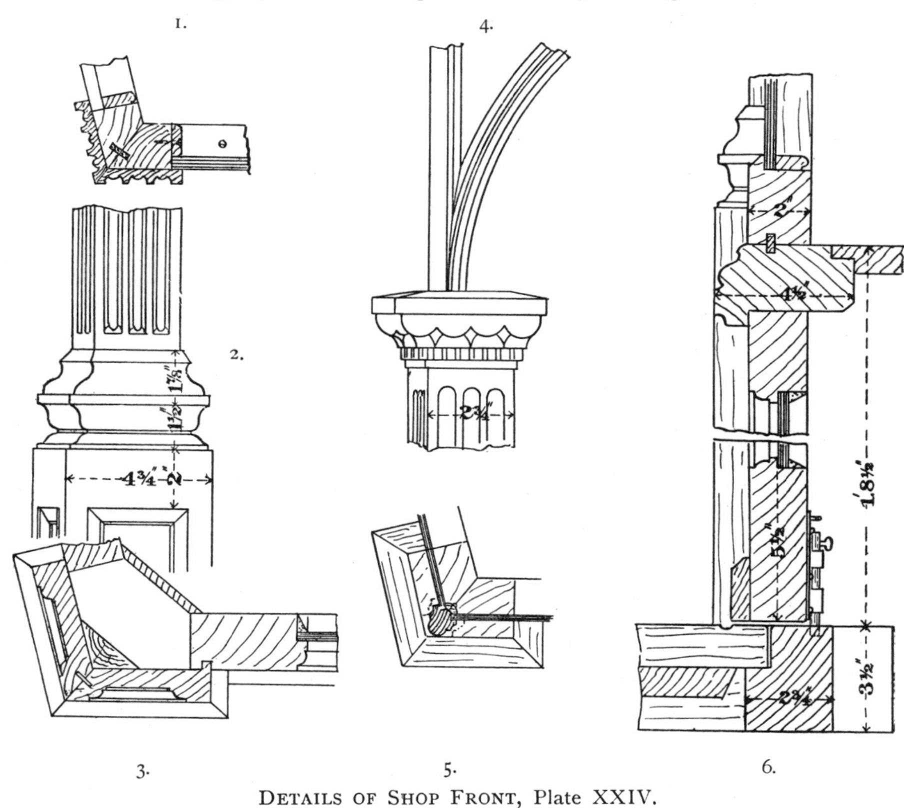

DETAILS OF SHOP FRONT, Plate XXIV.

1. Section at *b b*, f. 3, Plate XXIV. 2. Enlarged Elevation of same.
3 Section at *a a*, f. 3, Plate XXIV. 4. Elevation of Angle Post, f. 2.
5. Plan. 6. Section through Stallboard and Flap.

of the rough brackets, leaving them projecting $\frac{3}{8}$ in., and making slots in the back of the fascia to slip over their heads, or by fixing pairs of undercut fillets across the back to form dovetail grooves, and sliding these down over dovetail tongues previously formed on the face of the brackets. When they are of considerable width, they should have stout battens buttoned across the back to prevent casting. $\frac{3}{8}$-in. iron rods driven through from edge to edge are sometimes employed for this purpose, but are much less effective. If a proper selection of suitable boards were made, there would be less need of preventive measures against casting. Boards that are cut tangential to the annual rings will warp freely, however they may be fixed, whilst those cut radial, *i.e.*, with the rings at right angles to the surface, will have scarcely any tendency to cast. Honduras mahogany is undoubtedly the best and most serviceable wood for fascias, but the back and edges should be painted to protect them from dampness, otherwise it will rot quickly.

Shop Sashes should be fixed at the sides, not at the top, lest any sagging of the breastsummer should throw weight on the frame, which might cause the glass to break. At least ½ in. clearance should be allowed above the head, and an equal amount at each side, to give room for the fixing wedges, which are driven

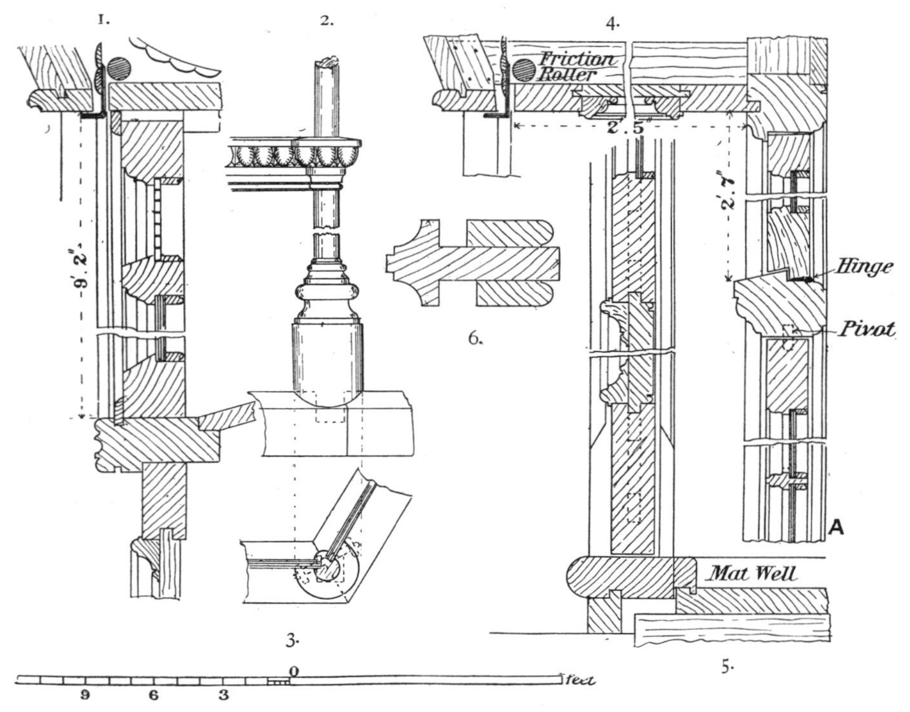

1-3. Details of Shop Front on Plate XXIV. 4, 5. Sections through Doorway.
6. Section of Bar.

in pairs opposite wood bricks or plugs in the walls, and are then secured by nails driven through the edges of the frame, the guard beads covering all over outside, and a lining inside. Heavy frames in superior class work are usually built into chases in the masonry. The upper edges of the rails are usually bevelled for the purpose of throwing off the rain; and bars, &c., should be scribed, not mitred, or housed in, as, in such cases, any shrinkage will allow of access of water to the interior. The bottom rail may be tongued to the stallboard, as in f. 6, p. 237, or the guard bead may be grooved into it, as shown in f. 1 above, to prevent water gaining entry

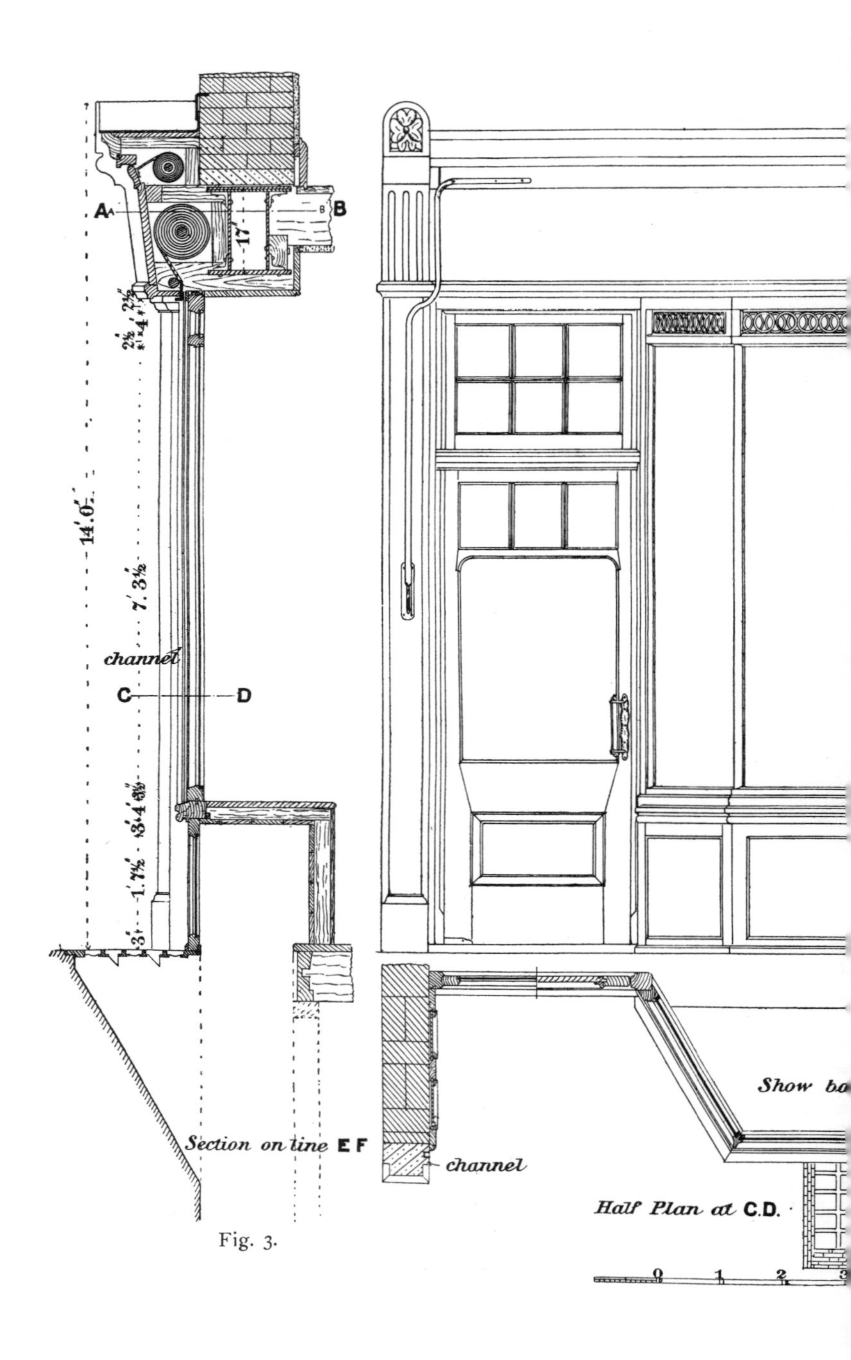

Fig. 3.

1-3. Design of a Shop Front suit

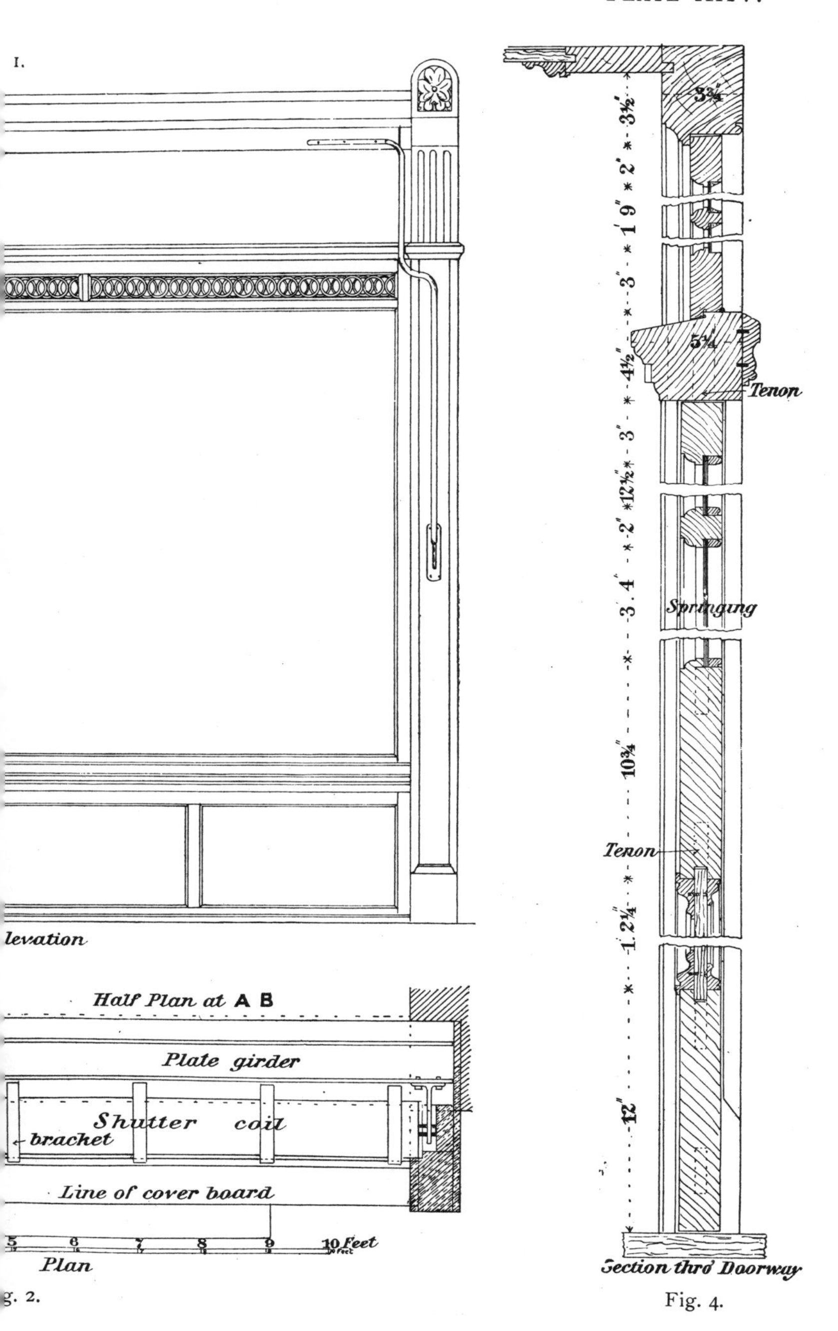

g. 2.

Fig. 4.

A GROCER, WITH SECTION AND DETAILS.

[*To face page* 239.

The Glass in shop fronts should always be fixed upon the inside, and is better bedded with wash leather than putty, which is liable to cause breakage when it hardens, if there is much vibration from the traffic.

Counters are generally arranged to suit the shopkeeper's requirements, and vary with the size of the establishment. Subject to these limitations, the following are average sizes for a few of the principal trades:—Drapers and Tailors, height 2 ft. 8 in., width 3 ft.; Grocers, height 2 ft. 10 in., width 2 ft.; Chemists, height 2 ft. 6 in., width 2 ft.; Publicans, height 3 ft. 1 in., width 2 ft. 3 in. Mahogany is the wood chiefly used for counter tops in the dry and clean trades, as it is of good appearance, polishes well, does not splinter, and can be obtained of large scantling. Tops that require frequent washing, such as those used by Butchers, Cheesemongers, Bakers, &c., should be made of spruce, lime, sycamore, or other light-coloured wood of compact texture. Public-house or restaurant counters are sometimes made of deal covered with pewter, also of Jarrah which takes a high polish and is not affected by the liquids used in these places. The top should be made to slope slightly towards the inside, and should overhang about 5 in. in front.

The Underfittings of shop counters are usually drawers of various sizes and stout shelving. Drapers require wide drawers, from 2 to 3 ft. wide, and 8 to 11 in. deep, paper lined. Grocers require a few deep drawers for dry fruit, &c., but most of the space is occupied by shelves with numerous divisions. Bakers have a few deep drawers, and grids or open shelving. Tailors, &c., require a raised base about 3 in. from the floor, carrying divisions about 3 ft. apart, with one or more wide shelves, the whole covered by sliding doors.

A **Front** suitable for a Grocer's or Baker's shop is shown in f. 1 to 3, Plate XXV., and an enlarged detailed section through the doorway in f. 4. The stallboard front is fitted with a glazed framing, which, with prismatic pavement lights, illuminate the basement. An opening is trimmed in the floor to correspond with the pavement light, to give further access of daylight. The shutter is arranged in this example, to show the method of fixing when there is not room beneath the girder, or provision in the wall has not been made for the same. Such cases arise when an existing building has to be converted into a shop, and in the present instance a heavy wall has to be carried over the opening. A wrought-iron box girder is used for the purpose, with a 3-in. York stone templet, to distribute the load of the brickwork. The shutter is hung in two lengths to wrought-iron brackets bolted to the girder, and the fascia and soffit are secured to a series of skeleton wood brackets fixed between the flanges of the girder. Above these are fixed another series of brackets to carry the cover board and cornice, and wood brackets wedged into the wall support the spring blind roller. The floor joists B are carried by a wood plate bolted on the inside of the girder.

The Shop Front shown in the photograph, Plate XXIII. (which has recently been fitted to a building in Queen Victoria Street, London), is the design of Mr George Walton, architect, High Street, Kensington, and is a very neat and effective one. The lower portion is constructed of oak, and, as will be seen by inspection of the sheet of working drawings here given, Plate XXVI, the sash, rising directly from a low marble step, is bow-shaped in plan, and the head and bottom rail are attached to the side door-jambs in the manner shown in the enlarged detail, f. 7. The bars, shown in section, f. 3, are chamfered off from the rebate to a small square in front, and they are tenoned and haunched into the sash rails, as shown in f. 4, also at f. 5 and 6, which are respectively a section and plan of the bottom rail; the mortise and haunching, which is stopped at the back and front, is shown hatched in f. 6, and its position indicated by dotted lines in f. 5.

In connection with the subject it may be useful to note that the provisions of the London Building Act require that no shop front shall project more than 5 in. from the external wall of the building wherein it is situate in any street, &c., not wider than 30 ft. In any street of greater width than 30 ft., such shop front may project 10 in., and the cornice may project 13 in. in the former and 18 in. in the latter case. Also, no part of the woodwork of any shop front shall be fixed higher than 25 ft. above the pavement level of the public footway, and no part of the woodwork shall be fixed nearer than 4 in. from the centre of the party wall or other wall of adjoining premises.

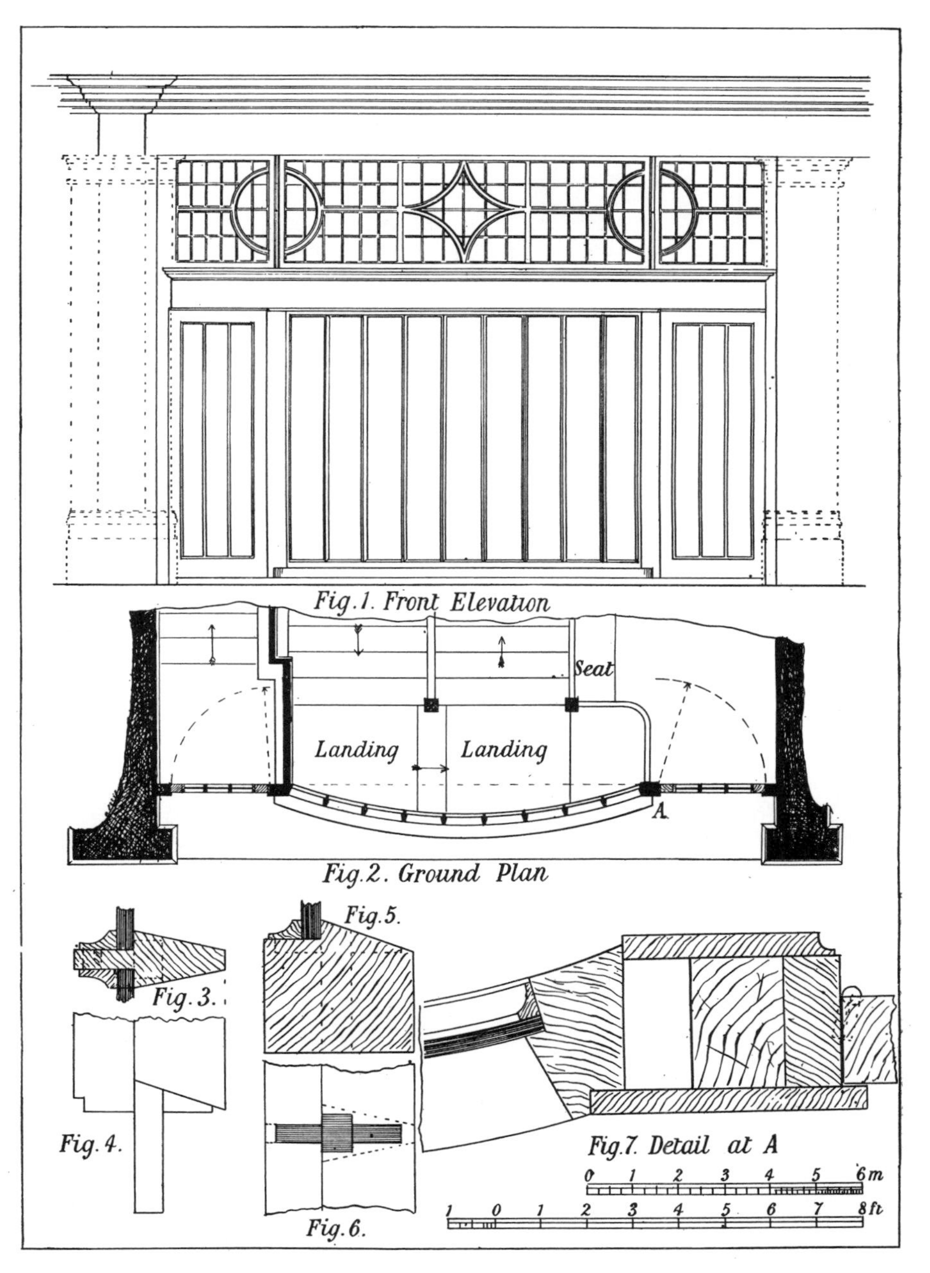

CONSTRUCTIONAL DETAILS OF SHOP FRONT, NO. 5 QUEEN VICTORIA STREET, LONDON.

George Walton, Architect

[*To face page* 240.

CHAPTER XV.

AIR-TIGHT CASE WORK.

Introduction of Air-tight Cases—Construction of a Counter Case—Requirements—Composition of Cements for Glass—Air-tight Joints, How Formed—Kinds of Wood Employed—Sizes of Parts—How Screws are Fixed in Ebony Bars—Fitting Up—Glazing—Construction of a Centre Case—How Joints are Rendered Air-tight—Arrangement of Parts for Removal—Method of Fitting Up—Ventilation, &c.

THE AIR-TIGHT SHOWCASE is a product of the latter part of the nineteenth century, and is a result of a high degree of expertness obtained by specialising in this branch of joinery, coupled with the precision of modern tools, both hand and machine. The examples herein given are confined to two typical varieties of showcase, but as the methods of forming the especial joints, which are the distinctive feature of this class of work, are the same in all cases irrespective of size or design, these will, it is hoped, be found sufficient for explaining the principles of their construction.

A Square Counter Case, such as is used by jewellers, tobacconists, &c., is shown in the two transverse sections, f. 1 & 2, next page, f. 3 being a sectional plan drawn to the same scale, but broken in length and width in deference to the exigencies of space. Fig. 4 is a back elevation broken in like manner, one side showing the door closed, the other open or rather removed. Next to the property of air and dust tightness, which is an essential feature in these cases, their most important qualification is, that the woodwork in them is reduced to the smallest possible dimensions that will keep the glass in position, so that an uninterrupted view of their contents may be had. This slenderness of the surrounding frame renders it necessary that the joints in it shall be made with the greatest accuracy and care, and that the glass panels shall fit the openings exactly. In the best work the joints between the wood and the glass are hermetically sealed by the use of a quick drying cement, which unites the two so firmly that they can seldom be separated without breaking. In a commoner quality of work an approximation to air-tightness is obtained by bedding the glass in a kind of soft putty made with white lead and boiled oil, which is also stained to match the wood of the case. The opening flap or door at the back or end, as the case may be, is rendered air-tight by forming small reeds upon the rebates in the frame, as shown in f. 5, and corresponding flutes upon the edges of the door which fit together accurately. These are worked by special planes made in pairs known as "air-tight joint" planes. For the shutting edge a reed and hollow are worked side by side, forming a miniature hook joint, f. 10.

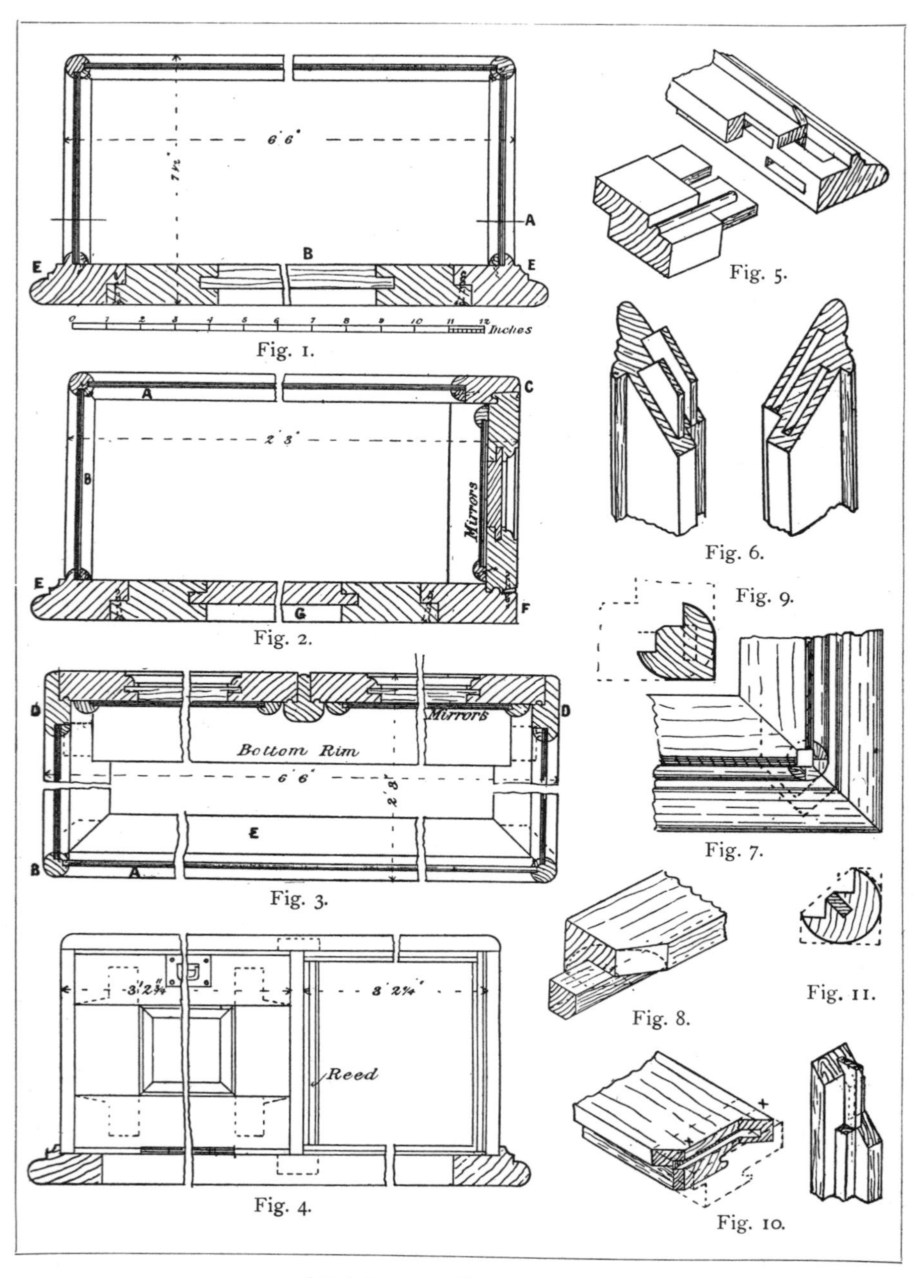

AN AIR-TIGHT COUNTER CASE.

1, 2. Sections. 3. Plan. 4. Back Elevation. 5-11. Sketches of Joints.

The Wood usually employed for these cases is clean straight-grained Honduras mahogany, of course perfectly dry, and polished both "bright" and "black." In superior class small work ebony also is used, but as the closest examination is required to detect the difference between the real and imitation ebony, it may be taken that the use of the former is very limited in consequence of the cost of working it. In common work, American basswood and birch are used, black stain covering a multitude of sins—of substitution. The top rim A and the standards B in the case illustrated in f. 2, previous page, are circular in section, and finish $\frac{3}{4}$ in. in diameter. The top back rail C and back standards D each finish 2 by $\frac{3}{4}$ in., and are rounded on the front edges, the back ones being worked respectively to a hook joint and a reeded rebate. The front and ends of the base E are $2\frac{3}{4}$ by $1\frac{1}{2}$ in., moulded and rebated for the glass, and also rebated inside to receive the bottom. The back rail F is $2\frac{3}{4}$ by $1\frac{1}{4}$ in., rebated and reeded for the doors and rebated at the bottom. The bottom is a panelled frame, so made to prevent casting and splitting; the panels are seldom required the full thickness, but must be flush on the inside. The bottom is fixed in the frame with screws, and is made removable to facilitate the glazing of the case; it may be made of pine, as it is usually covered either by a velvet lining or by a movable tray. The various joints used in connecting the parts of the frame are shown in isometric projection in f. 5-11. Fig. 5 is the joint between the back and end rails of the base. $\frac{3}{16}$-in. double tenons, stubbed in as deep as the section of the moulding will allow, are used. The outer tenon is kept $\frac{1}{16}$ in. nearer the bottom so that additional depth may be obtained for the mortise for the standard. In cutting the shoulder on the back rail the reed must be allowed to run over to meet the rebate in the standard. The whole of the seating, however, is not taken out until the base frame is glued up, otherwise the end of the top mortise would probably break through, neither is the mortise for the standard made until the glue is dry. Fig. 6 is the joint in the mitre of the base. The moulding and rebating is done first, and when gluing up, shaped blocks with their outer edges parallel with the mitre are glued on the moulding near the ends of the mitres, and these are drawn together with a hand-screw. Fig. 7 is a plan of the same joint after it has been glued up, and showing the mortise and mitreing for the standard. Fig. 8 is the end of the standard to larger scale showing the tenon, the dotted lines indicating the finished section. All the mitres in the standard and rim are made before rounding the edges, and the inner edges only are rounded before the case is glued up. Fig. 9 is a full-size section of the rim, the full lines showing its section when ready for gluing up; the outer angle is rounded off after the glass is in; the dotted lines indicate the position of the front and end rim in plan, and the mortise and tenon in the mitre. Fig. 10 shows the joint at the back angle of the rim. The dotted lines indicate the method of setting out the back rail; the width of the end rim is marked off, then divided equally by the line X X, the central part of which is cut in and becomes the joint shoulder; from the intersection of the outside lines with the front and back edges mitres are drawn with the templet. The width of the back rim being set off on the end piece, a gauge set to half the width of the latter is run on to obtain the joint, and mitres are marked in like manner from the sight lines. A similar mortise to that shown in f. 7 is made in the top rim after it is glued up for the tenons of the standard, but previously a thin piece of hardwood should have been glued over the top of the

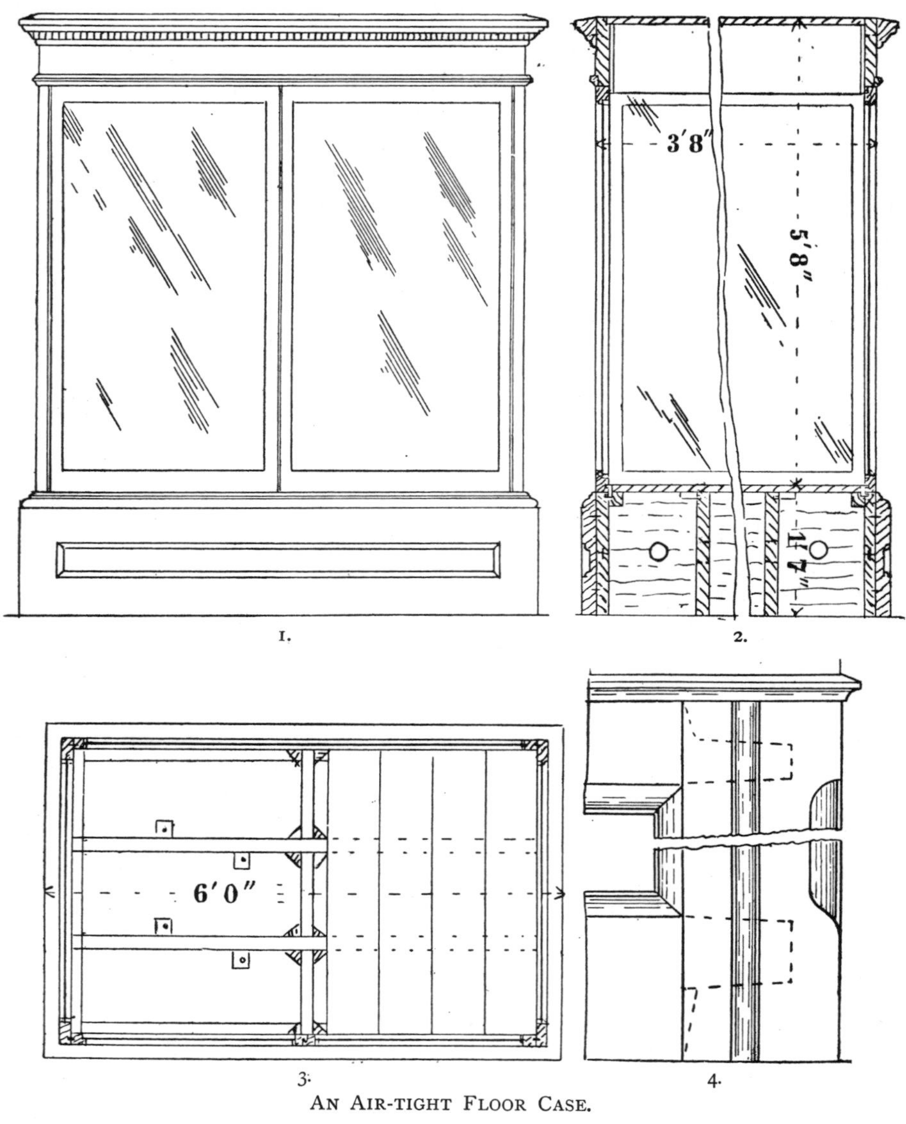

AN AIR-TIGHT FLOOR CASE.

1. Elevation. 2. Section. 3. Plan. 4. Enlarged Detail of Sash Stile.

mitre to strengthen it whilst it is being mortised and fitted. The base rim should be glued up, before the bottom frame is made, so that a good fit may be ensured. The doors are framed together with stub mortises and tenons, the laying members running through and the upright ones framed between them, as shown in f. 4, p. 242, so that no end grain should come in the hook joint.

To Fit the Glass in, the case is turned upside down upon a glazing frame—that is, an open frame with a true surface somewhat larger than the case, and raised about 3 in. from the bench by bearers so that the hand can be introduced beneath to manipulate the glass—straight strips are tacked around the rim to keep it in position, and the top plate is bedded in upon a cement made with red and white lead, knocked up in oil and japanner's gold size or terebene, with lamp black substituted for the white lead for black work. The edges of the glass and the rebates should be blacked to prevent reflection. The end plates are next bedded and inserted, being kept up with a light strut whilst the

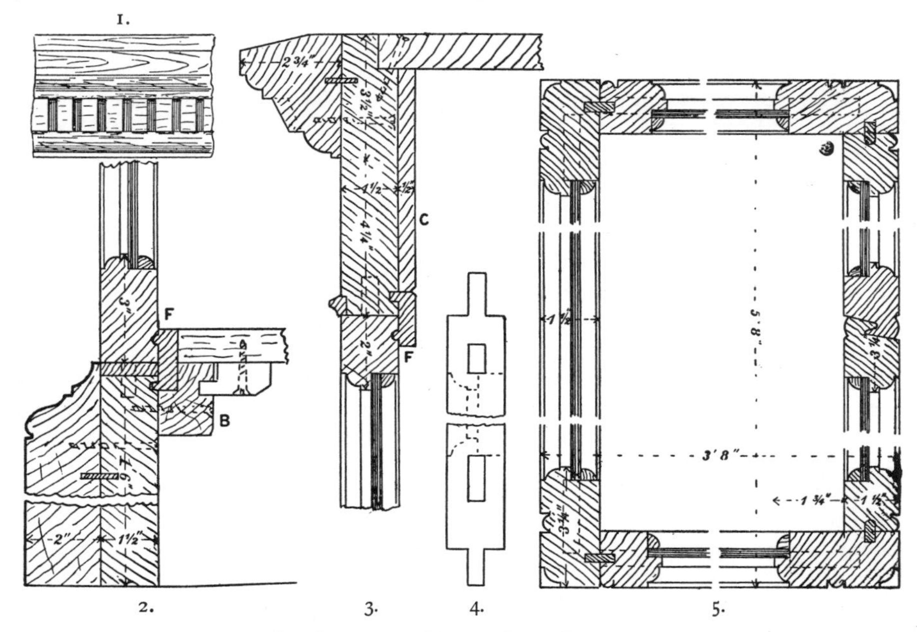

AN AIR-TIGHT FLOOR CASE DETAILS.

1. Elevation of Cornice. 2. Section of Bottom of Case. 3. Section of Top of Case. 4. Corner Standard. 5. Enlarged Section of f. 1, p. 244.

front plate is got in; then the quadrant fillets are screwed in, securing all in place. To facilitate the fixing of these fillets when ebony is used for the case, screwing slips of soft mahogany are glued into grooves made in the internal angles, as shown in f. 11, p. 242; the dotted lines show the preparation of the stuff for ploughing. The doors are sometimes lined with velvet inside, but frequently have mirrors fixed on them, as shown in the sections.

A Centre or **Floor Case** is shown in elevation in f. 1, section in f. 2, and plan in f. 3, previous page. Enlarged details are shown in f. 1-5 above. These cases are intended to be stood in the centre of the floor, so that their contents may be viewed from all sides, and they are provided with deep plinths or bases to bring the goods within easy view, and also to protect the glass from breakage. As they are usually too large to pass through doorways when complete, they are generally made in several pieces, as shown, and fixed together *in situ*.

The Base and the **Cornice** are each complete frames, into which the three glazed frames, forming the side and ends of the central part of the case, are fixed by means of tenons cut on the ends of the stiles or standards, as shown in f. 4, previous page, the joints at top and bottom between them being made air-tight by the tongued and reeded hardwood fillets F, f. 2 & 3, which are glued around inside the base and fascia. The projecting portions of these fillets are of course kept free from glue. The joints between the three fixed sides of the case are grooved and cross tongued (sometimes double tongues worked in the solid are used), and are fitted very tightly together. The two portions of the frames are further secured by means of screws sunk in the rebates, where they are hidden by the glazing slips.

The Joints at the Hanging Stiles of the doors are rendered air-tight by inserted beads, glued into the end frames, as shown in f. 5. The Meeting Stiles are furnished with a hook joint and beaded. The Base is made of deal, edged with mahogany; the boards forming it are ploughed and tongued together, and dovetailed at the angles, thus forming a complete frame to which the plinth is fixed. An inner frame B, f. 2, is screwed to the inside of the base to provide a fixing for the bottom; the bottom is made up of 1 by 10 in. pine boards, tongued and glued together, and buttoned to the frame B and to the rough bearers shown in the plan, f. 3, p. 244. The air-tight fillet F must be mitred in and fixed before the base is put together, and the mahogany plinth is fixed afterwards with glue and screws; two or three cross tongues should be inserted as the width may require, and a tongue should also be placed in each mitre. Two lines of holes should be bored through the rough bearers in the base, and an ornamental grating inserted at each end for ventilation. The Fascia is mitred at the angles, and cross tongued and rebated for the top. The cornice moulding is glued and tongued on, and screwed from inside, the holes being hidden by the cover piece C, which may be of pine, stained to match the outside work. The dotted lines in the sections, f. 2 & 3, indicate the tenons formed on the ends of the stiles of sashes. In framing the sashes, the mortises are stopped, as shown in f. 4, p. 244, when the ends of the tenons would show on the face; the mortises should be made tapering in depth, and the tenons cut to fit them tightly, so that when forced up by the cramps they will be very firm; the edges of the sashes that are not seen can be mortised right through and wedged. The hook joint stiles in the doors should be double tenoned. Fig. 4 also shows the stops at top and bottom of the ovolo moulding on the angles of the case.

CHAPTER XVI.

FITTINGS FOR BANKS, MUSEUMS, LIBRARIES, AND CHURCHES.

Bank Fittings—Design—Suitable Woods—Construction of a Double Slope Desk—A Pedestal Fitting—A Paying Counter—A Cashier's Desk—Height of Counters—Fittings for Cash Drawers—Museum Fittings—Design—Materials—Requirements—Bolts and Locks—Construction of a Floor Case—Standard Dimensions—Dust-proof Joints—Internal Arrangements—An Interchangeable Drawer Case—Entomological Drawers and their Fittings—A Museum Wall Case—Library Fittings—A Wall Case for a Public Library—Method of Construction—Fittings for Shelves—Size and Spacing for Shelves—Movable Folio Divisions—Hanging Doors with Butterfly Butts—Church Fittings—Mediæval and Modern Work Compared—Examples of Mediæval Screens—Description of the Construction—Stalls—Pulpits—Example from Westminster Abbey—A Modern Example—Method of Construction—Plan of Stairs—Method of Making the Opening Joint in the Door Capping.

THE fittings of public institutions provide examples of the highest class of modern joinery. Cost being usually a secondary consideration, the work is executed in the best possible manner to ensure permanent efficiency. Hardwoods are invariably employed, and great precautions are taken to neutralise the effects of our changeable climate upon the material. In **BANK** and **OFFICE FITTINGS** the designs are massive and imposing, but severe; as befits the business operations they are intended to serve. The effects are obtained by the free use of bold substantial mouldings, raised panelling, and the rich colouring or markings of the woods employed, which are chiefly Teak, Walnut, and Mahogany. The former wood is more suitable for those parts exposed to the weather, such as entrance doors, vestibule screens, and linings; the latter for the interior fittings; the Mexican and Spanish varieties for the more ornamental parts; and Honduras for the bulk of the less important work. Walnut is, in consequence of its sombreness, more sparingly used. For economical reasons yellow deal and even basswood is sometimes employed for the hidden or internal parts; and if placed judiciously, will serve equally well as the more expensive hardwoods. Basswood, however, must be used with caution, as it varies constantly in size with the state of the atmosphere, and warps persistently. White deal or spruce should not be used in the constructive parts, as its varying texture prevents it retaining a true surface. Its use should be confined to bearers, blocks, fillets, &c. All working parts, such as drawer runners, &c., should be of oak.

Carving is but sparingly introduced in this class of work, and **Veneering** is chiefly confined to curved parts in which the sweep is so sharp that the abrupt change in the direction of the grain becomes unsightly. All the material must be exceptionally dry, and all joints are grooved and tongued together in various ways.

Mouldings are secured by slot screwing, and wide surfaces by buttoning.

A Double Slope Pedestal Desk, such as is used by ledger clerks in banks, is illustrated on Plate XXVII.; f. 1 is a transverse section, f. 2 the half elevation, and f. 3 a sectional plan. The desk is composed of two plain ends of 1-in. mahogany shaped to section, the usual angle of slope being 15 degrees. They are rebated at the lower edges to receive the bottom,

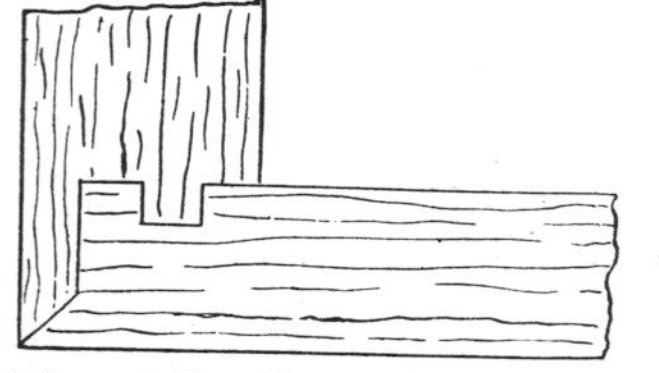

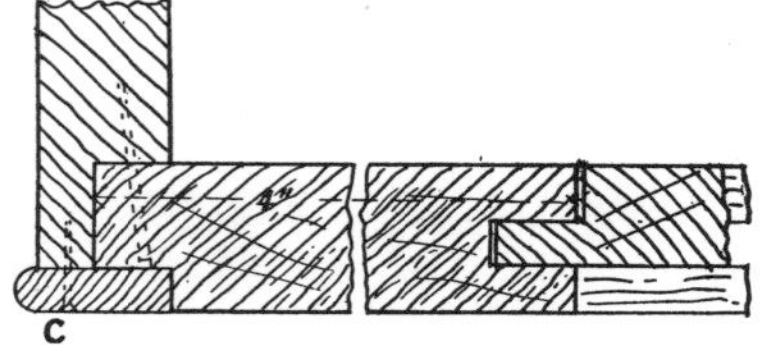

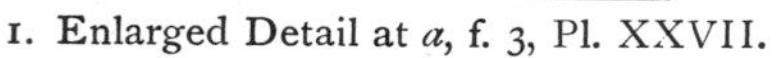

1. Enlarged Detail at *a*, f. 3, Pl. XXVII.

2. Detail at E, f. 1, Pl. XXVII.

and lipped and mitred and tongued to the fronts, as shown in the detail, f. 1. The two front rails are of $\frac{3}{4}$-in. mahogany, rebated similarly (see detail, f. 2), and have the openings for the drawers cut in them. The cross divisions, lettered D throughout the drawings, are of 1-in. deal, of similar shape to the ends, but the thickness of the bottom less in depth (see f. 1, next page). They are fixed to the fronts in a dovetail groove, as shown in the plan, and are half notched under the longitudinal division, as shown in sketch, f. 3. The bottom is a deal panelled frame, with the exception of the rails R, which coming under and forming runners for the drawers, should be of oak or birch. The panels are flush on the top side, and should have $\frac{1}{16}$ in. clearance all round to allow for swelling, as shown in f. 2. The tenon ends and wedges should be cut back $\frac{1}{16}$ in., so that they shall not project should the rails shrink. The bottom is rebated all round for the cock-bead C, f. 2, which both breaks the joint and strengthens the front rail under the drawer openings, and this bead is glued in at the last moment when fitting up. The desk top is composed of the central flat H, and six wings or rails I framed into it, with three bevelled tenons in each, and these tenons are indicated by the letter T in the plan.

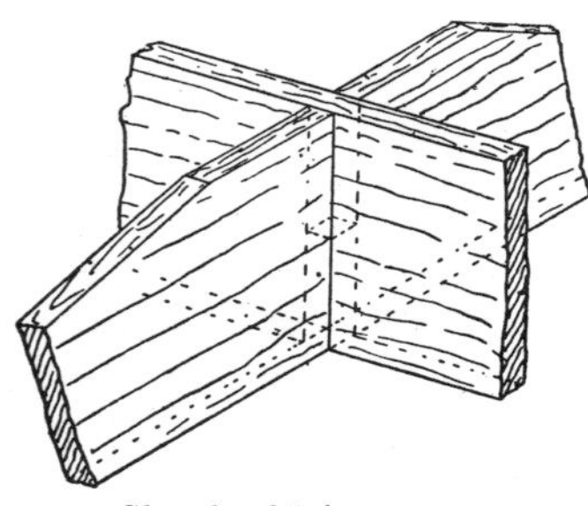

3. Sketch of Joints at Intersection of Divisions.

The Method of Cramping Up these rails is shown in f. 1, next page. A jack or saddle cleat is hooked on the edge of the flat opposite the rail to form a square pull for the cramp, and a shaped block is used at the nosing of the rail (which it is more convenient to work first) to prevent injury by the cramp. The flaps are mitre clamped, as shown in f. 2, and the mitres stopped back, to prevent end grain showing on the edge. It is necessary to run the grain of the flaps in a transverse direction to that of the rails, otherwise they would have to be hung

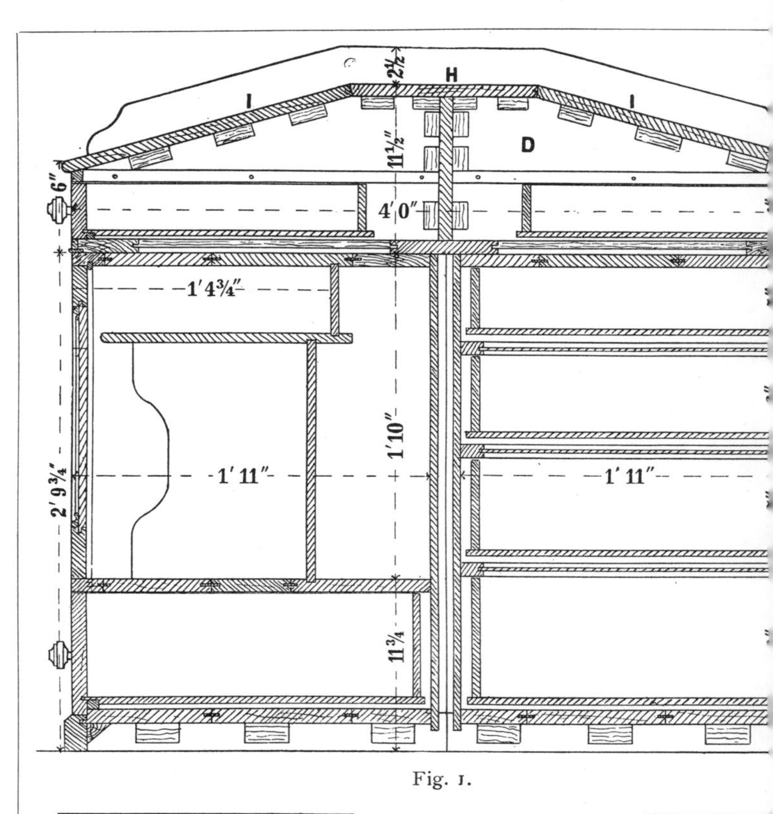

Fig. 1.

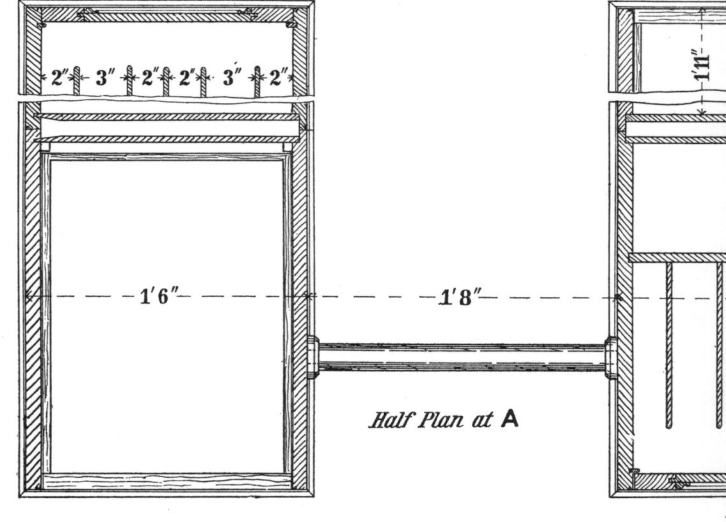

A Double Slope Pedestal

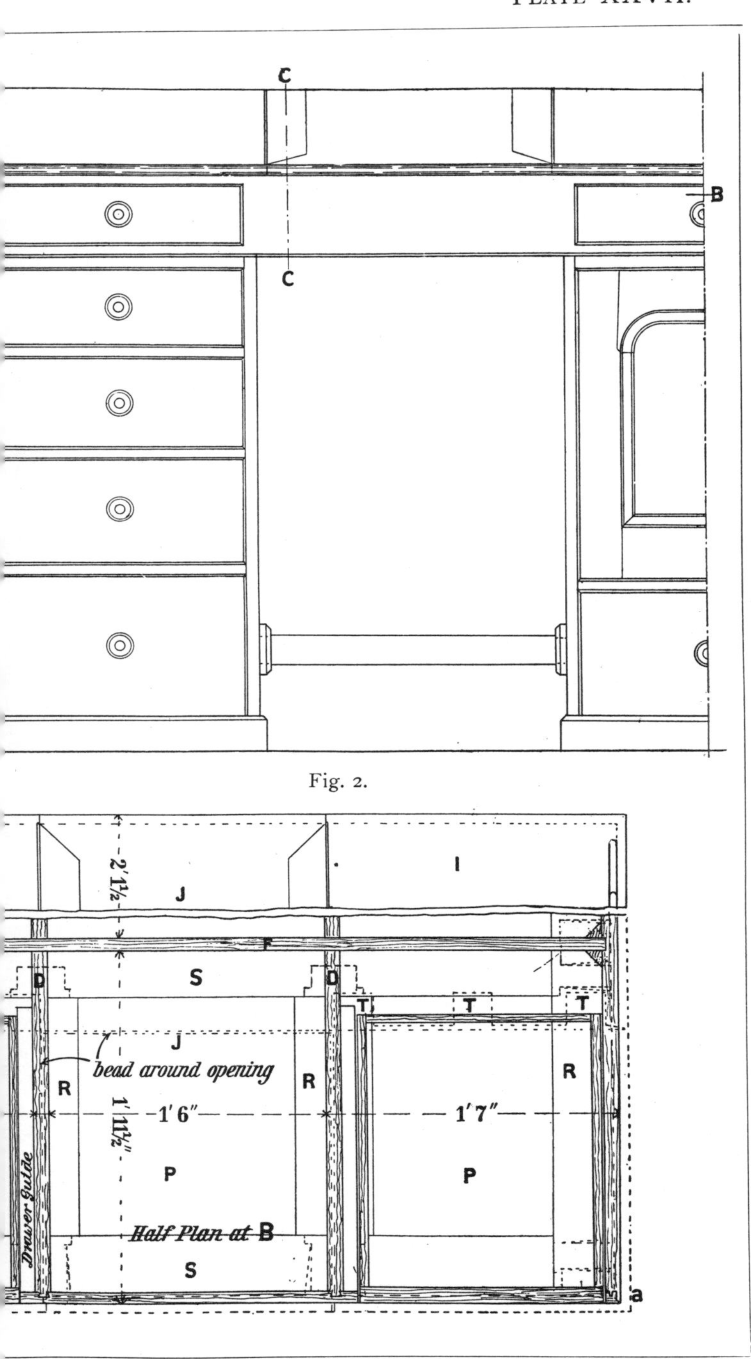

ONS, ELEVATION, AND PLANS.

[*To face page* 248.

by the clamps, a bad arrangement, as their shoulders would soon be started by the jarring. The flaps are fitted with a toothed rack and spur, as shown in f. 1, the latter being screwed on the side of a division. These divisions are,

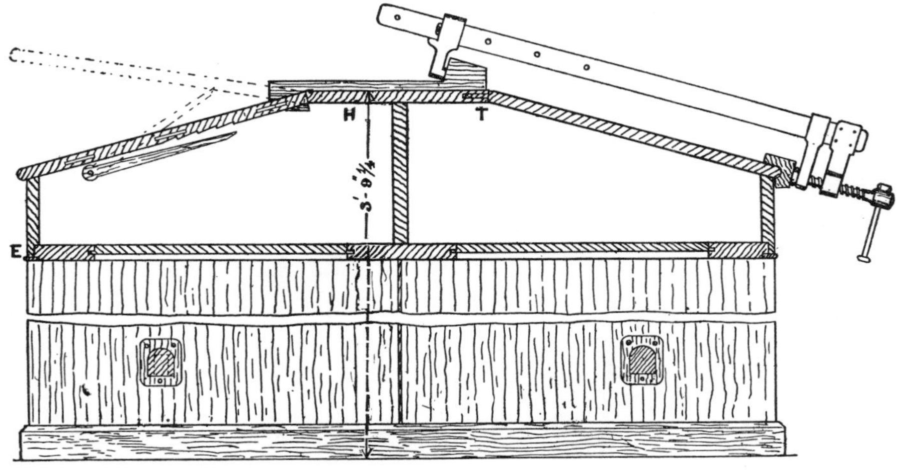

1. Section on C C, f. 2, Plate XXVII.

as will be seen by reference to the plan, kept ½ in. within the opening, to form a rebate for the flap. When the desk is without flaps, technically a "slope," the wings or slopes run in the same direction as the flat, and are connected thereto by ploughed and tongued joint. A shaped plinth cut in one piece is fitted at each end of the desk top by groove and tongue joint with the shoulder outside.

The Pedestals are fitted alternately with book cupboards and drawers. They are made in pairs, but separate, for convenience of handling, and are eventually connected by the side plinths. The tops are lap dovetailed, as shown at f. 1, next page. The bottom, which should run the same way of the grain as the sides, so that one shall not interfere with the shrinkage of the other, is housed, blocked, and nailed through the sides (see section, f. 2, next page).

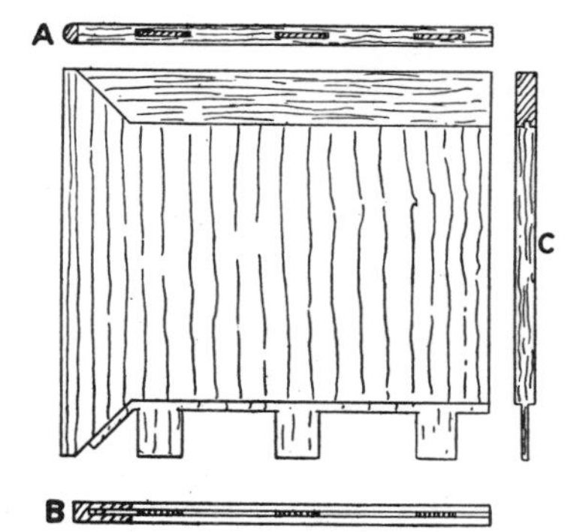

2. Method of Clamping Flaps, f. 1.

A. Outside Edge of Clamp.
B. Inside Edge of Clamp.
C. Back Edge of Flap.

The drawer divisions are panelled frames, housed into the sides dry. The front rail is stub tenoned with double tenons (see A, f. 3, p. 250); this rail is mahogany, the back one pine, the sides or runners oak, and the dust-board pine. A ¼-in. bead is stuck around the openings in the case, which renders it necessary to fix the rails in this manner. In the succeeding example another method is shown. The front rail and plinth are in one piece, grooved and tongued to the bottom (see f. 6, p. 250), and lip

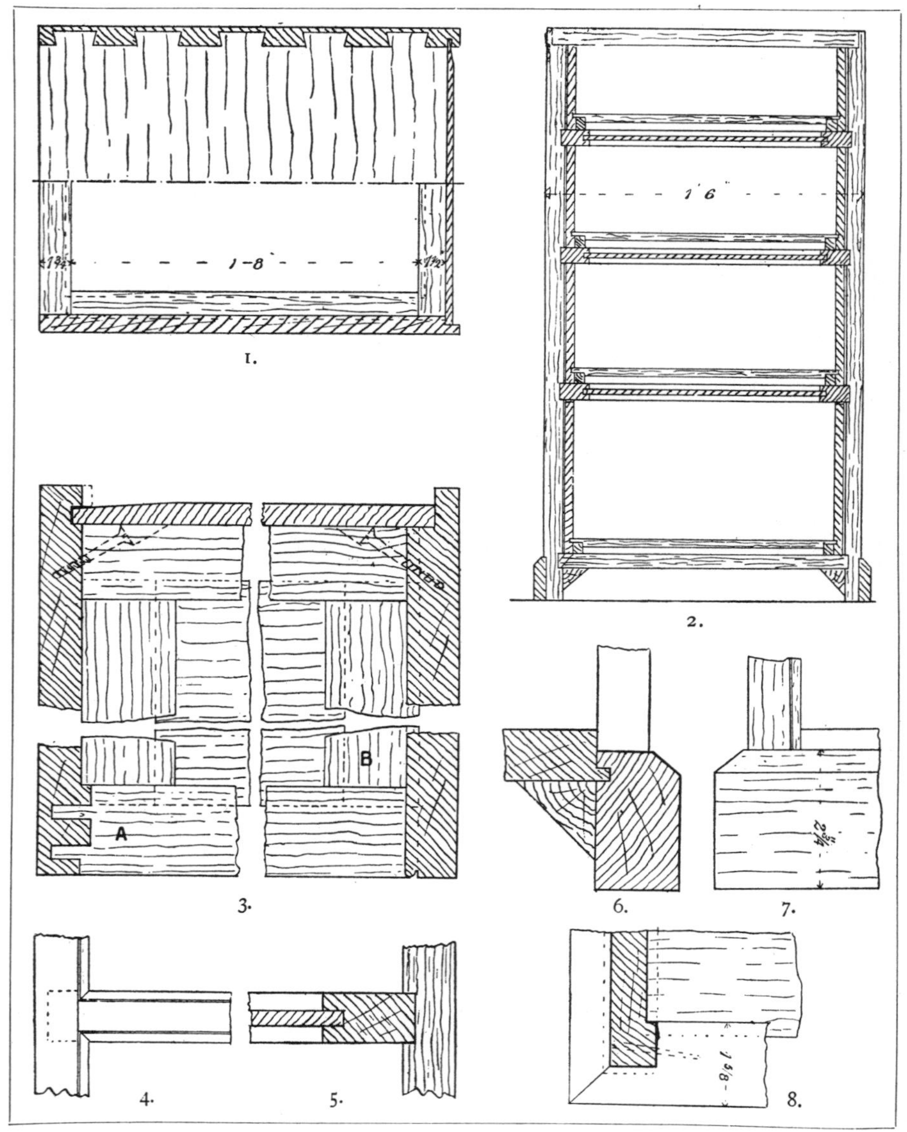

1, 2. Plan and Section of Pedestal. 3. Method of Framing Panelled Divisions.
4-8. Details of Joints.

mitred across the front to meet the side, as shown in f. 7 & 8 above; this mitre should be made before the case is put together.

The Drawers are made in the ordinary way, which has been described in an earlier chapter, as shown in the side elevation on next page, the bottom being

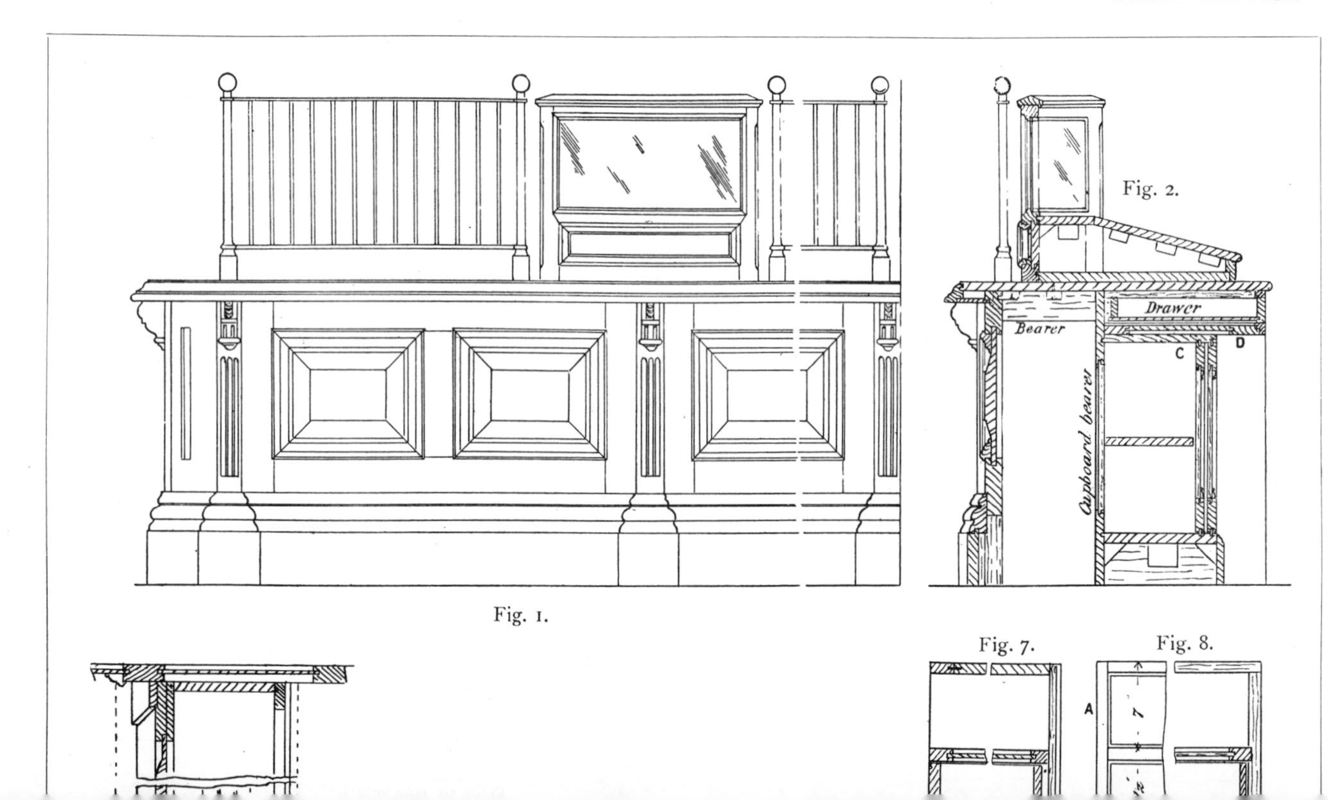

Fig. 1.

Fig. 2.

Fig. 7.

Fig. 8.

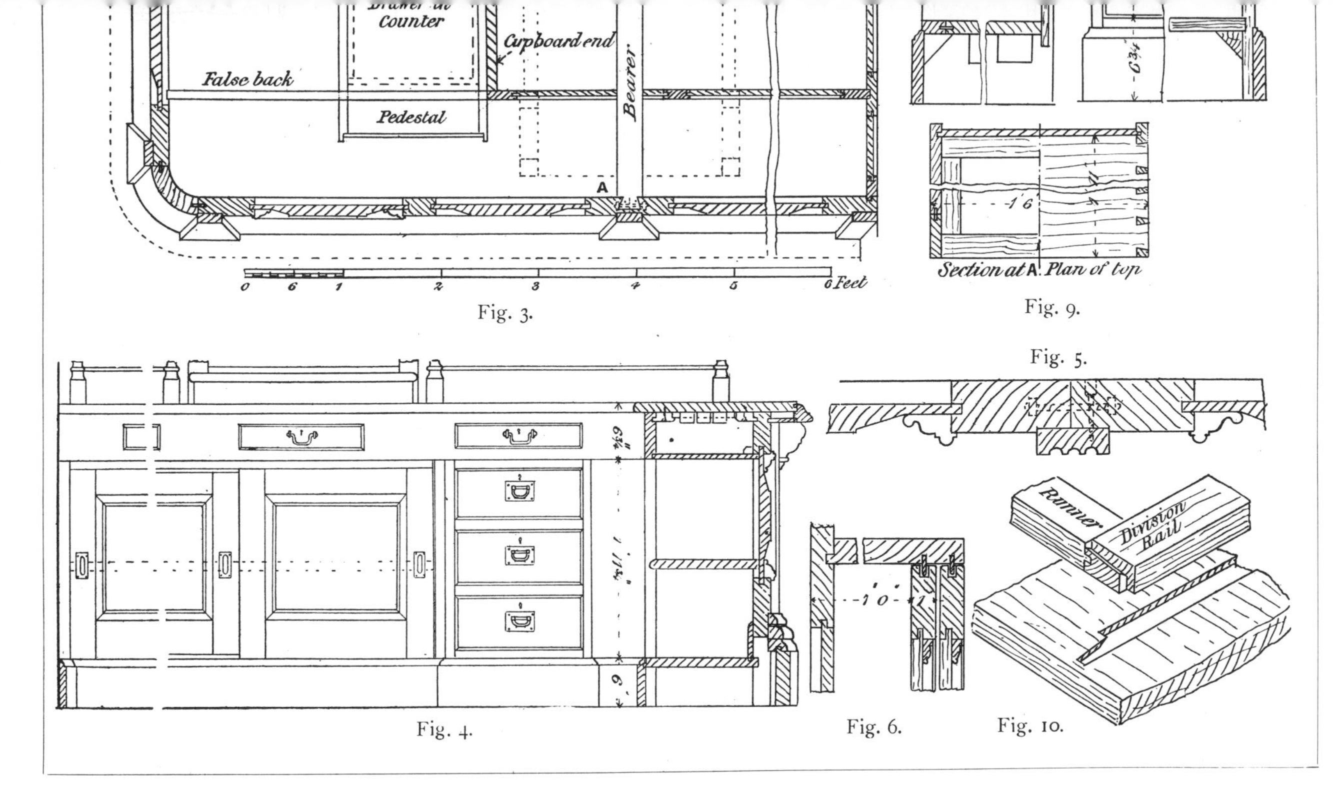

Fig. 3.
Fig. 9.
Fig. 5.
Fig. 4.
Fig. 6.
Fig. 10.

A Bank Counter, with Cashier's Desk, Pedestal, &c.: Elevations, Sections, Plan, and Details.

[To face page 251.

rebated out to receive ½-in. square oak blockings, which should be glued to the sides and not to the bottom. The bottom is secured to the back with two screws in slots, so that it may be able to shrink without splitting.

A Bank Counter, with a cashier's desk, pedestal, and enclosed underfittings, is shown on Plate XXVIII. Fig. 3 is the general plan, broken in length, but showing all the essential parts; one end abuts against a wall, the other is framed into a return counter somewhat narrower than the main one. There are generally no openings in bank counters, access to the enclosure being obtained through doors in screens or offices contiguous. Fig. 1 is a front elevation, and f. 4 a corresponding back elevation. Fig. 2 is a vertical section through the cashier's desk. The height of bank counters range from 2 ft. 10 in. to 3 ft. 6 in., 3 ft. being perhaps the more general height. The widths of paying counters range from 3 ft. to 3 ft. 6 in. Subsidiary counters vary as the requirements.

The Counter Top should overhang the front framing from 4 to 6 in. to permit of persons standing close without kicking the skirting. The top is seldom made thicker than 1 in., although its apparent thickness can be increased as desired by the use of edge mouldings, and in the present instance a thin soffit lining is used to fill in the void between the moulding and the front. The front panelling is made up in a series of frames, with the joints hidden by the pilasters. Apart from the greater ease in fixing and transport, this is constructively a better method than making the rails continuous, as the horns of the stiles can run down to the floor for fixing and support; the frames are connected by two handrail bolts in each joint, as shown in the detail, f. 5. The pilasters are sunk in ⅛ in. and glued on one edge, the other being eventually secured with screws when the front is fixed. The mitres of the mouldings are screwed and the frames fixed by slot screwing.

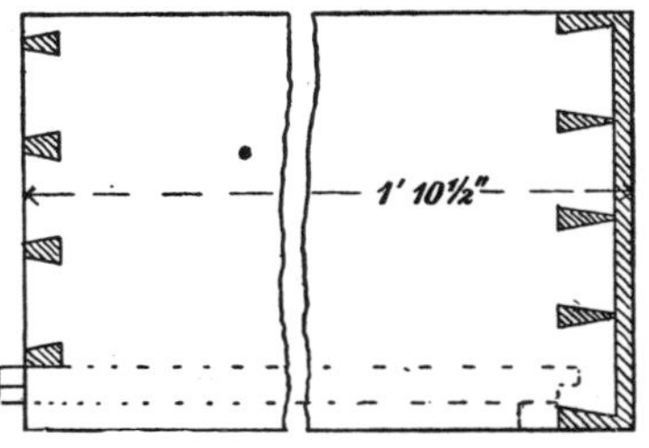

Side Elevation of a Drawer.

In the underfitting the pedestals are spaced about 4 ft. apart, the interspaces being fitted as book cupboards with sliding doors. A section through the top rails of these is shown in f. 6. The back of the cupboards is framed into a continuous top rail between the pedestals, this rail forming a back for the drawers, one of which is indicated by the dotted lines in the plan. A framed dustboard, D, f. 2, runs the entire length of the inner compartment of the counter, for the purpose of providing runners for the drawers, and to prevent any of their contents escaping.

The cashier's desk is arranged to stand between the pedestals, as the setting back of the cupboard, leaves room for the feet of the user. The drawers are fitted with locks *en suite*, one key fitting five locks—three cash and two paper drawers. Flush handles are fitted to the drawers in the pedestals, these being less liable to catch in the clothing than projecting handles; the counter drawers, being more out of the way, have stout drop handles, as these permit the lock being placed in the middle of the drawer, and are also more readily grasped.

The Construction of the Pedestals is shown in f. 7 & 8, and is similar to that already described on p. 249. In this case, however, the beads are planted around the drawers instead of being worked on the case, and the front rails are fixed as shown in f. 10: to make this joint, cut the division rail off to the neat length between the housings, stop the housing at the width of rail back from the edge, cut the remaining portion up to within $\frac{1}{4}$ in. of the edge, slightly tapering as shown, and also dovetailed under, then drive the rail into the housing, and mark the shape of the socket on its front edge. Next transfer the marks to the back edge by squaring, and gauge on the end of the rail the amount left on for the front stop. Measure the width of the socket at its front end, and mark this width on the gauge line, and draw a line from this point to the bevel at back edge, and cut down this line to the shoulder. These rails should be glued in first, the runners entered next, but dry, to allow the sides to swell or shrink, then the panel knocked in, and the back rail screwed on, as shown in f. 3, p. 250. If the sides of the pedestal are jointed up, the joints should break line with the top and bottom joints. All drawers are stopped with a $\frac{1}{2}$-in. square fillet or block glued at the back of the case. This is better than stopping them in front, as the shrinkage of the front does not then affect the surface, the side of the drawer in this method striking the stop endways. Cash and other drawers with heavy contents are fitted with friction rollers, as shown in f. 1, 2, 3. Small steel rollers are inserted, two in the drawer and two in the case, so arranged that they will thus just clear each other. Those in the drawer should be at the back, as shown in f. 1. Strips of brass are also inserted, both in the drawer bottom and the case runners, as shown in M, f. 3, for the roller to run upon. A plan of **The Cashier's Desk** shown in the elevation, Plate XXVIII., is drawn to larger

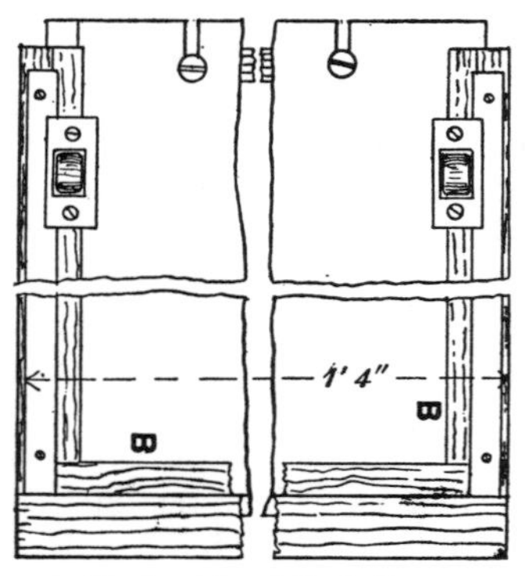

1. Underside of Cash Drawer.

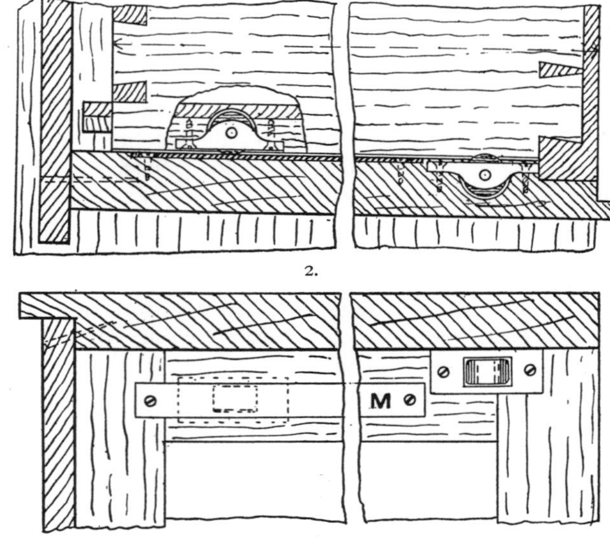

FITTINGS FOR CASH DRAWERS.
2. Sectional Elevation of Drawer showing Rollers.
3. Plan of Drawer Runners in Pedestal.

scale in f. 1, with enlarged details in f. 2 to 4. The middle rails of the screen are tenoned and fox-wedged to the posts, and the top and bottom rails dovetailed. The desk frame is tongued at the mitres, blocked and dovetailed to the posts. The bottom is rebated in; strips of baize are glued around the edges of the bottom to prevent marking the counter top. All the parts should be fitted together dry, and the whole cleaned off, then knocked apart and polished

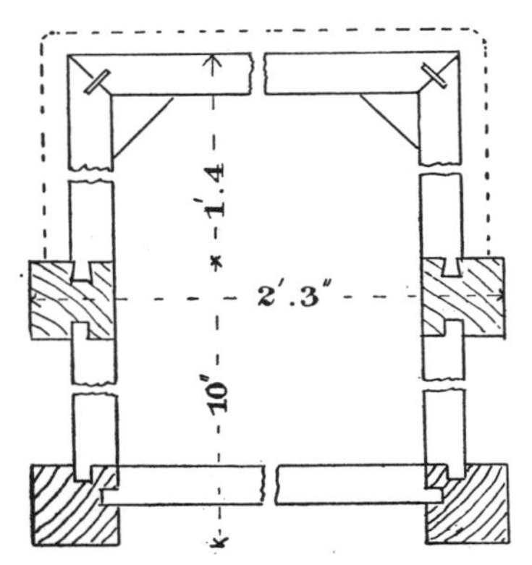

1. Plan showing Construction of Cashier's Desk, Plate XXVIII.

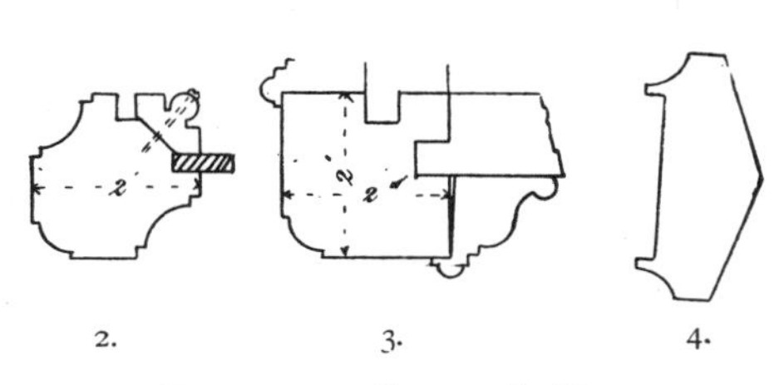

DETAILS OF CASHIER'S DESK.

2. Section of Standard above Middle Rail.
3. Section of Standard below Middle Rail.
4. Section of Capping.

before gluing up. The joint between the two counter tops is made as shown in f. 5 below; the nosing is mitred at the back edge; the joints are ploughed and cross tongued, and brought up with counter keys as described in the chapter on joints; the moulded edging is fixed after the tops are in position, and runs through to the quadrant corner to which it is dowel-jointed.

MUSEUM CASES are generally designed in a plain, substantial, and unobtrusive manner, being merely intended to act as settings or frames for their more important contents. The material employed is chiefly plain straight-grained mahogany, either stained to a uniform tone, or ebonised and polished; sometimes both of these methods are combined for effect. Pitch pine is occasionally used, but though its rich warm tones provide an effective contrast when the contents are dull and heavy in appearance, it is not to be recommended for the purpose. The nature of this wood renders it eminently unsuitable for small work, or where smooth true surfaces are required, as until all the superabundant resin in its tissue has evaporated, which takes many years, its surface continually alters

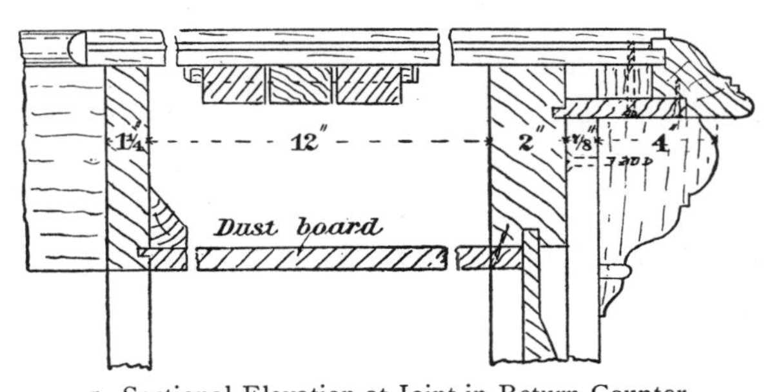

5. Sectional Elevation at Joint in Return Counter.

with the amount of warmth in its neighbourhood, and the polish or varnish has yet to be discovered that will resist the solvent action of its juices.

The Chief Requirements in this class of work are a maximum of lighted or visible area, absolutely dust-proof joints, great strength and rigidity in the framing, and perfectly secure fastenings. The first is obtained by making the

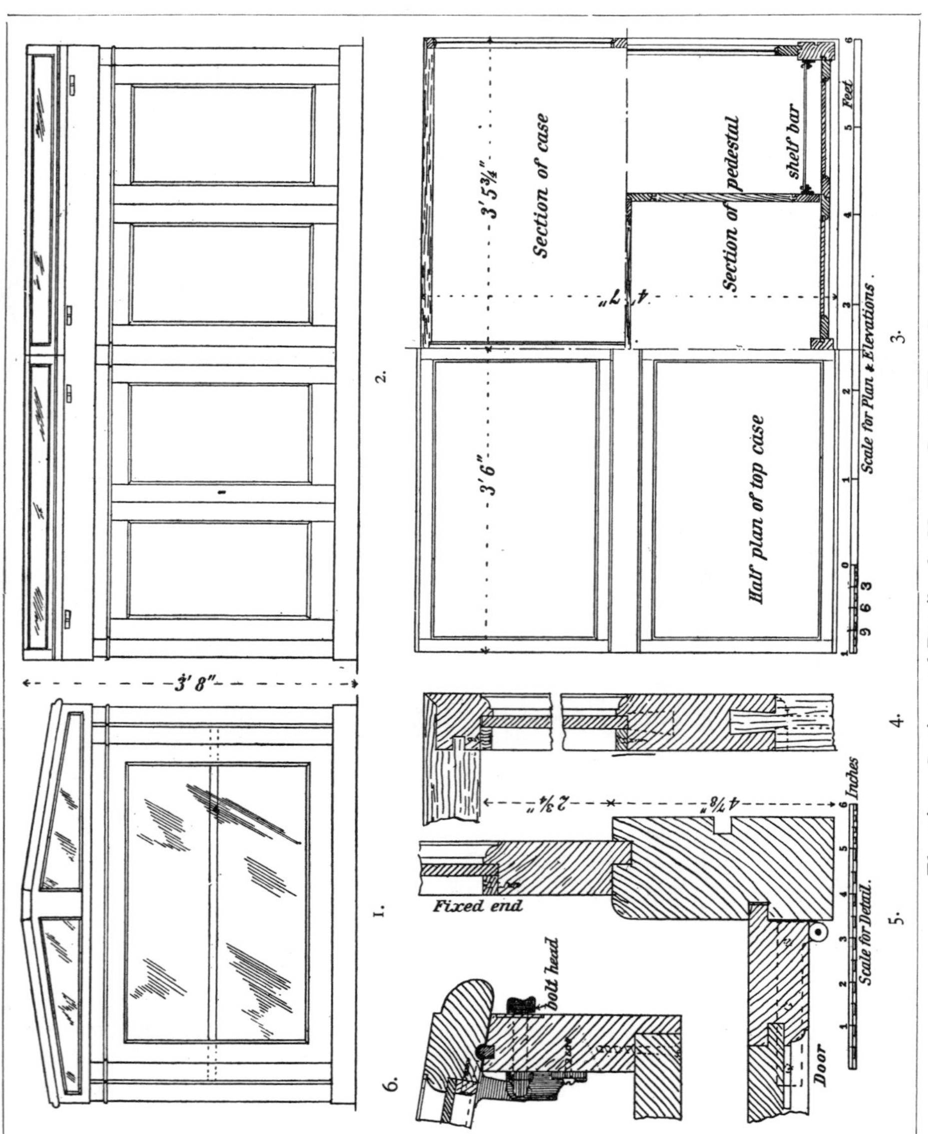

Elevations, Section, and Details of a Museum Centre Floor Case.

frames and sashes as small and light as possible, strengthening the weak parts with metal plates; and the second by means of velvet or rubber covered reeded joints, and the accurate fitting and bedding of the glass, which latter is usually done with wash leather. Strength is obtained by the use of substantial framing, and the stiffening of weak parts with metal bars; and security by the use of specially designed hinges, bolts, locks, &c., some of which are illustrated herewith.

A Centre or **Floor Case**, such as is used in the national museums, is shown in side elevation in f. 2 opposite, and end elevation in f. 1. Fig. 3 is a sectional plan, one half showing the top of the show-case, the other half sections through the case and pedestal respectively, and various enlarged details are shown in f. 4 to 6. These cases are seldom made more than 7 ft. long, although they are frequently placed end to end in long galleries, but when so arranged the sashes are made square-edged and flush at the ends, as shown, with a thumb mould on the front edge only; when isolated, they are usually moulded on the ends also. In the former case the rail tenons are stopped and fox-wedged; in the latter, taken through and wedged, as shown in f. 1 above, the stile moulding being tongued on and fixed after the sash is wedged up. The stile is of course ploughed before fitting, and the rail moulding left long enough to mitre. Fig. 4 opposite, which is an enlarged section through the end of the case, shows how the end frames are connected to the front rails; the end rails are tenoned through the front stiles, as the ends of the tenons are hidden by the lipping, and they are dovetailed into the central mullion at top, as indicated by dotted lines, and the latter is tenoned through the bottom rail. The sashes are hung to the top flat by a brass barrel hinge running the full length of the sash, one wing being sunk flush in the edge of the light; the other, with the whole of the knuckle, is sunk in a hollow worked on the edge of the flat. The sashes are fastened by locks in the middle, and are secured at each end by bolts that are connected with the lock in such manner that both must be moved simultaneously to open or close the case. A sketch of the bolt and its connection with the lock, as seen from inside the case, is given in f. 2 above; a small stud attached to the bolt passes through the front of the case for the purpose of actuating it; these are shown in the elevation, f. 2 opposite, just under the edge of

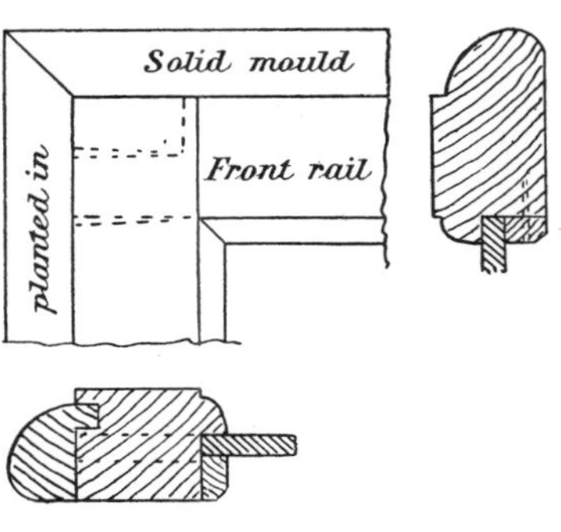

1. Plan and Section of Joint at Angle of Top Sash of Floor Case.

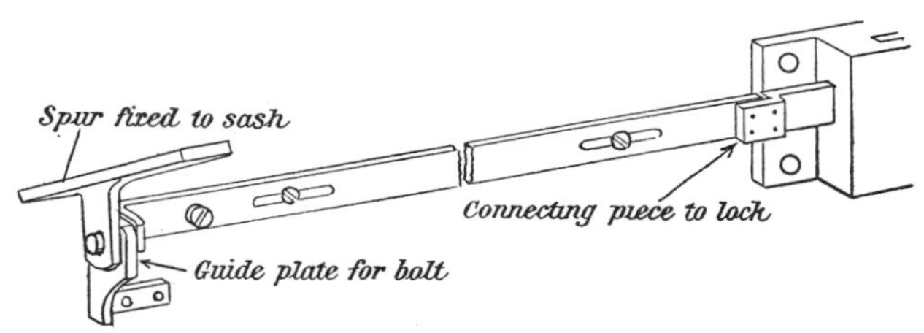

2. Bolt for Fastening Floor Case.

the sash. The reed shown in the detail, f. 6, p. 254, is ploughed in, all round the top of the frame, and a strip of silk velvet is pasted over it and the top edge of the case, the corresponding hollow in the sash being made to fit accurately over

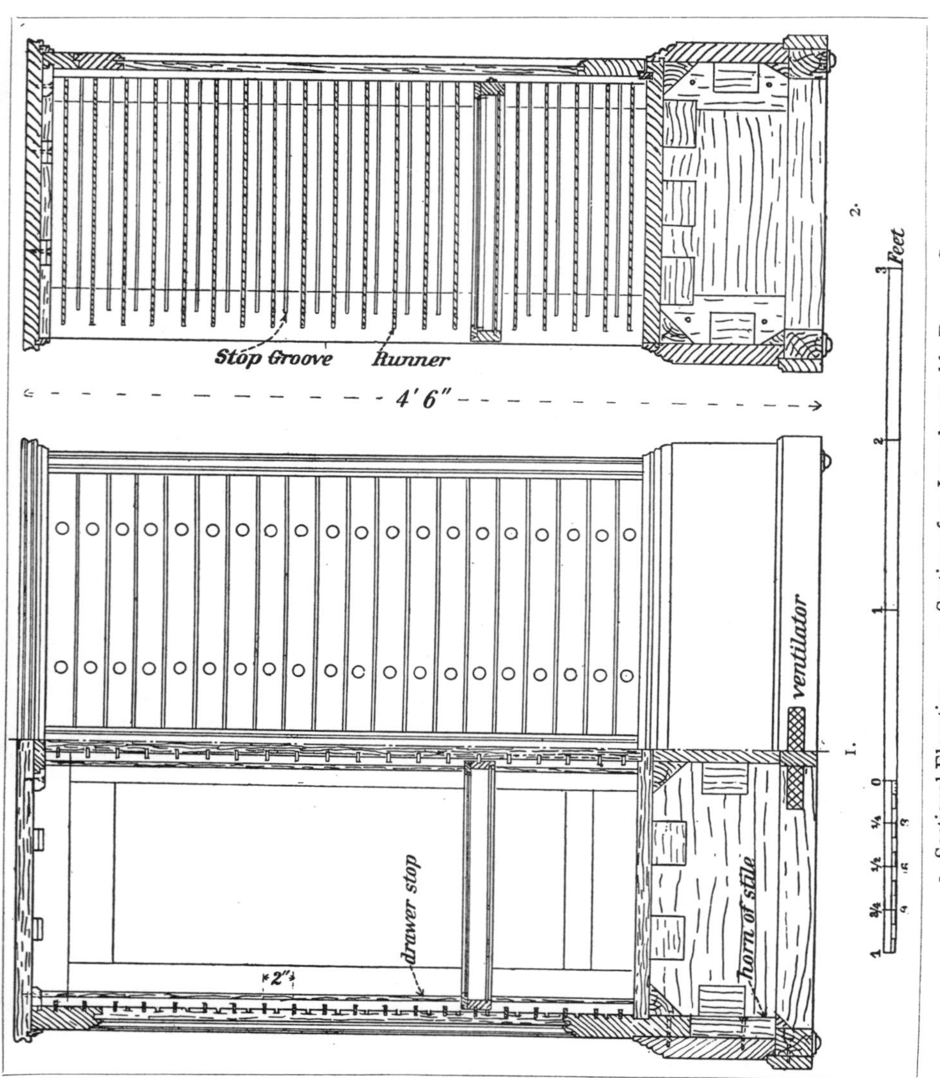

1. Sectional Elevation. 2. Section of an Interchangeable Drawer Case.

it; this hollow is best made after the sash has been hung, when the exact position can be marked. The hollow must be stopped at the ends. The arrangement of the interior fittings of the pedestals vary with the requirements. In the present case, the end compartments are glazed and contain shelves resting on bars bolted to T iron standards. The central part is arranged as a storage cupboard, with panelled deal divisions. One of each pair of doors is hung by projecting knuckle hinges, so as to fold back flat on the pilaster, as shown in f. 5. The other door slides on metal runners similar to f. 6, Plate XXVIII.

An Interchangeable Drawer Case, used for exhibiting entomological specimens, is shown in f. 1 & 2 opposite. The sides of the case are made up of panelled and moulded frames, moulded on the front edges and rebated on the back, as shown in the detail, f. 1 annexed, which is divided into two compartments by a central solid division, and has an open front. The base is a deep one blocked and screwed to the frame. The top is flat, built up and moulded on the edges, as shown in the detail, f. 2 below. The drawers are all made exactly the same size, and as there are no special openings, they can be inserted anywhere, or in any similar size case. Instead of the ordinary division rails for the drawers to slide upon, thin oak slips or cross tongues are fixed in grooves accurately spaced in the sides of the case and projecting about $\frac{3}{16}$ in. Corresponding grooves are made in the sides of the drawers, as shown in f. 1, next page, and in these the drawers work freely; between the runners another set of grooves is made, and stopped 2 in. back from the front edge. A small bolt attached to the back of the drawer (see f. 1, p. 258) working in this prevents the drawer being drawn entirely out. When it is wished to remove a drawer, the back of the case is opened (it is framed and hung as a door for this purpose), a set screw turned and the bolt shot back, when the drawer can be drawn completely out. The drawers have thin glazed tightly fitting covers screwed to the tops to protect the contents. A section of the frame is shown in f. 3, p. 258, and a sketch of the angle joints in f. 2, p. 258. They have also thin slabs of cork glued to the bottoms for pinning the specimen in, and in some instances cells are formed at the back and front, as shown in f. 3, p. 258, to contain an insecticide or a moisture absorbent, the division being pierced with fine holes as airways. Two square slips A, f. 1 above, with rubber or leather pads attached,

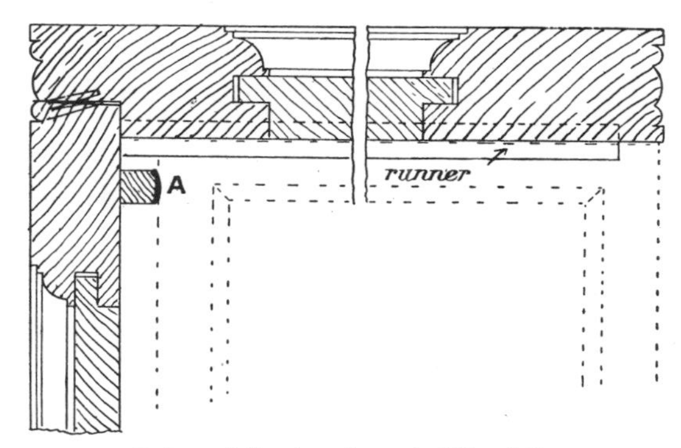

1. Enlarged Section through Side of Case.

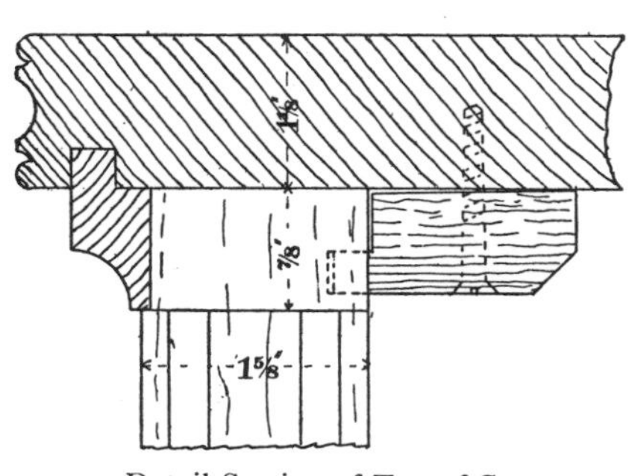

2. Detail Section of Top of Case.

are fixed at the back of the case to form buffers for the drawers, to prevent their contents being injured when shutting, and the drawer sides are well rubbed with powdered French chalk to make them run sweetly.

A Large Wall Case is shown in section on the next page, having pairs of doors in the front, a fixed glazed frame at one end, and a wood panelled frame at the other. These cases being very lofty, in some instances as much as 12 ft. in height, with the doors necessarily of slight substance, are difficult to make dust-proof. The methods employed at the British Museum are shown in the drawing. A specially made soft felt roll, is fixed in a groove on the edge of the hanging posts, and runs the whole height of the opening, except where

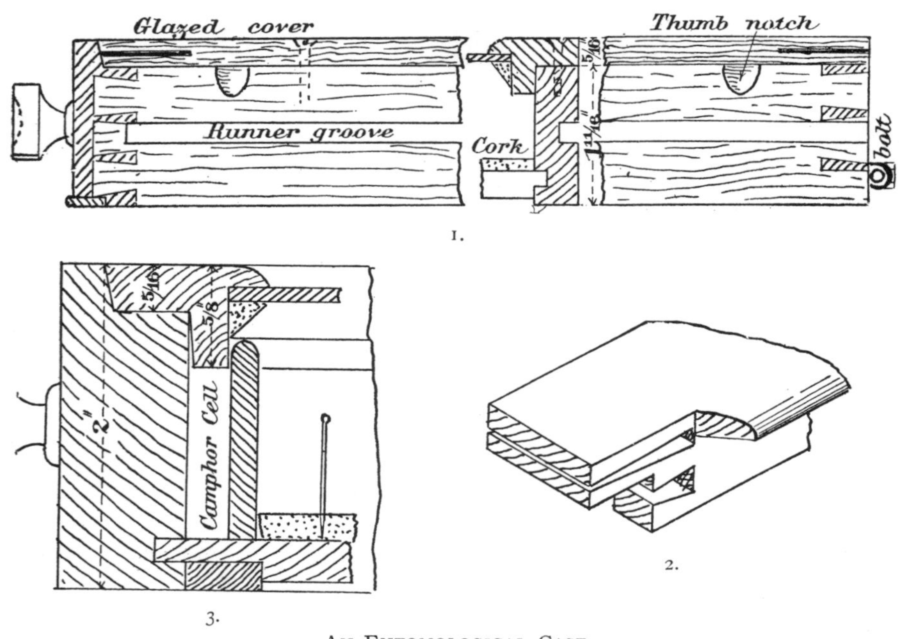

AN ENTOMOLOGICAL CASE.
1. Side View and Section of Drawers. 2. Joint in Air-tight Covers.
3. Enlarged Section of Drawer Front.

interrupted by the hinges, which, however, are themselves dust-tight when closed ; in addition to this, a strip of velvet, which is sewn in such manner that two parallel hollow pipes are raised on its surface, is attached to the edge of the rebate, and the rolls stand out when the door is open in the manner shown on the left of the sketch. When the door is closed both of these pads are nipped tightly and spread out flat between the joint, effectually filling any interstice that may exist, as indicated on the right of the drawing ; the stiles of the doors are stiffened by having brass plates screwed on the entire length of their edges, and four strong hinges shaped to the contour of the moulding prevent the aforementioned padding from forcing off the stiles.

A Library Bookcase.—Fig. 1, next page, is a part front elevation of a wall case suitable for a public library. Neatness, strength, and utility are the chief requirements in work of this class. The two portions, pedestal and shelf-case, are made separate and complete in themselves; as the upper part is entirely open in front, the back should be framed and well fitted to prevent the case "racking."

It should be sunk in rebates in the end standards, and well screwed to the top and divisions; the carcase is dovetailed at the angles, and the divisions housed into, and tenoned through the top and bottom. The cornice is fixed to the case and to cross-pieces over the divisions, which act also as bearers for the cover board, and supports for the blind rollers. The latter are

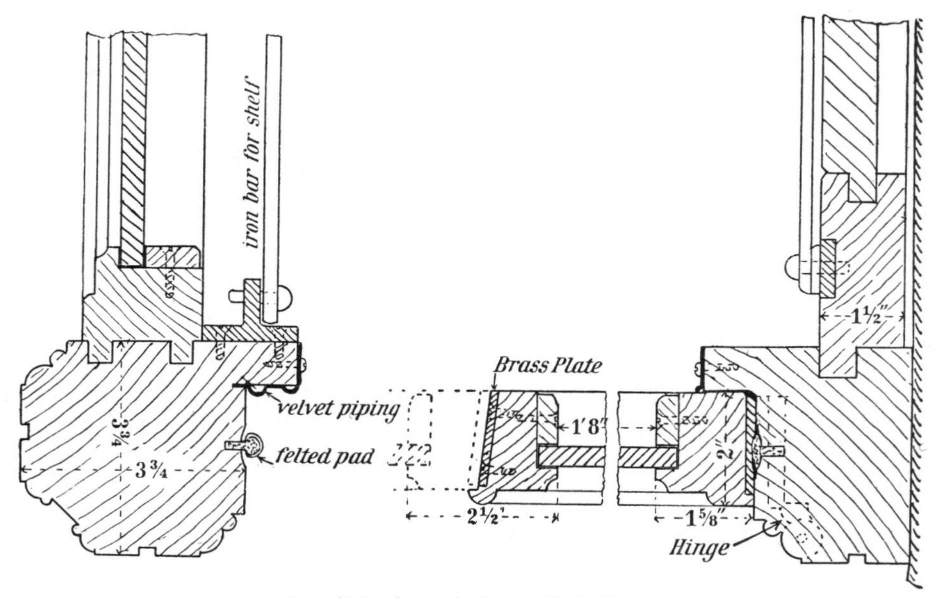

Detail Sections of a Large Wall Case.

spring rollers, and the stretcher of the blind, which is of metal, travels in grooves in the standards, as shown in the detail, f. 3, p. 261. The shelves are loose and interchangeable, the spacing of the divisions being alike throughout.

Two methods of fixing the shelves are shown in f. 3. The one on the front edge is known as "Tonk's fittings," and consists of a narrow iron bar perforated with a series of rectangular slots about $\frac{3}{4}$ in. apart; this is sunk in flush in the standard, to which it is fixed with screws. Fig. 5 shows the clip or shelf-bracket used with the bars. These clips have a lug or projection at the rear end, standing at right angles to the face, of a size to go easily through the slots in the bar. To enter them, they are turned upright, the top edge of the lug entered and pushed home in the slot, then the bracket is turned down level, the lug resting behind the bar and preventing its accidental slipping out. A shallow mortise must be made at the back of each slot to allow of the lug turning round, and a small spur on the front end of the bracket enters the shelf and prevents the latter slipping forward. The fitting shown *in situ* at the back edge of f. 3 is known as the library-bureau shelf-pin. It is shown to a larger scale in f. 4, and consists of a small square brass casting with a

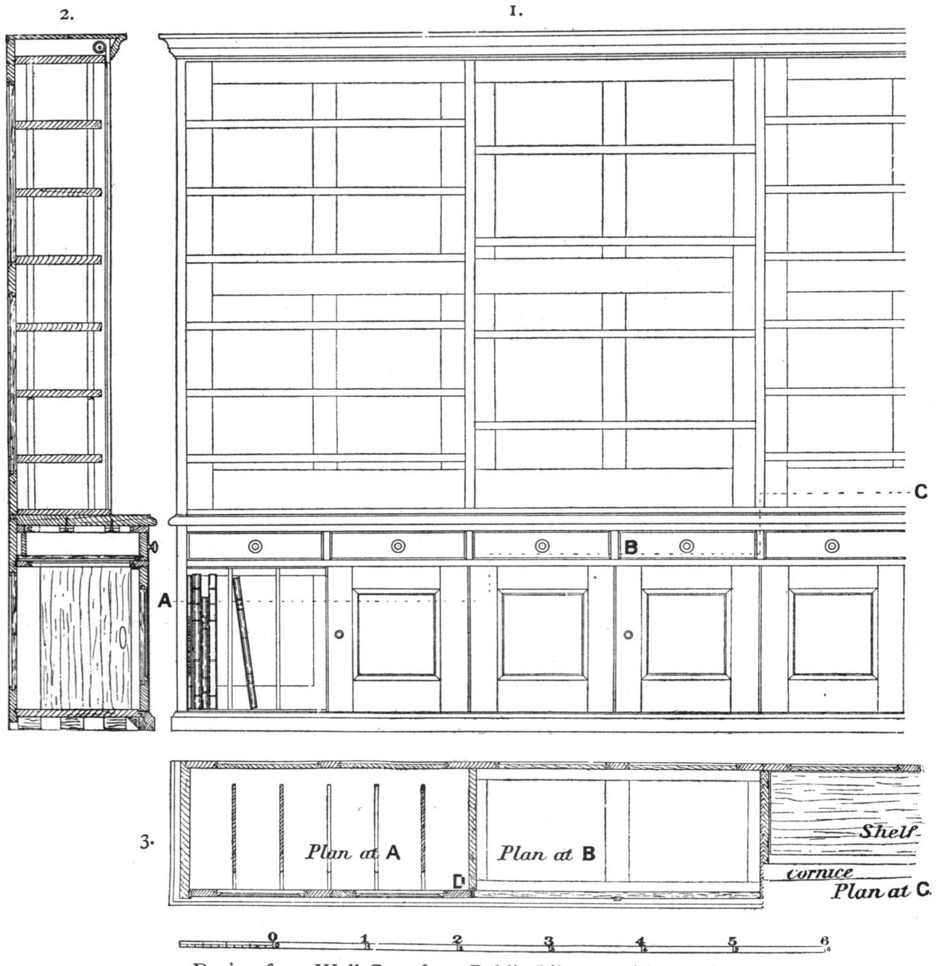

Design for a Wall Case for a Public Library, with Details.

stud at the back about $\frac{1}{4}$ in. long. This stud fits into a series of holes bored in the standard down the centre of a groove in which the back of the clip fits accurately. This pin has a second adjustment, made by turning the clip upside down. A third form, called a finger pin, is shown in f. 6. These are brass castings consisting of a short pin attached to a circular head, flat on each side, with tapering edges. The head contains a hole for the insertion of the finger for the purpose of withdrawal. These heads are sunk flush into the under side of the shelf and cannot slip out.

Shelves, if made out of 1-in. stuff, should not be longer than 3 ft. without support. The edges may be taken off or nosed. Shelves that are intended to receive books with expensive bindings should be covered with cowhide, and two pads or rolls of leather inserted in grooves in the standards, to protect the sides of the end books.

The Pedestal of the case illustrated is fitted with drawers, trays, and cupboards covered with doors hung folding. The drawers are useful for loose papers, circulars, indicator slips, cash, &c. The trays, the openings for which are covered by dummy drawer fronts, as shown at B, f. 1 & 3, opposite, are suitable for newspapers, maps, extra large folios, &c. The cupboards are fitted with sliding divisions to accommodate large folios and other volumes too large for the ordinary shelves. The top and bottom of the cupboard portion have a series of $\frac{3}{8}$-in. grooves cut about 6 in. apart, extending to within $2\frac{1}{2}$ in. of the back. The loose divisions slide easily within these grooves, and may be placed where required. Two of these are shown in position in the elevation on the left hand, where the door is supposed to be open. The doors are in pairs and hung folding, *i.e.*, they each fold back flat on the face of the adjacent door. The method by which this is accomplished without intervening divisions is shown in the detail and sketch, f. 1 & 2. The cupboard division or standard is kept back flush with the insides of the doors, which are separated from each other by a $\frac{1}{4}$-in. bead grooved into the standard. This bead lies flush with the face of the doors. Special butts are used called butterfly hinges. These consist of an ordinary brass butt hinge mounted on a pivot, which is fixed between two lugs projecting from a brass plate screwed to the edge of the division, as shown in f. 2. The butt moves freely round the pivot, the two leaves opening out back to back like the wings of a butterfly, hence the name. One wing is screwed to each door, and of course either door may be reversed upon its neighbour. The pivot of the hinge is continued through the projecting lugs about $\frac{3}{8}$ in. and is turned to a sharp point. These points enter the ends of the dividing bead, which is exactly the size of the hinge knuckle, and keep it in position, thus showing an unbroken line from top to bottom of the door. The tray compartment is shown at B in f. 3 opposite, the tray itself being omitted for the purpose of showing the construction of the carcase. The flap, which is made to match the drawer fronts, is hung at the lower edge, and falls over on the doors, to allow the tray to be withdrawn. The dustboards in these compartments should be flush on the top side, and in the cupboards where the folio divisions are used. They should also be flush underneath, as shown in the vertical section,

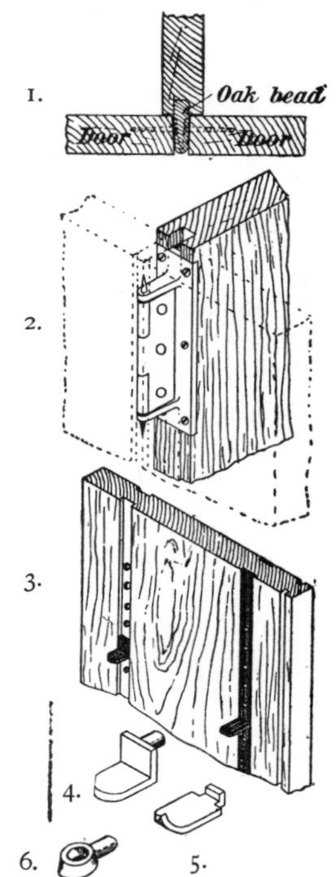

Bookcase Fittings.

f. 2. The standards should be placed directly under each other throughout, and a corresponding rough bearer under each across the plinth.

Spacing for Shelves.—When shelves are fixed, which method makes the stiffest cases, the distance between the shelves should be 10 in. for octavo books; 7¾ in. for small octavo; quarto, 13½ in.; large folio, 20 in.; small folio, 15½ in.

CHURCH FITTINGS comprise PULPITS, LECTERNS, FONT COVERS, STALLS, and SCREENS. Mediæval examples of these are interesting, in consequence of the great variety in design and the painstaking care and individuality of the execution, the craftsman of that period evidently possessing an originality, with the freedom to exercise it, which is not displayed by, or permitted to, his less fortunate successor of to-day. Modern church work, perhaps more than any other branch of Joinery, is based on, and rather closely follows, both in design and construction, the old methods, as will be observed by comparing the fourteenth-century examples given on pp. 263 & 264 with the photographs of recently executed work on Plates XXIX. & XXX.

The Parclose (a special form of screen used to partly enclose a tomb or a chapel within a church) illustrated opposite, is situate in Waltham Abbey, Essex, and is a good example of the earlier Decorated * period of Gothic architecture, massive in its proportions and vigorous in treatment, the pierced tracery being bold and flowing in design. An analytical sketch of the upper part of the tracery is shown to the right of the elevation. **The Chancel Screen** shown on p. 264 is another example of Decorated work to be found in Geddington Church, Northamptonshire. The upper part consists of open traceried panels of quatrefoil design, divided by a chamfered mullion at C, and vertical bars B, the latter being integral parts of the design. The lower panels are composed of 1 by 6½ in. boards, grooved and tongued together; the tongue is formed by chamfering off one edge of the boards on each side; the muntings dividing these panels are 5 in. wide, and are chamfered on one face, and struck with a cavetto mould on the other. The sill is square, and all the joints are stub tenoned and pinned together. The cornice is crenellated and pinned to the transom, as shown in the enlarged detail. The door is rebated in the solid upon each edge, and is hung with a pair of plain plate hinges.

The Stall, shown on p. 265, is a beautiful example of early Perpendicular * work in Lavenham Church, Suffolk. A part front elevation is shown at A, and an end elevation at B. A vertical section through the panel marked A in the elevation is shown in the lower left-hand corner of the plate, and various enlarged details to the right; these have all letter references, and will be easily identified. Lambs and other animals are carved as finials to the bench ends; the tracery is planted on the panels.

The Modern Stalls shown in the photographs, Plates XXIX. & XXX., are respectively fitted in St Michael's Church, Wendron, Cornwall, and the Chapel Home of the Community of the Epiphany, Truro. Both were made, entirely by hand, in the workshop of Messrs Harry Hems & Sons, the well-known ecclesiastical sculptors of Exeter, to whom the author is indebted for the photographs, which were taken whilst the fittings stood in the workshops.

---

* For explanation of this term refer to Glossary.

EXAMPLE OF MODERN CHURCH FITTINGS—CHANCEL STALLS AT ST MICHAEL'S, WENDRON, CORNWALL.

[*To face page* 262.

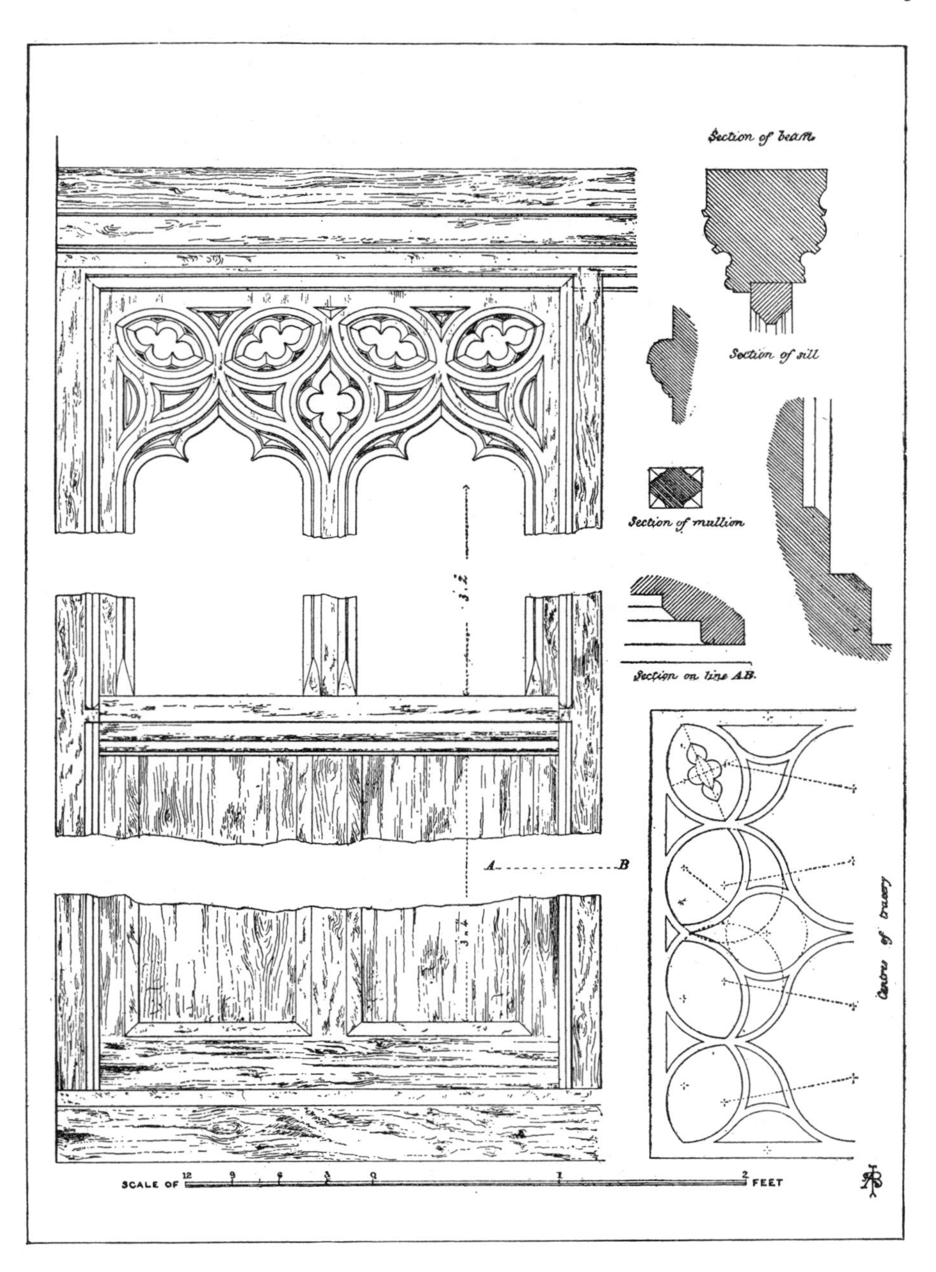

PARCLOSE SCREEN, WALTHAM ABBEY, ESSEX.

The Pulpit, shown on p. 266, is a typical example of late Perpendicular work, with its "linen-fold" panels, marginal fillets to the framework, and

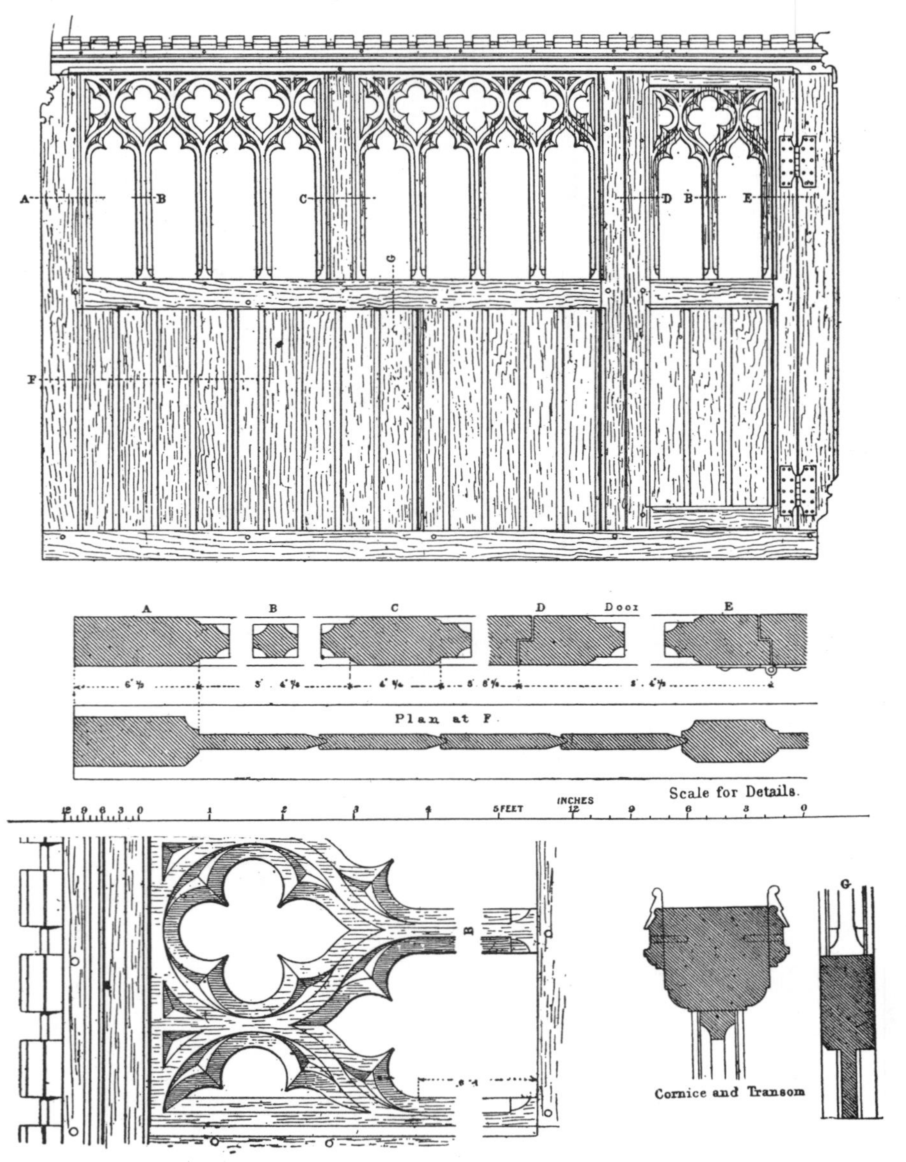

Elevation and Details of the Chancel Screen at Geddington Church, Northants.

shallow cut mouldings. This example is to be found in Henry VII.'s Chapel, Westminster Abbey.

A Pulpit in the modern style, given chiefly as an example in construction, is shown on p. 267. The pulpit proper, is constructed of a series of panelled frames, tongued into solid pilasters placed at the angles; these have each three

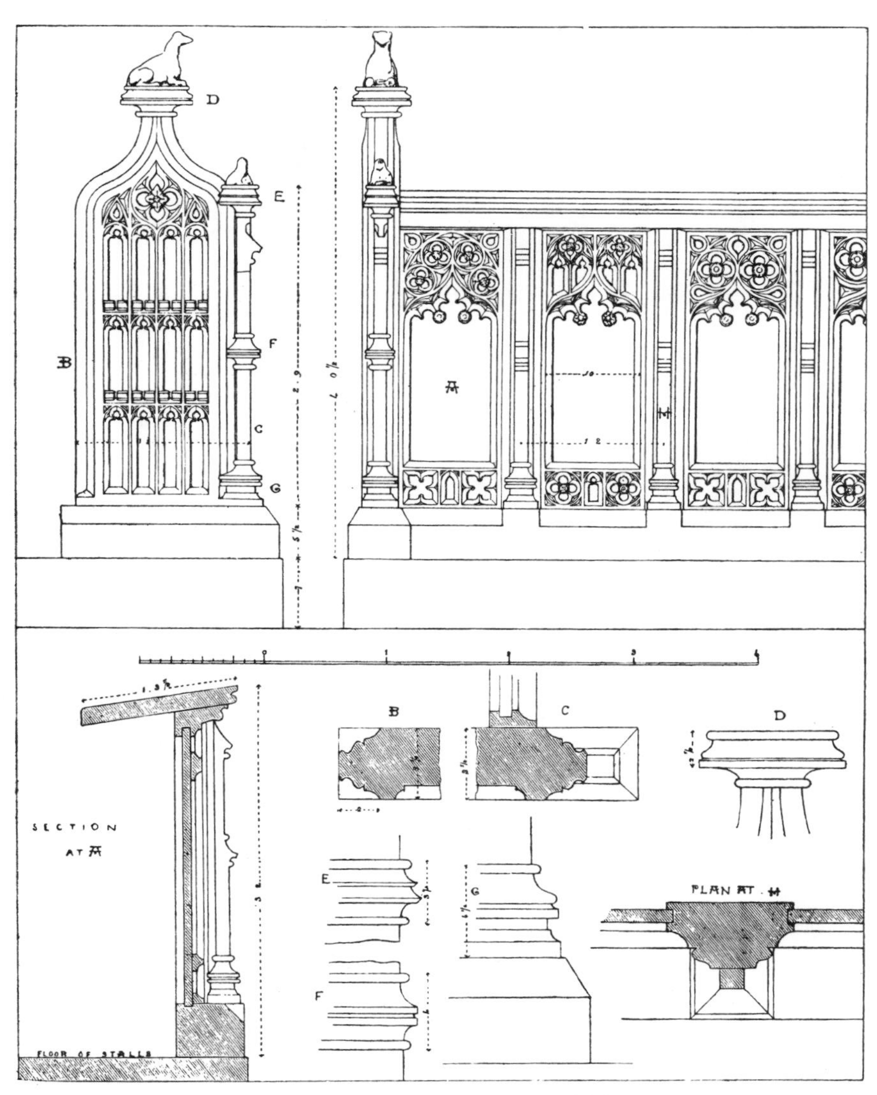

Chancel Stalls, Lavenham Church, Suffolk.

V-shaped grooves or sinkings, and are supported by moulded and mitred plinths. Upon the top of the framing rests a boldly moulded capping, the chief member of which is enriched with strapwork carving; mouldings of different projections

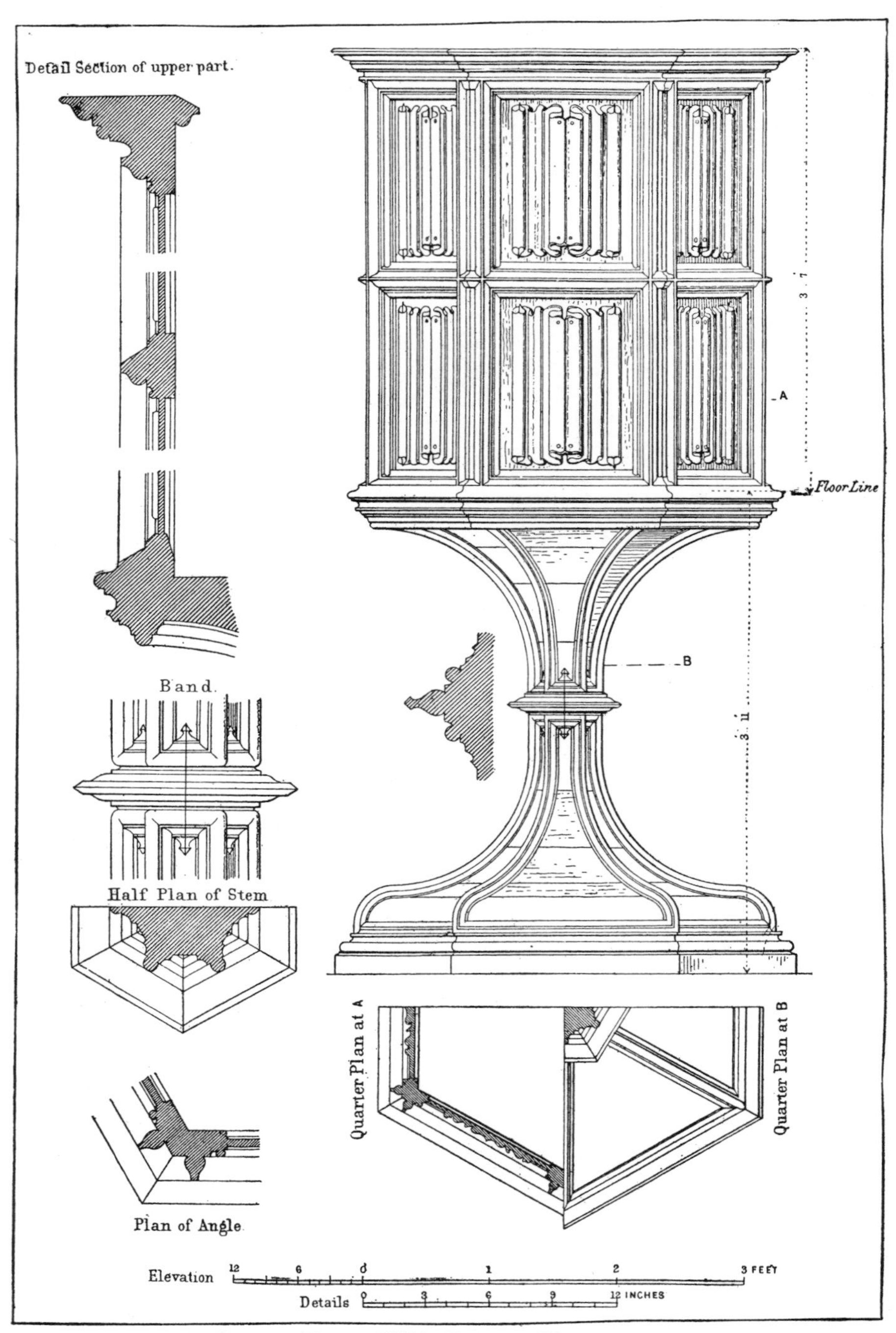

PULPIT, HENRY VII.'S CHAPEL, WESTMINSTER.

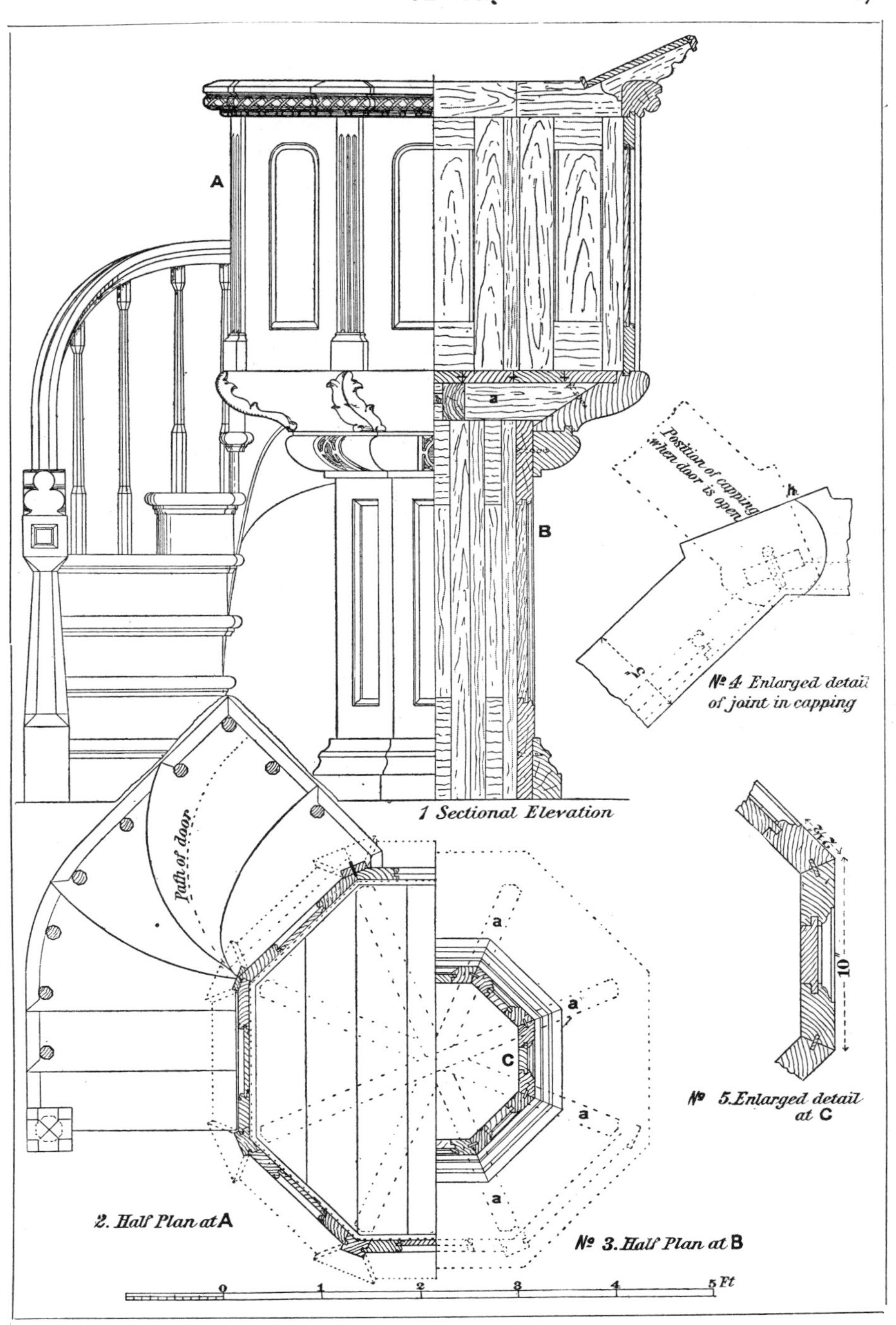

PLAN, SECTIONAL ELEVATION, AND DETAILS OF A MODERN PULPIT.

are required over the pilasters and framing to ensure correctness in the mitres. The panelling is supported by a heavy ogee and ovolo moulded surbase enriched with carving, and this in turn rests upon an octagonal framed pedestal, as shown in the half plan No. 3. The panelling forming the pedestal is ploughed and tongued together at the angles, as shown in the detail No. 5, and is surrounded by a solid moulded skirting sunk and screwed from the back, as shown in No. 1, the mitres of which are cross tongued. The floor of the pulpit is carried on four $4\frac{1}{2}$ by 2 in. fir joists radiating from the centre, as shown at *a a* in the plan. The one shown at *a* in the elevation runs through uninterruptedly, and the others are tenoned into it; a central piece of quartering resting on the floor supports the middle (this is not shown), and the outer ends rest on the framing of the pedestal, the ends being cut to fit the back of the surbase mouldings, to which they are screwed. The inner edge of this moulding is rebated to receive the ends of the flooring. A small stair of six steps, three of which have curved fronts, afford access to the pulpit. 2-in. cut strings are provided, and the newel and balusters are wrought to octagonal sections in consonance with the rest of the design. The method of making the joint in the capping over the hanging stile of the door is shown in No. 4, the full lines showing the capping when the door is closed, and the dotted lines the same when open; the door below is also indicated by dotted lines. The capping must be shaped and mitred in the solid, and a circular path cut with its sides vertical; the cut is struck from the centre of the hinge joint, and should afford a slight clearance at *h* when the door is wide open, as indicated by the dotted lines.

Example of Modern Church Fittings—Stalls for the Community of the Epiphany, Truro.

[To face page 268

CHAPTER XVII.

SHAPED, CURVED, AND BEVELLED WORK.

Methods of Forming Curved Surfaces by "Cutting in the Solid"—Bending in the Solid—Building Up—Veneering—Staving and Kerfing—Description of Bending Apparatus—Cylinders, &c.—Preparation of Moulds in Work of Double Curvature—Methods of Preparing same without Moulds—Use of the Falling Compasses—Construction of Circle-on-Circle Door Frame with Radial Jambs—Ditto with Parallel Jambs to be Machine Worked—A Semi-Head Sash Frame Circular in Plan, Obtaining the Moulds—A Framed Splayed Lining with Circular Soffit—Circular Pew with Inclined Bookboard—Methods of Developing Curved Surfaces, &c.—A Louvre Bull's Eye Frame; how to Set Out the Grooves, Obtaining Shape of the Boards, &c.—Methods of Obtaining True Bevels in Work Splayed in Plan and Elevation, as Window Linings, Hoppers, Trays, &c.

THE methods employed in the construction of curved surfaces in Joinery may be classified under five heads, viz., BENDING IN THE SOLID by the aid of steam or boiling water; VENEERING on shaped cores, or the backing of veneers first bent to shape; CUTTING IN THE SOLID; BUILDING UP in sections; and SAW KERFING. Of these, undoubtedly the first method is the best, both in respect of appearance and strength, as to bend a piece of wood in its entirety will ensure continuity of fibre, natural appearance of the figure, and freedom from alteration due to unequal shrinkage—effects that cannot be obtained by either of the other methods. In an ordinary builder's shop, however, the bending in the solid is necessarily confined to stuff of small scantling, in consequence of the difficulty of thoroughly steaming large pieces, which can only be done satisfactorily in large strongly constructed tanks, and under a high pressure of steam which softens the wood throughout equably and rapidly, when it may be bent readily and without impairing its strength. On the contrary, if stuff of considerable substance is steamed in a small tank, with a low-pressure steam, it must be subjected to the steam some hours before the latter will permeate the interior parts, and meanwhile the outer layers of tissue are being dissolved and the cell contents washed out, and when the pressure is applied these collapse, causing cripples in the curve, and of course greatly reducing the strength of the timber. Boiling the wood in water, as is frequently done, is to be deprecated for the same reason. No arbitrary rule can be laid down for the length of time pieces of specific scantling should be steamed, as the heat and pressure of the steam and the age and texture of the wood must all be taken into consideration. Generally speaking, the heating should be done quickly and the bending slowly. Perhaps the best procedure in work of importance is to subject a small piece of wood from the plank to be bent to the steam, noting the time of its immersion, then bending it, observe its behaviour under the strain, and deal with the bulk accordingly. When the wood has been made thoroughly pliable, it is bent around a centre or "drum" prepared to the requisite curve, and fastened thereon until it has acquired a permanent set.

When the curve is other than a portion of a circle, a "**Centre**" and a **Caul**, or reverse, must be prepared, and the piece gripped between them with hand-screws; but when the curve is part of a circle, a **Drum** similar to f. 1 & 2

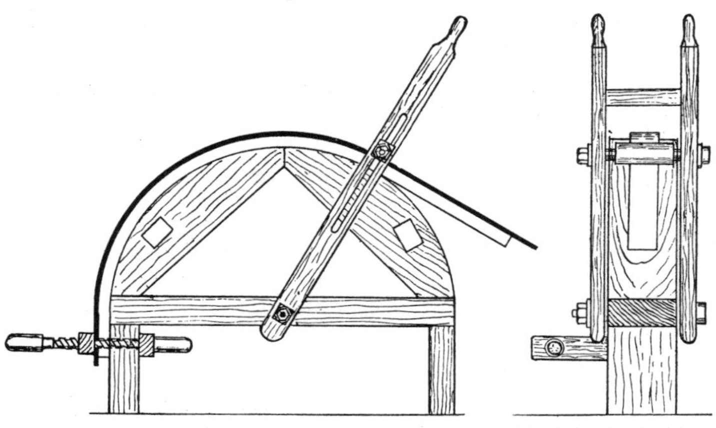

1, 2. Side and Edge Views of a Drum for Bending Work in the Solid.

is used. This is provided with a saddle lever pivoted at the centre of the curve and carrying a small adjustable roller between its arms, which is arranged to keep the work pressed tightly to the periphery of the drum. The piece to be bent is taken quickly from the steam chest. One end is secured by a hand-screw to the drum, as shown in the sketch, a length of galvanised hoop iron being introduced between the wood and the roller. The lever is then brought around steadily and not too rapidly, because if the fibres are stretched suddenly they will rupture. The piece should be left in position until it is thoroughly dry, and the pores on the outside should be filled in with hot glue. Any moulding, such as a bead that does not differ greatly in the thickness of its parts, may be worked in the straight and then bent, but otherwise the safer plan is to bend first and work to section afterwards.

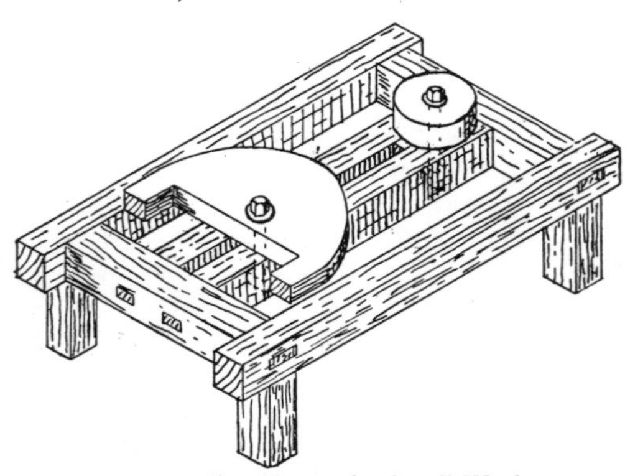

3. A Bending Frame for Small Work.

The Bending Frame.—Another form of apparatus for bending beads and similar small work is shown in f. 3. This consists of a stout frame fixed on legs, and with a disc or wheel mounted at one end upon a fixed spindle, so that it can revolve freely. A groove or slot is formed down the length of the frame in line with its axis, and "centres" of any size can be mounted on a flanged screw-bolt which travels in the slot, so that the "centre" may be adjusted at a suitable distance from the wheel and there fixed. The piece to be bent is then introduced between them, the centre turned around until one of its ears is parallel with the piece, which is secured to it with a hand-screw, and the centre is then turned back until the other end is reached, which is secured in like manner.

The Building Up process is illustrated in f. 1, 2, & 3 hereunder, which show the method as applied respectively to the rails, stiles, and panels of framing, &c. In the first, comparatively narrow pieces are cut to the required curve or portions of it, and fastened together edge to edge with glue and screws until the necessary width is obtained. The heading joints may be either butted or bevelled, the latter being the stronger, and they should be cross tongued. Fig. 2 shows the method when a long stile or similar piece curved in the direction of its width is required. The pieces are stepped over each other to suit the mould, and though shown square edged, are usually cut bevelled, as by reversing them two may be cut out of a batten. Panels and quick sweeps for similar purposes are obtained as in f. 3 by jointing up narrow boards edge to edge at a suitable bevel to contain the desired sweep—the internal curve is frequently worked approximately before gluing up. The numerous joints incidental to these methods limit their uses to painted or unimportant work.

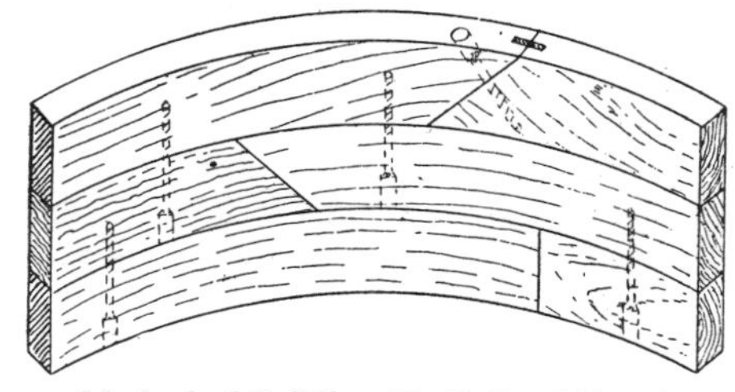

1. Method of Building Up Rails of Framing.

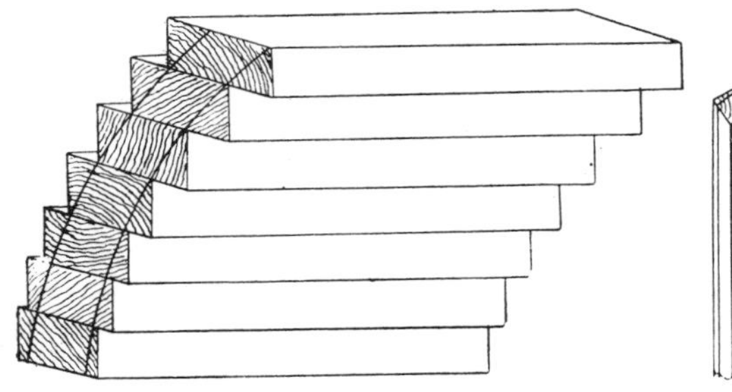

2. Building Up a Long Curved Surface.

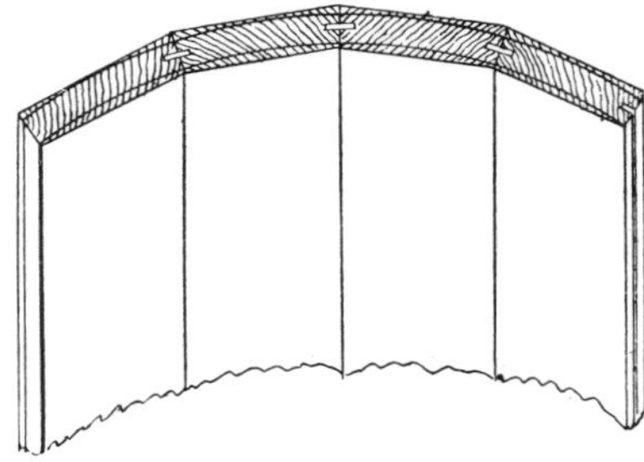

3. Building Up a Wide Curved Surface.

When the work has to receive a higher finish, such as polishing or varnishing, the **Veneer Method**, shown in f. 1, next page, is adopted. This consists of cutting a "core," usually in yellow pine of suitable shape, and which may be built up in sections to the required width, and covering one or both of its sides with a veneer of hardwood. Veneers are of two kinds, known as knife-cut and saw-cut. The former run from thirty to forty to the inch in thickness, and are shaved off the surface of the log, which is fixed in a kind of a lathe and revolved in front of a fixed knife; this method produces a very handsome figure upon the surface of the veneer, but there is obviously little wear to be obtained from wood of such slight thickness, and its use is mainly confined to cabinet work.

Saw-cut Veneers vary from four to sixteen to the inch, and are cut from the sides of balks or planks with large thin circular saws or horizontal frame saws. The veneer is prepared by first toothing and then well wetting it with boiling water, and bending it over a cylinder or "centre," where it is allowed to dry, when it will retain a shape very near to that of the mould. It is then made very hot, as is also the core, either by ironing with a hot iron or passing them over gas jets. The veneer is then attached to the core with glue, and fixed in position with hand-screws; if the hand-screws cannot be made to reach the middle, Cauls or reverses must be employed; these should be made very hot and well soaped before applying, and forced lightly down on the work in a veneer press or with saddle cleats.

Knife-cut Veneers are laid with a veneering hammer (f. 2). This consists of a stout piece of steel about 6 in. long, with a straight smoothly rounded edge mounted in a deal headstock, having a short stout handle fixed in

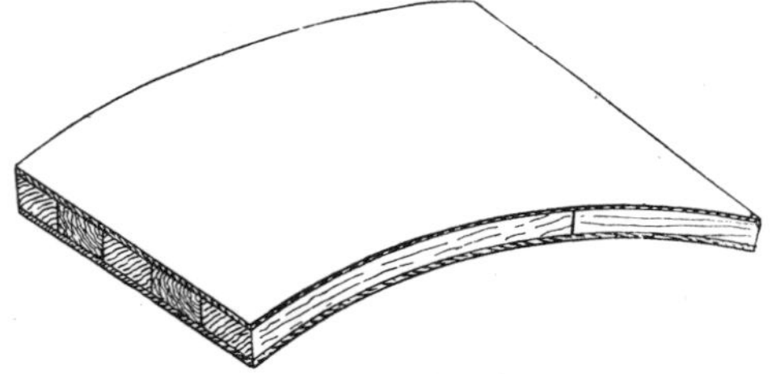

1. A Veneered Rail.

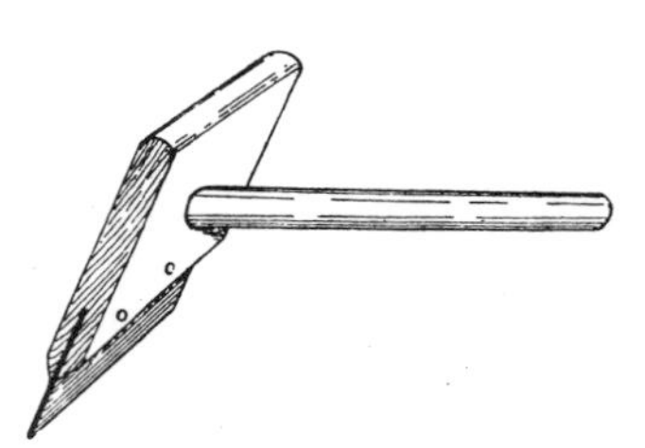

2. A Veneering Hammer.

it at an obtuse angle. It is used as follows:—The veneer having been cut a trifle larger all round than the core, is toothed on the back, then wetted on the face with hot water, to counteract the swelling caused by the glue on the back. The face of the core, previously "toothed," and the back of the veneer, are plentifully and rapidly covered with boiling glue. The veneer is then carefully placed in position, pressed down with a short straight-edge laid across the grain to prevent splitting, and the superfluous glue squeezed out with the hammer. The latter is drawn to and fro with a scraping action, commenced at the centre and proceeding towards the outside, the glue being driven out before it. The rubbing must be done very quickly and continuously with as heavy a pressure as possible. The tool must not be allowed to rest for a moment after commencing, otherwise the glue will set where the stroke has been stopped, and cause a ridge in the veneer. The veneer once pressed down to the core will not, as a rule, rise from it; but if this should happen, a hot iron should be immediately applied to the spot with a piece of wet flannel interposed, the steam from which will penetrate the wood and soften the glue, and pressure should be applied to the spot for a short time, either by an assistant, or the application of a hand-screw and caul. The surface of the veneer should be well soaped to form a lather in which the hammer may work freely. Oil, though it is sometimes used, is not to be recommended, as it will penetrate the veneer, and prevent the glue setting. After the work is finished, lightly sponge the surface with hot water to remove the glue that has oozed through, which if left on will make the cleaning off very difficult.

Staving or backing is another method of using veneers to form curved surfaces. In this method a centre or "cylinder," as it is usually termed, is prepared to the contour that the face of the veneer is desired to assume, and the latter, which in this case should be sawn, is steamed or wetted and bent around, and left till dry, when narrow staves, about an inch in thickness, and as wide as the curve will admit of, are glued across its back, as shown alongside. The backing is much strengthened by having canvas glued over its surface after the latter is cleaned off. This is the method generally used to form elliptic-headed and circle-on-circle linings, and similar surfaces of complex curvature.

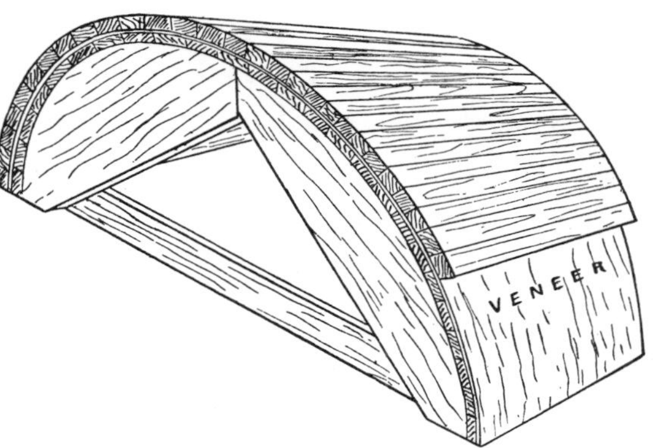

Method of Staving a Veneer on a Cylinder.

Kerfing is a very crude and inferior method of obtaining a curved surface, discreditable alike to the workman's skill and his ingenuity. It consists in making a number of saw cuts in the piece to be bent, nearly through to its face, and spaced apart according to the quickness of the desired sweep; then filling these with hot glue, bending the piece to the sweep, and securing it in position. The face will then have assumed a rough curve made up of short flat facets which are rubbed down to some degree of regularity with glass-paper; but there is neither beauty in the curve or strength in the work.

CURVED WORK.

Work of Double Curvature, *i.e.*, work curved both in plan and elevation, is perhaps the most difficult, as it is undoubtedly the most interesting and beautiful branch of the art of Joinery. Its proper execution necessitates considerable manual dexterity and a trained eye to realise its complex curves; and the difficulty of execution is added to in many instances by the limitations which the nature of the material or its economical use impose. For instance, the solid head of a door frame of this description is more easily executed than the cased head of a sash frame, or a handrail wreath than the framed winding soffit to the stair, for in both cases where the work is wrought in the solid the preliminary outlines drawn on the material may be varied or corrected as the work proceeds, according as the eye dictates, whilst in the case of work built up in numerous pieces, there is little to guide the judgment in their preliminary preparation but experience. The method usually advised for the production of surfaces of compound flexure is to make developments of the curved surfaces on the flat, and so obtain moulds which when bent around one of the surfaces will give the outlines of the other. In theory the process is correct, but its application in practice is not always the most economical or satisfactory in result, as the templets require extreme care in applying them to the work, the slightest deviation from the correct position showing itself in a crippled curve; and also

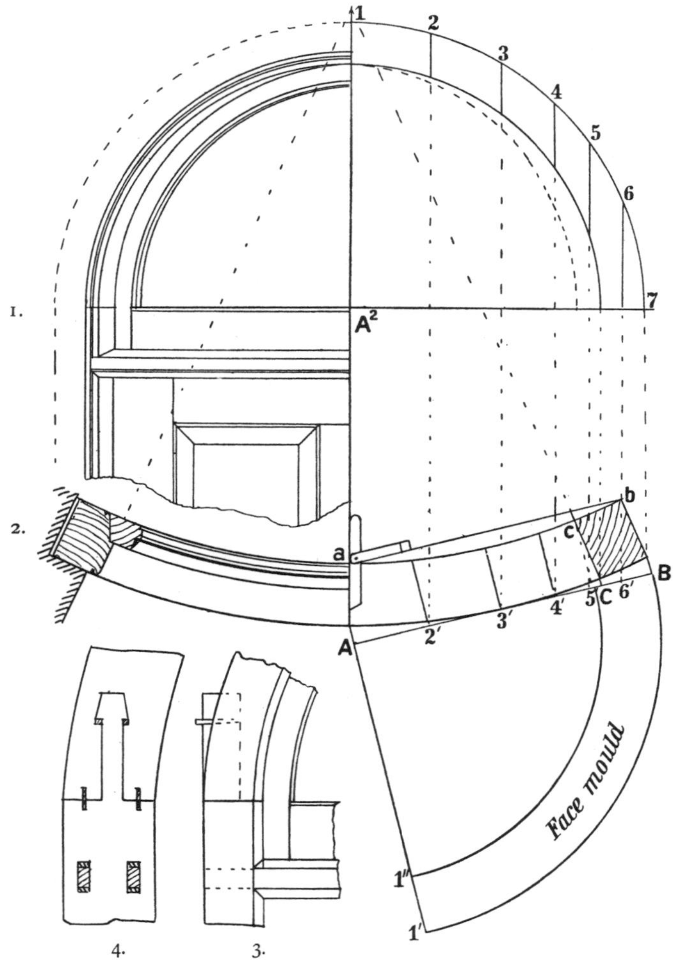

A Circle-on-Circle Door Frame, with Radial Jambs.

a wood templet has inherent faults that limit its usefulness, for should it be wide it will shrink or swell unequally, in consequence of being cut across the grain in places, and if narrow it will spring; in either case falsifying its curves, which require constant correction. For this reason sheet zinc is sometimes used for moulds, but the trouble of preparing it confines its use to important work. The author himself has in practice, whilst not discarding moulds entirely, used them only in conjunction with the methods described below which have given satisfactory results.

A Circle-on-Circle Door Frame with radial jambs, and soffit level at the crown.—Fig. 1 opposite shows in one half the outside elevation and in the other half a line diagram of the head. Fig. 2 is a half plan above the transom and a half plan diagram. Figs. 3 & 4 are enlarged details of the joints at the springings. The head is made in two pieces, jointed at the crown with a handrail bolt.

To find the thickness of the piece required to cut each half of the head out of, draw the chord line *a b*, f. 2 opposite, joining the two extremities of the head in plan; draw A B parallel with this and touching the plan curve outside; the distance between these lines measured square across gives the thickness. **To find the face mould** for the head in the "square": draw lines perpendicular to A B from the angles of the jamb *c* and B, and the crown joint A; then with A as centre, and B and C as radii, describe the arcs B 1′ and C 1″; draw the line A 1′ perpendicular to A B, which will give the joint line on the face. This mould is applied to the face of the stuff for marking it for the rough cutting out, and the lines are sawn out square from the surface. Strictly speaking, the edges of this mould should be quarter ellipses, but the difference between these and the quarter of a circle, is so slight as to be immaterial in the case of this rough mould. Having cut the two pieces out full to the lines, make the centre joint; using the

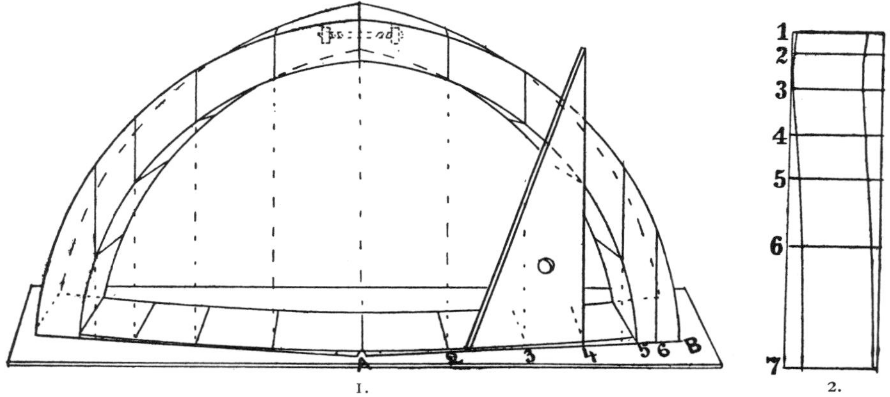

1. Marking Plan Curves on Head of Circle-on-Circle Frame.
2. View of Edge of Frame when marked.

mould to mark the joints on the face, and using the Bevel, as shown in the plan, to mark the edge of the crown joint (note that the stock of the bevel must be held level when set as shown in the plan); and a Square for the edge of the springing joint. When the pieces are jointed they should stand over the parallel lines in the plan A B and *a b*, and similar ones on the opposite side which are not shown. Divide the top edge of frame in elevation into any number of equal parts, and drop perpendiculars from them into the plan, cutting the line A B in 2′, 3′, 4′, 5′, 6′; draw lines from these points at right angles, or "square" with the line A B: these are shown full in f. 2 opposite. Having done this, stand the head upon the plan, as shown in the perspective sketch, f. 1, p. 275 (which represents a rod showing the plan set out as described), and transfer the lines to the face with the aid of a set square; also square them over on both edges with a try square, then with a pair of dividers prick in upon these lines the distances from the face that the corresponding line cuts the curves in the plan, then with a thin lath bent around the head over these points, mark the curves as shown in f. 2, and the same on the under side. Unbolt the joint and work the face off to the lines on the edges, keeping the surface flat by applying a straight-edge perpendicularly or parallel with the lines drawn on the face. Having worked the convex side, bolt up and test over the plan with a set square, and when correct, gauge the inside curve parallel and work it off in similar manner.

To mark the soffit or elevation curve, attach a stretcher to the head at the springing, as shown in annexed sketch. Fix a block in the centre of this equal in height to the head at the crown. Prepare a radius rod or "falling compass" with one edge shot straight and one end cut square with it; this end is grooved to receive a pencil, and the opposite end reduced to receive a bradawl which is fixed as shown, at a distance from the pencil point equal to the radius of the elevation.

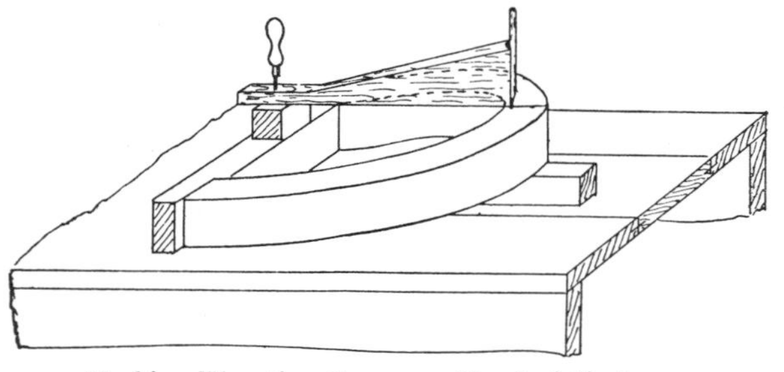

Marking Elevation Curves on Head of Circle-on-Circle Frame.

The rod is then swept around the head, keeping the lower edge level, and allowing the pencil to rise and fall in the notch as required; the soffit is produced squaring from the front, the stock of the square to be kept perpendicular. The rebates and beads are worked parallel from the edges.

A Circular Head Door Frame in a Circular Wall, with parallel jambs and soffit level at the crown.—It is assumed that this head is to be machine worked, and geometrical methods of obtaining the moulds and joint bevels are described; the drawings are also to be used in connection with those shown on p. 115, describing the methods of cutting out the head with the band saw. The elevation is a semicircle upon the line *x x* in plan, between the reveals in the wall; the edges of the frame are made parallel to this arch. The head is to be made in two pieces, with joints at the crown and springings. Proceed to enclose the plan and elevation curves within straight lines as C C, *a b*, f. 1, and A B, *b* S, f. 2; the resulting rectilineal figures show the width, thickness, and length of the piece required to cut one-half of the head.

To obtain the face mould.—Divide the outside curve in elevation, into a number of parts as 1, 2, 3, 4, 5, 6. Draw projectors from these points, perpendicular to the springing line S S, and continue them across the plan to cut the face line A B: from the intersections B II, III, IV, V, VI, erect perpendiculars to A B, and make them equal in length to the height of the corresponding points above the springing line in the elevation; through these points trace the curve. These lines can be utilised to obtain points for the soffit curve, as next described.

To obtain the edge mould.—This could be developed upon the plan, but to avoid a confusion of lines a separate drawing is made for the purpose in f. 3, which is a repeat of one-half of the plan, f. 2, and an auxiliary elevation of the top edge of the plank *a b′* is made parallel with the face of the plank on the line S¹ S². This line, which is the real length and inclination of the line *a b′*, f. 1, is obtained by revolving the top end of the line, point B in the plan, into the

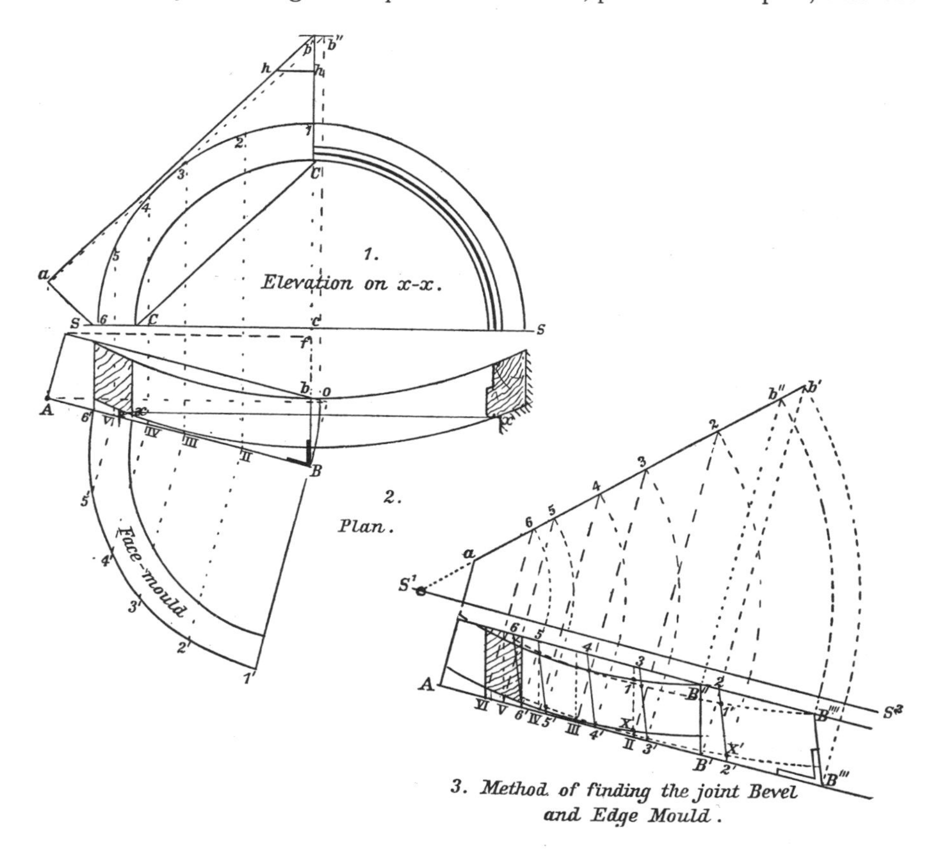

same plane as the lower end point A. Therefore with A as centre, and A B as radius, describe an arc, cutting a line drawn parallel to the elevation plane in point O. Project this point into the elevation, cutting a horizontal projector from the top end of the line *a b′* in point *b″*. Join *a b″*, which is the true length and inclination of the line *a b′*. Transfer this line to f. 3, making its ends the same height above S¹ S² that they are in f. 1. Produce the line *a′ b′* to cut the ground in point S¹. The dotted lines II, III, IV, V, VI, in f. 3 represent the same lines and points as in f. 2; these are projected into the elevation, cutting line *a′ b′* in points 2, 3, 4, 5, 6. These are next revolved into a horizontal plane on point S¹ as centre, as shown by the concentric arcs drawn from that point. On reaching the ground line they are projected to A B, perpendicular to that line, intersecting it in points 2′, 3′, 4′, 5′, 6′. The corresponding points on the other face can be obtained the same way, but it is quicker to draw lines from the first points, parallel with the joint line at B‴ (see next paragraph), as shown in full lines. Having found the true position of these lines upon the top edge of the plank, prick off upon them the distance from the face that the curves intersect the corresponding dotted lines in plan; for example, make 2′ X′ I′ equal to II X I, and so on, and draw the curves through these points.

To obtain the joint bevel, shown at B‴, f. 3.—Project points B′ and B″, where the joint line in plan cuts the faces of the plank, into the elevation, cutting the top edge of the plank in points *b′ b″*; revolve these horizontal on point S^1, then project them into the plan in points B‴ B⁗. Join these points, and the true bevel is disclosed for application to the back edge of the plank. But in such cases as shown in f. 1, where there is a waste part above the mould, the bevel may be taken direct from the plan, as shown at B, f. 2. This must be applied, however, in a line parallel with the ground. To do this cut off a portion of the waste as at *h h*, f. 1, square to the vertical joint line, and apply the edge bevel on this surface.

A Semi-head Sash Frame Circular in Plan is shown on Plate XXXI. So far as the parts below the springing are concerned, this frame is constructed precisely as a square head frame circular in plan would be. The pulley stiles must be made parallel, otherwise the edges of the outside linings would have to be removable to get the top sash in and out of the frame. The face edges of the sash stiles and the bars radiate from the centre of the plan, so that the mouldings, &c., can be worked square from the face. The pulleys, if ordinary ones are used, must be kept near the face of each sash, as shown, and the outside linings dished or bevelled to give room for the weights. Purpose made pulleys with the face plates at a suitable angle to the axle box are sometimes used, and these can be placed in the usual position. The inside lining is set square with the face of the pulley stile, as this position considerably reduces the thickness of stuff required for the head linings; but if it were required to be parallel with the outside lining, it would not affect the method of obtaining the moulds. The head of the frame may be cut out in the solid as described for the semi-head frame, p. 205, but is more economically constructed by bending a veneer of suitable wood over a cylinder, and staving the back, as shown in f. 6, 7, & 8.

To obtain the soffit mould for marking the veneer, f. 5, divide the elevation of the lower edge of the head, f. 1, into a number of equal parts, as A, B, C, D, E, S, and drop projectors from these points into the plan, cutting the chord line A″ E″ in A″, B″, C″, D″, E,″ S″. Draw the line S′ S′, f. 5, equal in length to the stretch out of the soffit in the elevation (the length of a curved line is transferred to a straight one by taking a series of small steps around it with the compasses, and repeating a like number on the straight line), and transfer the points A B C, &c., as they occur thereon, repeating them on each side of the centre line. Erect perpendiculars at the points, and make each of these lines equal in length to its correspondingly marked line in the plan, as A′ *a′ a′*, f. 5, equal to A″ *a a*, f. 3, these letters referring respectively to the chord line

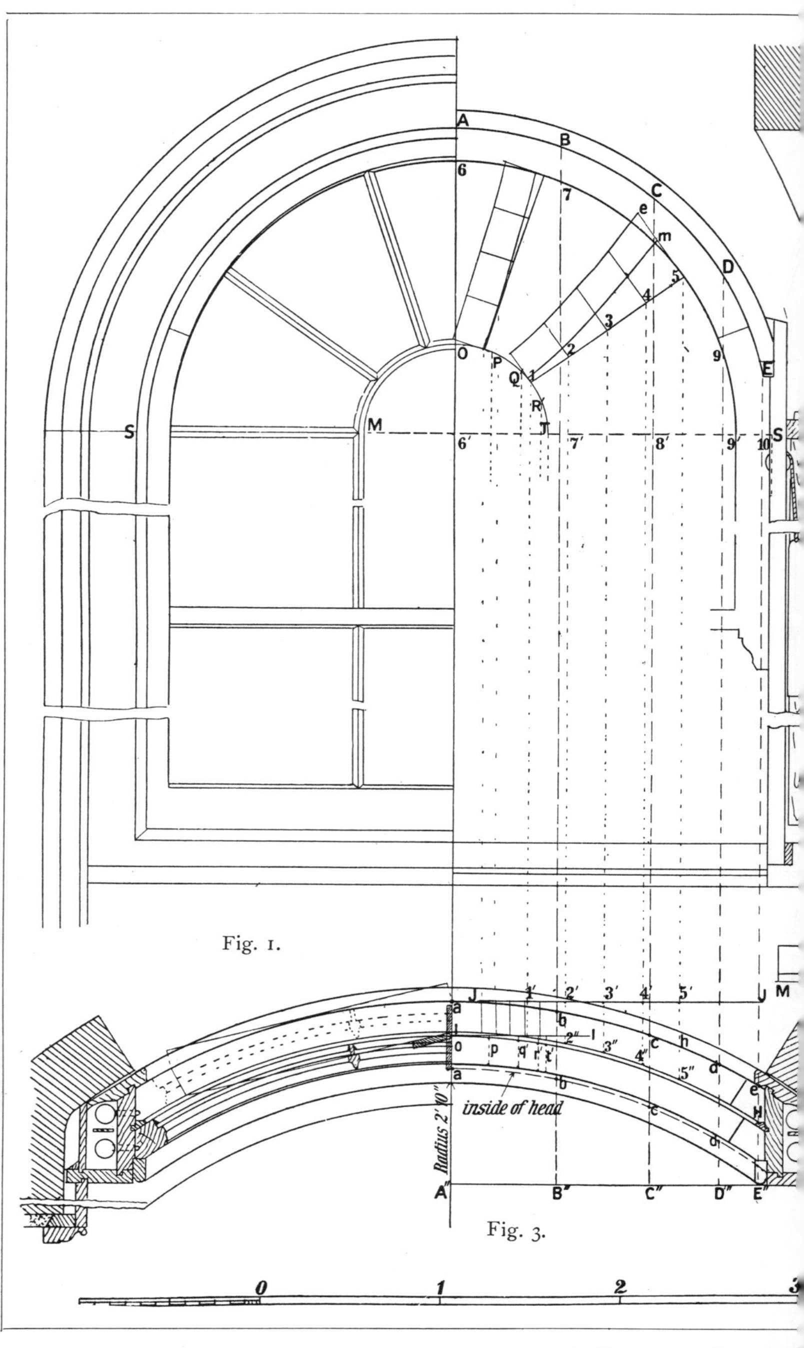

Fig. 1.

Fig. 3.

A Circle-on-Circle S

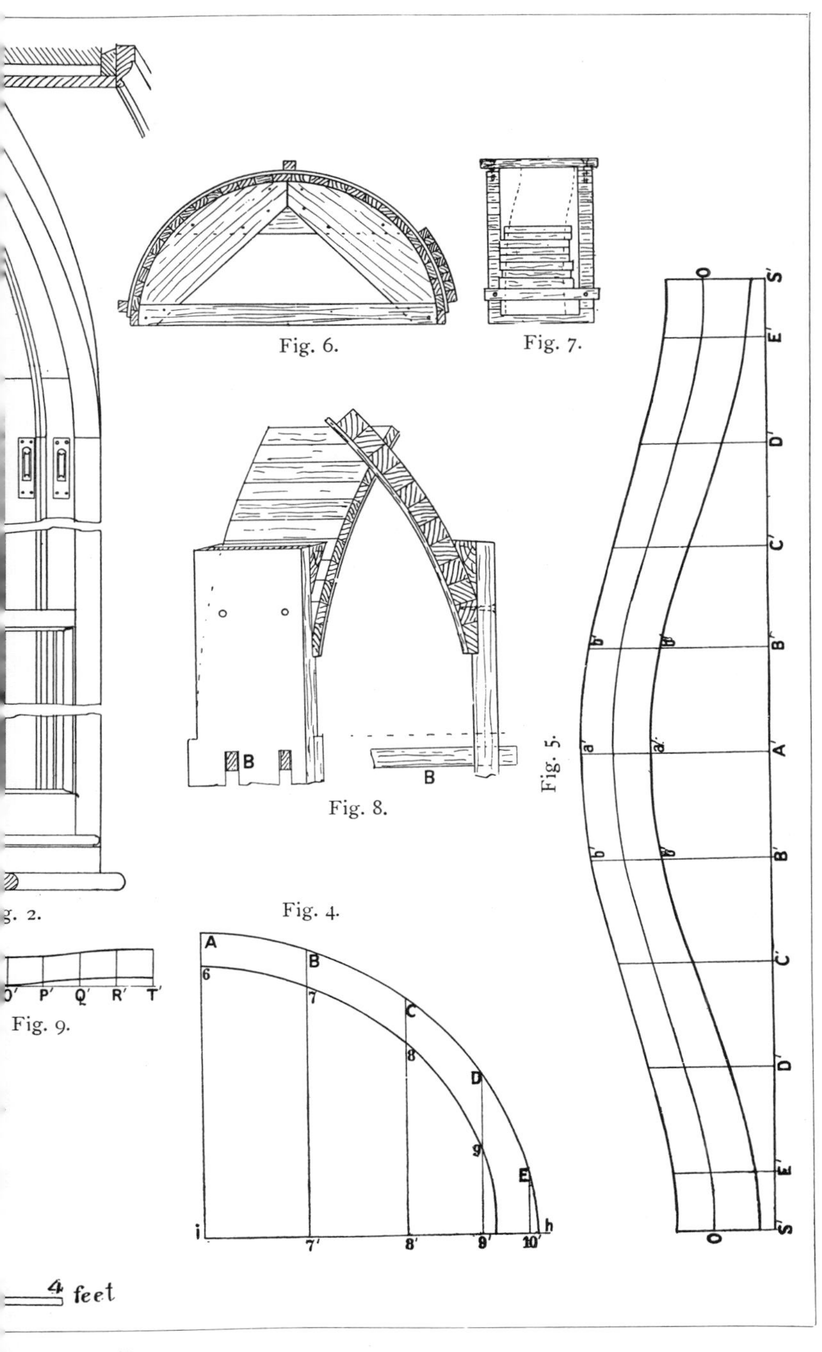

Fig. 6. Fig. 7. Fig. 8. Fig. 5. Fig. 4. Fig. 9. g. 2.

AME, WITH DETAILS.

[*To face page* 278.

and the inside and outside edges of the head. Draw the curves through the points so found. As will be seen by reference to f. 5, the mould is wider at the springing than at the crown; this is in consequence of the pulley stiles being parallel. If they were radial their width would be the same as the width of the head at the crown, and the head would be parallel; the gradual increase in width from the crown to the springing is also apparent in the sash-head and the beads, as indicated by the line O O, f. 5, which is the inside of the sash-head, and the outside of the parting bead; this variation in width renders it impossible to gauge to a thickness from the face, or the groove in the head, from its edges.

To form the head.—Having prepared the cylinder, f. 6, to the correct size, prepare a number of staves to the required section, which may be obtained by drawing one or two full size on the elevation, as shown to the right in f. 8. The staves should be dry straight-grained yellow deal, free from knots and sap, and not be so wide that they require hollowing to fit. If the veneer is pine, it will probably bend dry, but hardwood will require softening with hot water. One end should be fixed as shown in f. 7, by screwing down a stave across it. Then the other end is bent gently over until the crown is reached, when another stave is screwed on, and the bending continued until the veneer is well down all round, and a third stave secures it until it is thoroughly dry, when the remainder may be glued on. It is as well to interpose a sheet of paper between the cylinder and the veneer, in case any glue should run under, which would then adhere to the paper instead of the veneer. The head should not be worked for at least twelve hours after gluing. If a band saw is at hand, the back should be roughly cleaned off and the mould bent round it, the shape marked, and the edges can then be cut vertically with the saw, by sliding it over the cylinder sufficiently for the saw to pass. When cut by hand, the mould is applied inside and the cut is made square to the face, the proper bevel being obtained with the spokeshave, and found by standing the head over its plan and trying a set square against it. When fitting the head to the pulley stiles, the correctness of the joints is tested by cutting a board to the same sweep as the sill, with tenons at each end, and inserting it in the mortises for the pulleys, as shown at B, f. 8. A straight-edge applied to this and the sill will at once show if the head is in the correct position, and if the edges are vertical as they should be. The Head of the Sash may be cut out as described for the door-head, f. 1, p. 274, and fitted to the head of the frame by scribing, and the faces worked to the soffit mould of the frame cut down for the purpose to the line O O. The face edge can be worked, by gauging it from the back or top edge; or if preferred, a developed face mould can be used, as next described.

To obtain a developed face mould—Make the line *i h*, f. 4, Plate XXXI., equal to the stretch out of the plan of the face of sash-head, viz., I H, f. 3. Transfer the divisions as they occur, and erect perpendiculars thereon. Make these equal in height to the corresponding ordinates over the springing line in the elevation, and draw the curves through the points so found. The groove for the parting bead can be marked by running a ⅜-in. piece around the inside of the sash head, this bead being generally put in parallel. It is sometimes omitted altogether, the side beads being carried up until they die off on the head.

To find the mould for the cot bar.—Divide its centre line in the elevation into equal parts, as O, P, Q, R, T. Drop projectors from these into the plan, cutting the chord line I I in *o*, *p*, *q*, *r*, *t*. These lines should be on the plan of the top sash, but are produced across the lower to avoid confusion with the projectors from the other bars. Set out the stretch out of the cot bar on the line M′ O′ T′, f. 9, and erect perpendiculars at the points of division, and make them equal in length to the correspondingly marked lines in the plan. The cot bar is cut out in one piece long enough to form the two upright sides as well as the arch. The straight parts are worked nearly to the springing, and the bar

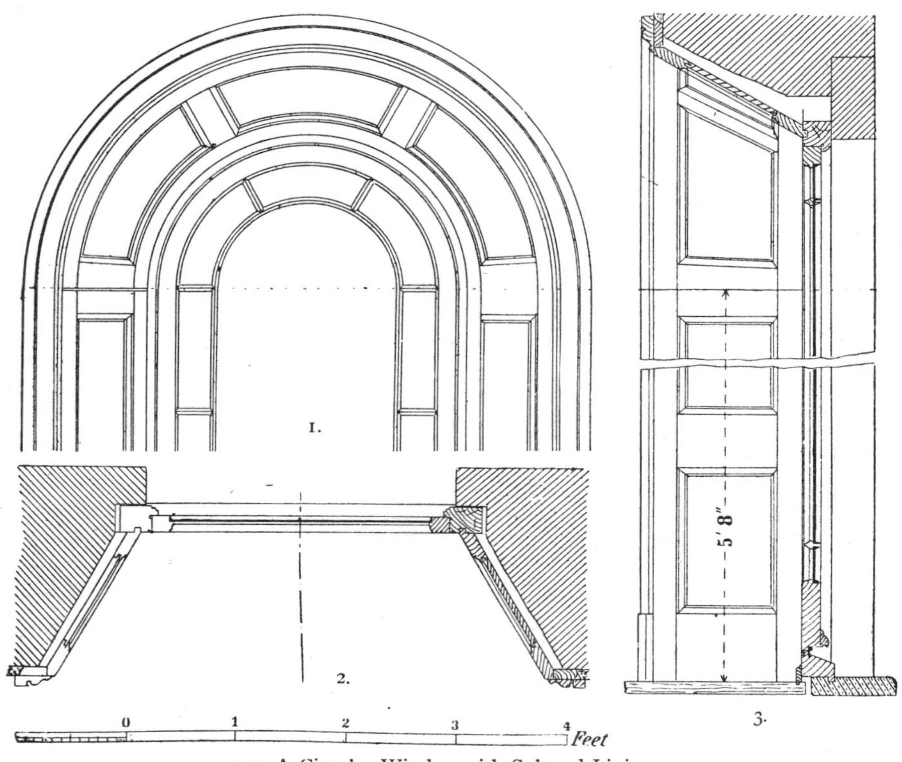

1-3. A Circular Window with Splayed Linings.

which is got out wider in the centre is then steamed and bent around a drum, and afterwards cut to the mould, f. 9, and then rebated and moulded. The arched bar should not be mortised for the radial bars, but the latter scribed over it and screwed through from inside.

To find the mould for the radial bars.—Divide the centre line of the bar into equal parts, as 1, 2, 3, 4, 5, f. 1, and project the points into the plan, cutting the chord line J J in 1′, 2′, 3′, 4′, 5′. Erect perpendiculars upon the centre line from the points of division, and make them equal in length to the distance of the corresponding points in the plan from the chord line, and draw the curve through the points so obtained. The other bar is treated in the same manner, the projectors being marked with full lines in the plan. The soffit moulds for the head linings are obtained in similar manner to those for the head mould, the width being gauged from the head itself.

A Framed Splayed Lining with Circular Soffit is shown in part elevation in f. 1, opposite page, plan f. 2, and section f. 3. The soffit stiles are worked in the solid in two pieces joined at the crown and springings. The rails are worked with parallel edges, their centre lines radiating from the centre of the elevation. The panels are veneered and staved on a cylinder similar to that on p. 273. Edge moulds only are required for the stiles, and a developed mould for the panels. The method of obtaining these has been shown on a separate diagram, annexed.

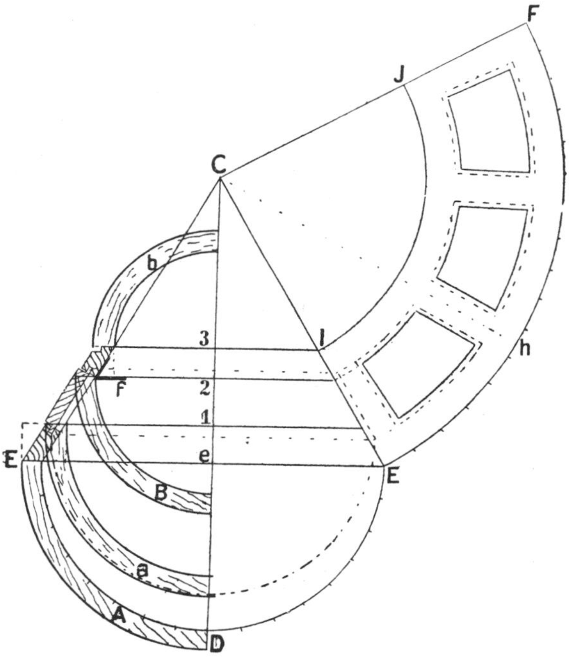

Method of Obtaining Moulds for Circular Splayed Linings.

To obtain the face moulds for the soffit stiles.—Draw the line E E, and produce the faces of the jambs until they meet in point C. From C draw the line C D perpendicularly to E E. This line will contain the centres of the various moulds, which are located by producing the plans of the edges of the stile across it, *e*, 1, 2, 3 being the respective centres, and the inside and outside faces of the jambs affording the necessary radii for describing the arcs A *a* and B *b*.

To apply the moulds.—Prepare the stuff equal in thickness to the distance between the lines *e* 1 and 2 3. Apply the mould A to the face of the pieces intended for the front stile, and cut the ends to the mould, and square from the face. Set a Bevel as at F, and apply it on the squared ends, working from the lines on the face, and apply the mould *a* at the back, keeping its ends coincident with the joints, and at the points where the bevel lines intersect the face. The piece can then be cut and worked to these lines, and the inside edge squared from the face. The outside edge is at the correct bevel, and only requires squaring slightly on the back to form a seat for the grounds. The inside stile is marked and prepared similarly.

The Development of the Conical Surface of the Soffit is shown on the right hand of the diagram on p. 281, and is given to explain the method of obtaining the shape of the veneer, but it is not actually required in the present construction, as the panel being necessarily constructed on a cylinder, its true shape is defined thereon, and its size is readily obtained by marking direct from the soffit framing when the latter is put together. Let the semicircle E D E represent a base of the semi-cone, and the triangle E C E its vertical section. From the apex C, with the length of one of its sides as radius, describe the arc

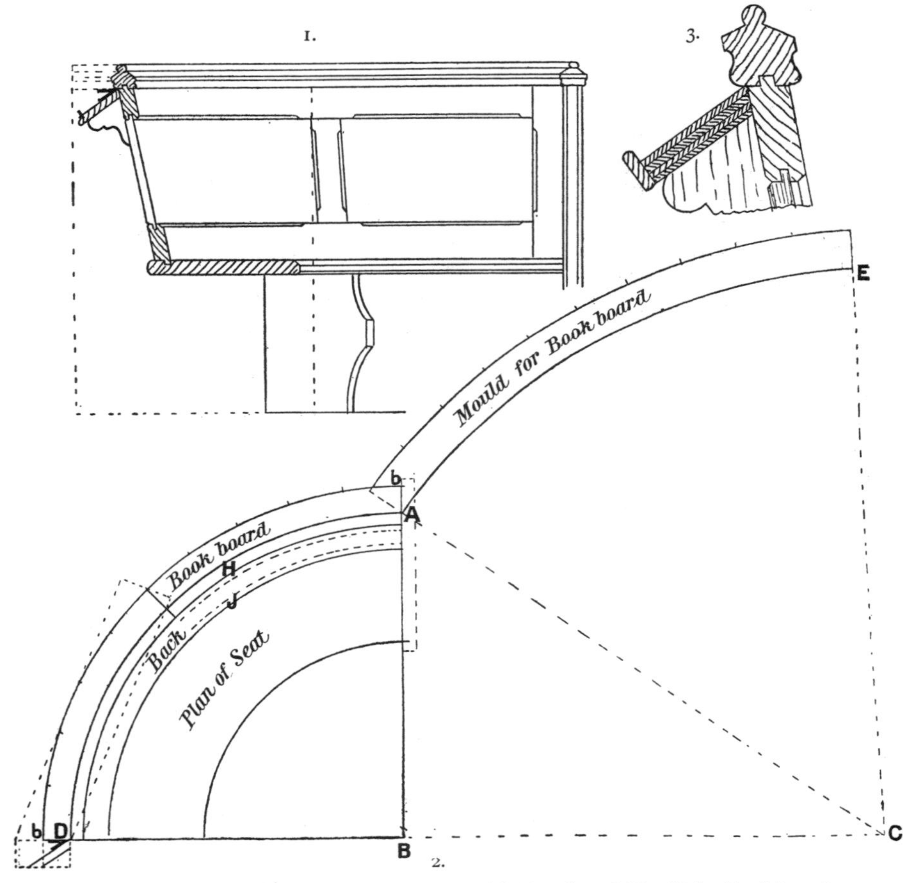

1-3. Plan and Elevation of a Circular Pew, with Developed Mould for Bookboard.

E F, which make equal in length to the semicircle E D E by stepping lengths as previously described. Join F to C, and E C F is the covering of the semi-cone. The shape of the frustrum, or portion cut off by the section line of the linings, is found by projecting the inner edge of the lining upon the side C E, and drawing the concentric arc I J; then E I J E represents the covering of the frustrum. Any portion of this, the panels for instance, is found in the same way. To draw the rails, divide their centre lines equally on the perimeter, and draw lines from the divisions to the centre as H C; make the edges parallel with these lines.

A Church Seat Circular in Plan, with inclined bookboard and back (see opposite page), affords another example of the development of the cone for obtaining moulds to line out complex curves. The bookboard may be cut out of the solid, and jointed in the middle with handrail bolts, or it may be built up with four $\frac{1}{4}$-in. veneers, cut to the shape of the developed mould shown, and bent around brackets fixed to the back of the seat, or upon rough brackets cut to the correct slope, and fixed upon the plan lines on the rod. In this case the first veneer is bent around and sprigged down to the brackets with headless sprigs that will draw through easily, then the next layer is well glued and fixed to it with handscrews until it is dry, and the successive layers fixed in the same manner; the edges are covered by the rim, which is steamed and bent to the shape on a drum (see p. 270), it may be fixed with brads or screws as shown in the section, f. 3.

To obtain the mould for the veneers.—Draw the plan of the bookboard full size, as at f. 2. From the inner edge of the board at A draw a line at an inclination to its end A B equal to the intended slope, as shown in the section, f. 1. Draw a line from D, the opposite end of the board perpendicular to A B, and produce it to cut the inclined line from A, and its point of intersection C is the apex of a cone, of which A B is the semi-diameter of the base, and the curve A D its elevation. From C as centre, and C A as radius, describe the arc A E; produce A C, and make the distance from A equal to the width of the bookboard, as shown in the section. Describe a concentric arc from this point, and make it equal in length to the stretch out of the plan, as described in previous examples. The moulds required to mark the stuff for Cutting in the Solid will simply be the plan curves of the inside and outside edges, as shown at D A and *b b*, f. 2: the width of the piece required to cut the board out of is shown by the dotted line enclosing the plan. The thickness is ascertained by drawing parallel lines touching the top and bottom edges of the board in the section, f. 1. The dotted rectangle at D in the plan represents the end of the piece turned up level with the top to show the application of the Bevel to the mould lines. The bevel is the inclination of the board to a level line as shown in the section. The mould *b* is applied on the bottom surface of the piece, as shown by the dotted lines across the turned up end. The edge cuts are made square from the face of the piece. The panels can be built up as shown at f. 3, p. 271, the component pieces being made tapering in length, or they may be veneered on both sides of a shaped core, as shown in f. 1, p. 272.

Plain Splayed Circular Head Lining Segmental in Plan.—The sketch on the next page shows the method of obtaining the shape of a veneer for the soffit of this, which is to be bent around a conical cylinder, and staved. The geometrical problem involved is, the finding of the envelope of a frustrum of a cone, formed by the penetration of two right cylinders perpendicular to the generating section, the generating section being the triangle A E B formed by producing the faces of the linings. Let A B and C D be the plan of the edges of the soffit, and A C and B D the plans of the faces of the jambs. Produce these to meet at E. Draw the chord line A B, and describe a semicircle upon it; divide this into any number of equal parts, as 1, 2, 3, &c. Draw perpendiculars from these points to the chord line A B, and from the intersections draw lines to the apex E. From E as centre, with the side E A as radius describe the arc A *e*, and make it equal in length to the semicircle A 9 B, as previously described. Set off upon it the divisions 1 to 17 as they occur on the semicircle, drawing lines from the divisions to the apex E. From the intersections of the plan curves with the radial lines, as at *a a*, &c., draw lines parallel with A B to cut the side A E. From the apex E describe arcs from these points, cutting the corresponding lines in the development. The intersections will be points through which to draw the curve, and the same process is repeated upon the edge C D.

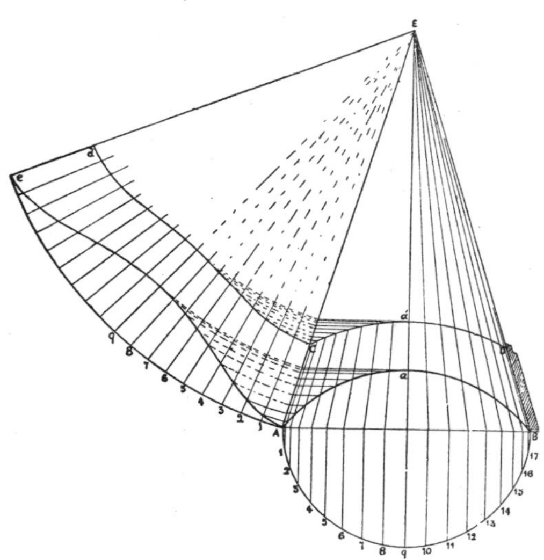

Method of Obtaining Shape of Veneer for Head of a Semicircular Splayed Lining Segmental on Plan.

A Louvre Bull's-Eye Frame. — The whole of the lines required for setting out this frame may be obtained geometrically, but instruments sufficiently accurate to produce the necessary developments are seldom to be found in the workshop, so a readier and more practical method is here described for obtaining the cuts upon the frame; the geometrical process will be explained elsewhere. The frame is first constructed in four segments, cut to shape and planed up, then jointed as shown, with handrail bolts and cross tongues. Usually these frames are beaded on the face side; if so, this should be done before jointing, otherwise the cross grain will break out at the ends when moulding. Having prepared the frame and put it together dry, also prepared the stuff for the louvres, which should be gauged to a thickness: prepare a rough frame of inch stuff, as shown in f. 1 opposite, just large enough to contain the bull's-eye, and equal to it in thickness; shoot the edges of two sides A A, and lay them on the section of the frame which has been set out full size on a rod, as f. 3. Strike up the sight lines of the grooves on each side of the frame; pair the other piece with the marked side, and strike over the lines on the second piece, then arrange the pieces in the form of a square box about the frame, with the marked sides exactly parallel to one pair of joints; nail the rough frame together, and also drive temporary nails into the bull's-eye to keep it in position when turning over. Next, by aid of a straight-edge, mark the groove lines across the face of the frame, as shown at *b b*. Having marked all on the face, turn the frame and its case over, and treat the back similarly. Then remove the rough case, take the bull's-eye to pieces, and fix one of the

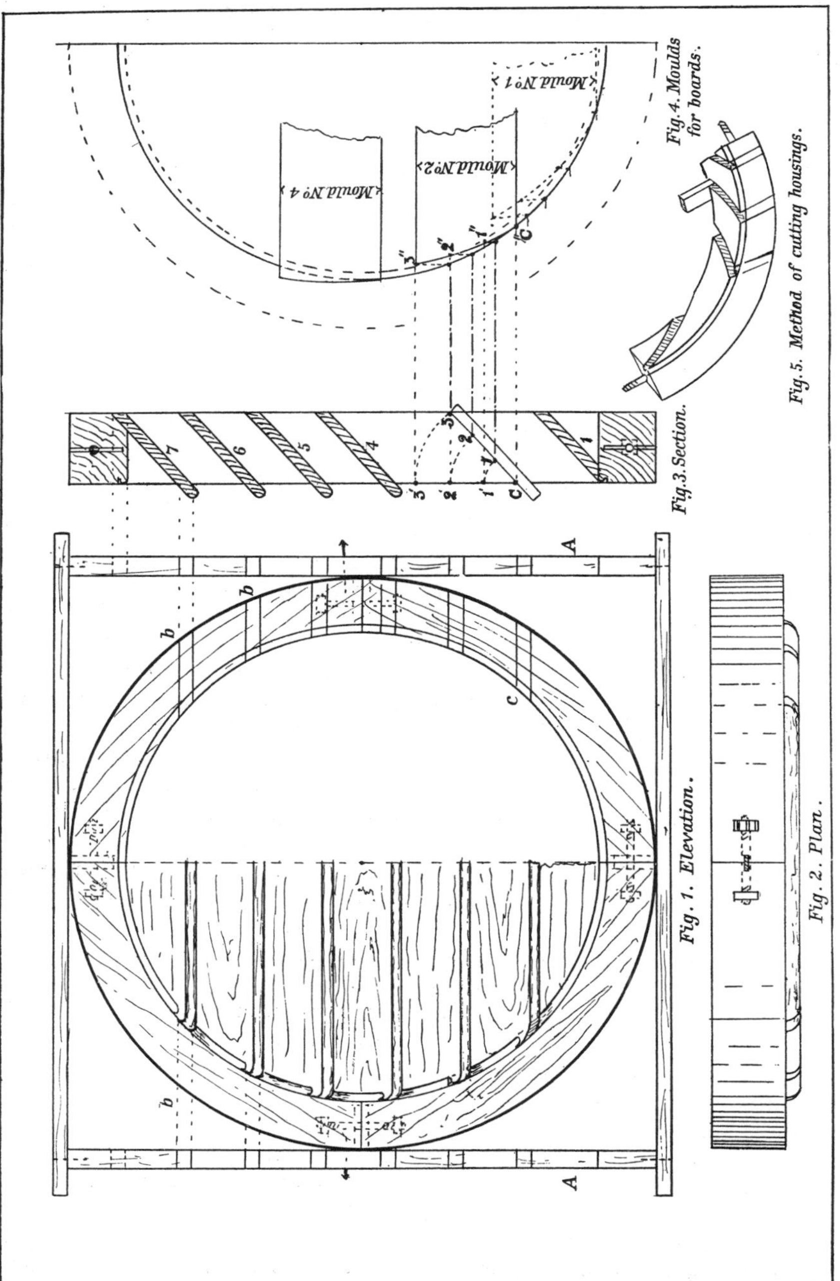

A BULL'S-EYE LOUVRE FRAME.

segments in the bench screw, as shown in f. 5, with a short strip of wood laying upon one of the offside lines ; this will act as a guide to the saw, the direction of the cut not being visible. The tenon or panel saw can now be run down to the quirk of the bead, and the process repeated until all are done ; the core can then be removed with a chisel, when it will be found that in the more oblique cuts the saw has barely gone beneath the surface in the middle of the thickness ; therefore the housing must be further sunk in with the chisel to a uniform depth. The housing for the bottom board marked No. 1 in the section should not be taken through at the front, otherwise the appearance of the bead around the frame will be spoiled.

The shape of the louvre boards may be obtained either by scribing or by geometrical development. The former method will be described first. Put the frame together in two halves, which may be permanently jointed with glue or red lead ; the two vertical joints must be left dry, but drawn up close. Mark the length of each board on the face edge between the bottom of the housings at opposite sides ; also lay a straight-edge on the vertical joints, and make a mark on the board where it crosses, square this over to the back edge, and work to this centre line in the next operation. Turn the frame over, and mark the corresponding lengths on the back edges ; draw lines across the faces of the boards from these points, which will give the shape of the ends roughly. Cut the boards off ½ in. longer than thus marked at each end, as they will require to be longer in the middle. Next open the frame on the two loose bolts sufficiently to insert the boards between the soffit, when the exact shape may be scribed with compasses; set them to the sight lines previously marked on the edges of the boards, and run them across the top and bottom faces, and cut with a bow saw to the two lines. The bevel for the front edge of No. 2 board is shown at *c*, f. 1. Each board should be marked out completely before commencing another, to avoid mistakes with the bevels. The process takes longer to describe than it does to execute, but is not difficult when the principle is grasped. Any shaped frame may be set out in this way.

The geometrical method, though easier to demonstrate, is more difficult in practice, and requires great accuracy in drawing. Fig. 4 is one side of the frame re-drawn to avoid confusion of lines ; the only line of the frame necessary is the quirk line of bead, which represents the depth of the housings. In the section, f. 3, the third board has been omitted to leave a clear space for the construction lines, and the process is shown complete on the second board only, though the developments of two others are indicated on f. 4.

Divide the face of the board, f. 3, into equal spaces in points C, 1, 2, 3, with C as centre ; turn these points into the vertical plane, as shown by the dotted arcs, cutting the face of the frame in points 1′, 2′, 3′. Draw horizontal projectors from points 1, 2, 3 on the face of the board, cutting the quirk line in the elevation ; these projectors are drawn in dot and dash line, to distinguish them from the others. Erect perpendiculars indefinitely from the points of intersection, then draw projectors from points C, 1′, 2′, 3′, intersecting the perpendiculars in points 1″, 2″, 3″, then a curve drawn through these points from the point C″ will give the true shape of the end of the board on its face side. A mould may be cut to this shape, and used for marking both ends of the board, its back edge being kept flush with the back edge of the louvre.

BEVELS.

A Window with a Splayed Soffit and Splayed Jambs is shown in part elevation, f. 1 below; plan, f. 2; and section, f. 3. The linings are grooved and tongued together, as shown in the enlarged section, f. 4. To obtain the correct bevel required for the shoulder of the jamb and the groove in the soffit, the lining must be revolved upon one of its edges until it is parallel with the front, when its real shape can be seen. This operation is shown in the diagram on the right-hand half of the plan and elevation. Draw the line *a b*, representing the face of the jamb in plan, at the desired angle. Project the edges into the elevation, and

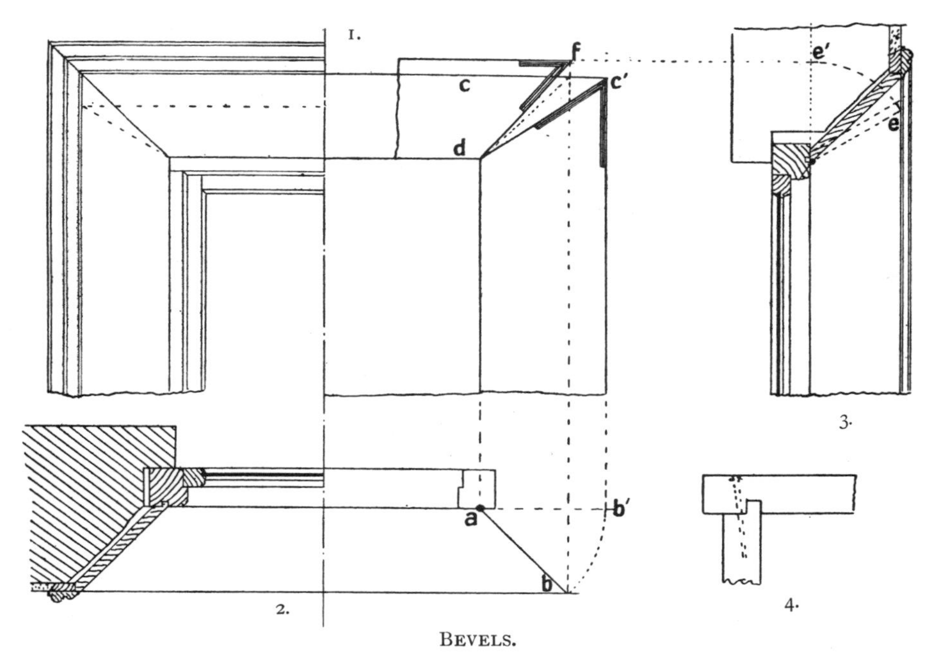

BEVELS.

1. Elevation of Linings with Splayed Jambs and Soffit. 2. Plan. 3. Section of f. 1. 4. Section through Joint in Linings.

intersect them by lines *c d*, projected from the top and bottom edges of the soffit in section. This will give the projection of the linings in elevation. Then from point *a* as centre, and the width of the lining *a b* as radius, describe the arc *b b'*, bringing the edge *b* into the same plane as the edge *a*. Project point *b'* into the elevation, cutting the top edge of the soffit produced in *c'*. Draw a line from *c'* to *d*, and the contained angle is the bevel for the top of the jamb. When the soffit is splayed at the same angle as the jambs, the same bevel will answer for both; but when the angle is different, as shown by the dotted lines at *e*, f. 3, then the soffit also must be turned into the vertical plane, as shown at *e'*, and a line drawn from that point to intersect the projection of the front edge of the jamb in F; join this point to the intersection of the lower edges, and the contained angle is the bevel for the grooves in the soffit.

The Hopper, shown in the sections, f. 1 & 3 below, and plan, f. 2, is another example of oblique or compound bevels similar to the splayed linings, f. 1, p. 287. The true shape of the sides and the bevels required for marking the joints are determined in the same manner—that is, by turning the inclined surfaces as upon a hinge until their opposite edges become parallel or in the same plane, and then connecting their ends by a straight line which will disclose the bevel. Two differing inclinations have been given to the ends of the hopper, to show that the method is applicable to any inclination, the plan and section being given.

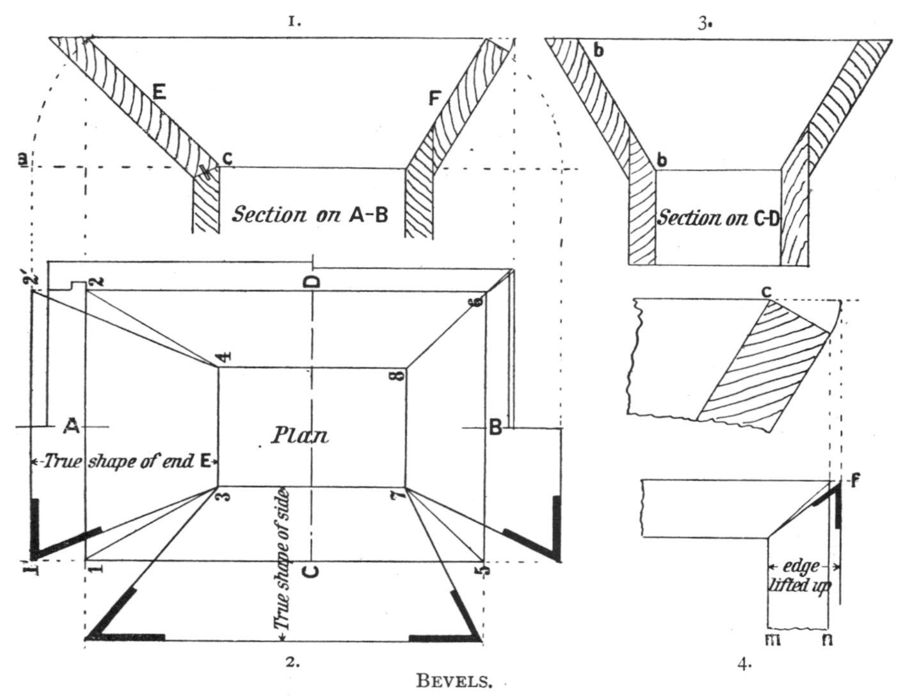

BEVELS.

1 & 3. Sections of a Hopper. 2. Plan of Hopper. 4. Development of Edge Mitre.

To Determine the Bevel for the Shoulders and the True Shape of the Sides.—Let 1 2 and 3 4 be the plans of the top and bottom edges of the side E. With the lowest point C, f. 1, as a centre, and the width of the side as a radius, describe an arc cutting a line drawn level from point C in *a*. Draw a perpendicular from this point into the plan, and draw projectors from the inside edges of the other sides to meet it, as shown by the dotted lines. Draw lines from the intersections 1′ and 2′ to the bottom edge at 3 and 4, and the figure enclosed within the lines 1′, 2′, 3, 4 is the real shape of the end E, within the clear of the adjacent sides, and either angle will give the bevel for the face cuts. The end F is determined in like manner. The shape of the side C is shown by drawing a parallel line to the bottom edge 3 7, at a distance equal to the width of that side as shown at *b b* in the section, f. 3, draw projectors from the top edges of the ends at 1 and 5, and from the intersections draw lines to the bottom edge 3 7.

The edge cut of the end E is square to the face, and if it were a mitre, as shown at the opposite end, it would be shown true in the plan when the edges are bevelled as shown, because the joint would then be drawn on a plane parallel with the ground; but the mitre shown in the plan of the end F is not the true bevel, because it lies upon an inclined surface. To find the true mitre, lift the lower arris of the edge to the level of the upper arris, and project it into the plan. To cut a projector from the corresponding arris of the side, join the intersection to the intersection of the inside arrises, and the angle will contain the true mitre. The parts have been drawn to larger scale in f. 4 to render the operation clearer.

For other methods of obtaining these bevels see p. 81.

CHAPTER XVIII.

MISCELLANEOUS FITTINGS AND FITMENTS.

Construction of Chimney-piece and Overmantel—Kitchen Dressers—Woods to Use for Kitchen and Pantry Fittings, &c.—Kitchen Tables—Advantage of Double Shoulders to Rails—Method of Cutting Boards for Table Tops—Gluing up the Frame—A Housekeeper's Closet—A Fitment for a Butler's Pantry—Bath and W.C. Enclosures—Removable Fixings—Setting Out W.C. Seats and Fitting Up Frames—Method of Fixing—Method of Setting Out Bath Top—Constructing the Framing, A Movable Panel—Pipe Casings.

A DESIGN for a **Chimney-piece** and **Overmantel** suitable for a drawing-room is given in f. 1-4 opposite. This design might well be executed in Weymouth pine and enamelled white, or the framework and mouldings prepared in Black Walnut with Huon pine or Sycamore panels. The chimney-piece is built on a cased frame of 1-in. stuff, with a wide rail under the mantel-board stub tenoned into the jambs, and the corbel mould screwed to its face. This mould is worked out of $2\frac{1}{4}$-in. stuff, as shown in the section, p. 292, with the return ends mitred and the joints cross tongued; the top edge is also tongued into the mantel-board. All the mouldings are glued and screwed from the back. The overmantel is framed complete in itself, and tongued into the top side of the mantel-board. The back is formed of 1-in. panelled and solid moulded frame, the centre panel under the cupboard being filled in with bevelled edged silvered plate glass. The upper shelf is continuous, with a moulded edge and circular ends, as shown in the enlarged section, f. 4. The ends of the cupboard and the brackets under are housed into it. The frieze rail is dovetail grooved into the cupboard ends, and the cornice mould sunk $\frac{1}{16}$ in. into and screwed to the frieze, p. 292. The cupboard doors are cut out of 1-in. stuff, the rails and stiles $1\frac{1}{2}$ in. wide, rebated $\frac{1}{4}$ in. for glass, and moulded with a $\frac{1}{4}$-in. ovolo. The bars are $\frac{1}{2}$ in. thick, and sunk the depth of the square from the face. The doors are also set back $\frac{3}{16}$ in. below the face line of the ends.

A Small Kitchen Dresser is shown on p. 293. These fittings are made with both open and enclosed pot-boards, each arrangement being shown in half elevation in f. 1. When the lower part is made open, it lightens the appearance of the legs to taper them slightly upon the insides, as shown. One or two bearers should be placed across under the pot-board to stiffen it. The drawers are generally made just under 9 in. deep, and a small bead is returned around the fronts to break the joint. The rail between the drawers and pot-board, known as the drawer rail, is, when no doors are used, made of the same width as the legs, and double stub tenoned into them. In the

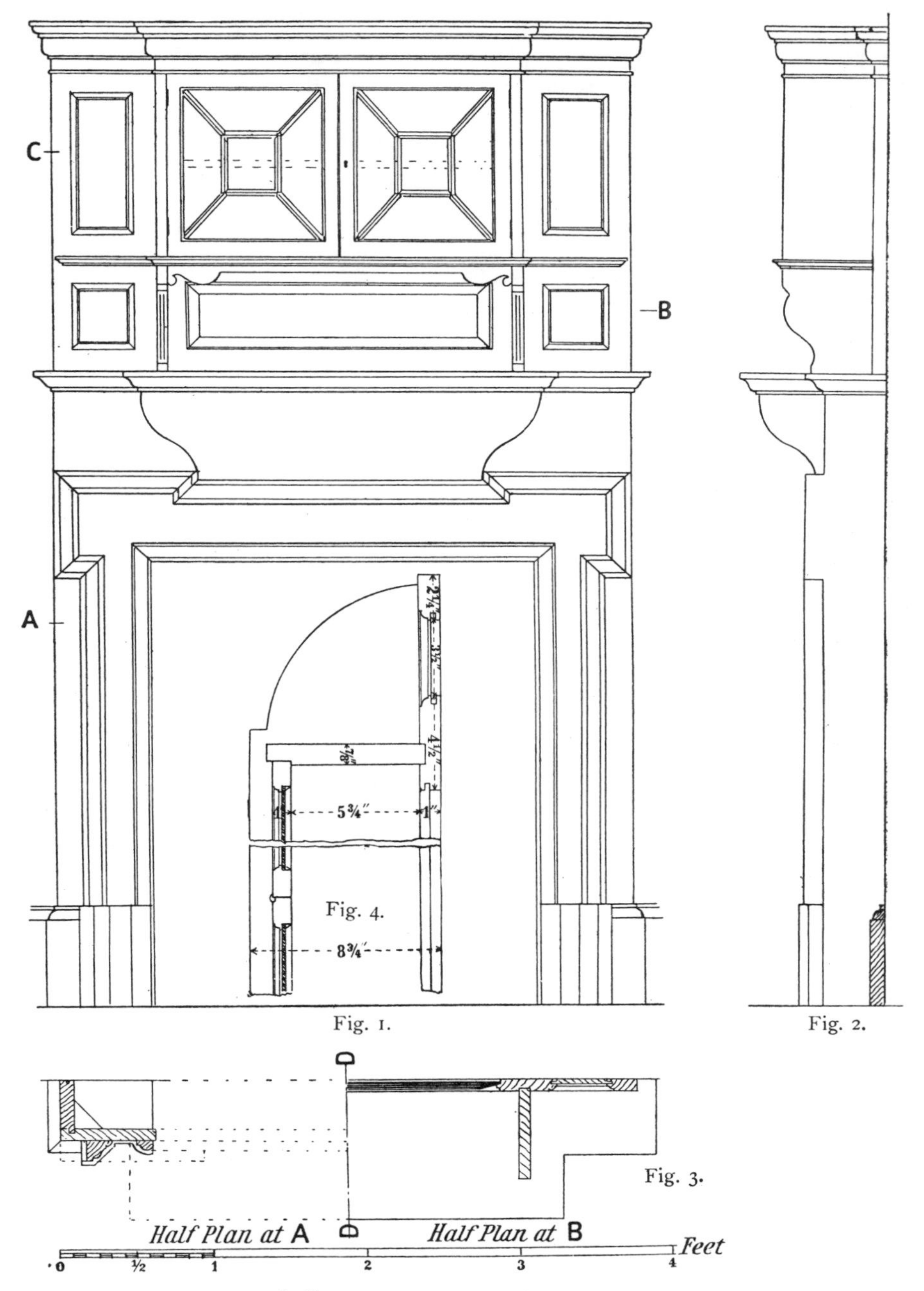

A Chimney-piece and Overmantel.

1. Elevation Front. 2. Elevation End. 3. Half Plans. 4. Enlarged Section at C.

section shown at f. 5, the rail is increased in depth and reduced in width to provide room for the door stop, and single tenons are used in this case as shown. The top rail is dovetailed to the legs, and the drawer divisions double tenoned through the rails. The back rail, which is an 11 by 1 in. board, is ploughed with a $\frac{3}{8}$-in. groove $\frac{7}{8}$ in. from the top edge inside for buttons, and is stub tenoned into the legs, the tenons mitreing with those of the end rails, f. 3 opposite. The drawer runners are tongued into the back and front rails, and the tilting pieces dovetailed. The top is 2 in. thick, buttoned to the frame and screwed to the front rail. The top should be of spruce or white deal, as this wood imparts no odour to food, &c., placed upon it, and being also free from turps and resin, it is very suitable for table tops, shelves, cupboards, &c. The shelf rack above the top is usually made separate and fixed in position with wall hooks. The back, when fitting against a plastered wall, is left open, as shown, with the exception of one or more standards, according to the length of the shelves, to support them in the middle. In other cases the back is match-lined. The shelves, f. 6, are housed into the end standards $\frac{1}{4}$ in., the housing fitting neatly around the plate groove. The ends of the standards are also housed into the dresser top, as shown in f. 5.

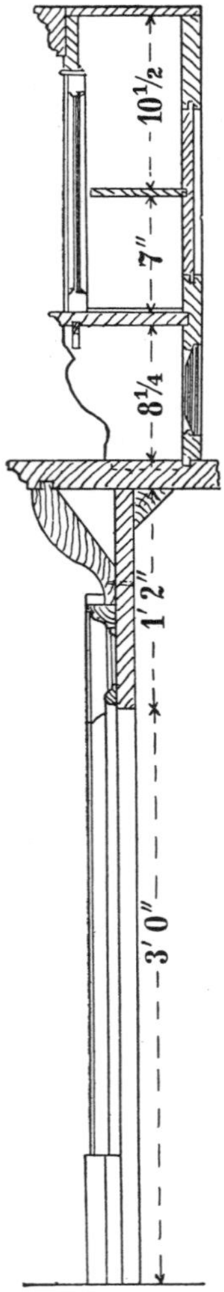

A Kitchen Table suitable for a large house is shown on p. 294. The dimensions vary with the requirements, the example is of medium size, but larger than this are sometimes required, in which case it would be necessary to introduce additional legs. These tables are necessarily heavy, and must be strongly constructed. The tops are generally of white deal or American birch, either of which stand frequent washing well, and they vary in thickness from $1\frac{3}{4}$ in. to $3\frac{1}{2}$ in. according to size. The top will keep a better surface if it is made up of several pieces, the width of which should not exceed 9 in. They should be grooved and straight tongued, the joints made with "Beaumontique" (a very thick paint composed of white lead, litharge, whiting, and boiled oil), and secured with two or more $\frac{3}{8}$-in. iron bolts with square heads at one end and screw nuts at the other, passing through the thickness from edge to edge. The top is sometimes screwed to the frame in pelleted holes, but is better fixed by buttoning, the centre board only being screwed. The framework is less likely to rack if the legs are placed raking as shown. The rails should be deep at the shoulders, but may be shaped out at the middle, as shown in the drawing, to provide knee space. Double shoulders are stronger than single, and have a more workmanlike appearance than barefaced tenons. The tenons, however, should be kept as near the outside as possible to obtain additional length. They should fit tight, taper in width, and mitre with each other, as shown by dotted lines in f. 5. The tenon is best

Section of Chimney-piece on D D, f. 3, p. 291.

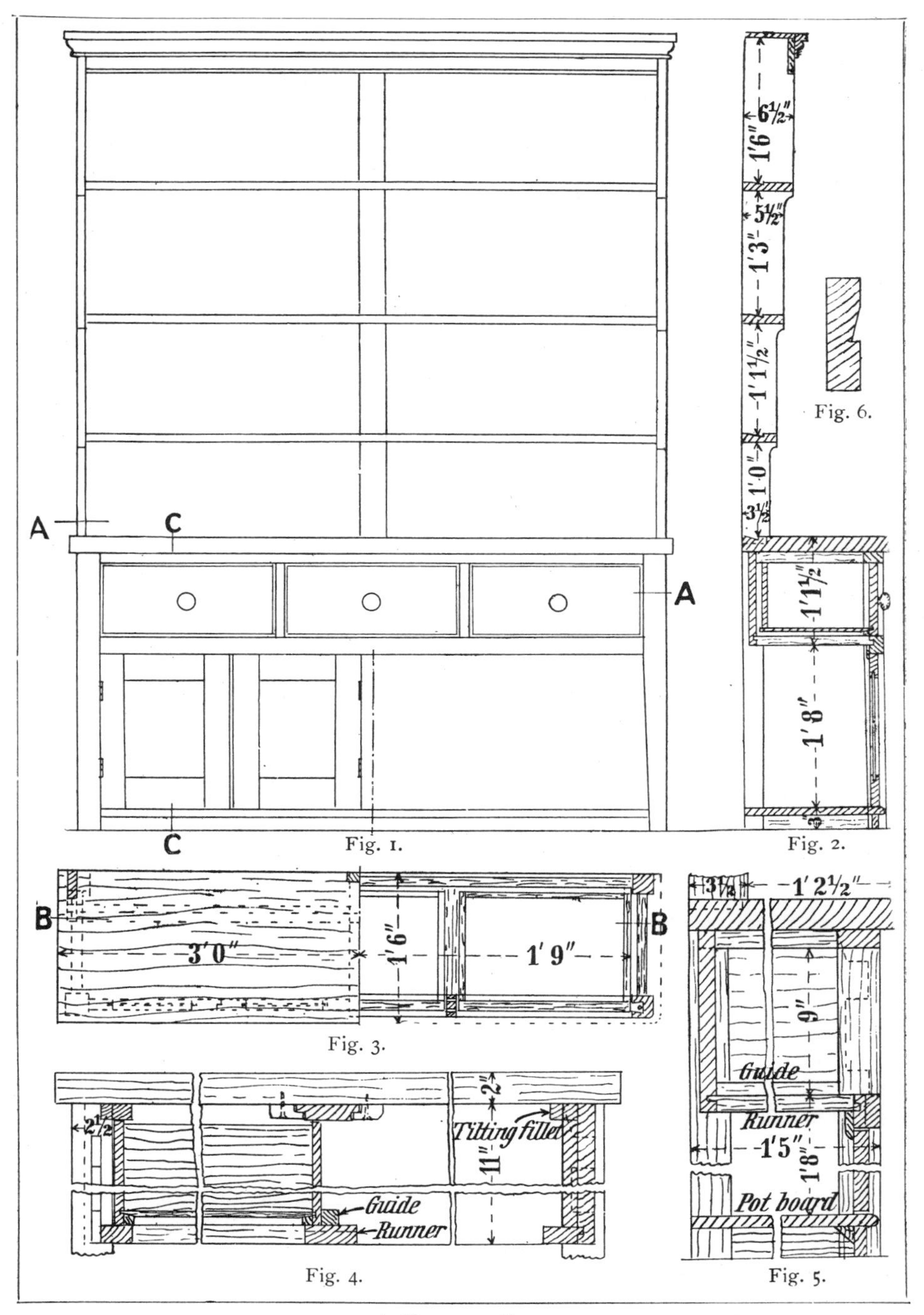

Fig. 1. Fig. 2. Fig. 3. Fig. 4. Fig. 5. Fig. 6.

A Kitchen Dresser.

1. Alternative Half Elevations. 2. Section on C C. 3. Horizontal Sections at A A. 4. Section on B B. 5. Enlarged Section of f. 2. 6. Section of Shelves.

fitted by knocking each rail up separately and scribing on the side of the tenon the inside of the adjacent mortise, which will indicate the short edge of the mitre. When fixing together, glue the end rails in the legs first, and let these dry before gluing in the side rails. The mortises rather than the tenons, should be well glued. If the tenons only are done, the glue will all be scraped off as the tenon is driven in. A couple of scores should be made with a bradawl on each side of the tenon, to let the superfluous glue escape, otherwise the shoulders will not come up close. The joints in the top should be ploughed from the back side, and the face cleaned off after it is buttoned down. The boards for table tops should not be cut tangential to the annual rings, *i.e.*, with the rings more or less parallel with the surfaces, as such boards

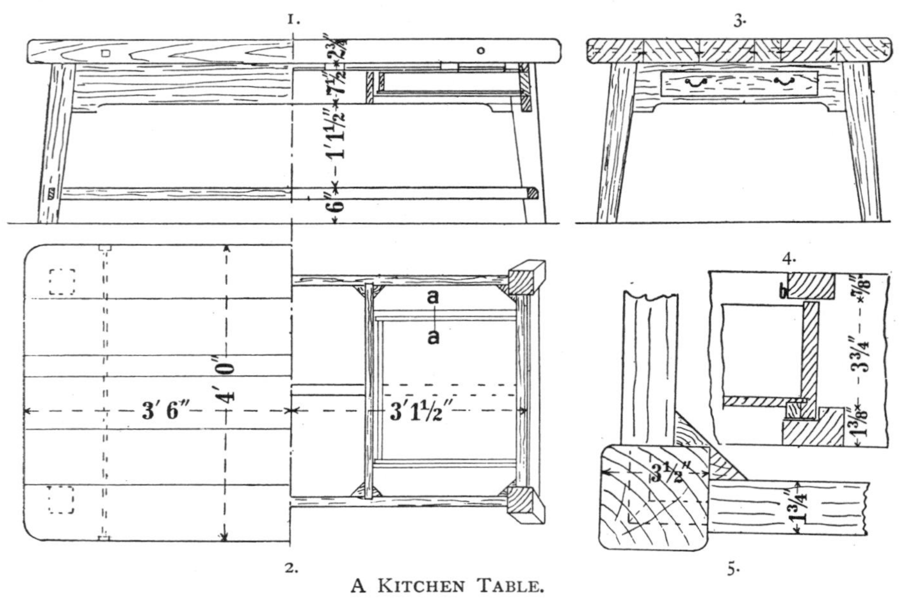

A Kitchen Table.

1. Sectional Elevation. 2. Sectional Plan. 3. End Elevation. 4. Section on *a a*. 5. Joint in Legs.

will invariably cast hollow, and make the surface irregular (see f. 3, p. 418). If, however, it is compulsory to use such boards, they should be placed with the convex side of the rings downwards, then buttons in the middle of each board will tend to counteract the subsequent casting. The correct cutting for the boards is indicated by the graining in f. 3. Drawers are often fitted at each end, as shown in f. 3; the openings to receive them are cut in the rail, and rebated runners are framed between the rail and division, as shown in f. 2 & 4.

A Housekeeper's Linen Closet is shown on opposite page. Fig. 1 is a front elevation with one pair of doors removed to show the interior. Fig. 2 is a broken vertical section, and f. 3 an elevation of the framed end. The plan, f. 4, shows an alternative treatment of the ends, that on the left hand being framed and fitted to a rounded corner post with rounded plinth and

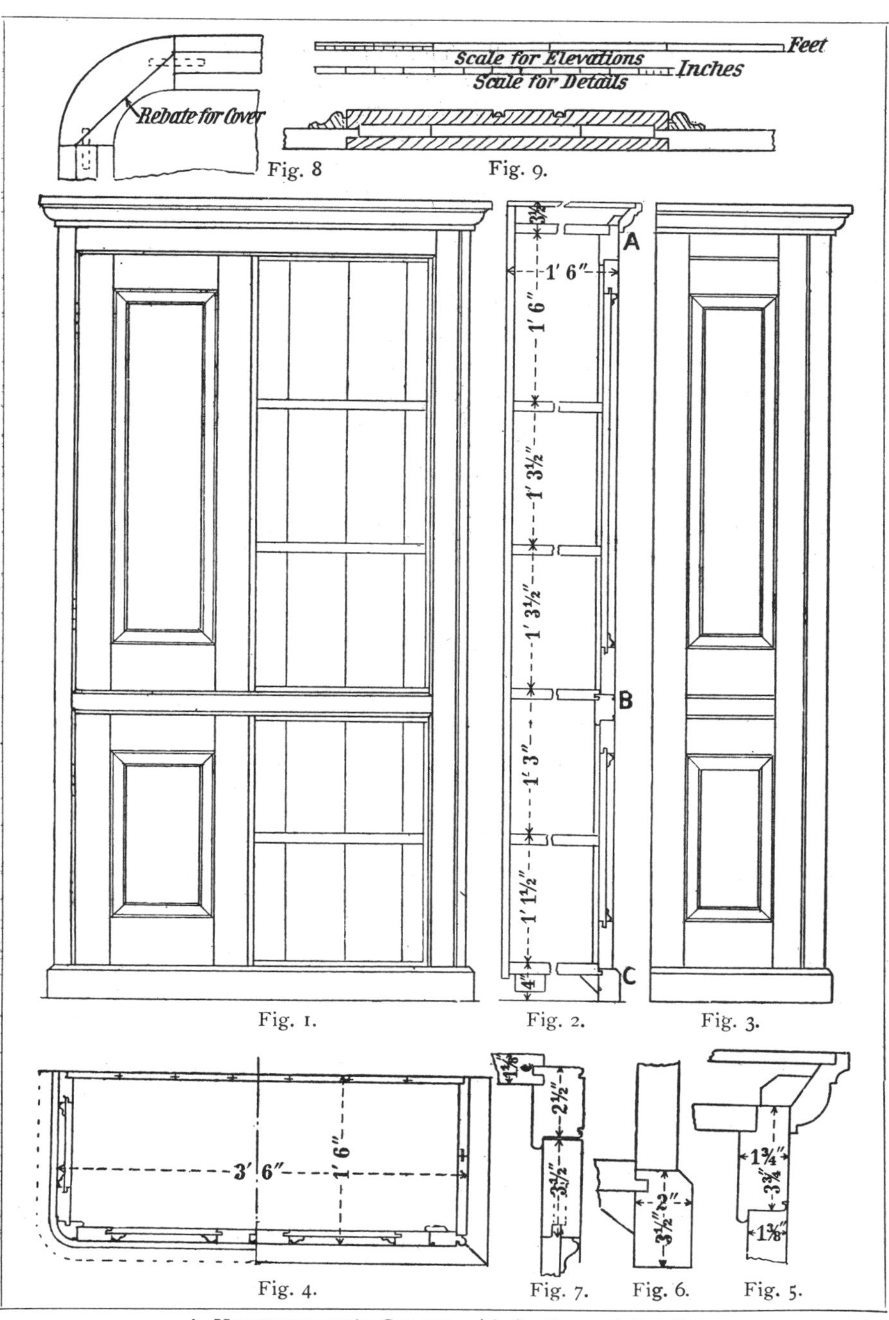

A HOUSEKEEPER'S CLOSET, with Section and Details.

cornice; this method is suitable when the closet stands near a doorway or projects from the middle of a wall. The right hand half plan shows a plain solid end rebated into the front with a staff bead on the angle, and is the usual treatment when the cupboard stands in a recess. The front frame is $1\frac{3}{4}$ in. thick, rebated with solid stops for $1\frac{3}{8}$-in. doors and stub mortised and tenoned together. The front plinth is worked in the solid on the bottom rail and tenoned into the stiles of the frame. The side plinths are planted on the faces of the ends which run down to the floor. Details of the various joints are given in f. 5-9. Fig. 8 shows the method of fixing the quadrant corner of the cornice. This is dowelled to the straight parts, and glued and bradded to the front. The cover board is dropped into a rebate cut straight across the corner. Fig. 9 is an enlarged section of the middle rail of the framed end showing the tenons, &c.; two $\frac{3}{8}$-in. beads are sunk across the rail to match the front. The mortises can be made through and the tenons wedged in this end, as the edges of the stiles are not seen, but they are better cut back $\frac{1}{8}$ in. on the rebated edge to prevent the joint being broken should the stuff shrink.

A Butler's Cupboard.—The drawing opposite shows a half elevation, transverse sections, and enlarged details of a fitment suitable for a butler's pantry. It consists of a pedestal fitted with a double set of drawers, and a cupboard above fitted with shelves enclosed by glazed doors. The two outer doors are hung on hinges in pairs, and the middle door slides between them, when they are open, upon hardwood runners. The detail at A shows the top runner, the bottom one is similar; a detail of the rebates in the meeting stiles is shown at B, and a corresponding section through the side of the cupboard and the hanging stile of the door is shown at C. The carcase of each part is made separate and afterwards screwed together. The top and bottom of the cupboard should be dovetailed to the sides, and if made to stand in a recess, as shown in the plan, the dovetails may be taken through, otherwise they should be stopped. A match-lining back is shown, which is quite sufficient if fixed in a recess, but if not, a panelled framing should be used. The cornice is simply cut in tightly between the walls and nailed to the top, it is rebated to receive the dust board, which is also supported by three cross bearers. The pedestal is constructed similarly to f. 1, p. 250, with the difference that here the beads are worked on the drawer fronts, consequently square shoulders are required on the division rails. This will be clear on examining the detail section D. The top in this case is secured to the top rail and ends by buttons and screws. The plinth is tongued, glued, and blocked to the bottom.

A W.C. Enclosure is shown on p. 298. The seat is carried on two deal bearers that are dry dovetailed to two legs in front, and fixed by wedging into the wall at back. The bearers are rebated out to receive the seat, and also a hanging rail for the same at the back. The seat is covered by a mitre clamped flap, hung to a frame, which is fixed, as shown in f. 4, to the seat rail. The riser is a panelled frame with a quadrant stile at the corner. The plinth and bottom rail are made in one piece, and the plinth portion is lipped across the wall stiles, as shown in the edge view, f. 7, being sunk about $\frac{1}{16}$ in. below the face to hide the joint. A tenon is formed on the back part to correspond with the top rail tenons. The tenon on the other end of the rail

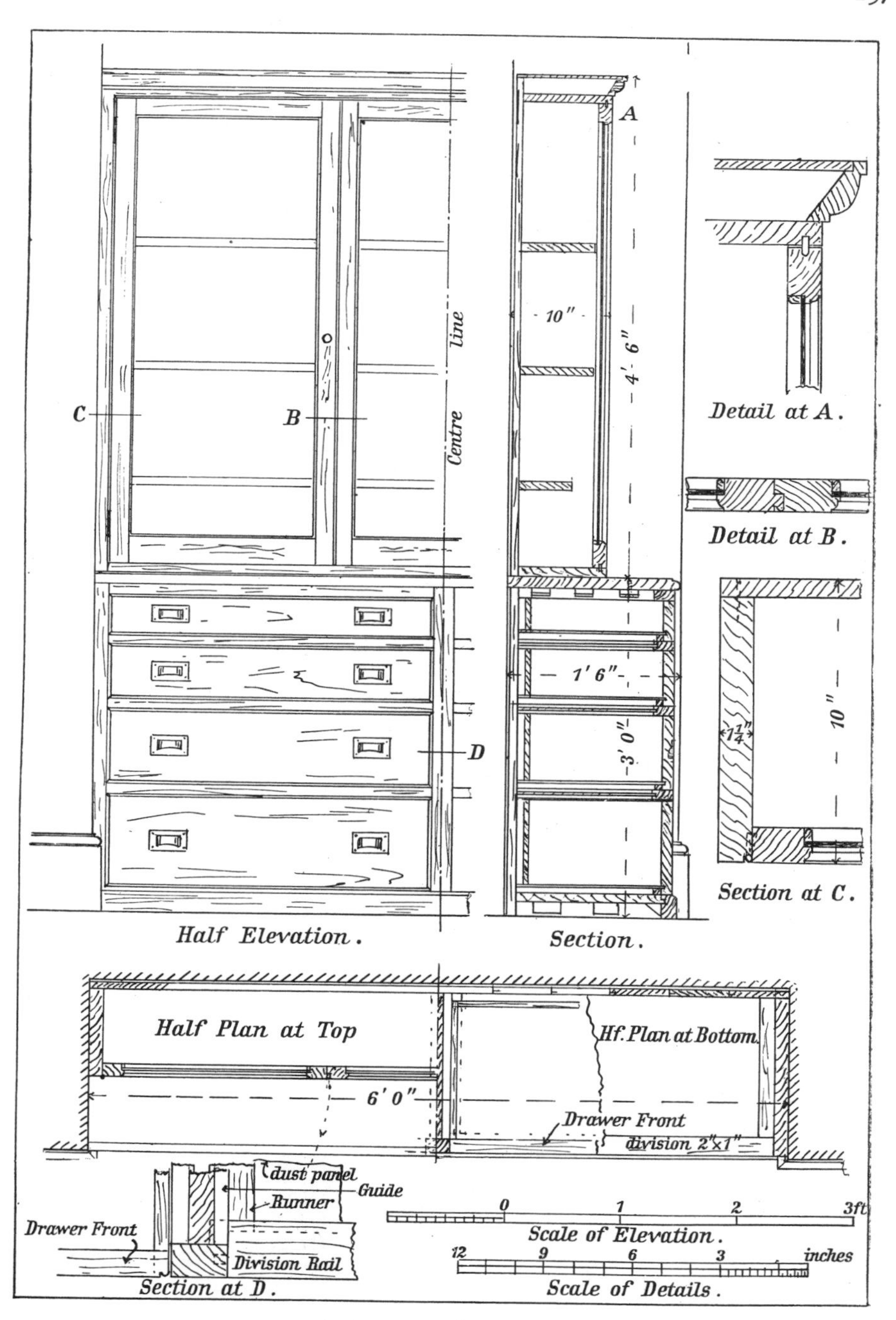

A BUTLER'S CUPBOARD.

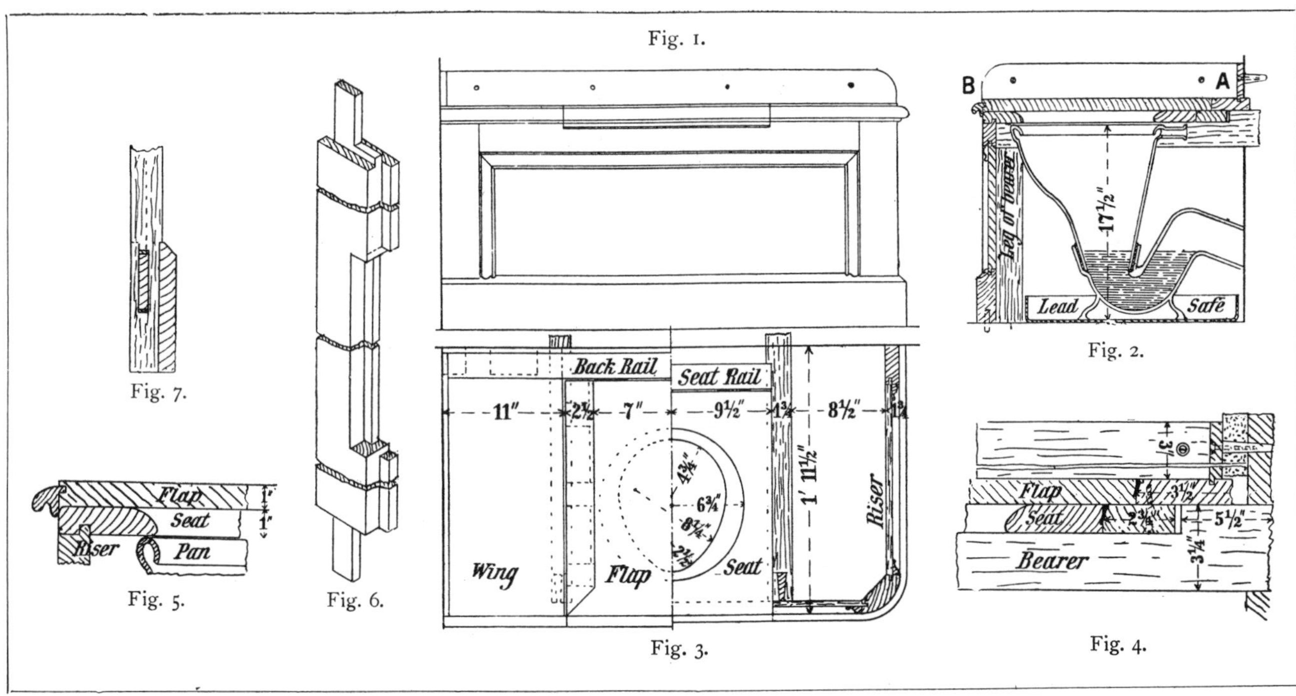

A W.C. FITTING, with Details.

is stubbed in as far as the rounding will allow and screwed from the back, a shoulder being made at the sight line of the stile, and the corner piece of plinth made to the thickness shown in section in f. 7, and shaped at the back to fit the sinking on the stile. The riser has two dowels fixed in the bottom rail which fit into holes in the floor, and its top edge is tongued both into the seat and the flap frame in the manner shown in the sketch of the top rail, f. 6. The skirting or seat border is also tongued into the flap frame, and is secured by cups and screws to plugs in the wall, thus fixing the whole enclosure, which is readily removed when necessary by withdrawing these screws. Seats are preferably made in one piece, but if jointed the joint should be dowelled, not tongued, and the dowels kept clear of the hole. The hole is usually cut of an oval shape, and struck from three centres, as shown in f. 3. The dishing is made circular, and struck with a radius an inch longer than half the major diameter of the hole. The section of the dishing is shown in f. 5. The grain of the seat should run parallel with the front, and two hardwood ledges should be slot screwed to the under side, or where there is not room for these, the ends of the seat may be ploughed, and a hardwood straight tongue inserted therein dry. The hanging rail of the seat must be kept forward sufficiently for the seat to lie flat back on the flap when both are open. The bead should be stuck just flush, and the knuckles of the hinges let entirely into, and flush with the bead. The moulded nosing shown (known as a "treacle mould" because of its supposed resemblance to that viscid substance when it is flowing over an edge), has two lips or tongues, the inner one preventing the fingers slipping between the edge of the flap and the seat, and getting nipped when closing it. This nosing is glued into the edges of the flap and wing rails in one length, and cut through at the joints after it is dry, the flap being first hung and shot off straight with the rails, and ploughed from the back side. To do this, place a screw cramp across the front of the frame, first inserting a chip in each joint to provide room to work the saw when cutting through the nosing. The flap should be fitted very close, with just sufficient joint to prevent it "speaking," and the chip should spring it open a little. Then when the cut has been made in the nosing with a fine saw, the ends will close up and form a good joint.

The Bath Enclosure, p. 300, is arranged for a full-size bath, which measures inside at the top 6 ft. long, 27 in. at the shoulders, 20½ in. at the foot, and 20½ in. deep. A circular end is introduced to show the different methods of jointing the top. The front is a panelled frame 1¼ in. thick with 1¼ in. planted ogee moulding. The rails and panel at the circular end are only shaped on the front, the back running straight across, as indicated by the dotted lines in f. 2. The joints in the rails are made with handrail bolts. The end panel at the foot opens on hinges, so that the trap and pipes can be reached without removing the whole of the enclosure; this is kept closed either by a lock or a small thumb bolt just under the moulding. Fig. 4 shows how the back is treated to prevent the panel casting or splitting. Oak rails 2 by ½ in. are fixed around the edges. Those running the same way of the grain as the panel are glued on, those running across it are slot screwed dry, the other pieces being tongued into them at the ends. A ½ by 1½ in. fillet is screwed around the opening to form a stop. The top is out of 1¼-in. stuff, and

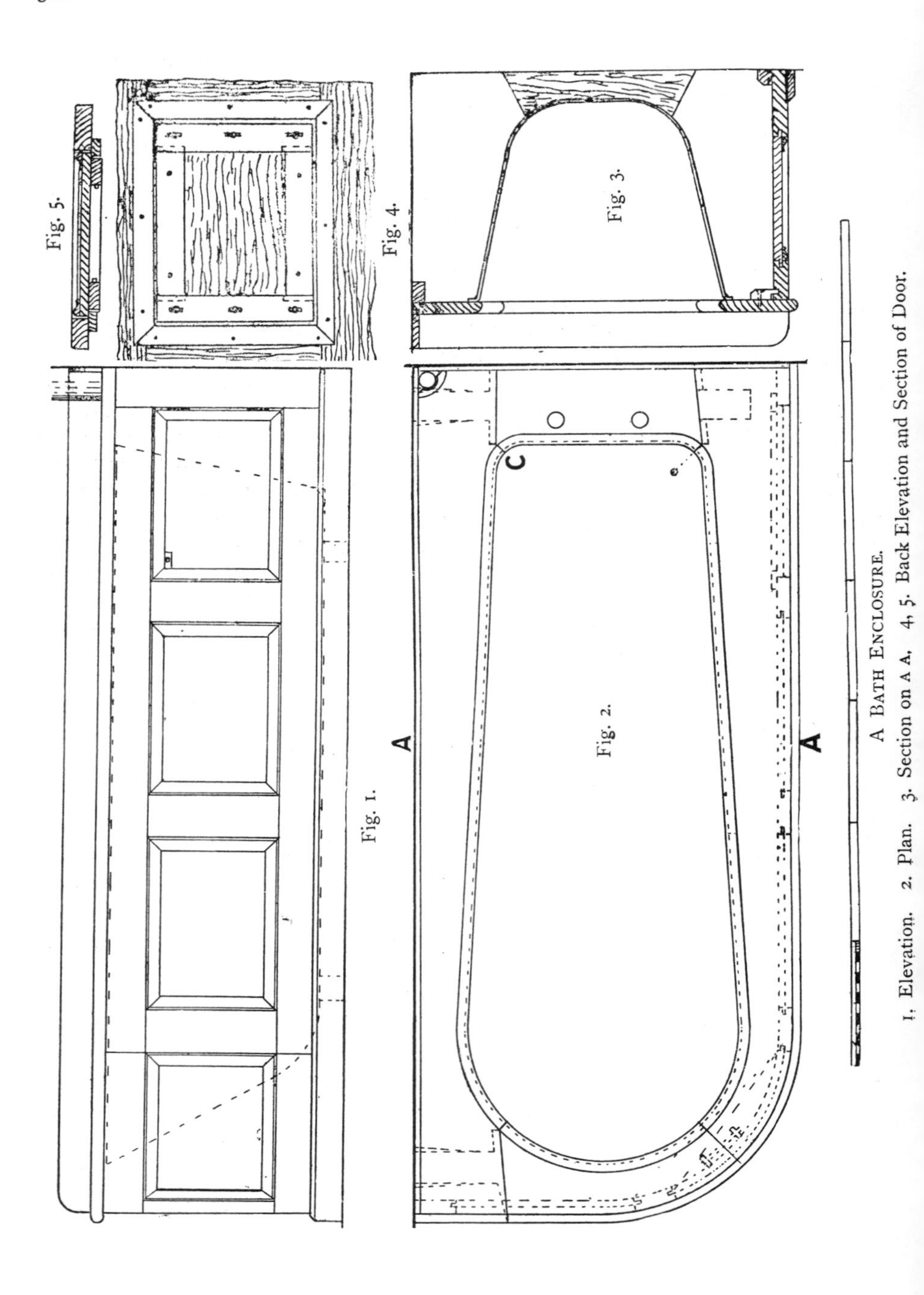

A BATH ENCLOSURE.

1. Elevation. 2. Plan. 3. Section on A A. 4, 5. Back Elevation and Section of Door.

should lie close down on the rim of the bath. In best work the edge is sunk in a rebate and bedded in a mixture of red and white lead. The tenons in the top are frequently only stubbed in about 2½ in. and screwed from the back, but where possible the tenons should always be taken through and wedged, as in moist situations stub tenons will always get slack and loosen the shoulders sooner or later. The tenons need be only ¼ in. thick, as there is no cross strain on them after the top is fixed, and they should be wedged so that they are tighter at the shoulder than outside. The shoulders are usually bevelled, as shown in the two opposite corners of the plan, under the impression that bevelled shoulders have a tendency to slip towards the inside when shrinking and so keep the mitre up, but personally the author has not noticed any difference when square shoulders as at C have been used. The mitres are drawn radial from the centres of the curves, the width of the stiles being kept equal at each end. The top is buttoned to the front, the latter being turned over for that purpose after both have been fitted into place, and they are then slipped down into position together, a fillet being fixed to the floor inside, to fix the bottom edge to (see f. 3 opposite). The shape of the bath top is usually obtained in practice by making a rough framed templet large enough to project the required distance over the end and front of the bath after the latter has been fixed, then running a pencil round inside the bath underneath, and working the inside of the templet off to the shape. A ¼-in. margin should be allowed inside this when lining the top out by it. The templet should be scribed round all pipes and projections, and the position of any tap connections, &c., coming through should be marked upon it, so that the tenons can be kept clear of them when setting out. Mahogany is

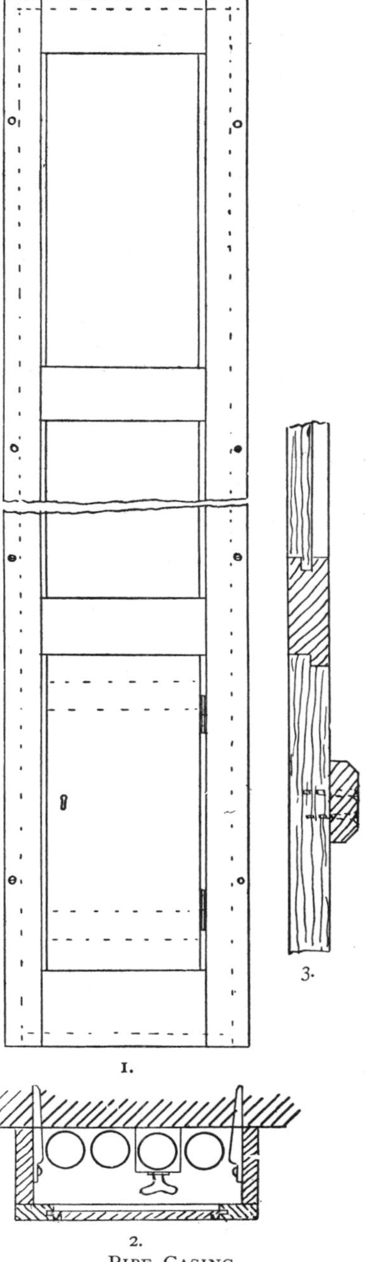

PIPE CASING.
1. Elevation. 2. Plan.
3. Detail of Door.

the most suitable wood for these and W.C. enclosures, though Birch is sometimes used for tops.

Pipe Casings up to 8 in. wide are generally made of $\frac{3}{4}$-in. boards rebated and nailed together in the form of a shallow box, and are fixed by screwing through the sides into plugs driven in the wall, the whole being taken down together when required; but when the pipes are numerous and the front board is required wider than 8 in., they should be framed, as shown in f. 1, p. 301, and f. 2, the sides fixed to wall hooks, and the front attached by cups and screws, or round-headed screws will answer, if the front is not likely to be removed very frequently. A small door should be framed in the front, or one of the panels be made to open opposite any taps or valves that require ready access. Fig. 3 shows the preparation necessary in the rails of the framing for a panel which is to open.

In fitting and fixing casings, discretion should be used as to which parts are fixed first. All elbow covers and bends or breaks around a set-off, should be fixed first, and the straight lengths cut between them, as in repairs, &c.; these lengths are the most likely to require removing. Casings should not be scribed over enclosure plinths, but should run down to the seat, &c., and the plinth be butted against the side of the casing.

CHAPTER XIX.

STAIRBUILDING.

Description of Stairs—Dogleg—Open Newel—Continuous—Circular Geometrical—Circular Newel—Self-supporting—General Remarks—Preparation and Planning—Size of Openings—Points to Avoid—Shape of Winders—Usual Sizes—Timbering—Cause of Creaking Stairs—Headroom, How to Obtain—How to Determine Rise and Going—Ratio of Rise to Going—Constructional Details—Taking Dimensions on Building—Preparing Story Rod—Preparation of the Working Drawings—Preparing the Pitch Board—Use of Pitch Board in Setting Out—Method of Working Strings—Making the Steps and Fitting Up the Stairs—A Special Clamp—Method of Marking Winders—Sunk Strings—Construction of Bullnose—Round End—Curtail and Commode Steps—Fitting Mitre Cap to Rail—Determining Section of Cap—Setting Out Dancing Steps—Construction of Cut and Mitred, and Cut and Bracketed Strings—Development of Wreathed Strings—How to Obtain a Good Falling Line—How to Fix Nosings and Balusters—How to Obtain Length of, and Set Out Balusters—Brackets, How to Fix and How to Obtain Reduced Patterns—How to Build "Laminated" and "Cored" Strings—Designs for Brackets and Handrails.

THE limits of this work do not permit of a large number of plans of stairs being given, but typical examples of each kind that are usually met with in practice are included, which it is thought will prove sufficient to explain the methods involved in the construction of stairs. The cognate subject of handrailing has been treated in a separate chapter for convenience; although the two subjects are so interwoven and interdependent, both in theory and practice, that cross references are unavoidable, and an acquaintance with each branch is necessary for a thorough mastery of the subject. The beginner should refer to the glossary for explanations of the terms used therein.

Description of Stairs.—Stairs are variously named according to their shape in plan, or the method of constructing the strings, and they may be roughly divided into two classes, newel and non-newel. In the first of these the newel post is an essential part of the construction, and this includes dog-legged, open newel, and spiral or circular newel stairs. In the second class the newel is usually dispensed with, or when used it is merely as an ornamental finish to the balustrade, and is in no sense a constructive part of the stairs. To this class belong the continuous string or geometrical stair, the elliptic, polygonal, and circular well stairs. All of these names refer to the disposition of the stairs in plan, and they are otherwise designated, in reference to the manner of treating the strings, as close or housed string, open or cut string, and bracketed strings.

A Straight Flight is one composed entirely of flyers, and differs only from a ladder in that the spaces between its steps are filled in with risers.

Dogleg Stairs (see p. 308) are those without wells or spaces between the outer strings, the return strings and rails being in the same vertical plane, and both are framed into the same newel post at the turns. These stairs occupy less space than any other variety with the exception of the spiral, and for this reason are the kind chiefly used in cottages and other small houses. They obtain their fanciful name from the supposed resemblance of the strings in elevation to the hind legs of a dog.

Open Newel (see p. 317), or as they are sometimes termed, open well stairs are those having rectangular plans with an open space or well between the strings of successive flights. These are, both from a constructive and an artistic point of view, the best form of stair there is. They are substantial, massive in appearance, and convenient to use. Their only drawback is the large amount of floor space required. Most of the mediæval and Renaissance stairs still in existence are of this type. The newel post is always a dominating and important feature of the design, often being very massive and elaborately ornamented ; one occurs at every change in the direction of the flight.

Geometrical or **Continuous Stairs** (see pp. 319, 326).—So called because the "setting out" of the strings and rails is based upon geometrical principles. In these stairs one or both strings and handrails run continuously from top to bottom of the successive flights. The well-holes are always curved at the ends, this being the chief characteristic of the type ; usually the curve is circular, and occasionally elliptic.

Circular Geometrical Stairs (see p. 328) are similar to the ordinary continuous string stairs, with this difference, that the space occupied by the stairs in plan is circular instead of rectangular, and there are no flyers—double width steps are used as landings. Occasionally, as in the example given, the stairs are placed within square walls, and the angles are cut off by octagonal strings, this method being more economical in construction than is a circular wall string. When these stairs stand clear of the surrounding walls, they are termed "independent" or "self-supporting."

Elliptic Geometrical Stairs differ only from the above in having their plans elliptic instead of circular or rectangular.

Circular, Newel, or **Spiral Stairs** (see p. 328) are composed entirely of winders radiating from a central newel post running through the entire height of the flight. They are usually built within a circular wall, but are sometimes treated as "independent," the steps being framed into the newel, and carried on bearers, fixed to the same as cantilevers.

General Remarks.

Before entering into the details of construction of the several kinds, a few instructions applicable to all varieties may be conveniently collected here for reference.

Stairs should be planned so that they may be well lighted, especially at the turns, and their arrangement should be duly considered when planning the house, so that strings may not have to be taken across windows, or floors cut away and trimmers rearranged when the staircase is inserted. Formerly stairs were built up *in situ*, a step at a time, but modern staircases are put together in the workshop, usually in flights, only leaving their connections at the turns to be made on the building. Considerable attention should be given to the size of the stairway and other openings that the staircase may have to pass through, so that not more than will permit of their passage shall be put together in the shop. Within these limits the more of the stairs that can be fitted on the bench the better, as all the necessary appliances are at hand, and the conditions are more favourable to good work. Stairs should not start or finish nearer than 12 in. from a doorway, as a person using them in the dark is liable to stumble, through the loss of support afforded by the wall, and doorways near the top of a flight are dangerous to the unaccustomed user. More than twelve steps in a flight without a landing should be avoided where possible, as they are fatiguing to mount.

Landings are to be preferred to winders for making a turn, as the latter are always more or less dangerous to descend. They are, however, often unavoidable when the total going is small in comparison with the rise. An odd number of winders should always be used in stairs having rectangular wall strings, to avoid the placing of a riser in the re-entrant angle, which forms a step difficult to keep clean or to fit coverings to. Dividing landings by a single step should be avoided, as these prove traps for the unwary. Landings should not be less in width than the stairs to which they are connected, but may be wider with advantage. No difference must be made in the rise of any connected flight. Common widths for dogleg stairs between the strings are, from 2 ft. 6 in. to 3 ft. 3 in.; open newel, 3 ft. to 6 ft.; geometrical, 3 ft. to 5 ft. There is, of course, no objection to making them wider if there is sufficient room, but the tendency is rather to cramp them than the reverse. Two people cannot pass with comfort on stairs less than 3 ft. wide, and stairs containing winders should be wider than corresponding stairs without.

Carriages are used for the dual purpose of supporting the stairs, and providing fixings for the laths to carry the plaster soffit; for the latter purpose they should be spaced not more than 15 in. apart, and the central one having to carry the most weight may be made stouter than the side pieces. They should be well secured to the trimmers or pitching pieces, and should be deep enough to prevent any movement in the stair. Weak carriages are often the cause of stairs creaking, but as an increase in their depth also causes an increase in the width of the strings, the stairs are frequently stiffened by rough bracketing instead of using a deep carriage. This consists of either pieces of board nailed upon the sides of the carriages, as shown in f. 2, p. 314, or triangular blocks fixed on top of the carriage underneath the step, as shown at D, f. 1, p. 321. Various ways of connecting the treads to the risers to form steps are shown in f. 1 to 4, next page, the method shown at *b*, f. 2 & 3, being probably the best, as the tread running under the riser, to which it is attached by slot screwing, is free to shrink without opening a joint or splitting. Tonguing the top edge of the riser into the tread is unnecessary when a scotia is used, which should be grooved in as shown in the enlarged detail, f. 1. Fig. 4 is an inferior

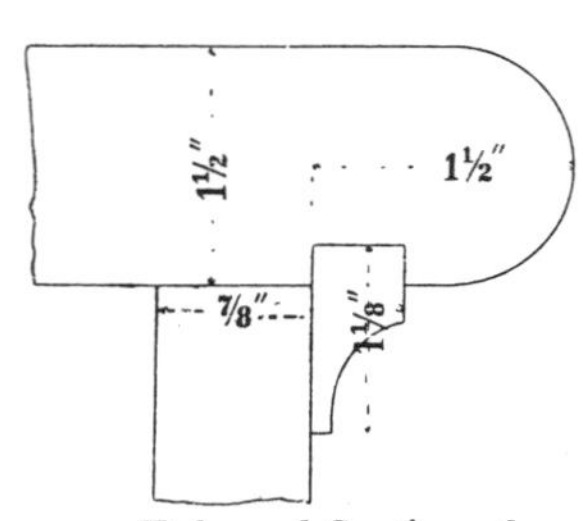

1. Enlarged Section of Nosing of Steps and Scotia.

method used in cheap stairs to economise material; by this arrangement a 9-in. board may be utilised for the tread.

The Nosings of steps generally project in front of the riser equal to the thickness of the tread (see f. 1).

Headroom must be provided over the stairs, and this sometimes presents considerable difficulty when the going is restricted, or the position of the trimmers has not been carefully considered when arranging the floor. A vertical height of 6 ft. 6 in. as a minimum must be provided between the line of nosings and the lower edge of well trimmer. The method of ascertaining if there is sufficient headroom in any given design is shown on opposite page. A nosing line is drawn on the elevation, and a line parallel with this, 6 ft. 6 in. distant, measured vertically; and unless all obstructions in the well, stand above this line, they will have to be cut further back, or the going altered.

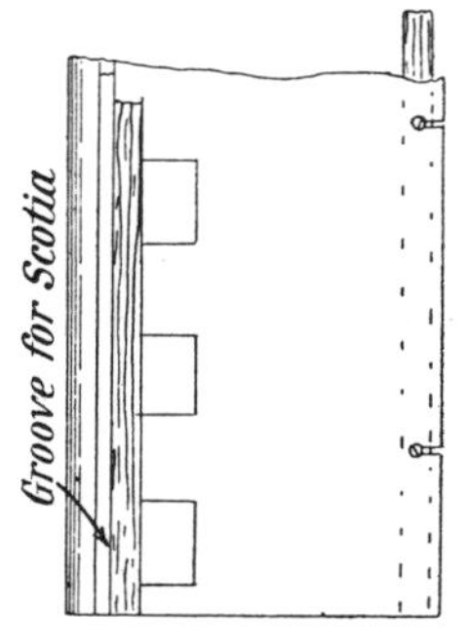

2. View of Under Side of Steps.

Landings in the quarter or half spaces are usually glued up in one piece and buttoned to the joists so that they may swell and shrink. Landings that are part of the general floor are cut back to the centre of the trimmer, and a nosing border inserted, as shown in f. 1, p. 314.

To Determine the Rise and Going of Stairs.—The amount of going and rise given to stairs depends chiefly upon the amount of floor space allotted to them, and upon the height of the story; but subject to these restrictions, there is room for considerable variation. To obtain a stair that shall not be fatiguing or awkward to ascend or descend, the going should bear a certain ratio to the rise. Various methods have been proposed by writers on the subject to obtain the ratio, of which the following are the best known and most practised:—

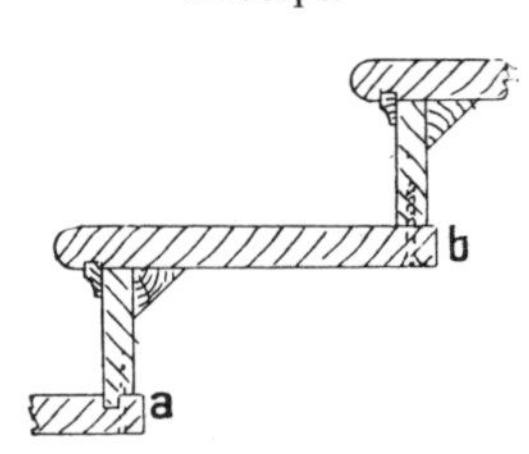

3. Proper Method of Fixing Risers to Treads.
(*a*) Tongued and Nailed.
(*b*) Slot Screwed.

1. It is assumed that the average length of step in walking on the level is 24 in., and that it is twice as difficult or fatiguing to climb upwards as it is to walk forward. From these premises it is deduced that one going or step forward, plus two rises or steps upward, should equal 24 in., which put in the form of a rule becomes—

TO FIND THE RISE WHEN THE GOING IS KNOWN.—Subtract the given going from 24 in., and divide the remainder by 2 for the rise.

TO FIND THE GOING WHEN THE RISE IS KNOWN.—Multiply the given rise by 2, and subtract the product from 24. The remainder is the proportionate going required.

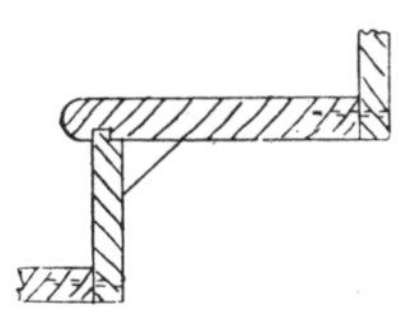
4. Inferior Method of Fixing Riser to Back of Tread.

2. The product of the going and rise multiplied together is to equal 66. Example—Going 11 in. × 6 in. = 66, and 7 in. rise × $9\frac{3}{7}$ in. = 66. Rule by this method—Divide 66 by the given rise or going to ascertain the proportionate going or rise.

3. Assume 12 in. going and $5\frac{1}{2}$ in. rise as a standard ratio. To find any other, for each addition of $\frac{1}{2}$ in. to the rise, subtract 1 in. from the going. Example—Rise 6 in., going 11 in.; rise 7 in., going 9 in. It will be noted that by this method the sum of 2 rises plus the going equals 23, which affords an easier stair than the first-mentioned method.

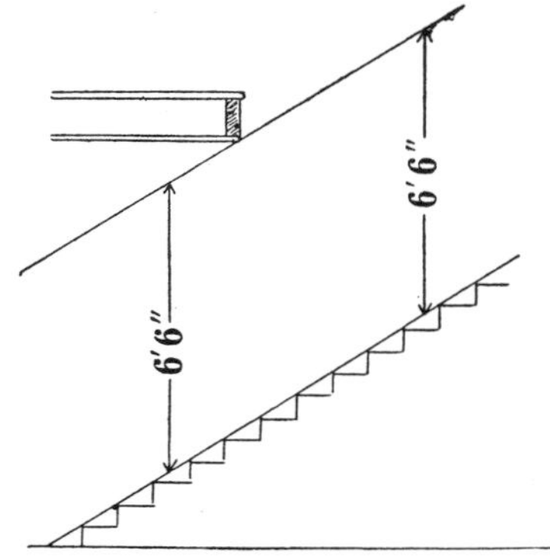

Method of Providing Headroom.

When the total rise of the stair is known, as shown by the story rod, and the approximate rise of the step is given, the exact rise is obtained by calculation thus:—Reduce the total height to inches, and divide it by the desired rise. If there is no remainder, the divisor will be the exact rise, and the quotient will be the number of risers required. If there is a remainder, again divide the sum by the quotient, discarding the fraction, and the result will be the exact rise. For instance, let the height of the story be 10 ft. 6 in., and the proposed rise $6\frac{1}{2}$ in.—10 ft. 6 in. = 126 in. ÷ $6\frac{1}{2}$ in. = 19 with 5 remainder; then 126 in. ÷ 19 = $6\frac{5}{8}$ in. full as the rise, and the proper ratio of going to this, as found by the first method, is $6\frac{5}{8} \times 2 = 13\frac{1}{4} - 24 = 10\frac{3}{4}$; but the exact going is found by dividing the plan into 18 equal parts, as there is always one less tread than the number of risers, in consequence of the landing acting as tread for the last riser. No arbitrary rule can be given for the treatment of the plan, which must be subject to circumstances. Every attempt should be made, however, to dispense with winders, which should only be introduced in case of necessity, when they are better placed at the top of a flight than at the bottom.

Constructional Details.

Dogleg Stairs.—Fig. 1, next page, is the elevation and f. 2 the plan of a two-flight dogleg stair, with a quarter space of winders and a quarter space landing, embodying all the characteristics of the type. The newel post at the return is taken down to the floor, which is a preferable arrangement when winders are used, as these prevent the throwing of a trimmer across the stairway; but when a spandrel framing is placed under the lower string board, the newel may be cut off flush with the under edge of the latter, unless a doorway is required.

Taking dimensions on the building.—A story rod about 1 to $1\frac{1}{2}$ in. square, and long enough to reach from floor to floor, is required. The nett height of the story from the surfaces of the flooring is marked upon this, the rod being held perpendicularly in front of the trimmer. This line should be marked in blue pencil, and the word *Up* written on it, as shown in f. 1, next page. The intervening space is later divided by compasses into the same number

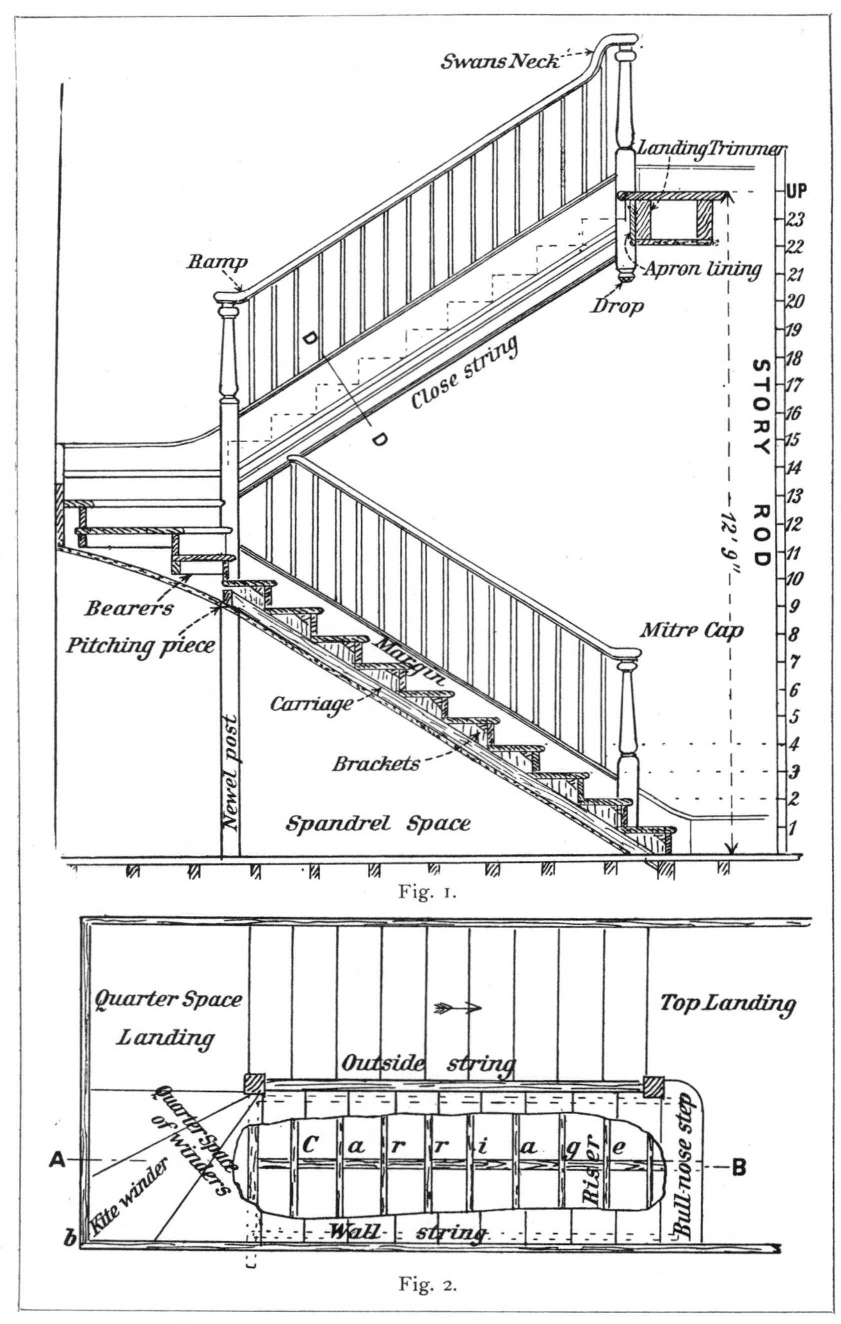

Fig. 1.

Fig. 2.

ELEVATION AND PLAN OF DOGLEGGED STAIRS.

of equal parts that there are to be risers. Write boldly on this side of the rod the word *Rise*. Next drop a plumb line from the face of the trimmer to the floor, and mark its situation thereon. Then lay the rod upon the floor with its squared end against the plaster upon the end wall, or allow 1 in. for this if the wall is not rendered, and transfer the trimmer mark thereto upon the unmarked edge. Also mark the position that it is intended the first riser should occupy. This point should be marked *Start*, and the edge of the rod inscribed *Going*. Take note of any windows or doors in or near to the stairway, marking their distance from one or other of the points on the story rod. With these data the stairs can be set out.

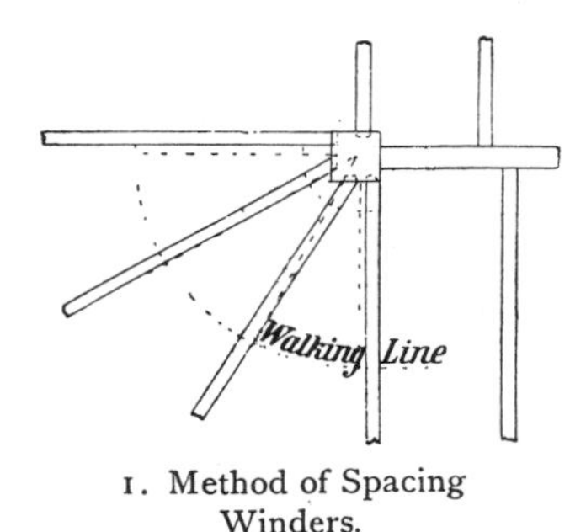

1. Method of Spacing Winders.

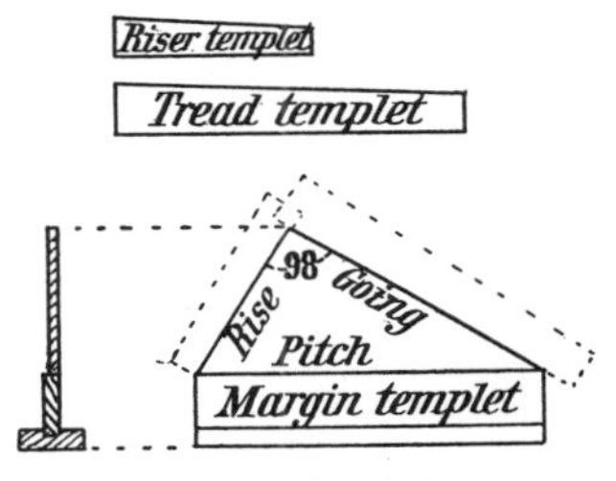

2. Pitch-board and Housing Templets for String.

The working drawings.—Prepare a plan to scale, 1½ in. to the foot is a very convenient one to use, drawing in the enclosing walls of the stairway, or rather the line of plastering, the trimmers, doorways, &c., as noted, and the wall strings. These are usually ½ in. thinner than the outside strings. Draw the outside string in the centre of the plan, and mark positions of first and last riser, and draw in the risers. The first landing newel must be arranged at an equal distance from each of the three wall strings. In the example given the bottom newel is placed so that the face of the second riser comes in its centre, but this may be varied as desired. The faces of the landing and the first winder risers also are arranged to come in the centre of the newel face. The remaining winders are spaced equally upon a line struck from the centre of the newel, with a radius of 18 in., and radiate to the centre through the points of division. This method of spacing is sometimes varied to obtain the winders wider at the narrow ends, as shown in the enlarged detail, f. 1 above, which is a repetition of part of the plan, f. 2 opposite, with the risers rearranged. The dotted lines indicate their position, as shown in f. 2. To set them out, draw the walking line as before, and place the first and last risers in the quarter space, either just beyond or in line with faces of newel. Describe a second arc just clear of the newel, and divide the risers upon the two arcs equally, drawing their faces through the points. The spaces between the first and last risers in each flight are next divided equally into the required number of steps as shown, this giving the going of the steps. Both tread and riser lines are shown in the drawings, but in making a working drawing only riser lines need be drawn, as the nosings are immaterial to the setting out.

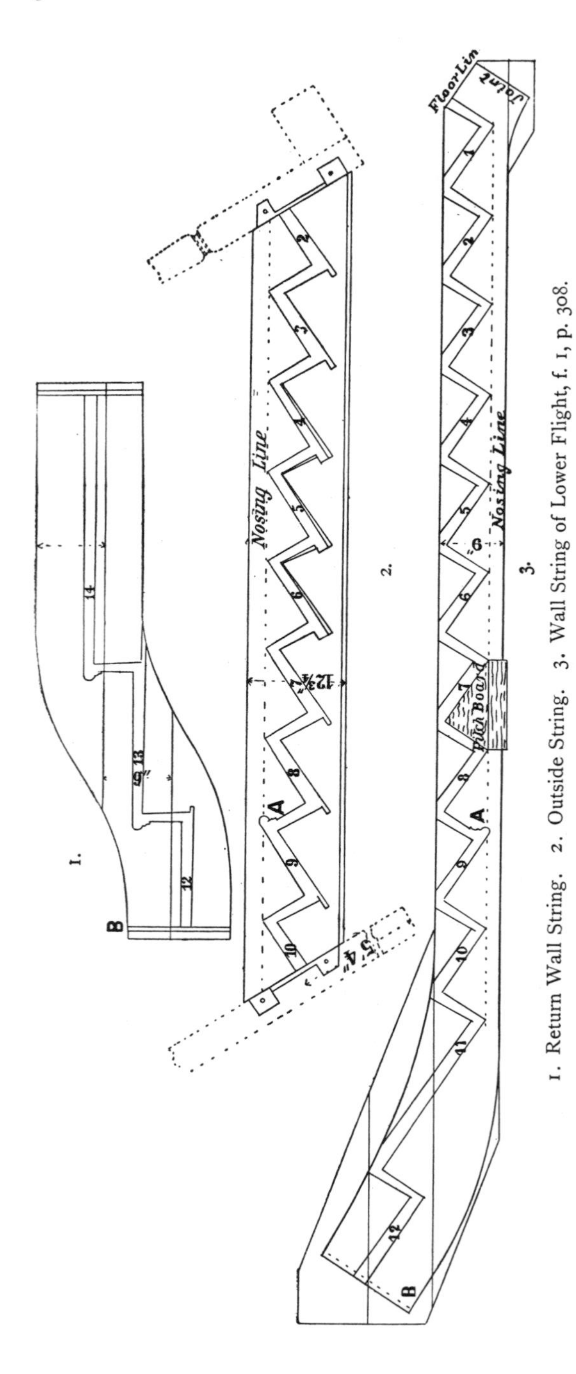

1. Return Wall String. 2. Outside String. 3. Wall String of Lower Flight, f. 1, p. 308.

In the above directions it has been assumed that winders have been decided upon before starting, but their necessity under the given conditions is found thus :—The total rise and going being given as 12 ft. 9 in. and 11 ft. 3 in. respectively, and suppose a suitable rise decided upon as about $6\frac{3}{8}$ in. 12 ft. 9 in. $\div$ $6\frac{3}{8} = 24$, this then is the number of steps required to get up. Now the landing will absorb 3 ft. of the lower going, leaving 8 ft. 4 in. available for the lower flight. A suitable going for the rise would be 10 in., and ten steps may be got in the space at this width, and nine in the upper flight, whilst the quarter landing will count one, thus totalling twenty, which is three short of the required number, and these three, under the given conditions, can only be supplied by winders, as shown in f. 2, p. 308. An elevation is not required for the setting out of the stairs, nor is it usual to set out a rod for the flyers, but the winders should be set out full size, as it is easy to make a miscalculation from a scale drawing when angles have to be measured. The first requirement for setting out the strings is a pitch-board. This, as has already been stated, is a set square made to the pitch of the stairs, as shown in f. 2, previous page. Three templets are required to use with it, as shown in the illustration, one for the margin, sometimes called the nosing templet. This is made with a fence at right angles to the margin, and acts as a gauge for the pitch-board. In close strings it is used from the top edge, and is made from 2 to $2\frac{1}{2}$ in. wide, according to the required margin. In cut strings it is used from the under edge, and is made equal to the width of the carriage and soffit. Two wedge-shaped templets are required to mark the housings—one for the tread and one for the riser. These are made in width to equal the thickness of tread and riser respectively, plus the amount for wedging, which should be about $\frac{1}{8}$ in. at the point and $\frac{3}{8}$ in. at the heel. Figs. 2 & 3 opposite show the outside and wall strings for the lower flight set out ready for housing, and with tenons and shoulders cut on the latter to fit the newels.

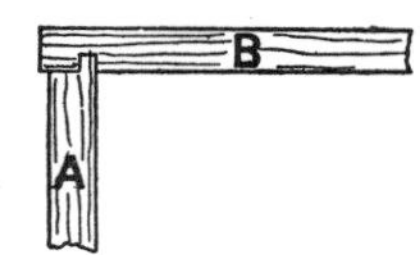

1. Enlarged Detail at *b*, f. 2, p. 308.

To set out the wall string, whose length may be found approximately by multiplying the longest edge of the pitch-board by the number of steps, adding the distance from the top riser to the return string, as shown on the "winder rod," and about 10 in. at the lower end to meet the skirting. If the board is longer it is better not to cut off until the steps are spaced out, when a bevel cut made as shown in the drawing will allow of the piece cut off being turned over, and jointed to form the easings without waste. Shoot the top edge of the board, and run down the nosing line. Set a pair of compasses exactly to the length of the pitch edge of the pitch-board, and run them along the nosing line, starting about 9 in. from the lower end. Mark off eleven steps and number them as shown. Next run the pitch-board along, keeping its ends to the compass marks and its edge to the templet, as shown in f. 3. Knife cut around its two edges; apply the wedge templets and pencil in the back lines. Next mark the twelfth riser at the distance from the eleventh shown on the winder rod, and the length to the shoulder line of the return string, allowing $\frac{3}{8}$ in. additional for a tongue. The faint lines in the drawing show where the board will have to be jointed to make up the necessary width to form the easings. These joints should be cross tongued. The easings are drawn by bending a thin lath to a gentle curve from the straight part, allowing it to rise over the twelfth riser from 4 to 6 in., as may be found suitable. When strings are worked by hand the housings are cut square to the nosing line at first, a few holes being sunk with a centre bit in the angle, the core cut out for about 2 in. with a chisel, then the side cuts run in with a tenon saw, the remaining core removed, and the bottom of the grooves levelled with a router, the trimming for the nosing and scotia being left until the steps are fitted individually; but in machine-wrought work the housing is generally worked right up to the nosings, and would require marking, as shown at A, f. 3 & 2, opposite, the templet used for marking the tread housing being cut to shape for the purpose.

2. Sketch of Centre Newel, f. 2, p. 308.

Housings are generally made $\frac{3}{8}$ in. deep. Fig. 1 opposite shows the return wall string, the method of building it up with 9-in. boards, and the setting out as obtained from the winder rod. In making the easings there is no necessity to keep the margin over the risers uniform with the flyers, but the heights over the corresponding treads in each string must be equal, as shown at B in f. 3 & 1, p. 310. The two strings are tongued together at the angles, as shown in f. 1, last page, A being the lower string and B the landing string. Fig. 2, p. 310, shows the outside lower string, which is set out in a similar manner to the wall string, but starts at the second step, the first being a bullnose, finishing on the outer face of the newel. The newel should be drawn on the wall string for a start, and its centre line taken as the face of the second riser. Its inside face gives the shoulder line.

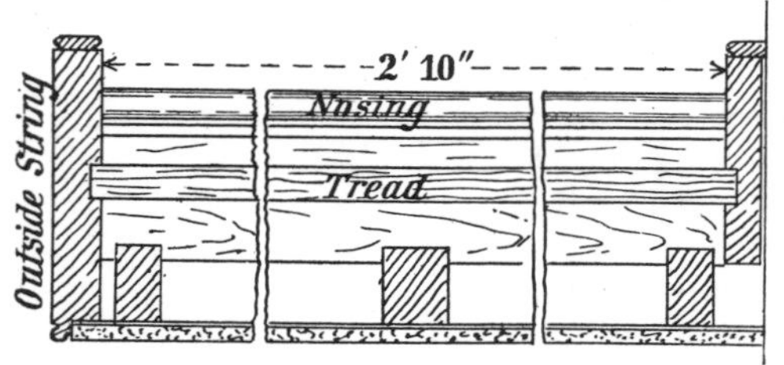

1. Enlarged Section across Stair showing Timbering, f. 2, p. 308.

In common work, a shoulder is cut on the inside of the string only, and a barefaced tenon formed on the outside; but it is much better to shoulder both sides, keeping the tenon about $\frac{1}{4}$ in. from the face of the string, as shown in the sketch of the newel, f. 2, previous page. The tenons are drawbore pinned from inside. The outside string must be wide enough to cover the carriages and also

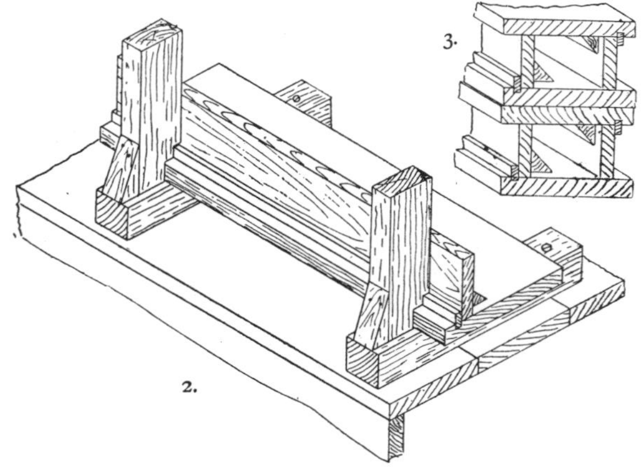

2. Sketch of Cradle for Gluing-up Steps.
3. Method of Stacking Steps to Dry.

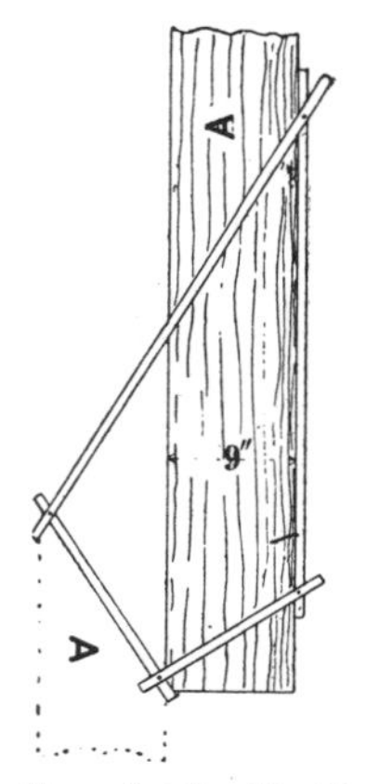

4. Templet for Marking Winders.

the plaster soffit, as shown in the enlarged section, f. 1 above. Fig. 2 shows the apparatus used in forming steps to ensure the "squareness" of the tread and riser. This is called a **cradle**, and consists of two brackets each formed with two pieces of quartering framed strongly together at right angles, and secured to the bench with screws, parallel and out of winding. The uprights are notched out to receive the nosing and scotia. To use the apparatus :—The treads, risers, and scotias being first squared off to correct length, the tread is laid face down on the base, and the riser shot to fit whilst it rests against the uprights. Then the scotia slip is glued in, the back of the scotia and front edges of the riser glued, and the latter rubbed until a joint is formed, then blocked inside and carefully lifted off the cradle, and placed in a pile on the bench or floor to dry as shown in f. 3. Afterwards the nosings and scotias are worked with the special planes described at p. 19.

Fig. 4 shows the method of marking the stuff for cutting winders by means of a skeleton templet made to fit the drawing on the winder rod. The piece cut off at A is reversed and jointed up, as shown by the dotted lines. The

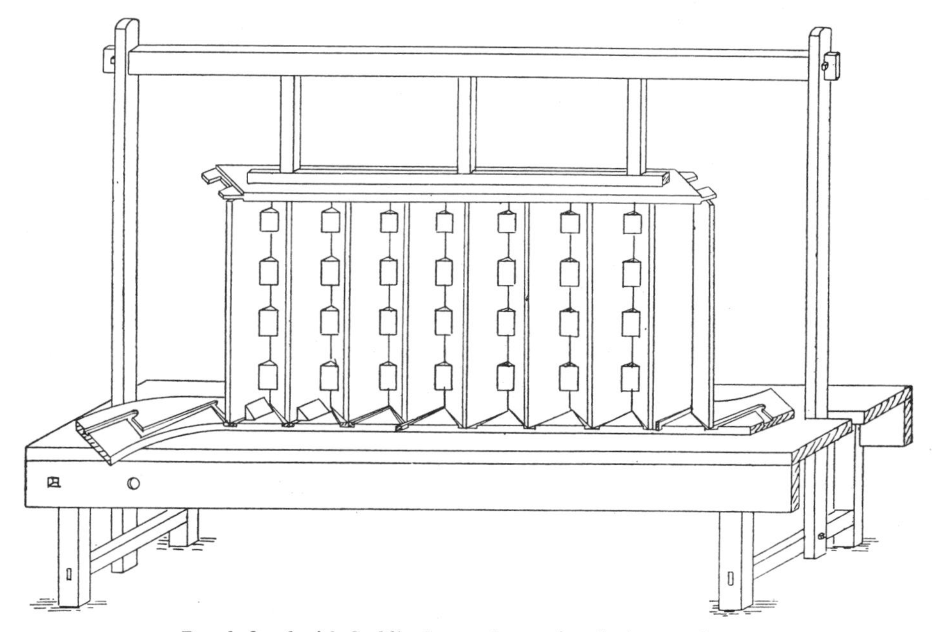

Bench fitted with Saddle Cramp for putting Stairs together.

winders should always be cut so that the front edge is parallel with the grain, and if they can be inserted in place in the shop, may be glued up in the cradle like the flyers.

The steps are next fitted into the strings separately and numbered. The width of each riser being accurately marked at each end with a chisel, and the superfluous wood shot off, they are then "put together," as illustrated above, which shows the lower flight of the stair in f. 1, p. 308, in course of construction on the bench. It is very necessary that the whole length of string should be brought hard down on the steps at once, so that it may be seen whether their front edges and the nosings are out of winding. This is frequently managed by strutting down from the roof of the workshop; but where this is inconvenient, the same result can be obtained by fixing a saddle cramp, as shown, to the ends of the bench by coach screws, and wedging down from the top bar.

The treads should be knocked up tight to the nosings, the wedges well glued and driven, then trimmed off flush, and the steps blocked to the strings. The various stages are all shown in the sketch. The backs of the treads are secured to the risers either by nails or screws, and when the blocks are dry the

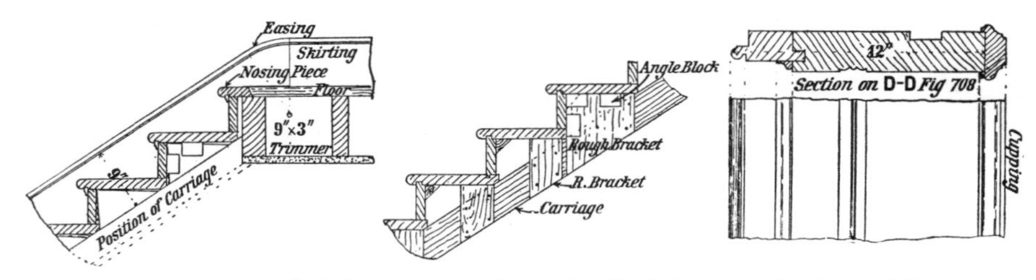

1. Detail at Top End of Wall String, f. 1, p. 308.

2. Rough Bracket Nailed to Carriage.

3. Section and Elevation on D D, f. 1, p. 308.

stair is turned up and the wall string nailed through from outside. For an extra good job, screws would be turned in to the outer string through the backs of the treads. The newels are next fixed—that is, if they can be passed through the openings, and in the present case the first winder; the others had better be left loose to fix upon the job. Fig. 2, p. 311, shows how the risers are fitted into the newel. These are inserted from the back and nailed. Kite winders are better not glued together until after they are fixed in position. An enlarged section of

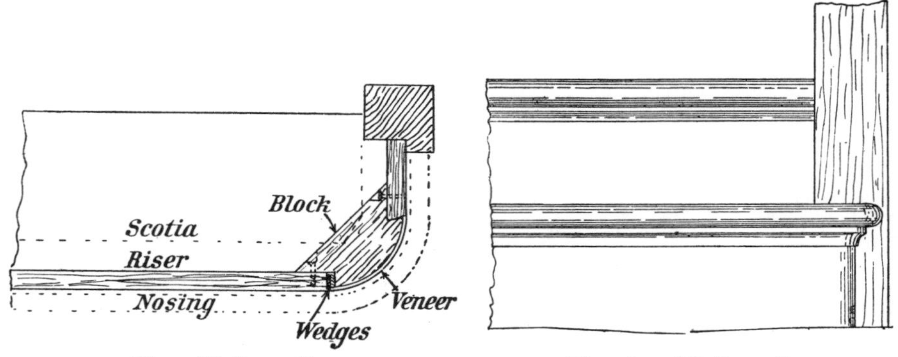

4. Plan of Bullnose Step.

5. Elevation of Bullnose Step.

the upper end of the top flight is shown in f. 1 above, to illustrate the method of easing the string into the skirting so that the moulding may be continuous. Fig. 2 shows two methods of stiffening stairs by means of rough bracket pieces nailed on the sides of the carriages. Another form of bracketing is shown in f. 1, p. 321. Here triangular blocks of quartering are cut to the pitch-board and fixed on the back of the carriage under the treads. These blocks, shown at D, are termed pitch-board brackets.

Sunk Strings.—Fig. 3 opposite shows the method of forming a close string which is too wide to be obtained from usual scantlings. The lower portion, called the sinking, is tongued into the main string and set back $\frac{3}{8}$ or $\frac{1}{2}$ in., and a moulding planted in the angle. This is further elaborated, especially in open newel stairs, by forming the string of three pieces, the centre piece forming a kind of sunk panel, the edges of the margin pieces being moulded in the solid.

Bullnose Step.—Figs. 4 & 5 opposite and f. 1 below are respectively the plan of riser, part elevation of complete step, and cross section to larger scale of the bullnose step commencing the flight in f. 2, p. 308. A perspective sketch of the block around which the riser is bent is shown in f. 2 below, and a part of the riser with a veneer formed near its end and the block fixed ready for bending round is shown in f. 3. The method of making the step is as follows:—Prepare the block to the shape of the plan, building it in three thicknesses, the laminæ with

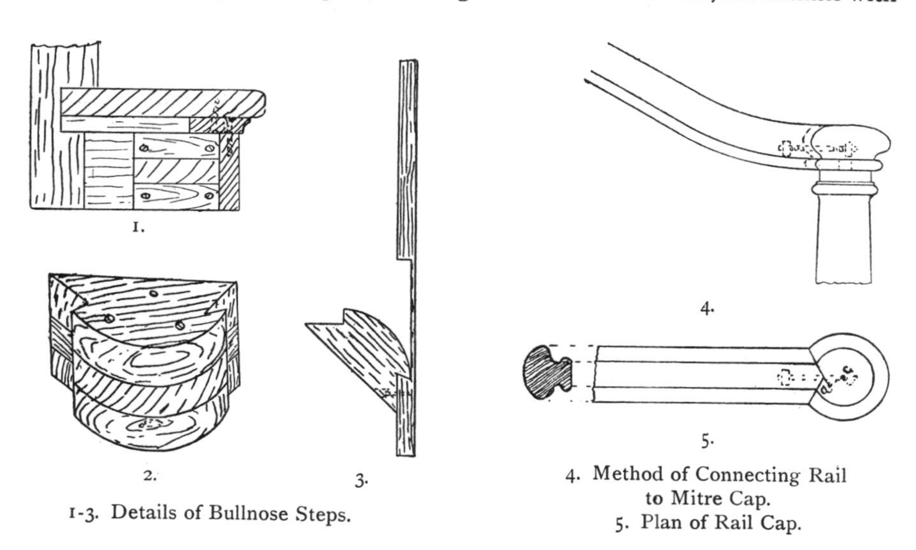

1-3. Details of Bullnose Steps.

4. Method of Connecting Rail to Mitre Cap.
5. Plan of Rail Cap.

the grain crossed as much as possible. Glue and screw them together. Make the seats square with each other, and undercut the newel end of the rebate as shown. Prepare the riser out of a piece of pine or sound dry yellow deal, free from knots. Mark off the distance of the springing from the newel end. Place the rebate in the block to this line, and roll it round carefully on the back of riser. Mark where the opposite rebate finishes, and allow an extra $\frac{1}{2}$ in. for wedge room. Gauge the veneer between these lines $\frac{1}{8}$ in. thick from the face, and remove the core with saw, chisel, and router by making two or three trenches across and routing them down to the gauge line. These will act as guides for the remainder of the surface. Tooth the veneer and face of block with either a toothing plane or coarse glass-paper. Prepare a pair of folding wedges to fit the recess in the block. Bore two holes at each end for screws square with the rebates. Have hot glue ready. Then well wet face of veneer with boiling water. Lay it on the bench and fix the block in the position shown in f. 3 above. Plentifully cover the veneer and block with glue, and roll the latter slowly along the veneer, pressing heavily down the while until the end is reached. Then glue and insert the wedges, which an assistant should tap in gently with two hammers. Drive equally and not too hard. The wedges are only required to bed the veneer firmly down on the block. When home, secure the block with screws. The scotia is worked on the edge and end of a piece of board $\frac{7}{8}$ in. thick, and afterwards reduced in width to about 3 in. This is screwed first to the riser and then to the tread, as shown in the section, f. 1, previous page. Cut strings are sometimes used with dogleg stairs. The construction of these will be found described under the head of geometrical stairs.

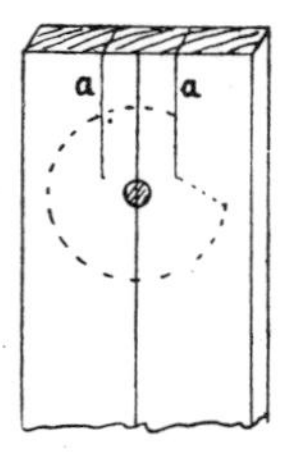

1. Block for Cutting Mitre Cap.

Fitting the Mitre Cap.—Fig. 4 overleaf shows the method of connecting the handrail to the mitre cap and the latter to the newel by a handrail bolt and a dowel pin respectively.

To obtain the mitre for the cap when its section is the same as the handrail. Draw the plan of the rail and cap as in f. 5, also draw lines on each, parallel with the edges, representing the greatest depth of the moulding. From the edge draw lines through the intersections of the curved with the straight lines, which will give the mitre.

To cut the mitre on the cap.—Prepare a cutting block, as shown in f. 1 above, out of a piece of 2-in. stuff, planed true on the face. Gauge a centre line upon it, and insert a dowel that will fit the hole in the cap tightly a few inches from the top end Next two saw cuts are made square through the block parallel with the centre line, as shown at *a a*, and at a distance from the latter equal to the square distance of the mitre line *a* from the centre of the cap *c*, f. 5, p. 315. The depth of the saw cuts below the edge of the cap, which is shown by the dotted line in f. 1 above, is made equal to the length of the mitre line, as shown in the plan, f. 5, p. 315. The width of the rail is marked upon the edge of the cap. The latter is then placed on the dowel and turned until one of the marks lies against one of the saw cuts. The saw is then run down to the bottom of the cut and the cap turned until the other scribe mark lies on the other cut, when the saw is again run in to meet the first cut, completing the mitre. The above-described method gives the common and best method of fitting the cap to the rail, but occasionally it is required

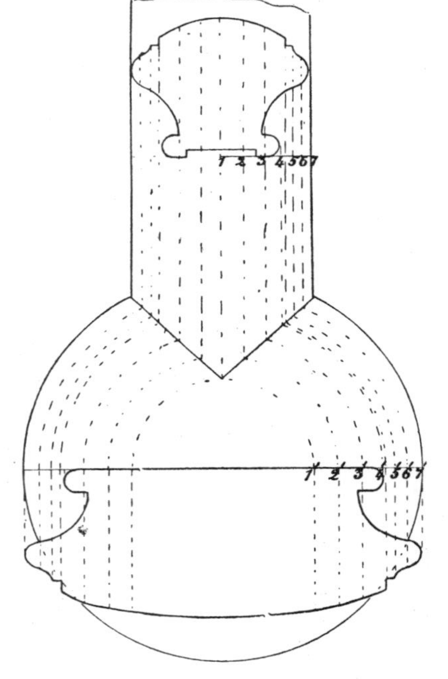

2. Method of Determining Section of Cap.

that the joint shall be a true mitre, necessitating different sections in the cap and handrail.

To determine the section of the cap upon a given mitre line (f. 2, previous page).—Draw the plan of the cap to the required size, also plan of part of the rail, and at some convenient point a section of the rail. Draw also from the intersection of the sides of the rail with the cap two joint lines at the required angle. In the present instance these are at 45 deg. with the side. Divide the outline of the rail into a number of parts, and draw lines from the points of division parallel with the sides of the rail. To intersect the joint lines, as shown by the lines 1, 2, 3, 4, 5, 6, 7, f. 2 opposite, draw a diameter line across the cap, and from its centre describe arcs passing through the points of intersection on the mitre and cutting the diameter in points 1′, 2′, 3′, 4′, 5′, 6′, 7′. From these points erect perpendiculars to the diameter, making them equal in height to the corresponding lines above the base of the section. The intermediate points where the ordinates cut the section can also be transferred, and these will give points through which the section of the cap can be drawn.

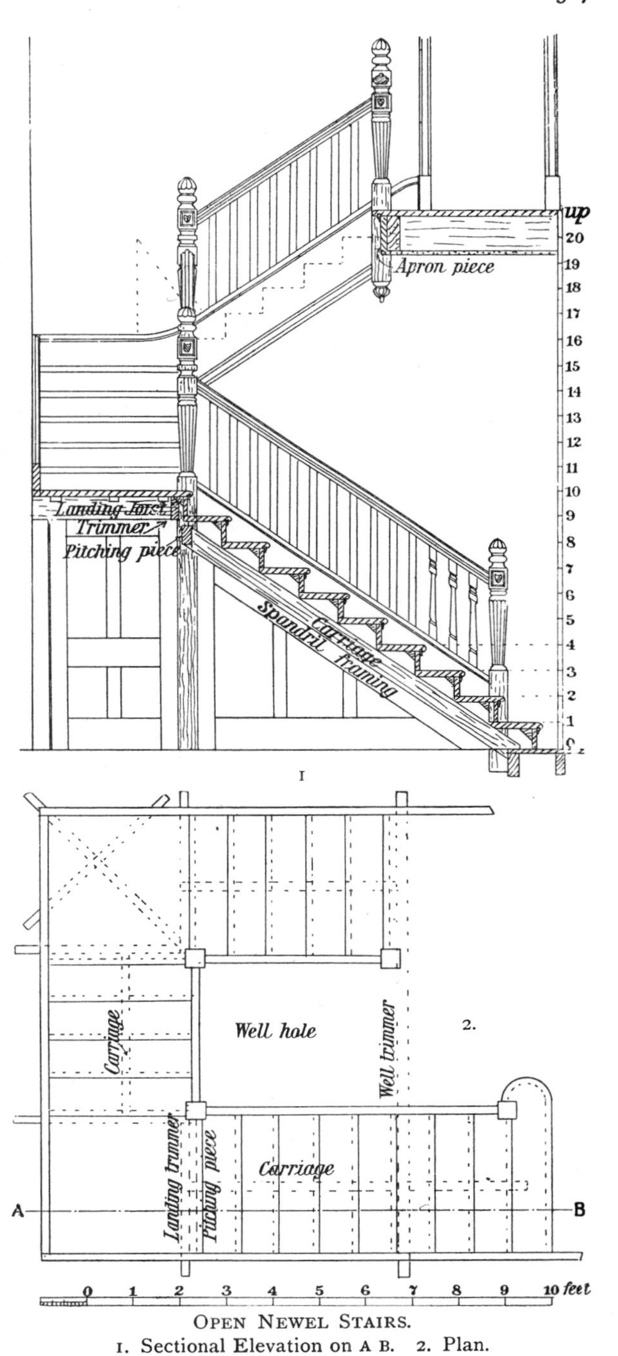

OPEN NEWEL STAIRS.
1. Sectional Elevation on A B. 2. Plan.

An Open Newel Stair with close strings, two quarter space landings, with a short flight between them, is shown in f. 1 & 2, last page. All the strings are "close," and the flight commences with a half-round step, usually termed a round-end step. The treads are shown in the plan by full lines and the risers by dotted. The trimmers of the first landing are framed into the newel post which runs down to the floor. The second landing newel finishes just below the string in a turned drop, and thus affording no support to the trimmers. Crossed joists are used to support the outer angle of the landing. The method of cogging these joists together is shown in f. 1, this page. The central carriages only are indicated in the plan to avoid confusion, but two others would be employed if the stairs had a plastered soffit. The first newel is carried below the floor and bolted to the face of the joist. Figs. 2 & 3 are plan and elevation to larger scale of the round-end step and newel. In the plan the position of the tread, scotia, newel, &c., is indicated by dotted lines; the riser and the block by full lines. The riser is formed similarly to that of a bullnose step previously described.

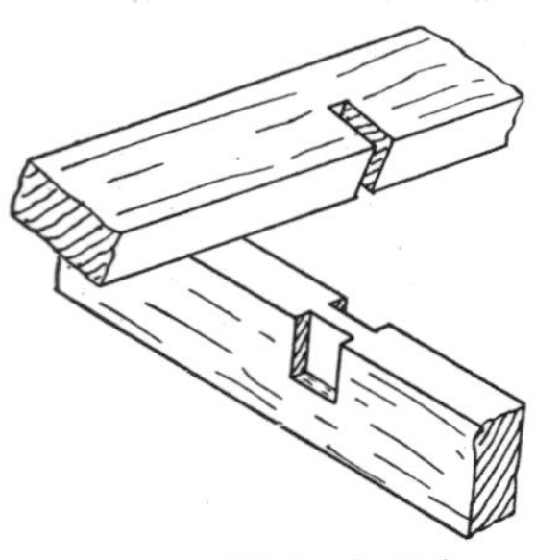

1. Detail of Joint in Joists, f. 2, p. 317.

Geometrical Stairs.—Fig. 2 opposite is the plan, and f. 1 is an elevation, of a cut string geometrical stair, having a quarter space of winders with two commode steps at bottom, and a quarter-space landing and quarter space of winders at the half height, the stairs finishing at a half-space landing. The lines shown in the plan represent the faces of the risers, nosings being omitted to facilitate explanations. This is a suitable stair to place in a wide hall or shop. The curved portion of the first wall string is treated as a cut string to correspond with the outer strings. The winders in each pace are made to "dance"—that is, to radiate from varying centres. This, by increasing the going of the steps, produces a better falling line in the strings and handrail, and makes the stair safer to use than they would be if the steps did radiate from a common centre.

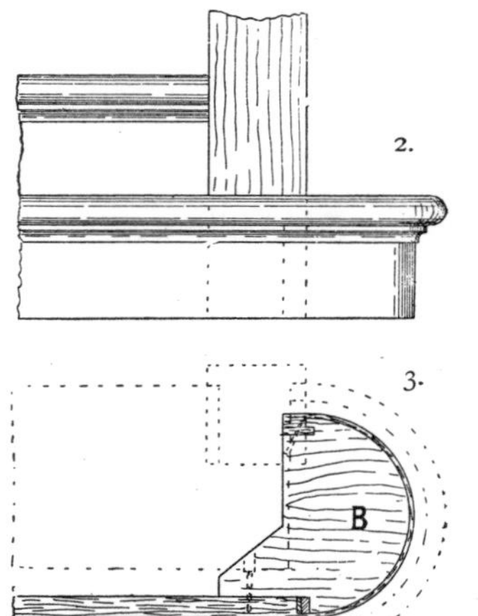

2, 3. Elevation and Plan of Round-end Steps and Newel.

To set out dancing steps.—Place the outer end of the first flyer riser 2 or 3 in. beyond the springing, as at *a*, f. 2 opposite—the exact amount depends upon the width of the flyer—and divide the face of the string from the point *a* to the landing riser *b* in the centre of the well into the required number of equal parts, in the present instance three. Next describe an arc from the centre of the well 18 in. from the face of the string, and divide the quadrant formed by its intersection with the risers *a* and *b* into a like number of equal parts. Join these points to the corresponding divisions on the string, and the lines will determine the position of the riser faces.

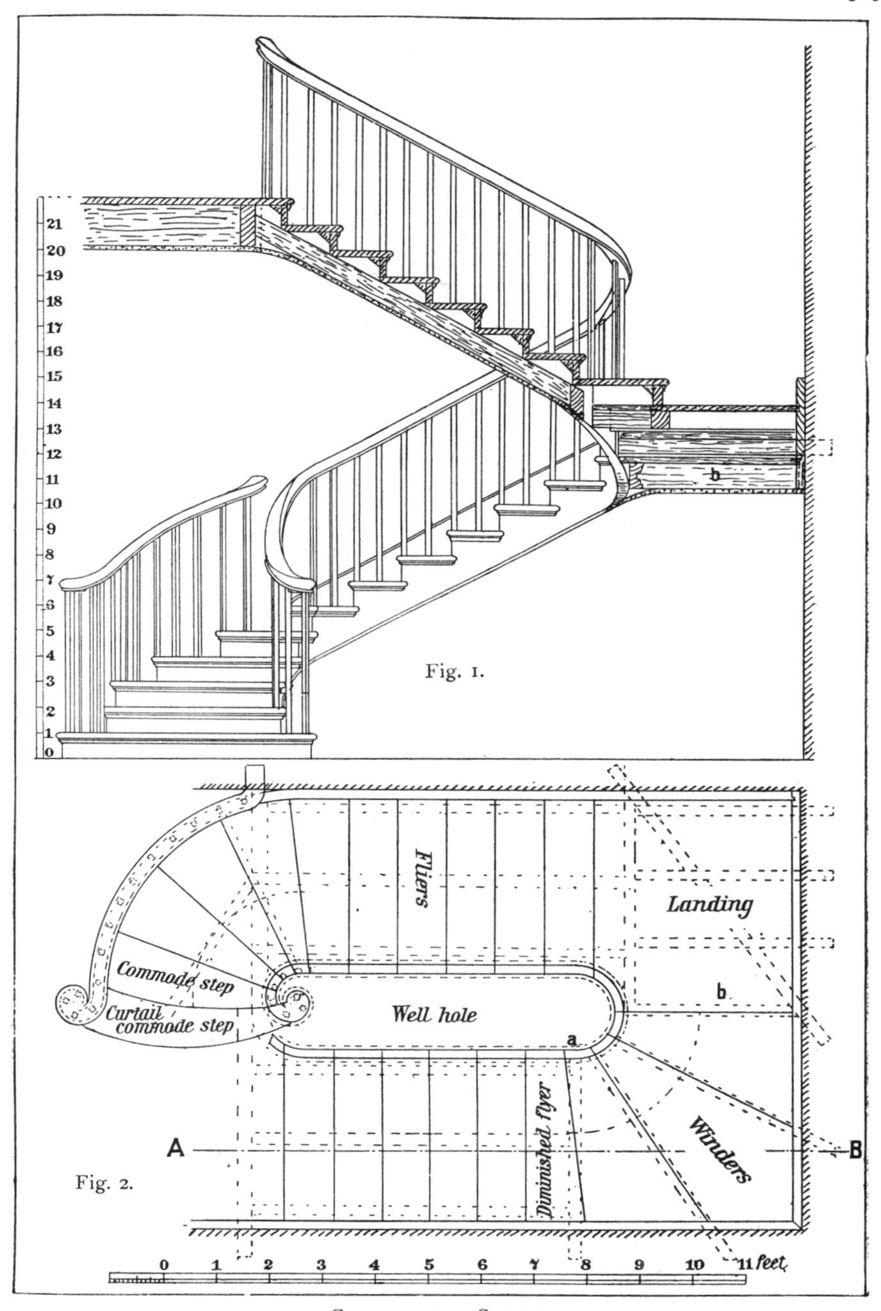

GEOMETRICAL STAIRS.
1. Section on A B. 2. Plan.

A Cut and Mitred String suitable for the above stairs is shown in different stages of construction in f. 1 below. At the upper end a portion of the tread and riser is shown with the return nosing removed to show the method of fixing the balusters. The two other steps show the preparation of the string to receive the risers. A third method sometimes employed is shown at *a*, f. 2 opposite, and in very common work the string and riser are both cut to a plain through mitre, a mitred fillet being glued at the back of the riser to fill the void formed through the difference in thickness of the two. The method of tonguing the string shown at *a*, f. 1 below, is chiefly applied to polished hardwood stairs where brads through the front would be inadmissible.

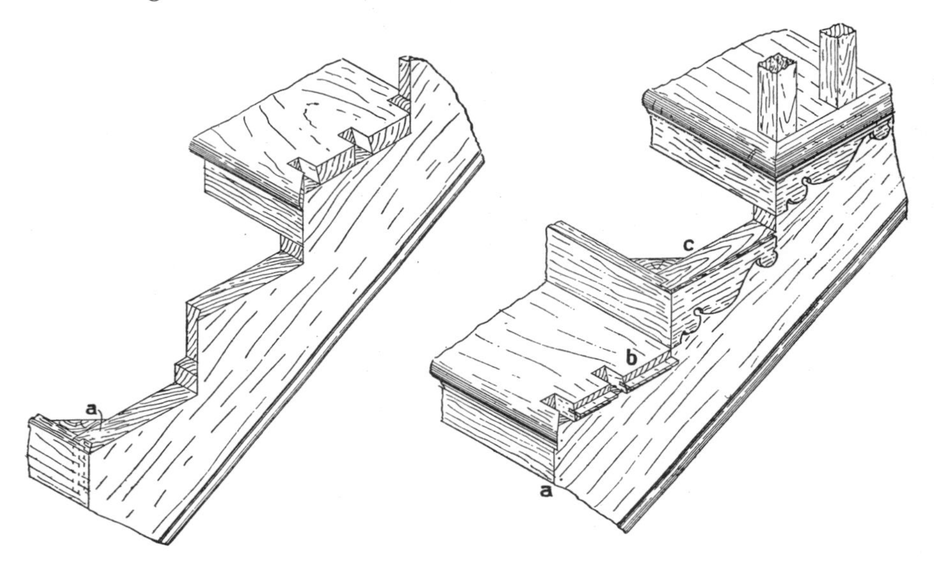

1. Sketch of a Cut and Mitred String. 2. Sketch of a Cut and Bracketed String.

A Cut and Bracketed String is shown in the sketch, f. 2, and a plan and elevation in f. 1 & 2 opposite. These show the difference in treatment required when ornamental brackets are planted on the face of the string under the return ends of the treads. The back end of the bracket mitres with the end of the riser which runs over the face of the string for that purpose, as shown at *b*, f. 2 opposite, and the string is merely rebated out for the riser to which it is secured by bradding through the rebate, the brackets eventually covering the brad holes. The tread must also project over the bracket, equal to its projection over the riser, and when the balusters are dovetailed in, a return nosing and scotia similar to f. 3 is cut to fit over the bracket and mitred to the tread nosing in front and mitred to a short return piece of itself at the back, as shown at *c*, f. 1 & 2 opposite. The dotted line at *d* in f. 1 shows the width of this piece. The return nosing should project beyond the risers at each end equally, and the brackets should be designed to fall within the nosing as shown. The nosing pieces are fixed either by groove and tongue joint as at f. 3, or by slot-screwing as at f. 4, and in common work simply by nailing through

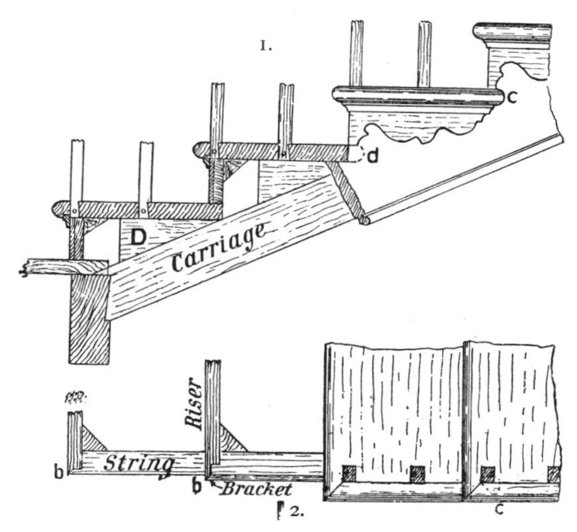

DETAILS OF A CUT AND BRACKETED STRING.
1. Elevation. 2. Plan.

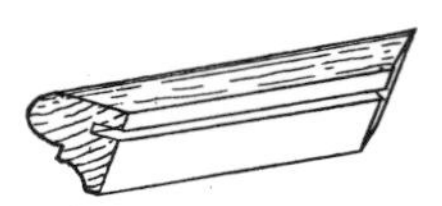

3.
Return Nosing
and Scotia.

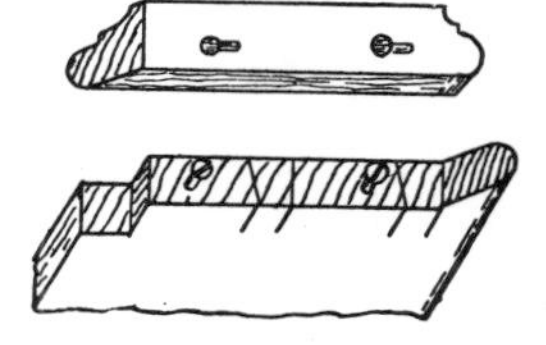

4. Method of
Fixing Nosing Piece with
Secret Screws.

the face. Whichever method is employed, they must of course be left free until the balusters are fixed. Brackets are usually from $\frac{1}{4}$ to $\frac{3}{8}$ in. thick. Their grain should run with the string; in common work they are cut below, not behind the return scotia.

Balusters, except in the case of very large examples, which are placed in the middle of each step, are usually placed in line with the face of riser and the face of string when plain cut strings are used, and in line with the brackets in bracketed strings. Intermediate ones are spaced according to taste, but it is not advisable to have them more than 5 in. apart. They are sometimes fixed into the steps by housing them into the treads $\frac{1}{4}$ in., in which case no nosing pieces are used, the tread running over the string and having the return nosing worked in the solid. When they are to be dovetailed the sockets should be sawn down in the treads before gluing them to the risers, but it is better not to cut the core quite through until the stairs are fixed, as the edges will get damaged and the sockets filled with dirt. The core, however, should be cut half-way through from the bottom, as shown

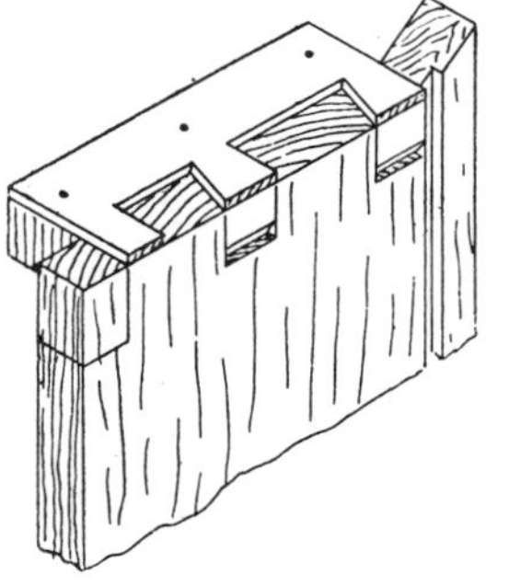

5. Method of Marking
Dovetails for Balusters.

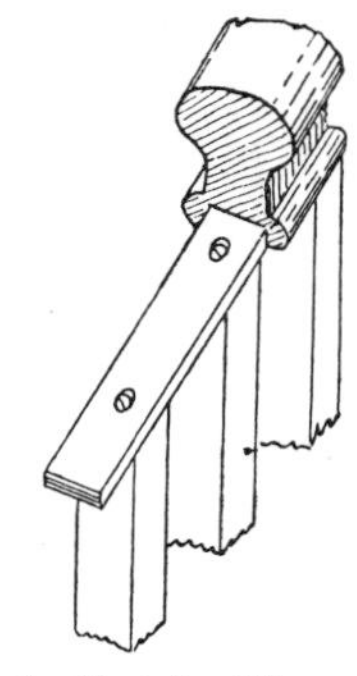

6. Sketch of Iron
Core in Handrail.

in f. 5, which also illustrates the method of marking the sockets upon the ends of the treads by means of a templet. The balusters are fixed at their top ends variously. Sometimes they are inserted in a groove in the rail, in other cases housed or tenoned to the rail, or simply cut to the rake of the stairs and nailed. In geometrical stairs they are frequently screwed to an iron core, as shown in the sketch, f. 6, the core being afterwards screwed into a groove in the rail.

Setting out length of balusters.—Two methods of arranging the moulded or turned part of the balusters are shown below. That at B is the more usual for wreathed strings, the turned parts being of uniform length. To set them out, draw the outline of a few steps and balusters full size, mark on any two front ones, as at *a* and *b*, the intended height of the square part, and draw the line *a b*. The intersection of this line with the faces of the balusters will show the lengths of the squares on the intermediate ones. A parallel line at the upper end will give points for squaring over the top part of the turning. The lengths being thus obtained, lay the whole of the balusters on the bench side by side, face up, in two lots, with a pair of each length on the outsides. Then the top and bottom members and the shoulder lines can be squared across, the bevel for the top shoulder being obtained from the pitch-board. The shoulders and dovetails should not be cut until after the turning has been executed, because the turner requires square ends to fix in the lathe. When iron balusters are used to strengthen the rail, one of these should be used as a pattern to set the wood ones out by. When the method shown at A is adopted, two pattern balusters will be required to mark them, as the squares are all of one length. The vertical height of the rail over the face of any riser is usually about 2 ft. 8 in., rising to 3 ft. over landings and winders.

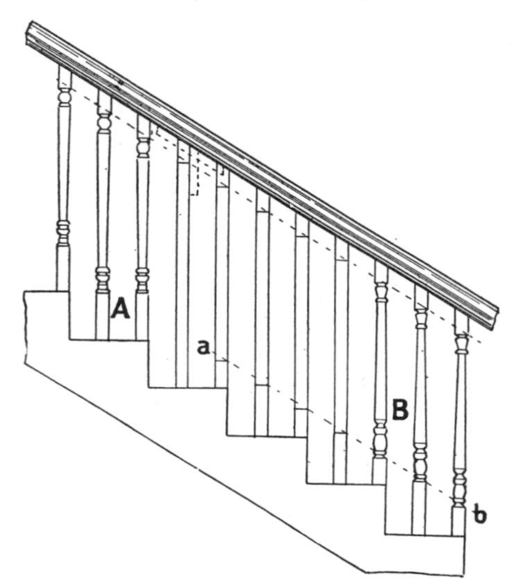

Method of Setting Out Turned Balusters.

The Wall Strings of Geometrical Stairs are set out in similar manner to the strings of dogleg stairs previously described, and the setting out of the cut string differs only in having the pitch-board run along the edge of the string without the intervention of the margin templet, and the use of a parallel tread templet equal to the thickness of the tread, instead of the tapering templet, f. 2, p. 309. This gives the line of cut for the underside of the tread, as in this case the tread rests upon the cut out portion of the string. See dotted line *a*, f. 1, p. 324.

The Wreath-piece of the outside string rising around the landing and winders at the half-space is formed by reducing a short piece of string to a veneer between the springings, and bending it upon a cylinder made to fit the plan, in the manner shown in f. 1, and when it is secured in position, filling the back of the veneer up with staves glued across it, and finally gluing a piece of canvas over the whole. The process has been more fully described in the chapter on Curved Work. The appearance of the wreath-piece after it has been built up and removed from the cylinder is indicated by f. 3; the canvas back has been omitted to show the staving, and the counter-wedge key used for connecting it with the string is shown. The wreath-piece is at this stage ready for marking the outlines of the steps.

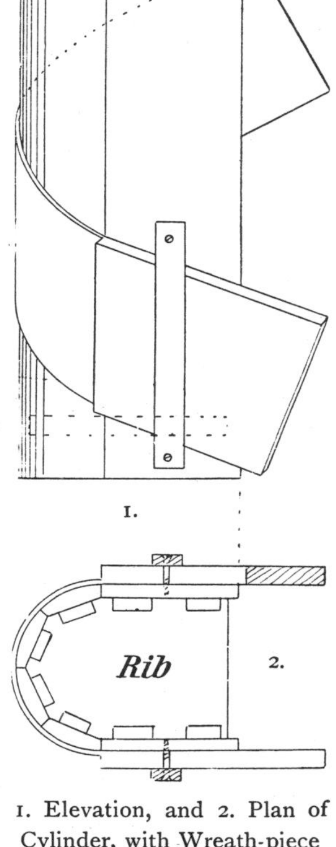

1. Elevation, and 2. Plan of Cylinder, with Wreath-piece Ready for Staving.

To obtain the mould for marking the veneer.—The circular part of the string, as shown in the plan, must be developed flat, so that its real shape and size can be ascertained. The method is illustrated in a separate drawing, f. 1, next page, a part of the plan, f. 2, p. 319, being reproduced to a larger scale. Draw the straight line T T tangent to the face of the string. Upon this set off a "stretch-out" of the face of the string between the springings. The stretch-out may be obtained in several ways; one is, to describe an equilateral triangle on the springing line, as shown by the dotted lines *e* C *e*, and produce the two inclined sides to meet the tangent line T T in points S S, whose distance apart will be equal to the length of the semicircle *e* 13 *e*; or the length may be obtained with more accuracy, and the points where the risers cut the strings ascertained at the same time, by making a templet of thin stuff, as shown in f. 2, next page, to fit the plan, and marking upon it the springing points, risers, &c., then rolling the templet, as shown in the drawing, carefully along the veneer or mould, and setting off the various points as they reach the veneer. Having obtained on the stretch-out the points S S, and the riser lines 11 to 16, erect perpendiculars upon these points, and intersect them by horizontals drawn from the six rises marked on the springing in the elevation; this will give the outline of the steps adjacent to the springing. Two of the ordinary flyers should be drawn on each side, to obtain the common pitch of top and bottom strings. Next draw the nosing line of the

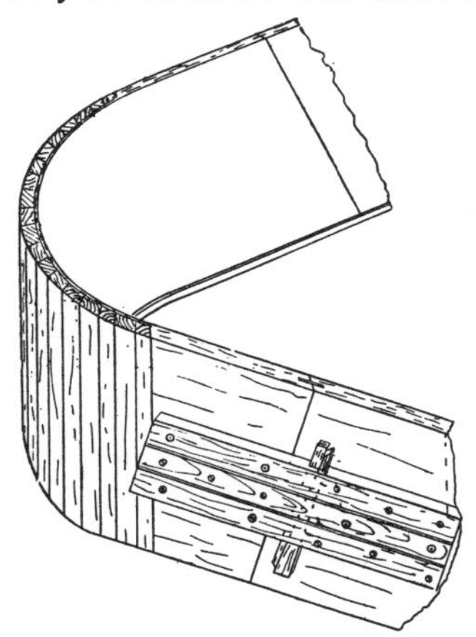
3. Wreath-piece Ready for Cutting, and showing Connection with String.

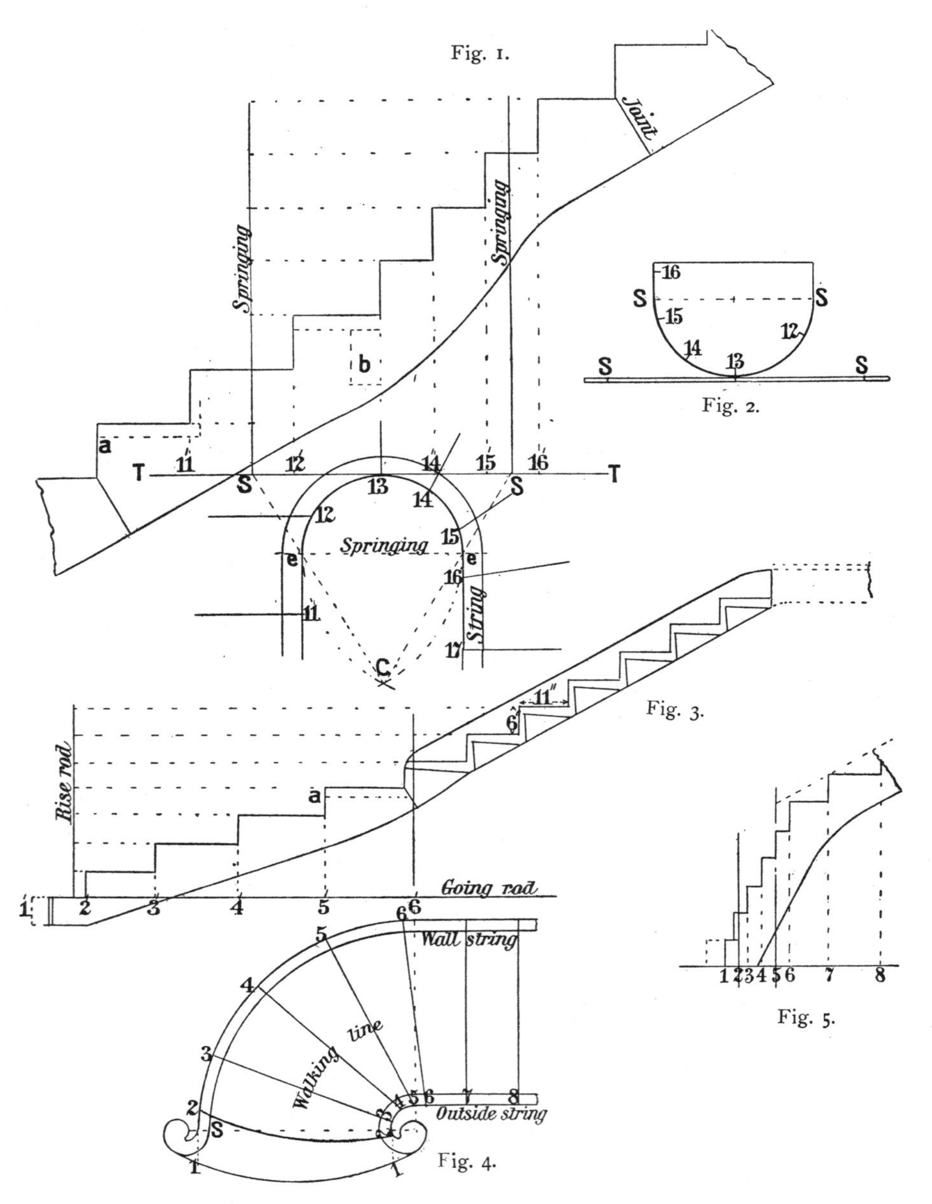

1. Development of Wreathed Piece for Plan, f. 2, p. 319.
2. Method of Setting Out Veneer with a Rolling Templet.
3. Development of Wall String for Lower Flight, f. 2, p. 319.
4. Plan of Winders in f. 2, p. 319.
5. Development of Outside String, f. 4 above.

flyers top and bottom, and the lower edges of the strings parallel with it, at the given width, which, as has been previously mentioned, must be wide enough to cover the carriages and soffit. Next draw under riser No. 13 the trimmer *b* which carries the landing. This must be kept below the step line a distance equal to the thickness of a tread, and a point 1 in. below the bottom of trimmer for the plaster line will represent the deepest point in the falling line of the edge of the string, which must be carried from it into the other two strings by gentle easings. The line so produced, whilst not unpleasing in the flat, has a rop in the centre that will cause it to appear somewhat crippled when bent to the plan curve, and this can only be obviated by making the flyer strings wider. This arrangement of the landing, which would have been better placed in the other space, has been purposely selected to illustrate the difficulty of obtaining a good falling line when the edge of the string is curved in elevation.

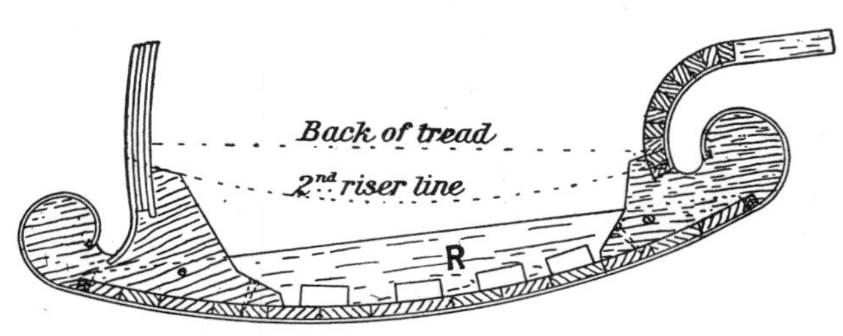

1. Enlarged Plan of Riser of Commode Step.

A Helical Curve depends for its beauty upon the regularity of its rise around its generating cylinder, and any departure from a straight line in the development will have a correspondingly ill effect upon the curve when seen in elevation. Therefore curving the edge of the wreathed string, though often unavoidable, should never be purposely introduced with the intention of im-

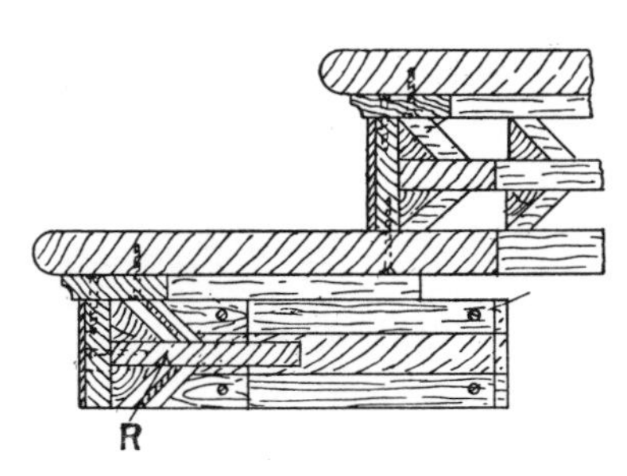

2. Transverse Section of Commode Steps, f. 1, to Larger Scale.

3. Method of Forming Riser and Block for Curtail Steps.

proving the falling line. The development, as shown in f. 1 opposite, may either be drawn directly on the face of the string before bending, or upon a piece of stiff paper or cardboard, cut out to shape, and applied to the face after the wreath-piece is built up. In cutting the templet it should be remembered that the development as drawn shows the top of the steps, and that the thickness of the tread must be allowed down, as shown by dotted lines at *a* and *b*. In cutting the lower edge of the string the saw should be held at right angles to the vertical face of the string, and radiating from the centre of the well. The joints must be made square with the pitch, about 1 ft. from the springings, and in the re-entrant angles of the steps: they should be grooved and tongued. A variation to the above-described method of forming the wreath-piece is sometimes adopted, especially in hardwood strings. The entire wreath-piece from joint to joint is formed of veneer, staved within the springing as described previously, and the straight pieces between the joints and springing backed up with solid pieces of pine glued on the same way of the grain. These pieces are called tangent backings.

The Development of the Wall String of the lower flight is shown in f. 3, p. 324, and the part of the plan redrawn to larger scale is shown in f. 4. The stretch-out in this case is found by the process of stepping, described on p.

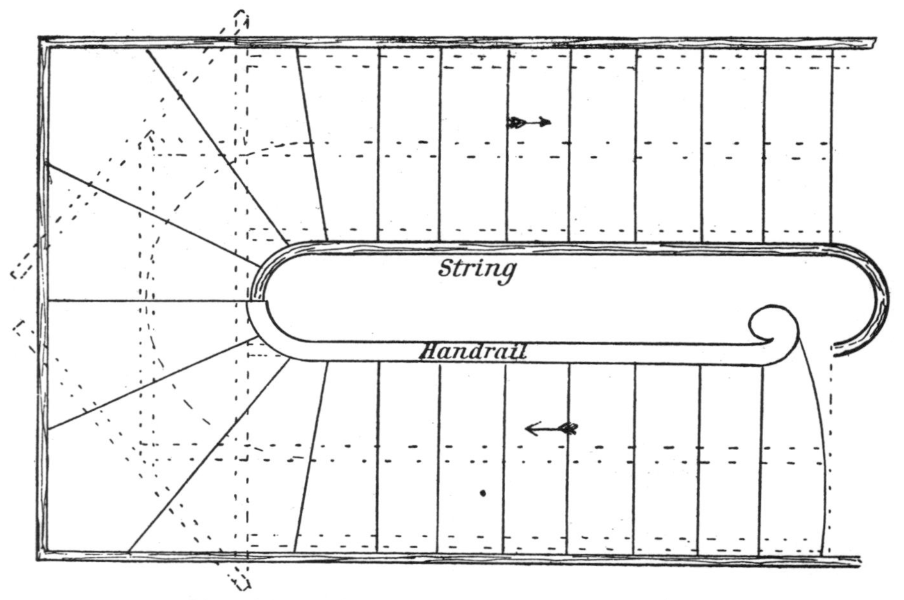

Plan of Geometrical Stairs with Half-space of Winders.

333, and the elevation of the steps, &c., is produced as described above. The wreathed part of this string is laminated or built up in four thicknesses. The inside board is steamed and bent around a cylinder, to which it is fixed until set; a few headless sprigs are then driven through it into the cylinder to keep it in position, these pulling through without damaging the surface when it is ready for removal; the other three laminæ are then glued on successively, each being allowed to dry before the next is applied. Fig. 5 is the development of the curtail portion of the outside string, and is obtained from the stretch-out of the curtail, as shown in the plan, f. 4, as above described.

The full width of the first **commode step** is indicated by the dotted line beyond the springing, to make the drawing clearer, but is not really required, as the string finishes in the neck of the curtail block, as shown to larger scale in f. 1, p. 325, which is a plan of the riser of the first commode step, showing the method of attaching it to the strings. Fig. 2 is a transverse section through the first and part of the second commode steps. R is a rib cut to the required sweep, and on its edge a number of pieces 6 in. long by 1 by 3 in. are nailed and blocked, and their edges glued together. These are afterwards worked off to a regular curve, and an $\frac{1}{8}$-in. veneer glued on their surface, the ends overhanging

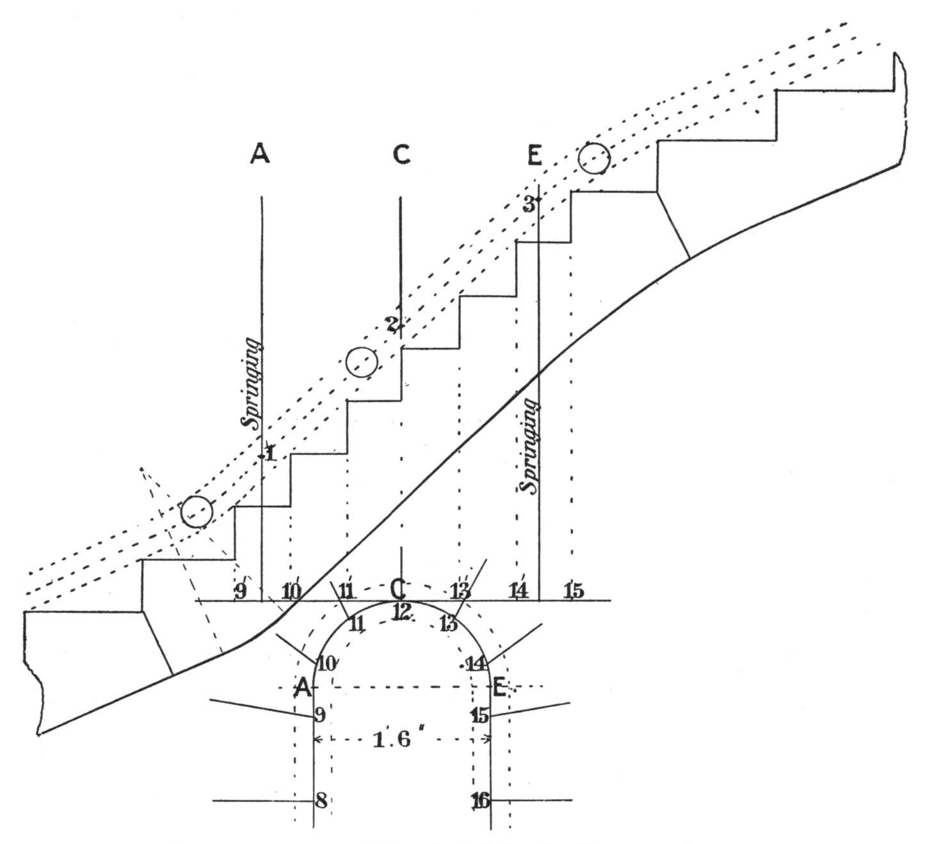

Development of Wreathed String for Plan, opposite.

sufficiently to reach round the curtail blocks, as shown in f. 1. The blocks are built up in a similar manner to that of the bullnose step described on p. 315. They are slotted out to receive the ends of the rib R, and secured to it with screws. In painted work the veneer is often dispensed with, the riser being merely kerfed at the back and bent around the rib, the marks of the kerfs being rubbed down with glass-paper. The ends are reduced to a veneer to bend around the curtail blocks, as described in the case of the bullnose step. In such work the curved portion of the outside string, adjacent to the curtail block, is made to form part of the block, no veneer being used, the surface of the block finishing flush with the face of string.

The construction of the riser of an ordinary **curtail step** is shown in f. 3, p. 325. The curve for curtail is described from the same centres that are used to describe the handrail scroll above it. The method of obtaining these is explained

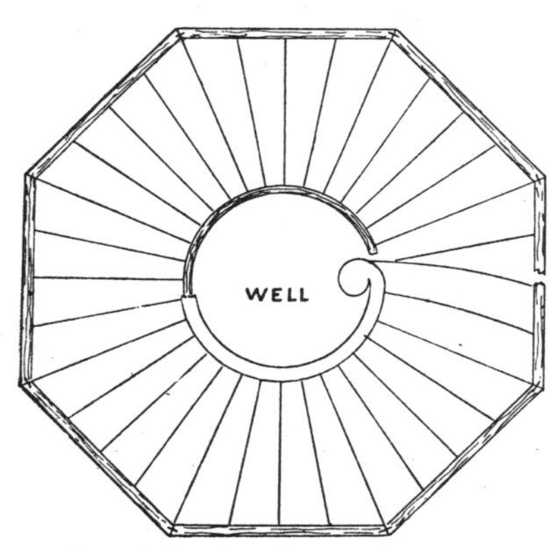

1. Plan of Circular Geometrical Stairs.

2. Plan of Circular Newel Stairs.

in the next chapter. The tail of the curtail is not continued into the eye as in the case of the scroll cap, because of the difficulty of keeping the space clean. The neck or narrowest part of the block must in all cases be at least as wide as the thickness of the balusters.

A Geometrical Stair with a half-space of "danced" or "balanced" winders is shown in plan on p. 326, and the Development of the Wreathed String, on p. 327. This string has a much better falling line than the one described at p. 324.

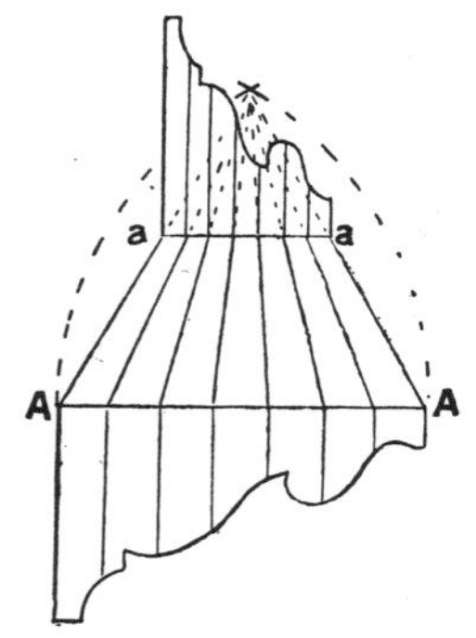

3. Method of Diminishing a Bracket.

Fig. 1 above is the plan of a **Circular Geometrical Stair** placed within an octagonal stairway.

Fig. 2 is the plan of a **Self-supporting Circular Newel Stair**. The wreathed strings for both of these varieties can be made in two ways, both needing a centre or cylinder for their construction, which is made to a portion of the plan of the concave face of string, and of sufficient length or height to build a given section of the string upon. The first or *Laminated* method has been described in a former paragraph, and need not be further entered into, except to mention that the joints or connections of the various sections of the string are made by toothing the pieces into each other, the ends of the laminæ being kept short alternately for that purpose. In the second or *Cored* process as it is termed, a stout veneer is first bent around the cylinder and secured. Then a shaped core made up of pieces of dry pine about 3 in. wide, and from 1 to 1½ in. thick, long enough to reach across the veneer, is glued to the back of the same, and when dry, gauged to a thickness and cleaned off, and a second veneer steamed and bent around, and when dry, glued to the face. The string in this method is made in sections of about a quarter of a revolution of the plan, and connected by grooved and tongued joint secured by handrail bolts, and may be further strengthened by screwing a $1\frac{1}{2}$ in. by $\frac{1}{4}$ in. iron bar which has been bent to the sweep inside the string under the steps.

Diminishing Brackets, f. 3 opposite, show the method of obtaining a reduced pattern for a bracket as required for the ends of winders. Upon the top edge of the bracket used for the flyers describe an equilateral triangle. Divide the contour of the bracket into a number of parts, and draw lines from the divisions

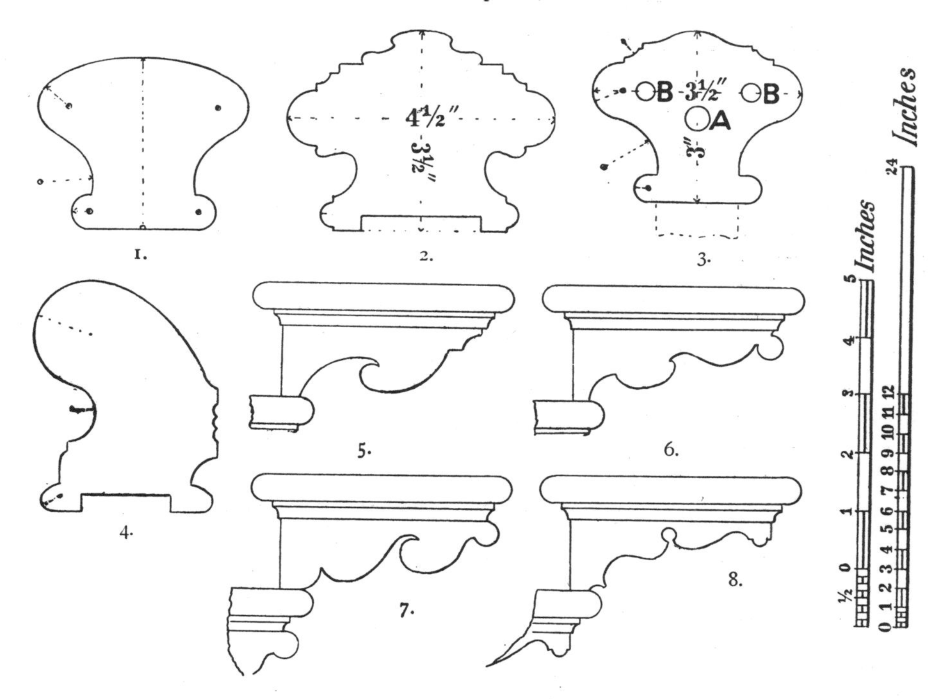

Sections of Handrails and Designs for Brackets.

perpendicular to the top or base of the triangle: from the intersections draw lines to the apex of the triangle. Next mark upon the sides of the triangle, from the apex, the length of the bracket required. Join these points by a line *a a* which is parallel with the base, and upon the points where this line cuts the lines drawn to the apex, erect perpendiculars, and make them equal in length to the corresponding lines drawn on the original bracket. Nos. 1 to 4 above show sections of handrails with the centres from which the various curves are drawn. Nos. 5 to 8 are designs for brackets.

CHAPTER XX.

THEORY OF HANDRAILING.

Requirements of a Handrailer—Definition of the Tangent System—Square and Bevel Joints—Explanation of Geometrical Terms—Theory of Handrailing Explained—Theory of the Tangent System, its Advantages—Classification of Tangents—Production of Face Moulds for a Quarter Turn—A Half-space Landing—A Quadrant of Winders—A Half-space of Winders—A Quarter-space of Winders in an Obtuse Angle—Wreath over Circular Winders—Scroll Shank for a Curtail Step—A Scroll Wreath over Winders—How to Obtain the Bevels—Theory of Bevels and their Use—How to Obtain Lengths of Balusters over Winders—How to Obtain Falling Moulds—When to Use them—Use of Projections—How to Project a Rail in any Position—To Trammel an Ellipse—To Discover the Direction of Elliptic Axes by Construction and Mechanically—To Find Thickness of Wreath, &c.

THE production of a continuous handrail for stairs curved in plan necessitates some acquaintance with the science of geometry, and also considerable manual dexterity on the part of the craftsman. This refers more particularly to the compound curved parts or "wreaths," the straight portions requiring no especial skill. The first of these requirements, involving what may be termed the theory of handrailing, is demonstrated in the following pages by means of sundry typical examples. The second can only be acquired by constant practice, and the intelligent observation of finished examples. To assist the beginner, we have, under the heading of practical work, described and illustrated the methods employed by expert workmen to realise their conceptions; and if these instructions are carefully followed, any workman of average ability will be able to produce rails of passable appearance; perfection in the art coming with experience. The writer having had considerable experience in the teaching of this subject, strongly advises beginners to work out exercises as large as possible—not less than half full size—and to finish them throughout, the moulding being the most difficult part to accomplish. It is a mistake to assume that the art has been acquired when the operator is able to produce the rail "in the square."

There are several so-called "systems" of handrailing, each having its supporters, but it may be safely laid down that there is no "best" system, the latter quality being largely a matter of individual opinion. They all resolve themselves into the method of obtaining the correct shape of an elliptic templet whose edge when placed at an appropriate altitude and inclination will be coincident with the plan curves of the rail. The system selected for description herein is that known as the **Square Cut and Tangent System**, not because its results are any better than that of others, but because it is more readily understood by those having little practical knowledge of geometry, and probably chiefly for this reason it is fast becoming the system in general use.

The term SQUARE CUT refers to the practice of cutting the wreath-piece out "square" to the surface of the plank, and incidentally to the method of making the joints, which are also square from the surface and square from the tangents. These are sometimes called BUTT JOINTS, in contradistinction from the vertical or BEVEL JOINTS of the earlier methods.

Tangent System refers to the method employed for determining the plane containing the section of the rail, by treating it as part of the surfaces of two or more planes which are tangent to the curve in plan. This will be more fully explained presently. The following preliminary observations will be found useful by readers unfamiliar with geometrical terms.

A Plane is an absolutely flat or even surface.

A Prism is a solid whose sides are parallelograms and whose ends are rectilineal figures (see f. 1) which represent in isometric projection a square prism at *a* and a pentagonal prism at *b*. If a square prism is divided longitudinally by a plane passing through its axis and two opposite edges, the resultant figures will be equal and similar triangular prisms. If the lengths of the three sides of a triangle are known, that triangle can be produced by drawing three lines of the given dimensions with their ends touching each other.

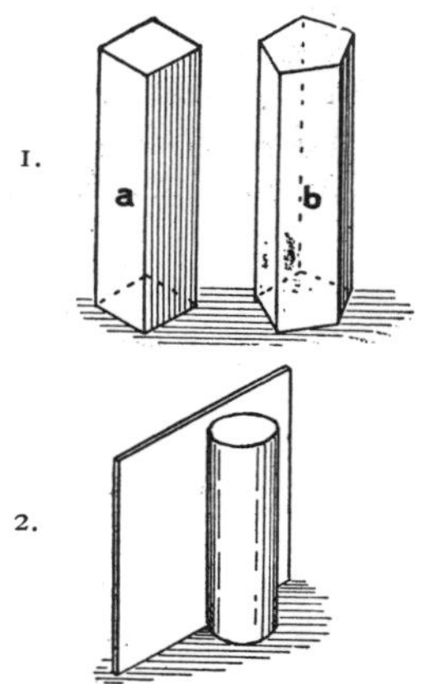

1. Examples of Prisms.
2. Example of a Tangent Plane.

A Cylinder is a solid with parallel sides and circular ends. The term cylinder is also applied in the workshop to any apparatus with a curved surface, whatever its section, if used for building strings or handrails upon. They are seldom used in the latter connection now, however, except for making joints in very large sweeps.

The Axis of a cylinder is a straight line passing through its centre from end to end.

A Tangent Plane is a plane, imaginary or real, touching the surface of a cylinder throughout its height, as shown in f. 2.

An Ellipse is a regular curved figure having all its four quarters exactly alike when divided upon its axes. In this it differs from the oval, which has only two of its quarters alike. The axes of an ellipse are two lines at right angles to each other, and passing respectively through the longest and shortest dimensions of the figure.

The Section of a Cylinder at right angles to its axis is a circle. Any section inclined to the axis will be an ellipse. The longer diameter or axis of the ellipse will vary with the angle of inclination of the section; the shorter or minor axis will always equal the diameter of the cylinder of which it is a section. **The Shape of the Ellipse** will not vary by changing the *position* of the section with relation to the sides of the cylinder so long as the inclination of the section remains the same in relation to the axis.

Ellipses can be drawn in several ways. One of the readiest is by trammelling, as shown in f. 1 below. Draw the two axes A B and C D at right angles to each other indefinitely; then upon any straight-edge or rod mark from one end half the length and half the width of the proposed ellipse. These points are indicated at 1 and *m* in f. 1, and are respectively the semi-major and semi-minor axes of the required ellipse. Next lay the straight-edge across the conjugate axes in such position that the two points 1 and *m* will be immediately over the lines. Then the end of the rod will be lying in the proposed curve, and a mark made at its end will be a point in that curve. Any number of these points may be found by moving the rod around, keeping it in the same relative position with regard to the axes, and the curve may be completed by freehand or by bending a thin lath around the points, whichever is more convenient.

Fig. 2 illustrates a method by which any required ellipse can be drawn when the *direction* of the axes is unknown, provided that the position of any two points in the curve, the length of one of the axes, and the position of the centre

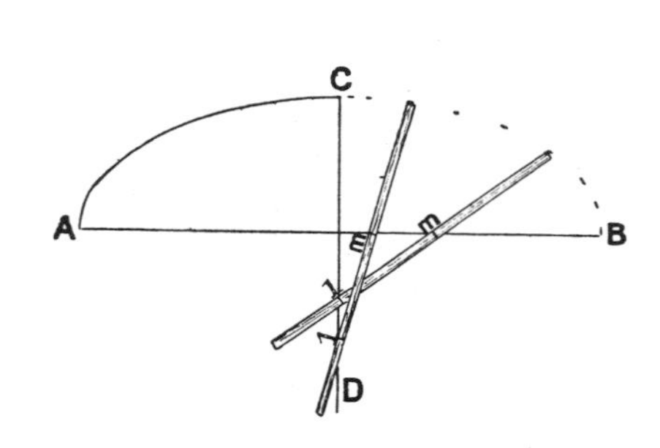

1. Method of Trammelling an Ellipse.

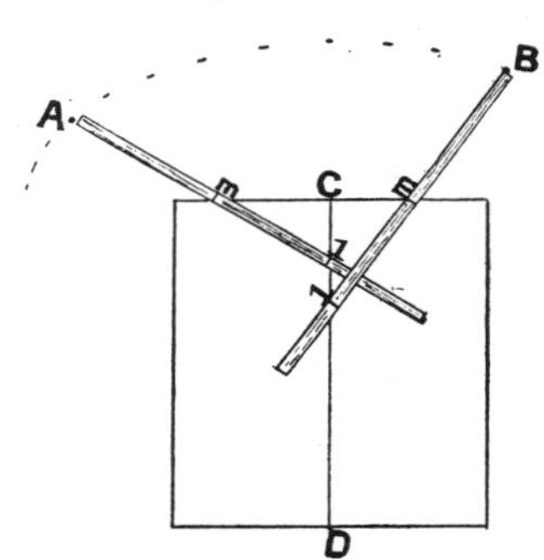

2. "Square" for Finding Direction of the Conjugate Axes of an Ellipse.

or axis of the generating cylinder are known. Provide a square templet, which may be of wood or paper, with one of its ends straight. Draw a line through its middle at right angles to the end, as at C D in the figure, and place the intersection of this line with the end, at the axis of the intended ellipse, shown at C in the figure. Arrange the end of the templet to lie in the *supposed* direction of the major axis. Let A and B be the known points in the curve. Mark on a rod the given length of the semi-minor axis, as at *m* in the figure, and place the rod with its end at one of the points A or B, and point *m* on the edge of the templet, and make a mark on the rod where it crosses the centre line as at 1 in f. 2. Next transfer the rod to the other point, and if the two marks *m* and 1 lie on the end and centre line of the templet respectively, then that end and centre line will be in the direction of the major and minor axes, and the curve can be drawn as described above. If, however, the marks will not lie as described, the templet must be moved around the centre until a position is found in which they will do so. A few trials may be necessary, and a 2-foot rule may be used in place of the rod, which will expedite the operation.

To obtain the length of a curved line, take a number of short steps or divisions around the line with a pair of dividers, and a corresponding number of like divisions set off along a straight line will equal the former in length. This line is then called the "stretch-out" of the curved line, and sometimes the development of the curve. The position of any point in the curve can be obtained on the stretch-out in the same manner.

The Development of a Solid is the drawing produced by the imaginary revolving of all its sides into one plane or surface parallel to the spectator, the object being that the real length and angles of inclined lines upon its sides may be seen. If these lines represent the edges of a section of the solid, and they are drawn in their relative positions, the shape of that section is determined. As the triangle is the only rectilineal figure of which the sides can only be arranged in one position with a given set of dimensions or lengths of sides, a prism of this shape is always used to determine the plane of section in handrailing. The application of the above theorems will now be explained.

THEORY OF HANDRAILING.—The **Plan** of the wreathed portion of a hand-rail may be considered as representing the plan of a hollow cylinder of like diameter, and the wreath-piece itself when in the "rough" as a thick slice or section of that cylinder cut with parallel surfaces to a given thickness, as illustrated in the sketch, f. 2, p. 335, which represents a cylinder in outline, the shaded portion representing a section made by two parallel planes passing through it from side to side. Fig. 1 is the plan of the same cylinder. If the section were cut through its longer diameter (the major axis of the section), it would be of the actual shape required for a wreath-piece to fit over the plan indicated by dotted lines in f. 1, and would be a "half-turn wreath," and the straight portions of the rails would be attached as indicated by the dotted outlines in plan and elevation, the upper rail showing a butt and the lower a bevel joint. It would not of course be practicable to cut a wreath-piece in this manner from an actual cylindrical piece, because the grain of the wood would be running in the wrong direction, and it is also not good workmanship to make such a wreath in one piece even when cut properly from the plank, in consequence of the grain running so much across the direction of the rails; but as the above-mentioned section shows the shape required at the top and bottom surfaces of the piece or pieces out of which the wreath is to be formed, we can use such a section as a mould or templet for the marking of the shape on the plank. A templet used for this purpose is called the **face mould** because it is applied to the face of the plank, and its edges will give the necessary curves to form two half-circles in plan when it is tilted up to the proper inclination. It will further be obvious that as the section in f. 2, p. 335, is cut throughout of equal thickness, both of its surfaces are exactly alike, and therefore if two moulds made exactly to its shape are applied, one on each side of a plank of parallel thickness, with their centres or any other common point so arranged in relation to each other that when the piece is set up to the required inclination these points shall be plumb or vertically over each other, the sides of the piece when cut to the two outlines will also be everywhere vertical, or in shop parlance, will "fit the cylinder." This cutting of the vertical sides of the wreath-piece is called **bevelling the wreath**, because the sides, although vertical in relation to the inclination or pitch of the plank, are at a bevel as compared with the *original* surface of the plank. This matter will be dealt with more fully presently, but we may here incidentally draw attention to the fact that the so-called "Bevels" of handrailing are really plumb lines, and are only bevels in relation to the inclination of the plank.

Having shown that a certain section of a cylinder will provide the shape of the face mould for a wreath that is to be inclined at a corresponding angle to that of the section, we have now to show how that inclination and section are determined from the plan of the rail by means of tangent planes.

The Theory of the Tangent System is that the imaginary cylinder of which the wreath forms a part, and of which its plan is a horizontal section, is enclosed within a hollow box or prism, the sides of which touch the cylinder at two places, called the springings, and that any plane of section or cut, passing through the sides of the prism, passes also at a like inclination through the cylinder; and as a section of a prism is easily determined, the ellipse forming the appropriate section of the cylinder can be drawn accurately within its boundaries, the section of the prism being obtained by unfolding its sides, measuring the lengths of the section as then disclosed upon each, and with these dimensions describing the section.

The Plan of a Rail forming a Quarter Turn, with its centre line enclosed by the plans of tangent planes *a b* and *b c*, is represented in f. 3 opposite. Two lines drawn parallel to these, from the centre or axis of the cylinder X, represent the springings, and the line *a c* joining these, is the plan of a plane called the "diagonal," converting the tangent planes into a triangular prism. It may be noted here that in all handrail drawings only the *centre line* of the rail is used as the basis of work, all measurements being made from this, joints made and bevels applied at the centre of the material. The necessity of this will be apparent when it is remembered that the surfaces are being continually altered in shape and position during the operations of fashioning the wreath. The rail in question is inclined in the direction of the tangent *a b*, and level in the direction of the tangent *b c*—in other words, it is a rail over a flight of stairs leading to a level landing.

To draw the plan, f. 3.—Draw the straight rails as shown over landing and stairs at right angles to each other, also their centre lines produced until they meet. Determine how much of the corner is to be cut off, *i.e.*, the size of the cylinder, in this case 6 in. Set this distance off from *b* on each centre line, as at *a* and *c*, and draw lines from these points at right angles to the centre line, meeting in X, the axis of the cylinder. From this point as centre describe an arc connecting the ends of the centre line of rail. The position of the risers is not considered in this example.

To obtain the elevation of the tangents—that is, the development of the edges of the section—project the parallel plane *b c* into the elevation, as shown in f. 4. Revolve the plane *a b* upon *b* as centre, until it is in the same line as *b c*, its path being indicated by the dotted line in the plan, then project it into the elevation at A. Draw the line of section A B parallel to the pitch of the stairs (in this case assumed to be 45 deg.), because the handrail is always parallel with the stairs over the flyers; and upon the elevation of the side *b c* draw the level line B C. These two lines then represent the edges of the section upon the two

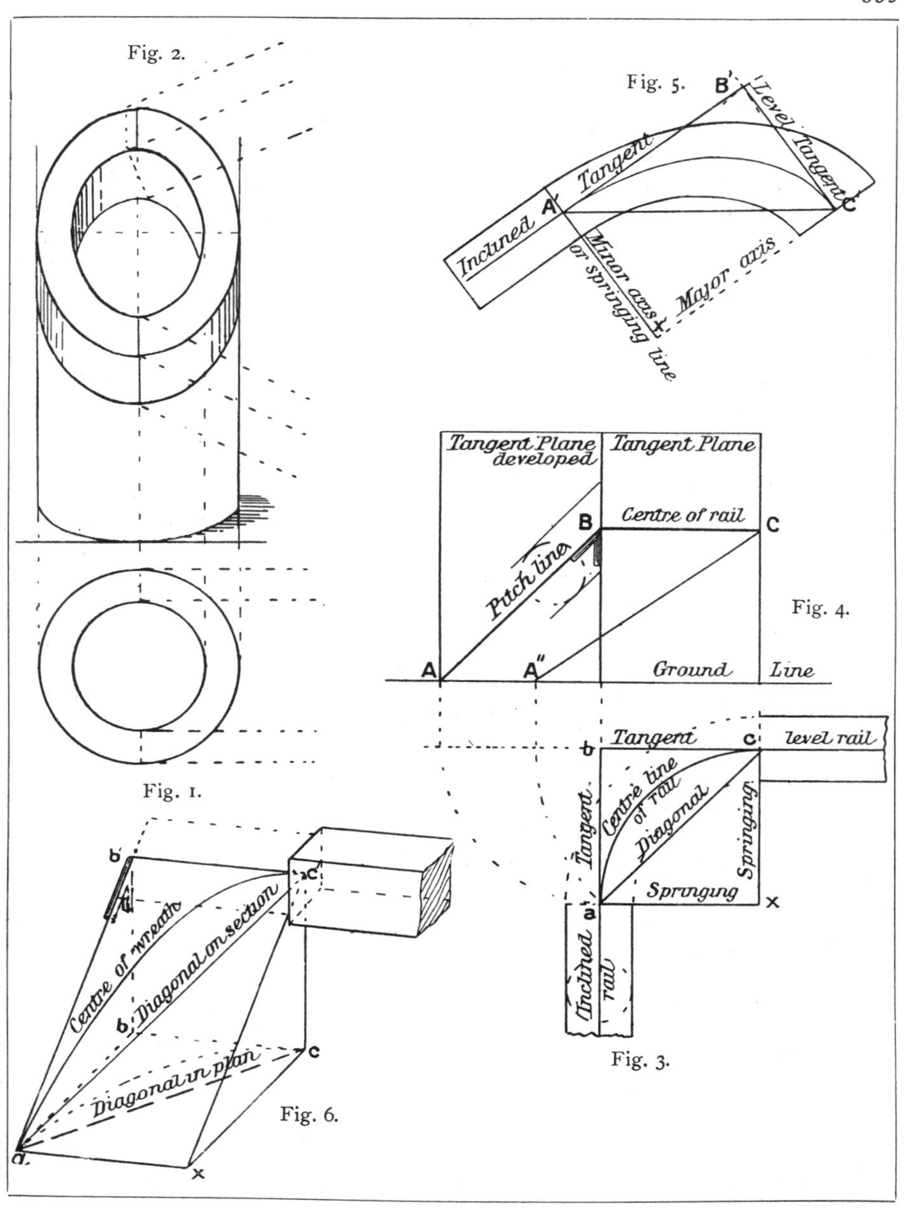

THE THEORY OF HANDRAILING.

1, 2. Plan and Elevation of the Section of a Cylinder.
3. Plan Diagram of a Wreath enclosed by Tangents.
4. Development of Tangent Planes in Elevation.
5. Face Mould Diagram.
6. Sketch of the Prism shown in Plan, f. 3 above.

tangent planes. To obtain the edge of the section upon the third plane formed by the "diagonal" *a c* in the plan, imagine it rotated upon the point *c* until it lies in the same plane as *b c*, then project its lower end into the elevation as at A″, which represents the level of the point A, and draw a line from A″ to C, which is the point where the section meets the vertical edge of the prism, and this line will be the length of the section upon the diagonal plane, because if the planes representing the prism were folded into the positions shown in the plan, and a saw cut made commencing on the horizontal line B C and kept level in this direction whilst it ran down the side *a b* at the angle shown by the line A B, it would pass through the side *a c* in a line running from C to A.

To draw the section of the prism.—Draw a straight line as A′ C′, f. 5, last page, equal in length to A″ C, f. 4. Then with A B, f. 4, as radius, and A′, f. 5, as centre, describe an arc; and with B C, f. 4, as radius, and C′ as centre, describe another arc intersecting the former at B′. From this point draw lines to A′ and C′, and the resulting triangle will be the true shape of the section of the prism *a b c* upon the given line of inclination.

To obtain the shape of the enclosed cylinder upon this section.—The position of the axis X, f. 3, must first be determined. On reference to the plan, it will be observed that the springing lines radiating from the centre or seat of the axis are parallel to the opposite tangents, and make with them a square which may be considered the base of a square prism. If this prism were cut to the same angle of inclination as the triangular one, its sides or the edges of the section would still remain parallel, and the shape of the section being a parallelogram (colloquially an "oblong"—that is, a four-sided figure having its opposite sides parallel and equal and its adjacent sides unequal), therefore to locate point X′ on the section, draw lines from points A′ and C′ parallel to A′ B′ and B′ C′. Their intersection at point X′ is the position of the axis of the cylinder upon the section, and the lines themselves the major and minor axes of the required elliptic curve. Note that the minor axis of the section is always equal to the radius or half diameter of the cylinder of which it forms part, but it is only shown on the section when one of the planes of section is horizontal, as in this instance. The elliptic curve can now be trammelled from A to C′ as described on p. 332, and the triangle and central elliptic curve, if placed over the plan, with the side B′ C′ horizontal, and the side A′ B′ inclined at an angle of 45 deg., would stand with these sides perpendicularly over the corresponding lines in the plan, and with the curved line over the central quadrant in plan.

To complete the face mould two similar ellipses must be drawn equally on each side of the centre one, representing the inside and outside edges of the rail. At the springing A′ they will be equal in width to the rail in plan, because the wreath is level across this direction, but at springing C′ they will be wider as the piece is inclined in this direction. To obtain the correct width, imagine the level portion of the rail cut off square and butted against the prism with its centre to the vertical edge upon the section, as shown in the sketch, f. 6, p. 335. Now if a mark were made at the outside of the rail upon the surface of the section, it would show on that section half the width required upon the inclined piece to equal the level piece when cut vertical, and as we have not a solid but only its projections to deal with, we draw upon the inclined edge of the section, f. 4, p. 335, half the width of the rail on each side of the section line as shown by the two arcs, and touching these and parallel to the section line two lines intersecting the vertical edge B, and the width shown on this line will be the width required for the mould at the top end. The reason of this will be apparent on referring to the sketch, f. 6. If we draw lines from the sides of the horizontal rail across the surface of the section to the opposite edge, as shown by the dotted lines, we shall obtain the necessary width of the inclined mould on that edge, and if we imagine the surface contained between the two dotted lines to be a plane capable of revolving on the line *b′ c″* as an axis, when its surface is vertical or parallel to us, we can see its real width, and this is what we do by construction upon the elevation as described above. Having found the width as at B, mark off the half on each side of the tangent at the springing C′, f. 5, and mark off on each side of the other tangent at the springing A′ half the width of the straight rail as it is shown in plan. These two points on each springing are points in the elliptic curves, and the lengths of each of them from the centre X will be the semi-major and semi-minor axes of the ellipses which may be drawn as described on p. 332. The mould is completed by drawing the sides of the shank parallel to the tangent A′ B′ and the joint square with the tangent.

The bevel shown in the sketch of the prism and also on its development, f. 4, p. 335, is the bevel used for locating the position of the face mould on the wreath-piece when making its sides vertical, and its application will be explained in the next chapter. I have entered somewhat minutely into the reasons of the various operations required for the production of the face mould, so that the reader may understand why the placing of a line in a given position brings about a certain result, in the hope that it will make the study of this difficult subject more interesting than the mere copying of the examples and the committing of the methods to memory would prove. I have purposely selected a very simple case as an illustration to render the explanations easier, but the principles involved are equally applicable to more complicated conditions, and as here laid down are followed throughout the succeeding examples. To avoid tedious repetition, only a bare outline of the above-described processes will be given with the following examples, the explanations being confined to any variations in detail or new principles that may arise. In dealing with handrailing from the point of view of the tangent system, the several positions which the handrail assumes under given conditions of inclination may be classified into four groups according to the relative positions occupied by the tangents under those conditions, and as this grouping renders it easy to remember certain characteristics in relation to the bevels and face moulds, we have selected examples that are types of the various groups rather than types of the various arrangements of stairs in plan. In this connection it must be understood that "tangents" mean lines either in plan or elevation representing the *edges* of the plane of section containing the centre line of rail.

CLASSIFICATION OF TANGENTS.—(I.) **When one Tangent is Level and the other Inclined**, examples f. 2, p. 339, and f. 2, p. 350, Properties of face mould and bevel. One end of the face mould equals the width of the rail in plan. The major and minor axes of the elliptic curves coincide with the springing lines on the section. The plank having one inclination only, one bevel only is required for locating position of face mould, a square being used at the other end.

(2.) **When both Tangents are Inclined Equally**, examples f. 2, p. 345, and f. 3, p. 349, Properties of mould, &c. The mould is wider at each end than the rail in plan, but the widths of the ends are equal. The major axis of the elliptic curves is parallel with the "diagonal" of the mould. The plank being inclined in two directions, two bevels are required to locate the face mould on the plank, but both are alike.

(3.) **When both Tangents are Inclined Unequally**, example f. 3, p. 342, Properties of mould and bevels. The face mould is wider than the plan at each end, and the ends are unequal. The direction of the axes of the elliptic curves is unknown, and must be discovered by construction or trial, as described in the examples. Two bevels are required, each different.

(4.) **When the Tangents are not at Right Angles in Plan**—that is, either obtuse or acute—examples f. 1, p. 348, and f. 1, p. 351. In these cases the axis of the cylinder or centre of the plan curves lies outside of the imaginary solid formed by the tangents and diagonal, and its position has to be located upon the section by a special construction, described in its place. The bevels depending upon the inclination of the tangents in elevation are subject to the same conditions mentioned in the other cases.

JOINTS are best made square to the tangents both in plan and elevation.

EXAMPLES.

A Pair of Wreaths for a Half-space Landing.—Fig. 1 opposite is the plan of part of a geometrical stair with a 10-in. well, and having a half-space landing connecting the return flights. The straight flights do not finish at the springings to avoid throwing the tangents out of level or inclining them in the wrong direction—that is, running in the opposite way to the inclination of the rail. This will occur under certain conditions of going and rise when the landing risers are set in the springings, and will prevent butt joints being used, as the joints cannot be made square with the tangents when the latter incline the wrong way. However, when such an arrangement of the plan is unavoidable, as would occur in stone stairs previously fixed, the difficulty may be got over by ramping the top rail up and kneeing the bottom rail down until the centres are level, and then proceeding as described in the present case. The going of these stairs is 9 in. and the rise 7 in.; the landing risers are placed at half a going from tangent points B and D. The full lines show the plan of the straight rails, the dotted lines the wreath. The centre line of the wreath is enclosed by the tangents A, B, C, D, E, found by producing the centre lines of the straight rails and connecting them by a line at right angles touching the wreath at its centre and the centre of the well. As before mentioned, a half-turn wreath should be in two parts to avoid cross grain and economise material: therefore, draw the direction of the joint from C to X, the centre of the curve, and draw the springing line A E through the centre, completing the plan of the containing solid. Draw two risers on each side to obtain the going. Next draw the development

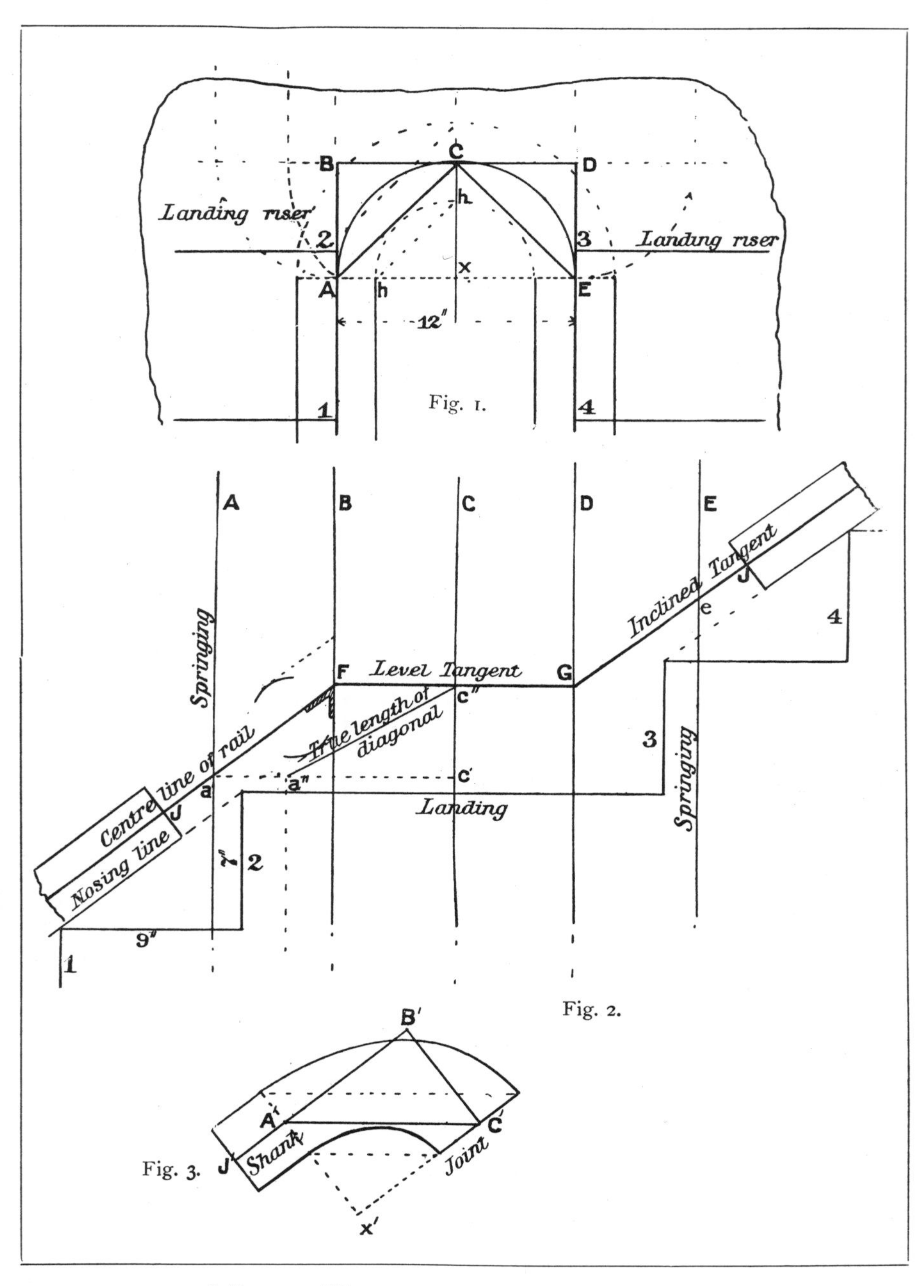

A PAIR OF WREATHS FOR A HALF-SPACE LANDING.
1. Plan. 2. Development of Tangent Planes. 3. Face Mould.

by unfolding the tangents in elevation, making the lines A, B, C, D, E, f. 2, previous page, equal in distance to points A, B, C, D, E, f. 1. Draw the horizontal line representing the landing in any convenient position, and the risers 2 and 3 at the same distance from the lines A and E that they are from the correspondingly marked points in the plan. Mark off 7 in. on these lines and draw the treads at right angles and 9 in. wide, adding another rise which will give the common pitch over the stairs. Next draw the nosing lines and the depth of the rails above them. Parallel with the nosing line draw the centre lines, cutting the tangents B and D in points F and G. Draw the level line connecting these points, and the lines *a*, F, G, *e* are the elevations of the tangents or edges of the section of the prism that contains the elliptical section of the cylinder whose plan is represented by the plan of the wreath. Draw the joint lines J J square with the centre lines at about 3 in. from the springing.

To obtain the face mould, f. 3, last page, the section of the prism must first be drawn. To do this, convert the base of the prism A, B, C, X, f. 1, into triangles by the line A C, and ascertain the length of this line upon the elevation. Draw the level line *a c′*, f. 2, from the point of intersection of the tangent A with the springing, and mark off from *c′* to *a″* equal in length to A C, f. 1. Join *a″* to *c″*, which will be the real length of the diagonal. Make A′ C′, f. 3, equal *a″ c″*, f. 2, and A′ B′ and B′ C′ equal *a* F and F *c″*, f. 2. The resulting triangle is the true shape of the section of the prism A, B, C, A, f. 1, and the lines A′ B′ and B′ C′ are the tangents of the mould. They should be square to each other, and if tilted up to the pitch of the stairs will stand immediately over the corresponding lines in the plan. To find the centre of the elliptic curves X′, draw the lines A′ X and C′ X parallel to the tangents.

To find the points from which to draw the elliptic curves. Mark off on the springing at A′ the width of the rail in plan equally on each side of the tangent. Draw lines from these points parallel to the tangent, thus forming the sides of the shank, and draw the joint line J′ at right angles with the tangent at a distance from A′ equal to J *a*, f. 2. The points at the other end of the mould are found by drawing a circle on the tangent *a* F, f. 2, equal in diameter to the width of the rail, and drawing lines touching the circle parallel with the tangent to cut the vertical line B. The width shown on this line is the width of the nould at joint C, and should be set off on the springing line, equally on each side of the tangent. The lengths of these points from the centre X′ are the lengths of the major and minor axes of the curves, and the springing lines are their directions, and the curves may be trammelled as described on p. 332.

The correctness of the width of the mould at the joint C′ may be tested by drawing lines parallel with the diagonal from the points of the curve on the springing A″ to the opposite springing, where they should coincide with the points found by construction, as shown. The reason is, that lines drawn from the corresponding points *h h* in the plan, f. 1, are parallel with the diagonal, and represent vertical planes passing through the plane of section. The method applies to all face moulds. Both halves of this wreath being alike, only one face mould is required. The use of the bevel and application of the face mould is dealt with in the succeeding chapter.

A Wreath for a Quarter-space of Winders.—In the foregoing examples the inclination of the tangents has been practically fixed by the inclination of the stairs, as the tangents over landings should always be level, except in the case of very small wells; but in rails over winders the inclination of the tangents will vary with the arrangement of the winders in plan, and the best arrangement of the tangents is a matter requiring some consideration. If the stairs are already made, or the disposition of the risers fixed, there is no alternative but to arrange the tangents so that the rail shall be as nearly parallel with their nosing line as possible; but if the position of the winders may be varied to suit the rail, a suitable falling line for the latter should be decided upon by making a stretch-out in elevation of its central plane, and the position of the steps made to conform to it; then the location of the risers upon the plan can be obtained from the stretch-out, and the stairs made accordingly. In the case shown, it is assumed that the position of the risers has already been determined, but it is still advisable to develop the centre line of the rail to determine the exact heights of the tangent points around the well, as any error in their inclination will cause a corresponding error in the bevels, and the wreath will consequently be thrown out of its true position, necessitating an alteration of the joints and lengths of the straight rails to correct it. This matter is often neglected both in text-books and practice, with the result we have mentioned, for although a skilled workman may by intuition immediately discern the most suitable arrangement of the tangents, a novice cannot of course expect to have that power. Fig. 1, next page, is part of the plan of a continuous flight of stairs, the upper part turning at right angles to the lower, and having four winders in the turn. The winders are "balanced," as explained in Chapter XIX. The centre line of the rail, really the only one wanted, is drawn in full, the sides of the rail being indicated by dotted lines, the centre line of the wreath being enclosed with tangents as explained in previous examples. Fig. 2 is the stretch-out of the centre line of the rail, showing the steps as they occur thereon.

To make the stretch-out of the rail.—Let the distance from A′ to C′, f. 2, equal in length the curved line A C in f. 1, which is ascertained by the process of "stepping" described in Chapter XIX., and transfer the positions of the risers 3, 4, 5, 6, as they occur in the plan *upon this line*. At A′ and C′ raise perpendiculars to represent the springing lines, and draw the outline of the steps as shown, with two flyers beyond each springing, to obtain the common pitch of the stairs. Draw the rails resting on the nosing lines of the flyers: if the lower rail is continued up to the springing line A′, it will be found that to connect it with the upper rail would necessitate the rail over the winders falling below the nosing line. This cannot be permitted, because it would result in the rail being lower over the winders than over the flyers, an arrangement of great danger to the users of the stairs, so the lower rail must be ramped up until it is high enough to permit of the wreath resting on the nosings of the winders. (In steep pitches it would be necessary to raise it still higher.) Then draw in the wreath on the nosing line, easing it off gently into the straight rails, as shown. The curves employed in the easings are purely a matter of taste, but when they are carried beyond the springings they affect the heights of the tangents, as will be presently shown, and the longer the easing the less abrupt the change in the falling line of the rail, but a long easing means extra thickness of stuff required

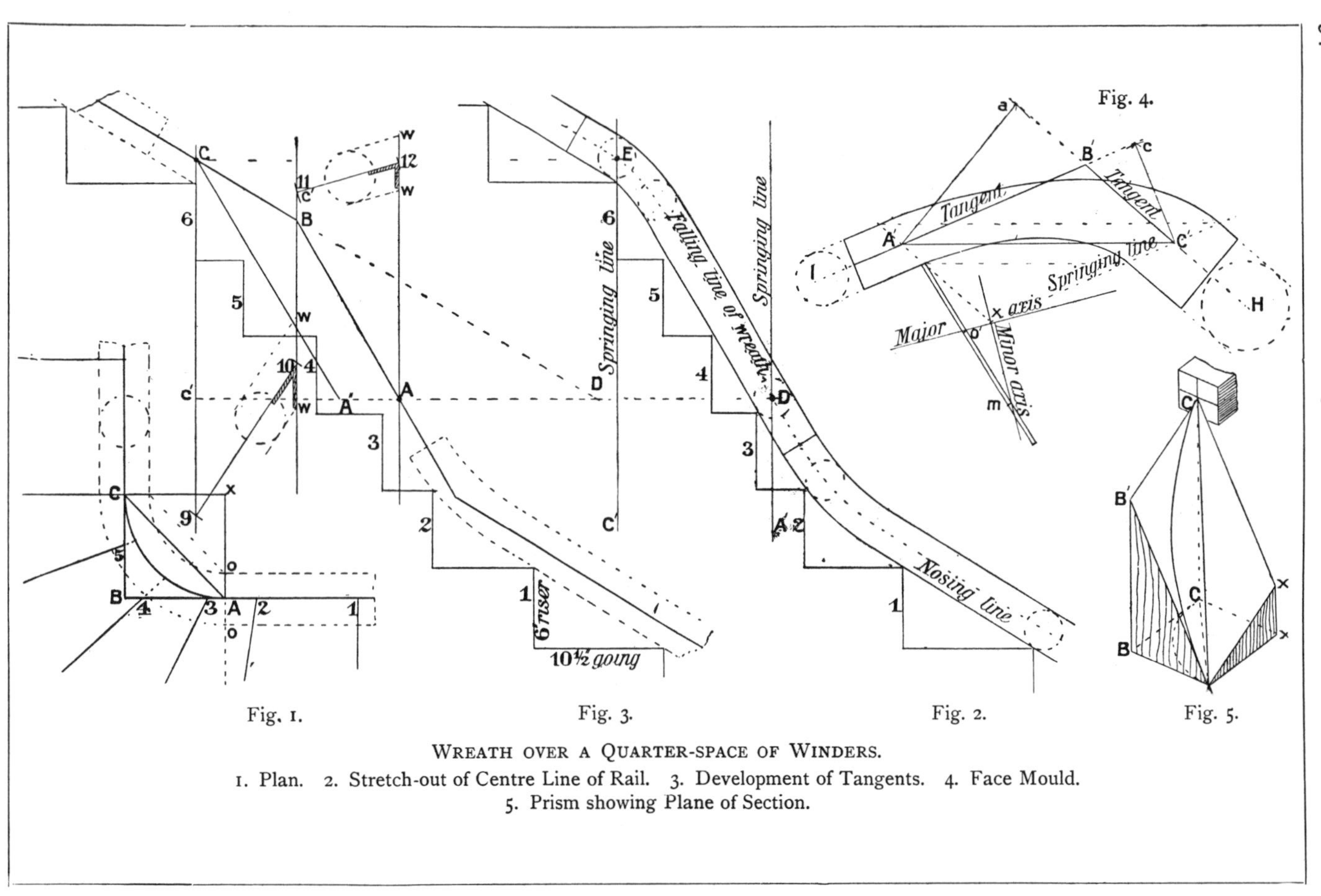

WREATH OVER A QUARTER-SPACE OF WINDERS.

1. Plan. 2. Stretch-out of Centre Line of Rail. 3. Development of Tangents. 4. Face Mould. 5. Prism showing Plane of Section.

to work the straight rails. Next draw the development of the tangent planes, as in f. 3 opposite, and draw the risers as they occur on the *tangents* in plan. Project the treads from the stretch-out, f. 2, and draw the straight rails resting on the nosing lines. Produce the centre line of the top rail to the vertical line B, and project point D in f. 2, where the centre line of the wreath intersects the springing, on to the same springing line A in f. 3. Draw a line from B through A to meet the centre line of the lower rail. Point C in f. 3 should coincide with point E, f. 2, where the centre line of the rail crosses the springing. Then the lines C B and B A are the lengths and inclinations upon two sides of the prism A B C, f. 1, of the section containing the centre line of the wreath, and will provide the "tangents" of the face mould.

To draw the section, the length of the third side, A C, f. 1, is necessary. Draw a level line from the lowest point A, f. 3, to the perpendicular from the highest point C. Upon this line from C′ set off the length A″ equal to the diagonal in plan, A C, f. 1. Draw A″ C, which is the real length of the diagonal. Make A′ C′, f. 4, equal A″ C, f. 3, and A′ B′ and B′ C′ equal A B and B C, f. 3, and through A′ and C′ draw lines parallel to A′ B′ and B′ C′, which will intersect at X′ the centre of the elliptic curves. Produce the tangents beyond the springings, and draw the joints square with them, at the distance shown on the development, from the springings. It may be here remarked that there is no fixed position for the joints that cannot be departed from as circumstances dictate. It is usual to work a shank or straight part on the wreath, for the two-fold convenience of making the holes for the bolts in straight wood, and also because the change of curve to straight can be better managed in the solid wood than on two separate pieces. 3 in. is the average length for shanks, but in some cases longer ones are advisable, as lessening the amount of easing to be worked on the straight rail. It is well to avoid placing a joint directly over a baluster, as the fixing of the latter may start the joint.

To find the widths of the mould and the bevel for each end.—In this case the plank from which the wreath-piece is cut will have to be tilted in two directions, therefore two bevels will be needed, and both ends of the mould will be wider than the rail in plan. To obtain the bevel for the top end, produce the tangent B′ C′, f. 4, and draw the line A′ *a* from the springing point A′, at right angles to the tangent B′ C′; draw a line equal in length to A′ *a* anywhere between the elevation of the tangents B C, as shown at the points 9-10 in f. 3. The bevel for that end is shown in the acute angle made between the line and the vertical, and if a circle equal in diameter to the width of the rail in plan is described on the line 9-10, and lines are drawn parallel to it touching the circle, it will show the necessary width of the top end of mould on the vertical line at W W. On the tangent B′ C′, f. 4, draw a circle equal in diameter to W W, as at H, and parallel with the tangent, draw lines touching this circle; where these lines cut the springing line will be the points from which to start the elliptic curves at that end. The width and bevel for the bottom end is obtained similarly by squaring a line from tangent A′ B′ to point C′, and laying it off on the development at 11-12, marking the rail upon this, and obtaining the width at W W. This is the diameter of the circle I, whose sides produced to the springing give the points for the curves at that end. The reason that the bevel is not shown by the angle of inclination of the tangent line in the elevation as in previous cases is that here, the joints where the **Bevel** has to be applied, are not parallel with the opposite tangents—a reference to the face mould will show this; and as the bevel must always be taken or drawn in the same position it is intended to be applied, if we are to obtain the correct angle we must draw a line parallel with the joint with which to take the bevel for that joint. For instance, the line A *a*, f. 4, p. 342, is parallel to the joint H, and this line laid off between the projections of the sides of the prism A X or B C, f. 3, will show the angle that line makes with the vertical. This explanation will perhaps be rendered clearer by an examination of the sketch of a prism, f. 1, p. 346. This prism refers to the next plan, but the principle applies to all similar cases. The dotted lines on the plane of section indicate the position occupied by the face mould or wreath, and it will be noticed that the joint at C does not lie in the same plane as the side of the prism that stands over the springing line, but is square with the edge or tangent B C. Therefore, if the bevel is taken at the angle B, as it occurs on the parallel side B A, it will be incorrect when applied at the joint C; but if a board or other flat surface is applied at the side of the prism which is at right angles to the joint, as shown in the sketch, and a Bevel is held with its stock at right angles to this board, and to the plane of section, with its blade in a line passing through the springing point A, that Bevel will be parallel with the joint C, and will be correct for drawing a plumb line thereon. The above is in effect what is done, when a line is drawn from the springing on the face mould, perpendicular to the opposite tangent, and then laying off a similar length between the projections of the tangent planes, as previously described.

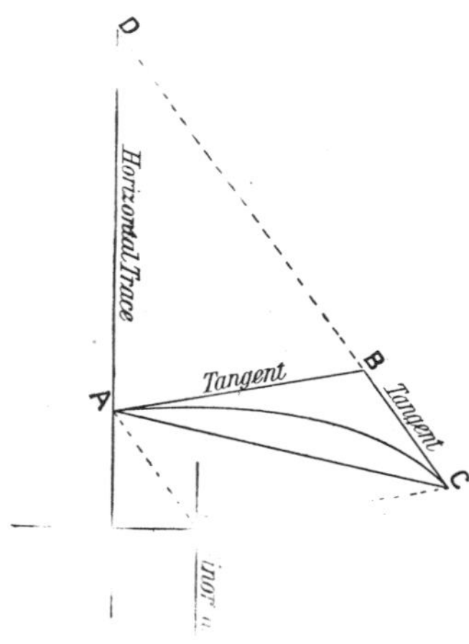

Method of Finding Direction of Elliptic Axes of Face Mould.

The axes of the elliptic curves in this example do not coincide either in length or direction with the springing lines, in consequence of the section having two pitches. The direction of the axes may be found by trial, as described on p. 332, taking the length of the minor axis from the plan, and obtaining the major from the "square"; or the direction of the minor axis may be determined geometrically by finding the horizontal trace of the plane of section, and drawing the minor axis parallel with this through the centre X. This method is shown in the separate drawing adjoining, to avoid confusion from the extra construction lines.

To find the direction of the elliptic axes by construction.—The annexed figure is a reproduction of the face mould, f. 4, p. 342, without its curve lines. Produce the tangent line C B, f. 3, downwards until it reaches the same level as point A, cutting the horizontal line therefrom on D. Add the length B D to the tangent C B in present drawing, and draw the level line, or horizontal trace D A, and a line drawn through point X parallel with this line will be the direction of the minor axis of the elliptic curves, and a line at right angles through the same point the direction of the major axis. The direction of the axes being thus found, the length of the minor axis of each curve is taken from the plan, f. 1, p. 342, measuring from X to o in each case, and the rod laid from one of the given points across the axes, as shown in f. 4, with the

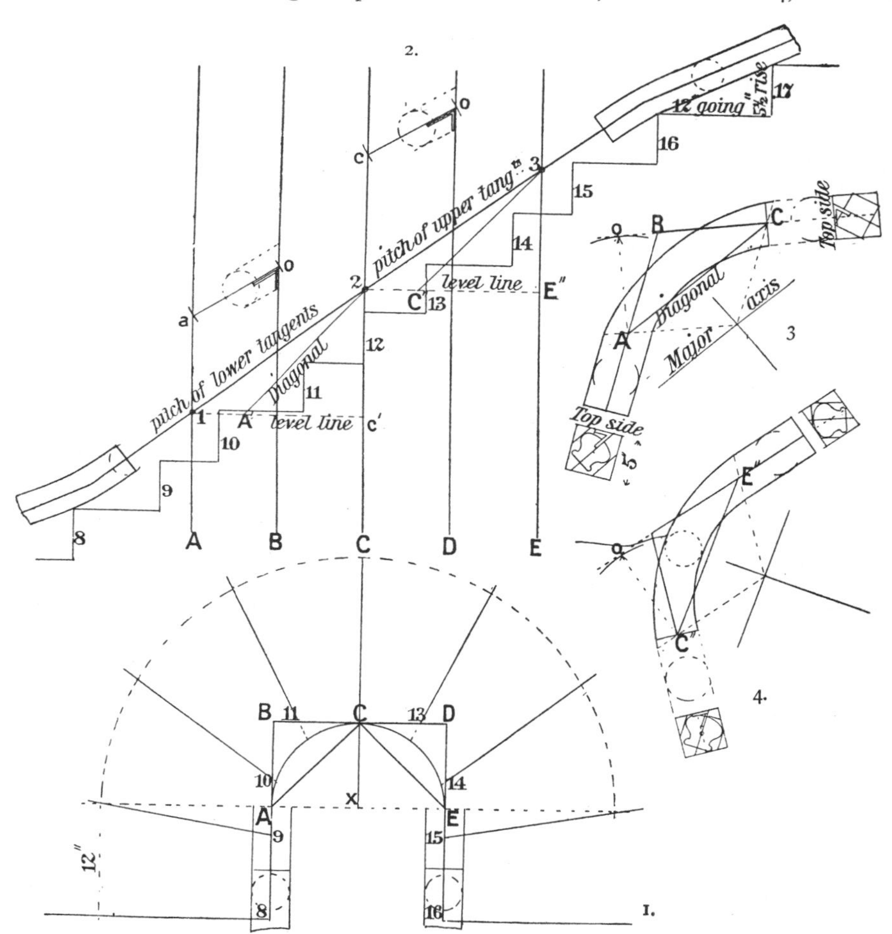

PAIR OF WREATHS FOR HALF-SPACE OF WINDERS.

1. Plan. 2. Development of Tangent Planes. 3. Bottom Face Mould. 4. Top Face Mould.

point o' resting in the direction of the major axis; then a mark made at point m where the rod crosses the minor axis will be the *length* of the major axis for that curve.

A Wreath for a Half-space of Winders.—To obtain the face moulds. —Fig. 1 is a plan of the stairs, on p. 326, prepared with tangents, diagonals, &c.,

and f. 2 is the development of the same in elevation. The development of the string on p. 327 has been utilised to obtain the falling section of the handrail from which the tangent points 1, 2, 3, f. 2, are obtained; this development is not strictly correct, as the face of the string and centre line of the handrail do not coincide, as will be seen by referring to p. 326, but the difference is so slight in a small scale drawing as to be immaterial.

To find the inclination of the tangents.—Draw the development of the tangent planes, f. 2, previous page, placing the steps as they occur on the tangents in plan. Draw the rails resting on the nosings, and produce their centre lines towards the springings. Set the points 1, 2, 3, on the tangent plane edges A, C, E, at the same heights over the steps beneath them that they occur in the development, p. 327. These are the points where the centre of the approved rail, as indicated by the dotted lines, cuts the springings, and must coincide with the section or tangent points of the enclosing prism. See the perspective sketch, f. 1 adjoining, where A and C on the plane of section are represented in the developments by points 1 and 2. Straight lines joining points 1, 2, 3, will be the tangents for each half of the wreath-piece: these are almost but not quite in a straight line in this case, consequently the pitches being different, the bevels for each half wreath will be different; but as the tangents in both cases are a straight line so far as each half wreath is concerned, the pitches are alike, and one bevel only will be required for each. These are shown in the elevation, as are also the widths of the moulds obtained in the manner described in the previous case, which need not be repeated here. Fig. 3 is the face mould for the bottom wreath, and the projections at the ends show the direction the Bevels are applied, to obtain the lines for sliding the mould, as fully described in the next chapter. Fig. 4 is the face mould for the top wreath. The pitches for these being alike, the major axis of the elliptic curves, runs parallel with the diagonal of the face mould, and the minor axis at right angles to it through the centre X. On this line the rail is level, as will be seen on reference to adjoining sketch, where the line B X represents the minor axis, and is obviously level, as the points are the same height from the ground. If the widths of the mould have been set out correctly, the mould across the direction of the minor axis will be equal to the width of the rail in plan, as shown by the dotted circle, f. 4, last page.

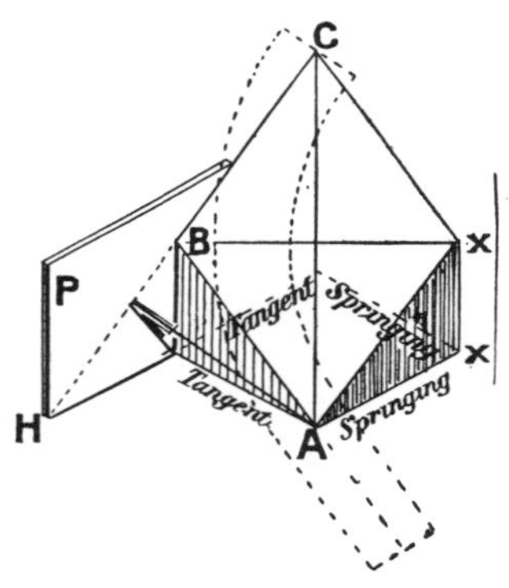

1. Illustration of the Theory of "Bevels" in Handrailing.

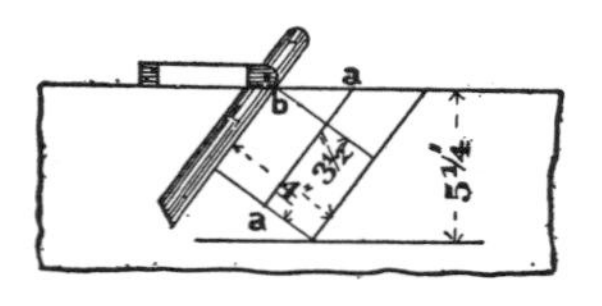

2. Method of Finding Thickness of Wreath-piece.

Particular attention must be given to the joints at C on each mould. These must not be made on the springing lines, but square with the tangent through the point where the tangent cuts the springing line.

To find the thickness required for the wreath.—Fig. 2 illustrates a general method. Draw on a board the line *a a* with the **Bevel** of greatest inclination: set off on each side of the line half the width of the straight rail, and draw parallel lines with the bevel. At point *b*, where the bevel side cuts the edge of the board, draw a line at right angles to the side and a similar line at a distance equal to the thickness of the straight rail, thus forming a rectangle. Through the lower corner draw a line parallel with the edge of the board, and the distance between them measured at right angles with that edge will show the thickness required to cut a wreath in the square at the given angle, which in the instance given is $5\frac{1}{4}$ in. to cut a rail $3\frac{1}{2}$ in. thick; but as most rails have their sections more or less rounded off at the corners, in practice a less thickness can be used, as shown in the projections at the ends of the face mould in f. 3 & 4, p. 345, the former showing the thickness required when the corners are left on, the latter when the rectangle is drawn just large enough to contain the section of the rail. This is shown clearer on p. 358, where the unshaded portion represents the section or templet, and the outlines the size of stuff needed to work it.

Wreath over Winders in an Obtuse Angle.—This is a case in which the tangents are not square in plan, and the wreathed piece of rail forms a small part only of a large cylinder. Fig. 1, next page, is the plan showing the risers around the well, the centre line of rail, and the tangents A, B, C. The rail is indicated by dotted lines. X is the centre, and X C and X A the springing lines of the curved rail. Fig. 2 is the development of the centre line of rail from which the elevation of the tangents is obtained, and f. 3 is the development of the tangents: both are obtained as previously described.

To draw the face mould, f. 4, make A′ C′ equal *a* C, f. 2, and A′ B′ and B′ C′ equal A B and B C, f. 2, which will complete the section. Produce the tangents beyond the springings to equal the length of shanks shown on the development, and draw the joints at right angles with the tangents. To find the position of the centre X, which, as will be seen by reference to the plan, lies outside the prism, draw the line B X, f. 1, crossing the diagonal A C at its centre, and draw a corresponding line on the section, f. 4, from B′ through the centre of the diagonal A′ C′. This line will pass through point X on the section, and its position on this line must be located. It will be found that the distance from B′ to the diagonal on the section is the same as the distance from B to the diagonal in the plan, and when this occurs it proves that the section is level in the direction B′ X, and therefore the distance of point X from B′ on the section will be the same as it is in plan. This is always the case when the tangents are inclined equally, and the reason will be clear on inspection of the prism opposite. Here the tangents are equally inclined, and the dotted line B X passing from corner to corner of the prism cuts the diagonal A C in its centre. That it is a level line is shown by the heights of the points B X being equal, and it would be directly over and equal in length to a line joining the corresponding corners in the plan B X. Having thus located the centre of the elliptic curves, the major axis is drawn through it, parallel with the diagonal and the minor axis at right angles. The remaining construction is the same as described in the previous example, with this difference, that to obtain the bevels, the distance of the tangent from the springing on the face mould is laid off on the plan, as shown in f. 1, instead of on the elevation, because in this case the develop-

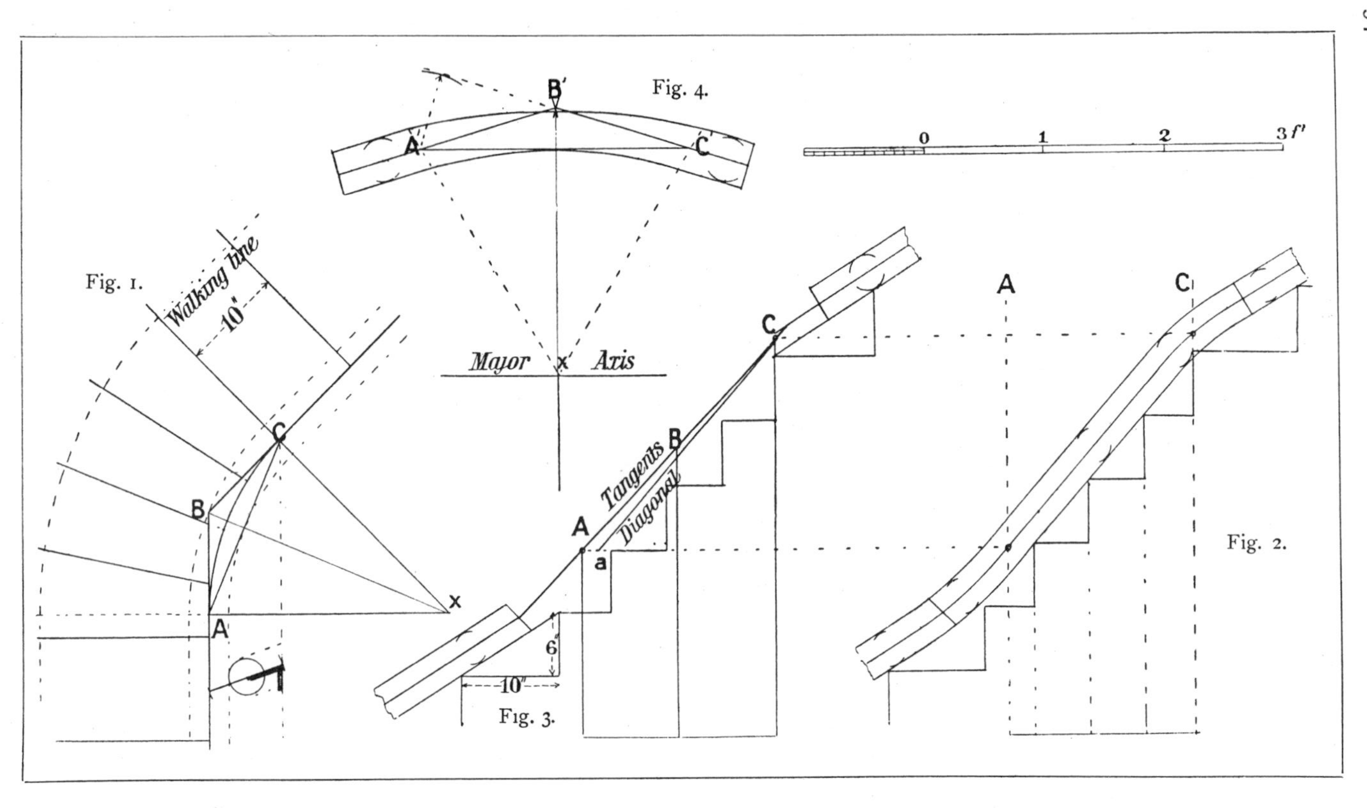

WREATH FOR WINDERS IN AN OBTUSE ANGLE.

1. Plan. 2. Stretch-out of Centre Line. 3. Development of Tangents. 4. Face Mould.

ment of the tangents do not show their actual distance apart when folded into position.

A Wreath over Circular Winders.*—This wreath is for the short rail shown on the left hand of the plan, f. 2, p. 319, and although a quarter of a circle only is dealt with, the method is applicable to a rail making a complete revolution. Fig. 1 hereunder is the plan of the centre of rail enclosed within tangents, and the risers radiating from them. Joints occur in the wreath at A and C, and the shank joint just beyond the springing E. The position of the joints locate the disposition of the tangents in plan, as the tangents must always pass square through the joints. The joint at A must be made where the curves

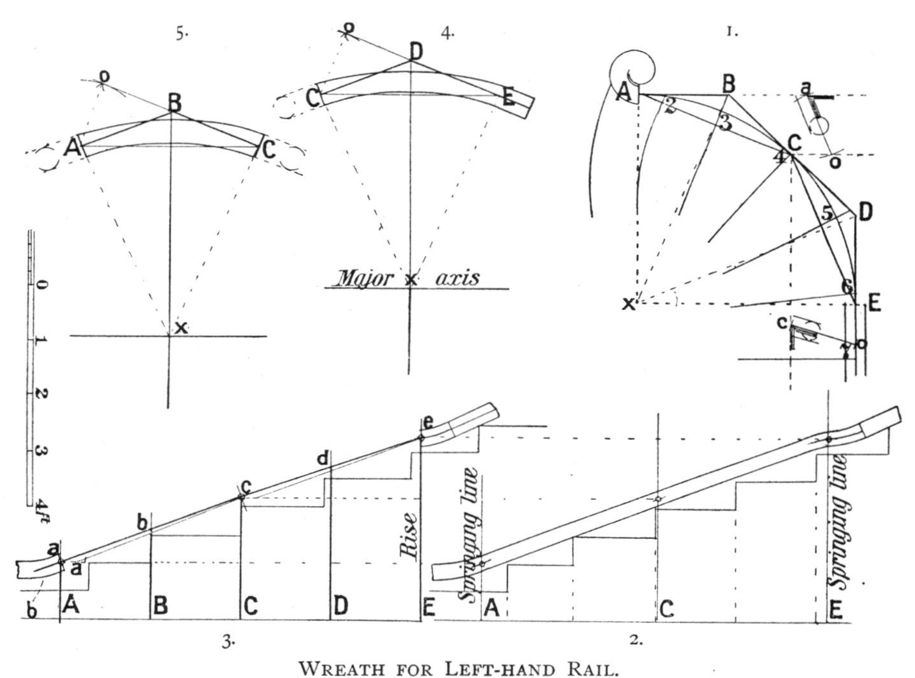

WREATH FOR LEFT-HAND RAIL.

1. Plan f. 4, p. 324, reproduced. 2. Stretch-out of Centre of Rail. 3. Development of Tangents. 4. Top Face Mould. 5. Bottom Face Mould.

of the rail in plan change their direction. The joint at C can be made anywhere near the centre, but is best arranged as shown when the sides of the tangent planes will be equal, which facilitates the working. Fig. 2 is the development of the centre line of rail, with the steps and springing lines as they occur thereon in the plan. The rail is lifted a little over the sixth riser, because if brought to meet the common pitch at the springing, the wreath would be lower over the

* A slight discrepancy will be noticed between the plan as here drawn and that at p. 319, the rail being here continued up the stairs for the purpose of showing the treatment required in such a case.

winders than the rail over the flyers, which arrangement, as before stated, must be avoided; so a ramp is here introduced, and at the bottom end the rail is placed so that the bottom of the scroll shall be 2 in. above the curtail step. This latter arrangement is a matter of taste, but it is usual to lift the scroll about 2 in. higher than the rail. The stretch-out of the rail thus becomes a straight line between the springings, which is the true geometrical shape of a regularly inclined line around a cylinder, the name of the curve so formed being a *Helix*.

Fig. 3, last page, is the development of the tangents in elevation produced as already described, points *a b* and *b c* being the points of the tangents for the lower half wreath, and *c d* and *d e* those for the upper half; *a c* and *g e* being the lengths of the diagonals A C and C E respectively of f. 1, found as previously explained. Fig. 4 is the top face mould, and f. 5 the bottom ditto, drawn similarly to the mould in the last example, f. 4, p. 348.

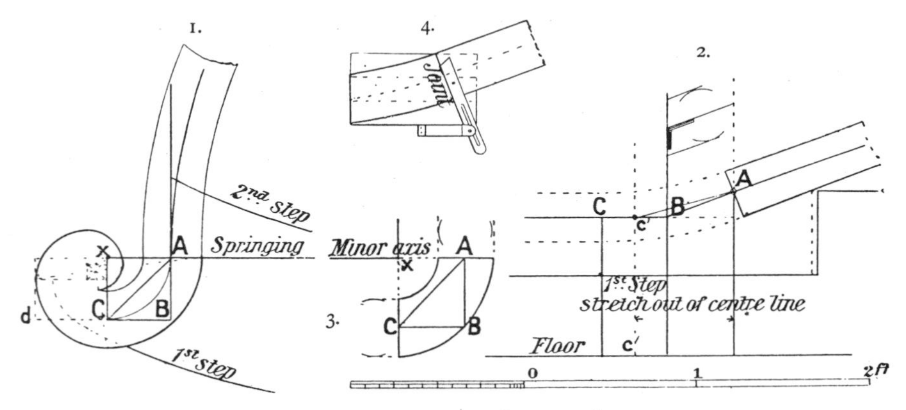

SCROLL WREATH FOR CURTAIL STEP.

1. Plan of Scroll. 2. Development of Tangents. 3. Face Mould.
4. Method of Making Joint in Scroll.

A Scroll Wreath for a Curtail Step.—This is the scroll completing the last wreath, and is drawn to larger scale in f. 1 above. The centres from which the scroll is drawn are indicated. The method is fully described on p. 362. The tangent A B is drawn in the direction shown in the plan, f. 1, last page, and the tangent B C at right angles with it, and touching the centre line of the scroll on the springing line drawn to the centre of the largest quadrant of the scroll. The diagonal A C completes the plan of the enclosing prism.

Fig. 2 is the development of the tangent planes. Superimposed on this in dotted lines is the stretch-out of the curved centre line. This is the usual workshop method, to save time. The previous examples have been drawn separately simply for the sake of clearness. The pitch of the rail above the springing is drawn the same as shown in f. 3, previous page, and the joint drawn through the springing square with the pitch. The scroll cap is usually lifted about 2 in. off the step, so that when fixed, the level part of the cap shall be

2 in. higher than the rail is, measured vertically over a riser, the latter height being commonly 2 ft. 8 in. to the back of the rail. The height being determined upon, draw the centre line level through the springing at C, f. 2, opposite page, and also the level tangent C B through the same point. The centre line of the upper rail is drawn down to meet this point, and determines the inclined tangent B A. Set off the length of the diagonal A C, f. 1, on the level line from B, f. 2, and draw the line C′ A for the diagonal of the face mould, f. 3. This is finished exactly as described in example 2. The length from C to X, f. 3, which is square with the tangent, is the length to lay off on the projection of the tangent plane to obtain the bevel as shown. The remainder of the face mould required to mark the cap beyond the springing line C X is cut exactly to the shape shown in the plan, and is jointed to the shank mould on the line C X.

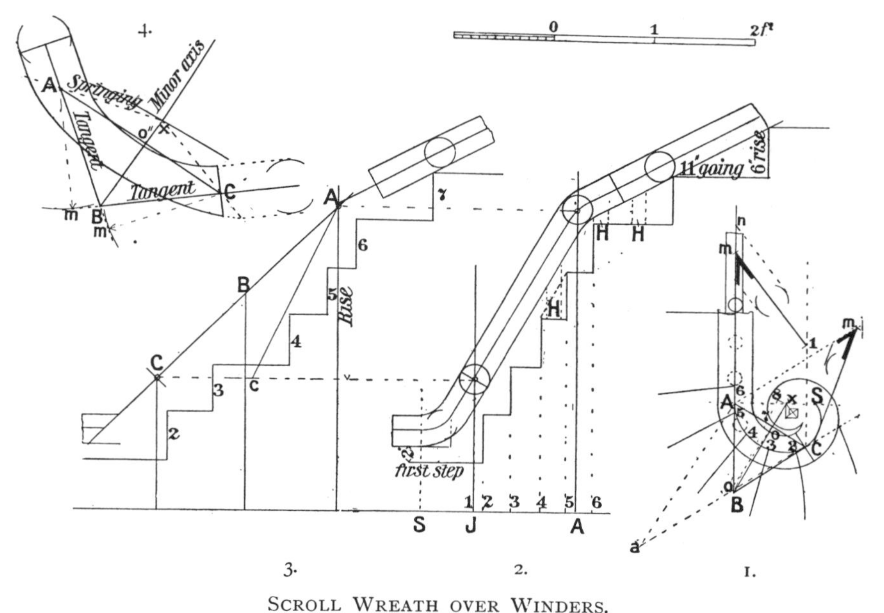

SCROLL WREATH OVER WINDERS.

1. Plan. 2. Stretch-out of Centre Line. 3. Development of Tangents. 4. Face Mould.

Fig. 4 opposite shows the method of finding the thickness required for the scroll block, the shank being drawn to the correct pitch. The width of the block measured on the plan parallel with tangent A B is drawn as shown. A level line is then drawn at the bottom of the pitch, allowing $\frac{1}{8}$ in. for fall into the level, and a similar line at the top touching the end of the shank at the joint. The rectangle shows the size of block required. The shank joint is made as shown by the bevel after the block is cut out square to the plan mould. The application of the face mould is referred to in the next chapter.

A Scroll Wreath over Winders.—This scroll is also for the stairs, shown in f. 2, p. 319, and is for the right handrail. Fig. 1 above is the plan of the scroll and risers radiating therefrom. The small circles represent balusters, and are introduced to show how the position of the handrail with reference to the string and curtail step is determined. A portion of the string beyond the sixth step is shown.

To determine position of the rail.—Draw the string and a baluster. If a plain cut string, the baluster must be flush with the outside of the string; if a bracketed string, flush with the face of bracket. Draw the centre line of the rail through the centre of the baluster, and mark half its width on each side of the line. The neck of the curtail step should equal in thickness the balusters, and the rail should overhang the curtail block equal to its overhanging of the string.

To draw the tangents. — Having drawn the scroll as explained on p. 362, and produced the centre line of straight rail, which will give the direction of one tangent, the position of the lower joint has next to be determined, as this governs the direction of the second tangent. This might be made on the springing line of the first quadrant, as in the previous example, but that as the fall is so much steeper, a very thick block would be required to cut the scroll, and also that the inside of the scroll and the outside of the shank would be very difficult to mould, so the joint is taken further into the eye, and made to radiate from the centre of that portion of the scroll; then the tangent is drawn through the centre square to the joint until it meets the other tangent at B. Fig. 2, previous page, is the stretch-out of the centre line of rail as far as the springing line S, f. 1. The line A represents point A in f. 1, and the line J the joint at C; the steps are drawn as they occur on the centre line. Two steps are drawn above the springing to obtain the pitch, the rail drawn on the nosing line, and the falling rail eased into it and into the scroll cap at bottom, where it is lifted 2 in. above the first step to correspond in height with the other side. Fig. 3 is the development of the tangents in elevation, their inclinations being determined from the stretch-out as in previous cases. The plane of section, f. 4, is obtained in the usual manner, the centre of the elliptic curves, point X, being determined by the following construction. Draw a line from B to X in f. 1, and as it does not pass through the centre of the diagonal A C, the place where it crosses the diagonal of the face mould must first be determined. Draw an elevation of the diagonal A C upon its plan, as shown by the dotted lines. To do this, draw A a perpendicular to A C. With C as centre, and the length A′ C′, f. 4, as radius, describe an arc cutting A *a* in point *a*. Draw the line *a* C, and from point *o*, where the line B X crosses the diagonal, erect a perpendicular cutting A C in *o*′. Measure the distance of *o*′ from *a*, and set it off from A on the diagonal at *o*″, f. 4, and draw the line B *o*″ indefinitely. This line will contain the centre X. To ascertain its distance from B, take the length B *o*″, f. 4, as radius, and with B, f. 1, as centre, describe an arc cutting the diagonal A C in point *7*: through this point draw a line from B, and produce it to meet a perpendicular raised from the centre X at 8. Then B 8 will be the real length of line B X on the section, and if set off from B′ on the line B′ *o*, f. 4, will be the position of the centre required. Lines drawn from A and C to this point will be the springing lines. The construction necessary to find the centre described above is shown more clearly in the sketch opposite, which is a reproduction to larger scale of f. 1 without the curve lines. The lettering is similar, and refers to the same points in each drawing. The bevels and widths of the mould are found in the manner described in previous examples, the lengths to obtain the bevels being set off on projections of the tangent planes in the plan, as described in the case of an obtuse angle, this example being the converse of that one.

To obtain the lengths of balusters over winders.—This may be done by aid of the stretch-out of the centre line of rail, as shown in f. 2, p. 351. Draw the balusters on the stretch-out as they occur in the plan in relation to the risers. The dotted lines H, H, H, indicate the positions of the three balusters shown in the plan, f. 1, and the vertical distance between the under side of the rail and the nosing line measured on the centre of the baluster will be the amount to add to the length of the flyer balusters, which are relatively in the same position. To ascertain the total length of each particular baluster over the winders in a regular fall, as shown in the example, one baluster only need be drawn, as the lengths of all are alike; but when the wreath does not follow the nosing line of the winders, then all must be drawn to ascertain their lengths. Measuring on the centre line is advisable, as the bevel on each side of the baluster is different, owing to the winding surface of the rail. These bevels can be obtained by projections from the plan, but are more accurately determined in practice by the application of a bevel to the edge of a plumb rule held in position on each side of the rail. In arranging the balusters in plan they need not be kept rigidly to the riser lines of the winders if any other situation will allow of their more regular spacing. This is shown in the example, where the baluster is made to stand in the middle of the fourth step. An equal spacing has the best appearance.

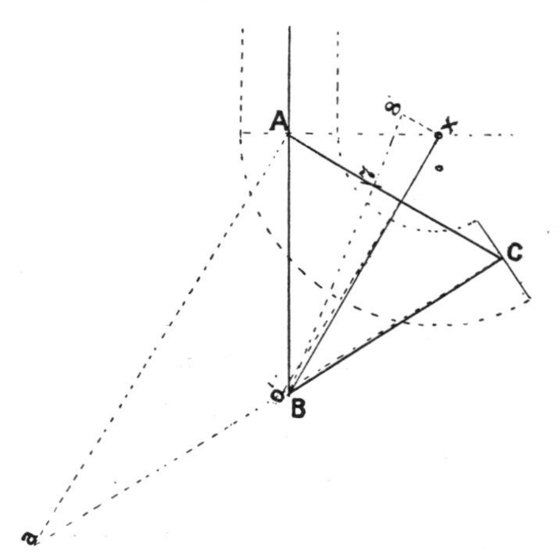

Plan Diagram of f. 1, p. 351, to Larger Scale, showing Method of Determining the Centre.

METHOD OF DRAWING FALLING MOULDS.—Falling moulds are the templets used to bend around the inside and outside of a wreath after it has been "bevelled," for the purpose of marking the top and bottom edges from which to "square" it. They are seldom used in these days by expert handrailers, except in the case of rails distorted or ramped out of their true geometrical curves, the workman in most cases being content to trust to his eye and hand for producing a good falling line; but as the ability to do this necessitates some experience, the tyro is often at a loss to know when his edges are right, so to assist him we have included an example of an inside and outside falling mould for the wreath shown in example p. 342, the method employed for obtaining them being equally applicable to other cases.

Fig. 1, next page, is a reproduction of the plan, f. 1, p. 342, and f. 2, next page, a reproduction of the stretch-out shown on the same plate. Divide the centre line of the wreath in plan into any number of equal parts, as 1 to 8; draw lines through the points radiating from the centre C; transfer these points by "stepping," as described on p. 333, to the stretch-out, f. 2, where they are numbered similarly, and draw perpendiculars from them cutting the centre falling line.

To obtain the inside falling mould.—Make a stretch-out of the inside surface of the rail in plan from 1′ to 8′, f. 1 below, as shown in f. 3, with the divisions as they occur thereon, numbering them similarly, and draw perpendiculars from them to intersect horizontal projectors from the correspondingly

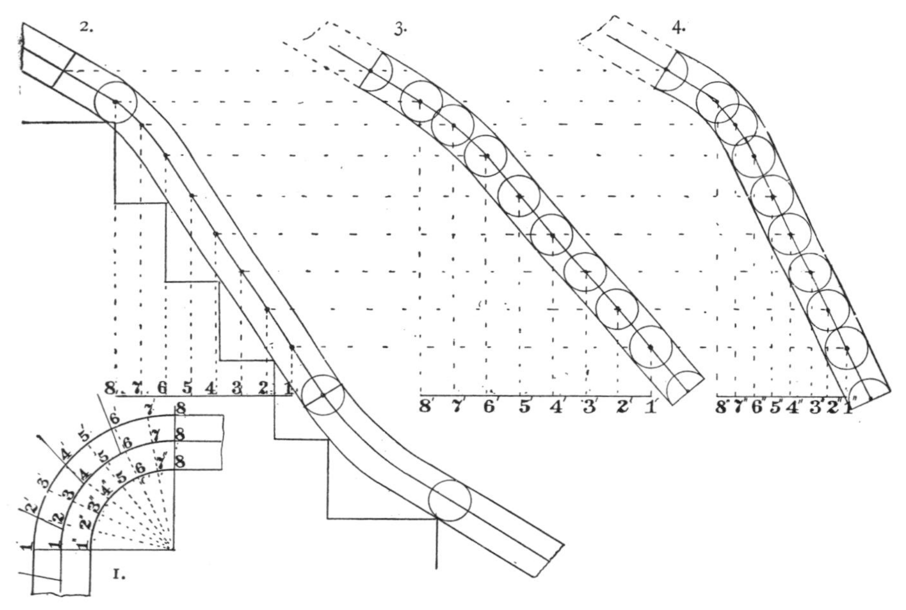

METHOD OF OBTAINING FALLING MOULDS.

1. Plan of Quarter-space of Winders. 2. Stretch-out of Centre Line of Plan.
3. Stretch-out of Inside Falling Mould. 4. Stretch-out of Outside Falling Mould.

numbered points on the stretch-out of the centre line. These intersections will provide a series of centres from which to describe circles equal in diameter to the thickness of the rail, and two parallel lines drawn touching their peripheries will give the outline of the falling mould, as shown in f. 3. The outside falling mould is obtained in like manner, as will be clear on inspection of f. 4. The application of these moulds is dealt with in the next chapter.

To obtain the projections of a wreath.—It is sometimes advisable, when the effect of a given falling line is doubtful, to make a projection of the rail in the "square," in one or more positions, when the appearance of the proposed rail may be better judged than it can be from the developed centre line. The method of doing this is illustrated in f. 1 to 4 opposite, where the wreath shown in f. 1, p. 342, is projected in two positions. Draw the plan, f. 1 below, which is reproduced from f. 1 opposite, and the same points of division utilised, but the number is immaterial; and also draw the stretch-out of the centre line of the rail, f. 2 below, which is the same as f. 3 opposite, although here reversed for convenience. Draw the projectors 1, 2, 3, &c., right across the falling mould to obtain corresponding points on the top and bottom edges; and draw projectors from the points of intersection of the radial divisions with the outside and inside curves on the plan, to meet correspondingly numbered projectors from the stretch-out, f. 2 below, and draw curved lines through the points of intersection so obtained which will show the appearance of the rail in the given position, as shown in f. 3. Note, it is advisable to project one edge only at a time, and

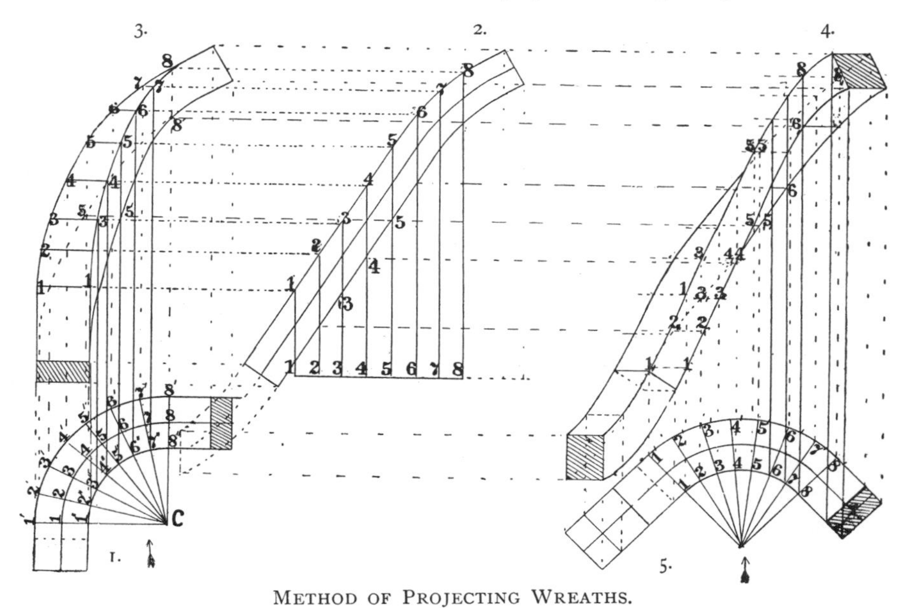

METHOD OF PROJECTING WREATHS.

1. Plan of Wreath, f. 1, p. 342. 2. Stretch-out of Centre Line of Rail in Elevation. 3. Elevation of f. 1. 4. Elevation in New Position. 5. Plan in New Position.

having obtained the necessary points, obliterate the projectors, and proceed with another edge in like manner. Only a few of the projectors have been drawn, to avoid confusion, but sufficient are given to show the method employed for the remainder. Full lines are used for the outside curve, and broken lines for the inside curve, to make them clearer. The projection, f. 4, is obtained similarly from the plan, f. 5, arranged in a new position, the view in each case being in the direction of the arrows.

CHAPTER XXI.

HANDRAILING—PRACTICAL WORK.

How to Prepare the Face Mould—Marking the Wreath—How to Make the Joints—How to Mark the Wreath for Bevelling—Use of the Tangents—Sliding the Mould—How to "Bevel" the Wreath—How to "Square" the Wreath—To Prepare the Section Templet—How to Apply Bevels—To ascertain Correct Direction of Bevel—Fitting the Bolts—Testing the Joints—To Obtain a Good Falling Line—Position of Square—Use of Callipers—Method of Moulding the Wreath—Tools Required—Thumb Planes, &c.—Dowels, when to Fit—Double Pitches—How to Slide the Mould—Precautions to Take—Use of Cylinder in Large Wells, &c.—Use of Falling Moulds—Cutting Wreaths by Machinery—Methods of Drawing Scrolls—Forming Easings.

IN the preceding chapter the methods of obtaining the various moulds and bevels required for given arrangements of tangents are fully explained, and this chapter is devoted to the explanation of the method of applying the moulds and bevels, and the practical work involved in the formation of wreathed rails by hand work. The latter part, being very much the same whatever the shape of the rail, the easy example shown in f. 3, p. 335, has been selected to illustrate these instructions.

Preparing the Face Mould.—This should be made of a piece of clean dry pine free from shakes and not more than $\frac{1}{8}$ in. thick. The shape may either be trammelled directly on the stuff or a tracing taken from the drawing and glued to the face. If this method is adopted, care must be taken not to stretch the paper, or the curves will be altered. The tangents and springing line must be drawn on each side of the mould carefully, so that they coincide, but the "diagonal" is not needed. The ends of the tangents and springings should be squared on to the edges of the mould, and a $\frac{1}{2}$-in. hole bored through each tangent. Only one mould is necessary. Where there are two shown in the drawings, it is merely to indicate two positions of the same mould.

To Mark and Cut the Wreath.—The mould being prepared, apply it to the face of the plank and mark its outlines, knife-cutting the joints. Cut the piece out with a compass saw, square through the plank. Face up the top side out of winding, again apply the mould in the same position as at first, and prick off the ends of the tangents on the face. From these points the tangents can be drawn on the piece, and the wreath will be "lined out," presenting the appearance shown by f. 1 opposite. Square the joints from the face and the tangents, using the square to the latter, as shown in the sketch. (Note, do not fall into the error of squaring the tangents from the joints, but always draw the tangents first and square the joints from them.) The joints being finished, square the ends of the

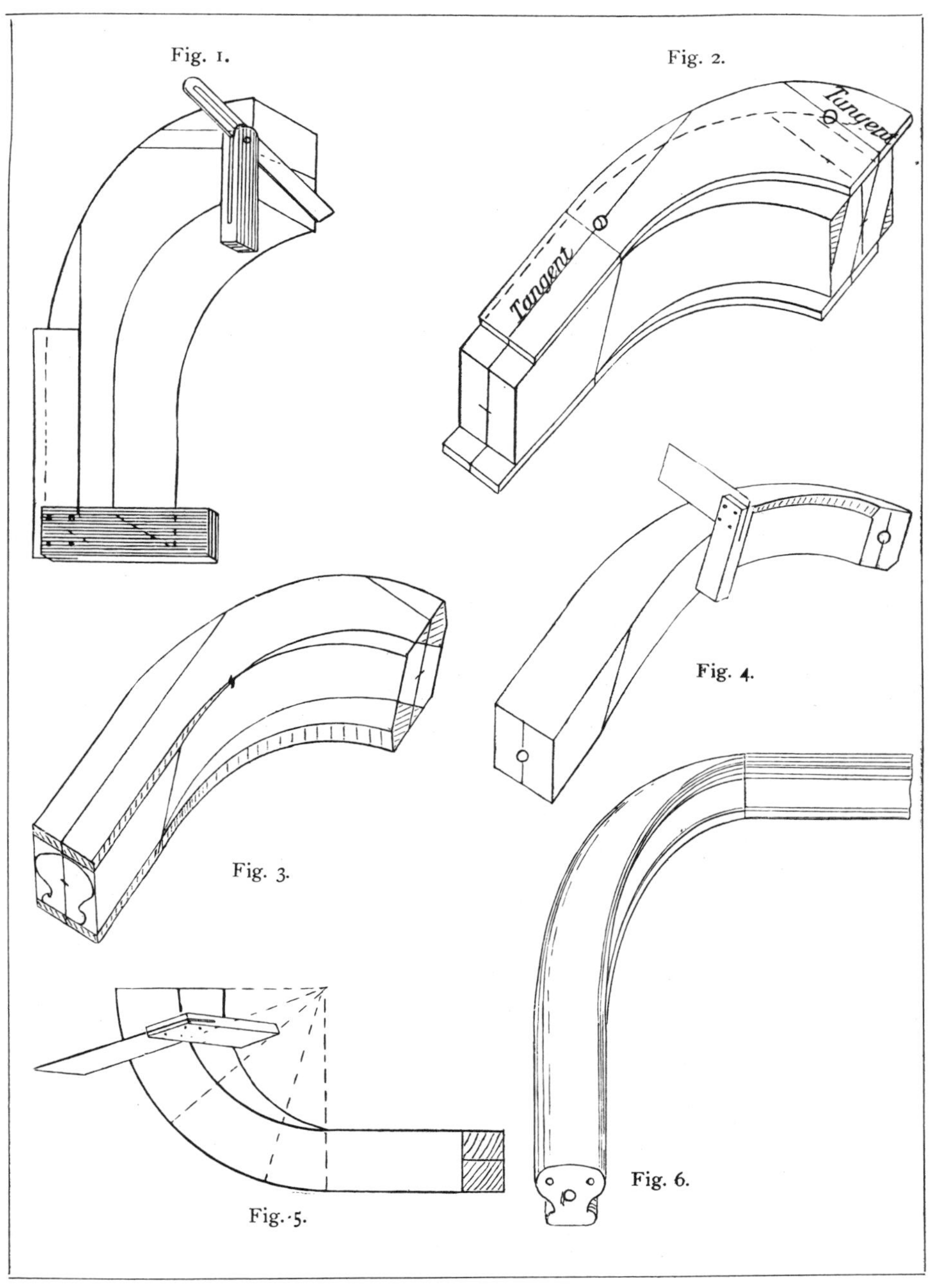

1. Making the Joints. 2. Marking for Bevelling. 3. Wreath Ready for Squaring.
4, 5. Position of Square in Squaring. 6. The Finished Wreath.

tangents across the joints, apply the mould on the under side, and mark the tangents on that side also. Next set a gauge to half the thickness of the wreath (we have shown on p. 346 how the exact thickness is obtained, but a rough-and-ready rule is to cut the wreath out of stuff as thick as the rail is wide, except when the rail is square or thicker than it is wide, when this rule will not work), and run it on the ends across the centre lines squared down from the tangents. Next apply the Bevel across the top or wide end, as shown in f. 1, last page, making the blade pass through the central point found by the intersection of the gauge and centre lines. Mark the bevel line on the end, and where this line cuts the top and bottom surfaces are the points from which new tangents are to be drawn parallel with the old ones. The new tangent on the face side is shown in the sketch. The old one should be crossed or rubbed out to prevent mistakes.

Marking the Wreath for Bevelling.—Apply the mould on the face of the wreath, as shown in f. 2, so that the tangents on the mould lie directly over the tangents on the wreath. The holes in the mould will enable it to be seen when they coincide. In this case the mould moves only in one direction, upwards, as there is only one inclination in the plank. This process is called "sliding the mould." Mark the outline of the mould upon the wreath, which can only be done on its concave side, and prick in with the marking awl the position of the springing line. Next apply the mould to the under side, this time sliding it downwards as shown in the sketch, and mark the outline of the convex edge. Prick in the springing as before, and draw a knife line on the outside, joining the springing points: this line is henceforward called the springing line. Fasten the mould to the wreath with two small screws or tin tacks.

Bevelling the Wreath.—Place the wreath in the bench screw and cut away the wood upon the concave or outside, between the edge of the mould and the line marked by the same on the top surface of the wreath; this part is shown shaded on the end of f. 2. The cutting may be done with a bow saw or chisels, as the operator pleases, and finish up the surface true with a spokeshave; applying the edge of a square to test the surface in a direction parallel with the bevel line at one end and the springing line at the other. Worked thus, the edges of the rail become parallel with the bevel throughout or vertical when set up to the proper inclination. Care must be taken when using the spokeshave not to take the edge off the mould. Having finished bevelling the outside, remove the mould and fasten it on the face in its original position, as shown in the sketch, and work off the inside of the wreath in similar manner, when the sides will be parallel and the wreath will be "bevelled" ready for the next stage, "squaring."

A Section Templet for Marking Ends of Wreath.

A Section Templet of the rail should be provided, which may be made of zinc, or of cartridge paper brushed over with thin glue, and allowed to dry, after cutting to shape. This templet should have three fine holes made in it, one in its centre for marking the bolt hole, and two others for the dowel holes, placed, as shown above, at a quarter of its greatest width from each edge, on a line drawn through its greatest diameter. A centre line should be marked on each side of the templet perpendicular to its bottom edge, and a face mark placed on one of its sides to identify it. When using the templet, as described below, pass a pin through the central hole, and enter its point in the centre of the rail, and then adjust the templet so that its centre line lies accurately over the bevel line on the joint, and prick in the dowel holes and mark round the outline, noting which side of the templet is up, and in applying it to the corresponding face of the other joint turn it over so that the former top side is now under. This reversing ensures the correspondence of the dowel holes, and if there is any difference in the outline it will be continuous throughout the rail. Note, the finer the holes in the templet, the more accurate will be the result.

To Mark the Wreath for Squaring.—Apply the templet as described in the former paragraph on each end of the wreath, with its centre line on the centre line of the wreath at the shank end, and on the bevel line of the wide end. Lightly mark its outline, but do not prick the dowel holes at this stage. Square lines on the joints from the outside of the wreath touching the top and bottom of the section templet; these lines are shown in f. 3, p. 357, and at the points where these lines cut out on the *original* surfaces of the wreath-piece, mark lines thereon by aid of the face mould, which must be slid backwards and forwards as required, until it coincides with the points, the other end being kept on the tangent, as in marking the outline in the first operation. It will only be possible to mark one line on each side thus, in consequence of the other end of each line falling on the *side* of the wreath. At the shank end draw the lines parallel with the top surface up to the springing line, when the rail will present the appearance sketched in f. 3, the shaded parts indicating the wood to be removed after the next operation.

Jointing the Rail.—Having marked the square lines on the ends as described, cut sufficient off the under side at the top end to obtain a flat surface for inserting the handrail nut. Do not cut quite up to the line, as a precaution against the bit running, when boring the holes for the bolt, an event which will necessitate a slight shifting of the wreath from the centre of the rail, when the additional stuff at the bottom may be useful. Keep the mortises for the nuts as small as possible, and bore the bolt holes in line with the tangents on the surface. Refrain from boring deeper than the bolts require. $4\frac{1}{2}$ by $\frac{5}{16}$ or $\frac{3}{8}$ in. are the usual size bolts, and for these the nuts can be placed at 2 in. from the ends. Bolt the wreath to the straight rail at the shank end, and apply the Bevel used throughout to the under side of the straight rail, and see if the springing line on the wreath coincides with the blade of the bevel; if not, the pitch is wrong, and the joint must be altered until it is correct. Generally it is better to make the alteration on the straight rail. See also that the sides of the shank are parallel with the straight rail; or better, draw a centre line on its back, and make this line range with the tangent on the wreath. The joint at the top end, if made carefully to previous instructions, should only require testing for square. Bolt up the rail, and with a 2-ft. square held against the inside, try if the blade is parallel with the shank: if not, alter the joint. Again apply the templet, mark the dowel holes and bore and fit in the pins, when the piece will be ready for squaring.

Squaring the Wreath.—The wood outside the rectangular section shown on the end of the rail in f. 3 has to be removed with saw and spokeshave in such a manner as to produce a rail of the appearance shown in f. 4, p. 357. The exact process is rather difficult to explain in writing, and will perhaps be best understood by first stating what it is that is wanted. The top and bottom of the rail should present regular curved surfaces from joint to springing, neither having lumps or hollows, and everywhere in a direction radial from the centre or imaginary axis, the surfaces should be straight across, not rounding, or expressed differently; if the rail were cut in two at any point, on a line radial with the centre of the cylinder, and parallel with the springing line on the side, its section should be a rectangle. Only the two lines previously mentioned as marked on the faces, can be used as guides in producing the surfaces, and the workman must trust to his hand and eye to obtain a regular falling line. Irregularities may be detected by drawing the palm of the hand slowly over the surface, and thus located, can be marked and removed. To obtain the true section a square must be applied to the concave side of the rail, as shown in f. 4 & 5, p. 357. The stock must be everywhere held parallel with the springing line or the bevel line on the joint, and the blade must radiate from the centre of the curve, as shown in f. 5, where the dotted lines indicate the direction of the blade of the square at different points. This really amounts to keeping the square plumb, when the wreath is pitched up at its proper inclination, and is the secret of the whole operation. It is not important which edge is done first,—personally the author prefers to work the under edge first, as this sometimes has to be fitted to an iron core, and if this edge is shaped correctly the upper surface can be partly gauged parallel with it. The use of a pair of callipers set to the required thickness helps to ensure the parallelism of the rail; they should also be applied radially. The wreath being squared, the next step is to mould it.

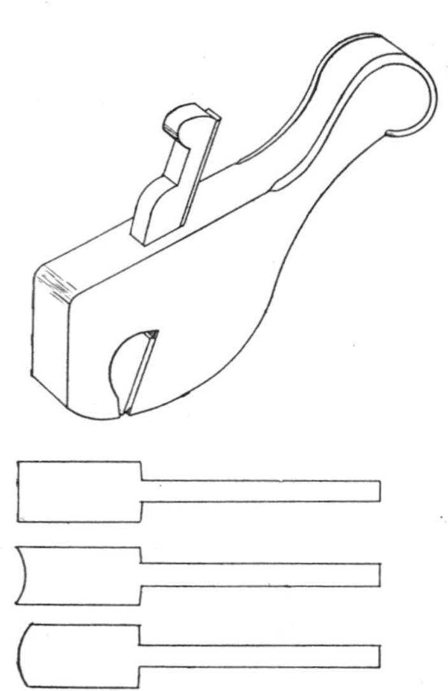

Thumb Plane and Irons for Moulding Wreaths.

Moulding the Rail.—For this purpose a great variety of tools are used by various workmen, including straight and bent gouges and chisels, quirk and bead routers, thumb planes of various sections as shown at side, shaped scrapers, rifflers or bent files, shaped cork rubbers and scratches (pieces of steel shaped to some part of the contour, and fixed in a kind of spokeshave). Of course these are not all required for the simple wreath we have been considering, nor would they all be used in any case by one man—each has his own particular fancy; but in every case the bulk of the "core" is removed with bent gouges or chisels, as most convenient, and as a preliminary proceeding, as many chamfers as possible should be worked around the edges touching the curves, as these will provide working surfaces or gauge points throughout. Bolt the wreath to the rail at one end, and carefully work with small chisels its section on the end of the wreath for about ¼ in. Remove the rail and fix on the other end, which treat likewise, then the straight rails need not be touched again until the final cleaning off. Next run in the quirk router to the deepest sinking, and so obtain a guide for other parts, remove the core roughly with gouges, and shape with the thumb planes, cleaning off with the rubbers. Cross-grain may be shaped up with the rifflers; the wreath should be bolted to the straight rail when papering up to finish off the joints. Fig. 6, p. 357, shows the appearance of the finished wreath, and the photograph hereunder, is that of a half full size model of a rail of similar section made to fit over a half-space of winders. Some of the tools mentioned above, and used in its production, are included in the photo, as is also the head of the bench screw, a special form used by handrailers to enable them to work on both sides of the wreath. As will be seen, it rises some 9 in. above the bench top. The above directions are for producing the wreath with face mould only, and is the usual method employed; but if it were desired to use falling moulds with which to mark the edges for squaring, they would be obtained as described on p. 353, and cut out of stiff paper or beech veneer, and bent around the wreath after bevelling, a centre line being drawn on the end of each, and this kept to the centre of the springing and to a line squared through the centre of the joint. The stuff would have to be thicker than that shown in the sketches, so that no "wane" would occur on the edges after bevelling.

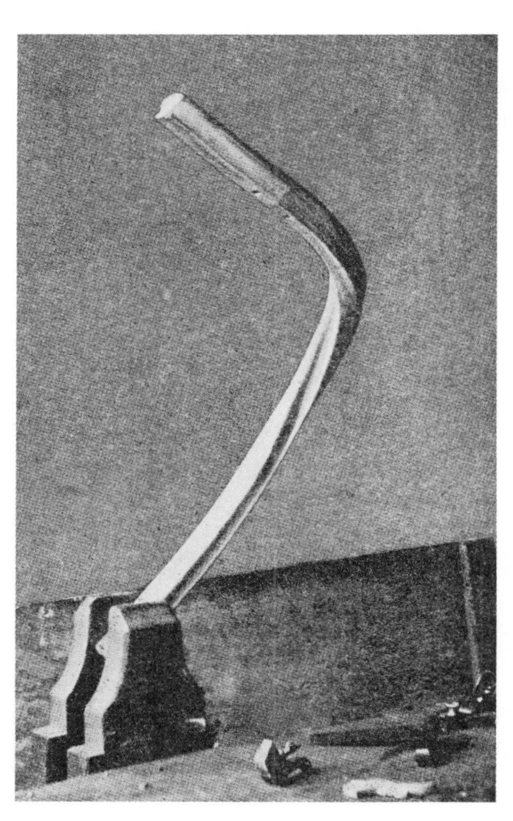

Model of a Finished Wreath for Half-space of Winders.

The Application of the Bevels and the moulds in the case shown on f. 1, p. 339, is similar in every respect to the foregoing, with the exception that in the upper half wreath, the mould slides towards the shank on the top side, and away from it on the bottom side, just the reverse of the lower half.

The Direction of the Bevels can readily be ascertained in any wreath, by tilting the rail to its approximate pitch, and applying the Bevel on the top side, when, if the bevel is in the right direction, the blade will be plumb, whilst if wrong, it will be horizontal. When making the centre joint in this and similar wreaths over half turns, two straight-edges should be fixed with clips upon the springing lines at the outside of each half, and their edges taken out of winding. This will ensure that the pitches are right, and if the tangents on the two pieces be in one straight line, the shanks will be parallel.

The wreath shown in f. 4, p. 342, requires two bevels, and care must be taken to apply them at the right ends. New tangents will be required at each end, and the mould slides upwards and outwards on the upper side, as shown in the sketch below. The wreath is here shown marked ready for bevelling, the mould being drawn in two positions. Remember that the tangent lines on the mould must lie over and be parallel with the tangents on the wreath: the fact that the ends of mould do not coincide with the joints is immaterial.

In Circular Stairs or other large turns, a light **Cylinder** should be made to fit the plan of the concave side of the rail, and long enough for one rise around the well. This should be stood on the rod, and the springing points marked on it, and vertical lines drawn on the surface therefrom. The amount that the wreath rises either around the well or in one revolution, as shown by the development

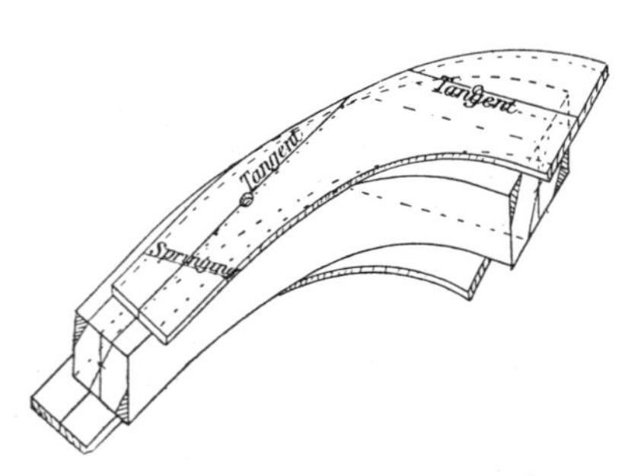

1. Method of Applying Face Mould for Bevelling Wreath of Two Pitches.

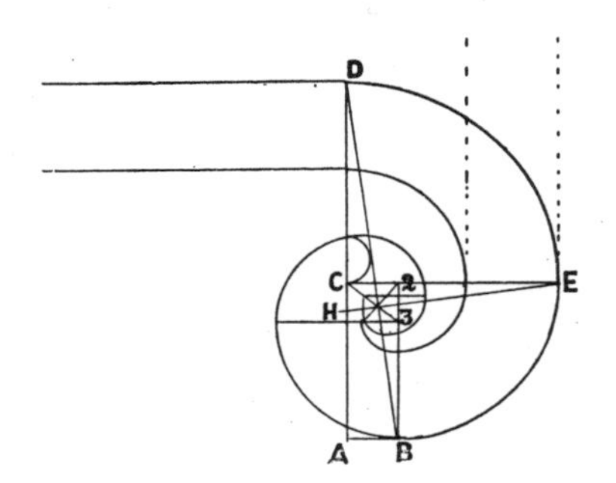

2. Method of Drawing a Scroll.

(see the distance marked "rise" in f. 3, p. 349, and f. 3, p. 351), should be set off on these springing lines, the one point perpendicularly over the other, and either the centres or the bottom edges of the wreath on each side kept to them when fitting the joints; this is a surer method than trusting entirely to the bevels to give the right pitch in such cases.

Machining Rails.—Any of the wreaths shown in the preceding chapter may be bevelled in the band saw by first cutting a prism or bed to the bevels given, as shown in f. 5, p. 342, and described fully in Chapter VII.; but the operation scarcely pays for itself unless there are more than one to the same pitch. The rail is better squared by hand, but when squared, can be readily moulded on a "spindle" machine. See p. 127.

Method of Drawing Scrolls.—The involute curve, known as the spiral scroll, is best drawn by winding a string around a cone, fastening a pencil in a loop at the end, and fixing the cone upon its point on the paper, unwind the string by passing the pencil around the cone, keeping the string taut and the pencil upright, and a continually and regularly increasing curve is the result.

This may be imitated to some extent by drawing the curve by means of a series of arcs of circles with continually diminishing radii, the latter being determined by an arbitrary arrangement of the centres. Fig. 2 opposite shows one method, f. 1 below being an enlargement of the series of centres shown therein. Having determined how far the scroll is to project from the stairs, draw the rail, and let A D, the springing line, equal the width desired. Divide this into eight equal parts (ten for large scrolls). Draw A B at right angles to A D equal to one of these parts. Join B D, and upon the centre of A D describe an arc touching A B and cutting A D in C. From C as centre and C D as radius describe the quadrant D E. Draw the line E H at right angles with B D. Join C E. Draw a line from B parallel with A D, cutting C E in point 2; this gives the second centre of the scroll. With 2 as centre and 2 E as radius, describe the quadrant E B. From C draw a line through the intersection of E H with B D; produce it to cut B 2 in point 3, which will be the third centre. With 3 as centre and 3 B as radius, describe the third quadrant. Draw a line from 2 through the intersection, cutting a line from 3 parallel with C E. This will be the fourth centre, and successive centres can be found by drawing parallels to the sides previously drawn, as shown. The same centres answer for the outside of the rail, and also for the curtail step beneath.

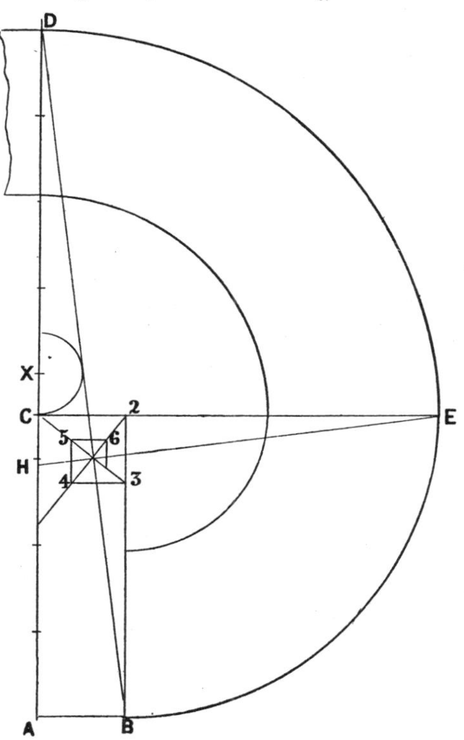

1. Centres for Scroll, f. 2 opposite.

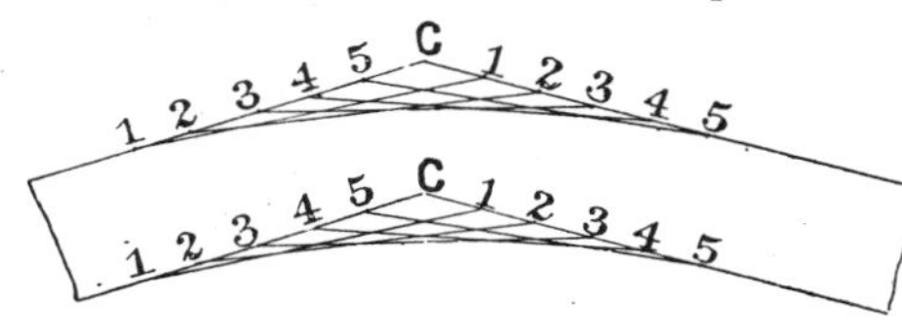

2. Method of Marking Circular Easings.

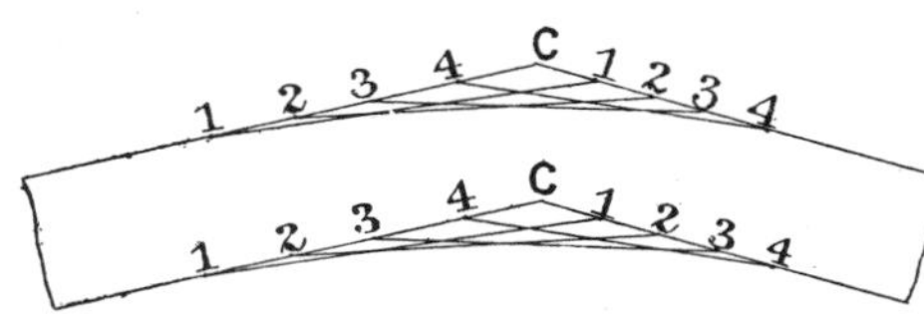

3. Method of Marking Parabolic Easings.

Easings of Rails.—The curves for these may be drawn by bending a thin straight lath to a suitable curve, touching the two straight parts to be connected, and running a pencil around it, or the curve may be produced by the intersection of a number of straight lines, as shown in f. 2 above. Set off on each side of the point of intersection of the directions produced (C, f. 2, previous page) the distance the curving is intended to go, and divide these spaces into the same number of equal parts, numbering them in reverse direction. Draw straight lines to the corresponding numbers, and their intersections will produce an approximately circular curve. When it is desired to have the easing to run further in one direction than the other, a parabolic curve is produced. Set off the desired distances on each side of point C, as in f. 3, and divide them both into the same number of equal parts, which will of course be of different lengths. Number these as in the former case, and connect them by straight lines in similar manner.

Easings may also be drawn by describing a circle which shall be tangent to the two straight lines it is desired to join up by a curve. Having drawn the two lines at the required angle, mark off from their point of intersection, as in the previous case, the distance the curve is desired to run, and from these points raise perpendiculars to the two lines: produce the perpendiculars until they intersect, and the point of intersection will be the centre, and its distance from either line the radius of a circle that is tangent to both lines.

CHAPTER XXII.

MOULDINGS.

Definition of Moulding—Design—Requirements of Position—Description and Methods of Drawing Mouldings of Various Types—Classic—Gothic—Renaissance—Modern—Their Characteristics—Methods of Copying—Enlarging and Diminishing Mouldings—Methods of Obtaining Sections of Simple Raking, Compound and Curved Raking Mouldings with, and without Drawings—To Obtain the Section of a Wide Moulding that will Mitre with a Narrower one—Methods of Mitreing Curved and Straight Mouldings Together — Sections of Angle Bars—Tracery, Description — Basis of Design—Method of Preparing Design—Method of Working Tracery Panels—Pierced Tracery—Chamfer Stops—Method of Working a Moulding.

ANY shaping of the rectangular edges of material, by which its contour is altered to a curved one, and that contour is continuous for some distance, is A MOULDING. It is this continuity of section which distinguishes a moulded, from a carved edge, the latter having a continually changing contour. A moulding with a portion of its surface carved is said to be "enriched." Mouldings depend for their effect mainly upon the shadows cast by their members, or the reflection of light from their surfaces, and therefore the position which a moulding has to occupy has a great influence upon its section; and in designing mouldings, less attention must be given to the production of a pleasing outline in the section than to the study of the *effect* of the outline when in position. A simple but convincing example of the difference in effect produced by a shadow is shown in the case of an edge merely rounded or nosed, and a similar edge with the round carried down into a quirk, as in a bead. In the first instance, when the surface is stood upright facing the light, only an indefinite edge will be visible; in the second, the quirk will provide a sharply defined margin of shadow throwing into high relief the brightly lighted round edge. Only the briefest outlines of the treatment of mouldings can be given in the limits of this chapter, as it would require a treatise to deal fully with the subject.

Mouldings that run horizontal, and are placed much above the level of the eye, should consist chiefly of deep coves or hollows, wide flats with narrow soffits at right angles: for the light reaching them from below, will throw all the soffits into grey shadows or half tones, and rounds in this position will appear as flats, unless bounded by deep quirks.

Vertical Mouldings should have rounds predominating, separated by small hollows and flats.

Horizontal Mouldings that are slightly below the level of the eye should have the upper members nearly flat, with narrow fillets between them. The under members should recede sharply and be of bold outline, chiefly rounds.

Horizontal Mouldings well below the eye should project most at the base, the mouldings separated by or surmounting wide flats. Ogee or other curves of contrary flexure should predominate, and the members be large and well defined. Sloped, rather than level edges should be employed, both to reflect the light and avoid the accumulation of dust thereon.

Architrave Mouldings act as a frame to an opening, and to be effective, should be large and dominant, forming a strong margin to the flats of the surrounding walls, and contrasting with the smaller mouldings of the enclosed framings. Bold projecting members separated by deep sinkings should be used. Fig. 2, p. 373, is a good example. A thin flat moulding like f. 3 is weak and ineffective, except as a border to a window, where the frame itself affords a substantial margin.

Panel Mouldings should be just the reverse of architrave mouldings in their effect, their office being to soften the abrupt changes of level in the framing, and break up the flat surfaces rather than emphasise them, and their best effect is obtained by the use of numerous small members, rounds in high relief and square fillets predominating. Deep hollows are unsuitable, as the various positions of the different portions cause them to throw shadows on the one side and high lights on the other, thus apparently breaking the continuity of the members.

Bolection Mouldings are exceptions to the above rule, as their purpose is to add to the apparent depth of the panel and thickness of the framing, seemingly reduced by the raising of the former. They approximate more nearly to architraves in their effect, subduing the preponderance of the panel, which would otherwise be obtrusive. They are meaningless and out of place on a sunk panel, giving the framing a spotted appearance. To be effective, they should be thick and comparatively narrow, with the dominant member outside.

Mouldings may be classified as Ancient or Classic, Mediæval or Gothic, Renaissance, and Modern.

CLASSIC MOULDINGS are the forms found ornamenting the remains of ancient Greek and Roman architecture, and these only exist in stone and marble. Whether they were applied to wood is doubtful, but the forms are largely used as a basis in the designs of modern mouldings. Altogether only nine distinct or definite forms have been discovered, and as these are always found in certain combinations, it is assumed that these were the only ones known. The chief difference in the mouldings of the two races is in the contour of the curves, not in their position or arrangement. The Roman curves all appear to be portions of circles, whilst the Greek curves are similarly portions of conic sections, viz., the ellipse, parabola, and hyperbola. No evidence exists that they were mechanically described, and it is the general opinion among archæologists that they were all drawn freehand ; but the curves can be reproduced approximately by means of these geometric sections. The first moulding or member shown on the plate of classic mouldings is the **Listel** or **Fillet**, f. 1 and 2 opposite. This is a narrow rectilineal band used as a finish to or separation of different members of a cornice. It is usually the crowning member.

The Astragal or **Bead**, f. 3 and 4.—The latter name is given because it is frequently carved into small balls or beads. In both varieties the profile is a semicircle. This mould is the prototype of the modern bead, which,

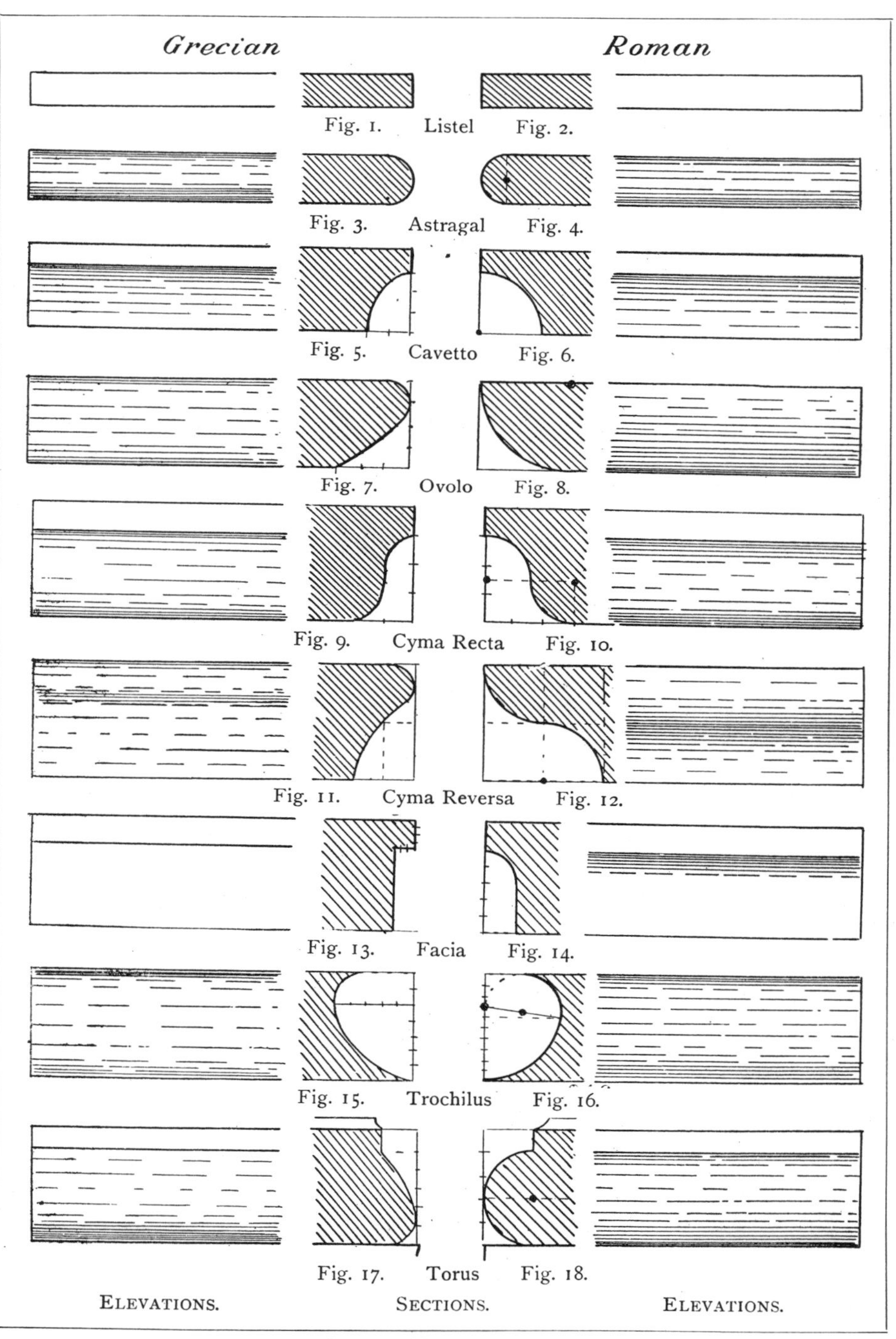

CLASSIC MOULDINGS.

however, is always made flush with or sunk below the surface to which it is attached, and used in conjunction with a quirk or sunk fillet. The original astragal always projects half its diameter.

The Cavetto, Cove, or **Hollow**, f. 5 & 6, has a section that is a quarter of a circle in the Roman variety, and a quarter of an ellipse in the Greek. The ratio of projection to height in the latter is as 2 is to 3.

The Ovolo or **Round**, f. 7 & 8, is the reverse of the cavetto, and is described in the Roman variety with a radius equal to its depth. The Greek moulding is a hyperbolic curve, the ratio of projection to height being as three parts are to four parts.

The Cyma Recta, **Wave,** or **Ogee**, f. 9 & 10, is a curve of contrary flexure, formed by two quadrants of circles in the Roman, and two quarter ellipses in the Greek form.

The Cyma Reversa, f. 11 & 12, is the above mould in reversed position in the Roman form. In the Greek the curves are parabolic and hyperbolic, quirked on the upper edge.

The Facia or **Corona**, f. 13 & 14.—In the *Greek* the height is divided into four equal parts, and the listel made equal to one of these, and its projection three-quarters of its height. In the *Roman* the listel equals two-sevenths of the height, and projects equally, being connected to the facia by a quarter hollow. The second name is used when the moulding is a member of a cornice, and projects considerably, in which case its soffit is recessed to form a drip around the front edge and ends.

The Trochilus or **Scotia**, f. 15 & 16, is a hollow moulding of unequal curvature, the upper part being a quicker sweep than the lower,—it is usually drawn as a rampant ellipse. To draw the Roman Trochilus, divide the height into ten equal parts; take seven of these as a radius, and with the seventh division from the bottom as a centre, describe an arc, and cut this by a horizontal line drawn from the sixth division. Join the intersection to the centre of the arc by a straight line, bisect this line, which will give the centre for the remaining portion of the curve. The modern adaptation of the scotia is made much shallower, and with a fillet at its upper edge.

The Torus, f. 17, is a parabolic curve, the projection being respectively one-fifth and two-fifths of the height.

The Roman Torus, f. 18, is similar to the astragal, but much larger, and is always used in conjunction with a fillet on its upper edge.

MEDIÆVAL MOULDINGS.—Several typical examples of these are shown on opposite page. Gothic architecture, of which mouldings form an important integrant, is usually divided into three distinct styles, which are typical of the periods during which they flourished in this country. The first of these styles is called **EARLY ENGLISH**, and commenced near the end of the twelfth century, lasting throughout the thirteenth, when it developed into the **DECORATED STYLE** of the fourteenth century. This style is considered the perfection of the architecture of the Middle Ages, which during the succeeding two centuries gradually deteriorated. The style in this latter period is termed the **PERPENDICULAR**, from the great predominance of vertical lines and members in the structures erected during this time. The style preceding the Early English is known as NORMAN or ANGLO-ROMANESQUE. The former name is given because it was introduced by the Norman conquerors of the country, and the

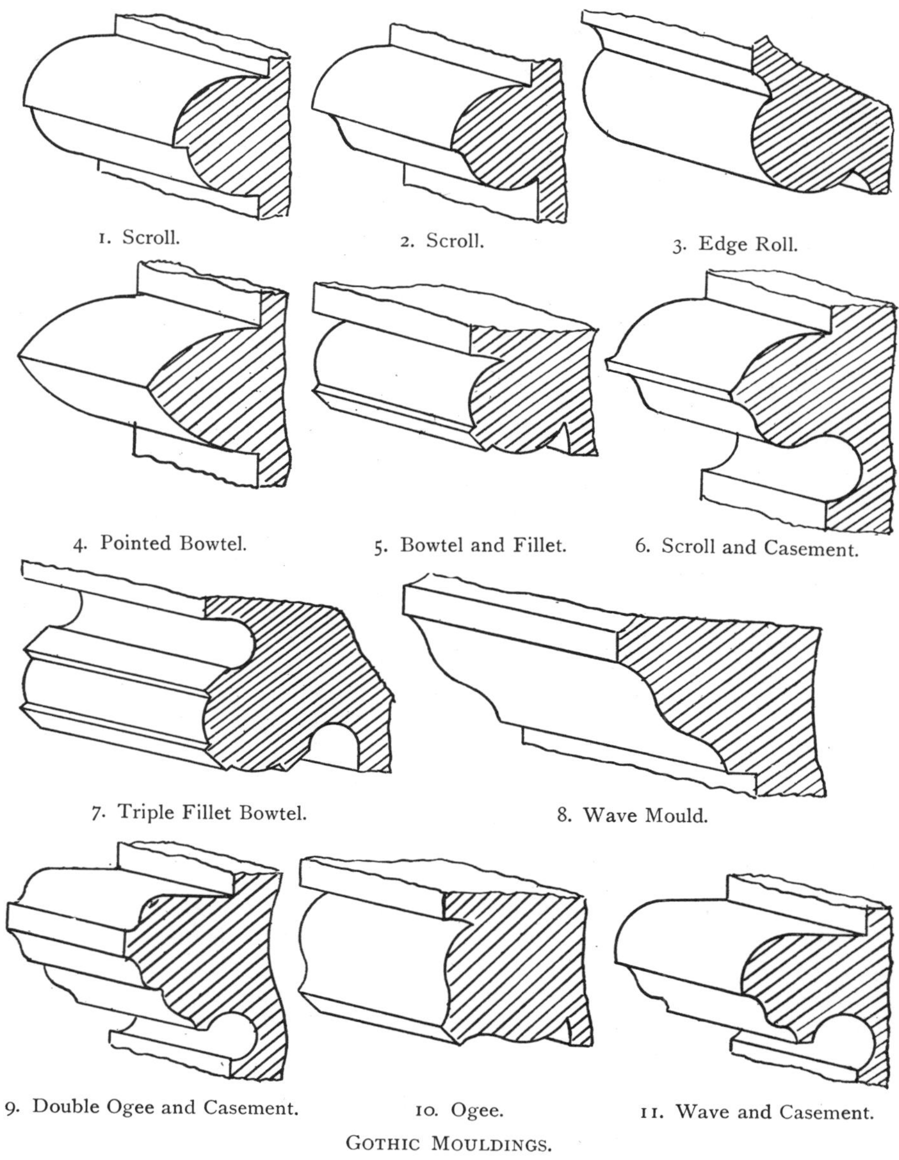

1. Scroll. 2. Scroll. 3. Edge Roll.
4. Pointed Bowtel. 5. Bowtel and Fillet. 6. Scroll and Casement.
7. Triple Fillet Bowtel. 8. Wave Mould.
9. Double Ogee and Casement. 10. Ogee. 11. Wave and Casement.

GOTHIC MOULDINGS.
1-6. Early English. 7-11. Decorated.

latter because of its resemblance to the architecture of the Romans. As the chief characteristics of its mouldings are a profusion of rude carvings unsuitable for woodwork, no examples have been included in this work. The mouldings shown in the drawings are chiefly from examples in stone, as these may be considered more truly representative of the several styles, the mouldings in wood, at least in the earlier periods, approximating more nearly to classic types.

The Edge Roll, f. 3, previous page, although used and developed during the EARLY ENGLISH PERIOD, was also much used during the Norman, and is probably a variant of the Roman Torus.

The Bowtel and **Fillet**, f. 5, is a development of the edge roll, and it is one of the typical mouldings of the thirteenth century. It was further developed during the succeeding period into the double and triple filleted bowtel.

Two other mouldings developed during the thirteenth century are the **Pointed Bowtel**, f. 4, and the **Scrolls** or Beak mouldings, f. 1 & 2.

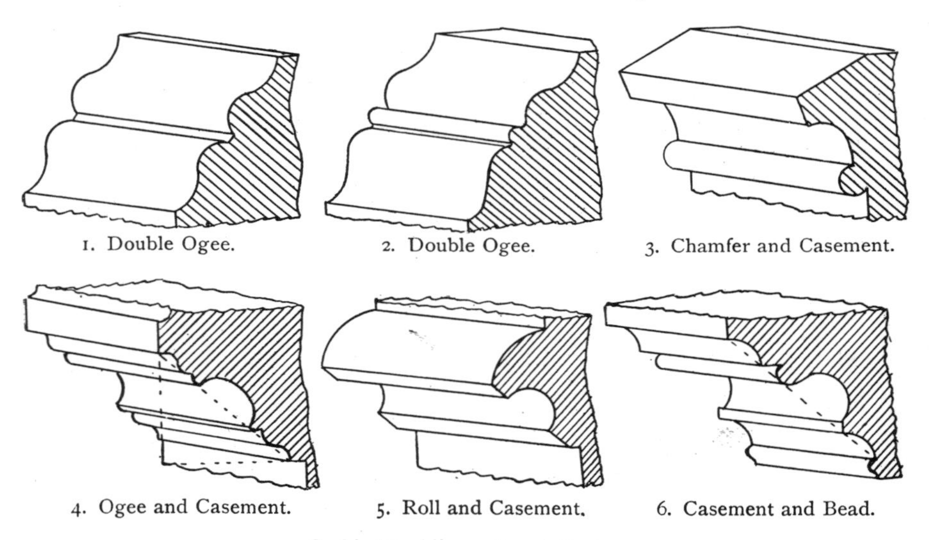

1. Double Ogee. 2. Double Ogee. 3. Chamfer and Casement.

4. Ogee and Casement. 5. Roll and Casement. 6. Casement and Bead.

Gothic Mouldings, Perpendicular.

The Casement or **Hollow**, shown in f. 6, was derived from the Norman, but was more deeply undercut during the Early English Period.

The Wave Mould, f. 8, is a further development of the bowtel, and, with the **Triple Bowtel**, f. 7, and the **Ogee**, f. 10, is especially characteristic of the DECORATED STYLE; whilst the **Double Ogee**, f. 1 & 2 above, and the **Chamfer** and **Casement**, f. 3, **Ogee** and **Casement**, f. 4, **Roll** and **Casement**, f. 5, and the **Casement** and **Bead**, f. 6, are common types of PERPENDICULAR MOULDINGS. "Casement" is a term applied to deeply cut hollows.

The chief characteristics of Mediæval mouldings may be briefly summarised as great complexity with boldness of outline, numerous members, deep undercutting, embellishment by carving often realistic rather than conventional.

RENAISSANCE MOULDINGS are shown on opposite page. This style came into vogue early in the sixteenth century upon the decline of the Gothic style. The term signifies a revival or return of the classic. The mouldings used in this style are derived from the ancient Greek and Roman examples, with variation and elaboration of treatment. Generally the arrangement of the members is similar to the classic, horizontal lines predominating; but mouldings originally designed for great heights and of massive proportions, in a material yielding quite different results from wood, are reduced in dimension and projection, to bring them within domestic requirements, and the inevitable weakness so induced is sought to be hidden by profuse ornamentation of a conventional type; "enrichment" it is termed, endless repetitions of a few typical classic ornaments, such as the egg and dart, acanthus leaf, beads, dentils, &c. Three examples

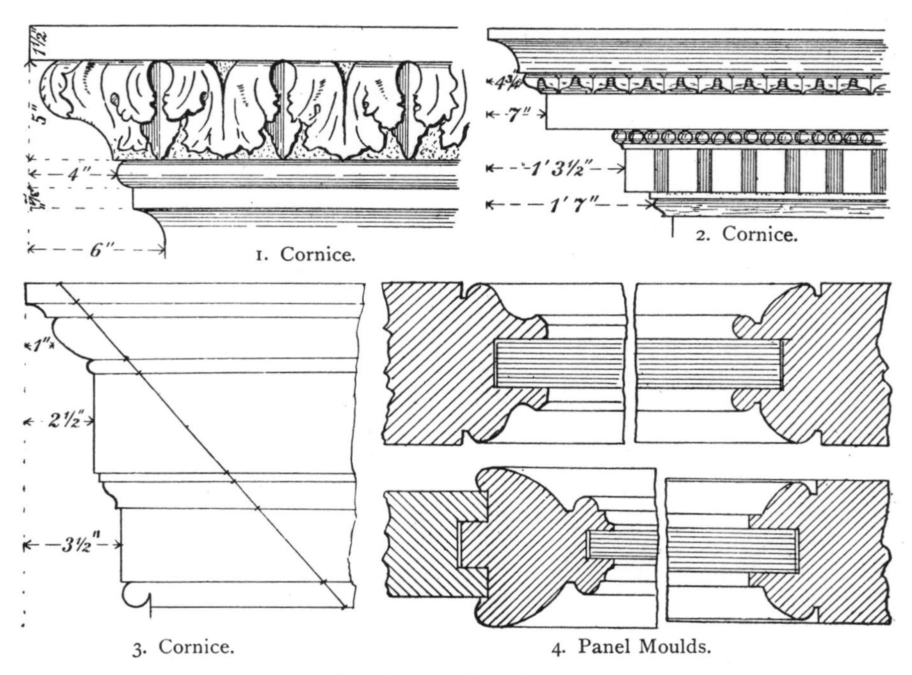

1. Cornice. 2. Cornice. 3. Cornice. 4. Panel Moulds.

Renaissance Mouldings.

of cornices are given in f. 1, 2, & 3. The facias and friezes of rooms are usually enriched with applique ornament in plaster, either festoons or garlands of flowers, fruit, &c., or strap work, and ribbons enclosing medallions painted with mythological or amatory subjects.

It was during this period that **Joinery** as an art distinct from **Carpentry** became fully established. Previous ages had gradually developed the skilled carpenter or constructor in wood, but little of that refinement of workmanship associated with joinery is seen until the fifteenth century begins to draw to a close, for beautiful and interesting as many of the earlier examples of woodwork are, it is evident that they owe their beauty to the art of the carver rather than to any skill shown in their construction, which is usually of the rudest description—precisely similar, in fact, to the methods employed in the constructional woodwork of the buildings; but with the increase in the wealth of the community, resulting from the opening up of new channels of trade and means of communication between this and foreign countries during the reigns of the last Henrys, and of Mary and Elizabeth, there arose a great demand for more luxurious fittings and dwellings than had formerly served, and in the supply of these the art of joinery was evolved. Typical panel mouldings of the later Renaissance or Georgian era, are shown in f. 4, previous page. These are sometimes carved in addition, and indicate the transition from the ornate treatment of the Jacobean era to the simpler, and it must be confessed, often meaningless assemblage of hollows and rounds that do duty for mouldings in our own times, thanks mainly to the exigencies of machine production.

MODERN MOULDINGS for various purposes are shown opposite. Figs. 1 to 6 are suitable for architraves; f. 7 is a cornice moulding; f. 8 a dado; f. 9 a frieze or picture rail mould; f. 10, 11, 12, skirtings; f. 13 to 16, sunk, panel mouldings; f. 17, 18, 19, raised, panel or bolection mouldings. The profiles of these mouldings are drawn in true section to half full size, the elevations are thrown in oblique projection, to show the effect of position upon the various curves.

THE ENLARGING AND DIMINISHING OF MOULDINGS.—The design of a moulding can be readily enlarged to any desired dimension by drawing parallel lines from its members, as shown in f. 3, last page, and laying a strip of paper or a straight-edge of the required dimensions in an inclined direction between the boundary lines of the top and bottom edges; and at the points where the straight-edge crosses the various lines, make marks thereon which will be points in the new projection, each member being increased proportionately to the whole. Projections drawn at right angles to the former from the same points will give data for increasing the width in like manner.

To diminish a moulding.—The method to be explained, which is equally applicable to enlargement, is based upon one of the properties of a triangle, viz., if one of the sides of a triangle are divided into any number of parts, and lines drawn from the divisions to the opposite point of the triangle, any line parallel with the divided side, will be divided in corresponding ratio. See f. 1, p. 374, where A B C is an equilateral triangle, the side A B being divided into six equal parts, and lines drawn from these to the apex C. The two lines D E and F H, parallels to A B, are divided into the same number of parts, and each of these parts bears the same ratio to the whole line that the corresponding part bears to A B, viz., one-sixth; the application of this principle will now be shown. Let it be required to reduce the cornice shown in f. 2 to a similar one of smaller proportions. Draw parallel projectors from the various members to the back line A B, and upon this line describe an equilateral triangle. Draw lines from the points on the base to the apex, then set off upon one of the sides of the triangle from C a length equal to the desired height of the new cornice as at G or H, and from this point draw a line parallel to the base line. At the points where this line intersects the inclined division lines, draw horizontal projectors corresponding to the originals. To obtain their length or amount of projection, draw the horizontal line b E, f. 2, at the level of the lowest member of the cornice, and upon this line drop projectors at right angles to it from the various members.

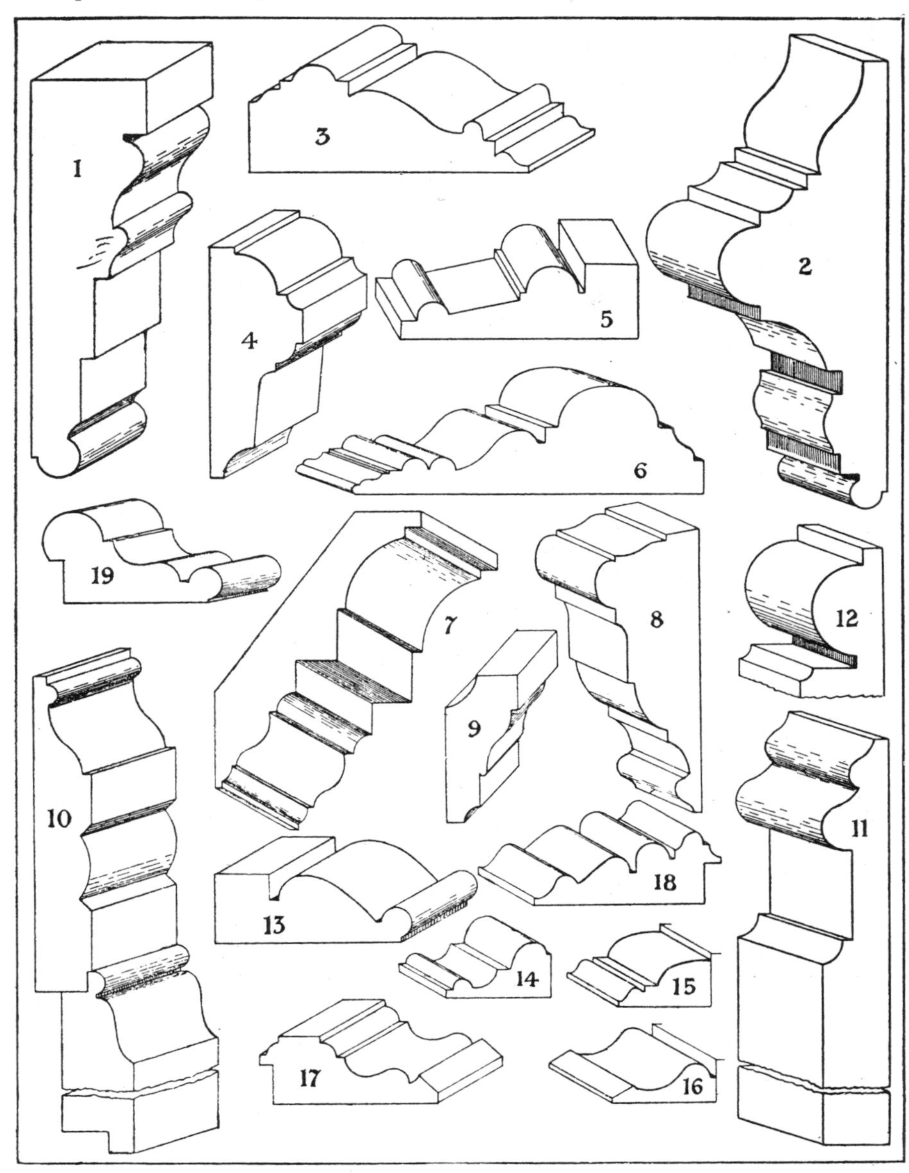

Modern Machine-struck Mouldings.

Describe the equilateral triangle b i E upon this side, and draw lines from the divisions to the apex i. To ascertain the length that shall bear the same proportion to *b* E that the line G H bears to A B, place the length of *b* E on the line B A from B to F, and draw a line from F to C; the portion of the line G H cut off from J to H is the proportionate length required. Set this length off parallel to *b* E within the triangle, as before described, and also draw the horizontal line L H, f. 3 below, making it equal in length to *a a*, f. 2. Upon this line set off the divisions as they occur on *a a*, noting that their position is reversed in the two figures. Erect perpendiculars from these points to intersect the previously drawn horizontals, and through the intersections trace the new profile. The frieze and architrave are reduced in like manner, M B, f. 2, representing the height of the original architrave, and N H. f. 3, the reduction. The cornice can be *enlarged* similarly by producing the inclined sides of the triangle, as shown

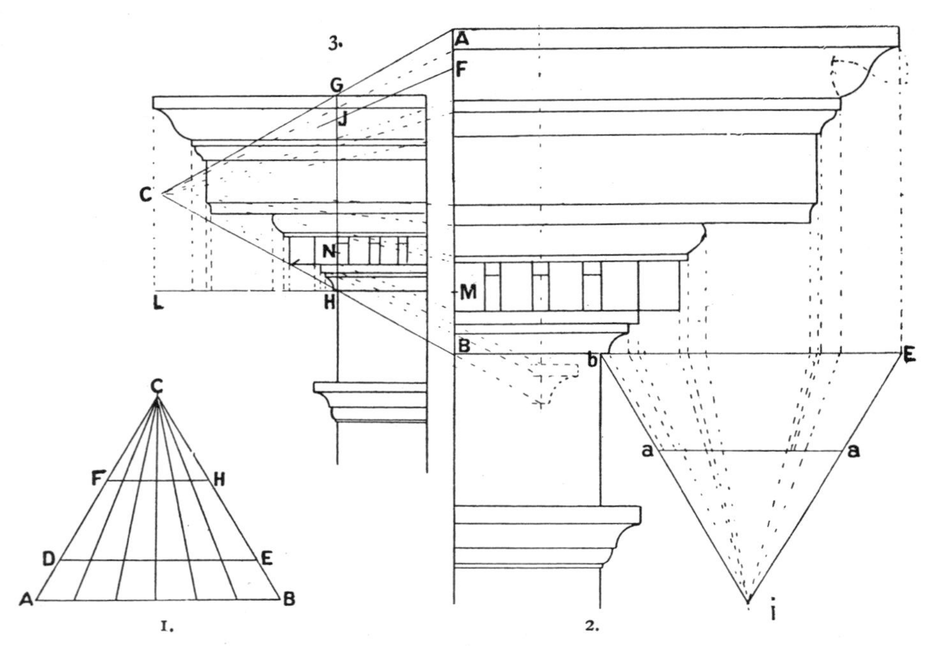

Method of Enlarging and Diminishing Mouldings.

by the dotted lines on f. 2, sufficiently to enable the required depth to be drawn within it, parallel to A B. One member has been enlarged to indicate the method, which should be clear without further explanation.

Raking Mouldings.—Fig. 1 opposite shows the method of finding the true section of an inclined moulding that is required to mitre with a similar horizontal moulding at its lower end, as in pediments of doors and windows. The horizontal section being the more readily seen, is usually decided first. Let the profile in f. 1 represent this. Divide the outline into any number of parts, and erect perpendiculars therefrom, to cut a horizontal line drawn from the intersection of the back of the moulding with the top edge, as at point 7. With this point as a centre, and the vertical projectors from 1 to 7 as radii, describe arcs cutting the top of the inclined mould, as shown. From these points draw perpendiculars to the rake, to meet lines parallel to the edges of the inclined moulding drawn from the corresponding points of division in the profile, and their intersections will give points in the curve through which to draw the section of the raking mould. When the pediment is broken and a level moulding returned at the top, its section is found in a similar manner, as will be clear by inspection of the

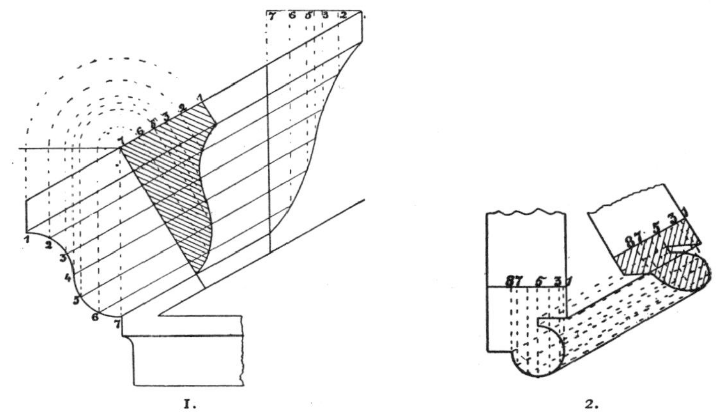

1. 2.

Methods of Obtaining Sections of Raking Mouldings.

drawing. If the section of the raking moulding is given, that of the horizontal mould can be found by reversing the process described above. Fig. 2 shows the application of the method in finding the section of a return bead when one side is level and the other inclined, as on the edge of the curb of a skylight with vertical ends, the plain outline is the level bead, the hatched section the inclined one.

Sprung Mouldings. — Mouldings curved in either elevation or plan are called "sprung," and when these are used in a pediment, require the section to be determined as in a raking moulding. The operation is similar to that described above up to the point where the back of the section is drawn perpendicular to the inclination, but in the present case this line E x, f. 3, is drawn radiating from the centre of the curve, and the projectors are drawn parallel to this line. The parallel projectors 1, 2, 3, 4, 5, are also described from the centre until they reach the line E x, when perpendiculars to this line are raised from the points of intersection to meet the perpendicular projectors.

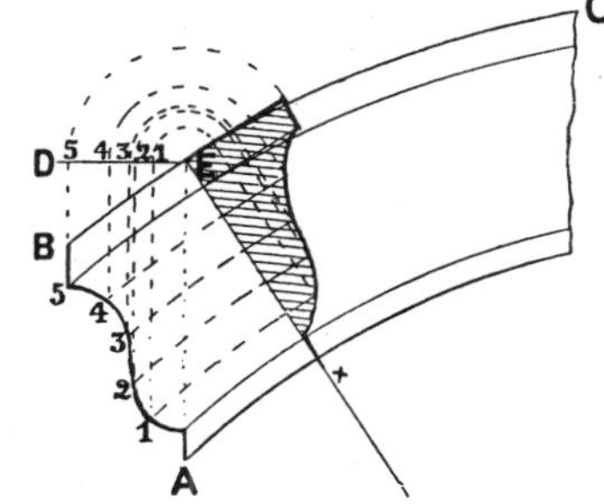

3. Method of Obtaining Section of a Sprung Moulding.

Oblique Raking Mouldings.—This term is applied to inclined mouldings attached to two surfaces that are not at right angles to each other in plan. On next page is the plan of such a case, in which one piece of moulding is horizontal and the other piece raking; and the walls or surfaces upon which they are fixed are at an angle of 135 degs.: A′ 8′ C′ is the wall, and *a b c* is the face line of the mouldings, the line 8′ *b* is the plan of the mitre. B is the given section of the level moulding, drawn perpendicular to its plan, and it is required to know the section of the raking moulding, whose *true* inclination is shown by the line A 8. Commence by dividing the profile of the given moulding into a number of parts, as 1, 2, 3, 4, 5, 6, 7, drop projectors from these points into the plan as indicated by the dotted lines cutting the mitre line in points 1′, 2′, 3′, 4′, 5′, 6′, 7′. Next revolve the plan of the raking piece on point 8′ until it is at right angles with the level piece, as indicated by the dotted outline *a*′, 8′, 1″. Again, with 8′ as centre, describe an arc from point 1′ cutting the face of the moulding in point 1″. Join 8′-1″ which gives the new position of the mitre line. Describe concentric arcs from points 2′, 3′, 4′, 5′, 6′, 7′, intersecting the line 8′-1″ in points 2″, 3″, 4″, 5″, 6″, 7″, and project these points into the elevation, as shown by the full lines. Intersect them by horizontal projectors from points 1, 2, 3, 4, 5, 6, 7, on the profile of B, obtaining the corresponding

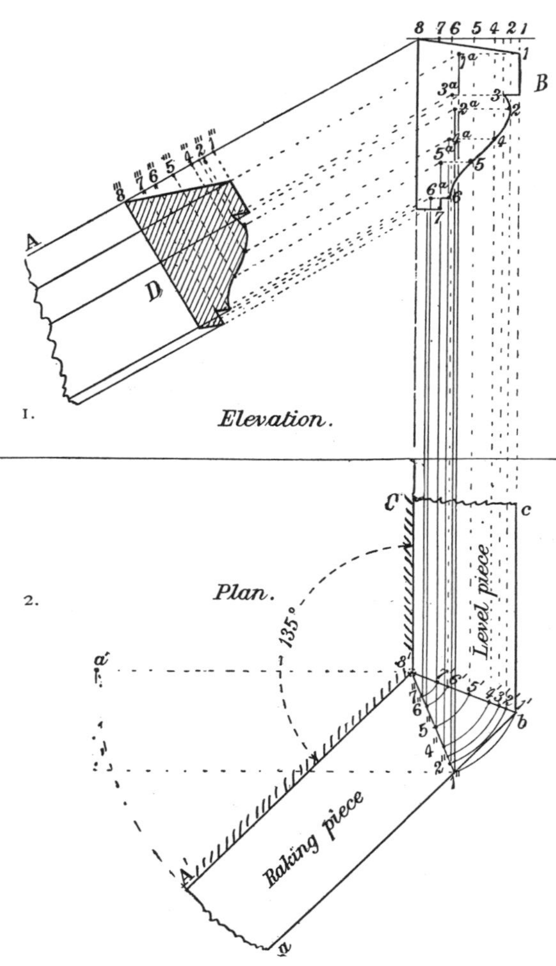

1, 2. Method of Finding Section of Oblique Raking Moulding.

points 1^a, 2^a, 3^a, 4^a, 5^a, 6^a, 7^a. From these points, draw lines parallel with the line A 8. Draw a line perpendicular to the raking line as 8‴, to represent the back of the moulding, and set off from it, on the raking line, points 7‴, 6‴, 5‴, 4‴, 3‴, 2‴, 1‴, equal to the corresponding numbered points on a line drawn at right angles to the back of the section B, which are projected upwards from the division points in the profile. Draw projectors from points 8‴, 7‴, 6‴, &c., perpendicular to the rake, intersecting the correspondingly numbered raking lines, and draw the profile of the required moulding D through the points so formed. It will be noticed in the above data that the true inclination of the raking moulding is given; this, however, may not always be known, it may have to be constructed from the drawing. In f. 1 & 2 is shown **a method of finding the true inclination of a line (or moulding)**, which is raking both in plan and elevation. The plan, f. 2, is similar to f. 2, p. 376, and is redrawn to avoid confusion of the lines. D C, f. 1, is the elevation of the line *d c′* in plan. Take any point in the elevation as A, and project it into the plan in point *a*. With *c* as centre, and *c a* as radius, revolve point *a* into the vertical plane (or to use precise terminology, into a plane parallel to the vertical plane), as *a′-c*; project this point into the elevation, and intersect it at A′ by a horizontal projector from the original point A. Draw a line from C through A′, and this line will be the true inclination of the line D C. Its use has been described in the preceding rule. The above described methods are applicable to plans of either acute or obtuse angles, and the most difficult position in plan has been selected for example, because it is the most usual one occurring in practice, although not the easiest to demonstrate. If, however, the raking member occurs parallel with the elevation, the construction is simplified. All that is then necessary is to draw a

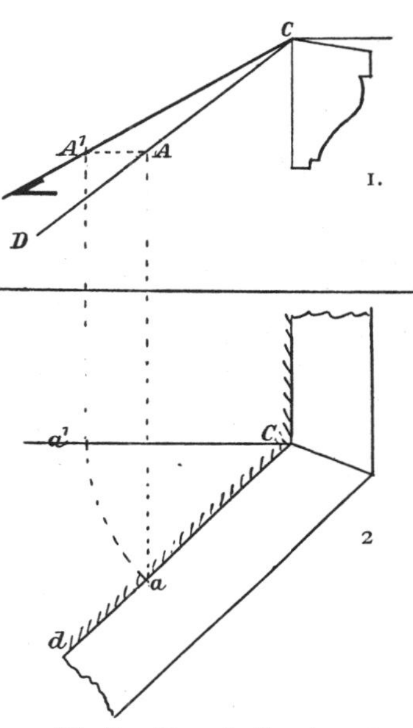

1, 2. Finding True Inclination of an Oblique Raking Moulding.

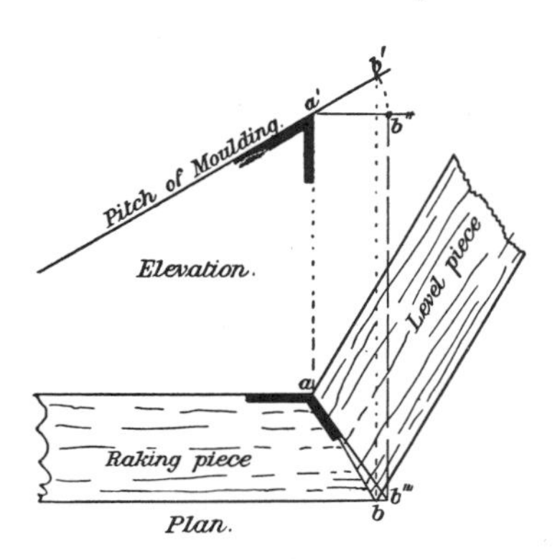

3. Method of Finding Bevels of Raking Mouldings.

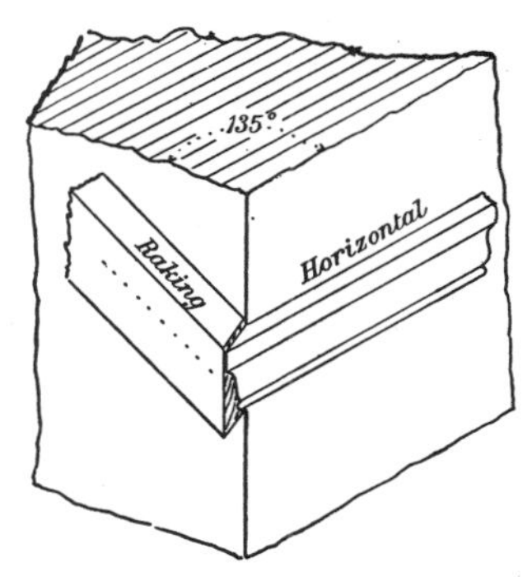

4. Method of Obtaining Section of Oblique Raking Moulding without Drawings.

section of the level piece on its plan and draw parallel projectors to the plan, cutting the mitre line, project these points into the elevation, and intersect them by corresponding horizontal projectors, which will give the profile of the moulding at the mitre. As both members are alike at the mitre, projectors drawn from these points parallel with the rake, will give points in the height of the raking moulding, and, if these are intersected by lines drawn perpendicular to the rake, at the same distance from the back line, that the corresponding lines are in the bevel moulding, points in the curve will be found.

The Method of Obtaining the Bevels for Raking Mouldings is shown in f. 3, p. 377. The true pitch of the raking moulding is drawn over its plan as shown, and the extremities of the mitre line *a b* projected into the elevation at *a′ b′*. Draw a horizontal line from *a′*; with point *a′* as centre, and *a′ b′* as radius, describe an arc cutting the horizontal line in point *b″*, from this point drop a projector into the plan, cutting the face line of the moulding produced in point *b‴*. Join *a b‴*, which is the true bevel for the inclined mitre; the mitre for the level piece will be as shown in the plan at *a b*. The bevel at *a′*, in the elevation, is the down bevel for the back of the moulding. These two bevels may be used for making the cuts in a mitre box in which to cut the moulding.

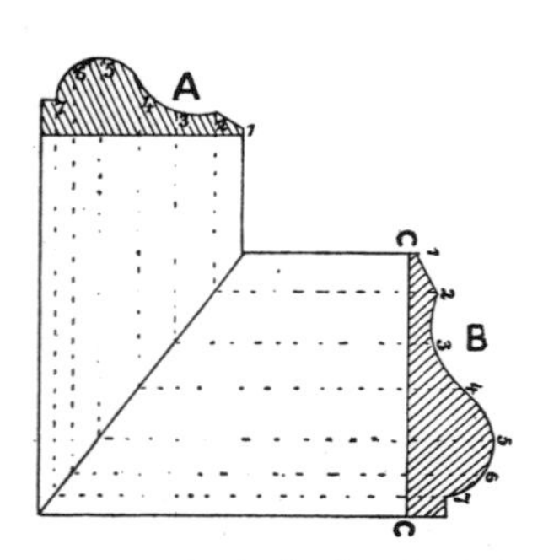

1. Method of Mitreing Straight Mouldings of Different Widths.

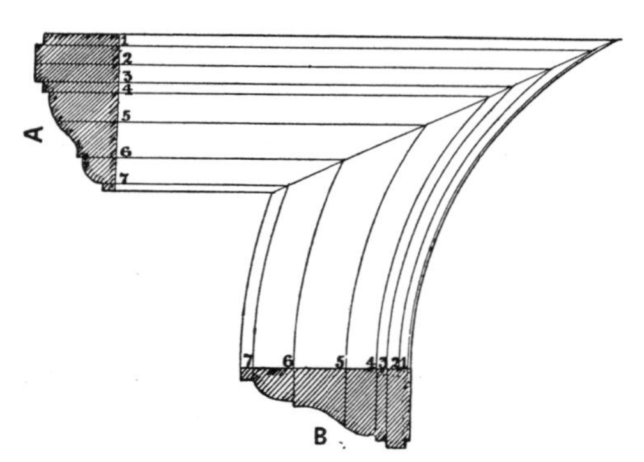

2. Method of Mitreing Straight and Curved Mouldings of Different Sections.

To obtain the section and mitre of a raking moulding without drawings.—When a single length of moulding only is required, it is hardly worth while to use the geometrical processes described above, to obtain the sections. Set off on the wall itself, or upon a block of wood cut to a similar angle as shown in the sketch, f. 4, p. 377, lines representing the top edges in their relative positions of two mouldings, mitre the horizontal piece in the usual way, *i.e.*, so that it bisects the angle in plan, and of course square to the edge in depth, then fix it temporarily in place as shown. Prepare a piece of stuff long enough for the required raking moulding of the same thickness as the other piece, and of a width sufficient to cover its mitred end. Apply the *back* of the stock of a Bevel, to the raking line, and adjust the blade to fit the mitred end of the level piece, this will give the edge cut. Apply the *side* of the bevel to the raking line similarly for the down cut; mark these on the piece, and cut off. Fix it as shown in the sketch, and pencil around the outline of the horizontal moulding upon the mitred end of the raking piece, remove it and work it off to the outline.

To mitre mouldings of unequal width.—Let A, f. 1 opposite, be the section of a moulding which it is required to mitre with a similar moulding of the width of B. Draw the plan as shown of each moulding at the angle they are required. The intersections of these boundary lines will give the mitre line. **To find the section of B.** — Draw the section A at right angles to the plan; divide the curve as shown from 1 to 7. Draw lines from these points parallel to the edges to intersect the mitre line. From these intersections draw lines parallel with the edges of the other moulding to cut a perpendicular line *c c*. Make each of these lines equal in height above the perpendicular to the height of the corresponding line in the section A, and draw the curve B through the points.

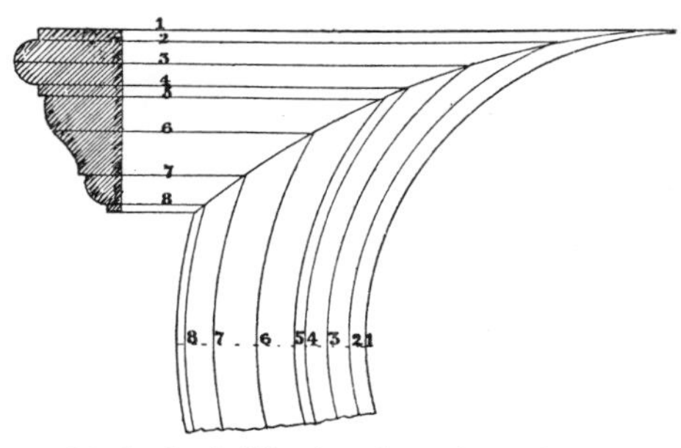

1. Method of Mitreing Straight and Curved Mouldings of the same Section.

Mitreing Straight and Curved Mouldings Together. — If a *straight* mitre is required, draw the plan of the mouldings, as in f. 2 opposite, and the section of the straight mould at right angles to its plan as at A. Divide its profile into any number of parts, and from them draw parallels to the edges intersecting the mitre line. From these intersections describe arcs concentric with the plan of the curved mould, and at any convenient point thereon draw a line radial from the centre. Erect perpendiculars on this line from the points where the arcs intersect it, and make them equal in height to the corresponding lines on the section of the straight moulding A, and these will be points in the profile of the curved moulding B. When it is required that the section of both mouldings shall be alike, a *curved* mitre is necessary, and its true shape is obtained as shown in f. 1 above. Draw the plan and divide the profile of the straight moulding as before, drawing parallels to

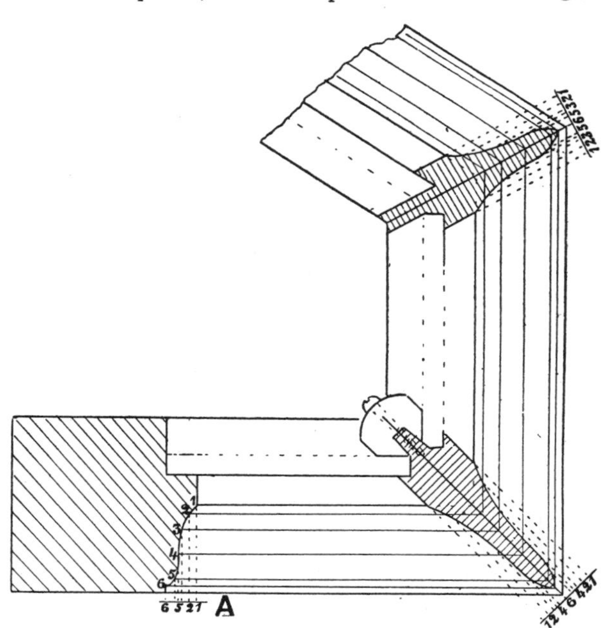

2. Method of Obtaining Sections of Angle-bars.

DESIGN FOR PANEL TRACERY.

the edge towards the seat of the mitre. Upon a line drawn through the centre of the curved moulding, set off divisions equal and similar to those on the straight part, as 1 to 8 in the drawing. From the centre of the curve describe arcs passing through these points, and through the points of intersection of these arcs with the parallel projectors, draw a curve, which will be the true shape of the mitre. Cut a saddle templet to this shape, and use it to mark the mouldings and guide the chisel in cutting.

To obtain the section of a sash bar raking in plan.—Fig. 2, p. 379, represents the plan of a shop front sash with bars in the angles. On the left hand is shown the section of the stile or rail into which the bars have to mitre. Divide the profile of moulding into a number of parts, as shown from 1 to 6, and from these points draw parallels to the sides of the rails intersecting the centre line of the bar. Also draw perpendiculars from the same points to any line at right angles to them, as at A. Draw a line at right angles to the centre line of the bar, and on it set off the divisions from 1 to 6 as at A. Draw projectors from these points parallel to the centre line of the bar, and where they intersect the correspondingly numbered lines drawn parallel with the sides of the sash will be points in the curve of the section of the bar. It will be noticed that there is no fillet or square shown on the bar, and that in transferring the points from the line A they must be reversed on each side of the centre. Should a fillet be required on the bar, additional thickness must be given for the purpose. Three methods of forming the rebates in the bar are shown, the screwed saddle bead being the best for securing the glass.

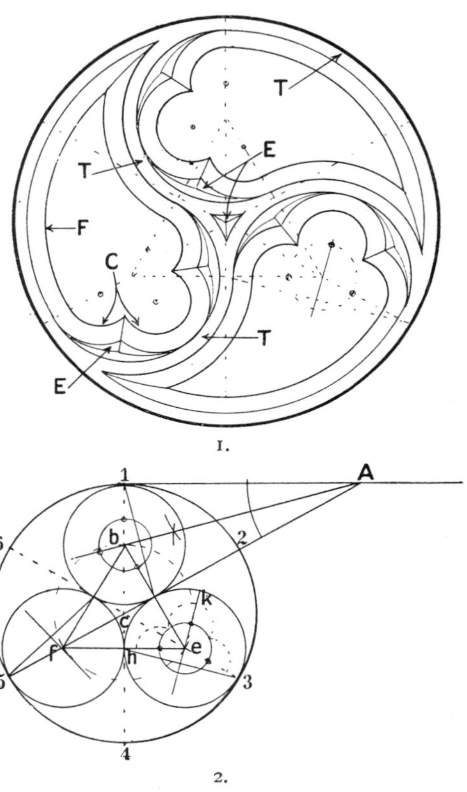

1. Design for Circular Tracery Panel.
2. Diagram of Construction, f. 1.

TRACERY.—This is a form of ornamentation peculiar to Gothic architecture. As applied to stone, it consists in the arranging of window bars or mullions into geometrical patterns, and the treatment of ribs and groining of vaults similarly. In its application to woodwork it is chiefly confined to panelling, and consists of the carving in the solid of geometrical patterns, recessed below the surface, or, as in the case of screens, &c., the recessed parts are entirely removed, leaving the pattern outlined in a delicate tracery or network of curves and bars. The latter description is distinguished as *Pierced Tracery.* Not infrequently the "solid" tracery was first pierced through a thin layer, and afterwards affixed to a solid backing, this latter method being usually adopted in the case of external doors of churches, &c. The patterns in this work are purely geometrical; circles, complete and in part, forming the principal curves; and it has been found that the triangle in its various forms, but more especially the equilateral, has been invariably used as a basis of construction in the designs of mediæval work. The upper part of the panel design, shown on p. 380, is constructed upon a triangle, as indicated by the dotted lines, the centres of the two main circles falling in the middle of the sides, and the centres of the subsidiary lozenges in the angles. The centres of the minor circles forming the trefoils also form a triangle. The centres of the circumscribing bars form a pentagon, which is a combination of five triangles within the larger triangle; further analysis of the design will show that not only all the centres but also all the ribs and backs of the cusping can be resolved into triangles. The design of the circular panel, f. 1, last page, is based upon the equilateral triangle, the centres of the bar tracery falling upon the angular points, and the centres of the foils forming isosceles triangles upon the corners of the main triangle. Fig. 2 is a skeleton diagram showing the preliminary construction to find the centres of the design. The first step is to inscribe three equal circles within the circles; to do this, first divide the circle into six equal parts by stepping the radius around the circumference. Draw dotted lines joining these parts as 1-4, 2-5, 3-6. From point 1 draw a line the tangent to the circle, and produce line 5-2 to cut the tangent in A. Bisect the angle so formed, cutting the line 1-4 in *b*. From the centre *c* with radius *c b* describe a circle cutting lines 5-2 and 6-3 in *f* and *e*. From *b f e* as centres, with radius *b* 1, describe three circles, and join their centres, forming an equilateral triangle. Next draw a line from point *h*, where the division line 1-4 passes through the centre of one side of the triangle to point 3, where the adjacent division line passing through the angle of the triangle cuts the circumference, and bisect this line as shown, *e k* being the bisector. Describe a circle around point *e* to the required size of the cusping, cutting the lines *e h*, *e* 3, and *e k*, which will be the centres of the trefoils, then each of the other angles are served similarly, and the centres are indicated in the design, f. 1. The technical names of the various parts of a tracery design are as follows. The portions of the design that are left on the original surface and form the groundwork of the design are called **Tracery Bars**, as T, T, T in f. 1. The large sweeps forming the interior edges of the design are **Foliations**, F. The small circles are called **Foils** when complete, **Featherings** or **Cusping** when incomplete or running into each other, as shown at C. The points which are formed between the featherings when the circles do not penetrate each other are called **Cusps** (see *c*, p. 380. The ridge down the centre of a cusp is its **Back**, and the sinkings between the cusping and the tracery bars are **Eyes**; these are marked E in the design. Foils are distinguished as trefoil, quatrefoil, cinquefoil, according as they have three, four, or five arcs.

In Working Tracery Panels to any symmetrical design similar to that on p. 380, copy one half of the design upon tracing paper with a soft pencil.

Prepare the surface of the panel with chalk well rubbed in. Set out the margin and centre lines, and pin the tracing in position, pencilled side down. Go carefully over the lines with a pencil sharpened to a round or dull point, and an exact copy will be produced on the panel. The other half may be reproduced in the same way by turning the tracing over and repeating. In this manner several repeats of the same drawing may be made, the lines being filled in on the wood as they get faint from continual use. If only one panel to a design is required, the latter may be drawn on thin paper and pasted on direct to the panel. When cutting the pattern, assuming that the work in hand is pierced tracery, cut the interior portions first with a bow saw or a jigger saw. Leave the lines showing, so that the edges may be cleaned up accurately with spokeshave or file. Work

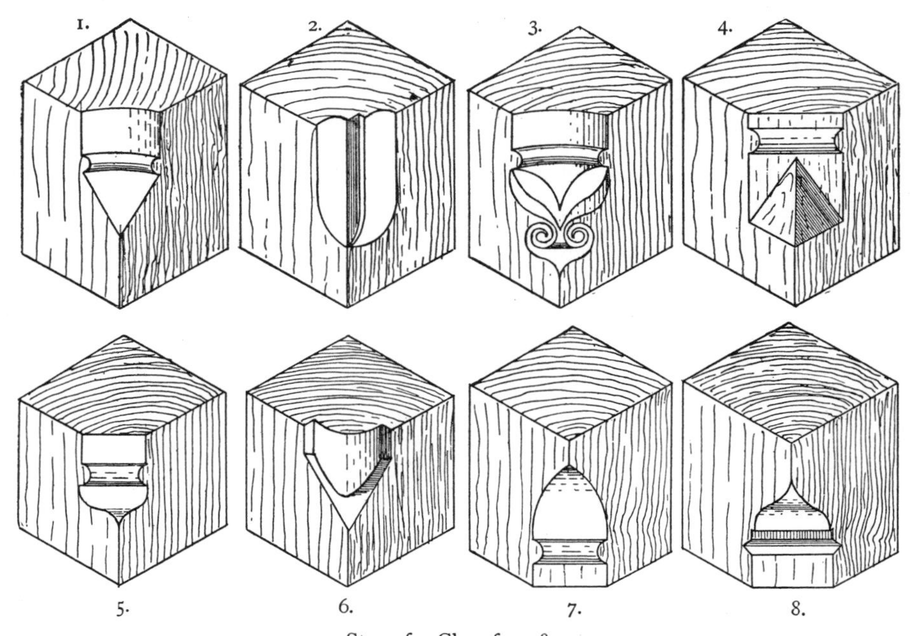

Stops for Chamfers, &c.

towards the outside, cutting the smaller and more intricate portions of the pattern first, leaving the large openings to the last, so that the wood may be better able to withstand the strain of the sawing. In cutting the eyes and forming the hollows in the cusping, bed the panel well by screwing it to a stout panel board. As a preliminary operation to working the hollows, form a chamfer all round the foliations, &c., with a spokeshave, first gauging the depth of the sticking from the back of the panel. The hollows are worked out with suitable bent gouges. The backs of the cusps must be worked up to a sharp arris hollow in its profile. All three sides of an eye should be worked down simultaneously, otherwise the mitres cannot be produced correctly; these will be curved when the intersecting curves are of contrary flexure, but straight when they are similar, as, for instance, a concave running into a convex curve will produce a curved mitre, but two convex curves intersecting will produce a straight mitre. This will be seen clearly on reference to the design, p. 380. A thin hardwood or zinc "reverse" of the mould is useful for testing the correctness of the working in the eyes; this should be applied "square" to the surface. The best class of work of this kind is left "from the gouge," no glass-paper is applied, and the individual characteristics of the workman are shown in his work. Some beautiful examples of such craftsmanship are shown in the photographs on Plates XXIX. & XXX., but influenced perhaps by the extreme regularity and uniformity of machine-worked moulding; many employers, or their clients, require a like uniformity in hand work; hence the following hints are given as a necessary concession to a demand. A small scraper should be made to the exact shape of the hollow, smoothed up on the oilstone, and its edges turned with the sharpener or a gouge. Work this round the hollows upright, and it may advantageously be secured in a small block of wood as a guide, for this purpose. Short pieces of cork should be filed up to fit the various curves, and pieces of glass-paper glued to them, which when dry must be trimmed off clean, as any overhanging edges will destroy the sharp arrises. Use these with a little chalk or whiting dusted over them, and finally burnish with shaped hardwood rubbers. Boxwood or beech make the best for the purpose. If the work is to be polished or varnished, "damp down" after scraping, with warm water. See Workshop Practice, p. 73.

Chamfer Stops.—In f. 1 to 8, p. 383, are given eight examples of stops for chamfered or moulded angles. These finishes are usually associated with Gothic work, but may suggest adaptations for modern door jambs, &c.

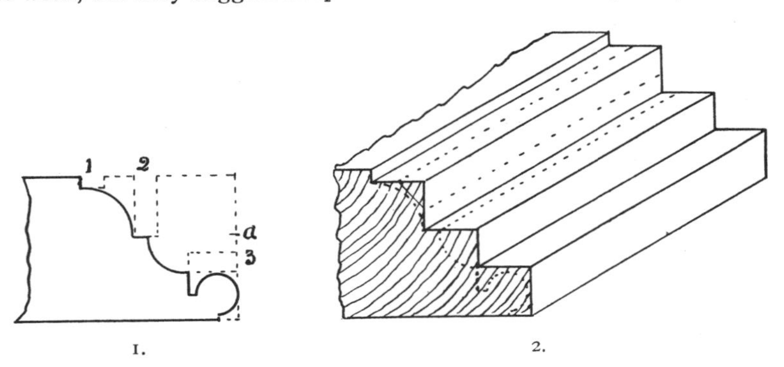

1. 2.

WORKING MOULDINGS BY HAND.—Mouldings are now generally stuck by machinery, but as occasionally a small quantity or a special section may be required which it is necessary to work by hand, a few hints as to the method employed may be useful. Almost every conceivable curve can be produced by skilful manipulation of the various Hollows and Rounds described on p. 18; aided by side Hollows and Snipes, when they are undercut; and corresponding thumb planes, as illustrated on p. 360, when the mouldings are on the sweep. In working the example shown in f. 1 above, prepare a templet to mark the outline of the moulding on each end of the board as shown, having first planed, shot, and gauged the board to thickness. Then from the front edge, run in two plough grooves, as shown by dotted lines Nos. 1 and 2, to the exact depth of the sinking and fillet respectively. Next run in the plough groove marked 3, from the back, and mark a gauge line along the front edge as at *a* to the same depth as groove No. 2. Proceed to chip away the core between groove No. 2 and the gauge line, and level off with a rebate plane. Run a second gauge line along the rebate in line with the bottom of groove No. 3, and cut away the core and level as before; in like manner form a rebate at point 1. The piece will now have the appearance of sketch in f. 2, consisting of a series of rebates parallel with the edge and face of the board. Take off the salient angle of the upper rebate, lining the chamfer as shown, and work the ovolo with a suitable Hollow. Next cut away the core over the cavetto with a gouge, and work this member with a Round; then with a "slipped" Bead of the requisite size work the return bead on the edge, sticking the member next the cavetto first; then turning the board up in the bench-screw, and tapping the iron slightly into the plane, so that it shall not cut twice on the outside, work the remaining half of the bead. Next rip the moulding off to the required width, which should be previously gauged upon the face. This back edge is easiest shot straight and square upon a "shooting board," as shown in the photograph, p. 54, by first nailing a straight strip of wood on the top of the board, parallel with the edge and at a distance therefrom, a little less than the finished width of the moulding. If the moulding is a panel moulding the back edge should be shot slightly "under," as shown on p. 155, and explained p. 157. To do this place a strip of veneer or similar thin stuff under the back edge when shooting.

Always make the more distant sinkings first, working outwards so that the outer edge is preserved intact as long as possible. The rebate in bolection mouldings should be sunk slightly less than the depth of sinking to the panel, so that it will sit tight upon the face of the work.

Curved mouldings are also worked as described above, the more rectangular sinkings that are made the easier and more correctly will the curves be produced.

CHAPTER XXIII.

FOREMAN'S WORK.

Method of Reading Plans—Meaning of Term Plan, Elevation, Section—Scales, How Described—Colouring of Plans—List of Colours—Copying Mouldings—Suggested Alterations—Working Drawings, What they are—Setting Out Rods—Various Methods of Setting Out—What to Do and what to Avoid—What a Rod should Represent—Points to Consider—Fixing Joints—Methods of Setting Out Curves—Uses of Templets—Setting Out Spaces equally—Storing, Labelling, and Indexing Rods—Examples of Rods for Door Frames—Sashes, &c.—Material Boards—Foreman's Tools—Instruments—Scales—Pencil Liners—Trammels, &c.

THE READING OF PLANS.—The drawings prepared in the architect's office for use in the workshop are usually tracings of the originals, and are called working drawings by architects, but are generally known in the workshop as "plans," this term being applied to all scale drawings indifferently; but to avoid confusion in the descriptions herein, the term plan is confined to its legitimate signification. *A Plan* strictly speaking is the shape or outline of any object upon the ground or other horizontal surface. In a wider and more general sense the term is also used to signify all horizontal sections which are identified by specific terms or reference letters, as, for instance, first floor plan, plan at A B, &c.

An Elevation is a geometrical drawing of an object, showing its height and width but not its thickness, and includes all that occurs within those dimensions upon the side shown. The particular portion of the object shown in the drawing is indicated by specific terms, as front, back, east, west, &c., elevations.

A Section is a drawing showing the interior parts of any object as they would appear upon a plane or cut passing through the object in any given direction. Sections are specified as horizontal or vertical, and their locality indicated by reference letters.

Working Drawings, as distinguished from architectural drawings, are representations to scale of the various *parts* of a building, whilst the latter are views of the building as a whole. The more important details of working drawings are generally drawn full size.

Drawing to scale means that the drawing is made to a proportionate reduction throughout of the object represented, and the ratio that the drawing bears to the object represented is called the fraction of scale. For instance, a drawing made to a scale of 3 in. to 1 ft. would show every part one quarter its real size, because 3 in. is a quarter of a foot, and the representative fraction would be $\frac{1}{4}$. Similarly a drawing of $\frac{1}{8}$ in. to 1 ft. is $\frac{1}{96}$ of the full size, and $\frac{1}{96}$ is the representative fraction of scale. The scale is sometimes expressed as being 4 ft. to 1 in., meaning that a line in the drawing 1 in. long represents

4 ft. in the object, and the same scale might be otherwise written as being a $\frac{1}{4}$ in. to 1 ft.—that is, each $\frac{1}{4}$ in. of the drawing will represent 1 ft. in the object, the fraction of scale being $\frac{1}{48}$.

To Draw a Scale.—Set off upon a straight line the length which is desired to represent 1 ft., and which may be anything from $\frac{1}{16}$ to 6 in., and as many repetitions of this length as may be required. Divide the first division on the left hand into twelve equal parts, and each of these will represent 1 in. to the scale. Scales of 1 in. to the foot and upwards may be further divided into quarter inches, and above 2 in. into eighths of an inch. The division subdivided into inches, &c., should be numbered 0, the next 1, and so on, as shown throughout this work.

COLOURING DRAWINGS.—Drawings are coloured, the better to identify the various parts and materials, but unfortunately there is not strict uniformity in the colours used in different drawing offices. The following list indicates the colours most generally employed to represent the various materials set against them, but as the recognition of the tint is of more utility for the reading of the plan than the knowledge of the name of the colour employed, especially as the former may vary, short descriptions of the tints employed are given in brackets.

LIST OF COLOURS USED IN WORKING DRAWINGS.

MATERIAL.	COLOUR IN ELEVATION.	COLOUR IN SECTION.
Soft woods (wrought)	Burnt sienna, pale (brownish yellow)	Same colour, full tint.
Soft woods (in rough)	Raw sienna or pale yellow ochre (lemon yellow)	Burnt sienna, dark (large sections in outline).
Oak	Sepia and yellow ochre, mixed (a warm drab)	Burnt umber (reddish brown).
Mahogany	Light red and burnt sienna mixed (a pale brownish red)	Crimson lake and burnt sienna.
Teak	Umber and yellow ochre (brownish yellow)	Same, but fuller tone.
Walnut	Crimson lake and sepia (brownish purple)	Same, with more red.
Ebony	Indian ink (grey black)	Dark green.
Existing timber (general plans)	Indian ink, pale (brownish grey)	Same, darker.
Proposed alterations (general plans)	Yellow ochre (warm yellow)	Same, darker.
Brickwork	Venetian red (light or pale red)	Crimson lake.
Soft stonework	Yellow ochre and sepia (pale brown yellow)	Dark sepia, also Prussian blue.
Hard stone and granite	Indigo (pale greenish blue)	Same, darker, or dotted with ink.
Concrete	Sepia and cobalt, pale (greenish brown)	Payne's grey and sepia (blue green splashed with brown).
Plaster or cement	Indian ink, pale (light drab)	Cobalt and Indian ink.
Slate	Payne's grey (green grey)	Same, darker.
Lead	Pale indigo and sepia (lead colour)	Indigo and sepia.
Brass	Gamboge, full (greenish yellow)	Indian ochre (bright yellow).
Steel	Indigo and crimson lake (warm violet)	Same, darker.
Wrought iron	Prussian blue, pale (very light blue)	Same, darker (bright blue).
Cast iron	Payne's grey, very pale (slaty blue)	Indigo, pale.
Zinc	Cobalt (azure)	Cobalt, dark.
Glass	Prussian blue and gamboge (mixed).	Hooker's green, No. 2.
Lines of section	Crimson lake (vivid red).	

Reading Plans.—If any discrepancy occurs between the scale drawings and the full-size details, adhere to the latter; but if it occurs between the drawings and the specifications, refer to the architect, or compare with the key plans. Pay great attention to the sections of any mouldings shown, as these are frequently designed with special reference to their position, and what may seem but a trifling departure from the drawing in section may quite destroy the intended effect; but on the other hand, a section may be given which it is impossible to produce except at a prohibitive price, but which may be brought within the capacity of the machines by slight alteration. In such a case make a tracing with the suggested alteration in coloured pencil, and submit it for approval (see that all such alterations are endorsed with the architect's "stamp" and the word "approved" before putting them in hand).

To Copy the Section of a Moulding.—Take a tracing of it with a medium or H pencil. Turn the tracing over, go over the outline with a soft or HB pencil, and lay this side upon the surface to which you desire to transfer the copy, and again pass the hard pencil over the outline; on lifting the paper an exact copy will be found which may be "lined in."

PREPARATION OF RODS.—Rods or staffs, in workshop parlance, are boards, square laths, or sheets of lining paper upon which full-size drawings are made of the finishings and fittings of a building, for the assistance of the workman in "setting out" the various parts. They are, in fact, a reproduction to full size of the scale drawings of the architect, and a translation of the "specifications"; they are exact sections of the work required. Elevations are but seldom used, except in the case of circular or other curved work, and framing of complicated design, the numerous parts of which might be difficult to show in sections.

Methods of Setting Out vary in different shops, unfortunately, both for employer and workman, tending much to the pecuniary loss of the former, and the mental distraction of the latter. In one the "rods" will be set out loosely, sections varying and requiring "allowances" to be made by the workman; in another they will be literal copies of the "plans," no discretion being used in the translation, with the result that the various parts of the construction will not "dovetail." Implicit reliance cannot be placed on "working drawings." Errors in construction creep in sometimes in the original, and more often in the tracing, which it is the duty of the intelligent foreman to eliminate. He should be able to distinguish between what is essential, as conveyed by the whole tenor of the specification, and what is merely suggestive. It must be remembered that the architect is not versed in the constructive details; he is a designer, not a craftsman, and his drawings are made to indicate, more fully and clearly than can be conveyed in the specification, what it is that he requires. To determine the correct way of obtaining the desired result is the duty of the foreman. Sundry deviations from given dimensions may be necessary to bring the work within marked sizes, &c., or to compensate for some authorised alteration. No definite rule can be laid down, common-sense must dictate when it is necessary to apply for authority to make some alteration.

A third method of setting out, and this is the one offered as an example for the embryo foreman to copy, is to set the rod out to the "tried up" sizes of the stuff—that is, the size the material is after having been planed up—and to make allowances for fitting and fixing, not upon the rod, but on the material. This will do away with the necessity of the workman making allowances, which is a fruitful source of error.

Before setting out a rod the foreman should well consider the construction of the particular piece of work he is dealing with, how it is to be put together, its conveyance, adaptability for fixing, &c. It is difficult to particularise in a matter so much subject to circumstances, but as an instance of what is meant, take as an example a large glazed screen with doors, such as are used in offices. This would probably be shown on the working drawings as one continuous framing, with details to half or full size of the more important parts. The first thing to ascertain would be whether the framing would have to match or correspond with any existing work, and if so, to take accurate dimensions of that as a guide in setting out the new, also to see whether any structural parts of the building would in any way interfere with the fixing or conveyance of the work as a whole to its situation; and further, should all these points prove satisfactory, whether its size and weight would preclude its being constructed and handled safely in one piece. Then if it is decided to make it up in sections, it must be determined where is the most suitable place to form the joints, and what kind to employ. Generally speaking, what may be termed fixing joints, as distinguished from joints of construction, should be as simple as possible, and of a form that do not require any extraneous aid to bring them up or keep them so. **A very Efficient Joint** of this class is made by ploughing and tonguing the edges of the two parts coming together to keep them flush, and securing the joint with wood screws or handrail bolts, the formation of the adjacent parts determining which is to be used. It might occur that there would not be a convenient pair of edges that could be thus united, and the junction would then have to be made at one of the constructional joints, as shown in f. 1, where piece A is joined to piece B in a line with the muntings. The muntings at the joint would

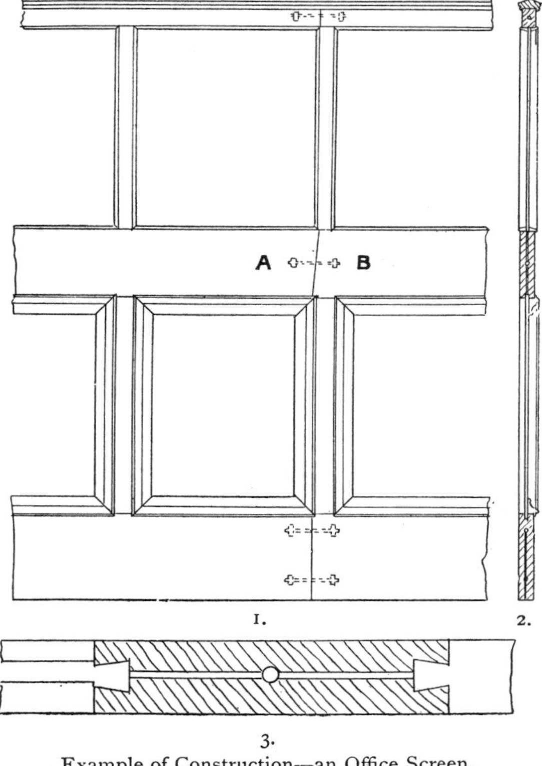

Example of Construction—an Office Screen.
1. Elevation. 2. Section. 3. Enlarged Joint.

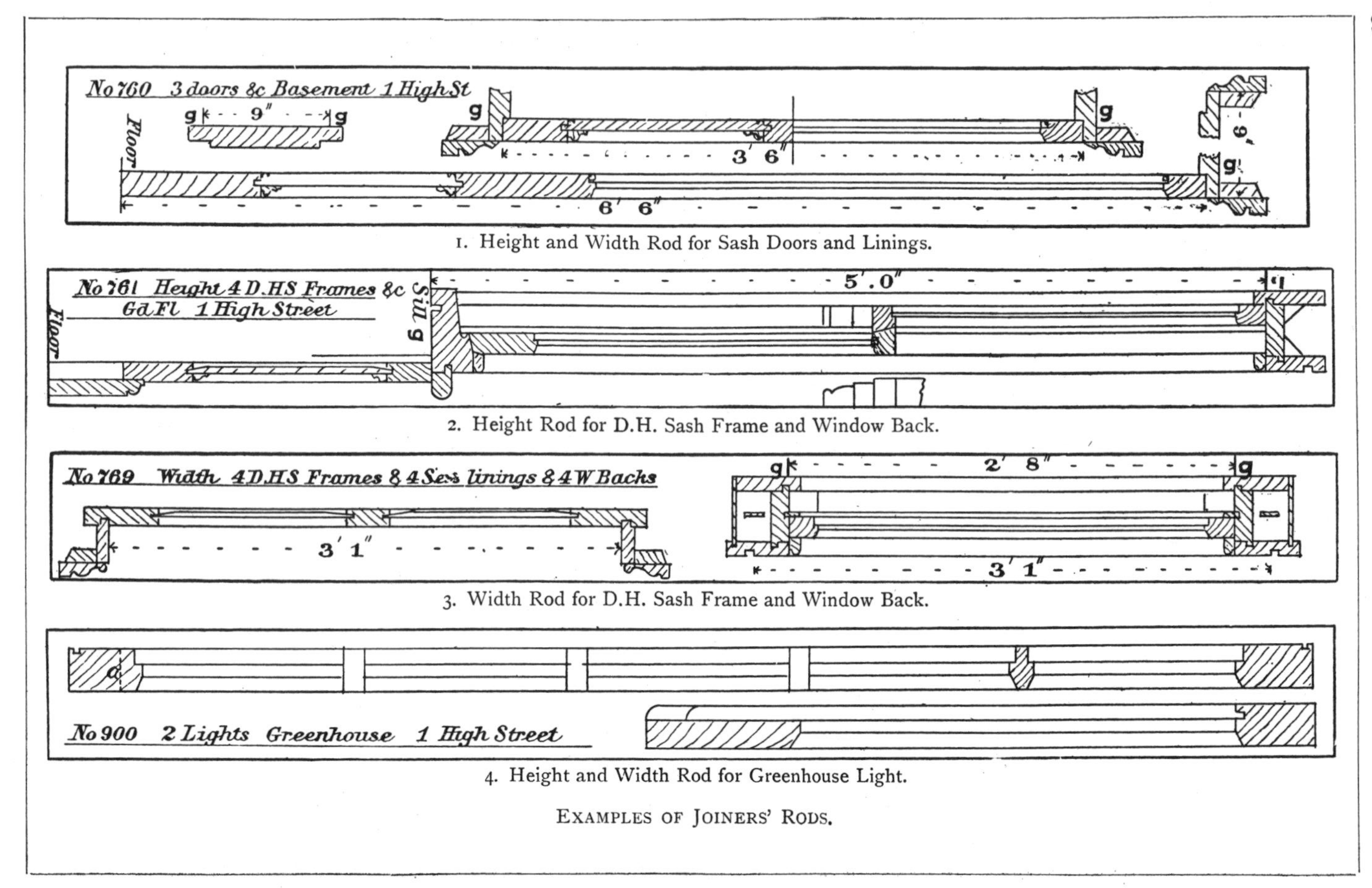

1. Height and Width Rod for Sash Doors and Linings.

2. Height Rod for D.H. Sash Frame and Window Back.

3. Width Rod for D.H. Sash Frame and Window Back.

4. Height and Width Rod for Greenhouse Light.

EXAMPLES OF JOINERS' RODS.

be better dovetailed, as shown in the enlarged detail, f. 3, than tenoned, and the rails butted, tongued, and secured with handrail bolts. The capping should be if possible in one length, or jointed elsewhere.

Rules for Setting Out.—It is not a good practice to superimpose one section on another, as, for instance, the sections of a sash door showing both above and below the middle rail in one drawing, but rather make two sections side by side, or if time and space is limited, show one half above the rail on one side of a centre line, and the other half below the rail on the other side of the centre line, as drawn in f. 1, opposite page. All parts that are really in section upon the line drawn should be hatched or sectioned with either straight lines drawn at an angle of 45 deg., or with freehand annual rings, as in the examples. This part of the work need not be elaborated, just sufficient to indicate that the shaded parts are in section is all that is needful. Any important parts that may be *above* the line of section, and any that may be *below*, but are hidden by some intervening part, should be drawn in dotted lines, all the other parts in full lines.

Glass and iron work are usually distinguished by drawing in blue pencil, brickwork in red, and stonework in yellow.

In Setting Out Frames or Linings for openings, the clear size of the opening should be ascertained and marked on the rod, as a guide to the arrangement of the fitting. In the given examples these lines are marked *g*, and should be marked in red pencil on the actual rod. It is advisable to figure in the chief dimensions, as shown. This is a precaution that will prevent many mistakes, especially in large jobs necessitating the use of several rods in lengths.

All Measurements above 10 ft. are better set off with a 5-ft. rule, inches or parts with dividers ; and where many repeated measurements are required, as in large light screens, &c., the various subsidiary measurements should be added together to see that they agree with the over-all dimension.

A Number of Small Divisions or Spaces, such as a number of bars in a light (see f. 4), are best spaced by adding the thickness of one bar or amount of one space to the total length to be divided—that is, the clear distance between the sight lines of the frame, as shown by the dotted line *a* in the figure—and dividing the total length by the number of intervals between the bars, which will be one more than the latter. Then set a pair of leg or beam compasses to the length of the quotient, and prick off the divisions, commencing at the mark *a*, and following on with the sight line as the second starting point.

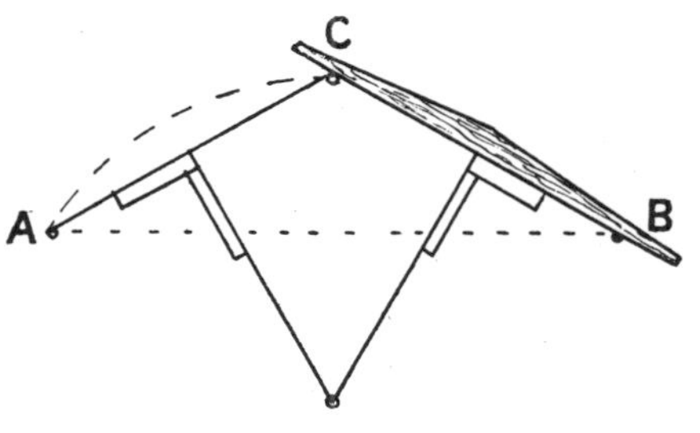

Method of Finding the Centre of a Circle.

The Curves for Circular or Segment-headed Frames may be struck with a radius rod, if the latter is not of inconvenient length, the centre being found by construction, as follows (see figure). Draw the springing line A B, and at its centre erect the perpendicular C, equal in length to the required rise. Draw lines from A and B to C, or rest two straight-edges upon the points as shown, and from their centres square out lines until they intersect. Their point of intersection will be the centre of a circle that will pass through the springing points A B and the rise C. When the centre is so remote that a radius rod would be unwieldy, the segment can be drawn by means of a triangular templet.

To Draw a Segmental Curve by means of a Templet, f. 1.—Let *a b* represent the spring, and *c* the rise of the required segment. Drive three wire nails in at these points. Take a piece of board, and cut off one end at any angle, but so that the edge is at least as long as the distance from *a* to *c*. Then

1. Templet for Drawing Segmental Curves.

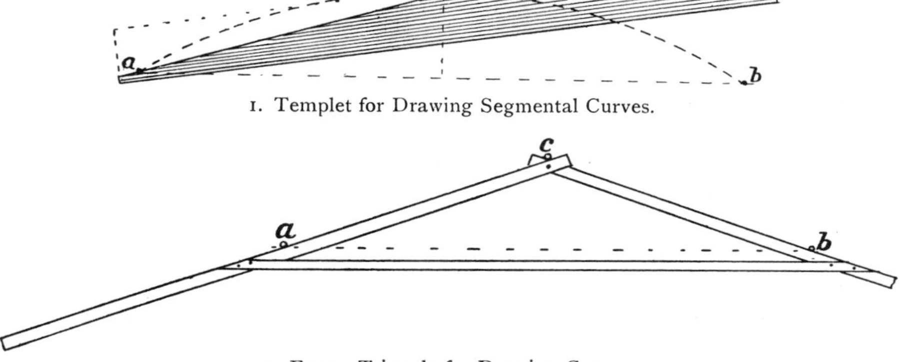

2. Frame Triangle for Drawing Curves.

place the cut edge against the nails, as shown in f. 1, and draw the line *c d* parallel with the springing. Cut the templet off to this line; then if it is moved towards *a*, whilst the edges are pressed to the nails at *a* and *c*, a pencil held at point *c* will describe the half segment, and if the templet is turned over, and moved towards *b*, the other half will be described.

When the rise is too much for the templet to be cut out of a single board.

The Frame Triangle for Drawing Arcs, f. 2, may be used, for arches of any size. The points *a b* represent the springings, and *c* the rise. Three nails are driven in at these points, and two laths of sufficient length (one of them should be twice the length from *a* to *c*) placed against them, and screwed together at *c*. A third lath, to act as a stretcher, is then fixed to the other two, forming the triangle. Remove the nail at *c*, and substitute a pencil. Move the triangle towards *b*, and the pencil will describe the required curve.

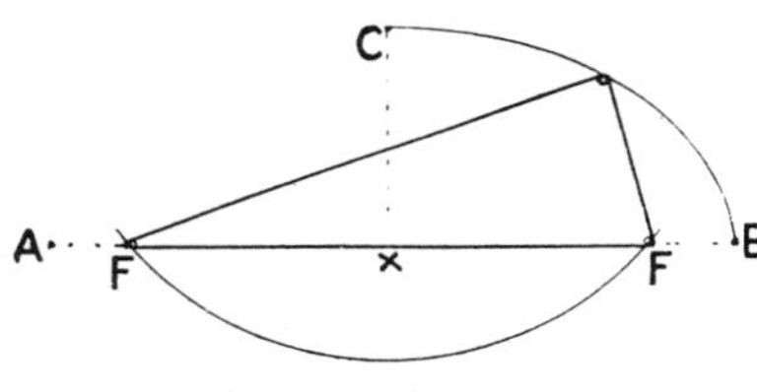

3. Drawing an Ellipse with String.

Elliptical Arches are best set out with the trammel, as described in Chapter II., although small ones may be struck with approximate truth by the string method, shown in f. 3. Let A B, f. 3, represent the springings, and C the rise of the arch. Draw a line from A to B and C *x* at right angles with it. With point C as centre and A *x* as radius, describe an arc cutting A B in points F F.

These are the focal points of the ellipse. Drive nails at F F and B. Take a thin pliable string, and tie it in a loop tightly around the nail at B, and the most distant one at F; then remove the nail at B, and placing a pencil in the loop, move it around from B to A, keeping the loop stretched tightly around the nails at F F, when an elliptic curve will be drawn.

When a solid frame has to be made to fit an existing arch, it is better to obtain a templet that has been fitted to the opening, and use this to set out by, as brickwork is frequently irregular.

Section Templets, f. 1 to 5.—Much time may be saved by the use of section templets of standard sizes for such things as door jambs, sash stiles, sills, pulley stiles, beads, &c. These should be made of zinc, filed accurately to shape, and the curved edges slightly smaller than the required outline, to allow for the thickness of the pencil.

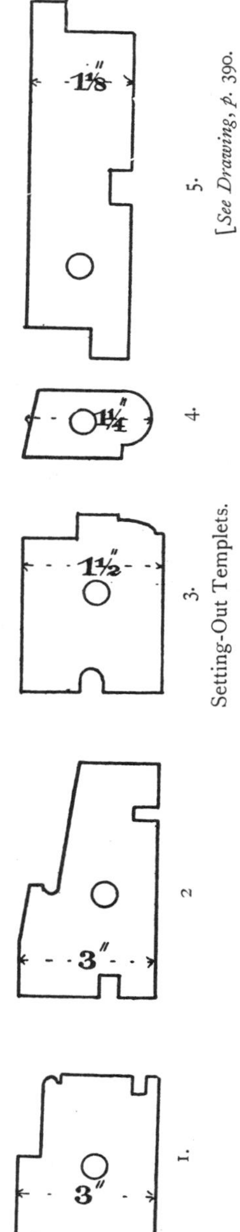

Setting-Out Templets.

[*See Drawing, p.* 390.

Preparing and Storing Rods.—Rods are usually made of pine boards, from 8 to 11 in. wide, ⅜ in. thick, and from 8 to 14 ft. long, planed both sides, and the edges shot. The faces may be rubbed with chalk, or better, covered with a mixture of whiting and thin glue, laid on with a brush whilst hot. Only just sufficient glue should be used to prevent the whiting rubbing up—too much will cause the surface to "glaze." When the covering is dry it should be rubbed smooth with worn glass-paper, and will then provide an excellent working surface. Erasions are easily made with a sharp knife, and the boards readily prepared for fresh work by giving them another coat. Wide rods are also required for elevations and large fittings. These should be made of stouter stuff, and be battened at the back, secured with slot screws. Every shop should have at least one large rod prepared with a true surface from which projections can be made on to the material when set up in position, and true bevels obtained.

Rods are usually stored in racks or on shelves, and should be labelled for identification on one edge, a strip of paper being glued on, as shown on next page, bearing a short reference to the job, the index number, and date. An index book should also be kept in which the entries are made at time of setting out, stating particulars of job, part of building, number of rod when put in hand, and to whom given out.

Where to Begin.—No hard-and-fast rule can be laid down about where to commence in setting out, further than what has already been stated about first laying down openings and sizes of spaces to be fitted, as so much depends upon special circumstances; but as indicating a general method of procedure, the following outline of the method of setting out a sash frame is given.

Setting Out a Sash Frame, &c., f. 2 & 3, p. 390.—Commencing with the "width rod," f. 3, draw in the width between the brick reveals *g g*, and run a line down near the edge of the rod for the outside of the frame. Then carry the reveal lines across faintly, and on the left-hand one set off the thickness of the outside lining, then the sash, parting bead, sash again, then the cover of the bead, usually $\frac{3}{16}$ in., and the thickness of inside lining, which will be the inside face of the frame. Run lines from these points between the two reveal lines, which can now be filled in, as shown, becoming the faces of the pulley stiles. Sink the parting beads $\frac{1}{4}$ in., and draw thickness of the pulley stile, with tongues $\frac{3}{8}$ in. square, on reverse edges. Make the linings $4\frac{1}{4}$ in. wide. Draw the back linings resting on the edges of the outside linings, and grooved into the inside ones. Next the parting slips in line, with parting beads. Draw the sight lines of the outside linings and guard beads, and the section of the sash stiles with a templet. The height rod is set out in similar manner, the height as given being always taken between top of stone sill and the soffit reveal, or if an arch, to the springing, the rise being added separately. When the rise does

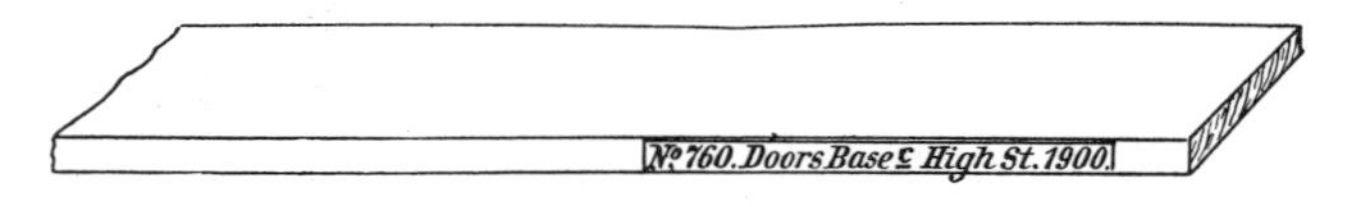

Method of Labelling Rods.

not exceed 1 in., a straight head lining is generally used, but in greater rises the lining should follow the arch.

In Setting Out Inclined Work such as the roof of a lantern, or a polygonal skylight, the true shape of the lights should be developed and set out on a wide rod, in addition to the section of the lantern, because such section will only show the length of the rails that are parallel, correctly; it will not show their true width or the true length of the stiles. The method of obtaining the true shape of the lights has been explained on p. 225, f. 7.

What is Required on a Rod.—The beginner is often at a loss to know what he is required to show on a rod. It may be taken generally that every piece of wood used constructively in the particular job must be shown at least twice—that is, in two sections—as in no other way can the length, breadth, and thickness of each piece be found by the workman. If then, on examining the rod after setting out, and taking each member seriatim, it is not found that its length is shown on one section and its width on another, and its thickness on both, then other sections are required; and when any portion of the work differs from the remainder, a section must be made of each part, and the place through which the section is taken must be indicated by reference letters.

When the work to be set out is too large in either of its dimensions to be drawn upon the rod in its appropriate position, a broken section is made, the dimension of the broken part figured in, and a complete section of the part made, either upon a separate rod or a convenient part of the first. A case illustrating this is shown in f. 1, p. 390, where the rod, not being wide enough to show the head of the jamb linings in its proper position, it is shown broken, and a full width at the other end of the rod.

Material Boards.—After the rod has been set out, a "cutting" and "planing list" should be prepared in shops where machinery is used. These are usually written on small boards about 6 by ⅜ in., of various lengths, bearing the same reference number as the rod, and containing entries in the "cutting list" of

No. 760.—Three Doors Basement, 1 High Street.							
No.	Name.	Length.	Width.	Thick.	Finished Sizes.		
		ft. in.	ft. in.	in.	ft. in.	ft. in.	in.
6	Stiles	6 9	0 4½	1½	6 9	0 4¼	1⅜
3	Mts.	1 6	0 4¼	1½	1 6	0 4	1⅜
3	T Rls.	3 7	0 3¼	1½	3 7	0 3	1⅜
3	M R	3 7	0 10	1½	3 7	0 9¾	1⅜
3	B R	3 7	0 10	1½	3 7	0 9¾	1⅜
6	Pan	1 4	1 4½	1	1 4	1 4	¾ full
1	Mld.	30 0	0 1½	¾	30 0	0 1⅜	to section.
6	Jambs	6 8	0 11¼	1½	6 8	0 11	1⅜
3	Soff.	3 11	0 11¼	1½	3 11	0 11	1⅜

A Material Board.

number of pieces required, their length, width, and thickness—this is for the use of the cutter-out—and a similar list for the machine shop, containing the finished sizes. Above is given an example of a material board for the rod shown in f. 1, p. 390. The amount to allow for wastage in planing depends upon the class of machinery used. Generally ⅛ in. in thickness and ¼ in. in width when both edges are planed will be found sufficient. Rather more should be allowed in hard wood than for soft, as the former springs more, and in some cases requiring reshooting. Mouldings should be traced, and the tracing attached to the board, the quantity being entered.

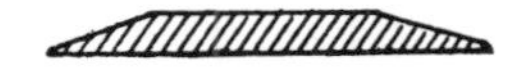

1. Section of Boxwood Scale.

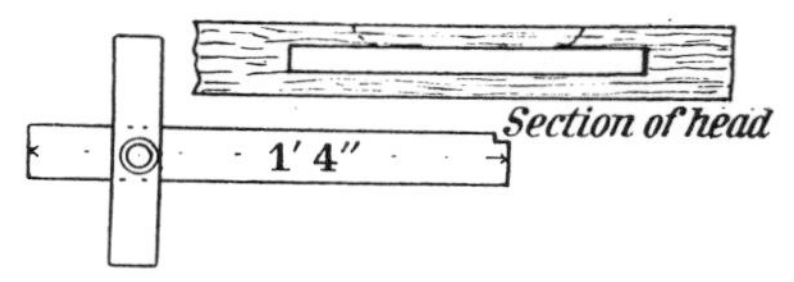

2. Pencil Liner for Setting Out.

The Special Tools Required for Setting Out are one or two pairs of pencil compasses, a pair of dividers, a set of trammel heads, and two or three "beams" of various lengths. A boxwood universal scale, containing scales divided from $\frac{1}{8}$ to 3 in. to the foot. The section, shown in f. 1, previous page, is the most convenient. 2 and 5 ft. rules, 12 and 26 in. wood squares, a large wood bevel with butterfly nut, a T square with divided head, and set screw, set squares of 60 and 45 deg. A pencil liner, f. 2. This tool is used, as its name implies, for running down lines parallel with the edge of the board. It is similar in shape to a T square, but the blade is movable, sliding somewhat stiffly through the head-stock. It is secured in any position by the pressure of the thumb through the hole shown in f. 2. A small notch is made at one end of the blade for a pencil rest, and the upper edge, as shown in the drawing, is sometimes divided into inches and parts. A 30-ft. tape measure is useful for taking dimensions on the building, also a pocket level. A few straight-edges of various lengths, one or two 5-ft. rules with bushed ends, sundry pencils, and coloured crayons complete a foreman's outfit, so far as his setting-out requirements are concerned.

CHAPTER XXIV.

FIXING JOINERS' WORK.

Requirements in Good Fixing—Special Tools—Various Fixings—Wood and Breeze Bricks — Joint Slips — Plugs—Firrings—Removable Fixings — Holdfasts—Rough Grounds—Fixing Framed Grounds—Fixing Jamb Linings—Fixing Architraves—Fitting and Hanging Doors—Open and Close Joint Hanging—Fitting Mortise Locks —Fixing Window Frames—Fixing Window Backs—Elbows and Linings—Fitting and Fixing Boxing Shutters—Fixing Skirtings—Single and Double Faced—Angle Joints—Secret Fixings—Fixing a W.C. Enclosure—Fixing a Counter.

THE FIXING OF JOINERY is, perhaps even more than its construction, subject to the law of circumstances, and therefore the following instructions can only be taken as a general indication of the course to be pursued in the given instances, which may have to be varied according to the quality of the work or the requirements of particular cases. The methods herein described are such as are used in what is termed "good class ordinary work." The chief requirements in fixing are that the work shall be secured firmly to the walls or other points of attachment, with its horizontal members or surfaces level, and the vertical ones plumb. The means of attachment should not injure the walls, and should be simple and economical in their application.

TOOLS AND FIXINGS.

The Spirit Level described in Chapter II., p. 29, is an indispensable tool in fixing, as is also the **Plumb Rule**, f. 1 next page. By the aid of these two instruments the work is placed level and vertical as desired. The plumb rule consists of a parallel straight-edge of convenient length, having an oval hole cut near one end, in which swings freely a leaden plummet or bob, which is suspended from the top end by a thin cord, so fastened that the cord lies immediately over a middle line on the rule, when the edges of the latter are vertical; and the distance this cord swings from the middle line, when the edge of the rule lies against the surface to be tested, shows the amount of its deviation from the vertical. The principle of the action of the plumb rule is that the heavy plummet, when freely suspended and at rest, is drawn by gravity towards the centre of the earth, and therefore the cord radiating from the point of support to that centre, is perpendicular to the earth at a point immediately under the centre of the plummet, and when the edges of the rule are parallel with the cord they also are perpendicular, or vertical to the horizontal surface of the earth at the same point. A useful combination tool that will do duty for a square, plumb rule, and level, is shown in f. 2, and consists of a 2-ft. wooden square with the blade fitted with a plumb bob. The same advantages, to a lesser degree, can be obtained by fitting the stock with a small spirit tube.

Fixings include wood bricks, breeze bricks, joint slips, wood plugs, iron holdfasts, rough grounds, backings, and firrings. Wood bricks are made of the same size as the clay bricks, and are built into the walls at suitable intervals to provide fixings for frames, &c. They are, however, fast going out of use, being superseded by solid bricks made of coke breeze that are fire and damp proof, but will take and hold nails. A later improvement of these is the pan-breeze brick. These are made of a mixture of coke breeze and clay, are burnt much harder, and are made hollow to receive plugs. They do not get red-hot so quickly as the ordinary breeze bricks. Joint slips are pieces of deal the size of a brick, and $\frac{3}{8}$ in. thick, built into the joints of the brickwork. Wood plugs are the most common fixing of any. These are small pieces of yellow deal, shaped as shown in the section, f. 3, and in the sketch, f. 4, and driven into the

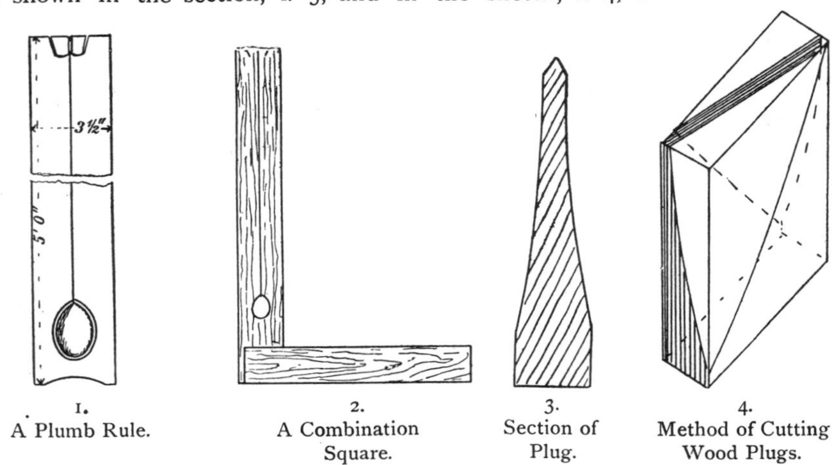

1. A Plumb Rule. 2. A Combination Square. 3. Section of Plug. 4. Method of Cutting Wood Plugs.

joints of the brickwork where required. They should be cleft, not sawn, to a rectangular shape, then bevelled off diagonally on each side to a knife edge as shown, the twisted surface giving them a better hold in the joint. The wall is prepared to receive them by making mortises in the joints with the plugging chisel, f. 14, p. 20, from 2 to 4 in. in depth, according to the nature of the fixing required, and the plug is then driven in steadily, and cut off flush or projecting, as may be desired. Plugs should not be driven into vertical joints near an opening, and when driven into a thin wall, an assistant should stand at the other side holding a heavy hammer against the joint plugged to absorb the force of the blows.

Stone Walls are plugged with dowels let into holes chipped or drilled in the stones, and fixed with cement, either Portland, or a mixture of red lead, oil, and litharge of lead.

A Good Removable Fixing is made by drilling a small hole in the wall, filling this up with Parian or other quick-setting cement, and bedding a screw therein, first well greasing the thread; they may afterwards be removed, and inserted through the work, when the screw will engage with the thread formed in the cement, and hold firmly. Holdfasts (see figure) are plugs of wrought iron from 2½ to 8 in. long, used for fixing solid frames, shelves, pipe casings, and the like in damp situations, or where there is a space between the work and point of attachment. They are driven into the joints of the brickwork in a vertical joint when the lug is wanted horizontal, and a bed joint when it is wanted vertical.

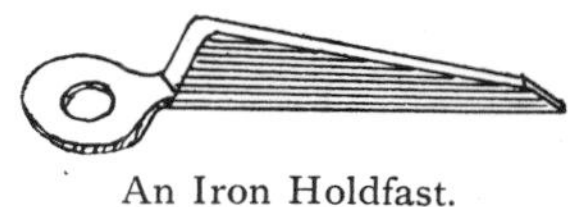

An Iron Holdfast.

Rough Grounds, f. 2, p. 407, are narrow battens from ¾ to 1 in. thick, unplaned, secured to plugs in the walls, and used for the purpose of providing a more continuous fixing than is afforded by the plugs alone. They are often used also as screeds for the plaster at the lower part of the walls, as shown in the figure, and in such cases are usually planed on one side and bevelled on the top edge to provide a key for the plaster. They are also at times grooved to receive bell and electric light wires, and in such cases the skirtings covering them are secured with brass cups and screws. These latter are also called wrought grounds, but generally all grounds not mortised and tenoned together are classed as "rough."

Firrings are small fillets or strips of wood nailed to any irregular surface to bring sundry points upon the surface into line, or to act as packings between the back of a finishing and the wall when there is a void between them.

Framed Grounds are used around the margins of openings, and have been fully described in Chapter XI. They must be fixed square in the opening and plumb on the face and edges. Double sets—that is, a set on each side of the wall—are connected by narrow strips called backings, and the edges of each set must be out of winding, and the heads level. They are generally made ⅞ in. thick, but whatever their thickness the plugs should be cut off so that the face of the ground is 1 in. from the wall. The *modus operandi* is as follows:—Drive a plug on each side of the opening about 3 in. from the floor. Lay a straight-edge on them at 1 in. from the wall and line. Then measure thickness of ground, and draw parallel lines on the plugs equal to the thickness within the former. Saw off to these lines and drive in remainder of plugs, one about every 2 ft. Rest the plumb rule against the cut ones and plumb, mark the edges of the plugs. Lay the straight-edge to the marks on opposite plugs, and line in. Cut off to these lines the plugs will be plumb and square, and the face of the ground will be similar when fixed. Drive two 2½-in. floor brads through into each plug on the skew. Do not punch them below the surface, or strike the ground after the nail is home, as the jar will start the plug. It is also an advantage to bore holes for the nails through the grounds, which will lessen the jarring. Before fixing both sides, try the opening with a rod diagonally for square, and if the floor is not level a line should be marked on each stile at an equal distance from the head, for the purpose of "squaring." When fixing a double set of grounds, in addition to the precautions mentioned above, see that they are parallel across the opening, and before fixing place both sets together, and mark the dovetail sockets for the backings, and cut these out. The backings are placed about 2 ft. apart.

Fixing Jamb Linings.—The grounds being duly fixed plumb and square, cut the soffit between their inner edges, and if there is any space between these and the wall, the ends of the soffit may be allowed to run on, the edges being notched out to clear the grounds. If the soffit is already grooved, arrange it so that the grooves are at equal distance from the edge of the grounds, and nail it up, having first reduced it to the proper width by planing flush with the grounds. If much has to be taken off, and due allowance has not been made in the rebates, these must be corrected before fixing. Next take the clear height between the floor and face of the soffit upon a rod, and apply this to the jambs. It should agree with the distance between the back shoulder line and the floor line on the edge. Test the floor for level through the opening. If it is not correct, set a bevel from face of ground, and apply this to the edge marked by the rod. Cut off slightly undercut, take the arrises off the tongues, enter them in the grooves, placing the jamb diagonally across the opening, and lever it into place with a wide chisel. In cases where the grounds are not fixed to the correct size of opening for the jamb linings, the jambs would first be cut off to the floor line as marked on them, the soffit nailed on, and the set squared, a brace being nailed on the edge to keep it in position. The lining would then be placed within the opening, equidistant from the grounds, and firrings cut in between the grounds and the linings. These would first be fixed to the grounds, and then the linings fixed to them.

To Fix a Set of Architraves.—Sets of architraves having foot blocks are usually framed up in the shop, and only require the latter to be scribed to the floor when fixing, so that the margin at the head is the same as at the sides. The amount of margin shown varies with the design, but commonly $\frac{1}{8}$ in. is shown when the edge of the architrave is square or chamfered. When beaded it should be kept practically flush with the rebate of the linings. To fit, place the set in position against the grounds, and pack up until the under side of head and soffit rebate are parallel. Measure the distance between the two edges, and set the compasses to it less the amount of margin desired. Run the compasses along the floor across the faces of the blocks, and cut off to the lines. Brad the architrave to the grounds at points where the plugs occur, driving the brads into quirks or hollows where they will be least noticeable. Both edges of the moulding should be fixed, and the nails should skew towards each other. Secret fixings of architraves have been described in a previous chapter.

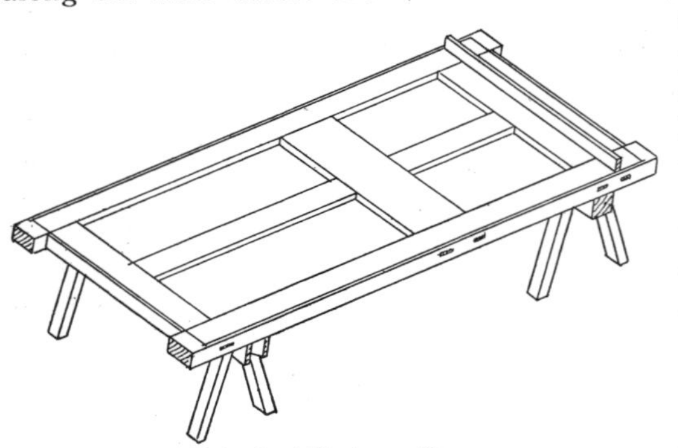

Method of Fitting a Door.

Fitting and Hanging a Door.—As a preliminary proceeding, test the opening to see if it is square in the usual way, by means of a rod applied diagonally. If "out of square," a bevel must be set for each end from the hanging jamb, but otherwise proceed as if the opening were correct. Cut two rods off the neat length between the rebates in width and rebate and floor in height. Lay the door room side up upon two sawing stools, as depicted opposite, and arrange the width rod at one end so that its ends oversail the shoulders equally, and mark its length. Next joint allowance must be made, and this will vary with the nature of the work and the time of year. Doors fitted in the summer time require more joint allowance than if fitted in the winter, as in the latter case they will shrink considerably in the following summer; whilst if fitted in the summer no further shrinkage will take place, but they will probably swell in the following winter. The allowance for ordinary four-coat painted work is a bare $\frac{1}{8}$ in. on each side, and for polished work $\frac{1}{32}$ in. on each side. The joint at top is usually half that at the sides, and at the bottom double that of the sides; but the latter is subject to circumstances, such as provision for thick carpets, uneven floors, &c. Having marked the allowance for joint in accordance with the requirements within the former marks, prick this length on the rod, as also the shoulder lines, and apply it at the other end, keeping the shoulder marks upon the shoulders (if a diminished stile, equidistant from the shoulders). Prick off the joint marks, and with a straight-edge applied at these points, line off the width as shown in the sketch. Next lay the height rod in the middle of the door, and arrange its top end to show the same margin as the stiles. Mark within its length the joint allowances, as described, and square (or bevel) lines through the points from the hanging stile, keeping the straight-edge on the line to square from. Next cut off to the lines all round and shoot the edges, the striking edge slightly "under" from inside of door. The ends should be shot first, then the edges, to prevent the latter being split out. Cut a V notch in the end of the sawing stool, as shown in f. 1, to hold the door whilst shooting. Take the sharp arrises off, and try the door in the opening. Set it at half and wide open to test the floor, and notice whether when resting squarely on the bottom rail its edge is parallel with the jamb, and if not, its want of truth may be corrected by easing the bottom or by arranging the hinges further out where required.

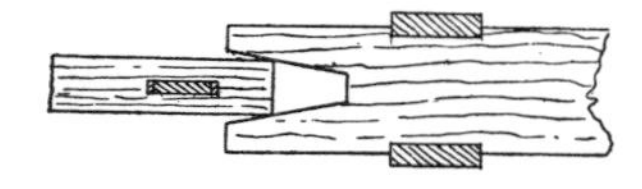

1. Method of Holding Door in Stool.

Hanging the Door.—There are several ways of hanging doors. Two of the best are explained below. These are known as open and close joint hanging. The former is generally used in painted work, the latter in polished.

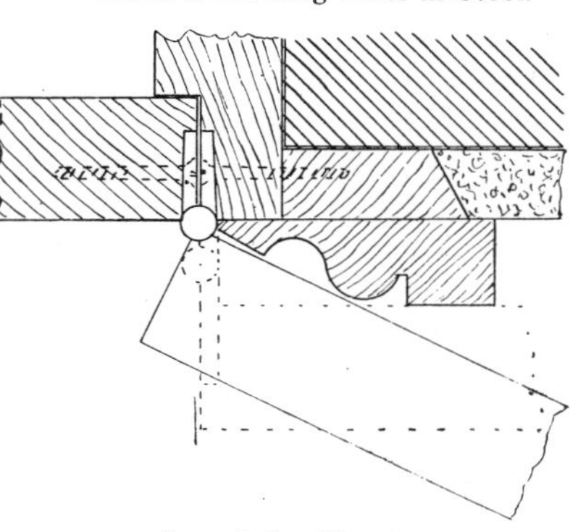

2. Open Joint Hanging.

Open Joint Hanging, f. 2. — This method is so-called because there is a space between the edges of the door and jamb when the door is open. This is more or less according as the door is required to open partly or wide back. Two positions are shown in the drawing—the full outline with the door resting against the moulding showing a gap of $\frac{1}{8}$ in. when open, the dotted line showing the door wide back leaves a gap of 1 in. between.

To Determine the Projection of the Hinge.—Place a straight-edge against the moulding in the position the face of the door is required to be when open back, and measure the distance of the straight-edge from the face of the jamb in a line with the rebate; half of this distance is the projection required from face of door to *centre* of the knuckle of the hinge. Set the door upon its front edge and mark the position of the butts upon the inside edge of hanging stile. Their outside ends should line with the panel moulding, and if a third is used, place it in the middle of the other two. Having determined the amount the hinge is to project, mark this amount across the end, as shown at *a a*, f. 1, and set a marking gauge from the edge of the wing to the line *a*. Run this gauge on the edge of the door from the inside, place the butts to this line, and knife in their ends. Set another gauge to the thickness of the wing close up to the knuckle, and run this one on the face of the door; then cut out the sinkings, not to a parallel depth but tapering upwards from the face, as shown in f. 2, because the wings are thinner on the outside edges. Screw the butts in, being careful to bore the holes square from the face of the butt. Next stand the door in the rebates, first opening the hinges wide back. Cut four small wedges; place two under the door near the edges, and one on each side near the top, and then by manipulating these the joints can be arranged exactly as required. Make the joint on the striking edge a trifle fuller than on the hanging edge, at the top, as the weight of the door has a tendency to pull over from the top hinge; then mark the top and bottom of the butts upon the jamb with a sharp chisel, and holding the chisel perpendicular to the jamb with its face towards the knuckle, draw it steadily down the knuckle, as shown in f. 3. Remove the door, and square in the end lines of the hinges on the rebate. Then set a pair of dividers to the distance of the edge of the hinge from the outside of the door, as shown at *a a*, f. 4, or rather slightly full of this to give clearance, and scribe down the rebate from the edge of the stop. Cut the sinkings out to these lines. Stand the door up on the wedges half open, and turn in one screw in each hinge. Try the swing, and if satisfactory, finish screwing. Should, however, it bind on the floor when anywhere about at right angles with the wall, let

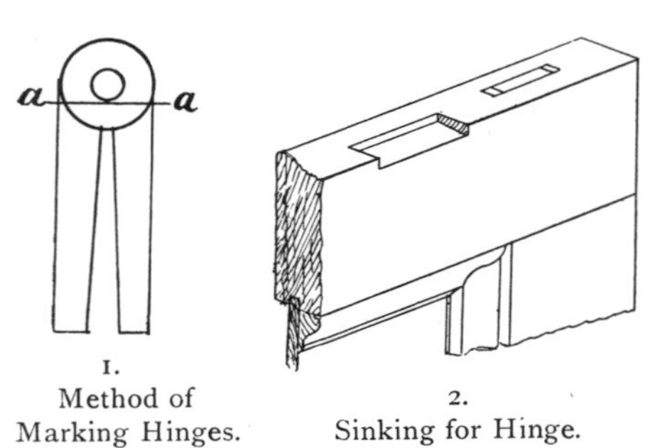

1. Method of Marking Hinges.

2. Sinking for Hinge.

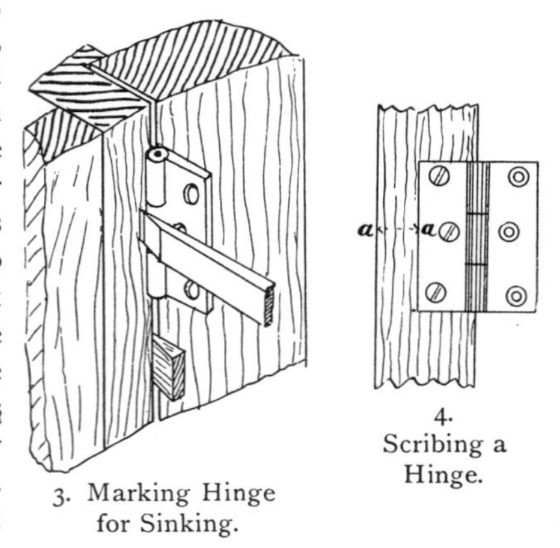

3. Marking Hinge for Sinking.

4. Scribing a Hinge.

the edge of the top hinge in a trifle nearer the stop; and if at the preliminary fitting there is any doubt about the door working free on the floor, it is best to make due allowance for this re-sinking by opening the dividers wider when scribing down from the stop. Should the door bind when either nearly closed or wide open, sink in deeper the top or bottom hinge respectively until it clears the high place on the floor.

Close Joint Hanging (see below).—In this method of hanging, which ensures a close joint at the hanging edge in all positions of the door, the knuckle of the hinge is flush with and sunk entirely into the bead either upon the jamb or the edge of the architrave moulding. The latter instance is illustrated here. The bead must be arranged to be of the same size as the knuckle of the hinge, $\frac{3}{8}$ in. in ordinary cases. Having fitted the door as described above, and in this case with the smaller joints, mark the position of the hinges thereon, and transfer the heights to the moulding. Hold the butt in position flush with edge of moulding, and mark around its outline, then cut away and sink in flush, as shown in the adjoining sketch. Turn one screw in each hinge and close them. Place the door in the rebate against the backs of the hinges, and with a bradawl scribe down its face upon the back of the hinges. Set a gauge to this mark from the front edge of the hinge, and run it on the edge of the door. Cut out the sinking from nothing at the face to the thickness of the wing at the outside, as shown, and turn in the screws square from the face of the butts. Place the door in position as before, and turn the screws into the jamb, and if the work has been carefully done the edge of the door will move around the bead without speaking. Should the door or the architrave be not perfectly straight, the edges will probably rub somewhere in the path. Whichever is in error should be corrected; or a slight chamfer may be made on the salient edge of the door, and a No. 1 round run down it, which will allow the edge to pass the high places.

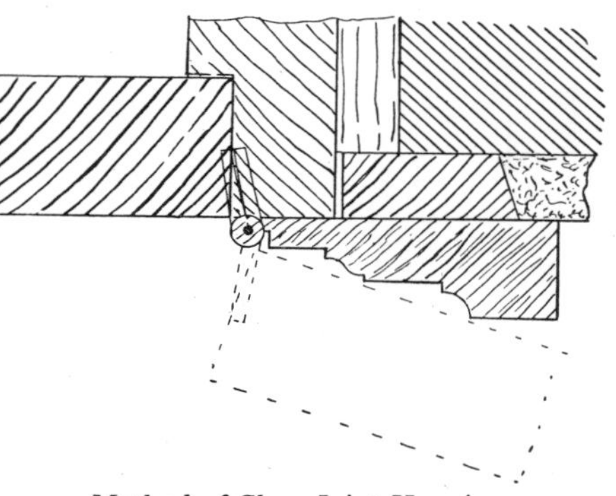

Method of Close Joint Hanging.

To Fit a Mortise Lock.—Mark the mortise upon the edge of the door, as in f. 1, next page, just large enough to allow the body of the lock to slide in stiffly. Bore a hole at one end of the mortise with a $\frac{5}{8}$-in. twist-bit to the requisite depth, and prepare a dowel to fit this hole and about 2 in. longer than its depth. Push this home into the hole and bore a second one close up to the dowel, which will prevent the bit running into the first hole. Repeat the operations until the whole core is bored out. This method dispenses with the use of the swan-neck chisel, and is much more expeditious than the latter. In some shops it is the rule to bore a hole in the lower edge of the lock rail before putting the door together at about 8 in. from the edge of the stile, as shown at *a* in f. 2 & 3 overleaf, but even this is not necessary if the above-mentioned method of removing the core is pursued. Place the lock in the mortise and mark around the face plate (if the lock is to be fitted into a polished door where great accuracy of fitting is required, test the edges of the flange with a try square, and if they are not undercut slightly, carefully file them so all round; the under flange will then pass easily into the sinking made for the face plate without damaging it). Mortise out the sinking, and level it with a router set to the thickness of the face plate and flange. Before trying on the lock, remove the face plate, and pass two screws outwards through the holes in the flange. These will provide a hold for withdrawing the lock, which is otherwise troublesome to remove. Place the lock upon the face of the door flush with the edge and on a level with the mortise, and with a marking awl scribe the key and spindle holes. Square the centres of these around on to the other side of the door, and bore them through from each side with suitable bits, finishing the keyhole with a pad-saw. Both holes should be of a size to permit the passage of the stems freely, but without rattle or side play. Of the two it is better to bore too small and to enlarge the holes with flat and rat-tail files. Next screw in the lock and replace the face plate, fitting this tightly and driving it in place with a piece of wood. Screw on the roses and escutcheons and fix the knobs on the spindle. Arrange the plug or fixing screws of the knobs underneath, and insert the striking plate in the rebate. The position for this can be ascertained by placing a little black oil from the oilstone upon the ends of the bolts and shooting them into the rebate when the door is closed.

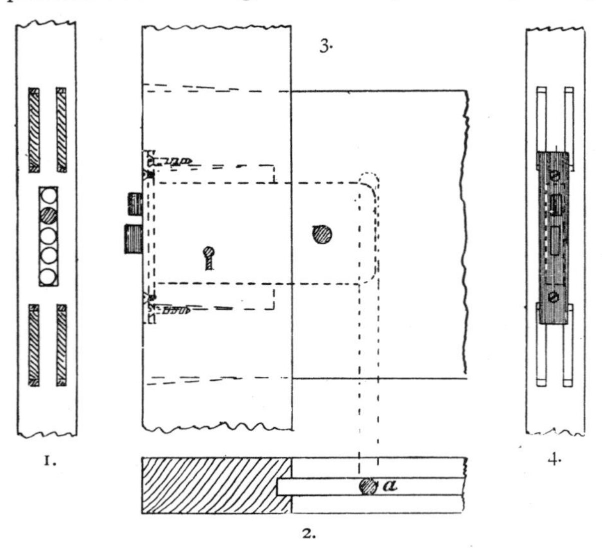

THE FITTING OF A MORTISE LOCK.
1. Edge of Stile showing Mortise. 2. Section of Stile showing Hole in Under Edge of Rail. 3. Outline of Lock on Door. 4. Edge of Door with Face Plate.

Fixing Window Frames.—In common work the window and external door frames are built into position as the walls are carried up, the sides being bedded in mortar on a bond timber, and the frames propped upright with stays from the joists, &c. Such methods are not advisable, as the dampness of the green walling sets up decay in the timber and causes the work to swell and warp, making good fitting impossible. For properly fixing window frames, fixing blocks should have been built into the brick jambs, or failing these, the jambs should be plugged about two or three on each side according to the height. The sill of the frame should be bedded level on the stone sill either in cement or white lead, and if an iron water bar is used, this should be painted and placed in the grooves, the frame arranged to show an equal margin on each reveal, and the face plumbed upright. The inside should be tried for square (unless the sashes are in, when the frame may be wedged so that these will slide properly) and wedged correctly, firrings cut in between the plugs and back of linings, and these skew-nailed to the plugs. If a good fixing cannot be obtained at the sides, blocks may be cut in between the head and the lintel over the pulley stiles. Next fix the jamb linings. These usually require firring out from the wall, or if the space is not much, the plugs may be left long enough to form a backing. The amount of projection is ascertained by entering the lining in the groove and marking its back on the frame, then with a square or bevel as required held to this line, mark the length of the plug or width of the firring piece, and cut them off to the line, then bring the lining to the required width and fit in each piece separately. Prop the soffit into position, square and tongue the top ends of the jambs, and holding them in position, mark their faces on the soffit. The latter can then be taken down and grooved, and the jambs cut off to the window-board or floor as required, the whole nailed together and slipped into place between the firrings to which they are nailed. The framed grounds are then nailed to the edge of the linings and the architrave to them. In cases where the grounds are previously fixed, the grooves in the sash frame should be made to line with the edges of the grounds, and the firrings made flush with the latter, and where the opening is fitted with a window back this would require fixing before the linings.

Fixing Window Backs.—These are usually fixed to rough grounds that are plugged to the wall under the window frame, but sometimes the top edge of the framing is made to rest against the oak sill under the nosing piece, and in such cases firring pieces only are required at the bottom to take the horns of the stiles. These are placed plumb under the face of the sill and nailed to the floor. Proceed to cut the framing so that it will stand between the brick jambs. Shoot the top edge and pack it parallel with the groove of the nosing. Measure its height above the bottom side of the nosing, and scribe that amount off the stiles at bottom. Arrange the frame in the opening so that the margins are equal, and mark down the lining grooves with a straight-edge from the sash frame. Plane the edges of the stiles parallel with these lines and plough in the grooves. Fix by screwing or nailing to the grounds, and glue in the nosing.

Fixing Boxing Shutters.—The method of fixing these finishings varies somewhat with the details of their construction, but the general procedure, subject to the above, is as follows :—Fix the window frame, back and elbow linings, as previously described. The position of the elbows is governed by that of the shutters with which they line, and must be obtained from the drawings. Plumb the face of the grooves in the frame down to the floor, and draw a line on the floor to represent the face of the elbow, splayed or square as required. Nail fillets on the floor parallel with these lines, but the thickness of the elbows behind them, and these with the bottoms of the boxings will provide fixings for the elbows. The soffit is next fixed to firrings from the lintel levelled down to bring it in line with the groove in the frame, and with the rails showing an equal margin on each side. If a back lining is used in the boxing, the groove to receive this must first be worked, its position being obtained from the groove in the window frame. Proceed to fit the back linings into the groove in the sash frame and soffit, and arranging them parallel with the elbows, mark their width to the back of the grounds, also mark the groove for the boxing bottom from that on the sill, and work these and also a tongue on the front edge if required, as in f. 2, p. 404. The linings are next propped up in position and the grounds fitted to them, then all the tongues glued and the work nailed together. The bottoms are next inserted and nailed to the elbows. Brad a $\frac{3}{8}$-in. bead temporarily to the soffit, and proceed to fit the front shutter into the opening and hang it to the sash frame out of winding with the grounds. The soffit bead may then be fixed permanently, flush with the shutter (see f. 1, Plate XXII). Next open both shutters upon the window frame, and ascertain whether they are parallel. If the boxings have been fitted plumb they will be; but if not, when the hinges are fixed to the flaps, they must be placed with the centres of the knuckles parallel with the *hanging* edge of the shutter, and plumb over each other, otherwise the flaps will bind at the bottom. Take the clear space between the edges of the shutters on a rod, then lay all the flaps together on the bench and mark their width with this rod : if there is much to come off, the amount should be distributed among the several rebates. Take down the shutters and lay them beside the flaps, arranging them exactly square with their middle rails in line, and mark across the ends with a straight-edge. Cut these off and shoot the edges square. The back flaps can now be fixed in line with the butts. If the flaps are less than $1\frac{3}{8}$ in. thick, simply screw them on the face; but if thicker, sink them in flush. Place the knuckles in the centre of the joint with the exception mentioned above, and see that the knuckles range in line over each other by applying a straight-edge to their sides. Each half set may now be rehung and tried in the boxings, and any necessary alterations made. The fixing of the drop-bar and the knobs as described in Chapter XII. will complete the job.

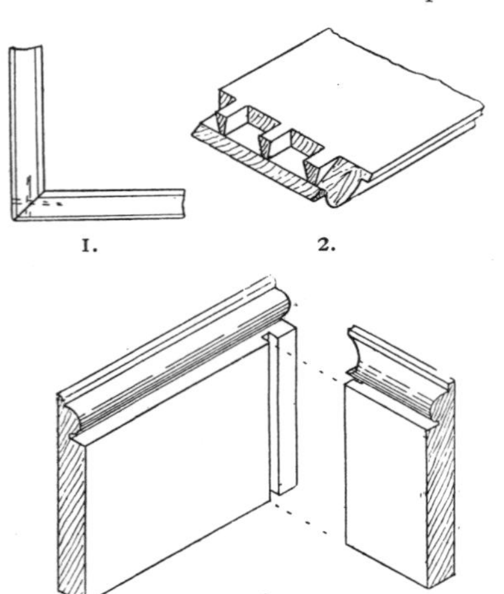

METHODS OF FORMING ANGLES OF SKIRTINGS.
1. Plain Mitre. 2. Dovetail Mitre. 3. Grooved and Scribed.

Fixing Skirtings.—Ordinary skirtings less than 10 in. deep are fixed to horizontal grounds, plugged to the walls or nailed to the studding in partition walls. Firrings are also nailed to the latter flush with the ground, and against the brick walls. Fillets are nailed to the floor to take the bottom edge of the skirting board. The latter should overhang the ground $\frac{3}{8}$ in. at the top, and be levelled and scribed to fit the floor at the bottom, as described under the head of "Scribing," Chapter IV. The external angles are plain mitres cross bradded, as shown in f. 1, or in the case of a polished skirting, mitre dovetailed, as shown in f. 2. The internal angles should always be scribed, and in good work are also grooved and tongued, as shown in f. 3. Very wide and double-faced skirtings are fixed to horizontal grounds and vertical backing pieces, and their lower edges are usually tongued to the floor instead of being scribed. Fig. 1 below is a section of a skirting so treated, and f. 2 is an isometric sketch of the same. The horizontal grounds are first fixed straight and level, and if these are done after the plastering is finished they must be made to lie flush with it. The best way to get all the plugs in a straight line is to drive the two end ones, and cut them off to the proper projection. Drive a nail in each, and stretch a chalk line between them; the intermediate plugs can then be cut off where the line crosses them. Backing pieces are next cut in between the ground and the floor, about 3 ft. apart, with two plugs behind each for fixings; drive them tight against the plugs, and hold a plumb rule against the face of the ground, marking its edge on the side of the backing piece. Another line can then be drawn parallel with this at the required projection to fit the lower part of the skirting, the height marked on, and the piece notched out to

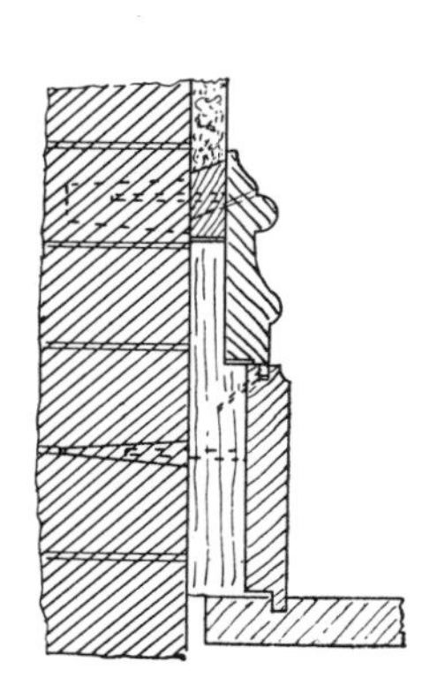

1. Section of Double-faced Skirting.

2. Sketch of Fixings for Fig. 1.

the lines, as shown in the sketch. When these are all fixed, cut the plinth in between the walls, pressing it close to the backings, mark its face upon the floor, then nail a fillet on the floor parallel to this line, but $\frac{3}{8}$ in. behind it, as a guide to the grooving plane, p. 16. When the groove is worked, the lower edge of the plinth is rebated and the tongue fitted. If the floor is out of level, the upper edge of the plinth should be levelled, and the lower edge made to fit the floor approximately before rebating it: generally it is best to fit the longer sides first; but when the plinth is grooved and tongued at the angles, as in f. 3 opposite, the groove should be put in the shorter piece, and this is necessarily placed in position first. The piece with the tongues on is then swayed out in the middle, the ends entered in the grooves and then sprung in, the whole driven down into the grooves and fixed. Where possible the external angles should be fixed together before placing in position, the dovetailed angles being glued up and allowed to dry before fixing. The moulded member is next fitted and fixed in a similar manner, the two opposite through pieces being placed in position first, and the clear distance between them taken on a rod which is cut in tightly between them. This rod gives the length of the sight lines on the scribed piece, and a mitre templet held to these lines will give the required mitre, which is afterwards scribed with saw and gouge. The top edges only of each portion are fixed by nailing, the lower edges being free to swell or shrink.

Secret Fixings.—The adjacent figure shows the method employed to secure hardwood skirtings with secret fixings, the various pieces being fitted as described above. Stout screws are turned into the backings with their heads left projecting about $\frac{3}{8}$ in., and corresponding slotted holes made in the back of the skirting, the slots running upwards, as shown. The screws must all be turned in to exactly the same depth, and this is best ensured by using a $\frac{3}{8}$-in. slip as a gauge; after they are all in, their heads are blacked with oil, and the skirting resting upon two $\frac{3}{8}$ in. slips is pressed against them, the imprint showing the places of the holes. An extra backing piece with screws turned in is used as a templet to fit the slots, and should be driven home in each one. All the skirtings containing internal angles must be knocked down together, and several men should assist, so that all parts may be driven equally, otherwise the fixings will be strained. A few sharp heavy blows are better than many light ones, and guard pieces should be used to prevent damage to the edges. After each blow down, the face of the skirting should be struck sharply opposite *every* screw, to jar them into position, as they may drag; each screw should be previously located by marks on the floor and plaster. In pitch pine or other sticky woods it is a good plan to rub the screws with tallow before driving.

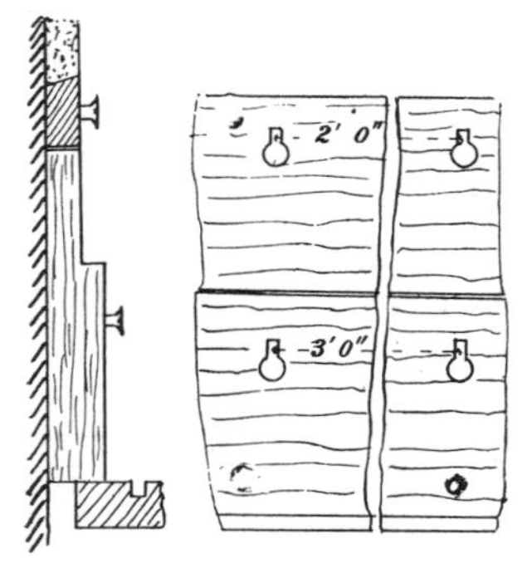

Secret Fixings for Skirtings.

To Fix a W.C. Enclosure.—Assuming the bracket bearers framed and fitted to the seat as usual, remove them and cut them off to such a length that the rebate to receive the seat, or if not rebated, the top edge of the bearer, will be about $\frac{1}{8}$ in. above the pan, first trying the pan by levelling and measuring from the highest part. Next cut two holes in the wall, large enough to receive the ends of the bearers, and about 3 in. deep. The lateral position will be located by the length of the seat, or if this is immaterial, place them about 1 in. clear of the pan: the height will be the same as that of the legs. Wedge the bearers in level in both directions, and parallel to each other, having first cut them off to length, which will be governed by the width of seat or position of riser. Have the holes stopped with cement. Next fit in the riser; if this runs from wall to wall, cut a straight-edge between the plaster, and level its bottom edge. Set bevels from this to fit the walls, and transfer length and bevels to the riser; working from its top edge, make the margins equal, and cut off, running the saw slightly under. If there is a skirting fixed, the riser must be cut over this, so place it in the opening resting on the skirting, and level its top edge; apply a straight-edge to the face of the skirting, and mark its edge on the riser. Set a pair of compasses to the distance between the top edge of the riser and the bearer, or that portion on which the seat rests, and scribe the bottom, also the skirting. If the moulding on the latter is at all intricate, it is best to fit a templet to it first, and use this to mark the riser with. Fasten the riser to the legs by buttons, bolts, or screws, as specified, and fit in the seat between the bearers, cutting the holes to suit the fittings as described at p. 296, f. 2. Cut the flap frame in the clear of the plaster, with the edges of the wings equidistant from the hole in the seat, or to coincide with the stiles of the riser frame, as the construction may indicate, and with the nosing back, tight up to the face of riser. This frame is sometimes screwed to the bearers, the screws being turned in from underneath through the hole, but is more usually secured by a plinth around the top; the wall is plugged, and the back plinth cut in and fixed with nails or cups and screws, then the side pieces fixed in similar manner.

To Fix a Counter similar to f. 1, Plate XXVIII., first shoot the top rails of the various portions to equal widths; place the first portion in its appropriate position with the stile against the wall. Level the top edge and plumb the face. Set compasses to the difference between the present and the required height, and scribe this amount off the horns of the stiles. Cut off waste, replace in position, and scribe the end to the wall. Prop up in position, and nail fillets to the floor, inside the stiles. Run these in line with a chalk line. Next level and scribe the return piece of framing shown in the plan, and work the tongue thereon; place it in position and mark its face on the inside of the front piece, take the latter down and run a groove to the mark. Set up and fix. Treat the second portion of the front similarly to the first, scribing it down to the level. Fit in the handrail bolts as shown; connect up, and nail the bottom ends of the stiles to the floor; the circular corner having been previously glued on. Take the length between the edge of the circular end and the screen upon a light rod; apply the latter to the return counter front, and mark at top and bottom; cut the end stile off to this mark, place the front in position, and scribe down to level of the remainder. Fix this in like manner. Next fit the false back into the grooves and scribe it down, taking the groove for the top C as a guide for amount. Having cut this, slide it down, and fit the underfittings into the grooves, scribing down the plinths, and blocking the back as shown in f. 2. Next fit in the framed dustboard and the drawer rail, making the former parallel and rebating the top edge: glue and nail the drawer rail to it. Dovetail in the cross bearers. Take the clear height between the drawer rail at D and the floor, and cut the pedestal down to this; place it in position against the end of the cupboard, fit in the return rail and dustboard, and the filling in piece against the pedestal. Fix rough bearers for the return counter bottom, and fit this; when the remainder of the plinth can be scribed down, and glue blocked. The pilasters in the front must be shot tightly into the grooves provided for them, glued in and screwed from the back, the skirting in each bay being mitred tightly between the pilasters, and scribed down to fit between the rebate and the floor; notched backing pieces are fixed behind to receive it. The fixing of the tops is explained on p. 253.

CHAPTER XXV.

NOTES ON TIMBER.

What constitutes Timber—Botanical Divisions of Trees—Growth of a Tree described—Why Sapwood should not be used—Structure of Wood described—Components of Wood—Cause of Wood being Hard or Soft—Cause of "Figure" in Oak and Plane Tree—Time of Maturity in Trees—Proper Time for Felling—Natural Seasoning—Wet Seasoning—Desiccation—Second Seasoning—Methods of Stacking Timber—Diseases and Defects of Timber—Wet Rot, Druxiness, Foxiness, Plethora, Doatiness, Dry Rot—Heartshake, Starshake, Radialshake, Cupshake, Rindgall, Upsett, Wandering Heart—Cause of Warping—Conversion of Softwoods and Hardwoods—Production of "Wainscot Oak"—Effect of Shrinkage upon Boards cut from Different Parts of a Log—Classification of Timber—Characteristics of Hardwoods and Softwoods—Minor and Local Classifications—Market Forms.

TIMBER is the term applied to trees that are suitable for building and similar purposes, and is confined to the class of trees that increase in size by successive layers or deposits of tissue upon the outside of previously formed layers. The botanical term for this class of tree is **Exogen**, signifying an outward grower, as distinguished from **Endogen**, an inward grower. These are the two grand orders into which the vegetable kingdom is divided, *i.e.*, omitting the fungi and cryptogams or non-flowering plants.

GROWTH OF TREES.—Described briefly, the method of growth of an exogenous tree is as follows. In the spring of the year, owing to the effects of the sun's warmth, vegetation, which has lain dormant during the winter, begins to awaken, buds begin to form, the roots absorb moisture from the earth. This fluid, chiefly composed of mineral salts in solution, ascends through the vessels of the tree, passing by way of stem and branches to the leaf buds, which expand under its influence, and are pushed forth, and the leaves develop. When this function of the sap is fulfilled, the leaves themselves take up their portion of the duty, and proceed to convert the watery sap supplied them by the roots into wood-forming matter. Most of the purely watery portion of the sap is evaporated from the surface of the leaves through innumerable microscopic orifices in the cuticle called **Stomata**, and the residue under the action of sunlight and air undergoes sundry chemical changes, becoming converted into starches, sugars, resins, &c., according to the requirements of the particular tree, and ultimately during the summer and autumn descends towards the roots again. This perfected or "proper sap" in its descending course forms a zone or ring of much denser tissue around the loosely formed zone produced by the ascending sap in the earlier part of the year, and this alternation of density and porosity in the successive layers of tissue produce the so-called annual rings observed in the cross section of a tree stem. As the tree ages, the interior zones of tissue become practically solidified, partly by deposit of the perfected sap within the intercellular spaces, and partly by the thickening of the walls or sides of the vascular tissue due to the absorption of the wood-forming secretions from the sap. This part of the wood, known as heartwood (or botanically as duramen, a word signifying durability), continues to acquire hardness and denseness for a considerable time, varying with the kind of tree and its situation, until it arrives at maturity, after which period it commences to decay; and the most suitable time to fell the tree for use is just at or before this period of maturity arrives, for then the bulk of the mass of the tree is at its most finished and endurable state, and the natural decay, if commenced, is at once arrested when the tree is killed.

The outer zones of tissue through which the sap circulates are known as sapwood or alburnum (in reference to its light colour), and this portion of the tree, in consequence of its loose spongy texture and the presence in its tissue of fermentable juices, should not be used for building purposes, as it is weak and inelastic and subject to rapid decay arising from its unfinished state.

Trees cut down before their prime or maturity display an inordinate amount of sapwood, and balks indicating this should be rejected, as such wood will be wanting in durability. The area of sapwood continues to lessen after the third or fourth year, up to which time the substance is all sapwood, until the period of maturity arrives, when, although the sapwood continues to increase, the heartwood begins to decay from the interior, and in the effluxion of time and the progress of decay, the tree in its senile state approaches again the state of its adolescence and becomes all sapwood.

Structure of Wood.—Wood is composed of cellular tissues of various kinds, each having separate and distinct functions to perform in the growth of the tree; but as their study belongs rather to the science of botany than to the art of woodworking, only slight reference need be made to them here. Those wishing to pursue the subject further should consult a small treatise on "The Oak" by Professor Marshall Ward,* where the subject is entered into fully. The bulk of the stem of a tree is made up of **Woody Fibre.** These are short, thick-walled hollow tubes with pointed ends overlapping each other. The character of these particular tissues differs somewhat in the wood of hardwood and soft-wood trees, the variant constituting the mass of softwood trees being termed **Tracheids.** The function of the true wood **Fibres** appears to be to add strength and toughness to the wood, as they are more numerous in the woods possessing these characteristics. Interspersed between the fibres are long tubular cells known as **Pitted Vessels.** It is through these elements that the sap circulates, and penetrating vertically between the rows of vascular tissue (bundles of vessels) are narrow bands or groups of cells which radiate from the centre or pith of the tree. These are the **Medullary Rays** or septa, and their function is to carry nourishment from the descending sap to the interior parts of the tree, and acting thus as conduits of the sap when highly charged with wood-producing

---

* Published by Kegan Paul, Trench, & Co., Charing Cross Road.

matter, the walls of these cells become thick and extremely hard by their absorption of lignine, and thereby act also structurally, as ribs, affording lateral strength and stiffness to the softer tissues. The medullary rays, due partly to their relative denseness, and partly to the contrast between their horizontally arranged cellular tissue and that of the vertical system of the rest of the tissue, afford in many woods a pronounced "figure" or pattern interwoven with the grain, when the surface of the boards cut from the tree lie parallel with the direction of their septa, familiar instances of which are the "silver grain" of the oak and beech, and the dappled or "partridge figure" of the plane tree.

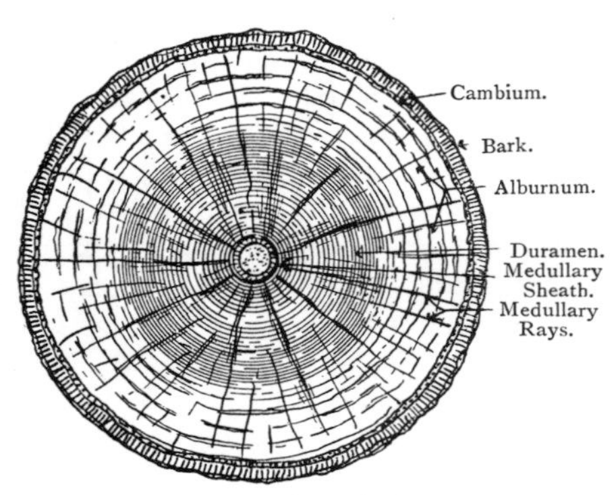

Transverse Section of a Tree Stem.

In the great majority of woods, however, including all the softwoods, these medullary rays are too fine to be perceptible to the naked eye, although they are always present. The adjoining figure is a diagrammatic transverse section of an exogenous tree stem, indicating the structure of the living tree: the **Pith Column** is shown in the centre surrounded by a thin membrane, the **Medullary Sheath**. This is the first formed ring of woody tissue deposited upon the pith or original shoot: the larger shaded portion encircling this, indicates heartwood, whilst the surrounding untinted zone is the sapwood, encircled in turn by the bark or cortex. A thin zone of tissue is shown lying between the bark and the sapwood; this is known as the **Cambium**, and is the ultimate ring of sapwood for the year in course of formation, the fundamental cellular tissue, from which is developed the various elements described above.

The hair-like lines radiating from the pith column are the large or **Primary Medullary Rays**, and the shorter ones interspersed between them are the small or **Secondary Medullary Rays.**

Felling.—Trees should be felled, as before mentioned, when they reach maturity, which period varies from 50 to 120 years, according to the species, and in the winter months, in cold and temperate climates when vegetation is suspended; the corresponding period in tropical countries being the dry season, as the less moisture there is in the substance of the tree the quicker it can be seasoned and used.

Seasoning, literally, is the submitting of green timber to the action of the weather for several seasons, during which period the natural juices gradually harden, and the moisture evaporates without any detriment to the timber. The term has, however, been rather loosely applied to all processes, whether natural or artificial, whose object is to remove the moisture of green timber, the latter process being more accurately described as drying.

Proper or natural seasoning has much the better effect in preserving the strength and ensuring the durability of the timber than any artificially accelerated process, as under the former influence all the natural secretions of the wood are preserved, only the uncombined watery matter escaping, whilst under the latter much of the partially solidified contents of the vessels are extracted with the watery sap. In preparing for seasoning, the trunks are lopped, barked, and hewn square as soon as possible after felling, then stacked in open formation, under cover, so that air can circulate freely around and between them. Protection from the direct action of the sun should be provided, and also from the rain. The adjoining figure, showing the end of a stack, indicates the usual manner of building the stacks. When individual balks are likely to be required from the interior of the stack, short blocks about 1½ in. square are placed under each balk at intervals, and by removing these a single balk may be extracted without disturbing the remainder. Timber requires from two to five years to season in this manner, according to its size.

1. Method of Stacking Balks for Seasoning.

Wet Seasoning.—In this method, which considerably shortens the time necessary for seasoning, the logs are submerged in a running stream, with their butts or root ends towards the flow of the tide, for from fourteen to twenty-one days, according to the size of the timber. The process considerably diminishes the strength of many woods.

Desiccation.—This is the submission of the wood in closed chambers to swiftly moving currents of dry-heated air, which absorbs the moisture contained in its pores; the moisture-laden air is drawn off from one end of the chamber, and replaced by a further supply of dry air at the other end. The temperature is kept constant at about 200 deg. Fahr. for about three days. This process has a tendency to render the wood brittle or "short," and reduces its tensile strength. It also bleaches to some extent the more highly coloured woods.

2. Method of Stacking Boards for Seasoning.

Second Seasoning is the further drying of timber after it has been seasoned in the log, then cut to scantlings, either planks or boards, the object being to dry out the sap of the interior parts that could not be reached in the first process. The sketch above illustrates the method applied to an oak log cut into boards.

The packing slips, which should be of the same kind of wood as the timber stacked, are placed about 4 ft. apart, and should be moved occasionally sufficient to expose the part covered by the slip, otherwise an indelible stain will result in the boards. Strips of wood or hoop-iron are also nailed on the ends of the boards to prevent them splitting whilst drying.

Diseases and Defects.

Growing trees are subject to sundry diseases and decay that affect the quality and quantity of timber that they yield.

Over-Maturity.—Trees reach their prime or perfect state in periods ranging from 50 to 120 years, according to their kind. When this stage is reached they begin to decay, at the heart first, and from the root upwards and outwards; in many cases where this decay has proceeded for a long time, the stem is found to be merely a hollow shell of sapwood supporting the bark, but continuing to put forth branches, leaves, and roots.

Wet Rot is a decomposition of the tissue of the tree set up by the access of water through wounds in the bark. This water, not entering into the circulation of the sap, becomes stagnant, and ferments the natural secretions, which are oxidised or burnt up in the process: the disease, progressing downwards, is chiefly confined to the bundle or group of vessels first attacked. The appearance of seasoned wood that has been so attacked shows a long deep furrow of greyish brown powdery remains, with blackened edges.

Druxiness or **Druxy Knots.**—This is the incipient stage of rottenness described above. A branch is broken off close to the stem, but irregularly, forming a receptacle for rain-water; in this water, floating spores of fungus are deposited and fructify, setting up decomposition. Appearance of the converted timber: light-coloured spots or streaks, which are soft and wanting in cohesion.

Foxiness is the term applied to reddish brown stains in oak and similar woods which indicate the commencement of decay, set up by over-maturity.

Plethora is a disease caused by an excess of nutriment being supplied to some portion of the tree at the expense of the other portions. Its causes are problematical, but its effects are that the wood is rendered less homogeneous, becoming hard and brittle in the over-supplied parts, and soft and pliable in the impoverished portions.

Doatiness is another form of incipient decay found in birch, beech, and oak. It is indicated by a dull white or greyish stain sometimes speckled with black.

Dry Rot is a disease of felled or dead timber due to the decomposition of its substance by a parasitic fungoid growth. There are several kinds of fungus which thus attack timber when the conditions are favourable, but the one most commonly found in buildings is the **Merulius Lachrymans**, or weeping fungus. In appearance it is a white cottony mass, with orange, red, or brown edges and fibres or tendrils. It covers and penetrates the vessels of the wood in all directions, searching for the secretionary matter which it absorbs, developing quickly into a thick unctuous mass, which in its early stage is covered with drops of water like grass on a dewy morning—hence its name. It eats up or destroys all the secretions of the wood, leaving it merely a mass of empty tubes that crumble under the touch. Its presence may be detected by a fœtid musty smell, also by brown spots or patches appearing on the surface of the timber. Damp, ill-ventilated situations are the most favourable to the propagation of this disease—a moist, warm atmosphere appears necessary for the growth of the fungus ; and conversely in cold, dry, well-ventilated situations, the wood is never attacked. The spores of dry rot fungus are said to be destroyed by corrosive sublimate, and when the disease has not progressed very far, it may be arrested

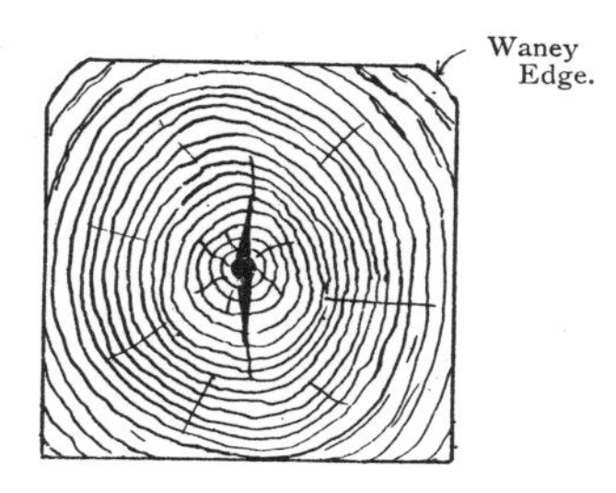

1. Balk with Slight Heartshake.

2. Balk with Bad Heartshake.

by washing all the parts attacked and the surroundings in hot limewash ; but thorough ventilation provides the most certain remedy.

Heartshakes are clefts starting from the heart or pith, and running towards the sapwood, as shown in f. 1 & 2. They are occasioned by the shrinkage of the interior parts due to age, and are a preliminary stage of decay. They are almost invariably found in trees past maturity, but an incipient heartshake may develop after a tree which is approaching maturity is felled, and is allowed to lie some time unbarked, which prevents the evaporation of the sap

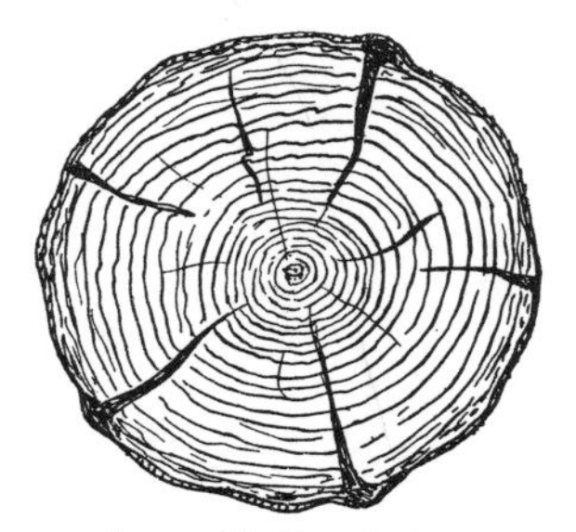

3. Log with Starshakes.

4. Log with Cupshakes.

from the outside, but allows it to proceed from the central parts. A heartshake straight across the trunk, as in f. 1, is not a serious defect in conversion, but when crossing at right angles, as in f. 2, it occasions considerable waste.

Starshakes (see f. 3) are numerous radial clefts in the tree commencing immediately under the bark, and running in the planes of the medullary rays towards the centre. They are chiefly confined to the sapwood, and their position is indicated on the living tree by ridges or ribs in the bark, due to the latter rising up from the crevice beneath. The starshake is not a sign of disease, but is the result of climatic disturbance during the growth of the tree. It may arise from severe frosts freezing the sap and causing the bursting of the tissue, or from fierce heat of the sun in dry seasons burning up and destroying the outer tissue, which shrivels up and recedes apart.

Radialshakes, somewhat similar to starshakes, occur in felled timber when the latter is exposed to the sun in seasoning. This causes the outer rings of tissue to dry relatively much faster than the inner ones, and the circumferential shrinkage set up results in numerous radial clefts in the substance. These may be identified from the true starshake by their number and fineness and also their irregularity, many appearing to start a few inches within the bark, and running a short distance towards the centre, will then follow the course of an annual ring for perhaps an inch, and again radiate towards the centre. This erratic course is due to the cleavage following the lines of the medullary rays, which is the line of least cohesion.

Cupshakes, f. 4, previous page, are clefts between the annual rings, and denote unequal growth at the parts of rupture, due probably to sudden increased moisture supplied by the roots after the heavy storms peculiar to tropical

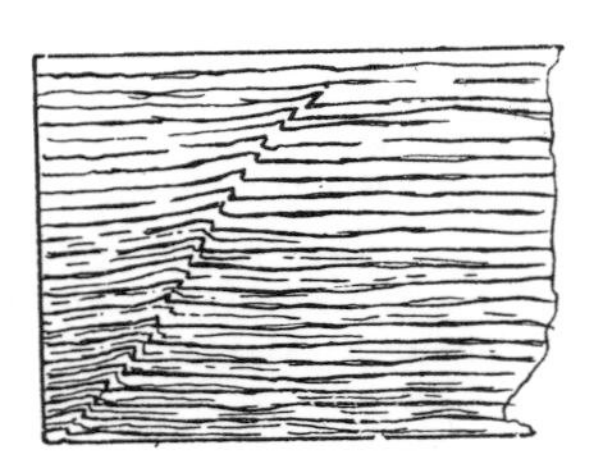

1. Balk Damaged by Upsett Fibres.

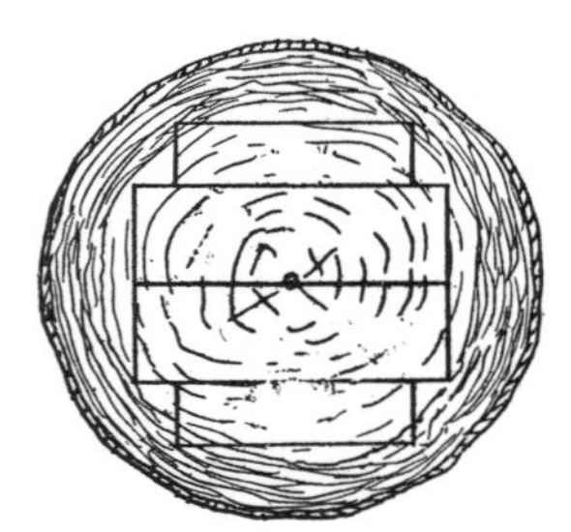

2. Conversion of a Log to Deals and Battens.

countries, these defects occurring principally in the trees of tropical climates. The theory that they are caused by the twisting of the tree in high winds has also been put forward, but there is little evidence to support this.

Rindgalls are protuberances upon the trunk of a tree, due to a branch having been broken off at the point and covered by successive layers of sap-wood, which, however, do not unite perfectly with the older wood, and either a cavity is formed, or the course of the fibres is so interrupted at the point that when the tree is converted it is wanting in strength and cohesion.

Upsetts, f. 1, are a crippling or buckling up of the fibres, due chiefly to unskilful felling; the trunk being thrown violently upon its end, crushes in the fibres as shown. When a tree that has met with this injury is converted, the parts so crushed will break through readily. A common defect known as **Wandering Heart** is due to the tree being either twisted or bent when young and the outside of the trunk assuming the natural straight direction later, to accomplish which unequal deposits of tissue are made on the crooked trunk. The result of cutting a plank or board straight through from end to end of the trunk is that it will pass through layers of different age and density and different direction of the grain, causing the board to warp and twist and be "cross-grained."

Conversion of Timber.

When a log or a balk is broken down with the saw it is said to be **converted.** Foreign timber is usually converted to market forms either in the forest or at the ports of shipment, the former being the usual American practice, and the latter the European. When timber once converted is reduced to smaller scantlings, it is said to be **re-sawn.** The method of cutting pine timber for deal and batten forms is shown in f. 2, p. 416. Cut thus, the pith comes upon the surface of the deal, where for carpentry purposes it may remain, and for joinery is disposed of in the re-sawing. If the deals are cut so that the pith remains in the middle of the deal, dry rot is very liable to attack them.

Fig. 2 shows another way of converting by which this danger may be avoided, a square piece sufficiently large to include the pith to cut away as waste,

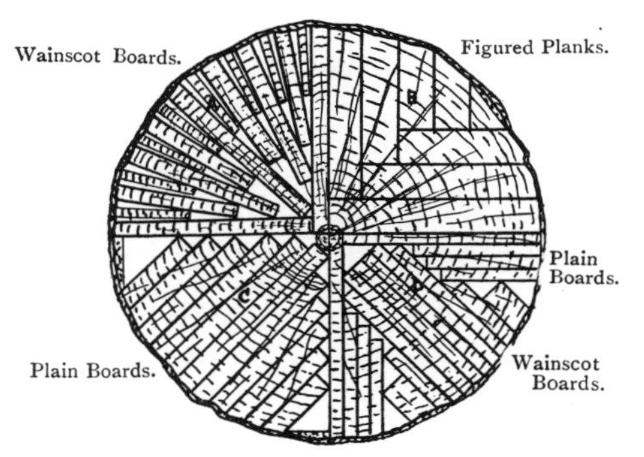

1. Method of Cutting "Quartered" Logs.

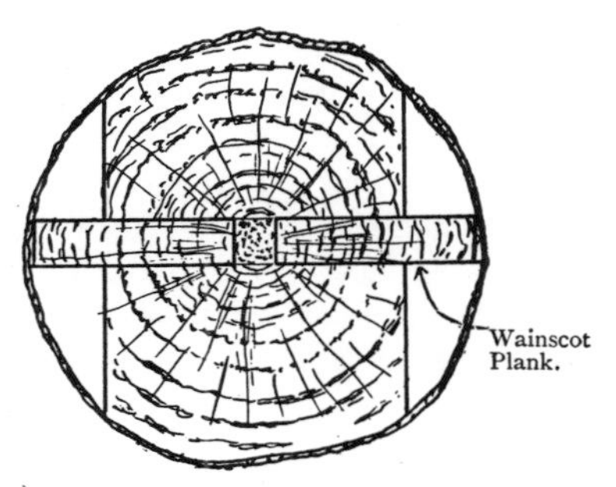

2. Method of Flitching Oak Logs.

and in the case of fir timber the two side pieces are converted into batten stuff or floor boards.

Quartered Logs (see f. 1 above).—This is a method of converting oak and other figured woods in which, as a preliminary operation, the log is cut into four quarters, which gets rid of the pith or centre, and opens up the log so that its defects can be seen, and the subsequent cutting arranged to minimise them. Four varieties of cutting are shown in the figure. The method shown at A produces what are termed **wainscot boards**—that is, boards showing figure or pattern of silver grain on each surface. It will be noticed that all the boards radiate from the centre, as do the medullary rays, and these cropping out on the surface irregularly, produce the figure. A few of the wider boards in the quarters marked C and D will also show some figure, the remainder will be plain. The quarter at B is cut into thick plank, and is the most economical, the white portions in each quarter showing the waste.

Fig. 2 shows how Russian and Austrian oak is cut for the English market. This method is termed **flitching**. In many woods, such for instance as pitch pine, the figure is produced by the contrasts in the annual rings, and the "feathering" occasioned by the rings being cut into at various depths. To obtain the largest amount of "feathering" upon the boards, they are cut tangential to the annual rings, as shown in f. 1 below.

The Effect of Shrinkage upon boards cut from different parts of the tree is shown in f. 3. Wood shrinks most in a circumferential direction, the medullary rays resisting lateral shrinkage. The sapwood shrinks much more

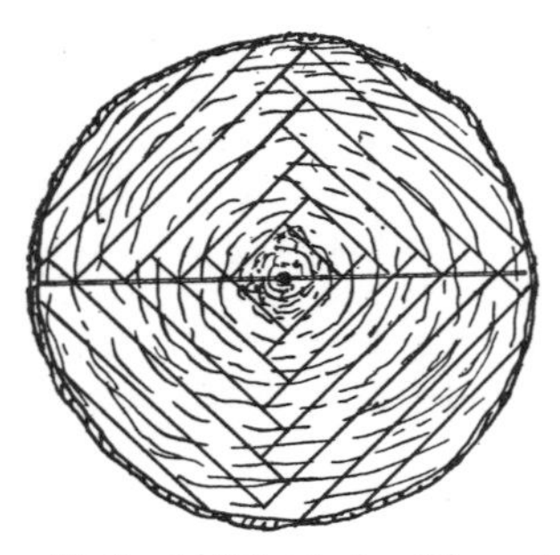

1. Method of Producing Figured Pitch Pine Planks.

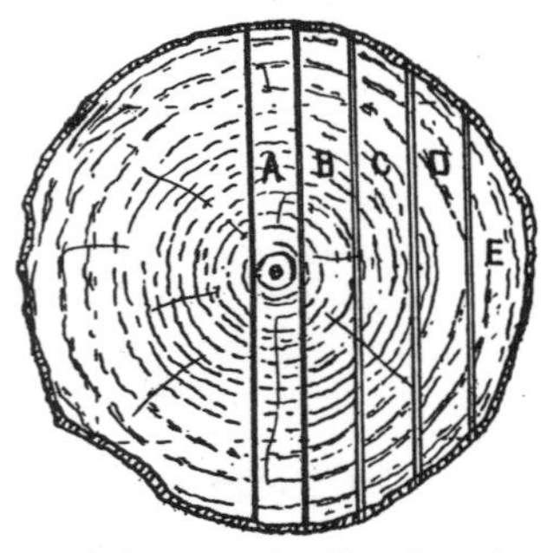

2. A Pine Log Cut into Boards.

than the heartwood, and consequently the nearer a board is to the sapwood the more it will shrink, and the effect of the shrinkage will be most pronounced on its outside face. A board cut across the centre of the tree, as at A, f. 2 above, will not shrink in width perceptibly, but will shrink in thickness at its edges, the two faces becoming "round," as shown at A, f. 3. The boards B C D E will shrink

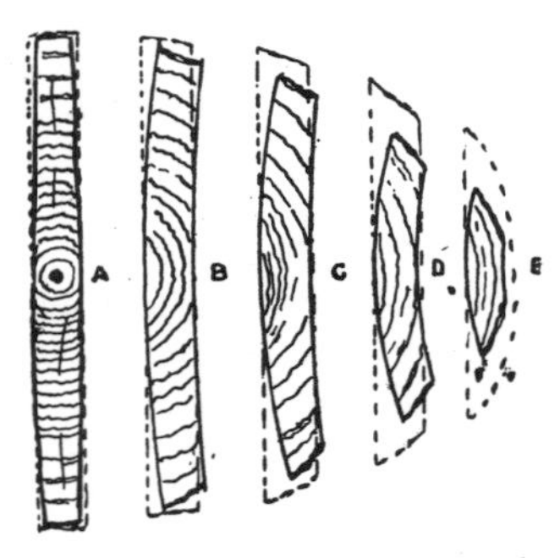

3. Effect of Shrinkage upon Boards.

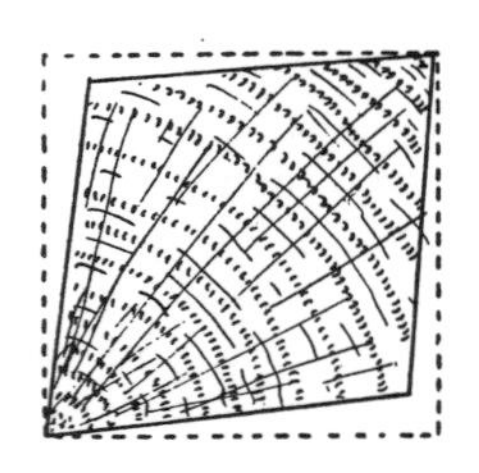

4. Effect of Circumferential Shrinkage upon Square Timber.

and cast, as shown in f. 3. The same effect of circumferential shrinkage on a square scantling is shown in f. 4, the dotted lines indicating the original shape of the piece. Boards that are desired to remain flat should be selected from those that have the annual rings parallel with the edges, or at right angles with the face, as at A, f. 2; these will remain true. Boards that have the rings nearly parallel with the face, as at C, will inevitably cast in drying.

CLASSIFICATION OF TIMBER.

Commercially all woods are divided into two classes, called respectively softwood and hardwood. The **Softwood** class comprise all the varieties of coniferous trees, *i.e.*, trees whose seeds are contained in a conical scaly shell called a cone; including the pines, firs, spruces, larches, cedars, yews, junipers, and cypress. The characteristics of this class are, the leaves are narrow, spinous or needle-like; the wood is resinous, of fragrant smell; most kinds secrete turpentine; the fibre is straight, and of soft regular texture, strong and tough in resisting cross stress, but weak and yielding in the direction of the fibres, or in resistance to shearing stress. The annual rings are very distinct, one side of each ring being soft, porous, and light coloured, the other hard, dense, and dark coloured, no visible medullary rays, and general colour of the wood pale tinted or light coloured.

The **Hardwood** class have broad leaves, are non-resinous, the secretions being either acid, astringent, or poisonous; the wood highly coloured and figured; annual rings indistinct; medullary rays in most are faintly perceptible, in some cases very distinct and producing an additional figure in the wood. The texture of the wood in this class varies very much with the species of tree, but all are harder and more difficult to work than the conifers, and are relatively much more resistant to shearing than to cross stress.

Minor Classifications. — There are several of these amongst timber merchants, woodworkers, and others, independent of the above and each other, with many purely local and arbitrary names which tend to make the identification of woods very difficult to all but experts. Some of these are dealt with below, and others in the particular description of various woods in the next chapter. Timber merchants call all timber belonging to the fir and pine species from European ports **Fir**, distinguishing it as red, yellow, or white, according to its appearance when converted. Similar wood from American ports is called **pine** and **American fir.** These woods, the product of varieties of the pine and spruce trees, are also named specifically after their ports of shipment, *e.g.*, Archangel, Dantzig, or Quebec fir, &c. Among woodworkers generally, the produce of the chief European conifers is termed **yellow** and **white deal**, and the American ditto, **pine** and **spruce**. The term **Baltic timber** is also applied to this wood when shipped from ports on the Baltic Sea, and **White Sea** goods when shipped from ports on that sea. **American Whitewood** is the product of the tulip tree, and **American Redwood** the product of a cypress.

MARKET FORMS OF TIMBER.

The following are the principal forms into which timber is converted for the market:—

A LOG is a trunk felled and lopped.

A BALK is the log squared by axe or saw.

A PLANK in hardwood is any cut stuff upwards of 9 in. wide and $1\frac{3}{4}$ in. thick; in softwood upwards of 10 in. wide and 2 in. thick.

A DEAL is over $2\frac{1}{4}$ in. thick and less than 10 in. wide.

A BATTEN.—Between $1\frac{1}{2}$ in. and 2 in. thick and less than 9 in. wide.

A BOARD.—Less than 2 in. thick and over 6 in. wide.

DIE-SQUARE STUFF.—Between 5 by 5 in. and 9 by 9 in.

WHOLE TIMBER.—Uncut balks.

A FLITCH.—Half of a balk cut in two lengthways.

QUARTERING.—3 by 3 in. to $4\frac{1}{2}$ by 4 in.

SCANTLINGS.—Miscellaneous cut stuff.

A SQUARE OF FLOORING or match boarding is 100 ft. superficial.

A HUNDRED OF DEALS is 120.

A LOAD OF TIMBER.—50 cubic feet.

A FLOAT OF TIMBER.—Eighteen loads.

ENDS.—Pieces of deals, planks, and battens cut off in the conversion of the latter to standard lengths, and shipped for firewood or lath rending, &c.

POLES.—Straight trunks, free from branches, greatest diameter not exceeding 8 in.

MASTS.—Ditto, exceeding 8 in. diameter.

Methods of Selling Timber.—Softwoods are sold—American goods, by the load of 50 cubic feet; European goods, by the standard hundred of deals. The two chief standards are the St Petersburg, which is 120 pieces 6 ft. by 11 in. by 3 in., or 720 ft. run of 11 in. by 3 in., or 165 cubic ft.; and the London standard, which is 120 pieces 12 ft. by 9 in. by 3 in. The former is the one in more general use, and to ascertain the number of feet run of any scantling required to make a St Petersburg standard, multiply 1.440 by 16.5, and divide the product by the sectional area of the scantling required. The quotient will be the number of feet run. For instance, a scantling of $2\frac{3}{4}$ in. $\times$ 10 in. $=27.5$ and $1.440 \times 16.5 = 23.7600 \div 27.5 = 864$, the number of feet run of 10 in. $\times\ 2\frac{3}{4}$ in. to equal a standard.

Hardwoods are sold variously by the foot super at 1 in. thick—the load of 50 cubic feet—by the log at per cubic contents, and in sundry valuable woods by weight per ton and per hundredweight.

CHAPTER XXVI.

DESCRIPTION OF WOODS USED IN JOINERY.

Place of Growth—Height, Size, and Appearance of Tree—Botanical Names—Common Names—Colour, Nature, and Texture of Wood—Purpose suitable for—Ports of Shipment—Market Sizes of Balks, Planks, and Battens of the following Trees—Softwoods—Northern Pine—Canada Pine—White Pine—Yellow Pine—Pitch Pine—Northern Pitch Pine—Kauri Pine—Huon Pine—Californian Pine—The Big Tree—Norway Spruce—American Spruce—Canada Spruce—Newfoundland Spruce—Oregon Fir—Larch—Cedar, Havannah, Australian, New Zealand—Basswood—American Whitewood—Hardwoods—Beech, English, Australian—Birch, English, American—Tasmanian Blackwood—Blue Gum—Ebony—Green Ebony—Greenheart — Jarrah — Karri — Spanish Mahogany — Honduras Mahogany — Mexican Mahogany—Tabasco—African Mahogany—Indian Mahogany — Oaks — English, Dantzig, Riga, American, Canadian, African—Padouk—Teaks, Indian, Burmese, Moulmein, Johore, Vindhyan, Karana, Siamese—Toon—Common Walnut—Black Walnut—White Walnut—Satin Walnut—Willow.

The common English designations are given in black letters, botanical names in brackets, and any local or variant names follow.

SOFTWOODS.

Baltic Pine (*Pinus sylvestris*).—Other names are yellow deal or fir, red deal or fir, Northern pine, Norway pine, Scotch fir, and European redwood. This tree is native in all the northern countries of Europe, and also Scotland. Its timber is exported in the form of balks, planks, deals, battens, and wrought flooring from Russia, Prussia, Norway, and Sweden. The general characteristics of the wood are—colour warm or honey-yellow; annual rings very distinct; the autumn growth brownish red and very resinous; the spring growth pale yellow and softer. The wood has a fragrant resinous smell when freshly cut, has large somewhat translucent and "live" knots, *i.e.*, the knots are firmly embedded in the tissue. The wood is strong, elastic, and very durable, works easily and freely. The sapwood is a pale blue.

The chief **Russian** ports of shipment are Archangel, Narva, Onega, St Petersburg, Riga, and Wyborg.

Prussian.—Dantzig, Konigsberg, Memel, and Stettin.

Swedish.—Gefle, Gottenburg, Holmsund, Stockholm, Soderham.

Norwegian.—Christiania, Dram, Holmstrand, and Frederikshald.

The largest **Balk** timber comes from Stettin; average sizes 18 to 20 in. square, 35 ft. long. Best balks from Memel; sizes 13 in. square, 30 to 35 ft. long. The longest balks, but of coarse strong timber suitable for heavy constructional work, are supplied by Dantzig; sizes 14 to 18 in. square, 40 to 50 ft. long. Riga balks are the smallest, but are of a clean mild description, lighter in weight and colour than Dantzig; sizes 12 to 13 in. square, 20 to 40 ft. long.

The best **Deals** for joinery come from the Russian ports Archangel, St Petersburg, Onega, and Wyborg. The best of the Swedish timber comes from the ports of Gefle and Soderham. Norway timber is generally small, coarse, tough, and only suitable for inferior work.

Canada Red Pine (*Pinus rubra* or *resinosa*).—This wood is known as Norway pine in Canada, where it is also called red pine from the colour of its bark. It is very similar in appearance to Memel or Dantzig pine, but is easier to work. It is tough, elastic, fairly strong, clean working, free from knots and sapwood, warps but little, takes glue well, and is much favoured for joinery purposes in America. It is a native of the north-west of Canada and Nova Scotia; usual sizes 10 to 18 in. square, 16 to 50 ft. long.

White Pine (*Pinus strobus*).—This is called **yellow pine** in England. It is also known as Weymouth pine, after Lord Weymouth, who is said to have introduced it in this country in 1708. The tree is a native of Canada and the Middle States of North America, and is chiefly exported from Quebec. The wood is soft, clean, straight-grained, and remarkably free from knots, which, however, when they do occur, are usually loose or "dead." The colour is a pale straw yellow, with fine annual rings not very distinct. This wood may be easily recognised by the presence of numerous fine open pores interspersed in the substance; these readily fill with dirt, and appear like short black hairs spread over the surface. The sapwood is brown. This is the softest and lightest of the pine species, and is a wood most extensively used for all kinds of interior joinery and cabinet work. It is largely used for mouldings, patterns, and cores for veneers, as it holds glue well. In this country it proves unsuitable for exterior work, rotting quickly when exposed to the weather, but in the drier air of America it is used largely for all kinds of exterior carpentry. It is a tall tree, often reaching 200 ft. in height.

Yellow Pine (*Pinus mitis*).—This wood, called also in America Georgia pine and short-leaved pine, is known in England as New York pine. It is a native of the Middle and Southern States of America. The wood is clean, fine-grained, strong, and durable, slightly resinous, and harder than the white pine. It, however, furnishes much smaller timber, the trees seldom exceeding 60 ft. in height and 18 in. in diameter. This timber decays quickly if painted. It is only imported to this country in small quantities.

Pitch Pine (*Pinus Australis*, *Pinus palustris*).—In America this tree is known as the long-leaved or southern pine, and also as the turpentine tree, in consequence of the quantity of turpentine obtained from it. It is a native of the Southern States, including Florida, Savannah, Carolina, and Georgia, and is a large tree of from 60 to 100 ft. in height, with a nearly parallel stem. Many of the trees have an average diameter of 2 ft. The leaves are of a brilliant green, and about a foot long. This is the hardest, heaviest, and strongest of the pine species, and is a most valuable wood for constructional purposes. The pitch pine, from its superabundance of resin, is very difficult to work, tools having to be kept bathed in oil or grease to enable them to work it successfully. The colour is a rich warm yellow, the annual rings red, very wide, and distinct.

The sapwood, which is comparatively small, is a greyish white. Much of the timber of this pine possesses a peculiar curly grain or figure upon one side of the tree; boards possessing this are in considerable demand for cabinet and ornamental joinery work. The wood, however, is not well suited for these purposes, as its shrinkage is great and long continued. It is also impossible to keep it with a good smooth surface, due to the unequal shrinkage of spring and summer growth of the annual rings, brought about by the gradual exudation of the liquid resin from the latter. This resinous fluid is also such a powerful solvent that it quickly destroys the surface either of paints or varnish. The tree is very subject to cup-shake. It is exported in the form of balk, plank, and wrought boards from the ports of Darien, Mobile, Pensacola, and Pascagoula. Sizes—Logs, 12 to 18 in. sq., 20 to 50 ft. long; planks, 3 to 5 in. thick, 10 to 18 in. wide. An inferior kind of wood, known as pitch pine in America, is the *Pinus rigida.* This tree grows in the Northern States of America. It produces a strong, tough, coarse-grained wood, but little exported, being mostly used for firewood.

Kauri Pine (*Dammara Australis*).—Known as the pitch tree of New Zealand; other names are cowry and cowdie pine: it is not a true pine although a conifer. It is a native of the province of Auckland, in the North Island of New Zealand, and is a stately tree, frequently rising 100 ft. without a branch. Some exceedingly fine specimens have been felled, averaging 90 ft. in circumference at the bole, and logs of 10 and 12 ft. diameter are of quite common occurrence: its total height averages about 140 ft. The wood is a light brownish yellow in colour, some specimens having a pink tinge, with a fine close and straight grain; it planes easily with a clean silky surface that takes polish and stain well. When polished, a very small but dense mottling appears, giving the wood some resemblance to plain satinwood. It shrinks little laterally, but is one of the few woods that shrink perceptibly lengthways of the grain. The author has had personal experience of this in preparing drawers of the wood, the fronts of which having been fitted tightly into the case on one day, have shrunk sufficiently both in length and width to fall out easily on the next; the wood casts also in length. The Kauri pine is shipped from Auckland and Whitanga in Mercury Bay in the form of balks and planks, the latter about 40 ft. long, 5 ft. wide, and 4 in. thick, often without a knot in them.

Huon Pine (*Dacrydium Franklinii*).—Is a native of Tasmania, and is found all over the island with the exception of the neighbourhood of the Huon River, from which it takes its name. Though originally dense forests bordered this river, it has now been entirely cleared by the settlers. The wood of this pine is a delicate yellowish brown, beautifully marked, and free from knots, except at the head, where they are small and numerous. It has a strong resinous smell. The grain is fine and close, planing easily with a lustrous satiny surface. It takes a high polish, and when used in conjunction with the native blackwood (*Acacia melanoxylon*) it is very effective. Huon pine wood is in such demand in the neighbouring colony of Australia that little of it reaches this country.

Californian Pine (*Sequoia sempervirens*).—Called also redwood and sequoia. This tree is a native of Northern California and the mountain ranges extending thence some 300 miles to the southern end of the Bay of San Francisco, and forming a thick belt of forest averaging 25 miles wide. It is one of the giant trees of the New World, being only surpassed by its sister tree the *Sequoia gigantea*, its mighty stem rising for 200 ft. without a branch, with gently tapering trunks from 10 to 15 ft. diameter. Many of these trees yield 10,000 ft. run (board measure) of marketable timber. Recently 4 acres of redwood trees yielded 1,000,000 ft. in board measure. The colour of the wood is dull red without lustre, with figure similar to Kauri pine. It is very soft, easy working, rather coarse grain, and very brittle or "short"; it polishes well. The trees are much subject to heartshake, but are so large that this defect is immaterial in conversion. It stands alternate wetness and dryness better than other pines. Its use in joinery is confined to panelling, shelving, and positions in which it can be used in the "solid," its peculiar shortness of grain precluding its use in framing. Although a cone-bearing tree and classed with the pines, it is not a true pine, but is allied to the cypress. The wood is shipped from Tacoma, Washington County, and Eureka, Humboldt Bay, in planks 3 to 7 in. thick, 11 to 24 in. wide, 12 to 40 ft. long.

The "Big Tree" (*Sequoia gigantea*) produces a similar wood, somewhat duller in colour, but otherwise indistinguishable from the better known redwood.

FIRS.—**Spruce** (*Picea excelsa*, or *Abies excelsa*), generally called **white deal**, is a native of the mountainous districts of Northern Europe. It is a tall tree, often reaching 100 ft. in height; its shape is pyramidal, having branches low down on the stem. This is the general habit of the fir genus, of which the spruces are varieties, and in which they differ from the pines, whose stems are free from branches for a considerable height. The wood is of whitish yellow colour, rather dull in comparison with the *Pinus sylvestris*, and much lighter in weight than the latter; it planes freely with a satiny lustre. The knots, which are numerous, are usually dead and extremely hard, and are dark coloured and opaque, which makes the wood easily distinguishable from "yellow deal," whose knots are always resinous and red. The hardness of the knots is the chief defect of this wood, limiting its usefulness, as few cutting tools will keep their edge upon them. The wood is odourless, and having a clean hard surface, is very suitable for shelving and receptacles for food. It is extensively used for "kitchen joinery," it swells and shrinks freely, and is not durable when exposed to the weather. Another characteristic defect is the presence of resin cells or "pockets" in the tissue, containing semi-liquid resin which slowly exudes when the timber is converted, leaving cavities in the wood. The Norway spruce is largely used for scaffold poles, boards, ladder sides, spars of ships, joists, and flooring. The best white deals come from Christiania in Norway, but the supply from this port is now very limited. By far the largest quantity of this timber is obtained from the Russian ports of Archangel, St Petersburg, and Riga, also Drammen and Frederikstadt.

American Spruce (*Picea nigra*).—Called the black spruce, from the sombre colour of its bark. Is a native of the colder regions of North America. The wood is strong, light, and elastic, less resinous than Norway spruce. It twists and shrinks freely. The knots shrink and fall out, leaving circular holes which have the appearance of having been bored out with a bit. The colour of the wood is a greyish white, much colder looking than Norway spruce, which in other respects it much resembles. The average height of the tree is 70 ft., and diameter at the base 24 in. Shipped from Quebec and Portland, Maine.

White Spruce (*Picea alba*).—Epinette, Brunswick Spruce; a native of Canada and Northern States, U.S. Commercially the wood is indistinguishable from the black spruce, the name referring to the light-coloured bark. The wood decays quickly. The tree is smaller than the *Picea nigra*, seldom reaching an altitude of 60 ft.

Newfoundland Spruce (*Picea rubra*).—The red spruce, so-called from the colour of the bark, is a native of Nova Scotia and the watershed of Hudson Bay. It is a softer and more resinous wood than *Picea nigra*, but is in other respects very similar. The wood turns brownish red with age. Sapwood greyish white. Exported from St John's, Newfoundland, and Parrsborough, Nova Scotia, also Puget Sound.

Oregon Fir (*Pseudotsuga taxifolia* and *Pinus Douglasii*).—Also called Douglas spruce or fir, Oregon pine, and Nootka fir. Is a native of the north-west coast of North America. Dense forests of this tree, associated with cedars and hemlocks, exist in Oregon, Washington, Columbia, and Alaska. It is a large tree, averaging 150 ft. high, with a conical trunk, often reaching 10 ft. in diameter at the base. There is a fine spar of this wood in Kew Gardens used as a flagstaff which is 159 ft. high. The wood is reddish yellow in colour, very similar in appearance to pitch pine, but with its annual rings much finer. The grain is fairly straight, but the wood is difficult to plane to a good surface in consequence of the great difference in texture of the spring and summer growth, which is sharply defined, and gives the boards when cut right across the trunk the appearance of light and dark yellow stripes upon the surface. It makes a fairly good carpentry timber for purposes where extra long scantlings are required. For joinery, however, it is inferior to European firs, with the exception of its freedom from knots. Shipped in balk and plank from Puget Sound, Washington County. Sizes—Balks, 30 by 30 in. by 120 ft.; planks, 11 to 40 in. by 12 to 40 ft. by 2 to 6 in.

Larch (*Larix Europæa*).—This tree differs from the pines and firs, which are evergreens, in shedding its leaves annually. It is a native of the mountain ranges of Central Europe and of Russia. The wood is a deep yellowish red colour, changing to nearly black with long exposure. It has a straight even grain, but is rather hard and tough to work. It shrinks excessively, and warps in seasoning. The wood will take a good polish, but has little figure.

Cedar [Havannah] (*Cedrela odorata*).—This tree is a native of Cuba, Jamaica, and Honduras; its wood is soft, porous, and brittle, with aromatic smell. Colour, dull reddish brown, used chiefly for cigar boxes and cabinet interiors. Exported in logs 20 ft. long, 1 to 2 ft. sq.

Cedar [Australian] (*Cedrela Australis*).—This tree is a native of Australia, where the wood is extensively used for all kinds of interior joinery. The wood is soft, porous, and open-grained in appearance, and colour not unlike plain Honduras mahogany, for which it is sometimes substituted.

Cedar [New Zealand] (*Libocedras Doniana*).—Native of the North Island. Colour a light red, straight-grained, easy working, durable wood used for house joinery, telegraph poles, railway sleepers, &c.

Basswood (*Tilia Americana*).—This wood is the produce of the American lime tree. In colour it is a yellowish brown with a slightly green tinge. The grain is straight, close, of even texture, light in weight and not very durable. It takes stain and polishes well, warps and twists freely.

American Whitewood (*Liriodendron tulipifera*).—Known in the United States as canary wood, and also as yellow poplar, is the produce of the tulip tree. It is a native of the Middle States of North America, where it reaches a height of from 70 to 100 ft., with trunks from 18 in. to 3 ft. in diameter. The wood is a lemon yellow in colour, and in some instances nearly white; the sapwood varies from white to pinky brown. The grain is fine, and the wood planes to a smooth surface, taking stain and polish well; it shrinks freely, and casts or warps each time it is planed. In the Northern States it is largely used for exterior work, clapboards, shingles, posts, &c., but in this country its use is confined to joinery, for which its great width and close texture make it very suitable. The wood is shipped from New York, Philadelphia, and Baltimore.

Hardwoods.

Beech (*Fagus sylvatica*).—Is a native of England and Central and South Europe generally. It is a straight tree, from 60 to 70 ft. in height, with a circumference of from 10 to 12 ft. The wood is of a reddish brown or yellow colour, hard, heavy, and close in texture, having a fine smooth surface when planed. Cut tangentially it has a "rowey" appearance, and the silver grain is very distinct in cross section, although it produces little figure in radial section. Beech is durable when kept either continuously wet or dry, but alternations of the weather soon cause it to decay. It does not absorb moisture readily, and consequently warps but little, for which reason it is almost solely used for the stocks of planes and other tools. Beech stains and polishes well, and is much used in cabinet work for constructive purposes, but little in building work.

Beech [Australian] (*Gmelina Leichhardti*).—A valuable timber for flooring, deck planks, piles, jetties, &c. Expands and contracts little, dark brown colour, straight grain, easy working, durable.

Birch (*Betula alba*).—Is a tough stringy wood of pinkish brown colour, curly grained and porous, washes well, used for kitchen joinery and furniture.

Birch [American] (*Betula lenta*).—Wood moderately hard, tough, and straight grained. Colour reddish brown, changes to warm yellow when polished. Figured examples highly valued for cabinet purposes. Exported from New York and Quebec.

Blackwood (*Acacia melanoxylon*).—This wood is a native of Tasmania and South Australia, and is largely used in Australia for all kinds of good joinery work and cabinetmaking in association with Huon pine. It is also cut into veneers for piano cases. The colour of the wood is dark reddish brown, which on exposure turns black. The figuring is similar to rosewood or Italian walnut. The wood is very dense, bends with facility when steamed, and polishes well.

Blue Gum (*Eucalyptus globulus*).—Is a native of Tasmania and South-

Western Australia. It is a very large tree, attaining a height of 300 ft., with a diameter of 20 ft. at the bole. This tree is reported to destroy the germs of malaria, and when planted in marshy districts quickly dries the land and eradicates fever from the neighbourhood. The wood is of a pale straw colour, very hard, heavy, and tough, with rather twisted or curly grain, subject to airshakes. It splits and warps excessively in drying, due to the immense quantity of green sap contained by the wood. The tree is said to absorb annually ten times its weight of water from the soil. Its chief use is for engineering purposes, as it has enormous crushing resistance, averaging 3 tons to the square inch. The wood is extremely durable under water, and is also good for railway sleepers and wood pavements. This wood gets so hard when dry that it has to be wrought in its green state. One of its chief uses is for piles in sea walls, &c., because being heavier than water, it sinks when broken off.

Ebony (*Diospyros ebenum*).—Is a native of Ceylon. The wood is jet black, with occasional streaks of brown and green: it is one of the densest and heaviest of woods, extremely hard, tough, and difficult to work. Planes with a beautiful glassy surface that only requires burnishing to put a fine gloss on it. The wood sinks in water, and it is subject to starshake. Used for high-class joinery and inlaid work in cabinetmaking. Indian ebony (*Diospyros melanoxylon*) is also imported under the name of Green Ebony.

Greenheart (*Nectandra Rodiæi, Sipirie bibiru*).—Is a native of British Guiana and the West Indian Islands. The wood is one of the strongest and most durable known, and having an extremely bitter secretion, is seldom attacked by insects. The colour of the wood is a dark yellowish green; the older parts of the heartwood are often a deep brown. The sapwood is equally as strong and nearly as durable as the heartwood, from which it is indistinguishable. The wood is very heavy, tough, and difficult to work. The grain is straight, and there is little figure. Viewed in cross or end section the substance appears to be very porous, similar to cane. The annual rings are indistinct, and the medullary rays invisible. One of the strongest woods in existence; crushing strength, 5 tons per square inch; weight, 68 to 70 lbs. per cubic foot. The uses are sills, warehouse floors, piles, jetties, and engine fittings. Exported from Georgetown or Demerara in Guiana.

Jarrah (*Eucalyptus marginata*).—This tree is a native of Western Australia. It is a rather large tree, averaging 120 ft. in height and 3 ft. 6 in. diameter at the base. Much larger specimens are, however, frequently met with. One recently felled measured 22 ft. in circumference at 5 ft. from the ground, and 80 ft. clear to the first branches, yielding twenty loads of sawn timber when converted. The wood is a warm brownish red in colour, not unlike Honduras mahogany, for which reason it is frequently called the Australian mahogany. There is scarcely any figure, and the wood being of porous texture, does not work up to a good face. It, however, varies considerably according to the soil in which it grows. The forests of Jarrahdale are said to yield the finest quality, this wood being of a much lighter colour than the other kinds. It is also free from gum veins, and has a mottled figure. It bends well, and is very durable under water. The weight of a cubic foot when fresh cut is 70 lbs. The principal uses to which Jarrah has been put are piles, wood-paving blocks, sleepers, bridges, general joinery, and furniture. Ports of shipment—Rockingham, Busselton, and Hamelin, in West Australia.

Karri (*Eucalyptus diversicolor*).—Is one of the giant trees of Australia, often reaching a height of 300 ft. with a girth of 30 ft., but the dimensions of average forest trees are 200 ft. high, 120 ft. to first branches, diameter 4 ft. at 5 ft. from the ground, yielding from 30 to 40 tons of marketable timber each. The Karri grows in close proximity to its sister, the Jarrah, to which its wood is very similar, being, however, somewhat redder, more dense and heavy, and its lateral strength or cohesion greater than that of Jarrah. It does not stand water so well as the latter. Its principal uses are, railway and bridge planking, wood blocks for paving, beams, &c. For ports of shipment see Jarrah.

Mahogany (*Swietenia mahogani*). — Several woods go by the name of mahogany commercially, but the true mahogany is a tree allied to the cedars, and is indigenous to the West Indian Islands and to Central America.

Spanish Mahogany.—Is the produce of a large tree growing in Cuba, Jamaica, and St Domingo, that from the island of Cuba being the largest and most valuable in consequence of its rich colour and handsome markings. The mahogany tree takes about two hundred years to arrive at maturity, and its wood is very durable. It swells and shrinks very little, and does not warp. Its colour is a rich brown red with darker markings. Many of these afford fine figure, which is variously known in the market as "veining," "watered," "festooned," "bird's-eye," &c. The pores of this variety contain a white gummy secretion which when freshly cut makes the wood difficult to work through the planes sticking to the surface, and later it dries hard and flinty, in which state it rapidly dulls the cutting edges of tools. This white flinty substance is a special characteristic of the island-grown timber, which distinguishes it from the variety grown on the mainland. The texture of the wood is dense and heavy, it faces up well and takes a high polish, but becomes nearly black with age. Market sizes—Cuba balks, 12 to 22 in. square, 10 to 18 ft. long; St Domingo average 11 in. by 11 in. by 10 ft.

Honduras Mahogany, also known as Baywood, is the produce of a large tree growing in the British colony of Honduras and the neighbouring States bordering the bay of that name. The wood is much softer and easier to work than the Cuban variety, is tough and elastic when freshly cut, but gets brittle and acquires a tendency to split as it ages. The colour varies from a dull brick red to reddish yellow, which changes to a bright golden red when polished. The figure is rather large and coarse, but occasionally a plank will show a finely mottled appearance with the play of light upon it. This figured baywood is in great demand for furniture, as the wood is worked easier and is much lighter in weight than Spanish: the grain is straight and rather porous, but faces up well. The wood shrinks and warps very little during seasoning, and when dry does not alter much. It will not, however, stand exposure to the weather, soon losing its colour and perishing, unless painted. For interior work, where it can be kept permanently dry, it is a most valuable and useful wood, and is more extensively used than any other hardwood for superior joinery. Baywood takes glue better than any other hardwood, and is very free from shake and sap. It has, however, a strong tendency to split in the direction of the grain if dried too rapidly, and boards will frequently split from end to end during cross-cutting unless all the fibres in one plane are severed at once, to accomplish which, the saw must be passed through it quite horizontal, and in ripping, it is also wise to cramp the end to prevent splitting. This wood is shipped chiefly from Belize, the chief port of British Honduras, which place exported in 1898, 6,832,546 ft., of which the United Kingdom received 6,476,443 ft. Market sizes—Balks, 2 to 3 ft. square, 12 to 15 ft. long; planks, 5 to 7 ft. wide, 3 to 6 in. thick.

Panama Mahogany.—Is similar to Honduras, but rather softer and of more woolly texture; inferior in durability.

Mexican Mahogany.—Is the produce of a very large straight-growing tree, very abundant in Mexico. The wood is similar to Honduras generally, although it varies in quality from the several districts. **Tabasco** supplies the best kind, which is a hard, brittle, dull red wood with very open pores and "rowey" figure, rather more difficult to work than Honduras. It is the largest of the mahoganies, being exported in balks 18 to 46 in. square and 18 to 32 ft. long.

African Mahogany (*Khaya Senegalensis*).—This is a hard and durable wood shipped from Sierra Leone. It is of a coffee-brown colour, without figure, extremely hard and durable, used chiefly for civil engineering purposes, and is very liable to heartshake and cupshake. Market sizes, 13 in. by 13 in. by 30 ft.

Indian Mahogany (*Soymida febrifuga*).—Is a large tree growing on the mountains of Central Hindostan. The wood is hard, heavy, and durable, of a red-brown colour, very plain. It is reputed to be the most enduring wood of India.

Oak (*Quercus*).—The oak is an almost universally distributed tree, growing freely in Europe, America, Asia, and Africa; over sixty distinct species are known to botanists. Commercially these distinctions are disregarded, and the wood is known either by the name of its place of shipment, or by the method in which it is cut or converted for use. After the pines, the oak is probably the most generally used and useful wood in existence, there being scarcely any situation in which it may not be used with advantage, as apart from the beauty of its markings, it unites in a remarkable manner the qualities of durability, strength, toughness, and stiffness. The general characteristics of oak are:—Colour, a cold greyish brown, varying, however, with the soil and climate grown in; hard, tough, fairly straight grained, rather difficult to work; does not "face" well on account of its open porous texture, which makes it less suitable for polishing than most hardwoods; warps, swells, and shrinks freely; is almost indestructible when kept either continually wet or dry; bends readily when steamed. All the European varieties secrete a pyroligneous acid which corrodes iron, and stains and decomposes its own tissue in the process. Viewed in cross section the annual rings are distinct, close, and dense on the one side, open or porous on the other. (The oak belongs to the ring-porous class of woods.) Numerous large and small medullary rays radiate from the centre of bright silvery appearance; when the wood is cut parallel with these medullary rays they crop out upon its surface in beautiful white marks called silver grain or overgrain.

English Oak (*Quercus pedunculata*).—Is the hardest and most durable of the oak species. Colour, a pale brownish yellow, which becomes nearly black after many years' exposure; hard, regular, straight grain; annual rings rather thick; overgrain or figure dispersed, giving large well-marked patterns; overgrain very white and hard. The best kind is grown in Sussex.

Dantzig Oak (*Quercus robur*).—Is grown chiefly in Poland. It is of a dark brown colour, with a straight close compact grain; bright overgrain, free from knots; bends easily; not so durable as English. It is classed for export purposes as crown and crown-brack qualities, marked respectively W and WW; imported in logs 10 to 16 in. square, 30 to 34 ft. long; planks, 9 to 15 in. wide and 2 to 8 in. thick.

Austrian Oak, also called Adriatic Oak, is shipped from Fiume and Trieste. It is soft, open, and porous, grained with good figure of somewhat mottled appearance. More easily worked and of lighter colour than English oak. It is largely used for fittings and furniture.

Riga Oak (*Quercus sessiliflora*).—Is grown in Southern Russia. It is rather a small tree, and only the best kinds are exported. These are cut through the middle, and are shipped as half round logs (see f. 2, p. 417). The wood is mild, of close texture, with the medullary rays or silver grain equally distributed, which gives this wood when quartered a pleasing mottled appearance: it is generally prepared as wainscot. One of the peculiarities of this wood is, that when polished the silver grain will show up as brown markings in certain positions· whilst appearing bright and silvery in others, therefore in using the wood attention should be paid to this effect in arranging its position.

American Oak (*Quercus alba*).—The **white oak**, so-called from the colour of the bark, is the most important variety of oak native to America. It grows in abundance from Mexico to the confines of Canada. The wood is redder than English oak, lighter, coarser grained, and less compact. It warps and twists freely, bends more readily than any other species. Prepared as wainscot, it is inferior in figure to most European kinds. Exported from Mobile, Baltimore, New York, and Quebec in logs from 25 to 40 ft. long, from 12 to 28 in. square; planks, 2 to 4 in. thick, 12 to 24 in. wide. This variety does not secrete pyroligneous acid.

Canadian Oak (*Quercus rubra*).—The **red oak** is a native of Southern Canada. The wood is coarse grained, very porous, and of woolly texture, very inferior to the white variety, and is not much exported to this country.

African Oak (*Oldfielda Africana*).—The Turtosa is a West African timber exported from Sierra Leone. It is a hard, tough, heavy wood, of very dense texture, planing with a highly finished "face"; it polishes well, but has little figure. The colour is a greyish brown inclining to red. Annual rings, clear, narrow, and well defined; large number of minute medullary rays; and, like all extremely hard woods, it has an immense number of small pores diffused equally throughout the substance. This wood is in appearance and structure more like teak than oak. It is very durable in wet situations, and is used for sills, keelsons of ships, guard strakes of jetties, &c., but its great weight and hardness preclude its extended use.

Padouk (*Pterocarpus indicus*).—A native of the Andaman Islands, and also of Burmah. The island wood is of a deep red-orange colour, hard, strong, and durable. The texture is dense, interspersed with porous bands, the wood being of the "ring-porous" order of which the oak is a type. "Faces" and polishes well. The Burmese variety is blood-red and much harder, planing with a clean glassy surface. Is used for good cabinet and joinery work.

Teak (*Tectona grandis*).—This tree is a native of South and Central India,

Burmah, Java, Ceylon, and Siam. It is of rapid growth, with a tall straight trunk averaging 100 to 150 ft. in height, and a girth of 10 ft. The wood is lighter in weight than oak, but is stronger and more durable under water. Its colour varies from a deep yellow to dark brown according to place of growth, the Siamese variety being the palest coloured, and the Burmese the darkest. The grain is open and porous, does not face well, and polishes indifferently. Teak is unrivalled for resisting the shipworm and the white ant. The Moulmein variety of Burmese teak is considered the best for constructional purposes, being straight grained, free from knots, shakes, and other defects. The Johore variety is the heaviest and strongest, and is used for piles, sleepers, beams, &c. The Vindhyan and the Karanee varieties are soft, easy working, and have a peculiar dark mottling in the figure which makes them much prized for cabinet work. Siam teak imports have recently largely increased, many thousand tons being shipped from Bangkok, the capital and river port of Siam, annually. Teak contains a resinous aromatic substance which has a preservative effect on iron. This secretion hardens in the pores and shakes in the wood, and is then very detrimental to the edges of tools. Teak is largely taking the place of oak for constructional purposes, being more economical to work, and withstanding the weather even better than the latter; it is also considered more fire-resisting. It is not, however, adapted for small or fine joinery work or mouldings, in consequence of its tendency to splinter and its greasy nature, which prevents it cleaning up well. The wood is found most suitable in joinery for doors and window frames, sills, stairs, and shop fronts, and in positions where clean sharp arrises are not required. It is less liable to alteration in size due to the state of the atmosphere than any other wood. Teak is rather a dangerous wood to prepare, as it splinters readily, and these splinters entering the flesh, are very liable to set up blood poisoning. It is also an unsatisfactory tree to convert, owing to the number of natural defects it is subject to. Only about 25 per cent. of the logs brought down to the ports are suitable for "first-class Europe grade." Teak in the market is classified according to size, not quality. Class A, logs of 15 in. side and upwards; Class B, over 12 in. and under 15 in.; Class C, under 12 in. Lengths average 20 to 40 ft.

Toon (*Cedrela Toona*).—Red cedar, sometimes called Indian mahogany, is a native of Bengal and Burmah. The wood is light and durable, resembling Honduras mahogany, but is of a looser, more open grain. It belongs to the same natural order as the mahogany tree. It is largely used in India for furniture and joinery, and in Assam for tea chests, as it is clean, strong, and odourless. The trees are of great size, often yielding 80,000 ft. of marketable timber. The wood takes a good polish, and is largely cut into veneer for cabinet work. Market sizes average 24 in. square, 30 ft. long.

Walnut (*Juglans regia*).—Is the common or Persian walnut. This tree is found in Persia, Asia Minor, China, Greece, and the forests bordering the Black Sea in Europe. The tree averages about 60 ft. high, and from 30 to 40 in. in diameter at the base. It is a quick-growing tree, arriving at maturity in fifty years. The colour of the wood is a rich warm brown with darker markings, the figure being wavy and somewhat coarse. The wood is fairly durable, easy working, faces and polishes well. Apart from its comparative scarcity, the walnut is too flexible for constructional purposes, and its use is confined to superior joinery and cabinet work. The most figured specimens are selected for veneer cutting, and are sold as Italian. Most of the wood imported to this country comes from the Black Sea ports, in balks of 10 to 16 in. side, and 6 to 9 ft. long.

American Walnut (*Juglans nigra*).—Black Walnut is a native of North America from New England to Florida. This tree is the largest and most important, commercially, of the walnut species, often reaching 100 ft. in height, and 6 to 7 ft. in diameter. The wood is strong, hard, durable, and easily worked, faces and polishes well. The colour when freshly cut is a purplish brown, which as it dries, changes to a very dark dull brown, or nearly black, hence its popular name. It is largely used in joinery and cabinet work, for framing, office fittings, shop fronts, furniture, &c. Exported in logs, planks, and boards chiefly from Baltimore, U.S.

White Walnut (*Juglans cinerea*).—Called also butternut. This is a large sea-coastal tree of North America, ranging from Canada to Virginia. Its height averages 50 ft., and 10 to 12 ft. in circumference. The wood is hard, dense of close texture, straight grain, faces well, takes a high polish, warps little stands well in wet situations. Colour, very pale pinkish yellow, not unlike birch. Viewed in cross section it is diffuse-porous, with numerous fine medullary rays. Annual rings indistinct.

Satin Walnut (*Hamonelideæ*).—This is a large tree of the Central and Southern States of North America, known also as the Red gum and sweet gum. The wood is of a salmon pink or reddish yellow colour, with a fine clean close grain; planes with a lustrous satiny finish, stains and polishes well. It has been extensively used both in this country and the United States for furniture purposes, and recently under the name of red gum, as wood block paving. It is a fine diffused-porous wood, with no visible medullary rays. Annual rings fine, compact, and distinct.

Willow (*Salix*).—This is a tree of moderate height, averaging from 30 to 40 ft., and is a native of all temperate countries. There are about 160 distinct species. The wood, although generally classed with the hardwoods, is not really much harder than pine; its colour is drab yellow. One variety, said to yield the most durable wood, is of a salmon pink tint. This wood was formerly much used in this country for flooring, but since the introduction of Swedish timber has gone out of use for the purpose. It is still used by the cabinet-maker, but chiefly by the waggon-maker and boat-builder. It is extremely durable in water, is very tough, does not splinter, will dent and bend rather than break when struck.

CHAPTER XXVII.

GLOSSARY OF TECHNICAL TERMS AND PHRASES CONNECTED WITH JOINERY.*

NOTE.—*Terms peculiar to machinery are prefixed by (M.).*

A.

ABACUS.—The uppermost member of the capital of a column. The "cushion" between the column and its load.

ABIES.—One of the orders or subspecies of the fir species of trees. The cones have round scales and the leaves are not clustered.

ABUT.—To touch or rest squarely against; usually contracted to BUTT (which see).

ABUTMENT.—The pier from which an arch springs.

ACUTE ANGLE.—An angle containing less than 90 degrees, *i.e.*, a right angle.

ALBURNUM.—The botanical term for sapwood, in allusion to its pale colour.

ANCHOR MOULD.—An ornament carved on an ovolo moulding in the shape of a dart or claw of an anchor.

ANCIENT LIGHT.—A window that has obtained a legal right in perpetuity of uninterrupted access of daylight.

ANGLE BAR.—The bar in the angle of a shop-front sash. See f. 2, p. 379.

ANGLE BEAD.—A bead upon the salient angle with a quirk on each side; otherwise a return bead or staff bead.

ANGLE BLOCK.—A small triangular block glued in the interior of an angle joint to strengthen it.

ANGLE BRACE.—A tie or strut placed in the corner of a framing to prevent alteration of shape.

ANGLE BRACKET.—A wood backing piece for a plaster cornice placed at an angle to the wall.

ANGLE OF REPOSE.—The angle of the bed joints in arch stones at which the stones will remain at rest without the aid of cements, supported only by friction.

ANGLE RAFTER.—Another name for a hip rafter. A rafter between the side and end of a roof.

ANNEX.—An addition to a building already in existence.

ANNUAL RINGS.—A term applied to the concentric rings of woody fibre shown in the cross section of a tree, each ring representing a year's growth of wood.

ANNULET.—A fillet or band encircling a column.

ANTEROOM.—A smaller room, or a passage attached to a principal apartment.

APPLIED ORNAMENT or APPLIQUE.—Ornament added or attached to the surface as distinguished from ornament worked in the solid material.

APRON or FLASHING.—Sheets of lead or zinc fixed to walls and overhanging the edges of gutters.

APRON LINING.—The vertical linings of a wellhole of a stair. See f. 1, p. 308.

APRON PIECE.—A false trimmer at the top of a flight of stairs.

---

* Terms strictly confined to Carpentry are dealt with in the companion volume upon that subject.

APRON RAIL.—The middle rail of a door having a raised piece worked or planted on its surface.

ARABESQUE.—Ornamentation in the Arabian style of conventional foliage and fruit, without animal forms.

ARCH BAR.—A semicircular bar in a sash ; otherwise a cot bar.

ARCHIMEDEAN DRILL.—A tool for boring holes in metal, worked in alternate directions by means of a bobbin and endless screw.

ARCHITECT.—A designer of buildings, and usually the chief supervisor during their erection. See Builder.

ARCHITRAVE.—In Classic architecture the lower division of the entablature ; also called the fascia in joinery.

ARCHITRAVES.—The moulded border, or frame, to a door or window opening.

AREA.—A small space between the surrounding ground and the walls of a building, provided to guard the wall against damp ; also the superficial contents of any large surface.

ARRIS.—The edge or line in which two plane surfaces meet.

ARRIS RAIL.—A rail of triangular section used for fences.

ASHLERING.—Perpendicular studding between the sloping roof and the floor of an attic chamber for fixing the plaster to.

ASTRAGAL.—A semicircular moulding projecting from the surface on which it is placed. When flush with the surface, it is termed a bead.

ATTIC.—A room formed in the roof of a building.

AUGER.—A wood-boring tool similar to a gimlet, but much larger, and worked with both hands.

AXIS.—An imaginary straight line joining the centres of the two ends of any solid ; also the line in a plane figure around which it is imagined the plane is revolved to produce the corresponding solid.

B.

BACK FLAP.—Any leaf of a folding shutter behind the front one ; also the name of the hinge used for these shutters.

BACK HEARTH.—That part of the hearth that is within the jambs.

BACK LINING.—That part of the weight box of a sash frame parallel to the pulley stile ; also the wall lining of a shutter boxing.

BACK OF A HANDRAIL is the upper side.

BACK OF A HIP.—The edge of a hip rafter made to range with the backs of the rafters on each side of it.

BACKINGS.—Fixing pieces nailed to the wall behind finishings.

BADGER.—A wide rebate plane with a skew iron. See f. 5, p. 12.

BAIL.—A wooden bar separating the stalls of a stable.

BALCONY.—A projecting platform before a window, enclosed with a railing, and supported on brackets or cantilevers.

BALL FLOWER.—An ornament carved in hollow mouldings of the Decorated period, resembling a ball within the open petals of a flower.

BALTIC TIMBER.—Timber of all kinds shipped from ports on the Baltic Sea, inclusive of Russian, Prussian, Polish, Swedish, and Norwegian timber.

BALUSTERS.—A series of small columns protecting the outer ends of stairs, and supporting the handrail.

BANKER.—A mason's bench.

BANNISTERS.—A corruption of the word baluster, used to indicate the slender rectangular sticks that do duty for a balustrade in common stairs. See p. 308.

BAR.—The interior division members of a sash ; also the serving compartment of an eating or drinking house.

BAREFACED TENON.—One with a shoulder on one side only.

BARGE BOARD.—Ornamental boards covering the ends of roof timbers projecting over gable ends of buildings.

BARREL or TOWER BOLT.—A straight bolt working in a barrel-like case.
BARREL ROOF.—One with the interior section semi-cylindrical.
BASE MOULD or SKIRTING.—A projecting moulding at the bottom of a wall.
BASEMENT.—That part of the building below the ground line.
BASIL.—The "grinding bevel" of any cutting tool.
BASILICA.—A Roman public hall. Some of these were afterwards converted into Christian churches.
BATTEN.—Converted timber of less scantling than 9 by 3 in. and more than 4 by 1 in.; applied also specifically to certain small scantlings, as slating battens 2 by 1 in., &c.
BATTEN DOOR.—One composed of boards or "battens" secured to cross ledges. See p. 153.
BAULK or BALK.—Timber roughly squared with the axe.
BAY.—Any portion of a building divided into equal compartments, as case and tail bays of roofs and floors, *i.e.*, the spaces between the beams or girders respectively near the middle or ends of the building.
BAY WINDOW.—A projecting window which rises from the ground. See Oriel.
BEAD.—A small moulding semicircular in section, generally used with a quirk or sinking. Without the quirk it is distinguished as a REED. When used with two quirks it is called a DOUBLE-QUIRKED BEAD; when projecting from the surface a COCK BEAD; when stuck below the surface a SUNK BEAD, and, when worked on the two faces of a salient angle, with a quirk on each face, a RETURN BEAD; when over 1 in. in diameter, a TORUS BEAD. Bead-butt and bead-flush are described on p. 156.
BEAM.—Any large piece of timber spanning an opening or space.
BEAM COMPASS.—An instrument for drawing large circles, comprising a thin lath or "beam," and two sliders, one carrying a pointed leg, the other a pencil. It is also incorrectly called a TRAMMEL (which see).
BEARERS.—Short pieces of quartering fixed under stairs to support the winders. See p. 308.
BEAT, otherwise "FIGURE."—The cropping out of the annual rings on the surface of a board.
BED.—The bearing surface of any portion of a building; also to place a thin stratum of cement between any two surfaces to make a good joint; generally, to fit solidly.
BED MOULD.—A moulding in a cornice or pediment, below the corona, or any moulding immediately below a deep projection.
BENCH.—A joiner's work table. See p. 35.
BENCH END.—The upright end of a church pew.
BENCH HOOK or JACK.—An appliance for holding boards, &c., steady whilst they are cut or worked on the bench top. See pp. 35 & 60.
BENCH PLANES.—The jack, trying, and smoothing planes; so-called because in consequence of their frequent use, they are kept constantly on the bench.
BENCH SCREW.—A wooden vice fixed at the fore end of a bench to hold the material operated upon.
BEST TIMBER.—A term used in specifications to denote good quality timber suitable for the work specified.
BEVEL.—Any angle other than a right angle or "square"; also a tool similar to a square, but with an adjustable blade.
BEVEL-CUT.—A system of Handrailing. See p. 331.
BIND.—A term used to denote that any moving part of a framework fits too tightly in its opening, as a door when it rubs against the edge of the rebate in the frame.
BINDER.—A thick joist in double and framed floors, used to carry the bridging joists.
BIRDSMOUTH.—An internal angle cut on the end of a piece of timber to enable it to fit upon the salient angle of a cross piece of timber.
BIT.—The generic name for various boring tools operated in a brace. See p. 25.
BLIND BOX.—A space or container for a blind to a window or shop front. See p. 214 and Plate XXV., 229.
BLIND STORY.—An apartment of a building having no windows; modern examples are chiefly confined to theatres.
BLOCK PLANE.—A substantially made smoothing plane with parallel sides for use with a block mitre shoot. See f. 4, p. 12.

BLOCK PLAN.—A plan of the site of a building and of the walls.

BLOCKS.—Pieces of wood fixed to the lower ends of architraves; also short pieces glued in interior angles to strengthen the joint.

BOARD.—Any converted stuff more than 5 in. wide and less than 2 in. thick. See Batten, also Plank.

BOARD AND BATTEN.—A method of forming the walls of wooden houses with a thick and thin board placed alternately. Usually the thin one is placed in a groove in the edge of the thicker.

BOLECTION MOULDING.—Moulding raised above the surface of the framing, and rebated over its edges. See pp. 155 & 373.

BOND.—Any arrangement of the component parts of a structure that causes the ends to overlap and so tie the whole together.

BOND TIMBERS.—Horizontal pieces of quartering built into walls for tying them together, or to provide fixings for the various finishings.

BONING.—Taking out of twist or winding. See p. 51.

BORROWED LIGHT.—A window in an interior wall transmitting light received from an outer window.

BOSS.—An applied ornament of circular shape on the intersection of ribs in vaulting, and flat ceilings.

BOW SAW, also called a Frame Saw, is a thin narrow blade fixed in a light frame for the purpose of making curved cuts. See f. 10, p. 9.

BOW WINDOW.—A bay window circular in plan. See Photograph, p. 234.

BOXING or FOLDING SHUTTERS.—Those which fold into a recess or boxing. See Plate XXIII.

BOXINGS.—A hollow enclosure to receive shutters or weights. See pp. 194, 229.

BOX STAPLE.—The iron case into which the bolt of a rim lock shoots.

BOX TENON.—An angle tenon in a corner post. See f. 7, p. 145.

BOXED PLANES.—Planes having slips of boxwood inserted in their working surfaces to increase their wearing capacities. See f. 5, p. 17.

BRACE.—A member introduced into a rectangular framing to triangulate it, and thus render it immovable; also a tool with a cranked shaft, used for revolving boring tools; called also a stock. The tools are called bits. See p. 26.

BRACKET.—A triangular-framed support for a shelf, &c.

BRACKETS.—Thin moulded pieces planted on the strings of stairs at the ends of the steps. See p. 320. Also ROUGH BRACKETS.—Pieces of wood nailed on the sides of the carriages to support the middle of the steps. See p. 314.

BRAD.—A small nail with its head flush with the sides.

BRADAWL.—A tool for making holes for brads and other small fastenings.

BRADDED.—Put together with brads.

BRAND.—A quality mark on timber either stamped or stencilled.

BRANDERING.—Small fillets nailed to girders to take the plastering laths, so affording a better key for the plaster.

BREAK.—Any recess or projection on the surface of a wall, &c., causing interruption to the uniformity of its surface.

BREAKING DOWN or BREAKING UP.—Dividing a balk into deals or boards.

BREAKING JOINT in flooring is when the butt joints of the boards are made to lie alternately on different joists.

BREAKING WEIGHT.—The exact weight required to break a beam when steadily applied.

BREAST LINING.—See Window Back.

BREASTSUMMER.—A stout lintel or beam spanning an opening in a wall and carrying the superincumbent mass. It differs from a girder in that the former merely carries a load, whilst the latter ties together also.

BRIDGING JOISTS.—The common joists carrying the floor boards.

BRIDGING or STRUTTING.—Pieces of wood placed between joists to stiffen them.

BRIDLE JOINT.—A joint the reverse of a mortise and tenon, chiefly used in carpentry.

BRIGHTS.—American timber that has had dry transit to the port of shipment.

BUCKLING (*M.*), as applied to saw blades, means that the tool has been suddenly bent, and has acquired a permanent set or cripple.

BUHL or BOULE WORK is the inlaying of veneer of wood with metal or tortoise-shell.

BUILDER.—The responsible erecter of buildings, and employer of the various craftsmen required. A technical and business expert, who carries out the designs of architects, &c. See Architect.

BUILT BEAM.—One composed of several pieces.

BUILT RIB.—One built up in several thicknesses.

BULL'S EYE.—A small circular window. See p. 216.

BULLNOSE.—A small metal rebate plane. See p. 14.

BULLNOSE STEP.—One with a quadrant end. See p. 314.

BUTT END.—The part of a tree nearest the roots.

BUTT HINGES.—Hinges with square or butt ends to be sunk into the wood, as distinguished from those planted on the surface. See p. 402.

BUTT JOINT.—A joint square to the length of the piece, the ends abutting.

BUTTON.—A small block of wood with a tongue on one end for securing wide surfaces, such as table tops, &c., to the framework, and allowing sufficient play for shrinkage. See f. 10, p. 148.

C.

CABIN HOOK.—A hook for securing a door or casement when open.

CAMBER.—The convexity given to the upper side of a beam to counteract the sagging produced by the load.

CAMES.—The metal bands or supports over small panes of glass, used chiefly to form or emphasise the pattern on stained glass windows. See p. 171.

CAMP CEILING.—The ceiling of an attic when all the walls are inclined equally.

CANT.—Any part of a surface that breaks out at an angle other than a right angle.

CANTILEVER.—A projecting beam supported at one end only.

CAP.—The head of a pier column or newel post.

CAPPING.—A moulding on the top of a screen, gate, or post.

CARCASE, applied to joinery, is the outer framework of cupboards, counters, and similar enclosures. Applied to carpentry, it is the structural timber work of a building. The term is also applied to the naked walls and rough timbering of a building before the joiner's work is applied.

CARRIAGES.—The timber framing supporting stairs. See p. 317.

CARYATIDES.—Female figures used as columns or supports.

CASE, CASED, or ENCASED BAY.—A division between two beams in a floor or roof. The outer bays between a beam and a wall are termed *Tail Bays*.

CASED FRAMES.—Sash frames whose sides are built up in numerous thin pieces, as distinguished from those having solid jambs.

CASEMENT.—A sash hung on hinges ; also a hollow moulding of the Gothic period.

CAST.—To wind or twist, generally applied to the distortion in length.

CATENARY CURVE.—One produced by a rope suspended at each end and hanging loosely.

CATHERINE WHEEL WINDOW.—A circular window with radiating mullions.

CAUL.—A piece of wood or zinc, the reverse of any curved surface that is to be veneered, which is heated and pressed on the face of the veneer after it is laid, to keep it in position until dry.

CAULKING.—Literally the corking or stopping of open joints in excavations with clay, in shipwrighting with oakum and pitch.

CAVETTO.—The Classic "Hollow" moulding. See p. 366.

CEILING JOISTS.—Small scantlings fixed to the under side of the floor joists to carry the plaster ceiling.

CEILING LIGHT.—A horizontal sash in a ceiling under a roof light.

CENTRES or CENTERING.—Wooden frames for building arches upon.

CHAIR RAIL or DADO RAIL.—A moulded rail fixed to the walls of an apartment to protect them from the backs of chairs.

CHANCEL.—That part of a church where the altar is placed.

CHANCEL SCREEN.—A dwarf screen or balustrade enclosing the chancel. See p. 264.

CHANNEL IRON.—A piece of iron of square U section forming a groove in which the ends of a revolving shutter move. See p. 230.

CHANTLATE or SPROCKET.—A piece of wood fastened to the foot of a rafter overhanging a wall, to tilt the slates up to a flatter pitch.

CHAMFER.—To take the salient angle off a piece of material, and leave a plane which is inclined to both surfaces.

CHASE MORTISE.—A mortise which runs out at one end to the surface in which it is placed, so that the tenon may be entered by placing the tenoned piece in a diagonal position across the opening.

CHECK.—A cabinetmaker's term for a rebate.

CHEEK.—The side of a mortise; also the portion of wood that is cut away to form a tenon.

CHIMNEY-PIECE.—An ornamental finishing to a fireplace.

CHISEL.—A cutting tool guided by the hand. See p. 20.

CHOIR SCREEN.—A partition between the choir and the public part of a church. In churches of the Gothic period these screens are usually of open tracery in the upper part, and richly carved panels in the lower. See p. 264.

CIRCLE ON CIRCLE.—A term applied generally to anything curved circularly, both in plan and elevation.

CLAMP.—A narrow strip of wood framed flush with, and across the grain of a larger piece, to prevent its warping.

CLAPBOARDS.—Pieces of cleft wood used to cover the sides and roofs of buildings in a similar manner to tiles.

CLEAR, IN THE.—Nett distance between any two points or surfaces.

CLEAR SPAN.—The horizontal distance between two walls bearing a beam.

CLEARANCE JOINT.—The amount of room given to a door or sash to allow it to open freely.

CLEADINGS.—The timber linings of a shaft or mine.

CLEAT.—A wooden cramp; also a chock or bearing block.

CLERESTORY.—The part of the walls of a church that rise above the side or aisle roofs, containing windows to light the nave.

CLEAN-UP, TO.—The finishing of the surface of joiner's work, with planes and scrapers or glass-paper.

CLOSE STRING.—The side of a stair in which the ends of the steps are entirely enclosed.

CLOUT NAIL.—One with a flat round head.

COACH SCREW.—A large wood screw with a flat square head.

COFFER PANEL.—One deeply recessed, usually in the soffit of an arch, or cupola, or in a ceiling.

COFFER DAM.—A timber structure filled with clay to form a watertight enclosure in waterways.

COGGING.—A form of joint in carpentry in which a central cog or tenon is cut in the bearing piece, and a corresponding slot or chase in the riding piece. See f. 1, p. 318.

COLLAR BEAM.—A horizontal beam in a roof truss placed at some distance above the feet of the rafters, usually at half the height between the wall plate and ridge.

COLUMN.—A long shaft carrying a load; made in one length of material.

COMMODE STEP.—One with a curved riser or front. See p. 325.

COMMON RAFTERS.—The smaller rafters of a roof carrying the covering.

COMMON PITCH.—The pitch of stairs above and below winders, *i.e.*, the pitch of the flyers.

COMPASS PLANE.—A smoothing plane with a circular sole. See p. 12.

COMPASS SAW.—A small hand saw with a narrow blade for cutting curves. See p. 9.

CONCENTRIC.—Having a common centre.

CONE.—A solid having a circular base and tapering to a point; also the seed vessel of an order of trees called coniferous, supplying the pine and fir woods of commerce.

CONSOLE.—An ornamental bracket or support to a shelf or head, having parallel sides, and of greater depth than projection, which d stinguishes it from a Modillion, which is of slight depth and considerable projection.

CONVERTING.—Re-sawing timber. See p. 417.
CORBEL.—A short projection from a wall, to carry a plate or frame.
CORE.—Waste cut out of a mortise, also the base or interior part of veneered work.
CORED SECTION.—(*M.*) A metal frame or other casting, which, appearing solid is really hollow, a "core" of sand having been inserted in the casting box. In machine frames this is done to reduce the total weight, and to place the metal in the right position to resist the stresses.
CORNICE.—The moulding at the top of a wall, column, or frame.
CORONA.—A flat overhanging member in a cornice, with a drip or recess in its soffit. Fig. 2, p. 371.
COT BAR or CRADLE BAR.—A semicircular bar in a sash.
COUNTER FLOOR.—An inferior floor fixed diagonally below a superior one, usually a "parquet."
COUNTER LATHING.—Laths nailed to the face of stout timbers to raise the sheet lathing above their surface to provide a space for the plaster to "key" into.
COUNTER SCREEN.—A dwarf glazed screen fixed on a shop counter.
COUNTERSHAFT.—(*M.*) A shaft carrying pulleys, interposed between the driving shafts and the machinery, to enable the speeds to be varied upon each machine.
COUNTERSUNK.—The head of a screw when let in flush with the surface.
COVE.—A large hollow moulding or half-arch in a ceiling.
COVER BOARD.—A board covering the top of a cornice on cupboards, &c.
COVER FLAP.—A hinged flap covering the boxings of shutters.
CRAB.—A movable winch.
CRADLE.—An apparatus used in gluing treads and risers together. See p. 312.
CRAMP.—A joiner's appliance for drawing the parts of a frame together. See p. 36.
CRAMPING UP.—The act of drawing the shoulders and parts of a frame up close with a cramp.
CROSS GARNET.—A T strap hinge fixed on the face of a door. See p. 153.
CROSS-CUT SAW.—A machine saw used for cutting across the grain; also a long saw with coarse teeth, having a handle at each end used for cutting logs.
CROSS-GRAIN.—Wood having its grain running in other directions than the length of the stuff; inferior wood. See Straight-grain Stuff.
CROWN.—The highest part of an arch.
CROWN POST.—A vertical support to a ridge board, usually resting upon the collar beam in a truss.
CUBE.—A rectangular prism having all its six sides equal in area.
CUPOLA.—A spherical roof; the interior surface of a dome.
CUPSHAKE.—A defect in timber in which the shake separates the annual rings. See f. 4, p. 317.
CURB.—A stout frame around an opening in a roof or floor, to carry a light or a trap door.
CURTAIL STEP.—One with its end spiral in plan; so called because the spiral is not taken to an eye—is curtailed. See p. 325.
CUSPS.—The points of the foils in Gothic tracery.
CUT STRING or OPEN STRING.—The side piece of a stair, which is cut out to receive the ends of the steps, thus showing their outline. See Close String.
CUT STUFF.—Deals cut into boards and other small scantlings.
CUTTER.—(*M.*) The knife or blade of a planing or moulding machine.
CUTTER HEAD or BLOCK.—(*M.*) A rectangular steel block, either cast upon or keyed to a revolving spindle for the purpose of attaching various knives or cutters. See p. 98.
CUTTING.—An excavation in a roadway.
CYLINDER.—A box or "centre" with a curved surface used for bending veneers around to form the wreathed parts of strings, &c. See p. 323.
CYLINDROID.—A solid having elliptic ends and parallel sides.
CYMA RECTA.—The ancient ogee moulding. See f. 9, 10, p. 366.

D.

DADO.—A framing around a room about breast high.
DANCING STEPS.—Winders that do not radiate from a common centre. See p. 318.
DATUM.—A fixed point in a building from which all levels are taken.

DEAD LOAD.—A permanent or settled load, the reverse of a moving load.
DEAD LOCK.—A lock with one bolt only that requires moving with a key.
DEAD SHORE.—An upright shore carrying a needle used for dead loads only.
DEAL.—A common term for all fir timber; also a market form of timber, size 3 by 9 in. or 3 by 10 in.
DECORATED STYLE.—The second and best style of Gothic architecture prevailing throughout the fourteenth century.
DEFLECTION.—The bending or sagging of beams under a load.
DENTILS.—Ornaments on a moulding consisting of rectangular blocks with narrow spaces between them.
DESICCATING.—The artificial drying of timber by means of dry heated air.
DETRUSION.—A force that acts by thrusting out; a bursting outwards.
DEVELOPMENT.—The laying out flat of any surfaces that are not in a common plane. See Stretch Out.
DIE SQUARE STUFF.—Equal-sided timber, less in scantling than a balk, and larger than quartering.
DISHED.—A shallow sinking in the shape of a dish, usually to receive the flange of a pipe.
DIVIDERS, SPRING.—A tool similar to compasses, but of lighter make. See f. 3, p. 22.
DOATINESS.—A speckled reddish appearance of timber beginning to decay.
DOG.—An iron cramp used for drawing up joints in joiner's and carpenter's work. Page 53.
DOG'S-TOOTH ORNAMENT.—A triangular fret, or a pyramidal carving in a moulding, the latter being an Early English characteristic.
DOME.—A semi-spherical roof. The term usually applies to the exterior. See Cupola.
DONKEY'S EAR.—A mitre shoot. See f. 5, p. 41.
DOOR HEAD.—A projecting canopy over an exterior door. See p. 180, A.
DORIC.—One of the five orders into which Classic architecture is divided.
DORMER.—A perpendicular window in a roof. See p. 220.
DOUBLE DOORS.—A pair of doors (see p. 165); also a door constructed in two layers with a fire-resisting material enclosed between them. See p. 173.
DOUBLE FLOOR.—One in which the joists are carried by binders (which see).
DOUBLE TENONS.—Two or more tenons side by side, *i.e.*, not in the same plane. See f. 4 & 5, p. 143.
DOUBLE-HUNG SASHES.—Frames in which both sashes are hung.
DOUBLE-MARGIN DOOR.—One prepared to look like a pair of doors. See p. 162.
DOUBLE-REBATED LININGS.—Those rebated on each edge of the jambs.
DOVETAIL KEY.—A key or slip in the back of a panel with its edges undercut. See f. 2, p. 149.
DOWEL.—A cylindrical wood pin used for securing joints or tenons.
DOWEL PLATE.—A steel plate with holes to suit "bits" in which dowels are made. Page 26.
DRAGGING.—The rubbing of an ill-fitted door or sash upon its lower part.
DRAGON TIE.—A cross beam at the angle of wall plates to receive the foot of a hip rafter.
DRAINING BOARD.—A wood fitting to a scullery sink on which plates are placed to drain after washing.
DRAPERY PANEL.—One carved to represent folds or rolls of linen; hence also called "linen-fold." See pp. 169 & 266.
DRAW-BORE PIN.—A tool used for drawing up shoulders in mortise and tenon joints, when cramps and wedges are not used. See f. 1, p. 70.
DRAW-KNIFE.—A double-handled chisel used horizontally. See f. 9, p. 20.
DRAWBACK LOCK.—One that is opened inside by sliding the knob back, and outside with a key.
DRAWER LOCK CHISEL.—A tool for making mortises in confined situations. See f. 8, p. 20.
DRAWING BOARD.—A board of standard size clamped or ledged to keep it true, used for attaching paper to for drawing purposes. The term is also applied to those "rods" in the workshop that are of considerable size.
DRAWING PEN (more correctly a Ruling Pen).—An instrument with a pair of parallel nibs used with a ruler for "inking in" lines in drawings.
DRESSER.—A kitchen fitment containing shelves and drawers. See f. 1, p. 293.
DRESSED STUFF.—Timber that is wrought or planed.
DROP.—The lower end of a newel projecting below the ceiling. See p. 308.

DROP HANDLE.—A drawer or door handle that drops down when not in use. See Plate XXVIII.

DRUM.—An appliance for bending wood upon; called also "a cylinder." See p. 273.

DRUXINESS.—A decay of growing timber caused by fungus getting access through a wound, generally at a broken branch, and indicated by spots and streaks of a bleached appearance.

DRY ROT.—This term is loosely applied to all forms of decay of dry or "dead" timber, in contradistinction to that of "Wet Rot," applied to the decay of living trees. There are numerous forms of this decay, the appearances varying with the immediate cause. They are chiefly due to parasitic attacks of various fungi which feed upon the secretions of the wood, and thus destroy its tissue. The one chiefly responsible for the decay of timber in buildings is the *Merulius lachrymans*, which is more fully described at p. 414.

DRY TIMBER.—Timber from which all moisture has been extracted. See Seasoned.

DRY WEDGED.—Temporary wedging up of framing without glue, for the purpose of fitting or keeping in condition whilst re-drying.

DUBBED OFF.—An abrupt or clumsy easing to a moulding, &c.

DURAMEN.—The heartwood of timber trees. See p. 412.

DUST BOARD.—Panelled divisions between drawers; also a cover board.

DWARF CUPBOARD.—One under 3 ft. in height.

DWARF DOOR.—One under 5 ft. 6 in. in height.

E.

EARLY ENGLISH.—The first of the pointed or Gothic styles of architecture. Flourished during the thirteenth century.

EASING.—The curve by which two straight or curved lines of differing directions are joined. The operation is termed "easing off"; also the gradual release of wedges sustaining loads, &c.

EAVES.—The lower part of a roof which overhangs the wall.

ECHINUS.—An ornament in the shape of an egg carved on an ovolo moulding.

ELASTICITY.—The power of returning to the original shape after a strain.

ELASTIC LIMIT.—A point in the straining of material which, if exceeded, a "permanent set" or distortion takes place, *i.e.*, the material does not return to its original shape when the load is removed.

ELBOW LINING.—The linings of window jambs below the window-sill. When the linings run all the way up they are called jamb linings.

ELEPHANT.—A machine of the "over-spindle" class for grooving and recessing generally. See f. 2, Plate VII.

ELLIPSE.—The inclined section of a cone cutting both its sides; also the inclined section of a cylinder.

ENDS.—Timber merchants' term for deals and battens less than 8 ft. long.

ENRICHMENT.—The carving of a moulding.

ENTABLATURE.—That part of a building in Classic architecture that rests on, or is carried by the columns, consisting of the architrave, frieze, and cornice.

ENTASIS.—The apparent swelling in the middle of a column caused by diminishing its upper part.

EQUILATERAL TRIANGLE.—One having all its sides of equal length and inclination to the others.

ESCUTCHEON.—A metal plate surrounding a keyhole.

ESPAGNOLETTE BOLT.—One used with French casements that fastens at the top, bottom, and middle with one movement of the lever or handle.

EYE.—The centre or finish of a scroll or volute.

EXOGEN.—A plant whose substance is increased by successive layers on the outside of previou layers, a characteristic of all timber trees.

EXTENSION BIT.—A centre bit with a movable cutter for making various sized holes.

EXTRADOS.—The exterior curve of an arch.

F.

FAÇADE.—The front or face of a building.

FACE MARK.—A mark made on prepared stuff to denote the face or working side. See p. 54.

FACE MOULD.—A curved templet applied to the face side of stuff to be shaped. For a special reference, see p. 333.

FACET.—The square or fillet between the flutes of a column.

FACING UP.—Preparing wood for setting out. See p. 51.

FALLING MOULD.—A thin templet used for bending around a wreath-piece of a handrail after it has been shaped to the face mould, for the purpose of marking the thickness; but seldom used by modern handrailers.

FAN.—The inclined boards guarding a scaffolding.

FANLIGHT.—A sash above a transom in a door or window frame.

FASCIA or FASCIA BOARD.—The broad flat surface between the sash and cornice in a shop front; also any wide flat band in a cornice. See f. 1, Plate XXV.

FATIGUE OF MATERIAL.—A deterioration set up by continuous overstrain.

FAULTS.—Shakes, large knots, sap, and decayed parts in timber.

FEATHER TONGUE.—A thin slip cut diagonally across the grain. See p. 54.

FEATHER-EDGED BOARDS are those cut thinner at one edge than the other, and are used overlapping for outhouses, &c. See Weather-boards.

FEATHERINGS.—The points or cusps in Gothic tracery. See f. 1, p. 381.

FEED.—(*M.*) The supplying of a machine with material to operate upon. There are many derivative terms from the word, as, Feeding End—the end of the machine in which the unwrought timber is placed (see Leading-off End); Rate of Feed—the speed with which a board is passed through a planer, &c.

FELT.—The medullary rays or "silver grain" in oak. See p. 417.

FENCE.—A guide piece on a plane, and the guide for the stuff passed through a machine.

FENDER.—A piece of timber placed at the base of a scaffold to protect it from vehicles.

FIELDED PANEL.—A raised panel with a wide flat surface. See p. 155.

FIGURE.—The pattern produced by the wavy grain in hardwoods, and in certain cases by the reflection of light from the cellular tissue. See p. 417.

FILLET.—A narrow strip of wood; also a rectangular moulding.

FILLISTER.—An adjustable rebate plane. See p. 17.

FINGER PLATE.—Ornamental plates of metal, glass, &c., fastened on the lock stiles of doors to prevent finger marks.

FINGER PLATE.—(*M.*) A plate filling the space in the table necessary for the insertion of a circular saw, containing holes at each end for insertion of the fingers. See p. 112.

FINIAL.—An ornament at the apex of a gable or spire.

FINISH or FINISHING OFF.—Terms used to denote the final preparation on the bench of joiner's work.

FINISHINGS.—Formerly this term covered all the joinery work in a building, but the increasing variety of work now placed in houses has made a distinction desirable, and the term is now applied only to the more permanent and ornamental parts of the fittings, such as architraves, picture rails, over-doors, and chimney-pieces, &c.; whilst removable fittings are termed Fitments, which see.

FIR TIMBER.—A generic term for all pine and spruce timber used in building.

FIRMER CHISEL.—A stout paring chisel. See p. 20.

FIRRINGS.—Fillets nailed to irregular timbers to bring them into one plane.

FITMENTS.—More or less removable fittings in houses, &c., such as dressers, cupboards, book-cases, tables, lifts, &c.

FITTING UP.—The general fitting, jointing, and trying together of joinery previous to gluing, &c.

FITTINGS.—A generic term, covering all the joiner's work in buildings, including hardware, such as locks, bolts, window fastenings, &c.; also special work, such as desks, screens, pedestals, pews, pulpits, &c.

FIXINGS.—Various appliances used for attaching linings, &c., to walls. See p. 398.

FIXTURE, A.—Legally something that cannot be removed by an occupier of a building.

FLAT PANEL.—One not raised or moulded in the middle.

FLIGHT.—A complete set of stairs reaching from floor to floor, or from floor to a landing.

FLOOR.—The horizontal divisions in a building.

FLOOR DOG.—A machine for cramping up the joints of floor boards.

FLOOR LINE.—A base or working line placed on all finishings that start from the floor.

FLOORING.—The covering or boards of a floor.

FLOORING, NAKED.—The assemblage of timbers supporting the covering.

FLUSH PANEL.—One level with the surface of the framing.

FLUTES.—Grooves of semicircular section in columns and pilasters.

FLYERS.—Steps that are of parallel width. See Winders.

FOLDED FLOORING.—Floor boards that are folded or sprung down, instead of having their joints cramped up with dogs.

FOLDING DOORS and SHUTTERS.—Those that are hung to each other and fold together. See pp. 161 & 229.

FORKED JOINT.—One in which the ends of the pieces are indented like a saw tooth.

FOX WEDGING.—A form of secret wedged tenon. See p. 150.

FOXINESS.—An incipient decay of oak producing a reddish tinge.

FRAME.—The thicker portions of joinery constructions which carry the panels, &c.

FRAME SAW.—See Bow Saw.

FRAMED FLOOR.—One in which binders are framed into girders ; sometimes called a double-framed floor. See Double Floor.

FRAMED GROUNDS.—Grounds formed with stiles and rails. See p 190.

FRAMED PARTITIONS.—A partition made with panelled frames ; and in carpentry, a partition that is trussed, and the parts connected by mortise and tenon joints.

FRAMED WORK.—Any constructive work whose joints or connections form an integral part of the members of the frame.

FRAMING.—The skeleton or substantial parts of a framed work.

FRANKING.—A form of reversed haunching. See p. 61.

FRENCH CASEMENT.—A sash hung and used as a door.

FRENCHMAN.—A sawing stool made with two boards in the shape of an inverted V.

FRIEZE.—That portion of a wall of an apartment immediately below the cornice and above the dado.

FRIEZE PANEL.—The top panel of doors having more than three rails.

FRIEZE RAIL.—The second rail below the top, in a door or other framing having more than three rails.

FURNITURE.—Applied metal and other fittings to doors, sashes, &c.

G.

GABLE.—The triangular end of a house under the roof.

GAINING.—(*M.*) The term used in America for housing or cross grooving.

GATE.—Usually a door of large dimensions used externally, and also a door with openwork framing.

GAUGE.—A tool for marking parallel dimensions. See f. 12 & 13, p. 17.

GIMLET.—A hand boring tool with a screw point.

GIRDER.—The principal beam carrying a heavy floor.

GLAZED.—Filled in with glass.

GLUE.—An adhesive cement made by boiling down skins, hoofs, and bones of animals.

GOING.—The horizontal distance between the face of one riser and the face of the next. GOING OF THE FLIGHT is the horizontal distance between the faces of the first and last risers in a flight of stairs.

GORE.—A board cut to a triangular shape with curved sides for the covering of a dome, &c.

GOTHIC.—A term applied to the architecture of the Middle Ages from the end of the twelfth century to the middle of the sixteenth.

GOTHIC ARCH.—A pointed arch. These were solely used in the architecture of the Middle Ages.

GRAIN.—The arrangement of the fibres and vessels of wood, particularly in their relation to the face of the material.

GRAINING.—An imitation of the natural figure of wood by painting, or drawing.

GRATING.—A wood lattice, used to protect gutters from snow.

GREENHOUSE.—A structure of wood and glass for preserving plants, not artificially heated.

GROIN.—The edge or angle of two intersecting vaults.

GROOVING.—A rectangular sinking in the surface of any material.

GROOVING SAW.—A thick saw with coarse teeth for making grooves by machinery.

GROOVED and TONGUED.—A form of joint in which the two edges of the joint are grooved and a loose slip or tongue placed therein.

GROUND FLOOR.—The floor of a house immediately above the ground line.

GROUNDS.—Strips and narrow frames to which the finishings of a building are affixed.

GUARDS.—(*M.*) Appliances placed over or around the cutting parts of machines to prevent injury to the operator. See p. 130.

GULLET.—A hollow or cavity at the root of a machine saw tooth to receive the sawdust whilst passing through the material, and so prevent the saw jamming in the work.

GULLETING.—(*M.*) The operation of forming gullets or spaces to receive the dust in circular saw teeth.

GUTTER BOARD.—A board for supporting the lead or zinc which is used to form gutters in roofs.

H.

HALF-LONG PLANE.—A plane 6 in. longer than a trying plane, formerly much used in making joints.

HALF-TIMBERED HOUSE.—One in which the walls are constructed of a framework of timber filled in with brickwork or plaster.

HAMMER-HEAD TENON.—One with a protuberance on each side of the exterior end against which wedges are driven to tighten the shoulder. See p. 184.

HAND OF A DOOR.—A term distinguishing on which side it is hinged. If when standing on the side of the door from which the knuckle of the hinge projects or is visible, the hinges are on the left side of the observer, then the door is hung "left-handed," and *vice versâ.*

HAND OF A LOCK is determined according to the side that the bolt shoots when held with the inside of the lock towards the observer; or stated otherwise, when facing the outside of the door. Thus a right-hand-hung door requires a right-hand lock.

HAND SCREW.—A wooden cramp actuated by a pair of threaded handle shafts.

HANDED.—Made in pairs, as door stiles, &c.

HANDRAIL.—The support for the hands at the sides of a stair.

HANDRAIL BOLT or SCREW.—A double-ended screw with a pair of nuts used in the joints of handrails, &c. See f. 3, p. 72.

HANGING SASH.—One sliding vertically and balanced by weights.

HANGING STAIR.—One in which the steps are built into the wall.

HANGING STILE.—The stile of a door on which the hinges are placed.

HATCH.—The cover of an opening in a roof or floor.

HAUNCH.—The heel or end of a tenon that has been reduced in width. See p. 143.

HAUNCHING.—A recess cut in a stile to receive the haunch of a tenon.

HEAD.—The upper horizontal member of a door or window frame.

HEADING JOINT.—Butt joints in flooring.

HEARTSHAKE.—Radial clefts in the heartwood of growing trees. See p. 415.

HEARTWOOD.—Perfected wood as distinguished from the sapwood which is in course of formation.

HELIX.—The curve formed by a regularly ascending line winding around a cylinder.

HERRING-BONE STRUTTING.—Crossed struts between floor joists.
HIP.—A rafter placed in the angle between the two inclined sides of a roof.
HOOD MOULD.—A projecting moulding over window and doorway in Gothic architecture; also called weather mould, dripstone, &c.
HOPPER WINDOW.—See Hospital Light, p. 217.
HORN.—A projection on the ends of a frame used as a fixing. See p. 184.
HOUSING.—Sinking the end of one piece of wood completely into another without reducing it in thickness. See p. 142.
HYPERBOLA.—A curve formed by the section of a cone parallel with its axis.

I.

INGLE NOOK.—A screen or enclosed seat in or near a fireplace.
INTERTIE.—An interior horizontal beam in a framed partition.
INTRADOS.—The interior surface of an arch.
IONIC ORDER.—One of the five orders of Classic architecture.
IRON.—The cutter of a machine or hand tool.
IRON CORE.—A thin iron bar sunk in the under side of a handrail to stiffen it. See p. 321.

J.

JAMB.—The side of an opening in a wall.
JAMB LININGS.—Thin finishings covering the jambs of an opening.
JIB DOOR.—One made to lie flush with the surface of the wall. See p. 168.
JOINT.—The abutting surfaces of two prepared pieces of wood.
JOINTING.—The act of making joints with planes. See pp. 52 and 54.
JOISTS.—The timbers carrying a floor or ceiling.
JOURNAL.—(*M.*) The bearings or hollow parts in which spindles and shafts move when running. They are made of highly polished and toughened composition metals, such as phosphor-bronze, Babbit's metal, &c. The journal and the part of the shaft it embraces are collectively, the bearings.

K.

KERF.—A saw cut.
KEY.—A piece of wood sunk across the grain in a board to keep it from casting. See p. 148.
KING POST.—A vertical timber in the middle of a roof truss supporting the tie beam.
KNEE.—A vertical curve in a handrail, convex on top. It is the reverse of a ramp. See p. 308.
KNUCKLE OF A HINGE.—The back, or part containing the pivot.

L.

LAGGING.—Strips connecting the ribs of centering, upon which the arch bricks rest.
LAMB'S TONGUE.—A very flat ogee moulding used for sashes.
LAMINATED WORK.—Boards built up of several thin pieces in shaped work.
LANDING.—Any step in a stair, wider than two flyers; also the floor at commencement and finish of a flight, and the intermediate floor between two flights. When the latter extends half way across the stairway or well, it is called a quarter space or pace landing; when stretching right across a half-space landing.
LANTERN.—A skylight with lights in the sides.
LATTICE.—Open work made by crossing strips of iron or wood.
LEADING OFF END.—(*M.*) The back end of a machine from which the wood issues after working; Am. outfeed.
LEAR BOARD.—Inclined boards at the sides of gutters.

LEDGED DOOR.—One formed by nailing boards on cross pieces or ledges.
LINE OF NOSINGS.—An imaginary line touching the rounded edges of the treads of stairs. See Nosing Line.
LINTEL.—A horizontal beam over the head of an opening in a wall.
LISTEL.—A square moulding or fillet. See f. 1, p. 367.
LIVE KNOT.—One so incorporated with the wood that it will not fall out.
LIVE LOAD.—A moving or changing load. See Dead Load.
LOCK RAIL.—The rail of a door in which the lock is placed.
LOG.—A tree in the round with branches lopped off.
LOG FRAME.—A machine containing a number of parallel saws used for breaking down logs into deals and boards.
LOUVRE-BOARDS.—The inclined boards in louvre-frames, which admit air.
LOUVRES.—Frames containing sloping boards designed to keep out rain, but permit the passage of air and light.
LYING or LAYING PANEL.—One with the grain running horizontal.

M.

MANDREL.—(*M.*) A fixed "Spindle," upon which a tool or block revolves within collars. In American phraseology mandrel or arbor is equivalent to spindle.
MANTELPIECE.—A shelf with moulded edges, over a fireplace.
MARGIN.—That part of a close string that rises above the nosings of the steps. See p. 309.
MARGIN LIGHTS.—Panes formed by bars near the sides of sashes and sash doors.
MARKING KNIFE.—A tool used for setting out shoulder lines. See f. 1, p. 24.
MATCH BOARDS.—Boards grooved and tongued on the edges, the tongue being formed in the solid. See p. 137.
MEDULLARY RAYS.—The radiating septa visible in oaks and beeches which produce "silver grain." See p. 412.
MEETING RAILS.—The two rails in a pair of lifting sashes that meet in the middle of the frame.
MEETING STILES.—The two middle stiles of a pair of doors.
MEZZANINE.—A story of small height introduced between the levels of the main floors.
MITRE.—A plane joint between two pieces of stuff, when the surface of the joint is at any angle other than a right angle. When at a right angle to the surface it is called a butt joint.
MITRE BOX.—A box or trough used to keep mouldings, &c., steady whilst cutting them to a mitre, and having guiding cuts for the saw in its sides. See p. 40.
MITRE CAP.—The moulded cushion on the top of a newel post, into which the handrail is mitred. See p. 316.
MITRE CLAMP.—A clamp with one or both ends mitred at an angle of 45 deg.
MITRE CUT.—A similar appliance to the mitre-box, but having a rebate on one side as a seat for the moulding, used for small work.
MITRE DOVETAIL.—A secret dovetail. See f. 7, p. 142.
MITRE SHOOT.—A box for holding mouldings, &c., at an appropriate angle whilst a mitre is planed. See p. 41.
MITRE SQUARE.—A tool similar to a try square, but with the blade at an angle of 45 deg.
MITRE TEMPLET.—A guide for a chisel in cutting small mitres. See p. 62.
MONKEY.—The slip-hook used for the ram in a pile driver.
MONKEY TAIL.—A vertical scroll at the end of a handrail.
MOP STICK.—A handrail of nearly round section.
MORTISE.—A rectangular sinking in a piece of wood to receive a tenon, called "stub" when it does not pass through the piece.
MORTISE LOCK.—A lock sunk in a mortise in the edge of a door. See Rim Lock also, p. 404.
MOULD.—A curved templet used for marking material to shape.
MOULDING.—See definition, p. 283.
MOUNT.—To fix on the face of any work.

MOUNTINGS or MUNTINGS.—The interior vertical divisions of doors and similar framing. See Stiles.

MOUTHPIECE.—(*M.*) A piece of wood inserted in the saw gap of a circular saw bench, when saws of a less diameter than the largest are in use. See p. 112. Also a piece of wood inserted in the sole of a plane when by wear, the mouth has become too large.

MULLION.—A vertical division in a window frame.

N.

NEAT SIZE.—A term expressing that no additions or allowances have been made to the dimensions.

NECKING.—Any small moulding near the top of a column or pilaster.

NEEDLE.—A piece of timber piercing a wall, used to support it during alterations.

NEUTRAL AXIS.—That part of a beam in bearing, where opposing stresses change or counteract each other, and their effect is *nil*.

NEWEL.—A post at the end of a flight of stairs, to carry the handrail.

NOGGINGS.—Horizontal slips built between the studding of a partition to tie in the bricks.

NORFOLK LATCH.—A fastening for a thin door, worked with the thumb, and also provided with a handle for closing it.

NOSING.—A semicircular projecting edge to any flat surface; the rounded edge of the tread projecting over the riser. See p. 306.

NOSING LINE.—A line drawn on the string of stairs touching the salient angles of the steps, used for setting the latter out. See p. 310. Not to be confounded with "Line of Nosings," which see.

NOTCH BOARD.—A cut string.

NOTCHING.—A joint formed by cutting a piece out of one timber equal to the thickness of the piece crossing it. See Cogging.

O.

OCTAGON.—A figure of eight equal sides and eight equal angles.

OGEE.—A moulding wave-like in section or consisting of reversed circular arcs. See p. 370.

OILSTONE.—A fine-grained stone used with oil to sharpen cutting tools. The common varieties are "Turkey," "Washita," "Arkansas," "Charnley Forest."

OLD WOMAN'S TOOTH.—A router or grooving tool. Fig. 12, p. 12.

OPEN JOINT.—A method of hanging in which the door shows an opening between the edge and the frame during its passage.

OPEN STRING.—Another name for a cut string, which see.

ORIEL.—A projecting window that does not rise from the ground.

OUT OF WINDING.—True or free from twist.

OUTER or OUTSIDE STRING OF A STAIR.—The one farthest from the wall.

OVER-DOOR.—A pediment or other fixture over a doorway.

OVERMANTEL.—The upper part of a chimney-piece.

OVOLO.—A moulding of quarter circular or elliptic section. See p. 367.

P.

PACE.—A contraction of "space" in stair nomenclature.

PACKING.—Pieces of felt, wool, or rope placed in the saw way of a circular saw to absorb the heat and prevent vibration; also synonymous with firrings, which see.

PAD SAW.—A small narrow saw fitted to a handle or pad into which it can be telescoped when not in use.

PAIR OF TENONS.—Two tenons lying in the same plane. See Double Tenon.

PAIRED.—Two of anything having their corresponding parts opposite.

PANEL.—Any surface sunk below its surroundings, particularly a thin wide piece inserted between the members of a framing.

PANES.—Panels of glass.

PARABOLA.—A section of a cone parallel with one of its sides.

PARCLOSE.—A general term for a wood screen in a church, which has not some specific name, as, for instance, a screen around a tomb, or one in front of a gallery. See p. 263.

PARLIAMENT HINGES.—H or shutter hinges which project considerably. See p. 232.

PARQUET.—Flooring formed into geometrical patterns with coloured hardwoods.

PARTING BEAD.—A thin slip separating sliding sashes.

PARTING SLIP.—A rough thin slip separating the weights in a sash frame.

PARTING TOOL.—A V-shaped bent gouge ; also called a veiner.

PARTITION.—A thin wall separating two apartments, both fixed and movable.

PATERÆ.—Small disc ornaments applied to any surface.

PATTERN.—A wood model of a metal casting.

PEDESTAL.—A short square base for a column or statue ; also a fitting to support desks, &c.

PEDESTAL CLOSET.—An "open closet," or one in which the pan, &c., is exposed.

PEDIMENT.—An ornamental head to a window or door triangular in elevation.

PENDENTIVE.—The triangular portion of a spherical vault formed by two intersecting vaults at right angles to each other.

PENTAGON.—A plane figure with five straight sides.

PERMANENT SET.—A permanent alteration of shape in a beam, &c., brought about by stressing it beyond its elastic limit.

PERPENDICULAR.—A line or surface at right angles to another.

PERPENDICULAR STYLE.—The last of the Gothic styles of architecture, flourishing during fifteenth and sixteenth centuries, and characterised by the predominance of vertical members in windows, &c., and the slenderness of the columns and piers.

PEWS.—Fixed and enclosed seats for the congregation in a church.

PIGEON HOLE.—A small compartment for holding papers in a fitting.

PILASTER.—An engaged pier of slight projection. See p. 181.

PILE.—A long stout timber driven into beds of rivers, &c., to carry superstructures or form dams to the water.

PILE DRIVER.—An engine or machine hammer for driving piles.

PINE.—Technically the wood of the tree *Pinus strobus ;* botanically the wood of all trees of the order *Pinus*.

PINNED JOINT.—One in which small wood dowels are used instead of wedges to secure a tenon.

PITCH.—The inclination of a roof or stair from the horizontal.

PITCH BOARD.—A thin triangular templet, one of whose sides equals the going, the other at right angles with it, the rise, and the third giving the inclination or pitch of the stairs. It is used for setting out the steps upon the strings. See p. 309.

PITCHING PIECE.—A piece of quartering fixed between the wall and newel under winders, and into which the carriages are framed. See p. 317.

PLAIN.—Free from ornament.

PLAN.—A geometrical representation of anything, parallel with the ground.

PLANCEER.—A soffit or ceiling, particularly the soffit of a shop front.

PLANE.—A perfectly flat surface ; also a cutting tool with a wide chisel fixed at a constant angle in a guide block.

PLANK.—Cut stuff, thicker than 1¾ in. and wider than 10 in.

PLANTED.—Mouldings inserted, as distinguished from those stuck in the solid on the framework.

PLINTH.—A small square member at the base of a wall or column.

PLOUGH.—An adjustable grooving plane. See f. 1 & 2, p. 17.

PLOUGHED AND TONGUED JOINT.—One in which the contiguous edges are grooved and a loose tongue inserted therein.

PLUGGING.—Small pieces of wood driven into the joints of brickwork for the purpose of forming fixing points for the finishings.

PLUMB.—Vertical.
POCKET.—The opening in a sash frame for the introduction of the weights.
POLISHED WORK.—Joinery with its surface polished, by rubbing on shell-lac dissolved in spirit.
PORCH.—A shelter to a doorway.
PORTICO.—A porch with its roof supported by columns.
POSTERN.—A small back gate or door.
POT BOARD.—The lower shelf of a kitchen dresser. See p. 293.
PUGGING.—Non-conducting material laid between floor joists to prevent the passage of sound.
PULLEY.—(*M.*) A wheel with a wide, slightly convex rim, to carry the belting that transmits motion from the driving shaft to the machine.
PULLEY STILE.—That part of a sash frame in which the axle-pulleys are placed. See p. 196.
PUNCH.—A tool for driving nails below the surface.

Q.

QUARREL.—A lozenge-shaped piece of glass in a leaded frame.
QUARTERED OAK.—Logs cut into four quarters, then cut radial; the method of preparing "Wainscot Oak." See f. 1, p. 417.
QUARTERING.—Square timber of small scantling from 3 by 3 to 4½ by 4.
QUATREFOIL.—Openings in tracery consisting of four intersecting circles. See p. 263.
QUIRK.—A narrow groove or sinking on the side of a bead.
QUIRK CUTTER.—A tool for cutting quirks in curved surfaces.

R.

RACKING.—Movement or distortion of a frame diagonally.
RAIL.—Any horizontal member in a framing.
RAISED PANEL.—One with its central parts higher than its sides. See p. 155.
RAKE or RAKING.—Anything whose position is neither vertical or horizontal, is described as raking.
RAM.—A large hammer driven by machinery.
RAMP.—A vertical curve in a handrail, concave on top, the reverse of a knee. See p. 308.
RANK STUFF.—Coarse fibred timber.
RAT-TAIL FILE.—A file of circular section used chiefly for enlarging round holes and forming gullets in saws.
RATCHET BRACE.—One fitted with a cogged stop that enables it to be worked in corners or confined situations, by completing a revolution in two or more stages.
REBATE.—Usually pronounced rabbit. A rectangular sinking on the edge of any board or framing. See pp. 56 & 75.
RENAISSANCE.—The style of architecture that followed the Gothic styles in the sixteenth century, professedly a return to the Classic forms as existing in the remains of Greek and Roman buildings.
REREDOS.—The screen behind a communion table or altar, usually fixed against the eastern wall of the church.
RETURN.—A term denoting that the thing in question is repeated upon two adjacent faces of the work.
RETURN BEAD.—A bead worked on the two faces of an angle.
REVEAL.—The outer sides or edges of a door or window opening in an outer wall; the interior sides of the opening are called jambs.
RIBS.—Arched or shaped rafters in roofs, and similar curved supports in "centres."
RIDE.—A term applied to a door when it touches a high place in its path.
RIDGE.—The uppermost horizontal member in a roof.
RIDGE ROLL.—A round piece of wood attached to a ridge as a finish and also to act as a core to the lead flashing. See p. 225.

RIFFLER.—A bent file or rasp used by carvers.

RIGHT-HAND LOCK.—See Handed.

RIM LOCK.—A lock finished with a rim or casing for affixing to the side of a door.

RIND GALL.—A defect in timber caused by the improper lopping of a branch.

RIP SAW.—A tool for cutting in the direction of the grain.

RISE.—The vertical height from the top of one tread to the top of the next, also the total height from floor to floor.

RISER.—One of the component parts of a step; its vertical front.

RISING HINGE.—One with a helical joint that causes the door to lift as it opens.

RIVER or RIVING KNIFE.—(*M.*) A mechanical contrivance to open the cut in timber, after it has passed the saw, to prevent it jamming the latter. See p. 90.

ROCOCO.—A style of ornamentation to furniture prevalent in the reigns of Louis XIV. and XV. of France, consisting of a profusion of shell-work and curls and flowers.

ROD.—A board upon which full-size sections of work are made to facilitate the setting out of the material.

ROLL AND FILLET.—A Gothic moulding. See p. 369.

ROMANESQUE.—A term applied to the architecture of the earlier period of the Christian era before the introduction of the Gothic styles; a debased form of Roman architecture.

ROSE WINDOW.—A church window circular in elevation and divided by radial mullions; sometimes called a wheel window.

ROSTRUM.—An elevated platform for public speaking.

ROUGH STRING.—A board notched out to fit the steps, and fixed to the inside of a cut string to strengthen it.

ROUND STEP.—One with a semicircular end. See p. 318.

ROWEY GRAIN.—Figure in wood consisting of minute dots similar to the roe of a fish.

RUN, TO.—A term applied to a saw cut when it deviates from the correct line.

S.

SADDLE CRAMP.—One made with deep shoes so that it may ride over considerable projections. Also one large enough to completely enclose the work. See p. 313.

SAG.—To bend or curve under a load.

SAPWOOD.—That portion of a tree through which the sap travels upwards; imperfect wood between the bark and the heartwood. See p. 411.

SASH.—The moving part of a window containing the glass.

SASH DOOR.—A door with the upper panels filled in with glass.

SASH SAW.—A long tenon saw formerly used in cutting sash shoulders.

SCALE.—The ratio or proportion which a drawing bears to the object it represents.

SCANTLING.—The sizes of timber; also rough irregular pieces cut off in the squaring of a log.

SCARFE.—A joint used in lengthening a piece of timber in which the two ends are cut to lap over and fit each other, chiefly used in carpentry.

SCOTIA.—The hollow moulding commonly used under the nosing of a step. See p. 306.

SCRAPE, TO.—A method of finishing hardwood after planing by removing extremely thin shavings with a piece of sharpened steel called a scraper. See p. 68.

SCREEN.—A partition in a public building or business office with the lower part panelled and the upper either open or glazed.

SCRIBING.—The fitting of the edge or end of a board or moulding to an irregular surface or the profile of another moulding.

SCROLL.—Anything with an involute curve—that is, a curve with a continuously diminishing radius.

SCROLL CAP or SCROLL.—The ornamental finish to a handrail over a curtail step, forming a volute or spiral in plan. See f. 1, p. 351.

SCUTCHEON.—A corruption of escutcheon; a cover for a keyhole.

SCUTTLE.—A small hatchway or aperture in a roof.

SEASONED.—Timber is said to be seasoned when all the green sap has dried out of it. See p. 412.

SECOND SEASONING.—The further drying of timber after it has been cut into smaller scantlings.

SECRET NAILING.—Nailing boards through their edges where the nail holes will not be seen.

SET.—The inclination from the plane of the blade given to saw teeth, so that they may cut a path wider than the thickness of the saw for its easy working (see f. 2, p. 9); also the cooling of glue in a joint which prevents its being further rubbed.

SET SQUARE.—A solid triangle in wood having one of its angles a right angle. See p. 28.

SETTING OUT.—Marking dimensions and joints on prepared stuff; working drawings on the rod.

SHAKE.—A longitudinal split or cleft in a board.

SHOOTING.—Planing the edge of a board straight and square.

SHOP FRONT.—All the woodwork comprised in the opening in the front wall of a building designed for a shop.

SHOULDER.—The abutting parts of a mortise and tenon joint. See p. 142.

SHOULDER PLANE.—A metal rebate plane for preparing shoulders accurately. See p. 14.

SHOW BOARD.—The board inside a shop front on which the goods are displayed.

SHOW CASE.—An air-tight glazed case for the display of goods.

SHUTTING STILE.—The stile of a door or casement on which the fastenings are fixed.

SIGHT LINE.—The interior edge of a frame; the working point in "setting out."

SILL.—The lowest horizontal member of a door or window frame.

SILVER GRAIN.—The figure of oak and beech wood produced by the medullary rays or septa cropping out upon the surface of the board; called also Overgrain and Felt.

SINGLE FLOOR.—One composed of bridging joists only.

SINGLE TENON.—One tenon irrespective of width. See Pair of Tenons.

SINKING.—A recessed part.

SKEW-NAILED.—Nails driven at an acute angle to increase their hold. See p. 70.

SKELETON FRAMING.—Any framing without panels. See pp. 188 & 190.

SKELETON OF A HOUSE.—A synonym for carcase, which see.

SKIRT.—To surround the extreme edge of anything.

SKIRTING.—The base board of a room having a moulded edge; without the moulding it is described as a plinth.

SKYLIGHT.—A sloping sash in the roof of a building.

SKY-LINE.—The profile of a building against the sky.

SLABBING.—Cutting the irregular faces from a balk to make it prismatic.

SLACK.—A term expressing an ill-fitted joint to a door or sash, or any portion of a frame that fits loosely.

SLEEPERS.—Thin floor joists resting on dwarf walls in a basement, and the lowermost members of a scaffold.

SLEEPY WOOD.—Timber over-dried by the desiccating process.

SLIDING DOOR or SHUTTER.—One that moves on a rail or in a groove horizontally.

SLIP FEATHER.—See Tongue.

SLIP or SLOT MORTISE.—One cut through the end of the piece so that the tenon can be inserted sideways.

SNIPES-BILL.—See p. 18.

SOCKET.—The aperture in a dovetail joint receiving the pin.

SOFFIT.—The head lining of any opening in a wall, and the framed or plastered under surface of the stairs.

SOLE or SOLE PIECE.—A piece of timber placed on the ground to receive the end of an upright and distribute its weight.

SOLID FRAME.—One with its sinkings, mouldings, &c., worked on the material, as distinguished from one formed with lining pieces.

SOLID PANEL.—Synonymous with flush panel, which see.

SOUND STUFF.—Wood free from shakes, loose knots, and other defects.

SPAN.—The clear distance between the faces of the supports of a beam or arch.

SPANDREL.—The triangular space beneath the spring of a stair, usually filled in with a spandrel framing.

SPANDREL FRAME.—Any triangular framing, particularly under a stair.

SPANDRELS OF AN ARCH.—The triangular parts between the springing and the crown, the intrados and the abutments.

SPECIFIC GRAVITY.—The ratio of weight of any substance to the weight of an equal bulk or volume of water.

SPILING.—A method of scribing. See p. 64.

SPINDLE.—(*M.*) In a general sense any straight cylindrical steel shaft or rod of a diameter less than 3 in. Above this size the term shaft is generally used. Specifically the arbor or axial rod, to which circular saws and other revolving cutters are attached in machines. (Am., mandrel.)

SPINDLE OF A LOCK.—The bar that connects the knobs or handles, and actuates the bolt.

SPINDLE MACHINE.—(*M.*) The irregular moulding machine is commonly so termed, presumably because the cutter spindle is visible for a greater portion of its length than in other machines, although *all* machines have spindles.

SPIRAL STAIRS.—Those consisting entirely of winders.

SPIRE.—An extremely tall roof relatively to its span, tapering in section.

SPLAY.—An acute or obtuse angle ; when of small extent it is called a chamfer.

SPLAYED GROUNDS.—Grounds with undercut edges to hold the plaster.

SPLAYED HEADING.—A floor joint in which one piece is bevelled under the other, the upper one when nailed holding both.

SPRIG.—A brad without a head and of round section.

SPRINGING.—The horizontal line from which an arch starts that part of a string, handrail or frame where the curves commence.

SPROCKET PIECE.—An addition to the foot of a rafter to alter the pitch of the eaves courses of the roof covering.

SPRUNG.—A term applied to a curved piece when it has become distorted or out of truth ; also a curved moulding. See p. 375.

SPURN WATER.—An angle fillet to throw off water, a term chiefly used in ship joinery.

SQUARE.—An arrangement of a framing in which the adjacent parts are at right angles to each other ; also the generic name of various tools and instruments for drawing lines at right angles to each other.

SQUARE-CUT.—The term applied to the modern method of forming handrail wreaths, to distinguish it from the older method called the BEVEL-CUT SYSTEM. In the SQUARE-CUT SYSTEM the pieces forming the wreath are cut out square from the surface of the plank, and the joints called butt joints are also square from the surface. In the BEVEL-CUT SYSTEM the edges of the wreath are cut to an appropriate bevel, and the joints are made to a corresponding bevel with the surface of the wreath piece, so that they are vertical when fixed. These are termed spliced joints.

SQUARE OF FLOORING.—100 superficial feet.

STAFF BEAD.—See Angle Bead.

STAIRCASE.—The complete construction comprised in one or more successive flights of stairs.

STAIRS.—A number of steps connecting two floors and closed in underneath, which differentiates them from a ladder, which is open between the steps.

STAIRWAY.—The aperture in the floors, and space between the walls, provided for the stairs

STALL.—An open seat fixed in a church. See p. 265.

STALLBOARD.—The thick rail supporting the sash in a shop front. See p. 234.

STALLBOARD FRAMING.—The woodwork below the stallboard in a shop front.

STAND, TO.—To rise slightly above the surrounding surface.

STANDARD.—A wide division placed vertically in any fitting.

STARSHAKE.—Radiating clefts in the outer parts of a log. See Heartshake.

STEEPLE.—A roof diminishing in dimensions towards the top by abrupt stages ; a series of reduced towers one above the other. See Spire.

STEP.—In joinery, a combination of tread and riser framed into strings.

STICK or STICKING.—The operation of working a moulding with a plane or a machine.

STILES.—The outer vertical members of a door and similar framing. See Munting.

STOCK.—See Brace.

STOP.—An ornamental finish or abrupt change to, or in a moulding. See p. 283.
STOPS.—Wide fillets planted on door jambs to form a rebate for the door.
STORY.—A floor or stage in a building.
STORY POSTS.—Posts placed under a floor or breastsummer to support it.
STORY ROD.—A staff used for taking the height between two floors, and on which the heights of the steps are spaced out. See p. 308.
STRAIGHT JOINT or SQUARE JOINT.—One without additional fittings, as tongues, dowels, slips, &c.
STRAIGHT STAIR.—One consisting entirely of flyers.
STRAIGHT-GRAINED STUFF.—Wood with its fibres free from curl or twist.
STRAIN.—A change in the shape of a beam induced by a stress.
STRAP HINGE.—Any long hinge fixed on the side of a door.
STRESS.—The effect occasioned by a load upon its supports.
STRETCH-OUT.—A drawing representing the unfolding into a plane of a curved surface. See Development.
STRIKE OVER.—To mark stuff all round in setting out.
STRIKING KNIFE.—See Marking Knife.
STRIKING PLATE.—The plate to receive the bolts of a mortise lock.
STRINGS are the inclined sides of a stair carrying the steps, and are of various forms. A CLOSE STRING is one whose edges rise above the steps and enclose their ends entirely. These are also called Housed Strings, see p. 308. A CUT STRING is one shaped to fit underneath the steps, the ends of which are returned across its face. These are also called OPEN STRINGS. When the string is mitred to fit the riser, it is called a CUT AND MITRED STRING; and, when ornamental frets or brackets are employed to cover the ends of the risers, the string is called a BRACKETED STRING, p. 321. A WREATHED STRING is one whose plan is part of a circle or of an ellipse, and, that is inclined in elevation. WALL STRING is the one adjacent to the wall. OUTSIDE STRING is the one near the centre of the stairway.
STRUT.—A post or brace thrown into compression by the stress.
STUB TENON.—One that does not pass through the piece mortised. See f. 9, p. 143.
STUD.—Small divisions in rough partitions.
STUFF.—Wood used in joinery; in carpentry termed timber.
STUMP TENON.—A tenon made thicker at its root to afford additional strength, chiefly used in carpentry. See f. 10, p. 143.
SUNK PANEL.—One below the surface of the framing, usually applied to the sham panels worked in the solid on newels, pilasters, &c.
SURBASE.—Mouldings midway in the height of a wall or at the top of a pedestal.
SWAN'S NECK.—A vertical curve in a handrail composed of a ramp and knee. See p. 308.
SWEEP.—A symmetrical curve, generally applied to a freehand curve.
SWING DOOR.—One mounted on pivots vertically and moving through the frame in both directions. See p. 164.

T.

TABLE.—(*M.*) The top of the frame of a machine upon which the work is placed to be manipulated.
TANG.—That part of a tool which is enclosed by the handle.
TANGENT SYSTEM is a geometrical method of obtaining the true section of a cylinder or other symmetrical solid when cut by an inclined plane (as in a handrail wreath), by enclosing the solid with tangent planes upon whose surfaces the inclinations of the rail are drawn. A plane containing these lines is the plane of section, and upon this plane the shape of the section is determined by methods described fully in Chapter XX.
TANGENTS are lines or planes touching a curved line or a curved surface, but not cutting into it.
TEMPLATE.—A piece of wood or stone placed under the end of a beam, to distribute its weight.

TEMPLET.—A pattern. See Face Mould.

TENON.—A projection formed by reducing the end of a piece of wood in thickness, which is inserted in a mortise to connect two members of a frame, &c.

TENSION.—The state of being pulled asunder in the direction of the length.

TEREDO NAVALIS.—A marine worm very destructive to timber.

THROATING.—A small groove worked upon the under horizontal surfaces of overhanging members of a frame to interrupt the flow of water and cause it to drop off. See f. 1, p. 225.

THUMB PLANES.—Small planes held between the fingers and thumb. See p. 360.

TIMBERING.—A term applied to the underframing or rough supports of a stair between the soffits and the steps.

TOAD'S BACK.—The section of a handrail with a flat curve on the top.

TOBIN VENTILATOR.—An open tube generally of wood, connected with an aperture in the lower part of the wall of a room, and rising to about half the height of the room, through which fresh air gains access. The tube contains a valve by which the supply of air is controlled.

TONGUES.—Slips of wood placed in grooves to form joints. There are three varieties—straight, cross, and feather. See p. 54.

TOOTHED.—Minute grooves made upon two surfaces to be glued together to afford additional grip to the glue.

TOOTHING PLANE.—A smoothing plane with a notched iron used to "tooth" flat surfaces.

TOP RAIL.—The highest rail in any framed work.

TOPPING-UP.—(*M.*) A makeshift method of sharpening circular saws, by filing the backs or tops of the teeth, and neglecting the gullets.

TORUS.—A large bead usually accompanied by a fillet. See p. 366.

TOTE.—The handle of a bench plane.

TOWER BOLT.—A straight cylindrical bolt sliding in a notched tube. See f. 6, p. 237.

TRACERY.—A form of decoration most highly developed in Gothic architecture; obtained by the disposition of the bars in windows, the ribs in vaulting, or the mullions and panels in parapets, screens, &c., into geometrical—as opposed to natural form—patterns; when, as in windows, the pattern is produced by the constructional members with the absence of groundwork it is termed Pierced Tracery; when, as in vaulting or panelling, the groundwork is recessed and the pattern is in relief, it is called Solid Tracery. See also subsidiary headings Cusp, Cusping, Foil, Foliation.

TRACING.—A copy of a drawing made upon semi-transparent paper or linen, which is laid over the original for the purpose.

TRAMMEL.—An apparatus for drawing elliptic curves. See f. 2, p. 23.

TRANSOM.—An intermediate horizontal member of a frame.

TRANSVERSE STRESS.—A stress applied across the fibres of a beam.

TRAVELLER.—The carriage of a winch mounted on a gantry.

TRAVERSING.—Preliminary planing across the grain of a wide surface to reduce it to an approximate flat before "trying up."

TREACLE MOULD.—A deeply undercut nosing to an overhanging flap. See f. 3, p. 298.

TREAD.—The board forming the top surface of a step.

TREFOIL.—In Gothic tracery a circle filled with three cusps.

TRIM.—To construct a framed opening in a roof or floor.

TRIMMER.—The joist carrying the intermediate joists forming a landing, or the ends of the joists cut to form the stairway.

TUDOR ARCH.—A very flat arch typical of the Perpendicular period described from four centres, two on the springing line and two below it. See p. 171.

TURN.—A horizontal curve in a handrail, as quarter turn, &c.

TURNBUCKLE.—A fastening for a cupboard door consisting of a knob and catch.

TUSK NAILING.—Skew nailing, which see.

TUSK TENON.—One having a projection (the tusk) upon the shoulder under the tenon to take the weight of the beam; properly a carpenter's joint.

TWISTED FIBRE.—Wood showing much cross-grain in the plank due to the fibres running up in a spiral direction. This renders the wood weak in cross-resistance.

U.

UNDERCUT.—A shoulder or other cut made out of square—that is, with its upper surface projecting beyond the lower. See f. 2, p. 60.

UPSETTS.—Timber injured by the fall in felling, causing the grain to be crushed together and destroying its tenacity. See p. 416.

V.

VENEER.—A thin piece of superior wood glued on the face of an inferior piece.

VENETIAN WINDOW.—One with three lights in one frame separated by mullions. See p. 207.

VENTILATING SLIP.—A piece of wood inserted upon the inside edge of the sill of a sash frame to enable the sash to be raised slightly without showing an opening at the bottom. See p. 195.

VERGE BOARD.—A barge board.

VERTICAL.—Perpendicular to the horizontal, or plumb.

VESTIBULE.—An enclosure at the entrance to a building, and within the building, in which it differs from a porch, the latter being exterior to the building. The door closing the interior side of the entry is called a Vestibule-door. See p. 165.

VOLUTE.—A spiral scroll.

W.

WAINSCOT.—Oak cut specially to display its "figure" or silver grain. So-called because originally used for the wainscotting or panelling of a room. See p. 417.

WALKING LINE.—An imaginary line generally placed 18 in. from the handrail, upon which the winders are divided. See p. 309.

WANDERING HEART.—A defect in timber in which the pith of the tree runs irregularly through the stem, causing boards cut from it to be "short" in the grain, useless for bearing purposes.

WARPING.—Twisting of a board in its width, due to unequal shrinkage. See p. 418.

WEATHER-BOARDING.—Stuff cut thinner at one edge than the other, used for fencing and covering of wooden houses.

WEATHERING.—The sloping of an exposed surface to throw off the rain. See p. 196.

WELL or WELL-HOLE.—The space in the centre of a staircase between two outside strings. See p. 319.

WELSH GROIN.—One formed by the intersection of two vaults of different heights. See Groin.

WET ROT.—Decomposition of the structure of a growing tree. See p. 414.

WHITE DEAL.—The wood of the spruce fir. See p. 424.

WHOLE TIMBER.—Uncut balks.

WINDER.—A step in the circular part of stair with one end wider than the other, also those which wind around a newel.

WINDING.—A twisted surface, the opposite of a "plane."

WINDING STRIPS.—Two parallel pieces of wood used by joiners to detect if a surface is in winding. See p. 31.

WINDOW BACK.—The lining to the wall under a window frame. See Plate XXIII.

WINDOW BOARD.—A narrow shelf between the window back and sill. See p. 229.

WING COMPASSES.—Those with a quadrant stay with a set screw attached. See p. 22.

WOOD BRICK.—A fixing built into the brickwork of a wall.

WOOD HINGE.—A form of hinge used in the brackets of tables in which the knuckles are of wood bored through to receive an iron wire which acts as a pivot.

WOOD SCREW.—Metal screw-nails having threaded stems, gimlet points, and slotted and bevelled heads, for use in woodwork. See p. 72.

WORKING DRAWINGS.—Sections full size or nearly so. See p. 393.

WORKING LOAD.—The weight that any structure will carry with safety, usually estimated at one-fifth the breaking weight.

WREATH.—That part of a handrail that twists round a curve in plan.

WREATH-PIECE.—That part of the string that is under the handrail wreath.

WREATHED.—A curve that rises as it revolves—a helical curve.

WROUGHT.—Planed up.

Y.

YORKSHIRE LIGHT.—A solid frame with one half glazed, the other fitted with a sliding sash. See p. 219.

Z.

ZONE.—Botanically, a specific band of tissue, encircling similar bands, as seen in the transverse section of a plant stem. See p. 412.

INDEX.

NOTE.—Distinct references to the index word are indicated by a comma between the numbers, continuous reference by a hyphen between the numbers. The chapter titles are printed in capital letters.

B

C

D

E

F

G

H

I

J

K

L

M

N

O

P

Q

R

S

U

V

Y

Z